Julio Licinio, Ma-Li Wong (Editors)

Pharmacogenomics

The Search for Individualized Therapies

Pharmacogenomics

The Search for Individualized Therapies

Edited by
Julio Licinio and Ma-Li Wong

Editors

Julio Licinio
Ma-Li Wong
Interdepartmental Clinical Pharmacology Center
UCLA Neuropsychiatric Institute
and School of Medicine
Gonda Center 3357 A
695 Charles Young Drive So.
Los Angeles, CA 90095-1761
USA
e-mails: licinio@ucla.edu
mali@ucla.edu

1st Reprint 2003

Library of Congress Card No.: applied for

British Library Cataloguing-in-Publication Data
A catalogue record for this book is available from the British Library.

Die Deutsche Bibliothek – CIP-Cataloguing-in-Publication Data
A catalogue record for this publication is available from Die Deutsche Bibliothek

printed in the Federal Republic of Germany
printed on acid-free paper

Composition K+V Fotosatz GmbH, Beerfelden
Printing Strauss Offsetdruck GmbH, Mörlenbach
Bookbinding J. Schäffer GmbH & Co. KG, Grünstadt

ISBN 3-527-30380-4

We dedicate this book to our parents
Áurea and João, and Yin Shen and Kwok Keung.

Acknowledgements

We acknowledge the exceptional role of Barbara A. Levey, M. D., in bringing us into the field of clinical pharmacology, of which pharmacogenomics is a new frontier. We are very fortunate to have worked with Karin Dembowsky at WILEY-VCH whose enthusiasm, expertise, and dedication greatly contributed to this project. Professor Boris Vargaftig's generous invitation for one of us (J. L.) to join the Scientific Committee of the Euroconference "Pharmacogenetics and Pharmacogenomics" (Institut Pasteur, Paris, France, October 12–13, 2000) greatly contributed to enhance our understanding of the state-of-the-art in pharmacogenomics.

Preface

In this book leading experts provide the state-of-the-art in the emerging and exciting field of pharmacogenomics. The multitude of ways that pharmacogenomics can be approached and applied reflects the possibilities brought about by the wealth of data generated by the Human Genome Project, in conjunction with parallel advances in bioinformatics and biotechnology. Procedures that are now routine were a decade ago thought to be impossible. We now study the simultaneous expression of thousands of genes and test thousands of discrete gene variations (single nucleotide polymorphisms) in one sample.

To clinicians and researchers pharmacogenomics is powerfully attractive. Individualized treatment is the Holy Grail of medical practice. However, unlike medieval knights who would leave family and country behind in their adventurous quest, we must stay firmly grounded in the reality of clinical practice and ethics as we search the rich minefields of genomic data for the sequences that will bring about a new era in individualized therapeutics. The quest for the Grail of pharmacogenomics is irresistible and enthralling. The promise of novel, individualized, more efficacious treatments with minimal or no adverse reactions is almost within reach. Given the fact that key diseases are complex and of unknown cause, the expectation of better treatments is extremely appealing. However, is such hope real or a mirage? Will the popular saying "too good to be true" apply to this new area of biomedical research?

Each chapter in this book contributes a piece of the puzzle that will reveal not only the possibilities, but also the complexities of the field. The final picture is still evolving, but our perception as we complete the editing of this volume is that we are witnessing the beginning of an eruption that will unleash revolutionary changes in patient-oriented research and in data processing and integration. It is hoped that this will directly impact on disease treatments, but to achieve such a goal there will be a need to overcome multiple challenges. We will need to develop analytical tools to deal with high volumes of data, data mining, and data integration. New strategies are required to bridge genomics and proteomics and new tools are needed to understand complex information, including behavioral data. Moreover, as the progress of pharmacogenomics is brought to the clinic, it becomes necessary to address increasingly complex ethical issues in patient-oriented research and in treatment design and delivery.

Pharmacogenomics represents a paradigm shift in medicine. In the 21st century the search to understand overwhelming complexity replaces the reductionistic approach to science that was a hallmark of the 20th century. Until recently scientific thinking led investigators to approach a topic by controlling conditions and studying one or few aspects of the problem. In this new century we fully acknowledge the complexity of biology, and the challenge is now one of feasibility. We need to discover what will the smallest unit of pharmacogenomic data be that will support final conclusions.

This book provides chapters on the latest updates on genomic science, related methodological issues, and the application of genomics to biological systems and to therapeutics of diseases that are public health problems worldwide. These are followed by chapters on ethical considerations. We conclude with a chapter on the role of vascular proteomics in individualized treatment. Many of the chapters start with a discussion of what pharmacogenomics is and its distinction from pharmacogenetics. Rather than edit out those – at times conflicting and redundant – paragraphs we thought it would be in the reader's best interest to leave those in, so that different individual perspectives could be presented. As this is such a new and emerging field, concepts and definitions are still evolving. While we personally believe that the term "pharmacogenomics" will eventually replace "pharmacogenetics," or be used interchangeably, others have a more strict view of the distinction between these concepts. We opted to have each author introduce the field in her/his chapter for readers to appreciate the diversity of views and the evolution of the field.

Our goal when we first conceptualized this project was to bring together in one volume the current level of development in pharmacogenomics. We are delighted that leading experts have participated in this endeavor and we are very grateful to them for having made that goal a reality.

Los Angeles, January 2002 | Julio Licinio and Ma-Li Wong

Contents

List of Contributors

*corresponding author

SIMON D. AHORVON
National Neuroscience Institute
11 Jalan Tan Tock Seng
Singapore 308433
Chapter 16

ISRAEL ALVARADO
Laboratory of Pharmacogenomics
UCLA Neuropsychiatric Institute
and School of Medicine
Gonda Research Center 3357A
695 Charles E. Young Drive South
Los Angeles, CA 90095-1761
USA
Chapter 19

MARGARET M. AMEYAW
Washington University School
of Medicine
Department of Medicine
660 South Euclid Avenue
Campus Box 8069
St. Louis, MO 63110-1093
USA
Chapter 24

WADIH ARAP
University of Texas
M.D. Anderson Cancer Center
1515 Holcombe Boulevard, Box 427
Houston, TX 77030-4095
USA
Chapter 26

VINCENZO S. BASILE
Neurogenetics Section, R-31
Clarke Site, Centre for Addiction
and Mental Health
250 College St., Toronto, Ontario
Canada M5T 1R8
Chapter 18

SAMUEL BRODER*
Celera Genomics
45 West Gude Drive
Rockville, MD 20850
USA
Chapter 2

EMMANUEL BROUILLET
URA CEA/CNRS 2210
Service Hospitalier Frédéric-Joliot
91401 Orsay Cedex
France
Chapter 18

MANUEL BRUN
INSERM U458
Hôpital R. Debré
48 Bd Sérurier
75019 Paris
France
Chapter 15

Guang Chen
Laboratory of Molecular Pathology
National Institute of Mental Health
Building 49, Room B1EE16
49 Convent Drive MCS 4405
Bethesda, MD 20892-4405
USA
Chapter 20

Philip Dean
De Novo Pharmaceuticals Ltd.
St. Andrews House
59 St. Andrews Street
Cambridge, CB2 3DD
UK
Chapter 7

Benoit Destenaves
In-Silico Biology
Genset S.A.
Genomic Research Center RN7
91030 Evry
France
Chapter 4

Jeffrey M. Drazen
Department of Medicine
Pulmonary Division
Brigham and Women's Hospital
and Harvard Medical School
181 Longwood Avenue
Boston, MA 02115
USA
Chapter 10

Jing Du
Laboratory of Molecular Pathology
National Institute of Mental Health
Building 49, Room B1EE16
49 Convent Drive MCS 4405
Bethesda, MD 20892-4405
USA
Chapter 20

Howard J. Edenberg
Department of Medical
and Molecular Genetics
Indiana University School of Medicine
545 Barnhill Drive
Indianapolis, IN 46202
USA
Chapter 21

Michel Eichelbaum*
Dr. Margarete Fischer-Bosch-Institute
of Clinical Pharmacology
Auerbachstr. 112
70376 Stuttgart
Germany
Chapter 8

Laurent Essioux
Departments of Biostatistics
Genset S.A.
Genomic Research Center RN7
91030 Evry
France
and
Valigen
La Défense
France
Chapter 4

Martin F. Fromm
Dr. Margarete Fischer-Bosch-Institute
of Clinical Pharmacology
Auerbachstr. 112
70376 Stuttgart
Germany
Chapter 8

Mike L. Furness*
Incyte Genomics Ltd.
Botanic House
100 Hills Road
Cambridge CB2 1FF
UK
Chapter 5

Paul Gane
De Novo Pharmaceuticals Ltd.
St. Andrews House
59 St. Andrews Street
Cambridge, CB2 3DD
UK
Chapter 7

David B. Goldstein
Department of Biology
Galton Laboratory
University College London
Wolfson House
4 Stephenson Way
London, NW1 2HE
UK
Chapter 16

Phyllis Griffin Epps
Health Law & Policy Institute
University of Houston Law Center
100 Law Center
Houston, TX 77204-6060
USA
Chapter 25

Siân Griffiths
Imperial Cancer Research Fund
General Practice Research Group
Oxford University
Department of Primary Health Care
Institute of Health Sciences
Old Road
Headington, Oxford OX3 7LF
UK
Chapter 22

Alex Gubanov
Neurogenetics Section, R-31
Clarke Site, Centre for Addiction
and Mental Health
250 College St., Toronto, Ontario
Canada M5T 1R8
Chapter 18

Philippe Hantraye
URA CEA/CNRS 2210
Service Hospitalier Frédéric-Joliot
91401 Orsay Cedex
France
Chapter 17

Michael M. Hoffmann*
Universitätsklinikum
Zentrale Klinische Forschung
Breisacher Str. 66
79106 Freiburg
Germany
Chapters 12 and 13

Thomas D. Hurley
Department of Biochemistry
and Molecular Biology
Indiana University School of Medicine
545 Barnhill Drive
Indianapolis, IN 46202
USA
Chapter 21

Federico Innocenti
Committee on Clinical Pharmacology
The University of Chicago
5841 South Maryland Avenue,
MC 2115
Chicago, IL 60637
USA
Chapter 14

Lalitha Iyer
Committee on Clinical Pharmacology
The University of Chicago
5841 South Maryland Avenue,
MC 2115
Chicago, IL 60637
USA
Chapter 14

PHILIPPE JAIS*
Medical Research
Genset S.A.
Genomic Research Center RN7
91030 Evry
France
Chapter 4

ELAINE JOHNSTONE
Imperial Cancer Research Fund
General Practice Research Group
Oxford University
Department of Primary Health Care
Institute of Health Sciences
Old Road
Headington, Oxford OX3 7LF
UK
Chapter 22

JAMES L. KENNEDY*
Neurogenetics Section, R-30
Clarke Site, Centre for Addiction
and Mental Health
250 College St., Toronto, Ontario
Canada M5T 1R8
Chapter 18

RICHARD B. KIM*
Vanderbilt University School
of Medicine
Division of Clinical Pharmacology
572 MRB-1
Nashville, TN 37232-6602
USA
Chapter 9

RAJAGOPAL KRISHNAMOORTHY*
INSERM U458
Hôpital R. Debré
48 Bd Sérurier
75019 Paris
France
Chapter 11

CLAUDINE LAPOUMÉROULIE
INSERM U458
Hôpital R. Debré
48 Bd Sérurier
75019 Paris
France
Chapter 11

TING-KAI LI*
Department of Medicine
Indiana University School of Medicine
Emerson Hall 421
545 Barnhill Drive
Indianapolis, IN 46202-5124
USA
Chapter 21

JULIO LICINIO
Interdepartmental Clinical
Pharmacology Center
UCLA Neuropsychiatric Institute
and School of Medicine
Gonda Center 3357A
695 Charles E. Young Drive South
Los Angeles, CA 90095-1761
Chapter 19

KLAUS LINDPAINTNER*
Roche *Genetics*
F. Hoffmann-La Roche
Bldg 93/532
4070 Basel
Switzerland
Chapter 6

WINFRIED MÄRZ
Abteilung für Klinische Chemie
Universitätsklinikum
Hugstetter Str. 55
79106 Freiburg
Germany
Chapters 12 and 13

Husseini K. Manji*
Laboratory of Molecular Pathology
National Institute of Mental Health
Building 49, Room B1EE16
49 Convent Drive MSC 4405
Bethesda, MD 20892-4405
USA
Chapter 20

Mario Masellis
Neurogenetics Section, R-31
Clarke Site, Centre for Addiction
and Mental Health
250 College St., Toronto, Ontario
Canada M5T 1R8
Chapter 18

Jeanette J. McCarthy*
Millennium Pharmaceuticals, Inc.
75 Sidney Street
Cambridge, MA 02139
USA
Chapter 3

Howard L. McLeod*
Washington University School
of Medicine
Department of Medicine
Room 1021 CSRB NT
660 South Euclid Avenue,
Campus Box 8069
St. Louis, MO 63110-1093
USA
Chapter 24

Urs A. Meyer*
Division of Pharmacology/
Neurobiology
Biozentrum of the University of Basel
Klingelbergstr. 50–70
4056 Basel
Switzerland
Chapter 1

Marcus Munafò
Imperial Cancer Research Fund
General Practice Research Group
Oxford University
Department of Primary Health Care
Institute of Health Sciences
Old Road
Headington, Oxford OX3 7LF
UK
Chapter 22

Mike Murphy
Imperial Cancer Research Fund
General Practice Research Group
Oxford University
Department of Primary Health Care
Institute of Health Sciences
Old Road
Headington, Oxford OX3 7LF
UK
Chapter 22

Markus Nauck
Abteilung für Klinische Chemie
Universitätsklinikum
Hugstetter Str. 55
79106 Freiburg
Germany
Chapter 12

Christian Néri
Laboratory of Genomic Biology
Fondation Jean Dausset – CEPH
27 rue Juliette Dodu
75010 Paris
France
Chapter 17

Matt Neville
Imperial Cancer Research Fund
General Practice Research Group
Oxford University
Department of Primary Health Care
Institute of Health Sciences
Old Road
Headington, Oxford OX3 7LF
UK
Chapter 22

Marie Hélène Odièvre
INSERM U458
Hôpital R. Debré
48 Bd Sérurier
75019 Paris
France
Chapter 11

Lyle J. Palmer*
The Channing Laboratory
Brigham and Women's Hospital and
Harvard Medical School
181 Longwood Avenue
Boston, MA 02115-5804
USA
Chapter 10

Renata Pasqualini
University of Texas
M.D. Anderson Cancer Center
1515 Holcombe Boulevard, Box 427
Houston, TX 77030-4095
USA
Chapter 26

Viswanathan Ramachandran
National Neuroscience Institute
11 Jalan Tan Tock Seng
Singapore 308433
Chapter 16

Mark J. Ratain*
Committee on Clinical Pharmacology
The University of Chicago
5841 South Maryland Avenue,
MC 2115
Chicago, IL 60637
USA
Chapter 14

Terry Reisine*
ActiveSite Biotech
10327 Rossbury Place
Los Angeles, CA 90064
USA
Chapter 23

Mark A. Rothstein*
Herbert F. Boehl Chair of Law
and Medicine
Institute for Bioethics,
Health Policy and Law
University of Louisville School
of Medicine
501 East Broadway, Third Floor
Louisville, KY 40202
USA
Chapter 25

Jean-Michel Scherrmann
Department of Clinical Pharmacy
and Pharmacokinetics
Laboratory of Neuropharmacokinetics
INSERM U 26
Hôpital Fernand Widal
200 rue du Faubourg Saint Denis
75475 Paris Cedex 10
France
Chapter 15

Eric S. Silverman
Department of Medicine
Pulmonary Division
Brigham and Women's Hospital
and Harvard Medical School
181 Longwood Avenue
Boston, MA 02115
USA
Chapter 10

G. Subramanian
Celera Genomics
45 West Gude Drive
Rockville, MD 20850
USA
Chapter 2

François Thomas
Medical Research
Department of Pharmacogenomics
and Medical Affairs
Genset S.A.
Genomic Research Centre RN7
91030 Evry
France
and
Bioserve Ltd.
Cambridge
UK
Chapter 4

Rommel G. Tirona
Vanderbilt University School
of Medicine
Division of Clinical Pharmacology
572 MRB-1
Nashville, TN 37232-6602
USA
Chapter 9

J. Craig Venter
Celera Genomics
45 West Gude Drive
Rockville, MD 20850
USA
Chapter 2

Robert Walton*
Imperial Cancer Research Fund
General Practice Research Group
Oxford University
Department of Primary Health Care
Institute of Health Sciences
Old Road
Headington, Oxford OX3 7LF
UK
Chapter 22

Scott T. Weiss
The Channing Laboratory
Brigham and Women's Hospital and
Harvard Medical School
181 Longwood Avenue
Boston, MA 02115
USA
Chapter 10

Heinrich Wieland
Abteilung für Klinische Chemie
Universitätsklinikum
Hugstetter Str. 55
79106 Freiburg
Germany
Chapter 13

Bernhard R. Winkelmann*
Universitätsklinikum Heidelberg
Kooperationseinheit Pharmako-
genomik/Angewandte Genom-
forschung
Im Neuenheimer Feld 221
69120 Heidelberg
Germany
Chapters 12 and 13

Ma-Li Wong
Laboratory of Pharmacogenomics
UCLA Neuropsychiatric Institute and
School of Medicine
Gonda Research Center 3357A
695 Charles E. Young Drive South
Los Angeles, CA 90095-1761
Chapter 19

Nicholas W. Wood
Institute of Neurology
University College London
National Hospital
Queen Square
London, WC1E 6BT
UK
Chapter 16

Edward Zanders
De Novo Pharmaceuticals Ltd.
St. Andrews House
59 St. Andrews Street
Cambridge, CB2 3DD
UK
Chapter 7

Color Plates

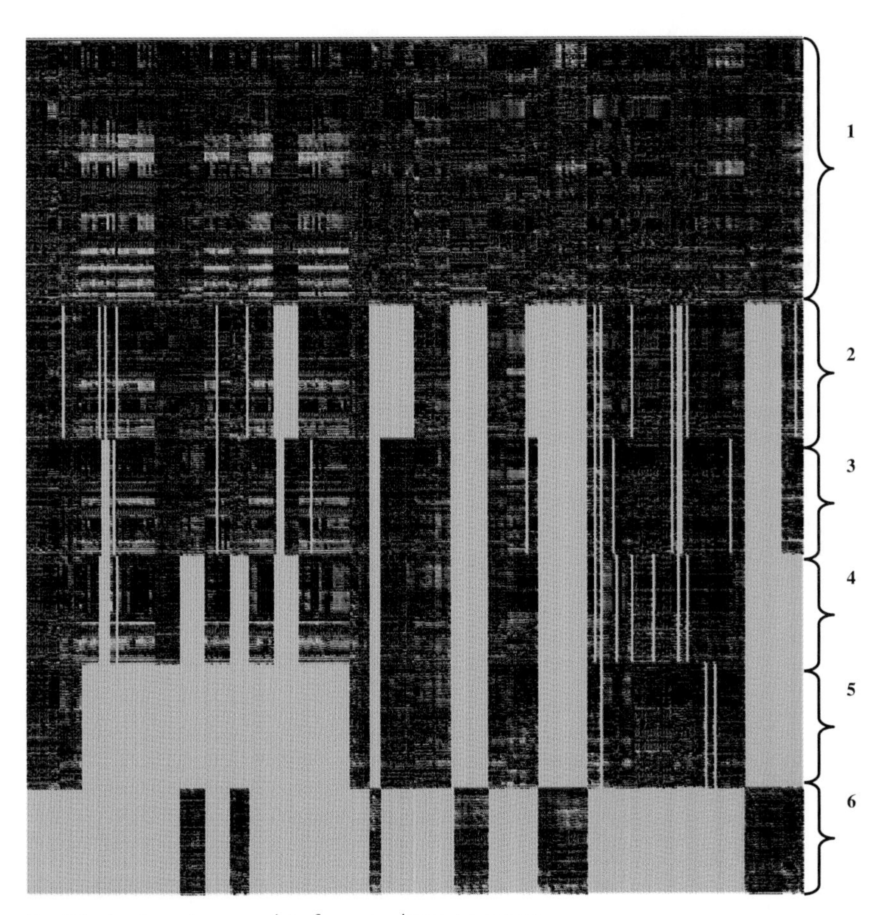

Fig. 5.2 *In vitro* microarray data from xenobiotic treatment.

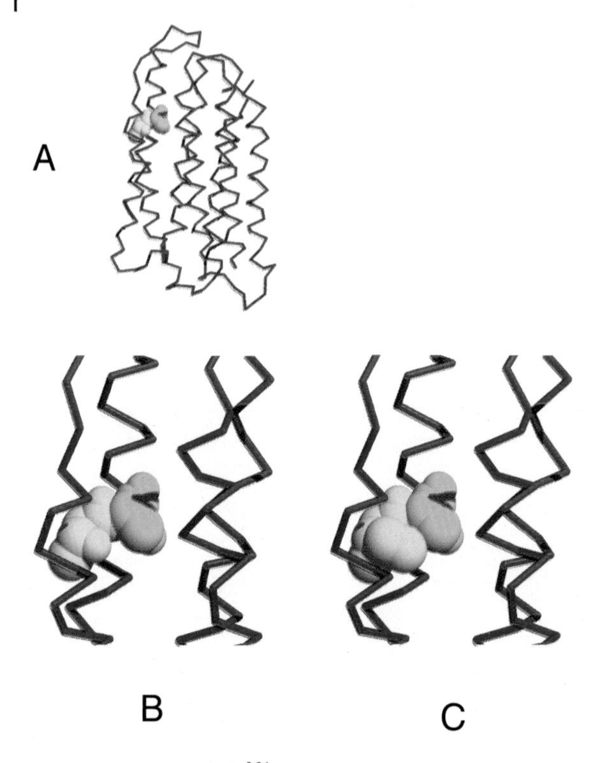

Fig. 7.1 Mutation of Thr^{164} to Ile (cyan) in the fourth transmembrane region of the β_2-adrenergic receptor leads to steric interference with Ser^{165} (green) within the active site of the receptor. (**A**) general domain structure and position of the residues, (**B**) wild type, (**C**) mutant showing interaction of Ile^{164} with Ser^{165}.

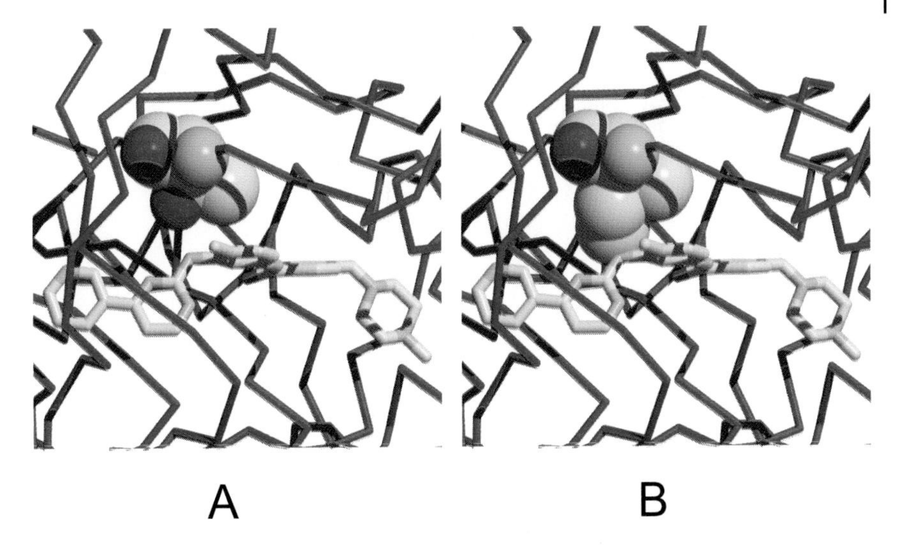

Fig. 7.2 Mutation of Thr^{315} (**A**) to Ile (**B**) in BCR-ABL interferes with the binding of STI-571.

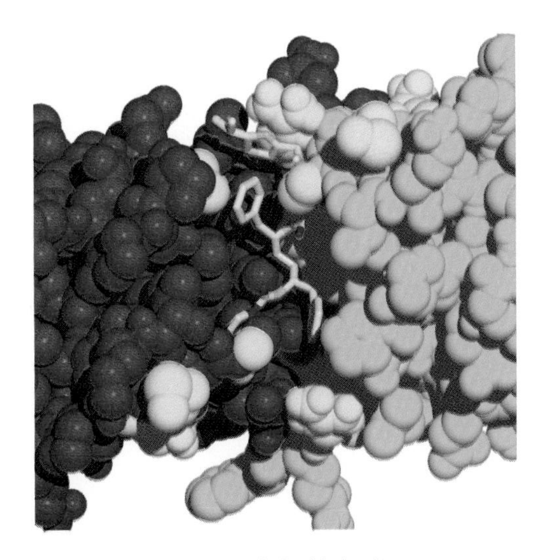

Fig. 7.3 Binding of the HIV protease inhibitor ritonavir. Amino acids highlighted in cyan are mutated in resistant strains of the virus and tend to occur at the extremities of the inhibitor.

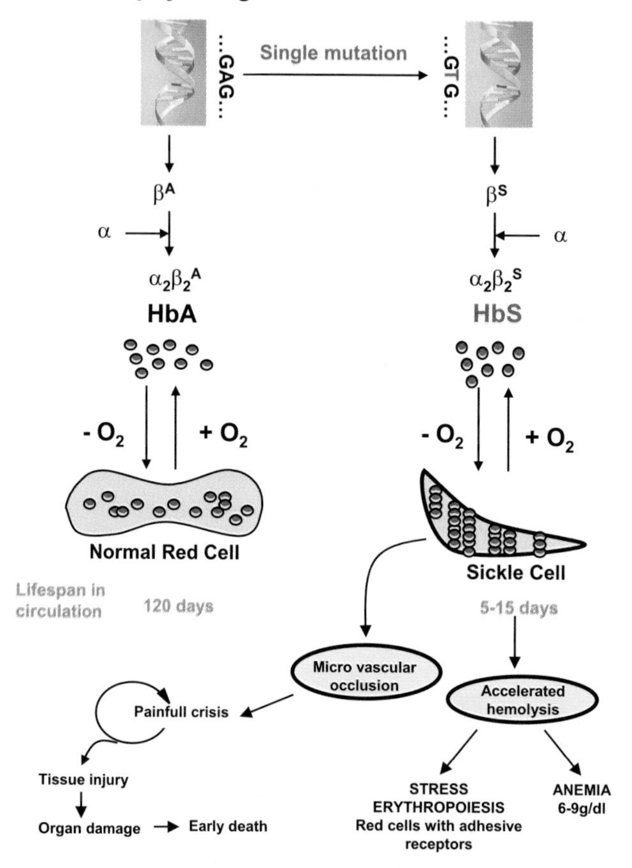

Fig. 11.1 Pathophysiologic scheme of sickle cell anemia.

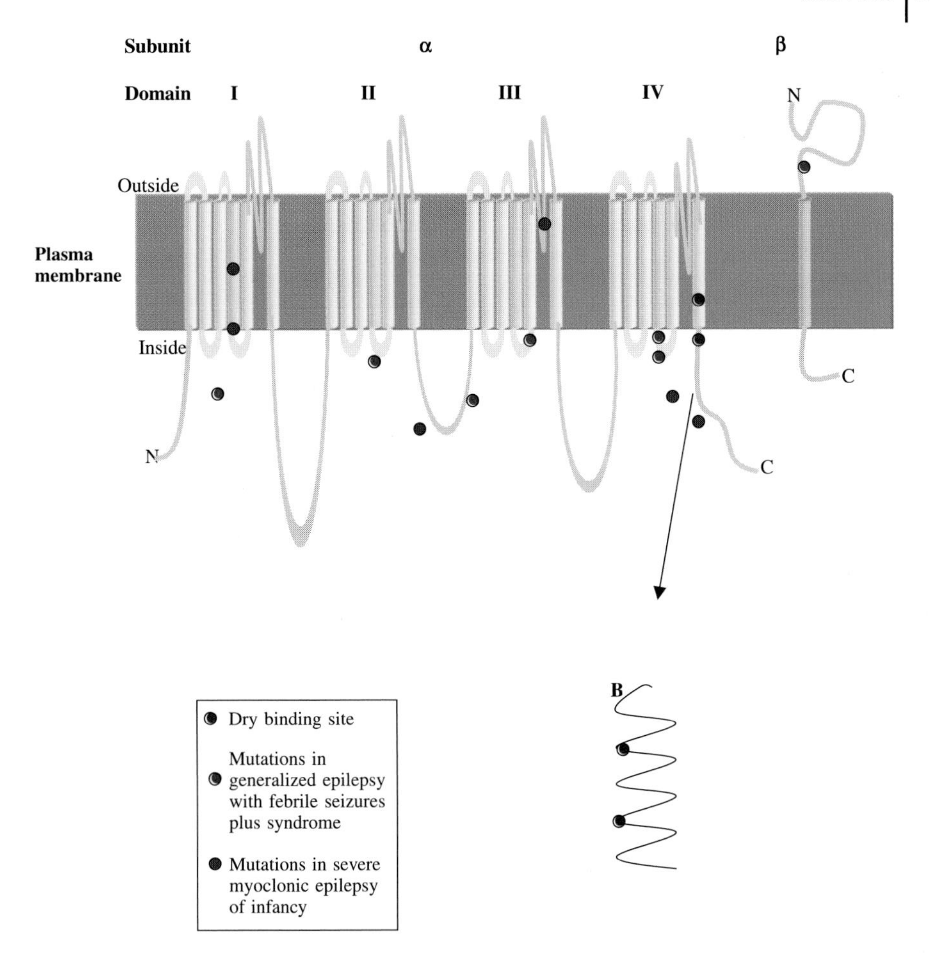

Fig. 16.1 Sodium channel structure. Schematic representation of the sodium channel subunits, α, $\beta 1$ and $\beta 2$. (A) The α-subunit consists of four homologous intracellularly linked domains (I–IV) each consisting of six connected segments (1–6). The segment 4 of each of the domains acts as the voltage sensor, physically moving out in response to depolarization resulting in activation of the sodium channel. The channel is inactivated rapidly by the linker region between III and IV docking on to the acceptor site formed by the cytoplasmic ends of S5 and S6 of domain IV. The β-subunits have a common structure, with the $\beta 1$ non-covalently bound, and $\beta 2$ linked by disulfide bonds to the α-channel (adapted from [4]). The S5/S6 and the segment linking them (P-loop) are believed to constitute the most of the pore of the channel. Specific mutations in the P-loop are associated with loss of selectivity of the channel. Mutations identified in generalized epilepsy with febrile seizures plus are denoted by red dots, while those in severe myoclonic epilepsy of infancy with black dots. The black dots denote the site of termination of the sodium channel. (B) An enlarged S6 segment of domain IV showing drug-binding site comprised of phenylalanine-1764 and tyrosine-1771 in human sodium channel $Na_v1.2$.

Saline Control

Valproate

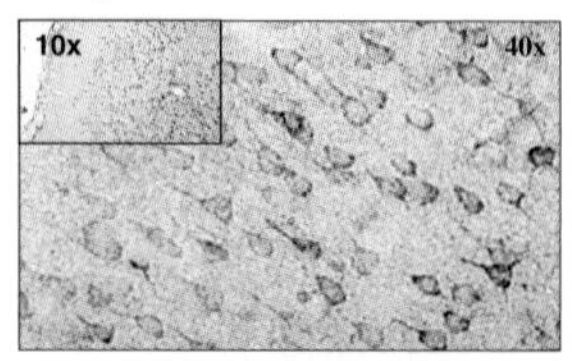

Lithium

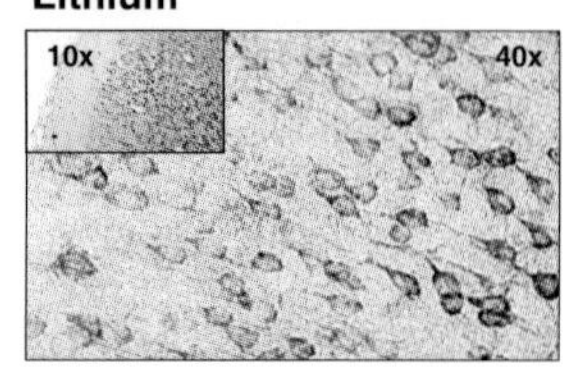

Bcl-2 Peptide Blocking

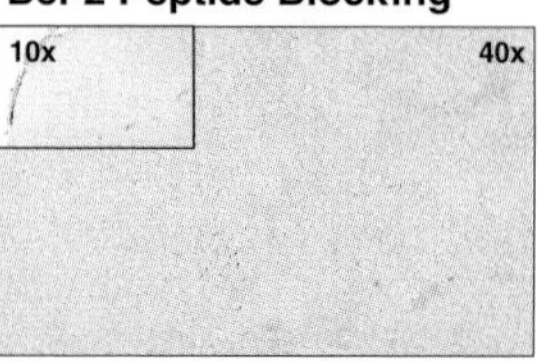

Fig. 20.2 Chronic lithium and valproate robustly increase bcl-2 immunoreactive neurons in the frontal cortex. Male Wistar Kyoto rats were treated with either Li_2CO_3, valproate or saline by twice daily i.p. injections for four weeks. Rats brains were cut at 30 μm; serial sections were cut coronally through the anterior portion of the brain, mounted on gelatin-coated glass slides and were stained with thionin. The sections of the second and third sets were incubated free-floating for 3 d at 4 °C in 0.01 M PBS containing a polyclonal antibody against bcl-2 (N-19, Santa Cruz Biotechnology, Santa Cruz, CA 1:3000), 1% normal goat serum and 0.3% Triton X-100 (Sigma, St. Louis, MO). Subsequently, the immunoreaction product was visualized according to the avidin-biotin complex method. The figure shows immunohistochemical labeling of bcl-2 in layers II and III of frontal cortex in saline-, lithium- or valproate-treated rats. Blocking peptide shows the specificity of the antibody. Photographs were obtained with 40× magnification. Modified and reproduced, with permission, from [40].

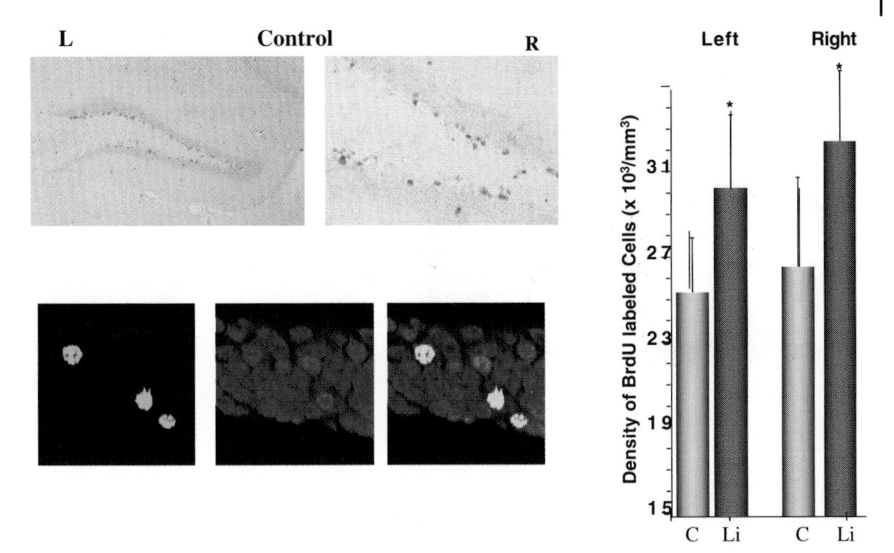

Fig. 20.3 Chronic lithium increases hippocampal neurogenesis. C57BL/6 mice were treated with lithium for 14 days, and then received once daily BrdU injections for 12 consecutive days while lithium treatment continued. 24 hours after the last injection, the brains were processed for BrdU immunohistochemistry. Cell counts were performed in the hippocampal dentate gyrus at three levels along the dorsoventral axis in all the animals. BrdU-positive cells were counted using unbiased stereological methods. Chronic lithium produced a significant 25% increase in BrdU immunolabeling in both right and left dentate gyrus (* $p<0.05$). (**a**) BrdU immunohistochemistry; (**b**) quantitation of BrdU-positive cells; (**c**) double labeling with BrdU and NeuN (neuron-specific nuclear protein, a neuronal marker). Modified and reproduced with permission from [45].

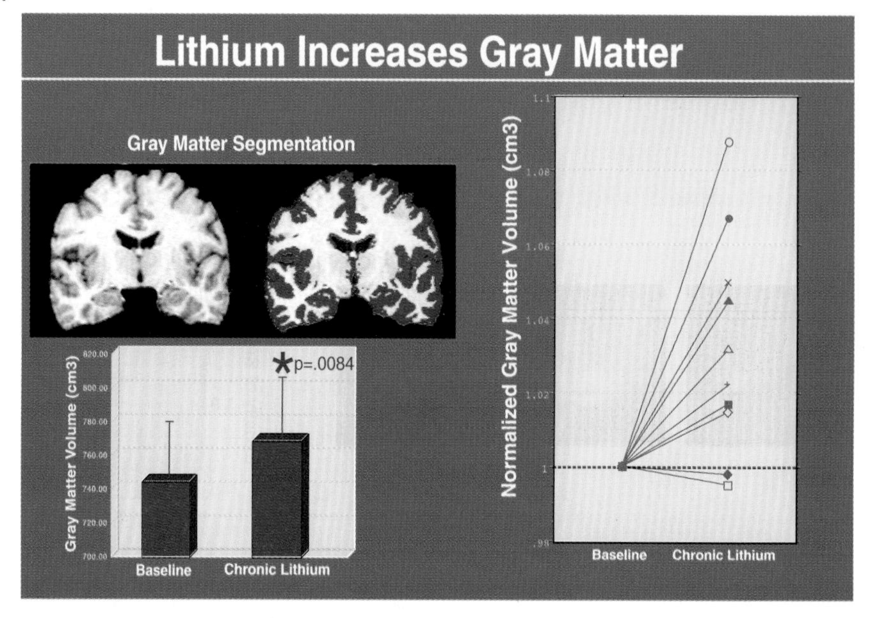

Fig. 20.4 Brain gray matter volume is increased following four weeks of lithium administration at therapeutic levels in BPD patients. Inset illustrates a slice of the three-dimensional volumetric MRI data which was segmented by tissue type using quantitative methodology to determine tissue volumes at each scan time point. Brain tissue volumes using high-resolution three-dimensional MRI (124 images, 1.5 mm thick Coronal T1 weighted SPGR images) and validated quantitative brain tissue segmentation methodology to identify and quantify the various components by volume, including total brain white and gray matter content. Measurements were made at baseline (medication free, after a minimum 14 day washout) and then repeated after four weeks of lithium at therapeutic doses. Chronic lithium significantly increases *total gray matter content* in the human brain of patients with BPD. No significant changes were observed in brain white matter volume, or in quantitative measures of regional cerebral water. Modified and reproduced with permission, from [53].

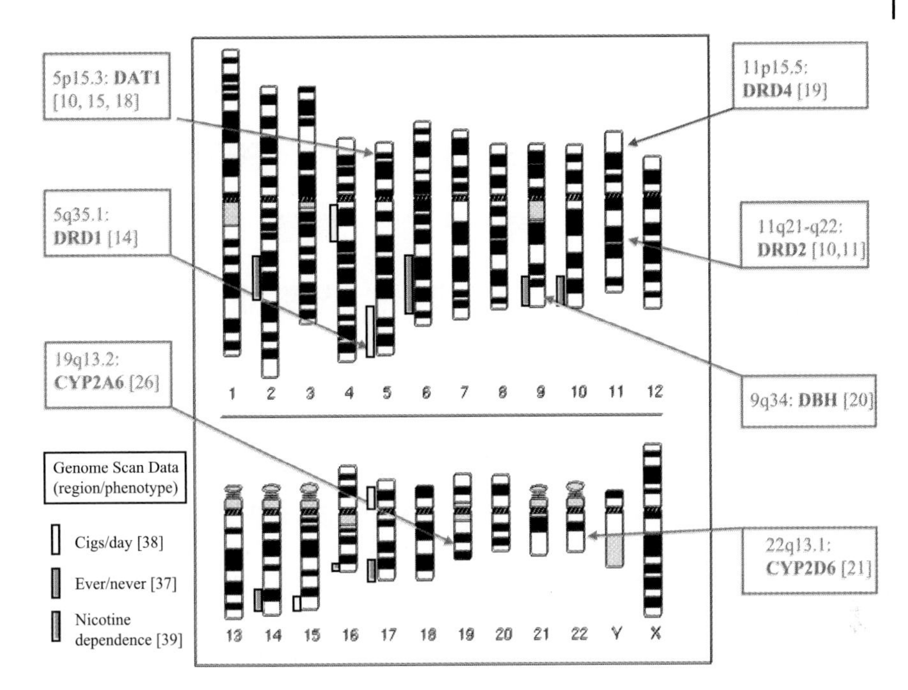

Fig. 22.2 Existing studies of links between polymorphisms in genes and their relationship to smoking behavior. *DAT1*, dopamine transporter; *DRD1*, dopamine D1 receptor; *DBH*, dopamine β-hydroxylase; *DRD2*, dopamine D2 receptor; *DRD4*, dopamine D4 receptor; *CYP2A6*, cytochrome P450 2A6; *CYP2D6*, cytochrome P450 2D6.

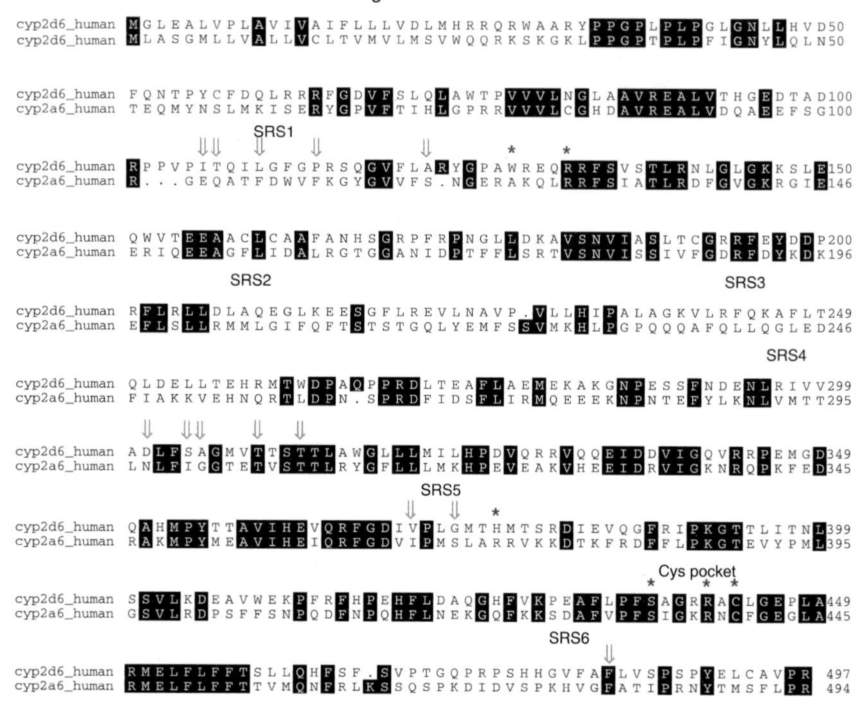

Fig. 22.3 Cytochrome P450 monooxygenases and nicotine metabolism. An alignment of the amino acid sequences of the enzymes 2A6 and 2D6. Occurrences of the same amino acid residue at the same position are shown in black. Putative substrate recognition sites (SRS1–SRS6) are shown by horizontal lines. Vertical arrows indicate amino acid residues predicted from modeling studies on cytochromes to bind to the enzyme substrate. The cysteine pocket contains key residues that bind to the heme cofactor, which is essential for enzyme activity. Inactivating amino acid changes for both enzymes are shown in red.

1
Introduction to Pharmacogenomics: Promises, Opportunities, and Limitations

Urs A. Meyer

Abstract

Pharmacogenomics leads to a better understanding of interaction of drugs and organisms. The promise of pharmacogenomics is that both the choice of a drug and its dose will be determined by the individual genetic make-up leading to personalized, more efficacious and less harmful drug therapy. The techniques of genomics and proteomics help to understand disease and to discover new drug targets. Finally, genomics allows to study the effects of drugs on gene expression. The limitations of pharmacogenomics are the complexities of gene regulation, of proteomics, of gene-environment interactions and also of the psychological complexities of interactions between physicians and patients.

1.1
Pharmacogenetics – The Roots of Pharmacogenomics

An ideal drug is one that effectively treats or prevents disease and has no adverse effects. However, a medication is rarely effective and safe in all patients. Therefore, when a physician determines the dose of a drug, it is always a compromise between "not too high" and "not too low" for this patient or group of patients. Dealing with diversity in drug effects is a major problem in clinical medicine and in drug development. The size of the problem is considerable. A meta-anaylsis of 39 prospective studies from U.S. hospitals suggests that 6.7% of in-patients have serious adverse drug reactions and 0.32% have fatal reactions, the latter causing about 100,000 deaths per year in the USA [1]. Of equal relevance is the fact that most presently approved therapies are not effective in all patients. For instance, 20–40% of patients with depression respond poorly or not at all to antidepressant drug therapy, and similar or even higher percentages of patients are resistant to the effects of antiasthmatics, antiulcer drugs, to drug treatment of hyperlipidemia and many other diseases (for review, see [2]).

The individual risk for drug inefficacy or drug toxicity is a product of the interaction of genes and the environment. Environmental variables include nutritional factors, concommittantly administered drugs, disease and many other factors including lifestyle influences such as smoking and alcohol consumption. These factors act in concert with the individual's genes that code for pharmacokinetic and pharmacodynamic determinants of drug effects such as receptors, ion channels, drug-metabolizing enzymes and drug transporters.

Pharmacogenetics deals with inherited variations in drug effects. It carries the promise of explaining how the individual's make-up of genes determines drug efficacy and toxicity. Pharmacogenetics had its beginnings about 40 years ago when researchers realized that some adverse drug reactions could be caused by genetically determined variations in enzyme activity [3, 4]. For example, prolonged muscle relaxation after suxamethonium was explained by an inherited deficiency of a plasma cholinesterase, and hemolysis caused by antimalarials was recognized as being associated with inherited variants of glucose-6-phosphate dehydrogenase. Similarly, inherited changes in a patient's ability to acetylate isoniazid was found to be the cause of the peripheral neuropathy caused by this drug. Genetic deficiencies of other drug-metabolizing enzymes such as cytochromes P450 CYP2D6, CYP2C9, CYP2C19 or methyltransferases were discovered later. Most recently, it was realized that drug receptors, e.g., the β_2-adrenoceptor, and drug transporters, e.g., the multidrug resistance gene MDR1, are subject to genetic variation. Adverse drug reactions in individual subjects and members of their families often were the clinical events that revealed genetic variants of these and other drug-metabolizing enzymes or drug targets (reviewed in [5–8]). All these observations dealt with variations of specific genes or polymorphisms. Genetic polymorphisms are monogenic variations that exist in the normal population in a frequency of more than 1% [9]. One reason for the pre-occupation of pharmacogenetics with single genes is that they were easier to study with the classical genetic techniques and many of them were clinically important. However, as will be discussed below, most differences between people in their reactions to drugs are multigenic and multifactorial.

Molecular genetics and genomics have transformed pharmacogenetics in the last decade. The two alleles carried by an individual at a given gene locus, referred to as the genotype, can now easily be characterized at the DNA level, their influence on the kinetics of the drug or a specific receptor function, the phenotype, can be measured by advanced analytical methods for metabolite detection or by sophisticated clinical investigations, e.g., receptor density studies by positron emission tomography. Molecular studies in pharmacogenetics started with the initial cloning and characterization of the drug-metabolizing enzyme CYP2D6 [5, 10] and now have been extended to numerous human genes, including more than 20 drug-metabolizing enzymes and drug receptors and several drug transport systems (*www.sciencemag.org/feature/data/1044449.shl*). Genotyping and phenotyping tests to predict dose requirements are now increasingly introduced into preclinical studies of drugs and into the clinical routine, e.g., in the choice and initial dose determination of antidepressants [11].

Another important aspect of pharmacogenetics is the realization that all pharmacogenetic variations studied to date occur at different frequencies among subpopulations of different ethnic or racial origin. For instance, striking cross-ethnic differences exist in the frequency of slow acetylators of isoniazid due to mutations of N-acetyltransferase *NAT2*, of poor metabolizers of warfarin due to mutations of *CYP2C9* and of omeprazole due to polymorphism of *CYP2C19*, and of ultrarapid metabolizers due to duplication of *CYP2D6* genes [5, 7, 12]. Some of the mutations of these genes indeed occur uniquely in certain ethnic subpopulations and trace the origins and movements of populations on this planet. This ethnic diversity, also called gene geography, pharmacoanthropology or ethnopharmacology, implies that population differences and ethnic origin have to be considered in pharmacogenetic studies and in pharmacotherapy.

Observations of person-to-person differences in the metabolism of drugs and consequently in drug kinetics and response led to the concepts of pharmacogenetics. The same principal concepts apply to the genetic variability in the reaction to food components (e.g., lactose intolerance) or to environmental toxins (e.g., carcinogens). These fields often are termed "ecogenetics" and "toxicogenetics".

1.2 Pharmacogenomics – It is Not Just Pharmacogenetics

Genomics involves the systematic identification of all human genes and gene products, the study of human genetic variations, combined with changes in gene and protein expression over time, in health and disease. Genomics is revolutionizing the study of disease processes and the development and rational use of drugs. Its promise is to enable medicine to make reliable assessments of the individual risk to acquire a particular disease, improve the classification of disease processes and raise the number and specificity of drug targets. In 2001, almost the entire human genome sequence became principally known and the information is increasingly accessible. Moreover, in association with the public and private efforts to sequence the human genome, a large number of techniques and bioinformatic tools have been developed. The term pharmacogenomics reflects the evolution of pharmacogenetics into the study of the entire spectrum of genes that determine drug response, including the assessment of the diversity of the human genome sequence and its clinical consequences.

There are three aspects of pharmacogenomics that make it different from classical pharmacogenetics.

1.2.1 Genetic Drug Response Profiles

Rapid sequencing and single nucleotide polymorphisms (SNPs) will play a major role in associating sequence variations with heritable clinical phenotypes of drug or xenobiotic response. SNPs occur approximately once every 300–3,000 base pairs if one compares the genomes of two unrelated individuals [13, 14]. Any two individuals thus differ by approximately 1–10 million base pairs, i.e., in < 1% of the approximately 3.2 billion base pairs of the haploid genome (23 chromosomes).

Pharmacogenomics focuses on SNPs for the simple and practical reason that they are both the most common and the most technically accessible class of genetic variants. For clinical correlation studies in relatively small populations SNPs that occur at frequencies of greater than 10% are most likely to be useful, but rare SNPs with a strong selection component and a more marked effect on phenotype are equally important. Once a large number of these SNPs and their frequencies in different populations are known, they can be used to correlate an individual's genetic "fingerprint" with the probable individual drug response. High-density maps of SNPs in the human genome may allow to use these SNPs as markers of xenobiotic responses even if the target remains unknown, providing a "drug response profile" associated with contributions from multiple genes to a response phenotype (see also [15], and *http://snp.cshl.org)*. The ability to predict inter-individual differences in drug efficacy or toxicity will thus be a realistic scenario for the future. Indeed, there is a rapidly growing effort to identify SNPs that will be useful for identifying patients who are at high risk to experience adverse drug reactions or to determine the best therapeutic approach in this particular patient [15, 16]. Thus, genotyping procedures will play an important role in future therapies. However, phenotyping methods will remain important to assess the clinical relevance of genetic variations, as discussed below.

1.2.2
The Effect of Drugs on Gene Expression

Genomic technologies also include methods to study the expression of large groups of genes and indeed the entire products (mRNAs) of a genome. Most drug actions produce changes in gene expression in individual cells or organs. This provides a new perspective for the way in which drugs interact with the organism and provide a measure of the drug's biological effects. For instance, numerous drugs induce their own metabolism and the metabolism of other drugs by interacting with nuclear receptors such as AhR, PPAR, PXR and CAR (for review, see [17, 18–20]). This phenomenon has major clinical consequences such as altered kinetics, drug-drug interaction or changes in hormone and carcinogen metabolism. Genomics is providing the technology to better analyze these complex multifactorial situations and to obtain individual genotypic and gene expression infor-

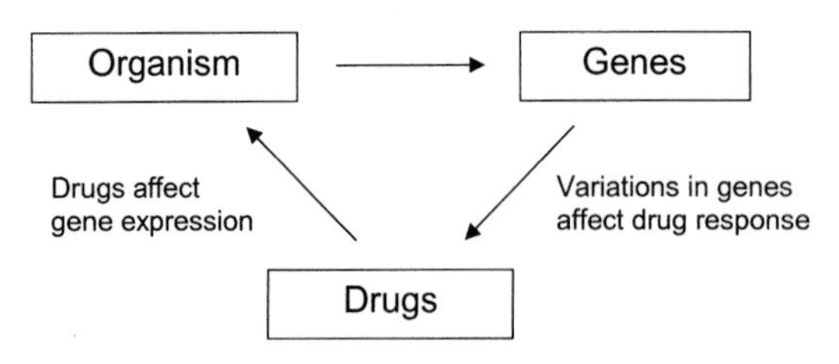

Fig. 1.1 Interaction of genes and drugs.

mation to assess the relative contributions of environmental and genetic factors to variation [21].

1.2.3 Pharmacogenomics in Drug Discovery and Drug Development

The identification of all genes, and the studies of ultimately all protein variants expressed in cells and tissues that cause, contribute to or modify a disease will lead to new "drugable" and "non-drugable" targets, prognostic markers of disease states or severity of disease information (for review, see [2, 22]. The pharmaceutical industry obviously has realized this potential, and chapter 6 in this book discuss this particular aspects of pharmacogenomics. Pharmacogenomic approaches and technologies for drug discovery and drug development have recently been reviewed [23]. It is obvious that the discovery of genes and proteins involved in the pathogenesis of disease allows the definition of new drug targets and promises to profoundly change the field of medicine in the future.

1.3 Pharmacogenomics – Hope or Hype?

At present, genetic testing is restricted to a limited number of patients or volunteers in academic institutions and clinical drug trials, although evidence is accumulating that prospective testing could be of major benefit to many patients [11]. The lack of large prospective studies to evaluate the impact of genetic variation on drug therapy is one reason for the slow acceptance of these principles.

Phenotyping tests require the administration of a specific marker drug or test drug, collection of urine, blood, or saliva for analysis of drug and metabolite concentrations. Gene expression analysis requires tissue, e.g., brain, liver, gut samples, that are accessible only under special conditions and are subject to major ethical constraints. Moreover, these tests are time consuming, expensive and subject to drug-drug interactions or other influences. Genotyping tests have the advantage of having to be done only once in a lifetime and providing unequivocal genetic information. However, gentoyping tests only indicate what "could" happen as they identify a group or category association (e.g., poor metabolizer, ultrarapid metabolizer) and do not predict the exact individual metabolic capacity or receptor interaction, because there still is considerable variation between individuals of the same genotype. Similar constraints apply to drug response profiles by SNP analysis, which do not consider epigenetic variables. These are serious limitations of the practical application of pharmacogenomics in medicine.

Studies in small patient populations have produced numerous controversies on gene-disease associations, e.g., the claimed association of a deletion variant of the gene for angiotensin converting enzyme (ACE) with the risk for myocardial infarction was inexistent in larger studies [24].

Are the promises of pharmacogenomics realistic expectations or just fantasies? It obviously depends much on the drug in question. The quantiative role of a drug-metabolizing enzyme (e.g., CYP2D6) or a drug uptake mechanism (e.g., influenced by the multidrug resistance protein MDR1) in the overall kinetics of a drug together with the agent's therapeutic range will determine how much the dose has to be adjusted in a poor metabolizer or ultrarapid metabolizer individual. The example of the CYP2D6 polymorphism again provides incontroversible clinical data for these concepts. The majority of patients ($\sim$90%) require 75–150 $mg \cdot d^{-1}$ of nortriptyline to reach a "therapeutic" plasma steady-state concentration of 200–600 $nmol \cdot L^{-1}$ [50–150 $\mu g \cdot L^{-1}$), but poor metabolizer individuals need only 10–20 $mg \cdot d^{-1}$, to reach the same levels. Ultrarapid metabolizers on the other extreme may require 300–500 mg or even >500 mg to reach the same plasma concentration [25]. Obviously, without knowing about the genotype or phenotype of the patient, poor metabolizers will be overdosed and be at high risk of drug toxicity, whereas ultrarapid metabolizers will be underdosed. Clinical observations have repeatedly confirmed these predictions.

Another situation is presented if the therapeutic effect depends on the formation of an active metabolite (e.g., morphine from codeine). Poor metabolizers will have no drug effect and ultrarapid metabolizers may have exaggerated drug responses [26]. The drug-related criteria that make a genetic variation clinically relevant thus are similar to those for drug concentration-drug effect monitoring, i.e., narrow therapeutic range or large inter-individual variation in kinetics or suspicion of overdose are the relevant factors.

A different limitation of pharmacogenomics is the requirement for behavioral changes to prevent disease or adverse reactions. Informing people about their genetic susceptibility may motivate them to change their life styles. However, recent analyses point to the fact that changing in behavior is very difficult. Providing people with personalized information on risk is not new. Situations such as obesity, smoking, inherited cancer are examples where intervention to induce change so far have largely failed. It is questionable if responses will be different if the information is based on DNA tests [27]. Pharmacogenomics and genomics in general has ethical, legal, financial and behavioral aspects that have not been analyzed. In spite of this, the promise of improved health care through personalized drug treatment remains a realistic scenario in many fields of medicine.

1.4 References

1 Lazarou J, Pomeranz BH, Corey PN. Incidence of adverse drug reactions in hospitalized patients: a meta-analysis of prospective studies. JAMA **1998**; 279:1200–1205.

2 Silber BM. Pharmacogenomics, biomarkers, and the promise of personalized medicine. In: Pharmacogenomics (Kalow W, Meyer UA, Tyndale RF, Eds.). New York, Basel: Marcel Dekker; **2001**; 11–31.

3 Motulsky A. Drug reactions, enzymes and biochemical genetics. JAMA **1957**; 165:835–837.

4 Kalow W. Pharmacogenetics: heredity and the response to drugs. W. B. Saunders, Philadelphia, PA; **1962**.

5 Meyer UA, Zanger UM. Molecular mechanisms of genetic polymorphisms of drug metabolism. Annu Rev Pharmacol Toxicol **1997**; 37:269–296.

6 Nebert DW. Polymorphisms in drug-metabolizing enzymes: what is their clinical relevance and why do they exist? Am J Hum Genet **1997**; 60:265–271.

7 Evans WE, Relling MV. Pharmacogenomics: Translating functional genomics into rational therapeutics. Science **1999**; 286:487–491.

8 Meyer UA. Pharmacogenetics and adverse drug reactions. Lancet **2000**; 356:1667–1671.

9 Meyer UA. Genotype or phenotype: the definition of a pharmacogenetic polymorphism. Pharmacogenetics **1991**; 1:66–67.

10 Gonzalez FJ, Skoda RC, Kimura S, Umeno M, Zanger UM, Nebert DW et al. Characterization of the common genetic defect in humans deficient in debrisoquine metabolism. Nature **1988**; 331:442–446.

11 Kirchheiner J, Brosen K, Dahl ML, Gram LF, Kasper S, Roots I. et al. CYP2D6 and CYP2C19 genotype-based dose recommendations for antidepressants: A first step towards subpopulation-specific dosages. Acta Psychiatr Scand **2001**; 104:173–192.

12 Ingelman-Sundberg M, Oscarson M, McLellan RA. Polymorphic human cytochrome P450 enzymes: an opportunity for individualized drug treatment. Trends Pharmacol Sci **1999**; 20:342–349.

13 The International SNP Map Working Group 2001. A map of human genome sequence variation containing 1.42 million single nucleotide polymorphisms. Nature **2001**; 409:928–933.

14 Kruglyak L, Nickerson DA. Variation is the spice of life. Nature Genet **2001**; 27:234–236.

15 Roses AD. Pharmacogenetics and the practice of medicine. Nature **2000**; 405:857–865.

16 Pfost DR, Boyce-Jacino MT, Grant DM. A SNPshot: pharmacogenetics and the future of drug therapy. Trends Biotechnol **2000**; 18:334–338.

17 Waxman DJ. P450 gene induction by structurally diverse xenochemicals: central role of nuclear receptors CAR, PXR, and PPAR[1]. Arch Biochem Biophys **1999**; 369:11–23.

18 Früh FW, Zanger UM, Meyer UA. Extent and character of phenobarbital-mediated changes in gene expression in liver. Mol Pharmacol **1997**; 51:363–369.

19 Bulera SJ, Eddy SM, Ferguson E, Jatkoe TA, Reindel JF, Bleavins MR et al. RNA expression in the early characterization of hepatotoxicants in Wistar rats by high-density DNA microarrays. Hepatology **2001**; 33:1239–1258.

20 Gerhold D, Lu M, Xu J, Austin C, Caskey CT, Rushmore T. Monitoring expression of genes involved in drug metabolism and toxicology using DNA microarrays. Physiol Genomics **2001**; 5:161–170.

21 Bailey DS, Bondar A, Furness LM. Pharmacogenomics – it's not just pharmacogenetics. Curr Opin Biotechnol **1998**; 9:595–601.

22 Ledley FD. Can pharmacogenomics make a difference in drug development? Nature Biotechnol **1999**; 17:731.

23 Kalow W, Meyer UA, Tyndale RF. Pharmacogenomics. New York, Basel: Marcel Dekker; **2001**.

24 Keavney B, McKenzie C, Parish S, Palmer A, Clark S, Youngman L et al. Large-scale test of hypothesised associations between the angiotensin-converting-enzyme insertion/deletion polymorphism and myocardial infarction in about 5000 cases and 6000 controls. International Studies of Infarct Survival (ISIS) Collaborators. Lancet **2000**; 355:434–442.

25 Bertilsson L, Dahl M-L, Tybring G. Pharmacogenetics of antidepressants: clinical aspects. Acta Psychiatr Scand **1997**; 96:14–21.

26 Fagerlund TH, Braaten O. No pain relief from codeine...? An introduction to pharmacogenomics. Acta Anaesthesiol Scand **2001**; 45:140–149.

27 Marteau TM, Lerman C. Genetic risk and behavioural change. BMJ **2001**; 322:1056–1059.

2
The Human Genome

Samuel Broder, G. Subramanian and J. Craig Venter

Abstract

The 2.9 billion letter nucleotide base pair sequence of the human genome is now available as a fundamental resource for scientific and medical discovery on many fronts [1, 2]. Pharmacogenomics (the science of individualized prevention and therapy) is no exception.

Some of the findings from the completion of the human genome were expected, confirming knowledge anticipated by many decades of research in both human and comparative genetics. Other findings (e.g., relatively low gene number, large segmental DNA duplications) were unexpected and even startling in their scientific and philosophical implications [1]. In either case, the availability of the human genome sequence is likely to have profound implications, in clinical pharmacology and pharmacogenomics.

2.1
Introduction

Our genomic sequence provides a unique record of who we are and how we evolved as a species, including the fundamental unity of all human beings [1, 3]. The knowledge fostered by understanding the genome might resolve which human characteristics are innate or acquired as well as the interplay between heredity and environment in defining susceptibility to illness. Such an understanding will make it possible to study how our genomic DNA varies among cohorts of patients, and especially the role of such variation in the causation of important illnesses and responses to pharmaceuticals [4–6], which in a sense is the focus of this entire book. We may also begin to ask new and fundamental questions regarding complex aspects of the human condition such as language, thought, self-awareness, and higher-order consciousness. The study of the genome and the associated protein content (proteomics) of free-living organisms will eventually make it possible to localize and annotate every human gene, as well as the regulatory elements that control the timing, organ-site specificity, extent of gene expression, protein levels, and the post-translational modifications that define health or illness. For any given physiological process, we will have a new paradigm for addressing its evolution, development, function, and mechanism in causing disease and in affecting the onset and outcome of disease.

We have also completed the mouse genome, and currently rat genome sequencing is underway. A number of novel genes have been discovered. These achievements, and the supporting computational biology developed simultaneously, will drive the discovery of new diagnostics and pharmaceuticals in ways unimaginable even a few short years ago. For the first time, we can utilize the reference DNA sequence for the entire human genome, and the entire set of protein coding genes, numbering roughly 30,000, a number smaller than expected. We will have an ever-growing body of genomic information from various model organisms essential to modern pharmaceutical discovery and development, and eventually we will have the tools to understand how human complexity, whatever one means by that term, is reconciled with relatively small gene numbers.

Target discovery will be accelerated through interactive programs of protein-based analysis at scale, proteomic analysis of cell compartments in tissues and standardized cell lines, evaluation of post-translational modification and proteolytic processing profiles, true exon-based RNA analysis, DNA variation analyses, RNA-editing profiles, high-throughput functional assays, and predictive/molecular toxicology (toxicogenomics) including protein surrogate markers of adverse reactions and efficacy. Sophisticated tools of computational biology now make it possible to examine gene classes and gene variations (polymorphisms) in broad terms, including the regulatory elements that govern the rate and tissue specificity of gene expression. Comparative genomics will allow dramatically more efficient prediction of gene structure and function, and perhaps even more important, will inform better use of animal models to define and validate targets for drug development and someday even help predict the outcome of clinical trials. Understanding the full range of gene duplications may make it possible to anticipate unintended or "non-specific" actions of what appear to be "specific" therapeutic interventions. All of these advances will enhance and expand the science of pharmacogenomics; indeed, perhaps transform is not too strong a term.

2.2
Expressed Sequence Tags (ESTs) and Computational Biology: The Foundation of Modern Genomic Science

The journey to the world of modern gene discovery and genomes was not linear, nor was it easy in any sense. Expressed Sequence Tags (ESTs) were a crucial starting point [7]. Since recombinant DNA techniques became available in the 1970s, scientists had developed an ability to use cloned DNA, representing a gene of interest, in a wide variety of molecular studies in biology clinical research. In the late 1980s, as the Human Genome Project was under discussion, a consensus emerged to seek a complete genome sequence and catalog of genes. It is astonishing how quickly access to the complete genome sequence of an organism has become an essential step for any new comprehensive research project. However, only a few years ago, this goal seemed very far away for all but a handful of viruses with very small genomes. Most genome sequence projects, prior to 1990, were unavoidably slow and tedious, and targets for achieving even intermediary goals were measured literally in decades. The EST technology was the first to unleash the full power of an automated random cDNA library sequencing strategy for rapid gene discovery [7, 8]. Other efforts lacked both scale and speed. By sequencing 300–500 base pairs each of a very large number of cDNAs from a variety of tissues, ESTs would help identify new genes. ESTs could also be used to help map the chromosomal location of genes, recover genomic copies, and retrieve complete cDNA clones for further analysis. Perhaps most important of all, ESTs contained enough information to identify an enormous number of genes by similarity searching of electronic databases. When the results were published, the scientific community instantly had the largest collection of human genes in the history of genomic research up to that point in time.

By the mid 1990s, increasingly large numbers of ESTs necessitated the development of computational methods to combine overlapping sequences in a way similar to contig (continuous stretches of DNA) assembly, but with orders of magnitudes more data. EST assembly served both to reduce redundancy (multiple copies of the same EST sequence) and to capitalize on it (to create consensus sequences representing up to the full length of the cDNA). The bioinformatic tools developed in consequence of those efforts, in turn, made it possible to explore the entire genomic sequence of a free-living organism.

The first fully-sequenced genome of a free-living organism belonged to *Haemophilus influenzae* [9], completed in 1995 by The Institute for Genomic Research. This is an exceedingly important pathogen in its own right, and an elegant model for much of microbiology. The plan was to randomly fragment the bacteria's genomic DNA into small pieces, repeatedly sequence the fragmented DNA until on average every nucleotide had been sequenced an appropriate number of times according to a Poisson distribution, and then apply very powerful computational assembly tools (combined with a directed effort to close the remaining gaps) to provide a final fully-assembled complete genome. Along the way, it became necessary to master the advanced automation, robotics and other features of industrial scale DNA sequencing. Since the 1995 publication of *H. influenzae* [9], many more genome sequences of free-living organisms have been determined [10]. The most effective and efficient approach is called whole-genome "shotgun" sequencing, and it formed the basis for our publication of the sequence of first the fruit fly [11] and then the human genome [1]. We have also recently used this approach to complete the mouse genome. We are in the process of sequencing the genomes of the rat and a variety of other model organisms, using these approaches. It is clear that the availability of entire genomic sequence information on any one species allows a global perspective and framework for future research difficult to achieve prior to the advent of whole genomic information. The capacity to study many genomes *simultaneously* adds a capacity to achieve unified knowledge heretofore impossible.

2.3
Microbial Genomics

The science of microbial genomics requires special attention, in part, because of its medical and economic importance. Even more so, microbial genomics provide important lessons for all of computational biology and comparative genomics, and therefore, microbial genomes are prototypes for the future of pharmacogenomics.

The clinical practice of infectious diseases has undergone a tumultuous cycle of transitions over the past century. The last century began with microbial infections accounting for much of human morbidity and mortality statistics. This dramatically shifted with the advent of modern concepts of sanitation, public health, and antibiotic interventions, only to shift yet again in the direction of a worldwide resurgence of virulent pathogens as microbes adapted to antibiotic pressure [12, 13]. These circumstances, terrible enough in their own right, have worsened beyond imagining with the AIDS pandemic. Thus, with infections and microbial drug resistance on the rise, we need a better understanding of the molecular determinants of microbial virulence and host susceptibility to these infectious pathogens. We urgently need anti-microbial agents that do not exhibit unpredictable toxicity after marketing, necessitating "black box" warnings on the product label or marketing withdrawal.

The biosphere includes ecological niches where microbes peacefully co-exist with their eukaryotic host; in the human this includes the concept of the "microbiome" [14], which is defined as the totality of microbial organisms that co-habit with human beings. On the other hand, microbial colonization of human mucosal surfaces or prosthetic devices, often results in the development of biofilms, with significant deleterious effects on human health [15]. These are some of the challenges in infectious diseases that reflect the need to maximally utilize genomic sequence information and related sciences to better control microbial disease in human populations, and to develop anti-microbial agents with a better therapeutic index.

It may be useful to highlight certain computational and experimental approaches used in genomic analysis, with examples of evolutionary adaptations that likely influence microbial virulence and the host response to microbial invasion. These concepts have applicability to the entire field of comparative genomics. The experimental strategies discussed include assays that provide a global picture of microbial biology, which we expect will enable the transition of genomic sequence-based technologies into clinical diagnostics and therapeutics. Recently, *B. anthracis* was adapted for use in a clear example of lethal bioterrorism in the USA. The "weaponization" of microbes is an ominous development adding even greater urgency to the study of microbial genomics and the genomics of host resistance. The computational analysis of whole genomes has moved from a purely scientific or medical undertaking into a crucial component of national defense.

2.3.1
Computational Analysis of Whole Genomes

The cornerstone for all genomic analysis is the availability of high-quality genomic sequence, computational programs for genome assembly and gene predictions that enable the functional analysis of the predicted protein set (proteome). *In silico* studies on the predicted proteome have been improved by the recent development of sensitive sequence/profile/analysis methods that allow objective detection and statistical evaluation of subtle sequence similarities. Methods include PSI-BLAST (position-specific iterated basic local alignment search tool) [16]; this uses a position-specific weight matrix obtained from the primary gapped BLAST search to iteratively search the database. Programs such as SMART [17] and Pfam [18] use a library of protein family and domain alignments represented as hidden Markov models to search the genomic sequence. The availability of over 50 bacterial, archeal and eukaryotic genomes (many sequenced by The Institute for Genomic Research) provides the computational biologist with a rich substrate of sequence information to infer evolutionary and functional relationships based on comparative patterns of sequence similarity across species. Orthologs are proteins from different species, encoded by genes that evolved by vertical descent (speciation), and typically retain the same function across species [19]. Paralogs are proteins from within a given species, encoded by genes that evolved by duplication, and may have evolved new functions. One approach for organizing such data includes the "COGs" (Cluster of Orthologous Groups of proteins) [19, 20], that is extracted from publicly available microbial sequence information, and comprises proteins that are either orthologs or paralogs. Computational programs available for prokaryotic genome analysis include the COGnitor (compares the predicted proteome against the COGs database) [20]. Alternatively, the LeK clustering program developed for eukaryotic genome analysis was recently used in the analysis of the *Drosophila melanogaster* (fly) [11, 21] and the human genomes [1].

2.3.2
Comparative Genome Analysis

Table 2.1 provides representative examples of pathogenic microbes and select eukarya to provide an appreciation for the genome size, the gene content and the fraction of genes that likely share common ancestry among these pathogens on the basis of computational sequence analysis. The wide range in genome size and gene number is reflective of the diversity in the ecological niches of these pathogens. Free-living organisms that need to survive diverse environmental conditions are invariably equipped with larger genomes with comprehensive biosynthetic (metabolic) pathways, transporters and signal transduction apparatus to enable efficient uptake and sensing of nutrients. Obligate parasites on the other hand tend to have smaller genome sizes with genomic adaptations that enable an existence, by definition, entirely dependent on the host. It is therefore, remarkable that genome size and a large

Tab. 2.1 Genomic features of representative pathogenic bacterial and eukaryotic genomes

Organism	***Genome size [Mbp]/ gene number [% in COGs]*** [1)]	***Clinical disease or syndrome***
Haemophilus influenzae	1.83/1695 (88) – This was the first genome of a free-living organism to be sequenced	Otitis media, epiglottitis, and meningitis, especially in children. Meningitis can be lethal, and survivors may have neurologic sequelae. Contains abundant simple sequence contingency loci
Escherichia coli K12	4.6/4288 (77)	Non-pathogenic laboratory strain
Escherichia coli O157	5.5/5416	Hemorrhagic colitis, hemolytic uremic syndrome (HUS); a not infrequent cause of industrial food poisoning
Helicobacter pylori	1.66/1578 (68)	Peptic ulcer, gastric cancer, gastric lymphoma (MALT)
Campylobacter jejuni	1.64/1634 (78)	Diarrheal disease
Pseudomonas aeuruginosa	6.30/5567 (75)	Severe opportunistic infections; can be lethal in immunocompromised patients and patients with extensive full thickness burns
Vibrio cholerae	4.0/3828 (71)	Epidemic and endemic fatal diarrheal disease
Neisseria meningitides	2.27/2081 (70)	Communicable meningitis; sometimes fatal. Contains abundant simple sequence contingency loci
Mycobacterium tuberculosis	4.4/3924 (63)	Tuberculosis – the consumption of an earlier era. This disease is staging a comeback
Mycobacterium leprae	3.3/1604	Leprosy
Chlamydia pneumonia	1.23/1053 (62)	Pneumonia
Chlamydia trachomatis	1.05/895 (71)	Trachoma, blindness
Mycoplasma pneumoniae	0.81/680 (62)	Community acquired pneumonia
Mycoplasma genitalium	0.58/471 (79)	Urinary tract infection
Borrelia burgdorferi	1.44/1637 (43) [2)]	Lyme disease – transmitted to humans by infected deer ticks. This disease is a growing problem
Treponema pallidum	1.14/1036 (68)	Syphilis
Rickettsia prowazekii	1.1/836 (81)	Epidemic typhus
Bacillus anthracis	Not published yet	Anthrax; first use of proven bioterrorism in the United States

Tab. 2.1 (continued)

Plasmodium falciparum	30/~6000 NR	Malaria – Roughly one million people die from this disease annually, many of them children in the African countryside
Drosophila melanogaster	120/14336 NR	Not applicable – the common fruit fly has been an important model in medicine and population genetics for much of the last century. It is suitable for many types of studies, including electrophysiology
Homo sapiens	2900/~ 27000 NR	Not applicable

1) COGs refers to the percentage of genes in a genome that show shared inheritance with other microbes

2) refers to the significant number of proteins encoded on plasmids. NR-COG analysis was not relevant

gene complement by themselves are not absolute requirements for evolutionary success or survival in the human host.

While up to 70% of all genes in some bacterial genomes have a shared ancestry, the molecular functions of about 40% of all genes in a typical microbial genome are yet to be elucidated [22]. In a sense, this is true for virtually all genomes (including the human) as they are first sequenced and annotated. Although significant advances have been made in our capacity to define core biological processes and parse meaning across species, defining the precise determinants of microbial pathogenicity will require experimental validation. This includes our ability to predict the physiological consequences of gene family expansions, and also the functions of the unique gene complements in each of these microbial species. As a detailed summary of the genomic features in these pathogenic microbes is beyond the scope of this chapter, we have provided a glimpse into only a few of the evolutionary forces at play in the microbes. In Table 2.2, we provide a few examples of expansions of gene families and gene loss or reductive evolution. In the case of the mycobacteria, the *Mycobacterium tuberculosis* genome provides some insights into the relationships between natural drug resistance of this microbe to antibiotics and the genomic expansions of enzymes involved in lipid metabolism and cell wall biogenesis, expected to affect permeability and transport of new drugs [23]. Equally interesting is the remarkable use of enzymes of the glyoxylate pathway that enable this organism to survive in lung tissue in the human [24]. Given the global resurgence of multi-drug resistant strains of the tubercle bacillus, understanding the functions of these genes is of paramount clinical importance in the diagnosis and treatment of tuberculosis. *Mycobacterium leprae*, on the other hand is an intracellular obligate parasite, which despite massive gene decay and severely stunted metabolic capabilities [25], still retains several en-

Tab. 2.2 Examples of functional attributes of representative genomes based on computational sequence analysis

Organism	***Genomic features and functional significance***
Mycobacteria (*M. tuberculosis/leprae*)	The genome of *M. tuberculosis* (the *tubercle bacillus*) shows a remarkable expansion of over 250 enzymes involved in fatty acid metabolism, a feature that accounts for the highly hydrophobic cell envelope. *M. leprae* is an extreme case of reductive evolution in a related obligate intracellular parasite. 50% of the genome is comprised of non-coding genes (pseudogenes and gene remnants), an estimated loss of more than 2,000 genes since the divergence of the mycobacteria. The *tubercle bacillus* is evolving multi-drug resistance in several portions of the world
Escherichia coli (K-12, O157:H7)	Extensive lateral gene transfer in the pathogenic strain O157, with 1,387 new genes that encode candidate virulence factors-type III secretion system and secreted proteins, iron uptake and utilization clusters, several toxins and non-fimbrial adhesins. Antibiotics that partially damage strain 0157 bacteria might be capable of releasing SHIGA toxins in the host. This is a potential explanation for the paradoxical clinical deterioration observed in some antibiotic-treated children, infected with this pathogen
Pseudomonas aeruginosa	One of the largest pathogenic bacterial genomes, with an expansion of nutrient transport proteins(~200) and two-component regulatory system proteins(~160); features that likely reflect adaptation to thrive in diverse environments and resist antibiotics
Rickettsia prowazekii	Reductive evolution in this obligate intracellular pathogenic bacterium is characterized by extensive gene loss and a remarkable similarity to the mitochondrial genome, with loss of genes involved in anaerobic glycolysis and preservation of bioenergetics genes
Spirochetes (*B. burgdorferi, T. pallidum*)	Related pathogens with reduced genome sizes, with drastic differences in their signal transduction apparatus, and unique sets of surface molecules
Plasmodium falciparum	Highly A+T rich genome with an unusually high proportion of low complexity genes. This fact lead some to (erroneously) conclude the genome would be exceedingly difficult to sequence. Large variant antigen gene families that comprise > 10% of the genome. Both these factors likely play a major role in the host immune response in malaria
Homo sapiens (compared to *Drosophila melanogaster*)	Large-scale gene duplications with substantial expansion of genes involved in acquired immune response (B cells, T cells, major histocompatibility complex genes, cytokines, chemokines and their receptors), plasma proteases (complement and hemostatic proteins), proteins associated with apoptotic regulation; and proteins related to neuronal network formation and electrical coupling

Tab. 2.3 Molecular determinants of microbial pathogenesis and host response

A. Pathogenic determinants in microbes and viruses

- Lateral gene transfer of DNA between different bacterial species. Examples include
  - (a) Pathogenicity islands (cassettes of genes ranging from 5–100 kbp) containing adhesins, toxins, invasins, protein secretion systems, iron uptake systems.
  - (b) Eukaryotic host derived extracellular domains that likely play a role in pathogenesis, e.g., von Willebrand A domain containing proteins in the spirochetes, perforin domain containing protein in Chlamydia.
  - (c) Plasmid mediated transfer of antibiotic resistance genes, and prophage mediated toxin production.
- Phase and antigenic variation in bacteria. These may be mediated by slipped-strand mispairing of repeat sequences that occur in these phase variable genes or by gene duplication events.
- Genome loss and specific host adaptive features in obligate intracellular pathogens
- DNA viruses modulate the host immune response, e.g., most likely the etiologic agent of Kaposi's sarcoma, KSHV (human herpes virus –8), has captured complement-binding proteins, three cytokines (two macrophage inflammatory proteins and interleukin 6), bcl-2, interferon regulatory factors, interleukin 8 receptor. Certain retroviruses (i.e., viruses that contain an RNA-dependent DNA polymerase, also called reverse transcriptase) can capture host oncogenes, which may then bring about malignant transformation in target cells, at least in animal models.

B. Selected determinants of host response or susceptibility to microbial infection

- TH_1 versus TH_2 cytokines
- TNF-alpha promoter polymorphisms and cerebral malaria
- Chemokine receptor (CCR5) polymorphisms and reduced susceptibility to human immunodeficiency virus (HIV) infection
- Toll receptor polymorphisms and bacterial sepsis
- Beta-defensins and urinary tract infections
- Activated protein C in the response of sepsis
- Complement deficiency and susceptibility to disseminated Neisserial infections
- Complement factor H and platelet-activating factor acetylhydrolase polymorphisms and hemolytic uremic syndrome due to *E. coli* 0157
- NRAMP-1 gene polymorphisms and tuburculosis susceptibility in The Gambia
- NADPH oxidase deficiency and catalase positive organisms in chronic granulomatous disease

zymes and species-specific genes not represented in the larger *M. tuberculosis* genomes. Another example is the recently published genome of the enterohemorrhagic *Escherichia coli* strain 0157:H7 [26], which poses formidable clinical challenges in its clinical presentation, diagnosis and paucity of therapeutic options. When compared to the K-12 strain [27], the expansions in several pathogenic determinants (Tables 2.2 and 2.3] provide key insights into the adaptations of this pathogen, highlighting the importance of complete genomic sequence in furthering our understanding of pathogenic microbes.

Detailed analyses of expansions of paralogous proteins are important not only from a diagnostic perspective (identifying members of gene families unique to specific organisms), but also from a therapeutic perspective (developing vaccines and drugs against microbial infections). The challenges in understanding obligate human parasites are equally important. Despite smaller genomes and severely restricted biosynthetic capacities, their ability to cause very distinct clinical disease reflects adaptive tissue tropism, with the enumeration of specific enzymes and surface molecules enabling infection in the human host. Examples include differences in the two *chlamydial* species that cause trachoma and pneumonia, respectively [28], and the spirochetes *Borrelia* and *Treponema*, which despite shared ancestry, have retained discrete genomic features likely to account for their distinct disease processes [29–31].

Even among the eukarya, computational analysis provides interesting insights into genomic differences, which may translate into novel therapeutic options. A striking example is the metal-dependent RNA triphosphatase protein family, members of which play a central role in mRNA cap formation and eukaryotic gene expression. The active site structure and catalytic mechanism of this protein family in *Plasmodium falciparum* and fungi are different from the RNA triphosphatase domain of the metazoan (including human) capping enzymes, and metazoans encode no identifiable homologs of the fungal or *Plasmodium* RNA triphosphatases [32]. Moreover, the structural similarity between the plasmodial and the fungal RNA triphosphatases raises the theoretical possibility of achieving antifungal and antimalarial activity with a single class of mechanism-based inhibitors.

Several recently developed computational approaches in comparative genomics go beyond sequence comparison. By analyzing phylogenetic profiles of protein families, domain fusions, gene adjacency in genomes, and expression patterns, these methods predict functional interactions between proteins that may help deduce specific functions for proteins sometimes leading to remarkable therapeutic applications [33, 34]. These developments mark a new era in which the benefits of comparative analysis of complete genomes will complement experimental approaches aimed at improving our understanding of microbial physiology and host-pathogen interactions. Perhaps the operative term here should be cautious optimism. It is our expectation that this area of study will have ramifications for many areas of pharmacology and pharmacogenomics, as well. Genomics provides exciting new opportunities, but it is important to take stock of how much remains to be done.

2.4 Genomic Differences that Affect the Outcome of Host-Pathogen Interactions: A Template for the Future of Whole-Genome-Based Pharmacologic Science

Microbial virulence is often the outcome of the complex interactions that take place as the pathogen establishes itself in the human host. The molecular determinants of pathogenicity include factors that cause damage to the host cell and those that help the microbe establish productive infection for survival [35]. The human host immune response counters the presence of these microbes with its acquired or innate immune response arsenal with outcomes that range from acute to chronic or latent infections. A clear definition of the host and microbial factors that result in microbial colonization or in disease pathology in the human host will almost surely play a major role in shaping new diagnostics and therapeutics in the field of infectious diseases [36].

Table 2.3 provides a summary of the major molecular determinants that contribute to microbial virulence as well as select examples of features in the host immune system that determine the outcome of microbial infection. Analysis of genomic sequence provides several clues towards predicting pathogenic potential in a microbe. The phenomenon of lateral transfer of genetic material in microbes is one of great interest and has been shown to contribute to genomic diversity in free-living bacteria and to a lesser extent in obligate bacteria [22, 36, 37]. Of interest is the identification of "pathogenicity islands" in bacteria [35]. Of comparable interest is the acquisition of eukaryotic host extracellular or signaling domains by pathogenic organisms, likely establishing productive infection or subverting the host immune response [35]. These observations have substantial implications for the development of new anti-microbial agents, and the selection of the best anti-microbial agents for any given host. This is a template for conceptualizing individualized medicine of the future.

Pathogenicity islands comprise large genomic regions ranging (in size) from 10–200 kilobase pairs (kb). These are present in the genomes of pathogenic strains, but absent from the genomes of non-pathogenic members of the same or related species. The finding that the G+C content of pathogenicity islands often differs from that of the rest of the genome, the presence of direct repeats at their ends and the presence of integrase determinants and other mobility loci, all argue for the generation of pathogenicity islands by horizontal gene transfer. Pathogenicity islands appear in pathogens such as *Vibrio, Helicobacter, Yersinia, E. coli* and others, often determining bacterial products such as adhesins, toxins, invasins, protein secretion systems and iron uptake systems, all of which constitute important determinants of microbial virulence. In the case of *H. pylori* the so-called *cag* island is a 40-kb genetic element that likely entered the genome after the *H. pylori* became a species. *H. pylori* organisms that are positive for *cag* induce higher levels of certain pro-inflammatory cytokines. Other classic examples of lateral transfer include the transfer of antibiotic resistance genes that are either chromosomally or plasmid-encoded [37].

Microbial genes whose products interact with the host immune response include unique classes of adhesive surface molecules that are often immunogenic. Mechanisms of phase and antigenic variation that are commonly seen in bacteria, or the presence of large variant antigen gene families as in plasmodia, represent some strategies for subverting the immune response by varying the surface molecule seen by the host defense system [38, 39]. Antigenic mimicry (likely to represent a form of convergent evolution), where the bacterial epitope is recognized as a self-antigen, is yet another ploy to escape immune surveillance. Of course, such a strategy can lead to autoimmune destruction of host tissue [40]. DNA viruses, on the other hand, often "capture" host immune molecules such as cytokines, chemokines or apoptotic regulators that help them establish their intracellular niche in the human [41, 42].

The human genome sequence has provided us with a clear understanding of some of the major features that distinguish the human from invertebrate genomes such as the fruit fly [1]. Newly sequenced genomes such as those of the mouse (a crucial model for biologic research) and *Anopheles gambiae* (the vector for malaria) promise to teach us even more. While many components of the innate immune system are shared (including the Toll receptors [43] and the small antimicrobial peptides called defensins [44], (which are essential for maintaining mucosal integrity), there are several prominent differences in the acquired immunity arm of the human. Table 2.3 provides representative examples of human genes and DNA polymorphisms in some of the genes that modulate immune response [43, 45]. (These polymorphisms are likely to be important in defining the best therapy to individual patients, and we will take up this issue in more detail later in the chapter.) Through the use of a candidate gene approach, population association studies, and linkage analysis, several genes such *Nramp1* (encoding a protein located in the lysosomal compartment of resting macrophages) [46], cytokines (such as tumor necrosis factor, interleukin-12 and interferon gamma receptor) [47, 48] and chemokines [49] and Toll receptors [43] have each been associated with susceptibility to a wide range of bacterial and viral infections, including human immunodeficiency virus (HIV) [49]. There are other genes, such as the plasma proteases including complement proteins (which link innate and acquired immune response) and protein C, important in modulating bacterial sepsis [34]. Hemolytic uremic syndrome (HUS) is yet another situation in which polymorphisms in genes affect clinical outcome. Thus, complement protein H [50] or platelet-activating factor acetylhydrolase [51] predispose to severe disease (HUS) in patients infected with *E. coli* 0157:H7.

Given the role of *H. pylori* infection in gastric cancer and duodenal ulcer disease, one must note that certain interleukin-1 gene cluster polymorphisms (suspected of increasing production of interleukin-1-β) are associated with a predisposition to hypochlorhydria and gastric cancer [52]. One of these host DNA polymorphisms involves a TATA box.

These are examples of a much larger principle. We are learning that many infectious diseases, autoimmune disorders, and cancers are heavily affected by alleles in genes encoding cytokines. Thus, future research must focus on DNA variation in genes encoding cytokines produced by the "pro-inflammatory" Th1 cells (producers of interleukin-2, interferon-gamma and lymphotoxin-beta) critical for cell-mediated immunity; or Th2 cells (producers of interleukin-4, interleukin-5, and interleukin-10) critical for humoral immunity [53]. DNA variation, often in the form of single nucleotide polymorphisms (to which we will return later), within regulatory regions, may explain many clinical syndromes or adverse reactions to therapies. Progress in pharmacogenomics will depend in part on harnessing how DNA variation in cytokine (and chemokine) genes contributes to the causation or outcome of common diseases. In this context, whole genome analyses at all phylogenetic levels coupled with the new technologies of proteomics and computational biology allow us to seek a unity or harmonization of knowledge never before possible. Thus, polymorphisms in host response genes have profound implications beyond microbial science *per se*, and represent a striking example of how studying the entire genomes of pathogens and their hosts can offer opportunities for significant advances in diverse areas of clinical pharmacology.

It may be worth focusing on the issue of "programmed" mutation or DNA variation, a topic which has considerable implications in many areas of biology. Genomes can respond to "unanticipated" challenges that in a strict sense are not encoded prior to the challenge. Indeed, one can argue that DNA polymorphisms in the broadest sense represent a method by which natural selection can respond to changes in circumstances, which might otherwise threaten a species with extinction. Some members of a species have DNA polymorphisms that provide a survival advantage in the new circumstances. This is true for prokaryotes and eukaryotes. Indeed, in one sense, cancer cells have genomes that differ from the host who gave raise to them. The "genomic" adaptations of a cancer cell to pharmaceutical or biologic interventions represent a response to an "unanticipated" challenge, at least from the cancer cell point of view. In a sense, both bacteria and cancer cells have confronted the issue of global mutation rates as in the natural selection of "fit" progeny. Thus, how various species deal with the problem of unanticipated challenges and adaptive evolution has broad implications for many disciplines.

The study of whole genomes in pathogenic bacteria has yielded considerable knowledge about how bacterial populations deal with scavenging essential nutrients or alter surface-exposed molecules in response to selective pressures seeking to contain or destroy the bacteria. Such organisms have evolved mechanisms of hypermutability, which can result from several mechanisms [54]. In this context, there is considerable interest in what are called simple sequence contingency loci [55]. These loci contain tandem DNA repeats (microsatellites), whose size generally varies from 1–8 base pairs. Whole genome analysis of *H. influenzae* and *N. meningitides* reveals an abundance of these simple loci, and this is a reminder of the unique power of whole genome sequencing approaches in generating knowledge, not otherwise available.

Such loci contain tandem repeats located within an exon or a promoter. Alterations in the (hypermutable) repeats may thus permit changes in the translational reading frame or changes in the rate of transcription. Random changes in these loci exert disproportionate effects on the selective advantage of a species in various environmental conditions. Bayliss et al. point out that there is both a hypermutability and reversibility to this process because expansion or contraction of the repeat tracts by one or more repeat units can readily occur [55]. Multiple "on" and "off" settings are possible. The gain or loss of perhaps one repeat unit could convert an "on" tetranucleotide repeat tract in a reading frame to an "off"; similar considerations apply to promoter regions. Because these changes (or mutations) can occur in a very small number of generations, these loci appear highly polymorphic, possibly even within progeny of a single clone. One area for future research might be interventions that block or retard the mutability of simple sequence contingency loci.

One of the biggest challenges in pharmacogenomics will be our ability to identify patients who are likely to show increased susceptibility to microbial infections as well as to identify determinants of pathogenic potential in microbes that are obtained from clinical specimens. The lessons we learn here will have broad ramifications in many medical disciplines. The availability of several million single nucleotide polymorphisms (SNPs) from the recent human genome projects (see below) will undoubtedly facilitate future efforts to help identify naturally occurring human variants among genes that modulate infectious diseases in humans [1].

It is worth concluding our discussion of microbial genomics with the following reflections. Only a fraction of organisms that might reside within various human compartments can be readily cultured by available techniques. A great many of common diseases might conceivably be ascribed to a microbial pathogen but for the inability of available axenic cultures and related technology to detect the causative pathogen. The etiologic role for *H. pylori* in gastric cancer was certainly a great surprise in the era of its discovery. There is no truly complete reference set of microbes for the purposes of basic research, clinical investigation, agriculture and the formation of strategies to counter threats of biologic warfare. In the future, we now have the tools to explore the feasibility of using sequencing and assembly algorithms to detect, identify, and characterize microbes in mixed flora in human body spaces. It is thus possible that computational biology can eventually supplement axenic culture in studying microbes and defining the human microbiome.

2.5 More Lessons from the Human Genome

With the lessons of microbial genomics in mind, we might now explore what broad lessons we have learned from the human genome. In the remaining portion of the chapter, we address the following major issues: number of protein-coding genes in the human genome and certain classes of non-coding repeat elements in the genome; features of genome evolution, including large-scale duplications; an overview of the predicted protein set to highlight prominent differences between the human genome and other sequenced eukaryotic genomes; and DNA variation in the human genome. In addition, we show how this information lays the foundations for ongoing and future endeavors that will revolutionize biomedical research and our understanding of human health.

2.5.1 Protein-Coding Genes

One of the most startling findings in the human genome is the relatively low number of genes [1, 2]. A gene, in this context, is defined as a locus of cotranscribed exons, which ultimately result in the production of a peptide or protein. There are a number of computational tools used to identify and enumerate genes within the genome of any organism, and these have been applied by several researchers, including us. The text and subtext of biology prior to the availability of the full sequence for the human genome was that the number of genes in an organism would in some fashion reflect its "complexity". There were expectations that the human genome would contain approximately 100,000 genes or more [57].

So what do we have? The fruit fly (*Drosophila melanogaster*) has approximately 14,000 genes [11]. The roundworm (*Caenorhabditis elegans*) has roughly 19,000 genes [58]. The mustard plant (*Arabidopsis thaliana)* has roughly 26,000 genes [59]. Those who might be tempted to use the number of genes for the exaltation of human complexity might then pause to consider that by this measure, the human with approximately 30,000 genes [1, 56], is approximately a fly plus a worm or the equivalent of a plant. So too, the expectation that new ways to prevent, diagnose, and treat illnesses would be driven by gene number according to a simple formula needs to be revised.

Surveying the landscape of the human genome leads to several other observations.

Only about 1% of the genome is spanned by exons, while just under 25% is contained within introns, and about 75% of the genome is contained in intergenic DNA. Thus, genes often exist in non-random clusters or gene-rich "oases", separated by what appear to be large "deserts" of millions of letters that do not appear to encode genes. There is no simple explanation for why natural selection has taken this path in the evolution of the human genome, but we believe it is premature to conclude that such "deserts" lack biological or medical importance.

2.5.2 Repeat Elements

The human genome is filled with blocks or "elements" of repetitive letters of code whose function is still a mystery. It has been known for many years, and amply confirmed with the completion of the genome, that human DNA contains large and complex families of such repeat elements [1, 2, 56, 60, 61]. These include the long interspersed repetitive elements (LINE) and short interspersed repetitive elements (SINE), which include Alu sequences that arose with the evolution of primates, including humans [60]. Alu sequences represent a distinct class of retrotransposon-amplified repeat DNA. During primate evolution these DNA elements could be replicated and transposed to new sites in the genome [60, 61]. They comprise approximately 10% of the human genome. Their biological function and role in natural selection has remained an enigma.

Yet in surveying the landscape of the human genome, a striking and non-random distribution of Alu sequences is evident. They appear to preferentially co-locate within gene-rich regions of the genome. One inference is that the biological role of these Alu sequences, the effects of nucleotide variations within such elements [61], and their ability to mediate recombination events [62], will be important in understanding their regulatory effects [61, 63, 64, 65] on gene function and disease. Further investigations are required to add to the known examples where Alu sequence variations have been shown to affect biology and clinical conditions. Such elements had previously been characterized as "selfish" DNA, with no direct impact on medicine or natural selection. The availability of the human genome suggests that this view should be revised.

2.5.3
Genome Duplication

The human genome reveals an astonishing level of genomic duplication [1]. Though the biological impact of gene duplication events in generating gene superfamilies is well established, the first comprehensive view of the genome-wide landscape has especially revealed the widespread impact of two distinct mechanisms of duplication.

These two forms of duplication are very different: one form mediated at the DNA level (segmental duplication), and another mediated at the RNA level (retrotransposition). Both mechanisms produce paralogs – a term for genes that make their appearance in more than one copy in the genome (albeit with possible modifications).

Let us first consider segmental duplication. The extent of the segmental duplications is 10- to 100-fold greater than that observed in the fly and worm genomes. There are over 3,500 genes in over 1,000 genomic blocks ranging in size up to chromosomal lengths, that have shown a duplication, with linear preservation of order on another chromosome [1]. In many cases, there is a disease-causing gene with a paralog on the duplicated segment, whose linkage to a disease is not currently recognized [1]. It is possible that an understanding of segmental duplication will provide new insights into the pathogenesis of disease. To be sure, every duplication event will not lead to a paralog that results in the same pathophysiologic consequences. However, it might well be possible to observe a unity for disparate diseases through the optic of genomics. It may also be possible to understand and explain adverse reactions and side effects of drugs through their previously unknown collateral activities against paralogs.

The other remarkable finding is the dramatic extent of duplication of genes that have resulted from retroviral-based transposition of gene transcripts. The ancestors of humans encountered retroviruses capable of transcribing RNA to DNA (reverse transcription). Indeed, such viruses are not extinct, as diseases such as AIDS amply confirm. The human genome carries the results of many such encounters. Gene duplication by this process in effect creates paralogs that lack introns and often occur in multiple copies scattered randomly over the genome. The medical implications of this form of gene duplication are similar to those that apply to segmental duplication. In addition, the degree of identity between the source gene and the retrotransposed gene is often very high thus leading to the possibility of confounding DNA or protein-based diagnostic tests. It is important to note that changes in coding or non-coding regulatory regions in these paralogs, leading to different functions or expression patterns, may be one way of providing an increased functional repertoire in the human genome.

2.5.4 Analysis of the Proteome

Earlier in this chapter, we discussed the unexpectedly low number of protein-coding genes. Does an analysis of the full set of proteins (also called the proteome) help us resolve the problem that human beings do not appear to carry many more genes than a fruit fly, a roundworm, or a plant? Indeed, we do note that the average human gene makes more proteins, and more complex proteins, than its invertebrate counterparts. A number of such features are worth detailing. These include the evolution of new protein domains, accretion of domains as well as greater combinatorial diversity in the human. In addition, certain genes produce more than one type of peptide utilizing alternative start and splicing processes.

Extensive domain shuffling is observed in the human proteome, and this would serve to increase or alter combinatorial diversity to provide an exponential increase in protein-protein interactions. Moreover, certain special genes show patterns for generating combinatorial diversity at the protein level. For example, immunoglobulins and the T cell receptors show DNA shuffling or re-arrangement to increase the immune repertoire, while the cadherins show exon trans-splicing (a form of RNA shuffling) to generate increased extracellular interactions [66, 67]. All of these factors taken together contribute to complexity not captured by examining gene number alone.

Many proteins (and protein domains) found in the human evolved early in the animal radiation and are hence present in invertebrate genomes. However, several noteworthy vertebrate-specific domains exist, especially within proteins involved in developmental, homeostatic and nuclear regulation. These proteins have profound implications in understanding human development, malignant transformation, and stem cell biology. In addition, proteins related to acquired immunity, complement fixation, and hemostasis are either unique or show a dramatic expansion in the human genome compared to known invertebrate genomes. Thus, we find several instances where evolution has harnessed "old" domains to provide novel distinct domain architectures in the human when compared to the fly or worm; i.e., "new" proteins created using "old" domains. Examples include the serine proteases which occur with a widely diverse set of protein domains in the plasma proteases (coagulation, complement and fibrinolytic systems) and the recruitment of the immunoglobulin fold into molecules of the acquired immune system, e.g., antibodies, major histocompatibility complex (MHC) and cell adhesion receptors (CD). Also, in concordance with the greatly increased neuronal complexity in the human compared to the fly and worm, there is an increase in the number of members of protein families involved in neural development, structure and function [1, 68, 69]. These include neuronal growth regulators, as well as classes of voltage-gated ion channels that play a vital role in neuronal network formation and in electrical coupling. Understanding how these components interact to generate the neuronal infrastructure in humans will have an impact on therapeutic modalities to address neuronal injury, as well as to provide insights into new ways to diagnose and treat neuropsychiatric disorders. Proteins involved in apoptosis or programmed cell death, a central effector mechanism that regulates cellular physiology, are also dramatically expanded in humans [1, 70]. The central role for this process in neurodegenerative diseases [71], malignancy and inflammatory conditions [72] related to extrinsic (pathogens) and intrinsic mediators (cardiovascular disease, inflammatory bowel disease, etc.) constitute areas of intense current investigation. Therapeutic interventions that can modulate the apoptotic process will likely have major effects on some of the most devastating clinical illnesses that afflict mankind [73].

However, a focus on genomic DNA sequence alone will not be sufficient to resolve all the important problems of medicine and biology. The availability of the human genome will dramatically enhance the power of proteomics (the study of the proteome) [74, 75]. In the near future, once the sequence of any "unknown" peptide is determined in any human fluid or cell culture (say, for example, by a technology dependent on mass spectroscopy for separation and identification of proteins), there will now virtually always be a "hit" or "match" between proteins and their genes. The applicability of the approach to better understand disease processes will undoubtedly increase as additional genomes of "model" organisms (such as mouse, rat, dog, etc.) become available. Such approaches will also enhance the capacity for detecting novel microbes and their protein complements, either pathogens or commensals, both of which have profound implications for enhancing microbial diagnosis and developing improved antimicrobial therapeutics [36]. It will be possible to link peptides and their post-translational modifications to the pathophysiology of illnesses [75, 76]. Post-translational modifications refer to the fact that proteins may undergo various types of modifications (phosphorylation, glycosylation, acetylation, covalent or non-covalent bonding between peptides from different genes, etc.) after they are synthesized. Many of these modifications likely affect the activity and disposition of proteins in health and disease. One special form of post-translational modification involves protein cleavage, which is essential to the activity of certain proteins, of which insulin and other hormones are classic examples, as well as those involved in the apoptotic process. Ultimately, the number, complexity, and modifications of proteins encoded by human genes all contribute to the complexity of human biology, and underscore that not all answers lie at the level of genomic information *per se*. Advances in proteomics will thus likely enhance the next generation of pharmacogenomic diagnostics as well as guide therapeutics in ways that were previously impossible or exceedingly difficult to do [74, 76].

2.5.5 DNA Variation

The study of the genome supports the fundamental unity of human beings throughout the world. We all share at least 99.9% of the letters of code (nucleotide sequence) in our genome [1, 77]. And yet, it is remarkable that the extraordinary diversity of human beings at the genetic level is encoded by less than 0.1% variation in our DNA. In a sense, this is the variation that is the basis for pharmacogenomics. In any physician's practice, patients are predisposed to different conditions, respond to the environment in variable ways, metabolize pharmaceuticals differently [4, 5], vary in the dose-response relationships for common drugs, and have a range of susceptibilities to adverse side effects from therapeutic agents (even when there is no obvious difference in individual pharmacokinetics or biochemical pharmacology) [4–6,). These are some challenges for the future of pharmacogenomics. In the past, the field of *pharmacogenetics* focused on genetic variation in genes controlling phase I (oxidative) and phase II (conjugative) metabolism and eliminations of drugs. The new field of *pharmacogenomics* will draw upon whole genomes and DNA variation across a broad front.

The most common form of DNA variation in the human genome is the single nucleotide polymorphism (SNP) [78, 79]. We briefly addressed some of these issues in the discussion of microbial genomics. The nomenclature defining a mutation (a disease-causing change) versus a SNP is arbitrary and relative. By convention, when a substitution is present in >1% of a target population, it is called a variant or polymorphism. We now have a genome-wide survey of several million SNPs, with precise nucleotide localization, in an ethnogeographically divergent group of individuals. There is on average 1 SNP per 1,250 letters (nucleotide base pairs) of genomic DNA. These SNPs can occur within exons, with synonymous or non-synonymous (results in a change in amino acid) attributes, or can occur outside exons within intronic or intergenic regions of the genome. Less than 1% of all known SNPs encode a direct amino acid change of the ultimate protein product of a gene [1]. Therefore, there are only thousands (not millions) of genetic variations that directly contribute to the structural protein diversity of human beings [1]. However, this in no way diminishes the importance of DNA variation in regulatory regions in the generation of diversity, predisposition to illness, or drug side effects.

While such changes are certainly important to medicine, this finding implies that future medical research will need to also focus on the contributions of polymorphisms in non-coding regions or intergenic regions of the genome, something that was previously very difficult or impossible to do. Thus, SNPs in proximity to various regulatory regions [46], some of which exist at a great distance from the regulated gene in either 5′ or 3′ direction, are likely to be important. By the same token, SNPs in introns may have an unexpected role in the causality of human disease [80, 81]. In addition, SNPs in genes whose final product is an RNA may also be of unexpected importance [82].

An understanding of the human genome and its DNA variation will allow a rapid expansion of medical applications of pharmacogenetics [4, 6]. There are a number of clear examples where DNA variation, primarily but not exclusively in the form of SNPs, has dramatic implications for medical practice and clinical pharmacology [4–6]. Thus, a brief list includes: angiotensin II type 1 receptor polymorphisms can affect the severity of congestive heart failure as well as the response to angiotensin-coverting enzyme inhibitors [83, 84]; β-adrenoreceptor variants may alter airway hyperreactivity and response to β-agonists [85]; and the apolipoprotein E4 allele impacts the differential response to anticholinergic agents in patients with Alzheimer's disease [86]. Also, the bioavailability of drugs is affected by polymorphisms, e.g., MDR1, a drug efflux pump on digoxin levels [87],

CYP2C19 on omeprazole metabolism [88], and CYP2C9 or 2C19 on tolbutamide and phenytoin metabolism [89]. Many of these issues are taken up in other chapters of this book. The availability of genome-wide data on DNA variation is thus likely to expand progress in prevention, diagnosis, and treatment customized to the needs of a specific patient, rather than a statistical average. In addition, SNPs provide a new tool for familial linkage and even more so for population-based indirect association studies to speed the identification of genes as targets for new diagnostics and therapeutics [4, 6, 78, 79]. In this context, it is now possible to integrate information on DNA variations in human populations with an understanding of entire networks of genes. Again, this would have been difficult or impossible prior to the sequencing of the entire genome. Since most common human diseases culminate from long standing interactions between many genes and environmental factors, predicting the contributions of genes in complex disorders will remain a challenge for medicine for many years to come.

2.6 Biological Complexity and the Role of Medicine in the Future of the Genome

The modest number of human genes means that we must explore mechanisms that generate the complexities inherent in human development and the sophisticated signaling systems that maintain homeostasis. Practicing physicians and clinical investigators have an active role in shaping our understanding of the genome. The transfer of knowledge does not flow simply from the laboratory to the bedside. Traditionally, very important insights at the basic research level are derived from clinical observations. This includes our attempt to understand the complexity of the human genome.

There are a large number of ways that the functions of individual genes and gene products are regulated. The key point is that certain observations at the clinical level provide unique opportunities to understand how the genome functions as an integrated system. Thus, the study of Mendelian disorders has lead to unique insights regarding the functions of over 1,000 genes [90]. However, many common disorders including cancer, asthma, type 2 diabetes, cardiovascular abnormalities, neuropsychiatric illness, and others cannot be generally explained on the basis of a single gene, i.e., they are polygenic in origin [91]. Other illnesses are manifestations of

[1] the process of creating triplet repeats [92, 93] (Huntington's disease, spinocerebellar ataxia, fragile X syndrome);
[2] a consequence of abnormalities of certain epigenetic phenomena such as gene imprinting [94–97]. (Prader-Willi syndrome, Beckman-Wiedemann syndrome);
[3] abnormalities of mitochondrial genes [98] (MELAS syndrome, Kearns-Sayre syndrome);
[4] the process of somatic mutation or mosaicism [99, 100] (McCune-Albright syndrome, paroxysmal nocturnal hemoglobinuria, cancer).

In addition, there is growing evidence for illnesses caused by genes whose product is an RNA molecule, not a protein *per se*. Also, the protean manifestations of RNA molecules and their modifications (including RNA editing) may contribute to our understanding of disease [82, 101–103].

The genome will provide practical benefit when there is integration of genomic information (genetic loci) with the phenotypes of clinical disease. We are in the midst of a major paradigm shift in biology and medicine [104]: the process of looking at genes in isolation has now shifted towards exploring "networks" of genes involved in cellular processes and disease, identifying molecular "portraits" of disease based on tissue or organ involvement, and ultimately defining the biochemical readouts that are specific to clinical conditions. The era of "personalized medicine" will evolve as a parallel process, wherein DNA variations recorded in human populations will be integrated into the above paradigm, to guide a new generation of diagnostic, prognostic and therapeutic modalities designed to improve patient care.

2.7 Conclusion

The availability of whole genome sequence information will transform pharmacogenomics and unify many disparate fields. The new era of pharmacogenomics can draw upon lessons from a huge number of genomes from the most ancient microbe to human beings. Target discovery, lead compound identification, biochemical pharmacology, toxicology, exhaustive literature annotation, and clinical trials can now be merged with the science of bioinformatics into a powerful and unified machine for discovering and developing the right products for the right patient. This unified process will provide new opportunities for rational small and large molecule design, including novel approaches for drug resistance in microbes and cancer cells, and the reduction of unexpected serious side effects.

Taken together, these advances will permit scientists and clinicians who are versed in the new world of information and computation to speed the identification of new pharmaceutical agents, eliminating products that are likely to display toxicity or poor efficacy, and reducing the formidable costs and risks associated with the current paradigms of drug discovery and development.

2.8 References

1 Venter JC, Adams MD, Myers EW et al. The sequence of the human genome. Science **2001**; 291[5507]:1304–1351.

2 The Genome International Sequencing Consortium. Initial sequencing and analysis of the human genome. Nature **2001**; 409:860–921.

3 Cavalli-Sforza LL. The DNA revolution in population genetics. Trends Genet **1998**; 14[2]:60–65.

4 Weber WW. Pharmacogenetics. Oxford University Press **1997**.

5 Broder S, Venter JC. Sequencing the entire genomes of free-living organisms:

the foundation of pharmacology in the new millennium. Annu Rev Pharmacol Toxicol **2000**; 40:97–132.

6 Roses AD. Pharmacogenetics and the practice of medicine. Nature **2000**; 405[6788]:857–865.

7 Adams MD, Kelley JM, Gocayne JD, Dubnick M, Polymeropoulos MH, Xiao H, Merril, CR, Wu A, Olde B, Moreno R, Kerlavage AR, McCombie WR, Venter JC. Complementary DNA sequencing: "Expressed Sequence Tags" and the Human Genome Project. Science **1991**; 252:1651–1656.

8 Adams MD, Kerlavage AR, Fleischmann RD, Fuldner RA, Bult CJ, Lee NH, Kirkness EF, Weinstock KG, Gocayne JD, White O, Sutton G, Blake JA, Brandon RC, Chiu M-W, Clayton RA, Cline RT, Cotton MD, Earle-Hughes J, Fine LD, FitzGerald LM, FitzHugh WM, Fritchman JL, Geoghagen NSM, Glodek A, Gnehm CL, Hanna MC, Hedblom E, Hinkle PS Jr., Kelley JM, Klimek KM, Kelley JC, Liu L-I, Marmaros SM, Merrick JM, Moreno-Palanques RF, McDonald LA, Nguyen DT, Pellegrino SM, Phillips CA, Ryder SE, Scott JL, Saudek DM, Shirley R, Small KV, Spriggs TA, Utterback TR, Weidman JF, Li Y, Barthlow R, Bednarik DP, Cao L, Cepeda MA, Coleman TA, Collins E-J, Dimke D, Feng P, Ferrie A, Fischer C, Hastings GA, He W-W, Hu J-S, Huddleston KA, Greene JM, Gruber J, Hudson P, Kim A, Kozak DL, Kunsch C, Ji H, Li H, Meissner PS, Olsen H, Raymond L, Wei Y-F, Wing J, Xu C, Yu G-L, Ruben SM, Dillon PJ, Fannon MR, Rosen CA, Haseltine WA, Fields C, Fraser CM, Venter JC. Initial assessment of human gene diversity and expression patterns based upon 83 million nucleotides of cDNA sequence. Nature: The Genome Directory **1995**; 377(Suppl):3–174.

9 Fleischmann RD, Adams MD, White O, Clayton RA, Kirkness EF, Kerlavage AR, Bult CJ, Tomb J-F, Dougherty BA, Merrick JM, McKenney K, Sutton G, FitzHugh W, Fields C, Gocayne JD, Scott J, Shirley R, Liu L-I, Glodek A, Kelley JM, Weidman JF, Phillips CA, Spriggs T, Hedblom E, Cotton MD, Utterback TR, Hanna MC, Nguyen DT, Saudek DM, Brandon RC, Fine LD, Fritchman JL, Fuhrmann JL, Geoghagen NSM, Gnehm CL, McDonald LA, Small KV, Fraser CM, Smith HO, Venter JC. Whole-genome random sequencing and assembly of *Haemophilus influenzae* Rd. Science **1995**; 269:496–512.

10 http://www.tigr.org/tdb/mdb/mdbcomplete.html

11 Adams MD, Celniker SE, Holt RA et al. The genome sequence of *Drosophila melanogaster*. Science **2000**; 287[5461]:2185–2195.

12 Cohen ML. Changing patterns of infectious disease. Nature **2000**; 406[6797]:762–767.

13 Ferber D. Antibiotic resistance. Superbugs on the hoof? Science **2000**; 288[5467]:792–794.

14 Blaser MJ, Berg DE. *Helicobacter pylori* genetic diversity and risk of human disease. J Clin Invest **2001**; 107[7]:767–773.

15 Costerton JW, Stewart PS, Greenberg EP. Bacterial biofilms: A common cause of persistent infections. Science **1999**; 284[5418]:1318–1322.

16 Altschul SF, Madden TL, Schaffer AA et al. Gapped BLAST and PSI-BLAST: a new generation of protein database search programs. Nucleic Acids Res **1997**; 25[17]:3389–3402.

17 Ponting CP, Schultz J, Milpetz F, Bork P. SMART: identification and annotation of domains from signalling and extracellular protein sequences. Nucleic Acids Res **1999**; 27[1]:229–232.

18 Bateman A, Birney E, Durbin R, Eddy SR, Howe KL, Sonnhammer EL. The Pfam protein families database. Nucleic Acids Res **2000**; 28[1]:263–266.

19 Tatusov RL, Koonin EV, Lipman DJ. A genomic perspective on protein families. Science **1997**; 278[5338]:631–637.

20 Tatusov RL, Natale DA, Garkavtsev IV et al. The COG database: new developments in phylogenetic classification of proteins from complete genomes. Nucleic Acids Res **2001**; 29[1]:22–28.

21 Rubin GM, Yandell MD, Wortman JR et al. Comparative genomics of the eukaryotes. Science **2000**; 287[5461]:2204–2215.

22 Fraser CM, Eisen JA, Salzberg SL. Microbial genome sequencing. Nature **2000**; 406[6797]:799–803.

23 Cole ST, Brosch R, Parkhill J et al. Deciphering the biology of *Mycobacterium tuberculosis* from the complete genome sequence. Nature **1998**; 393[6685]:537–544.

24 McKinney JD, Honer zu Bentrup K, Munoz-Elias EJ et al. Persistence of *Mycobacterium tuberculosis* in macrophages and mice requires the glyoxylate shunt enzyme isocitrate lyase. Nature **2000**; 406[6797]:735–738.

25 Cole ST, Eiglmeier K, Parkhill J et al. Massive gene decay in the leprosy bacillus. Nature **2001**; 409[6823]:1007–1011.

26 Perna NT, Plunkett G, 3rd, Burland V et al. Genome sequence of enterohaemorrhagic *Escherichia coli* O157:H7, Nature **2001**; 409[6819]:529–533.

27 Blattner FR, Plunkett G, 3rd, Bloch CA et al. The complete genome sequence of *Escherichia coli* K-12. Science **1997**; 277[5331]:1453–1474.

28 Kalman S, Mitchell W, Marathe R et al. Comparative genomes of *Chlamydia pneumoniae* and *C. trachomatis*. Nature Genet **1999**; 21[4]:385–389.

29 Fraser CM, Casjens S, Huang WM et al. Genomic sequence of a Lyme disease spirochaete, *Borrelia burgdorferi*. Nature **1997**; 390[6660]:580–586.

30 Fraser CM, Norris SJ, Weinstock GM et al. Complete genome sequence of *Treponema pallidum*, the syphilis spirochete. Science **1998**; 281[5375]:375–388.

31 Subramanian G, Koonin EV, Aravind L. Comparative genome analysis of the pathogenic spirochetes *Borrelia burgdorferi* and *Treponema pallidum*. Infect Immun **2000**; 68[3]:1633–1648.

32 Ho CK, Shuman S. A yeast-like mRNA capping apparatus in *Plasmodium falciparum*. Proc Natl Acad Sci USA **2001**; 98[6]:3050–3055.

33 Aravind L. Guilt by association: contextual information in genome analysis. Genome Res **2000**; 10[8]:1074–1077.

34 Bernard GR, Vincent JL, Laterre PF et al. Efficacy and safety of recombinant human activated protein C for severe sepsis. N Engl J Med **2001**; 344[10]:699–709.

35 Hacker J, Kaper JB. Pathogenicity islands and the evolution of microbes. Annu Rev Microbiol **2000**; 54:641–679.

36 Wren BW. Microbial genome analysis: insights into virulence, host adaptation and evolution. Nature Rev Genet **2000**; 1[1]:30–39.

37 Ochman H, Lawrence JG, Groisman EA. Lateral gene transfer and the nature of bacterial innovation. Nature **2000**; 405[6784]:299–304.

38 Gardner MJ, Tettelin H, Carucci DJ et al. Chromosome 2 sequence of the human malaria parasite *Plasmodium falciparum*. Science **1998**; 282[5391]:1126–1132.

39 Hoffman SL, Rogers WO, Carucci DJ, Venter JC. From genomics to vaccines: malaria as a model system. Nature Med **1998**; 4[12]:1351–1353.

40 Bachmaier K, Neu N, de la Maza LM, Pal S, Hessel A, Penninger JM. *Chlamydia* infections and heart disease linked through antigenic mimicry. Science **1999**; 283[5406]:1335–1339.

41 Russo JJ, Bohenzky RA, Chien MC et al. Nucleotide sequence of the Kaposi sarcoma-associated herpesvirus (HHV8]. Proc Natl Acad Sci USA **1996**; 93[25]:14862–14867.

42 Senkevich TG, Koonin EV, Bugert JJ, Darai G, Moss B. The genome of molluscum contagiosum virus: analysis and comparison with other poxviruses. Virology **1997**; 233[1]:19–42.

43 Lorenz E, Mira JP, Cornish KL, Arbour NC, Schwartz DA. A novel polymorphism in the toll-like receptor 2 gene and its potential association with staphylococcal infection. Infect Immunol **2000**; 68[11]:6398–6401.

44 Li P, Chan HC, He B et al. An antimicrobial peptide gene found in the male reproductive system of rats. Science **2001**; 291[5509]:1783–1785.

45 Weatherall D, Clegg J, Kwiatkowski D. The role of genomics in studying genetic susceptibility to infectious disease. Genome Res **1997**; 7[10]:967–973.

46 Bellamy R, Ruwende C, Corrah T, McAdam KP, Whittle HC, Hill AV. Variations in the NRAMP1 gene and susceptibility to tuberculosis in West Africans. N Engl J Med **1998**; 338[10]:640–644.

47 Knight JC, Udalova I, Hill AV et al. A polymorphism that affects OCT-1 binding to the TNF promoter region is associated with severe malaria. Nature Genet **1999**; 22[2]:145–150.

48 Dorman SE, Holland SM. Interferon-gamma and interleukin-12 pathway defects and human disease. Cytokine Growth Factor Rev **2000**; 11[4]:321–333.

49 Littman DR. Chemokine receptors: keys to AIDS pathogenesis? Cell **1998**; 93[5]:677–680.

50 Ying L, Katz Y, Schlesinger M et al. Complement factor H gene mutation associated with autosomal recessive-atypical hemolytic uremic syndrome. Am J Hum Genet **1999**; 65[6]:1538–1546.

51 Xu H, Iijima K, Shirakawa T et al. Platelet-activating factor acetylhydrolase gene mutation in Japanese children with *Escherichia coli* O157-associated hemolytic uremic syndrome. Am J Kidney Dis **2000**; 36[1]:42–46.

52 El-Omar EM et al. Interleukin-1 polymorphisms associated with increased risk of gastric cancer. Nature **2000**; 404[6776]:398–402.

53 Bidwell J et al. Cytokine gene polymorphism in human disease: on-line databases. Genes Immun **1999**; 1[1]:3–19.

54 Moxon et al. Adaptive evolution of highly mutable loci in pathogenic bacteria. Curr Biol **1994**; 4:24–33.

55 Bayliss CD et al. The simple sequence contingency loci of *Haemophilus influenzae* and *Neisseria meningitides*. J Clin Invest **2001**; 107:657–666.

56 Cooper DN. Human gene evolution. BIOS Scientific Publishers Ltd. **1999**.

57 Dickson D. Gene estimate rises as US and UK discuss freedom of access. Nature **1999**; 401[6751]:311.

58 Genome sequence of the nematode *C. elegans*: A platform for investigating biology. The *C. elegans* Sequencing Consortium. Science **1998**; 282[5396]:2012–2018.

59 Analysis of the genome sequence of the flowering plant *Arabidopsis thaliana*. Nature **2000**; 408[6814]:796–815.

60 Hamdi H, Nishio H, Zielinski R, Dugaiczyk A. Origin and phylogenetic distribution of Alu DNA repeats: irreversible events in the evolution of primates. J Mol Biol **1999**; 289[4]:861–871.

61 Hamdi HK, Nishio H, Tavis J, Zielinski R, Dugaiczyk A. Alu-mediated phylogenetic novelties in gene regulation and development. J Mol Biol **2000**; 299[4]:931–939.

62 Saikawa Y, Kaneda H, Yue L et al. Structural evidence of genomic exon-deletion mediated by Alu-Alu recombination in a human case with heme oxygenase-1 deficiency. Hum Mutat **2000**; 16[2]:178–179.

63 Rohlfs EM, Puget N, Graham ML et al. An Alu-mediated 7.1 kb deletion of BRCA1 exons 8 and 9 in breast and ovarian cancer families that results in alternative splicing of exon 10. Genes Chromosomes Cancer **2000**; 28[3]:300–307.

64 Sharan C, Hamilton NM, Parl AK, Singh PK, Chaudhuri G. Identification and characterization of a transcriptional silencer upstream of the human BRCA2 gene. Biochem Biophys Res Commun **1999**; 265[2]:285–290.

65 Norris J, Fan D, Aleman C et al. Identification of a new subclass of Alu DNA repeats which can function as estrogen receptor-dependent transcriptional enhancers. J Biol Chem **1995**; 270[39]:22777–22782.

66 Wu Q, Maniatis T. A striking organization of a large family of human neural cadherin-like cell adhesion genes. Cell **1999**; 97[6]:779–790.

67 Wu Q, Maniatis T. Large exons encoding multiple ectodomains are a characteristic feature of protocadherin genes. Proc Natl Acad Sci USA **2000**; 97[7]:3124–3129.

68 Ranscht B. Cadherins: molecular codes for axon guidance and synapse formation. Int J Dev Neurosci **2000**; 18[7]:643–651.

69 Missler M, Sudhof TC. Neurexins: three genes and 1001 products. Trends Genet **1998**; 14[1]:20–26.

70 Aravind L, Dixit VM, Koonin EV. Apoptotic molecular machinery: vastly increased complexity in vertebrates revealed by genome comparisons. Science **2001**; 291[5507]:1279–1284.

71 Yuan J, Yankner BA. Apoptosis in the nervous system. Nature **2000**; 407[6805]:802–809.

72 Krammer PH. CD95's deadly mission in the immune system. Nature **2000**; 407[6805]:789–795.

73 Nicholson DW. From bench to clinic with apoptosis-based therapeutic agents. Nature **2000**; 407[6805]:810–816.

74 Broder S, Venter JC. Whole genomes: the foundation of new biology and medicine. Curr Opin Biotechnol **2000**; 11[6]:581–585.

75 Pandey A, Mann M. Proteomics to study genes and genomes. Nature **2000**; 405[6788]:837–846.

76 Banks RE, Dunn MJ, Hochstrasser DF et al. Proteomics: New perspectives, new biomedical opportunities. Lancet **2000**; 356[9243]:1749–1756.

77 Sachidanandam R, Weissman D, Schmidt SC et al. A map of human genome sequence variation containing 1.42 million single nucleotide polymorphisms. Nature **2001**; 409[6822]:928–933.

78 Chakravarti A. Population genetics-making sense out of sequence. Nature Genet **1999**; 21[1 Suppl):56–60.

79 Risch NJ. Searching for genetic determinants in the new millennium. Nature **2000**; 405[6788]:847–856.

80 Horikawa Y, Oda N, Cox NJ et al. Genetic variation in the gene encoding calpain-10 is associated with type 2 diabetes mellitus. Nature Genet **2000**; 26[2]:163–175.

81 Kishi F, Fujishima S, Tabuchi M. Dinucleotide repeat polymorphism in the third intron of the NRAMP2/DMT1 gene. J Hum Genet **1999**; 44[6]:425–427.

82 Ridanpaa M, van Eenennaam H, Pelin K et al. Mutations in the RNA component of RNase MRP cause a pleiotropic human disease, cartilage-hair hypoplasia. Cell **2001**; 104[2]:195–203.

83 Andersson B, Blange I, Sylven C. Angiotensin-II type 1 receptor gene polymorphism and long-term survival in patients with idiopathic congestive heart failure. Eur J Heart Fail **1999**; 1[4]:363–369.

84 Benetos A, Cambien F, Gautier S et al. Influence of the angiotensin-II type 1 receptor gene polymorphism on the effects of perindopril and nitrendipine on arterial stiffness in hypertensive individuals. Hypertension **1996**; 28[6]:1081–1084.

85 Johnson M. The beta-adrenoceptor. Am J Respir Crit Care Med **1998**; 158[5 Pt 3]:146–153.

86 Poirier J, Delisle MC, Quirion R et al. Apolipoprotein E4 allele as a predictor of cholinergic deficits and treatment outcome in Alzheimer disease. Proc Natl Acad Sci USA **1995**; 92[26]:12260–12264.

87 Cascorbi I, Gerloff T, Johne A et al. Frequency of single nucleotide polymorphisms in the P-glycoprotein drug transporter MDR1 gene in white subjects. Clin Pharmacol Ther **2001**; 69[3]:169–174.

88 Furuta T, Shirai N, Takashima M et al. Effect of genotypic differences in CYP2C19 on cure rates for *Helicobacter pylori* infection by triple therapy with a proton pump inhibitor, amoxicillin, and clarithromycin. Clin Pharmacol Ther **2001**; 69[3]:158–168.

89 Inoue K, Yamazaki H, Imiya K, Akasaka S, Guengerich FP, Shimada T. Relationship between CYP2C9 and 2C19 genotypes and tolbutamide methyl hydroxylation and S-mephenytoin 4'-hydroxylation activities in livers of Japanese and Caucasian populations. Pharmacogenetics **1997**; 7[2]:103–113.

90 Antonarakis SE, McKusick VA. OMIM passes the 1,000-disease-gene mark. Nature Genet **2000**; 25[1]:11.

91 Peltonen L, McKusick VA. Genomics and medicine. Dissecting human disease in the postgenomic era. Science **2001**; 291[5507]:1224–1229.

92 Usdin K, Grabczyk E. DNA repeat expansions and human disease. Cell Mol Life Sci **2000**; 57[6]:914–931.

93 Lieberman AP, Fischbeck KH. Triplet repeat expansion in neuromuscular disease. Muscle Nerve **2000**; 23[6]:843–850.

94 Feinberg AP. DNA methylation, genomic imprinting and cancer. Curr Top Microbiol Immunol **2000**; 249:87–99.

95 Cui H, Horon IL, Ohlsson R, Hamilton SR, Feinberg AP. Loss of imprinting in normal tissue of colorectal cancer patients with microsatellite instability. Nature Med **1998**; 4[11]:1276–1280.

96 Ohta T, Gray TA, Rogan PK et al. Imprinting-mutation mechanisms in Prader-Willi syndrome. Am J Hum Genet **1999**; 64[2]:397–413.

97 Engel JR, Smallwood A, Harper A et al. Epigenotype-phenotype correlations in Beckwith-Wiedemann syndrome. J Med Genet **2000**; 37[12]:921–926.

98 Zeviani M, Tiranti V, Piantadosi C. Mitochondrial disorders. Medicine (Baltimore) **1998**; 77[1]:59–72.

99 Aldred MA, Trembath RC. Activating and inactivating mutations in the human GNAS1 gene. Hum Mutat **2000**; 16[3]:183–189.

100 Gottlieb B, Beitel LK, Trifiro MA. Somatic mosaicism and variable expressivity. Trends Genet **2001**; 17[2]:79–82.

101 Wang Q, Khillan J, Gadue P, Nishikura K. Requirement of the RNA editing deaminase ADAR1 gene for embryonic erythropoiesis. Science **2000**; 290[5497]:1765–1768.

102 Grosjean H, Benne R. Modification and Editing of RNA. ASM Press, Washington, D.C.; **1998**.

103 Eddy SR. Noncoding RNA genes. Curr Opin Genet Dev **1999**; 9[6]:695–699.

104 Vidal M. A biological atlas of functional maps. Cell **2001**; 104[3]:333–339.

3
Turning SNPs into Useful Markers of Drug Response

Jeanette J. McCarthy

Abstract

The efforts of the Human Genome Project have resulted in, among other things, the identification and mapping of hundreds of thousands of single nucleotide polymorphisms (SNPs) for use in large-scale association studies. Turning these SNPs into useful markers of drug response is the goal of researchers in the field of pharmacogenomics. The two main approaches taken to uncover pharmacogenomic markers include whole genome linkage disequilibrium mapping and candidate gene studies. Among the challenges of pharmacogenomics is sorting through the vast number of SNPs available and deciding how many and which ones to analyze based on location, frequency, and type. The existence of linkage disequilibrium between markers located close together affords an opportunity to examine combinations of SNPs, or haplotypes, within genes. Analytical issues that arise in SNP association studies include the effect of linkage disequilibrium on relative risk and the need to correct for multiple hypothesis testing. Pharmacogenomics examines the interaction between genes and a drug in affecting a disease outcome. Complex paradigms of not only gene-drug interaction, but the combined effects of several SNPs in different genes should also be evaluated. The field of pharmacogenomics is still in its infancy. Associations must not only be uncovered, but replicated and their utility evaluated before they are put into use. Ultimately, researchers will turn SNPs into markers of drug response that promise to dramatically alter the practice of medicine and drug development.

3.1
Introduction

With the complete human genome sequence, an expanding catalog of the genetic variation between individuals and the technologies to query thousands of DNA variants, we find ourselves on the edge of major advances in genetic research that may have substantial impact on the future practice of medicine. Genetic research to date has led to the identification of hundreds of genes where mutations harbored in the gene have been pinpointed as the cause of a disease. The majority of these are rare, single-gene disorders such as Huntington's disease, cystic fibrosis and neurofibromatosis. Until now, DNA variants in only a handful of genes have been unequivocally associated with more common multifactorial diseases such as cancer, atherosclerosis and diabetes. More recently, attention has turned to using genetic approaches to identify markers of drug response. The most common type of genetic variation, single nucleotide polymorphisms (SNPs), are being exploited in these efforts. Over the next several years, the numbers of SNPs found to be associated with drug responses will grow at an unprecedented rate. Sorting through the relevant SNPs and demonstrating clinical validity and utility of these SNPs as pharmacogenomic markers is a challenge that lies ahead.

3.2 Two Approaches for Employing SNPs in Pharmacogenomics

The two basic approaches which exploit SNPs to uncover markers of drug response include candidate gene and random whole genome linkage disequilibrium mapping. While these approaches differ in their underlying principles, they represent two complementary strategies that together may provide the greatest chance of success.

3.2.1 Candidate Gene Studies

The candidate gene approach utilizes experimental approaches or *a priori* knowledge of the drug pathway, metabolism or disease pathogenesis to identify genes with possible relevance to drug response. SNPs identified in these genes are then assessed in populations of patients exposed to the drug of interest and tested for statistical association or correlation with drug response. If associated, these "susceptibility genes" are hypothesized to directly influence an individual's likelihood of responding to the drug. One of the keys to the success of this approach lies in the ability to identify relevant candidate genes. Candidate genes for pharmacogenomic analysis may include the drug target and pathway genes, drug metabolizing enzymes or disease genes. The selection of these genes is facilitated through understanding of the disease pathogenesis and the mechanism of drug action. To this end, experimental approaches can be employed to characterize genes involved in drug metabolism and the primary and secondary targets of the drug in the preclinical phases of drug development. Biological paradigms to identify markers of drug efficacy or toxicity may include the use of comparative gene expression profiling of tissues or cell lines with *in vitro* or *in vivo* exposure to therapeutic compounds. Numerous examples of pharmacogenomic markers discovered using the candidate gene approach exist.

Drug Metabolizing Enzymes

Perhaps the oldest and most well-studied class of pharmacogenomic markers are the drug metabolizing enzymes. These include variants in genes which encode proteins involved in absorption, distribution, metabolism and excretion of a drug. These mechanisms are known to show inter-subject variability and, therefore, make good candidates for pharmacogenomics. Variants in a number of drug metabolizing enzymes have been linked to adverse drug reactions. Indeed, some of markers have made it into drug development and clinical practice as a means of identifying poor or extensive metabolizers of therapeutic compounds. An example is thiopurine methyltransferase (*TPMT*), an enzyme which catabolizes thiopurines. Thiopurines including mercaptopurine, azathioprine, and thioguanine are used to treat patients with cancers, autoimmune disorders and transplant recipients. Individuals with diminished *TPMT* activity are poor metabolizers of thiopurines and thus accumulate several-fold increased concentrations of toxic 6-thioguanine nucleotides. The genetic basis for deficiency in *TPMT* activity has been well characterized [1]. Genotyping to identify poor metabolizers before initiation of thiopurine therapy can identify individuals who require much lower doses, thus avoiding toxicity and failed therapy.

Drug Targets and Pathway

Present day drug therapies are developed around a relatively small number (<500) of targets primarily including cell membrane receptors and enzymes [2]. In addition, genomic approaches are being employed to discover novel gene targets to exploit in drug discovery. Because these targets interact with the therapeutic compounds to affect disease, they are thought to be good pharmacogenomic candidates. Genetic variation in the regulatory region of the target may affect transcription, thereby increasing or decreasing the amount of target available to the drug. Genetic variation in the coding region of a gene target, which changes amino acids of the resulting protein, may affect the efficiency with which a compound can bind the protein target. A survey of SNPs in known drug targets has found an average of 16 SNPs per gene after screening a series of 20 drug targets in a Japanese population [3]. The identification of SNPs in drug targets, such as these, may provide a useful set of pharmacogenomic markers to explore for their association with response to existing therapies. An example of a drug target where SNPs were associated with response to a therapeutic directed against that target is 5-lipoxygenase (*ALOX5*). Failure of asthma patients to respond to a 5-lipoxygenase inhibitor was associated with variation in the *ALOX5* gene [4]. While the genetic variation seems to be highly penetrant (the probability of non-response in carriers of variant alleles was 100%), only 6% of asthma patients have a variant allele. Nonetheless, the implication for these findings (if they are indeed real and reproducible) is that they may begin to identify subsets of patients who would receive no benefit from this specific asthma therapy, and who may, therefore, benefit from an alternative therapy.

Disease Genes

Advances in genomic technologies such as transcriptional profiling are facilitating the classification of disease at the molecular level. Clinical phenotypes previously thought to be one disease will be subclassified by a new genomic taxonomy. Recent discoveries in the molecular pathology of cancer have highlighted important and clinically significant differences in the gene expression patterns of a variety of tumors. Studies have found that breast cancers caused by mutations in either *BRCA1* or *BRCA2* have a distinct molecular taxonomy from each other, as evidenced by differential expression of over 170 genes [5]. It is plausible that a tumor's response to therapy may differ based on this taxonomy and that molecular classification in general may identify individuals with differential response due to an underlying difference in disease pathogenesis. An example of a disease gene in which SNPs have been correlated with response to therapy is the Alzheimer's disease gene, apolipoprotein E (*APOE*). Carriers of the E4 allele of the APOE gene, a risk factor for developing Alzheimer's disease, have been shown to respond differently to several cholinesterase inhibitors [6–8] compared to non-carriers. While these results have not been unequivocally confirmed [9], they represent a valid hypothesis of how different molecular classification of disease may influence response to therapy.

3.2.2
Whole Genome Linkage Disequilibrium Mapping Studies

An alternative to the candidate gene approach is a whole genome analysis of random SNPs using linkage disequilibrium mapping. This approach relies on linkage disequilibrium (LD), or non-random association between SNPs in proximity to each other. Tens to hundreds of thousands of anonymous SNPs need to be identified and their location in the genome mapped. While these anonymous SNPs may fall within genes and in fact be susceptibility SNPs, most will be found in the vast amount of non-coding DNA between genes and play no obvious role in drug response. Through LD, associations found with these anonymous markers can identify a region of the genome that may harbor a susceptibility gene without any *a priori* assumptions about what or where the susceptibility gene is. Additional significant efforts are then required to develop and genotype dense SNP maps covering the region in order to narrow down the precise location of the causative SNP and define combinations of SNPs that mark the underlying susceptibility.

Linkage Disequilibrium

The power of LD mapping depends on the SNP allele frequencies and the extent of LD between SNPs [10]. Genome-wide SNP LD mapping is predicated on the assumption that LD exists between SNPs. The extent of LD occurs as a consequence of many factors including population admixture, genetic drift, mutation and natural selection [11]. For genetic distances measured in kilobases (kb) of DNA, LD tends to decline with larger distance between SNPs in the 10–100 kb range. Over shorter genetic distances the degree of LD is highly variable from one genomic region to the next. In some genomic regions, LD extends over several thousands of kilobases [12] whereas in other genomic regions surrounding single genes, LD can be quite small [13]. Understanding the *average* extent of LD is useful for estimating the number of markers needed in a SNP map and the strength of the association that the markers are capable of detecting.

Theoretical estimates of the average extent of LD in the human genome have varied widely, ranging from <100 kb [14, 15] to <3 kb [16]. With the completion of the human genome sequence, the true extent of LD throughout the genome is being uncovered empirically. Several recent studies which examined LD between markers located <500 kb apart are summarized in Table 3.1. These reports suggest that similar to what has been found over longer distances, LD tends to be inversely correlated with distance between markers with stronger LD found between markers that are close together [18, 19, 21]. Nonetheless, the genomic location of the

Tab. 3.1 Studies which examine multiple genomic regions for the extent of linkage disequilibrium

Author	*Genomic region*	*Mean D′ by distance*	*% Markers with useful/ significant LD*	*Population*	*Conclusions/ comments*
Huttley et al. [12]	whole genome		4%	Europeans (CEPH families)	Distribution of LD is nonuniform; nine genomic regions with extensive LD
Goddard et al. [17]	33 genes on 16 chromosomes	D′ ~ 0.50 in US Caucasians	82%	US Japanese, US Chinese, US African, US Hispanic, US Caucasian	Mean LD similar among Asian populations and higher than Caucasians, African Americans, and Hispanics
Dunning et al. [18]	13q12–13, 19q13.2, 22q13.2	D′ = 0.68 for markers < 5 kb apart; no significant LD for markers > 500 kb apart	50% at < 5 kb none at > 20 kb	East Anglican, Afrikaners, Ashkenazi, Finnish	Useful LD does not extend beyond 5 kb
Abecasis et al. [19]	14, 13, 2	D′ = 0.08–0.69 for markers < 50 kb apart	50% at < 50 kb	British	Useful LD may extend to 50 kb for 50% of markers; mean D′ < 1 even for closely linked markers
Reich et al. [20]	19 genomic regions		50% at < 60 kb	US Caucasian	Useful LD may extend to 60 kb for 50% of markers

kb = kilobases of DNA; D′ = measure of linkage disequilibrium markers may be more important than simply the distance between markers [12, 22]. While LD may be quite similar among European populations, it may vary significantly between Europeans, Africans and Asians. Studies have shown stronger LD occurring in populations of Asian decent versus European or African [17, 23].

D′ is a measure of linkage disequilibrium [24] that ranges in value from 0 (no disequilibrium) to ±1.0 (complete disequilibrium). The formula for calculating D′ and a brief explanation can be found in a recent review [25]. Three published studies [18,19,21] which examined polymorphisms located < 500 kb apart in several chromosomal regions reported mean values of D′ ~ 0.70. Despite these consistencies, there is wide discrepancy between estimates of how much useful (D′> 0.33 or average D′> 0.50) LD exists for SNPs located < 50 kb apart. One study found half of all markers at < 50 kb to be useful [19] while another study found that only when the markers were much closer together, at distances < 5 kb apart, were they as useful [18]. Another more exhaustive study recently examined 19 randomly-selected genomic regions and found half of markers at 60 kb apart to be useful [20]. This order of magnitude difference in the distance over which substantial useful LD exists suggests that more investigation is required to map out the exact patterns of LD across the genome and in various ethnic groups in order to construct the most useful maps of human variation. Some genomic regions where LD is weak will require very dense SNP maps, while those regions with extensive LD may require fewer SNPs.

LD mapping has been employed successfully in a limited fashion in genetic linkage studies which take advantage of families with multiple affected individuals to uncover genes for monogenic diseases [26–28]. LD mapping is now being considered in the context of association studies [14, 29]. However, to date no studies have successfully utilized this strategy in the evaluation of unrelated individuals for either diseases or drug response. One published study [30]demonstrated the potential utility of LD mapping through post-hoc analysis of a known disease-causing gene. Nonetheless, data from another study [13] suggests that this same approach would be unreliable. As analytical methods are improved to accommodate large numbers of SNPs, this approach may become more feasible.

3.2.3
Comparison of Candidate Gene and Whole Genome LD Mapping

Whole genome LD mapping and candidate gene studies are both valid methods for use in pharmacogenomics. The major drawback of the candidate gene approach is that it relies on current knowledge of a disease or drug response to choose which genes to examine, whereas a whole genome approach makes no assumptions about what or where the underlying gene(s) are. The whole genome LD mapping approach, on the other hand, will require very large sample sizes and low cost genotyping before it becomes feasible. Even then, since this approach relies on being able to detect an association with an underlying causative SNP by querying a surrogate marker in LD, there is a chance that association will be missed.

LD mapping and candidate gene studies are two methods aimed at finding genes that account for differences in drug response; however, the two are complementary strategies that together may provide the best overall strategy. One way in which the approaches can be used in complement would be to first carry out a whole genome random SNP analysis with the goal of pinpointing large regions of the genome that may harbor susceptibility genes. This can be followed up with a large-scale candidate gene study, choosing genes that lie in those very chromosomal regions identified through the random search. SNPs in these "positional candidate genes" would be thoroughly evaluated to determine their association with drug response. The challenge of this approach will be in identifying only a few strongly-associated regions of the genome. If many weakly-associated regions are found, following each of them up makes this a less efficient strategy.

Another way to incorporate both whole genome and candidate gene methods would be to carry out a whole genome study, focusing on all ~30,000 genes rather than random genomic regions. Because LD within a gene is thought to be much stronger than between larger genomic regions, evaluation of a few to several dozen evenly spaced common SNPs across each gene may be an efficient strategy to pinpoint causative SNPs without having to find and test every SNP in the gene directly. The advantage to this approach is that, if we believe that SNPs within and around genes are going to be the most important, the chance of missing a true underlying association is diminished. However, it may be several years before this approach can be put into practice. The existence of ~30,000 genes in the human genome is based on predictions. Experimental evidence verifying gene expression, alternate splice forms and genomic structure (intron/exon boundaries) for all genes is currently incomplete. This knowledge will be critical for designing comprehensive SNP discovery experiments and genotyping assays across genes.

3.3 How Many SNPs are Needed and What Kind are Useful for Pharmacogenomic Studies

The 3.2 billion base pairs of the human genome are estimated to harbor on the order of 11 million SNPs where the minor allele frequency is at least 1% [31]. These SNPs are the result of mutations that occurred over time at an estimated rate of 1 nucleotide per 1,331 bp of DNA [32, 33]. The identification and mapping of SNPs has been the focus of the publicly funded Human Genome Project and the SNP Consortium led by the Wellcome Trust. As a result of these efforts, public databases are overflowing with SNPs throughout the genome, including some within and around genes. Determining which of these SNPs to evaluate in pharmacogenomic studies requires an understanding of what makes a SNP useful.

3.3.1
Location

Most SNPs in the genome fall within the vast amount of DNA between genes (~99% of the genome) and may play no obvious role in disease or drug response. These random SNPs may be useful as markers for linkage disequilibrium mapping studies. However, optimal SNP maps should be constructed to take advantage of our knowledge of genomic organization. The distribution of genes is not random and, therefore, SNPs focused in gene-rich regions may be more useful than those in gene-poor regions. Furthermore, since an estimated 43% of the genome is comprised of repetitive elements [34], genotyping SNPs in these regions may be problematic. Finally, depending on patterns of LD in the genome, denser maps may be required in regions of low LD while sparse maps may suffice for regions of high LD.

For candidate gene studies, the most biologically relevant SNPs are thought to be those that fall within genes: either the coding region where changes can affect the structure of the resulting protein product, or in the regulatory regions where changes can affect the amount of the protein product. While the International SNP Map Working Group (ISMWG) has found on average 2 cSNPs per gene (*in silico* discovery) [32], other studies taking a more focused approach have found ~4 cSNPs per gene [35, 36]. If for any given drug response, 10% of the estimated 30,000 genes in the human genome are involved, then 12,000 candidate cSNPs should exist. The ~4,800 (40%) of these which are estimated to change amino acids, in addition to yet unknown regulatory and non-coding SNPs, may be the most promising SNPs to examine. Without *a priori* knowledge of which genes are involved, a survey of the estimated ~50,000 cSNPs which would change amino acids in all 30,000 genes may be feasible.

Upstream and downstream genomic segments that regulate a gene's expression may be important regions to examine for SNPs. SNPs that occur in these regulatory regions may influence binding of regulatory factors and the ultimate production of the gene product. The exact location of the regulatory elements of most genes is unknown but can be determined in some cases through computational approaches, including inter-species genomic sequence comparisons and genome-wide expression profiling [37]. In the absence of knowing a gene's regulatory region, a strategy which examines several thousands of base pairs upstream and downstream of the gene may be employed. However, regulatory regions occurring further away from the gene will be missed.

SNPs that occur in intronic gene regions may also have functional consequences, affecting gene splicing or gene expression, and should be considered. SNPs that lie close to intron-exon boundaries are good candidates for affecting splicing. However, splice sites may also be found further within the intron. An example of splice site variants that affect a disease phenotype are the several splice site mutations in the *SURF1* gene which result in loss of *SURF1* protein and development of Leigh syndrome [38]. Intronic sequences can also play a regulatory role as in the *BRCA1* gene where an intronic sequence has been demonstrated to have transcriptional repressor activity [39].

3.3.2
Frequency

The minor allele frequency of the most useful SNPs for pharmacogenomics studies depends on whether the genetic basis of drug response is due to strong effects of uncommon variants or due to the interactive effect of common, weakly-associated variants. The SNPs most likely to have a direct impact on the protein product of a gene will be coding region SNPs (cSNPs) that change amino acids and SNPs in gene regulatory regions which control protein levels. Surveys of SNPs in candidate genes have revealed that 75% of cSNPs which change amino acids have minor allele frequencies less than 15% [35], the average being about 7% [36].

To accommodate the possibility of both strong, uncommon and weak, common SNP models one may need to examine SNPs with a variety of frequencies. The successful evaluation of uncommon SNPs [1–10%) will depend in part on the ability to find these SNPs. Most publicly available SNPs have been discovered as a by-product of large-scale sequencing efforts where the genomes of a small number of individuals are compared [40, 41]. Table 3.2 illustrates the likelihood of finding SNPs of various frequencies as a function of the number of individuals screened. Because of the small numbers of individuals screened, the methods employed to find publically-available SNPs are likely to miss SNPs of low frequency which may be most relevant. Therefore, the identification of uncommon SNPs will require directed efforts at screening larger numbers of individuals.

The successful evaluation of the association between uncommon SNPs in candidate genes and drug response will rely on more powerful studies. Sample sizes will need to be much larger to have sufficient power to detect even relatively large effects of uncommon variants. Finally, the decision to evaluate uncommon SNPs depends in part on the likelihood that they will be medically useful. Even those low-frequency SNPs that have strong effects may prove to have little clinical utility as general screening markers for patients receiving drugs.

Common SNPs in candidate genes that appear to have functional consequences have been the subject of most investigations to date with the hypothesis that common SNPs with small effects may underlie common outcomes, such as drug re-

Tab. 3.2 Likelihood of finding SNPs of various frequencies as a function of the number of individuals screened

No. of individuals		***SNP frequency***			
	>1%	*>2%*	*>5%*	*>10%*	*>20%*
2	4%	8%	19%	34%	59%
5	10%	18%	40%	65%	89%
10	18%	33%	64%	88%	99%
20	33%	55%	87%	99%	>99%
40	55%	80%	98%	>99%	>99%

sponse. These SNPs will be more easily detected by screening only a few individuals and will require smaller patient populations to demonstrate and validate their association with drug response. Common SNPs in coding, regulatory and intronic regions that show no evidence of being functional may nonetheless prove useful in identifying pharmacogenomic markers through haplotype analysis.

3.3.3
Haplotype Analysis

Individual polymorphisms located closely together on a chromosome and in strong linkage disequilibrium are inherited together as a unit referred to as a haplotype. These haplotypes, through their proximity to a causative SNP, may themselves have no effect on drug response, but rather act as markers of the underlying cause of the drug response. The evaluation of haplotypes across candidate gene regions will, in theory, allow the identification of associations between genes and drug response without requiring the discovery of the causative variant first. For example, LD was used to identify a major genetic determinant of hereditary hemochromatosis. Long before the gene implicated in hereditary hemochromatosis (*HFE*) was identified, associations were found between the disease and certain Major Histocompatibility Complex (MHC) genotypes [42]. The MHC is located on chromosome 6 where many genes involved in immune regulation are found. The MHC region exhibits extensive LD and as a result, several distinct ancestral haplotypes are found in the population. The MHC genotypes associated with hereditary hemochromatosis were part of an extended ancestral haplotype which included the *HFE* gene. Subsequent analysis of the chromosomal region implicated *HFE* as the cause of hereditary hemochromatosis.

Alternatively, specific haplotypes may themselves be responsible for the variation in drug response and be a far better marker than any one of their component SNPs. A recent pharmacogenomic example is described in a paper by Drysdale et al. [43] where 13 variable sites within a 1.6 kb contiguous segment of DNA encompassing the 5′ upstream region and open reading frame of the β_2-adrenergic receptor gene (*B2AR*) were examined. Twelve haplotypes were identified and tested for their association with response to the anti-asthma compound albuterol. The authors found that mean responses varied by >2-fold for different haplotype pairs. While haplotypes were significantly related to response, the individual SNPs comprising the haplotypes were not. The pairs of haplotypes in question were subsequently shown to correlate with *in vitro* transcript and protein expression levels of the β_2-adrenergic receptor, thus lending biological plausibility.

Haplotype Determination

Figure 3.1 illustrates the concept of haplotypes. The most accurate methods for determining haplotypes are experimental ones involving laborious and expensive laboratory analysis [44–46]. Alternatively, accuracy can sometimes be achieved by genotyping additional family members [47]. In pharmacogenomic studies where

SNP1 and SNP2 are located close together on a chromosome. SNP1 has alleles A and G. SNP2 has alleles G and T. For an individual who is heterozygous at each locus (i.e. having genotypes AG for SNP1 and GT for SNP2), with no additional information, the haplotypes are ambiguous resulting in two possible combinations of SNPs found on their chromosome pair.

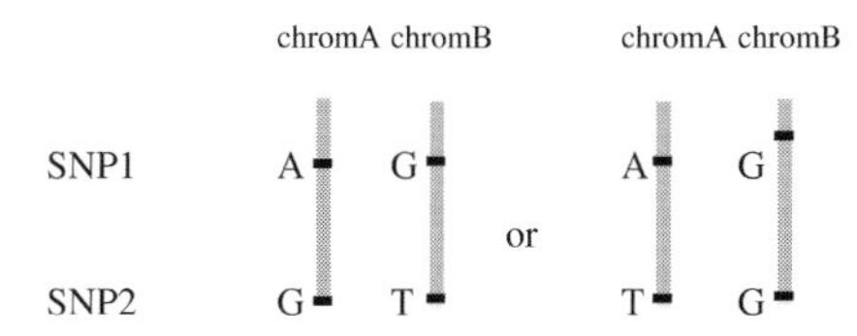

Using experimental approaches, family data or stastistical methods, the correct pairing of SNPs can be determined. In this example, if SNP1 A and SNP2 G alleles are in positive LD, and SNP1 G and SNP2 T alleles are in positive LD, then the chromosomes will look like this:

SNP1 A G

SNP2 G T

If carrying SNP1 alleleA is associated with poor drug response, and this association becomes even more pronounced in carriers of both SNP1 allele A and SNP2 allele G, then the haplotype defines the risk factor associated with poor drug response. Further work is needed to determine if the combination of variants SNP1A-SNP2G are directly responsible for the observed association or rather, are part of a more extended haplotype which contains another SNP further awsay that is the true underlying variant associated with drug response.

Fig. 3.1 Haplotypes are combinations of SNPs inherited together on a chromosome.

the focus is not on families, but on population data, the most cost-efficient approaches are statistical. The most popular method in use is maximum likelihood, which is implemented in the expectation-maximization (EM) algorithm [48]. In addition, other statistical approaches have been developed [49, 50]. The EM algorithm estimates haplotype frequencies in the population and its performance, when assessed for small numbers of biallelic markers, has been deemed reliable [51], although less accurate than haplotypes derived from families [19]. Using statistical methods for initial haplotype determination, followed by confirmation of selected haplotypes by experimental approaches may be the most powerful strategy.

Because LD between SNPs (even those within the same gene) is not perfect, haplotype analysis may result in false negatives where some important associations may be missed. Therefore, at some point it may be necessary to evaluate all SNPs within a gene directly. This would first require screening the gene to identify SNPs, including those that are uncommon. Because this would require additional significant effort, its value must be weighed with respect to risk of missing associations as well as cost. A strategy to carry out comprehensive SNP discovery on a limited number of high-priority candidate genes, such as the drug target, may be worthwhile.

3.3.4
Number of SNPs Required for Whole Genome LD Mapping Studies

The size, or density, of a useful SNP map will be in part determined by the patterns and strength of LD throughout the genome. LD mapping requires that a susceptibility allele be detectable with a marker that lies within the interval afforded by the SNP map density. Given the 3 billion base pair size of the human genome, a minimum of 30,000 evenly spaced SNP markers would be needed to have a marker every 100 kb, i.e., the maximum estimated average extent of LD. As this estimated extent of LD is the best case scenario, 30,000 represents the minimum numbers of markers needed. As patterns of LD in the human genome are determined empirically, the number and distribution of SNPs needed to carry out whole genome LD mapping will be optimally determined. Smarter SNP selection could be achieved if blocks of LD were first defined.

3.4
Study Designs for Pharmacogenomic Analysis

3.4.1
Challenges Unique to Pharmacogenomics

In contrast to studies of disease which can take advantage of familial inheritance, homogeneous populations and relatively straight-forward ascertainment of affected individuals, options for pharmacogenomic analyses are limited. Since drug response is a trait whose expression is mediated through the administration of a therapeutic compound, "responders" can only be identified after they receive the drug. This makes ascertainment of individuals from a general population setting difficult and precludes the use of families except in the rare instance where multiple family members are given a drug. Whereas studies of disease can take advantage of homogeneous, isolated populations to increase LD and hence the likelihood of finding the disease gene [52], clinical trials are likely to be the main source of patients for pharmacogenomic studies. Although drugs on the market may be distributed worldwide, most clinical trials of new therapies are performed in Caucasian Americans or Europeans. Pharmacogenomic studies will most likely be carried out in these genetically heterogeneous clinical trial populations or in case-control or cohort association study designs employing either candidate gene or LD mapping approaches.

3.4.2
Clinical Trials, Case-Control and Cohort Studies

Pharmacogenomic studies can either be carried out in the context of interventional studies, i.e., clinical trials or in observational studies such as case-control or cohorts. Clinical trials are advantageous because they control many more parameters than observational studies. Extraneous variables such as the indication for treatment, dose, timing and compliance can all influence the efficacy of a drug. Clinical trials are much more reliable and powerful than observational studies since these factors are controlled for, thus eliminating their possible confounding effects. Furthermore, since clinical trials employ a placebo group, inference can be made on the independent effects of the gene, the drug, and gene-drug interactions on the course of disease. Ongoing trials can be leveraged to not only accomplish their primary goal (understanding the effect of a drug on the course of disease), but to assess the possible relevance of a person's genotype on that response as well. By evaluating the placebo group, the effect that the genotype may have on disease progression can be assessed. In this way, one could distinguish between simple markers of disease progression and true pharmacogenomic markers, whose effect on disease progression is only seen in the presence of a drug.

For drugs on the market without access to phase IV clinical studies, case-control or cohort studies may be employed. Cohort studies follow a group of patients taking a drug to see how many develop a specific outcome, which may be some discrete or continuous measure of efficacy or toxicity. Frequency of the outcome is compared in the genotype groups to see if there is an effect (relative risk) of the genotype on the outcome. Case-control studies, on the other hand, retrospectively identify a group of patients who took the drug and experienced a specific outcome (cases) and a group who took the drug but did not experience the outcome (controls). Genotype frequencies are compared between the cases and controls to see if there is an effect (odds ratio) of genotype on the outcome.

The choice of the study design is determined by the frequency of the outcome and the time to develop the outcome. While cohort studies are preferred since incidence of the outcome can be measured directly in the different genotype groups, case-control studies may be the only practical design for some outcomes. Pharmacogenomic studies of rare side effects, such agranulocytosis or drug-induced hepatitis, may be more amenable to a case-control design. Since these outcomes occur in $<5\%$ of patients who take certain compounds, a cohort design

Tab. 3.3 Examples of study designs used for pharmacogenomic analysis

	Case-control	*Cohort*	*Clinical trial*
Basis of selection of study population	Cases: patients who took drug and responded; Controls: patients who took drug and did not respond	Group of patients who took drug during a specific timeframe; response to drug unknown at time of selection	Patients randomly assigned to Treatment group: given drug or Placebo group: not given drug
When outcome (response) measured	prior to selection of study population (necessary to define cases and controls)	following selection of study population, either prospectively or retrospectively (med. records)	following selection of study population, prospectively
Effects measured	Gene-drug interaction; no independent effects of gene or drug can be measured	Gene-drug interaction; no independent effects of gene or drug can be measured	independent effects of gene, drug and gene-drug interaction
Major drawback	Dose, indication, timing may not be standardized and therefore could confound associations	if prospective, then same drawback as clinical trial; if retrospective, same drawback as case-control	Prospective study requiring follow-up time until endpoints develop to measure response
Major strength	Most efficient design for rare outcomes (e.g., toxicity such as hepatitis)	Selection of patients not based on outcome; therefore more appropriate than case control when response is continuous	Most accurate; controls for dose, indication, timing of treatment and other possible confounders

would need to enroll at least 5,000 people taking the drug in order to have 250 with the outcome of interest. Case-control studies or perhaps retrospective cohort studies may be the most practical choice for examining the long-term efficacy of a compound where 10–20 years of follow-up would otherwise be required to measure the outcome. Table 3.3 summarizes various study designs used in pharmacogenomic research.

3.5 Analytical Issues is Pharmacogenomic Studies

3.5.1 Effect of LD on Sample Size

Sample sizes required for discovering SNPs associated with drug response depend on a number of factors, among which are the SNP allele frequencies, the number of SNPs being tested and, for LD mapping studies, the strength of LD. A marker

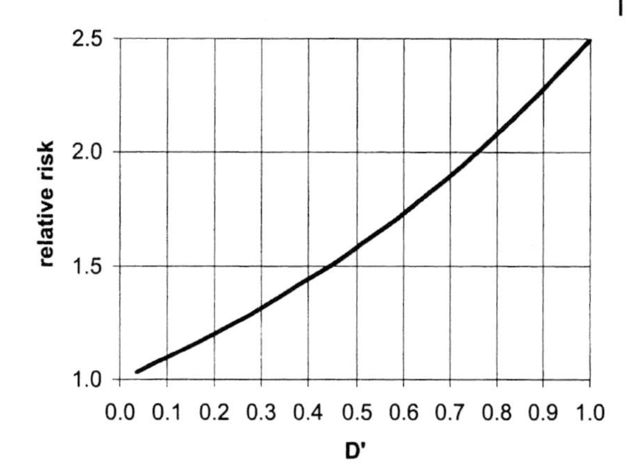

Fig. 3.2 Graph demonstrating the effect of varying linkage disequilibrium (D′) on relative risk. The underlying susceptibility SNP ($D'=1.0$) yields a relative risk of 2.5 and the marker and susceptibility SNPs have equal frequencies.

in LD with a susceptibility SNP will yield a relative risk that is smaller than if the susceptibility SNP were tested directly. Figure 3.2 illustrates the effect that varying LD has on relative risk. In this example, a SNP marker in complete LD ($D'=1.0$) with the susceptibility SNP and having a similar frequency yields a relative risk of 2.5. As D′ declines to 0.5, the detectable relative risk is only ~1.6. The weaker the LD between marker and susceptibility SNPs, the smaller the relative risk and the more difficult the association will be to detect unless the sample size is increased proportionately. To determine the sample size required for various values of D′, the formula below can be applied:

$$N/(D')^2$$

where N is the required sample size for detecting the association when $D'=1.0$.

For example, if 400 patients were required to find an association by measuring the causative SNP directly, then 816 patients would be required to find the same association using random SNPs in LD with the causative SNP where D′ is 0.7, and 1,600 patients if D′ is 0.5. This situation becomes even more complex when large differences in allele frequencies exist between markers and susceptibility SNPs. If marker allele frequencies are substantially different from the susceptibility allele frequency, then the required sample size, the number of markers, or both will need to be dramatically increased [53].

3.5.2
Multiple Hypothesis Testing

As the number of SNPs evaluated in an association study increases, a concern arises with false positives. This is true for candidate gene studies, but moreso for whole genome random SNP studies where tens to hundreds of thousands of SNPs will be evaluated. The traditional cutoff for assessing statistical significance ($p<0.05$] by definition can result in 5% false associations occurring simply by chance. The difficulty becomes distinguishing the true associations from these false associations. To account for the multiple hypothesis testing that occurs when many SNPs are evaluated, a simple correction can be applied to the *p*-value called Bonferroni's Inequality. This correction uses the formula [54]:

$$P^* = 1 - (1 - P)^n$$

where P* is the overall *p*-value, taking into account the observed *p*-value, P, and the number of hypotheses tested, n.

The major drawback of this method is that Bonferroni's Inequality is a conservative correction, especially if some of the hypotheses being tested are not independent. When many SNPs in the same gene are evaluated, for example, and are in LD with each other, the Boneferroni correction would not be appropriate, resulting in the possibility of false negatives or failure to detect a true association. A better approach would be to test the true level of significance directly through simulations.

Another option would be to not apply a correction at all but rather, to require any association, whether in the context of a single hypothesis or several thousand, to reach a level of genome-wide significance, similar to what is done in linkage analysis. Genome-wide significance may mean achieving a *p*-value on the order of 10^{-7} or 10^{-8} to account for 100,000 to one million multiple comparisons. Regardless of the correction method used, sample sizes will have to be substantially augmented to reach a reasonable level of statistical significance.

3.5.3 Gene–Drug Interaction

Pharmacogenomic studies are often carried out in the context of clinical trials which, by virtue of having both a treatment and placebo group, are able to assess the association between a gene and disease progression independent of therapy as well as interactive effects of genotypes and therapy on disease. Several of the phar-

Tab. 3.4 Structure of data to perform stratified analysis to assess gene-drug interaction

Pattern	*Genotype*	*Drug*	*Responders*	*Non-responders*	*Odds ratio*
1	+	+	*a*	*b*	*ah/bg*
2	+	–	*c*	*d*	*ch/dg*
3	–	+	*e*	*f*	*eh/fg*
4	–	–	*g*	*h*	1.00

+ present; – absent; italicized letters are numbers of patients in each category; Odds ratios are calculated relative to Pattern 4.
If the odds ratio for pattern 1 (joint effect of genotype and drug) is significantly greater than the product of odds ratios for patterns 2 (independent effect of genotype) and pattern 3 (independent effect of drug), then there is evidence for statistical (multiplicative) interaction. This analysis can be carried out in the context of multiple regression analysis by the inclusion of an interaction term.

macogenomic markers reported in the literature are also correlated with disease progression among the placebo group [55–57]. This indicates that the SNPs may simply be acting as a surrogate for disease prognosis (i.e., aggressive versus indolent disease) and not directly associated with the drug response at all. To test whether the gene and drug are having an interactive effect on disease progression, a stratified analysis can be undertaken [58] as illustrated in Table 3.4.

If there is no evidence of interaction, and yet there is evidence of an association between genotype and disease progression, the SNP still may be useful. Identifying fast disease progressors up front may be a mechanism to enrich clinical trials, decreasing the time needed to reach progression-related endpoints.

3.6
Development of Pharmacogenomic Markers

One of the biggest challenges that faces the field of pharmacogenomics is demonstrating that an observed genetic association is both real and medically useful. Many studies generate provocative hypotheses, but follow-through and confirmation are often lacking. So far, efforts to link SNPs to complex diseases have raised more questions than they have answered. The biggest concern raised is the pervasive lack of consistency in associations reported in the literature. The causes most often attributed to this are technical ones related to study design such as lack of statistical power or population stratification (confounding by race). The inability to consistently reproduce an association between a SNP and disease is more the rule than the exception. Assuming that the association is not a false positive, the inconsistency may be due to more fundamental issues related to differences between study populations. Diseases can be defined in an infinite number of ways, interactions with other genes or non-genetic factors may not be obvious and finally, the choice of a comparison (control) group will define what association is being tested. A general paradigm should be hypothesis generation followed by replication or validation in additional populations. Validation should answer the questions: Is the association real? Is it useful? Is it generalizable?

In the field of disease gene associations, many associations have been reported but follow-up studies fail to consistently replicate the initial finding. Several articles reviewing specific controversial gene-disease associations appear in the literature [59–61]. These articles reveal a common theme. Taken as a whole, the body of literature suggests that the individual polymorphisms in question are indeed associated with the disease, but have only a very small effect. The magnitude of association found in these studies is typically less than 2-fold. The effect of a genetic variant on disease predisposition or drug response may vary depending on other genetic or non-genetic factors. It is already well-recognized that substantial differences in SNP allele frequencies can occur between ethnic groups [17]. This implies that any individual SNP which predisposes to disease or drug response in one ethnic group may have little relevance in another. In addition, the SNP under investigation may not be the underlying cause of the association, nor the best marker. In order to improve the predictive value of a SNP, one must identify both the exact SNP or combination of SNPs that confer risk and the subset of patients for whom the SNP is most relevant.

Further complexities such as gene-gene or gene-environment interaction will become more evident as we decipher the genome. For example, a recent study has suggested that risk of Alzheimer's disease in carriers of the APOE4 allele is modified by a variant (TNFα-t) in the tumor necrosis factor alpha gene promoter. Carriers of both APOE4 and TNFα-t alleles had a 4.6-fold increased odds of Alzheimer's disease compared to non-carriers, whereas APOE4 carriers who lacked the TNFα-t allele had only a 2.7-fold increased odds of Alzheimer's disease [62]. As with disease markers, it is likely that the most useful tests for predicting drug response will be those that combine a number of genetic and non-genetic factors together in order to achieve sufficient predictive power in a subset of patients. This is illustrated well in one study where six polymorphisms in four genes used in combination produced the best test for predicting response to clozapine in schizophrenic patients [63].

3.7 Conclusion

A number of challenges exist today in turning SNPs into useful markers of drug response. The identification and mapping of hundreds of thousands of SNPs is only the beginning. With a plethora of SNPs available, choosing which SNPs to examine becomes a daunting task. The extent of LD and the resulting need to detect associations with very small effect is the pivotal issue that will determine for which situations SNP LD mapping could work as a global discovery tool. While LD mapping is appealing in that it is an unbiased approach and allows for a comprehensive genome-wide survey, the challenges and limitations are significant. Alternative methods, such as the candidate gene approach, offer several advantages: the candidate gene approach is proven; currently genotyping a limited number of candidate SNPs is economically feasible; no assumptions are made about LD; and the required sample sizes are consistent with current clinical trials.

Regardless of the method chosen, large, well-designed clinical study populations, cost-effective genotyping and analytical strategies all need to converge in the process of discovering markers of drug response. Pharmacogenomic markers should be judged not only on the strength of their association with drug response, but the ability to reproduce these associations in independent populations. More sophisticated analytical techniques may need to be developed to identify combinations of genes and other factors that together produce the most accurate pharmacogenomic marker profile. The ultimate challenge of pharmacogenomics will be the application of these SNPs to clinical practice and drug discovery and the ultimate improvement of the health of individuals.

3.8 References

1 Weinshilboum R. Thiopurine pharmacogenomics: clinical and molecular studies of thiopurine methyltransferase. Drug Metab Dispos **2001**; 29[4 Pt 2]:601–605.

2 Drews J. Drug discovery: A historical perspective. Science **2000**; 287:1960–1964.

3 Iida A, Sekine A, Saito S, Kitamura Y, Kitamoto T, Osawa S et al. Catalog of 320 single nucleotide polymorphisms (SNPs) in 20 quinone oxidoreductase and sulfotransferase genes. J Hum Genet **2001**; 46:225–240.

4 Drazen JM, Yandava CN, Dube L, Szczerback N, Hippensteel R, Pillari et al. Pharmacogenomic association between ALOX5 promoter genotype and the response to anti-asthma treatment. Nature Genet **1999**; 22:168–170.

5 Hedenfalk I, Duggan D, Chen Y, Radmacher M, Bittner M, Simon R et al. N Engl J Med **2001**; 344:539–548.

6 Poirier J, Delisle MC, Quirion R, Aubert I, Farlow M, Lahiri D et al. Apolipoprotein E4 allele as a predictor of cholinergic deficits and treatment outcome in Alzheimer disease. Proc Natl Acad Sci USA **1995**; 92:12260–12264.

7 Farlow MR, Lahiri DK, Poirier J, Davignon J, Schneider L, Hui SL. Treatment outcome of tacrine therapy depends on apolipoprotein genotype and gender of the subjects with Alzheimer's disease. Neurology **1998**; 50:669–677.

8 Richard F, Helbecque N, Neuman E, Guez D, Levy R, Amouyel P. APOE genotyping and response to drug treatment in Alzheimer's disease. Lancet **1997**; 349:539.

9 Rigaud AS, Traykov L, Caputo L, Guelfi MC, Latour F, Couderc R et al. The apolipoprotein E epsilon4 allele and the response to tacrine therapy in Alzheimer's disease. Eur J Neurol **2000**; 7:255–258.

10 Muller-Myhsok B, Abel L. Genetic analysis of complex diseases. Science **1997**; 275:1328–1330.

11 Hartl DI, Clark AG. Principles of Population Genetics. Sinauer Associates. Sunderland, MA, **1990**.

12 Huttley GA, Smith MW, Carrington M, O'Brien SJ. A scan for linkage disequilibrium across the human genome. Genetics **1999**; 152:1711–1722.

13 Clark AG, Weiss KM, Nickerson DA, Taylor SL, Buchanan A, Stengard J et al. Haplotype structure and population genetic inferences from nucleotide-sequence variation in human lipoprotein lipase. Am J Hum Genet **1998**; 63:595–612.

14 Lai E, Riley J, Purvis I, Roses A. A4-Mb high-density single nucleotide polymorphism-based map around human APOE. Genomics **1998**; 54:31–38.

15 Jorde LB, Watkins WS, Carlson M, Groden J, Albertsen H, Thliveris A, et al. Linkage disequilibrium predicts physical distance in the adenomatous polyposis coli region. Am J Hum Genet **1994**; 54:884–898.

16 Kruglyak L. Prospects for whole genome linkage disequilibrium mapping of common disease genes. Nature Genet **1999**; 22:139–144.

17 Goddard KA, Hopkins PJ, Hall JM, Witte JS. Linkage disequilibrium and allele frequency distributions for 114 single-nucleotide polymorphisms in five populations. Am J Hum Genet **2000**; 66:216–234.

18 Dunning AM, Durocher F, Healey CS, Teare D, McBride SE, Carlomagno F et al. The extent of linkage disequilibrium in four populations with distinct demographic histories. Am J Hum Genet **2000**; 67:1544–1554.

19 Abecasis GR, Noguchi EM, Heinzmann A, Traherne JA, Bhattacharyya S, Leaves NI et al. Extent and distribution of linkage disequilibrium in three genomic regions. Am J Hum Genet **2001**; 68:191–197.

20 Reich DE, Cargill M, Bolk S, Ireland J, Sabeti PC, Richter DJ et al. Linkage disequilibrium in the human genome. Nature **2001**; 411:199–204.

21 Eaves IA, Merriman TR, Barber RA, Nutland S, Tuomilehto-Wolf E, Tuomilehto J et al. The genetically isolated

populations of Finland and Sardinia may not be a panacea for linkage disequilibrium mapping of common disease genes. Nature Genet **2000**; 25:320–322.

22 Taillon-Miller P, Bauer-Sardina I, Saccone NL, Putzel J, Laitinen T, Cao, A et al. Juxtaposed regions of extensive and minimal linkage disequilibrium in human Xq25 and Xp28. Nat Genet **2000**; 25:324–328.

23 Kidd KK, Morar B, Castiglione CM, Zhao H, Pakstis AJ, Speed WC, et al. A global survey of haplotype frequencies and linkage disequilibrium at the DRD2 locus. Hum Genet **1998**; 103:211–227.

24 Lewontin RC. The interaction of selection and linkage. I. General considerations; heterotic models. Genetics **1964**; 49:49–67.

25 Jorde LB. Linkage disequilibrium and the search for complex disease genes. Genome Res **2000**; 10:1435–1444.

26 Friedman TB, Liang Y, Weber JL, Hinnant JT, Barber TD, Winata S et al. A gene for congenital, recessive deafness DFNB3 maps to the pericentromeric region of chromosome 17. Nature Genet **1995**; 9:86–91.

27 Houwen RHJ, Baharloo S, Blankenship K, Raeymakers P, Juyn J, Sandkuyl LA et al. Genome screening by searching for shared segments: mapping a gene for benign recurrent intrahepatic cholestasis. Nature Genet **1994**; 8:380–386.

28 Puffenberger EG, Kauffman ER, Bold S, Matise TC, Washington SS, Angrist M et al. Identity-by-decent and association mapping of a recessive gene for Hirschsprung disease on human chromosome 13q22. Hum Mol Genet **1994**; 3:1217–1225.

29 Risch N, Merikangas K. The future of genetic studies of complex human diseases. Science **1996**; 273:1516–1517.

30 Zubenko GS, Hughes BHB, Stiffler JS, Hurtt MR, Kaplan BB. A genome survey for novel Alzheimer disease risk loci: results at 10-cM resolution. Genomics **1998**; 50:121–128.

31 Kruglyak L, Nickerson DA. Variation is the spice of life. Nature Genet **2001**; 27:234–236.

32 The International SNP Map Working Group. A map of human genome sequence variation containing 1.42 million single nucleotide polymorphisms. Nature **2001**; 409:928–933.

33 Przeworski M, Hudson RR, DiRienzo A. Adjusting the focus on human variation. Trends Genet **2000**; 16:296–302.

34 Li WH, Gu Z, Wang H, Nekrutenko A. Evolutionary analyses of the human genome. Nature **2001**; 409:847–849.

35 Cargill M, Altshuler D, Ireland J, Sklar P, Ardlie K, Patil N, et al. Characterization of single-nucleotide polymorphisms in coding regions of human genes. Nature Genet **1999**; 22:231–238.

36 Halushka MK, Fan JB, Bentley K, Hsie L, Shen N, Weder A et al. Patterns of single-nucleotide polymorphisms in candidate genes for blood-pressure homeostasis. Nature Genet **1999**; 22:239–247.

37 Pennacchio LA, Rubin EM. Genomic strategies to identify mammalian regulatory sequences. Nature Rev Genet **2001**; 2:100–109.

38 Pequignot MO, Desguerre I, Dey R, Tartari M, Zeviani M, Agostino A et al. New splicing-site mutations in the SURF1 gene in leigh syndrome patients. J Biol Chem **2001**; 276:15326–15329.

39 Suen TC, Goss PE. Identification of a novel transcriptional repressor element located in the first intron of the human BRCA1 gene. Oncogene **2001**; 20:440–450.

40 Venter JC, Adams MC, Sutton GG, Kerlavage AR, Smith HO, Hunkapiller M. Shotgun sequencing of the human genome. Science **1998**; 280:1540–1542.

41 Buetow KH, Edmonson MN, Cassidy AB. Reliable identification of large numbers of candidate SNPs from public EST data. Nature Genet **1999**; 21:323–325.

42 Online Mendelian Inheritance in Man, OMIM (TM). Johns Hopkins University, Baltimore, MD. MIM Number:23520: May **2001**. World Wide Web URL: http://www.ncbi.nlm.nih.gov/omim/

43 Drysdale CM, McGraw DW, Stack CB, Stephens JC, Judson RS, Nandabalan K et al. Complex promoter and coding region 2-adrenergic receptor haplotypes alter receptor expression and predict *in*

vivo responsiveness. Proc Natl Acad Sci USA **2000**; 97:10483–10488.

44 Woolley AT, Guillemette C, Li Cheung C, Housman DE, Lieber CM. Direct haplotyping of kilobase-size DNA using carbon nanotube probes. Nature Biotechnol **2000**; 18:713.

45 Yan H, Papadopoulos N, Marra G, Perrera C, Jiricny J, Boland CR et al. Conversion of diploidy to haploidy. Nature **2000**; 403:723–724.

46 Sarkar G, Sommer SS. Haplotyping by double PCR amplification of specific alleles. Biotechniques **1991**; 10:436, 438, 440.

47 Sobel E, Lange K. Descent graphs in pedigree analysis: applications to haplotyping, location scores and marker sharing statistics. Am J Hum Genet **1996**; 58:1323–1337.

48 Excoffier L, Slatkin M. Maximum-likelihood estimation of molecular haplotype frequencies in a diploid population. Mol Biol Evol **1995**; 12:921–927.

49 Clark AG. Inference of haplotypes from PCR-amplified samples of diploid populations. Mol Biol Evol **1990**; 7:111–122.

50 Stephens M, Smith N, Donnelly P. A new statistical method for haplotype reconstruction from population data. Am J Hum Genet **2001**; 68:978–989.

51 Fallin D, Schork NJ. Accuracy of haplotype frequency estimation for biallelic loci via the expectation-maximization algorithm for unphased diploid genotype data. Am J Hum Genet **2000**; 67:947–959.

52 Terwilliger JD, Zollner S, Laan M, Pääbo S. Mapping genes through the use of linkage disequilbrium generated by genetic drift: "drift mapping" in small populations with no demographic expansion. Hum Hered **1998**; 48:138–154.

53 Cox NJ, Bell GI. Disease associations: chance, artifact or susceptibility genes? Diabetes **1989**; 38:947–950.

54 Fisher L. Biostatistics: a methodology for the health sciences. John Wiley & Sons. New York, **1993**.

55 Jukema JW, van Boven AJ, Groenemeijer B, Zwinderman AH, Reiber JH, Bruschke AV et al. The Asp9 Asn mutation in the lipoprotein lipase gene is associated with increased progression of coronary atherosclerosis. Circulation **1996**; 94:1913–1918.

56 Kuivenhoven JA, Jukema JW, Zwinderman AH, de Knijff P, McPherson R, Bruschke AV, et al. The role of a common variant of the cholesteryl ester transfer protein gene in the progression of coronary atherosclerosis. N Engl J Med **1998**; 338:86–93.

57 de Maat MP, Kastelein JJ, Jukema JW, Zwinderman AH, Jansen H, Groenemeier B et al. –455G/A polymorphism of the β-fibrinogen gene is associated with the progression of coronary atherosclerosis in symptomatic men. Arterioscler Thromb Vasc Biol **1998**; 18:265–271.

58 Yang Q, Khoury MJ. Evolving methods in genetic epidemiology. III. Gene-environment interaction in epidemiologic research. Epidemiol Rev **1997**; 19:33–43.

59 Schunkert H. Polymorphism of the angiotensin-converting enzyme gene and cardiovascular disease. J Mol Med **1997**; 75:867–875.

60 Altshuler D, Hirschhorn JN, Klannemark M, Lindgren CM, Vohl MC, Nemesh J et al. The common PPARγ Pro12Ala polymorphism is associated with decreased risk of type 2 diabetes. Nature Genet **2000**; 26:76–80.

61 Zmuda JM, Cauley JA, Ferrell RE. Molecular epidemiology of vitamin D receptor gene variants. Epidemiol Rev **2000**; 22:203–217.

62 McCusker SM, Curran MD, Dynan KB, McCullagh CD, Urquhart DD, Middleton D et al. Association between polymorphism in regulatory region of gene encoding tumour necrosis factor a and risk of Alzheimer's disease and vascular dementia: a case-control study. Lancet **2001**; 357:436–439.

63 Arranz MJ, Munro J, Birkett J, Bolonna A, Mancama D, Sodhi M et al. Pharmacogenomic prediction of clozapine response. Lancet **2000**; 355:1615–1616.

4
Association Studies in Pharmacogenomics

Laurent Essioux, Benoit Destenaves, Philippe Jais and François Thomas

Abstract

Single nucleotide polymorphisms (SNPs) are common genetic variations present in DNA that are thought to account for most of the genetic variations that occur between individuals. The near completion of the sequencing of the human genome and the development of high-throughput genotyping technologies, provide the means to assemble SNPs into maps that can be used for the dissection of complex genetic traits. This approach, reflected in the branch of genetics known as pharmacogenomics, can be used to identify gene variations that determine individual drug responses and has the ability to elucidate the gene networks that determine drug efficacy and associated toxicity in individuals. This chapter focuses on the methodologies that use SNP maps for the identification of genetic factors modulating drug responses. With this information, new drugs can be developed that target individual genetic profiles to achieve maximal efficacy and minimal side effects.

4.1
Introduction

In a large number of instances, the optimal dosage regimen of clinically useful drugs is limited by variability in the way individuals respond to these drugs, both in terms of drug efficacy and side effects, i.e., adverse drug reactions (ADRs). The latter are a major cause of non-compliance and failure of treatment, particularly for chronic pathologies. For instance, the daily doses required to treat patients vary by 20-fold for the antithrombotic drug warfarin, by 40-fold for the antihypertensive drug propranolol, and by 60-fold for L-dopa, the standard treatment for Parkinson's disease [1]. Furthermore, both the duration of treatment and the progression of Parkinson's disease require increasing amounts of L-dopa and as more drug is given, side effects increase. Other drugs have clinical utility in a subset of patients with a given pathology, e.g., antipsychotics that are ineffective in 30% of schizophrenics [2], suggesting the fact that such drugs are only effective in patients with specific disease etiologies.

Inter-individual variability in drug response and ADRs are major public health problems. In the United States in 1998, costs related to hospital admissions for underdosing, overdosing and prescription of unnecessary drugs exceeded US$ 100 billion a year [3]. Furthermore, this 1998 report indicated that ADRs are responsible for approximately 7% of hospitalizations with fatalities occurring in 0.3% of cases, making ADRs the fourth to sixth leading cause of death in the United States. As a whole, each year ADR-associated hospitalizations cost from US$ 30 billion to US $ 150 million [3]. Thus, a significant proportion of the drugs prescribed for disease treatment does not either result in the desired outcome or produce side effects that limit drug utility. The ability to understand by pharmacogenomics the reasons for variations in drug effects at the individual level has the potential to improve responses, to prevent ADRs and make both the practice and cost of medicine less of a societal burden. In addition, there is considerable discussion in the pharmaceutical industry regarding the use of pharmacogenomics to reduce both costs and risk in the drug development process.

4.2
Variability and ADR in Drug Response: Contribution of Genetic Factors

A number of non-genetic factors can cause drug response variability, including renal, pulmonary, cardiovascular or hepatic system status, age and gender, drug-drug interactions, environmental and nutritional factors, pathogenesis and severity of the disease. In addition to the causal relationship of such factors in determining drug response and ADR variability, a growing body of evidence also demonstrates that multiple genetic factors contribute to this variability. An example of such multiple interaction is illustrated in patients undergoing warfarin therapy, for which a number of genetic and non-genetic factors have been shown to affect response (Tab. 4.1).

Early evidence that genetic factors are associated with drug response variability is based on the phenotypic differences of drug metabolizing enzymes in individuals showing ADRs. The reduced activity of a phase II liver metabolism enzyme and the frequency of neurotoxicity of the antitubercular drug isoniazid having an almost identical geographical distribution has provided the first evidence that a genetic factor was involved in ADR [4]. More recently, the reduced activity of the phase II liver metabolism enzyme has been shown to be due to a homozygous inactivating polymorphism of the enzyme *N-a*cetyl *t*ransferase *2* (NAT2]. Other examples include: hemolysis due to the antimalarial drug chloroquine in patients with glucose-6-phosphate dehydrogenase deficiency, prolonged muscle relaxation caused by suxamethonium in patients with plasma cholinesterase deficiency, and more recently, lupus-like syndrome induced by procainamide in patients with mutations in the liver enzyme CYP2D6. Familial studies have provided further evidence for the role of genetic factors in drug variability. For instance, a high concordance rate of chloramphenicol-induced aplastic anemia occurs in identical twins [5], as well as a high concordance rate of lithium response in bipolar first-

Tab. 4.1 Known causes of individual variability in warfarin therapy in thromboembolism

Causes of variation	***Consequence on warfarin therapy***
Extra-genetic factors	
Association to miconazole, thyroid hormones, and others	Unknown
Association to allopurinol, amiodarone, fluoroquinolones, carbamazepine, phenobarbital, rifampicin, and others	Inhibition of CYP activity by the drugs
Association to cholestyramine, and sucralfate	Inhibition of digestive absorption
Interaction with non-steroidal anti-inflammatory drugs, androgens, fibrates, sulfamethoxazole, and others	Binding of drugs to albumin that increases warfarin free-circulating level
Food intake	Increase of alimentary vitamin K1 intake, resulting in decreased anticoagulant effect
Alcohol intake	Induction of liver-metabolism during chronic alcohol intake
Hepatic insufficiency	Decrease of liver metabolism, leading to increased efficacy of warfarin
Renal insufficiency	Decrease of renal elimination, leading to increased efficacy of warfarin
Genetic Factors	
Albumin	Reduced warfarin affinity of some albumin variants (18)
α1-acid glycoprotein	Variable warfarin affinity of some α1-acid glycoprotein variants (19)
CYP2C9	Multiple allelic variants of CYP2C9 may account for the occurrence of poor metabolizers and increased efficacy of warfarin (84)
CYP2C19	Effects of CYP2C19 polymorphisms on the metabolism of warfarin are not yet elucidated, although this enzyme is known to metabolize warfarin (85)

degree relatives of bipolar probands (e.g., they often have a similar response pattern) [6].

These examples have opened a challenging area of pharmacology, called pharmacogenetics, which focuses on the discovery of the genetic factors responsible for the variability in drug responses. Polymorphisms in the genes coding for proteins involved in drug absorption, distribution, metabolism and excretion (ADME), as well as in drug response can significantly influence the *in vivo* response to drugs. More recently, technologies enabling high-throughput analysis of genetic polymorphisms and expression have led to a new era for pharmacogenomics. Nowadays, the differences between pharmacogenetics and pharmacogenomics is somewhat subtle and arbitrary, such that these terms are often used interchangeably [4].

4.3 Multiple Inherited Genetic Factors Influence the Outcome of Drug Treatments

4.3.1 Background

The *in vivo* response to drug treatment is a complex and highly dynamic process which involves multiple factors. As many as 50 proteins, e.g., carrier proteins, transporters, metabolizing enzymes, receptors and their transduction components take part in pharmacodynamic response to a drug. Many genes coding for such proteins contain polymorphisms that alter the activity or the level of expression of the encoded proteins. Thus, the response of a given drug in a given individual reflects the interaction of multiple variable genetic factors that cause important variations in drug metabolism, distribution and action on its target.

4.3.2 Liver Metabolism Enzymes

Although hydrophilic drugs can be eliminated unchanged by passive filtration or excretion, other drugs, due to their lipophilic nature, cannot. Thus liver metabolism, which involves the synthesis of more hydrophilic polar residues, is a major pathway for the elimination of drugs in the bile. Such liver metabolism enzymes, mainly found in the endoplasmic reticulum of hepatocytes, can be classified into two groups: phase I or phase II. A number of these display wild type and other phenotypes, the latter resulting from genetic polymorphisms. Poor metabolism results in the accumulation of drug substrate and is typically a codominant autosomal trait due to deletions, null mutations or inactivating mutations in metabolizing enzymes. Ultra-extensive metabolism resulting in increased drug metabolism is usually due to gene duplication and results in an autosomal dominant trait [7]. However, there is no absolute correspondence between phenotype and genotype. This can be due to incorrect phenotype assignment by co-administration of drugs that may change gene expression, confounding effects of disease, and allelic heterogeneity due to the second allele product.

Phase I enzymes are responsible for modification of functional groups, through hydrolysis, oxidation, reduction or hydroxylation. Phase I metabolism is mostly carried out by the monooxygenase heme–thiolate protein superfamily, referred to as *cytochrome P450* isoenzymes (CYP). Approximately 80 different forms of P450 have been characterized in humans, each of which have a distinct catalytic specificity and unique regulation [8]. Because of this diversity, the high frequency of CYP genetic variations may be explained by their redundant or dispensable nature. In humans, it appears that approximately half a dozen CYP enzymes are responsible for metabolizing the vast majority of prescribed drugs. Among these, three (CYP2D6, CYP3A4, CYP2C19] are polymorphic and responsible for most ADRs demonstrated to date [8]:

1) CYP2D6 is involved in the metabolism of more than 30 drugs. At least 16 different genetic variations including point mutations resulting in early stop codons and amino acid substitutions, microsatellite nucleotide repeats, and gene amplifications or deletions are the cause of differences in the enzymatic activity of CYP2D6, which ranges from complete deficiency to ultrafast metabolism. The CYP2D6 poor-hydroxylator phenotype, the so-called sparteine-debrisoquin allele, is caused by homozygous inheritance of inactivating alleles. The frequency of this trait varies extensively among ethnic groups, ranging from 1% in Arabs to 30% in Chinese [9]. Consequently, clinical responses to standard doses of drugs metabolized by CYP2D6, such as haloperidol, chlorpromazine, codeine, and antiarrhythmics, vary from increased risk of ADRs at recommended doses in poor metabolizers, to therapeutic failure in ultrafast metabolizers [9].
2) CYP3A4 is involved in the metabolism of the largest number of drugs and influences their intestinal absorption. Although wide phenotypic variations have been found in human liver CYP3A4 activity [10], possibly caused by a common polymorphism in the promoter region [11], the resulting changes in drug responses remain to be elucidated. CYP2C19 contains several inactivating mutations, whose frequencies vary among ethnic populations [9].
3) Polymorphisms of CYP2C19 cause differences in metabolism of omeprazole, a proton pump inhibitor used for treatment of gastroduodenal ulcers or reflux esophagitis. Such polymorphisms result in resistance to treatment at a standard dose regimen in nearly 20% of European Caucasians, and in an even higher percentage of Asians [12].

Phase II conjugation of drugs can occur alone or after phase I metabolism. Phase II enzymes link large endogenous polar moieties to drug molecules in order to enhance their excretion in urine or bile. Several of these proteins have an important role in interindividual variations of drug response, such as *u*ridine diphosphate *g*lucuronosyl*t*ransferase 1A1 (UGT1A1], NAT2 and *t*hio*p*urine S-*m*ethyl*t*ransferase (TPMT). Firstly, UGT1A1 detoxifies various lipophilic chemicals and endogenous substances including bilirubin. This gene contains a promoter polymorphism that alters the level of expression of the encoded enzyme, resulting in a wide variation of drug metabolism [13]. Secondly, the homozygous occurrence of NAT2 inactivating mutations results in a slow-inactivator phenotype. This phenotype is responsible for increased dose-dependent toxicity due to the accumulation of non-metabolized drugs such as hydralazine-induced lupus, isoniazid-induced neuropathies, and sulfonamide-induced hypersensitivity reactions in some ethnic groups [4]. Thirdly, some polymorphisms of TPMT reduce biotransformation of thiopurine drugs, such as azathioprine and 6-mercaptopurine, and can lead to potentially fatal hematopoietic toxicity in some patients [14].

4.3.3 Transporters

Drug disposition does not only depend on liver metabolism and passive glomerular filtration, but also on absorption and excretion across tissues that have a barrier function such as blood-brain endothelium, as well as biliary, intestinal and tubular renal epithelia [15]. The total number of transporters is still unknown despite important efforts dedicated into their identification. The best characterized transporter remains the human multidrug resistant protein (MDR1], a member of the P-glycoprotein family, that is known to be polymorphic [16]. This transporter pumps a large number of structurally diverse drugs out of the cell and back into the intestinal lumen, e.g., chemotherapic agents, cyclosporine, HIV protease inhibitors and other CYP3A4 metabolites [16]. Although there are no clear structural features defining MD1R substrates, most of these molecules are lipophilic and/or amphipathic, and contain one or more aromatic rings. A polymorphism in exon 26 of MDR1 correlates with reduced absorption of digoxin, and is, therefore, proposed as a possible cause of inter-individual response variability and ADRs [17]. Other less characterized transporters include a second P-glycoprotein coded by the human MDR3 gene (also known as MDR2 in mouse), the canalicular multidrug-resistant protein (cMRP, also known as cMOAT or MRP2], and the organic anion tranporters, OATP and OAPT2. Despite the functional role of transporters in drug disposition, their polymorphic nature and eventual involvement in drug response variability remains to be elucidated.

4.3.4 Plasma Binding Proteins

Several physiological or pathological conditions including renal failure, cirrhosis, and inflammatory states, can modify the circulating level of plasma binding proteins, mainly serum albumin and α1-acid glycoprotein, resulting in drug response variability. While a number of polymorphisms have been identified in these plasma binding proteins, only a few studies have related these to variations in drug response. For instance, some albumin variants such as Canterbury (Lys313Asn) and Parklands (Asp365His), cause a decrease in high-affinity binding of warfarin, salicylate, and diazepam [18]. Genetic variants of α1-acid glycoprotein can affect binding of imipramine, warfarin and mifepristrone [19] and, consequently, their free-concentration in the plasma.

4.3.5 Drug Targets

Most drugs interact with specific target proteins in order to exert their pharmacological effects. These include membrane receptors and enzymes, such as the β_2-adrenoceptor, insulin receptor, various serotonergic receptors, angiotensin converting enzyme (ACE), and HMG CoA reductase. More rarely, drugs link to signal transduction molecules such as the *p*eroxisome *p*roliferator *a*ctivated *r*eceptor γ-2 (PPAR-γ-2]. Most of these drug targets exhibit polymorphisms that can alter the sensitivity to specific drugs [4]. Polymorphisms can play a major role in response variation when minimal inter-individual variation in plasma levels of a drug and its metabolites are seen, but major pharmacodynamic differences occur. Additional complications exist when multiple targets participate in the overall drug response. Arranz et al. [20] suggest that at least six receptors can influence the response to the atypical neuroleptic clozapine. While these findings have not been replicated to date [21], they represent a first approach to understand the genetics of multiple targeting of one of the most important psychiatric drugs available today.

Target genes involved in drug response can also be involved in disease susceptibility. For example, polymorphisms of the *c*holesteryl *e*ster *t*ransfer *p*rotein (CETP) are associated with both coronary atherosclerosis and response to pravastatin [22]. In addition, β_2-adrenoceptor polymorphisms are associated with severity of asthma [23] and variability in antagonist responses used in treatment [24]. ACE polymorphisms are associated with both susceptibility to essential hypertension [25] and response to therapeutically effective ACE inhibitors [25].

In most cases, ADRs result from the drug interaction with a target distinct from the therapeutic one. For example, polymorphisms in the 3'-untranslated region of the prothrombin gene result in enhanced risk of cerebral vein thrombosis and embolism in women receiving oral contraceptives [26]. Furthermore, the malignant hyperthermia induced by anesthetics such as halothane or succinylcholine, results from multiple mutations in the ryanodine receptor whose consequences segregate in certain families as an autosomal dominant trait [27]. In contrast, other ADRs can be caused by interaction of a drug with the target involved in drug therapeutic response. For instance, a polymorphism of the dopamine D3-receptor is associated with an increased risk of tardive dyskinesia induced by typical neuroleptics [28], this gene being also associated with response to this family of drugs [29].

Genetic variation of pathogenic agents in microorganism-caused disorders can be an additional source of individual drug sensibility. Since this article only focuses on human polymorphisms, we will illustrate this concept with one example. The severity of chronic hepatitis C and its complications is influenced by variations in host genes e.g., TAP2 polymorphism (*t*ransporter *a*ssociated with antigen *p*rocessing) [30], TNF-α promoter variants [31], TGF-β1 polymorphisms [32], and HLA class II and III alleles [33]. Moreover, drug response to interferon-α therapy in chronic hepatitis C patients also depends on variations in host genes including polymorphisms of IL-10 [34], interferon (IFN)-inducible MxA protein [35], and mannose-binding lectin [36] genes. Furthermore, polymorphisms in the sequence of the hepatitis C virus, e.g., the "interferon sensitivity determining region" (ISDR) of HCV-1b genotype [37] and the complexity of the hypervariable region 1 (HVR1] [38], can also modulate the efficacy of interferon-α-2 therapy. This situation is even further complicated by taking into account the evolution of the hypervariable region of hepatitis C virus quasispecies in response to interferon-α therapy, that likely causes profound changes in virus-host interactions [39].

4.4 Association Studies in Pharmacogenomics

4.4.1 The Principles of Association Studies

Association studies are used to identify common alleles more frequently associated with phenotypic traits such as drug response or ADRs. These associated alleles are called "causal polymorphisms" or "functional polymorphisms" and are risk-conferring factors of the trait at the population level, while, at the individual level, they impact either on the cellular level of the gene product or its function.

All polymorphisms that can be identified by such an approach are common. This population-based approach thus postulates that common factors explain a substantial genetic component of the trait. Numerous reports show that such polymorphisms do exist in pharmacogenomics (Tab. 4.2) and that this is a valid model [40]. If a more complex allelic structure is assumed where the effect of a gene on a trait is mediated by a multitude of rare functional polymorphisms (a model such as BRCA1 for breast cancer), population-based approaches have no power to identify causative genes. Whether this model is more frequent is currently unknown due to a paucity of knowledge and understanding of human genetic diversity [41].

If the polymorphism tested in an association study is a disease risk factor, its frequency is expected to increase with the studied trait. If the polymorphism tested is silent, but is correlated to a nearby causal polymorphism, an increase of frequency is still expected, although the effect will be milder, reflecting an indirect association of the tested polymorphism to the trait. The correlation between polymorphisms, known as *l*inkage *d*isequilibrium (LD), is a cornerstone of indirect-based association studies. The stronger the LD between the polymorphisms and the risk-conferring mutation, the stronger the effect found by an association study would be. This population-based parameter quantifies the concordance of two alleles of physically linked markers from an individual. Its intensity depends on biological processes, such as crossing-over and recurrent mutations, which tend to lower it. Linkage disequilibrium is also influenced by population factors including geographic migration, population size, evolution, and biological selection. Thus, LD is specific for a given population. Due to the multiplicity of factors listed above, all of which are critical to quantify, LD is extremely difficult to model and predict.

4.4.2 Study Design

The main potential drawback of genetic association studies is the bias resulting of the lack of ethnic matching between the groups under study, which can lead to spurious results [42, 43]. This bias is the best known and acute problem of this design and affects the interpretation of the results. When comparing two groups

Tab. 4.2 Examples of drug targets associated to disorder susceptibility, and therapeutic response

Gene	***Drug***	***Association with drug response and/or ADRs***	***Association with disorder***
5-hydroxytryptamine 2A receptor (HTR2A)	Clozapine and other neuroleptics	Variation of clozapine (20) and typical neuroleptics (86) efficacy in the treatment of schizophrenia	Increased susceptibility to schizophrenia (87)
Angiotensinogen converting enzyme (ACE)	ACE-inhibitors (enalapril, lisinopril, captopril)	Variation of ACE efficacy in the treatment of hypertension (88)	Susceptibility to essential hypertension (89)
Angiotensin I (angiotensinogen, AGT)	Salt intake and ACE inhibitors	Variation in the efficacy of reduced sodium intake (90) and of ACE inhibitors in the treatment of hypertension (25)	Susceptibility to essential hypertension (25)
Apolipoprotein E (APOE)	Cholinesterase inhibitor (Tacrine)	Variation in cognitive function improvements with tacrine in the treatment of Alzheimer's disease (68)	Susceptibility to Alzheimer's disease, increased by the epsilon-4 allele, and decreased by epsilon-2 allele (91)
Cholesteryl ester transfer protein (CETP)	HMG-CoA reductase inhibitor (pravastin)	Efficacy of pravastatin in the treatment of coronary atherosclerosis (22)	Susceptibility to coronary atherosclerosis (22)
Beta-2-adrenergic receptor (ADRB2)	Beta-2-adrenergic receptor antagonists (salbutamol, formoterol)	Beta-2-adrenergic receptor polymorphisms affect airway responsiveness to beta-2-adrenergic receptor antagonists in asthmatics (24)	Susceptibility to lower airway reactivity in asthmatic (23) and severity of asthma (92)
Peroxisome Proliferator activated receptor (PPAR-GAMMA-2)	Insulin	Variation of insulin efficacy in the treatment of diabetes (93)	Susceptibility to type II diabetes (93)
Sulfonylurea receptor 1 (SUR1)	Sulfonylurea (tolbutamide)	Decreased tolbutamide-stimulated insulin secretion in healthy subjects with sequence variants in the high-affinity sulfonylurea receptor gene (94)	Susceptibility to type II diabetes (95)
Vitamin D receptor (VDR)	1,25 dihydroxyvitamin-D3	Vitamin D response in patients affected with rickets (96)	Susceptibility to osteoporosis (97) and autosomal dominant rickets disease (96)

(e.g., responders/non-responders), the two collected samples originating from different source populations with different ethnicities, the result of the association study might simply reflect this inherent difference. Most of these issues could then be overcome by the analysis of individuals recruited for pharmaceutical trials, as such controlled studies take into account major confounding factors. A less obvious case is encountered when the study is designed in a mixed population, as different proportions of ethnic groups may be present in the analyzed population [44].

During recruitment of patients, the ethnic background can be taken into consideration, but this process may be inadequate for association studies. In recent years, methodological developments have focused on the assessment of the genetic homogeneity of the collected samples [44, 45] using sets of random markers (markers that *a priori* are not associated with the trait under study) to quantify the stratification effect between the groups under study. To some extent, this homogeneity study guarantees that the association found using this type of sample could not be ascribed to a major difference in ethnic background. These approaches are easily applied in genetic case control studies due to the availability of marker and genotyping capabilities, illustrating how genomic information can improve the quality of genetic study design. Other approaches, such as detection of genetic outliers [46] or the study of the genetic diversity of samples [45], can also provide improvements to the assessment of genetic homogeneity in the near future [47].

4.4.3
Direct Approach: A Hypothesis-Driven Strategy

Historically, association studies in pharmacogenetics have used one or a few polymorphisms per gene due to the lack of documentation regarding gene polymorphisms and access to genotyping facilities. Such studies focused primarily on non-conservative exonic markers or markers previously suspected to have a functional effect. In doing this, if an association exists, it is more likely to be directly mediated by the studied polymorphism, which explains why this has been called "direct strategy" or "hypothesis-driven strategy". Several compelling examples of this approach have been obtained and are detailed in Tab. 4.2. One critical point is the choice of the marker to studying and compelling the data that supports its functional role. For example, Kuivehoven et al. [22] have studied the effect of an intronic polymorphism of the CETP gene on atherosclerosis and cholesterol-lowering therapy. Although this polymorphism is intronic, it was chosen because previous studies showed it to be associated with lipid transfer activity and CETP plasma concentration, making it a plausible functional polymorphism. After reporting an allele effect of the polymorphism in patients undergoing cholesterol-lowering therapy, Kuivehoven et al. concluded that they could not ascribe this effect to the studied polymorphism or to another non-genotyped polymorphism in LD with the studied marker. The same concerns can be applied to all direct association studies, even those involving non-conservative polymorphisms.

4.4.4
Indirect Approach: A Hypothesis-Generating Strategy

With the availability of the sequence of the human genome and increased genotyping throughput, new approaches can be used based on the study of several polymorphisms across one gene that are no longer selected on their putative functionality. The indirect approach is based on the assumption that alleles of these unselected markers are in LD with the causative polymorphism(s) and may thus act as surrogate markers of the unknown causative polymorphisms and indirectly reflect the association. The feasibility of this approach has been illustrated, e.g., by the identification of a polymorphism in the 5' promoter region of the 5-lipo-oxygenase gene (ALOX5], which influences the response of asthmatics to the 5-lipo-oxygenase inhibitor, ABT-761 [48].

By not pre-defining a model of association, the indirect approach is less restrictive than a direct approach and can better be viewed as a hypothesis-generating, rather than a hypothesis-testing approach linking a gene to an observed trait and not merely identifying risk-conferring alleles. This strategy completely alters the way in which pharmacogenomic association studies are designed. After a positive association is revealed, other studies can be undertaken in order to refine the associations and to replicate the previous results in other populations.

Compared to the direct approach, the indirect approach requires detailed documentation of polymorphisms and their relation to the genes studied. In the last two years, numerous studies have been published on this topic. As pointed out by Cambien et al. [40], common polymorphisms are frequently in nearly complete LD in Caucasian populations. This has a major impact on haplotype diversity in the gene [49, 50] and thus on the study design. In such genes, all common polymorphisms do not have to be typed, as a subset is sufficient to provide an accurate image of the common diversity of the variation of the gene. The feasibility of this approach has been successfully validated for the ATM gene (*a*taxia-*t*elangiectasia *m*utated) [51]. Assembling SNPs into haplotypes has demonstrated that haplotype-based association studies for candidate genes have significant potential for the detection of genetic background that contribute to disease. Nevertheless, the extreme variability of LD has to be taken into account when designing indirect association studies. Consequently, the pattern of haplotype diversities or LD may first have to be described before choosing the proper set of markers to use. In addition, these patterns are highly population-dependent as illustrated by Peterson et al. [50] in a study which provided a good description of the worldwide genetic subdivision for the ALDH2 gene (*al*dehyde *deh*ydrogenase *2*) and demonstrated that a common polymorphism is only present in individuals of Asian origin.

Many statistical methods have been developed for association studies [52, 53] that mostly consist of testing each polymorphism separately with the disease. It has been shown that the power of such methods decreases rapidly when the markers are in low or even moderate LD with the risk-conferring polymorphism [54]. One way to overcome this defect is to use haplotype-based tests [47, 55, 56] that combine different alleles of different markers. Haplotypes are likely to capture more adequately the overall diversity of one chromosomal segment thus increasing the power to detect variations between groups. This was illustrated by an association study of ApoE (*apo*lipoprotein *E*) with Alzheimer's disease where it was found that a set of markers, none of which are significantly associated in univariate analysis, are associated when using an haplotype-based test [57].

In addition, haplotype-based methods can increase the power of association studies since the allelic architecture of the risk factor is unknown. Recent examples of association studies suggest that haplotypes can be the responsible factors [58]. The Apo E4 "allele", which is associated with Alzheimer's disease, is a possible example since it results from the substitution of two of non-conservative polymorphisms encoding for residues 112 and 158 [59].

4.5 SNP Assembly into Maps

Association studies are based on organized maps of non-repetitive genomic sequence variations, among which single nucleotide polymorphisms (SNPs) are by far the most common in the human genome. SNPs, the simplest genomic variations, are considered as the major source of genetic inter-individual variation. Other less frequent sequence variations due to insertion/deletion of one or more bases are not formally considered as SNPs, but can be detected by most of the methods used for SNP analysis [60]. An additional interesting source of polymorphisms are VNTRs (*v*ariable *n*umber of *t*andem *r*epeats), but their relatively rare occurrence in the human genome (only several thousand are supposed to occur throughout the entire genome) does not allow the development of an association strategy. It is estimated that SNPs occur on average every 1,330 bases when two human chromosomes are compared [60]. Therefore, a few million bases out of the 3 billion nucleotide pairs making up the human genome are thought to be polymorphic, and are thus present at a sufficient density for association studies [61, 69]. In addition, SNPs are less prone to mutations than microsatellites, as their rate of mutation is thought to be about 10^{-8} changes per nucleotide per generation. This relative stability of SNPs is well-suited for indirect association studies permitting a better maintenance of haplotypes across generations [61].

Construction of genome-wide SNP maps is an important step in characterizing and correlating genes with complex traits such as drug responses or ADRs. The SNP Consortium, a joint effort of 10 leading pharmaceutical companies under the aegis of the Wellcome Trust as well as the Human Genome Sequence program, have facilitated the construction of such SNP maps. The current release of the SNP Consortium coupled with the analysis of clone overlaps by the International Human Genome Sequencing Consortium has anchored by "*in silico*" mapping of 1.42 million SNPs to the human genome working draft, providing an average density of one SNP every 1.91 kb with only 4% of the genome having gaps between SNPs greater than 80 kb [60]. Furthermore, the intensive integration of public data to privately owned genomic information has allowed Celera Genomics to develop a high-density SNP map containing more than 2.1 million SNPs [62]. Overall, nucleotide frequency appears to be roughly homogenous across chromosomes, except for sex chromosomes in which diversity is much lower – but is much more variable at a finer scale – suggesting that these polymorphisms do not occur by random and independent mutation [60, 62]. However, 93% of gene loci are predicted to contain at least one SNP and 85% of the exons are within 5 kb of the nearest SNP [60], making it possible to analyze in detail the coding regions of the genome by indirect association studies. Interestingly, it is thought that only a very small proportion of SNPs (<1%) are able to impact on protein function, and that only one or two thousand genetic variations contribute to the diversity of human protein structure [62].

Although the simple structure of SNPs is well suited for automated high-throughput genotyping, their scoring remains a bottleneck for association studies. Most current methods are based on PCR technology, analyzing either directly the variable nucleotide polymorphism (single nucleotide primer extension methods, PCR-RFLP, allele-specific oligonucleotide hybridization, and pyrosequencing), or PCR products as a whole (dHPLC, SSCP and DGGE). High-density oligonucleotide microarrays and MALDI-TOF-MS (*m*atrix-*a*ssisted *l*aser *d*esorption *i*onization *t*ime-*o*f-*f*light *m*ass *s*pectrometry) have, in the past two years, been considered for use in high-throughput genotyping. All of these technologies have their advantages and disadvantages, as detailed in Tab. 4.3. Although most of these technologies are robust, their price per genotype (US $ 0.50–$ 1.5], however, limits their extensive use.

Other promising SNP scoring methods have been developed to fill the deficiencies of the currently used technologies. For instance, some methods have been developed for direct determination of haplotypes, in contrast to technologies that only estimate haplotype frequency. These direct haplotype scoring methods, such as single molecule [63] or allele-specific amplification [64] and carbon nanotube probe technologies [65], should achieve maximum power in haplotype-based indirect association studies. Furthermore, pooling methods have been proposed as an alternative to reduce the cost of genotyping. By mixing DNA from different individuals, this method allows the simultaneous study of a large number of samples for a given polymorphism. Despite significant economic advantages and the robustness of the method, the use of DNA pooling is hindered by the fact that it prohibits haplotype and subgroup analysis [66].

4.6 Strategies for Pharmacogenomic Association Studies

4.6.1 Candidate Genes

The candidate gene approach directly tests the association of selected genes with drug response or ADR. The major sources of candidates which code for receptor, enzyme or ADME targets, are obtained through literature surveys. Most pharma-

Tab. 4.3 Examples of current SNP genotyping methods. Information given in the table are based on the manufacturer's data provided in Q4 2000

Technology	*Throughput*	*Cost per assay*	*Accuracy*	*Robustness*	*Operational flexibility*
Invader assay (Third Wave)	Could be high, depending on implementation	High	Untested	Untested/low	High, DNA use
Mass Spectrometry (Sequenom)	High but costly to implement	High, decreased by possible multiplexing	Very high	High	Moderate
TaqMan (PE Biosystem)	Moderate	Very high	Moderate	High	Integrated system
GBA (Orchid)	Midlevel	High with current detection scheme	High	High	Low
Fluorescence LJL Biosystems	High	Moderate	Potential for high accuracy	High	Multistep system
DNA Chips (Affymetrix)	High number of SNPs analyzed, but low number of samples	High setup cost	Low	Low	Very low
Sequencing	Low/moderate (loading & profile analysis)	Very high	High	Robust	High

cogenomic variations validated so far have been identified by this approach [4]. Data mining of genomic and cDNA sequence databases has increased the number of candidates by identifying genes that possess similar sequences to established candidates [67]. Another important class of candidates are those genes associated with disease susceptibility, as some of them may also participate in drug response (Tab. 4.2). For instance, polymorphisms of the β-2-adrenoreceptor are associated with asthma susceptibility, as well as with variations in response to β-2-adrenergic agonists [24]. However, in some cases, no pathophysiology link has been established between the disease susceptibility gene and the mechanism of drug action. This is the case for the association between response to the cholinesterase inhibitor tacrine and variants of the ApoE gene in Alzheimer's disease [68]. Another source of candidates involves construction of protein-protein interaction maps for drug receptors and effectors. As a whole, choosing candidate genes is strongly limited by our present knowledge of disease pathophysiology, mode of action, metabolism and distribution of the drug.

Gene expression profiling appears to be an extremely promising technology for the identification of novel candidate genes by identifying genes that are over-/under-expressed in response to drugs studied in various cellular or animal models. Although limited by their inability to detect novel genes for which EST (*e*xpressed *s*equence *t*ag) information is not available, microarrays are efficient for high-throughput gene expression profiling [69]. Several other technologies offer the opportunity to discover novel genes that are differentially expressed, including differential display, RDA (*r*epresentational *d*ifferential *a*nalysis), SSH (*s*uppression *s*ubtractive *h*ybridization) and GeneCalling. The use of such technologies in pharmacotoxicology models has led to the identification of differentially expressed genes that can be tested as putative targets for both drug response and ADRs by association studies [70, 71].

4.6.2
Genome-Wide Scan

Genome-wide scans provide the means to identify many trait-related genetic factors and, therefore, appear to be useful tools for genetic studies [67]. A critical factor in such a strategy is the profile of LD throughout the genome, which influences the number and choice of markers to be genotyped. Although population-based studies and large simulation studies have been carried out to evaluate the genome-wide scan strategy [72], the estimation of the suitable number of markers still remains at the level of "best guesses" or is driven by economic constraints (e.g., the genotyping cost). Taking a mean density of one marker per 30 kb, the number of markers to be genotyped should be roughly around 100 000. Consequently, with an average cost of US$ 1 per genotype (without pooling the individuals), the cost of this approach would be around US$ 100 million for a sample size of 1 000 individuals, which is currently prohibitive in cost [73].

In a genome-wide scan study, a large proportion of the genotyped markers fall into intergenic regions, and are likely to be not relevant for association studies.

This limits the value of high-density SNP maps in genome scans. Gene-wide scans can be proposed as an alternative to genome-wide scans. This strategy is based on analysis of marker sets lying in functionally interesting regions, i.e. genomic regions surrounding predicted or annotated gene sequences. To date, between 32 000–40 000 genes are estimated to lie in the human genome [62, 74]. However, alternative approaches based on assemblies of ESTs estimate the total number of genes to range between 84 000 and 120 000 [75, 76]. Furthermore, among the already fully identified genes, only 60% of them belong to a predicted family for which a function might be anticipated by their sequence [62, 74]. Thus, a large number of genes remains to be discovered or annotated [77], and strategies based on current predictions may miss some genes that are still unknown. Ultimately, developing a SNP map organized around genes should be considered for association studies. This approach will document the extent of LD for each gene in a given population, which is a valuable information for further studies.

4.7 Expected Benefits of Pharmacogenomics in Drug R & D

4.7.1 Background

Pharmaceutical drug discovery and development of small molecules is a risky, expensive (around US$ 800 million per marketed drug) and time-consuming process of about 10–12 years [78, 79], as summarized in Figure 4.1. While the time from discovery to market is relatively stable, the number of new drugs that receive New Drug Application (NDA) approval is decreasing [78, 79]. Thus the pharmaceutical industry is searching for ways to reverse this productivity trend (e.g., decreased time of development and/or decreased cost of development and/or increased number of drugs that obtain marketing approval). This is why new technologies such as high-throughput screening and combinatorial chemistry have been widely adopted. The integration of pharmacogenomics in the drug development process has the potential to reduce the amount of time for drug discovery and development, as well as to increase the odds of compounds that will ultimately obtain marketing approval.

In order to understand how pharmacogenomics can be used in the drug research and development process it is necessary to understand what the aims of each phase of this process are, and are what the questions each phase is trying to answer.

4.7.2 Identification of New Targets

Pharmaceutical drug development of small molecules is initially based on the selection of candidate targets – mainly enzymes, receptors or circulating proteins that are currently targeted by 45, 28 and 11% of marketed compounds, respec-

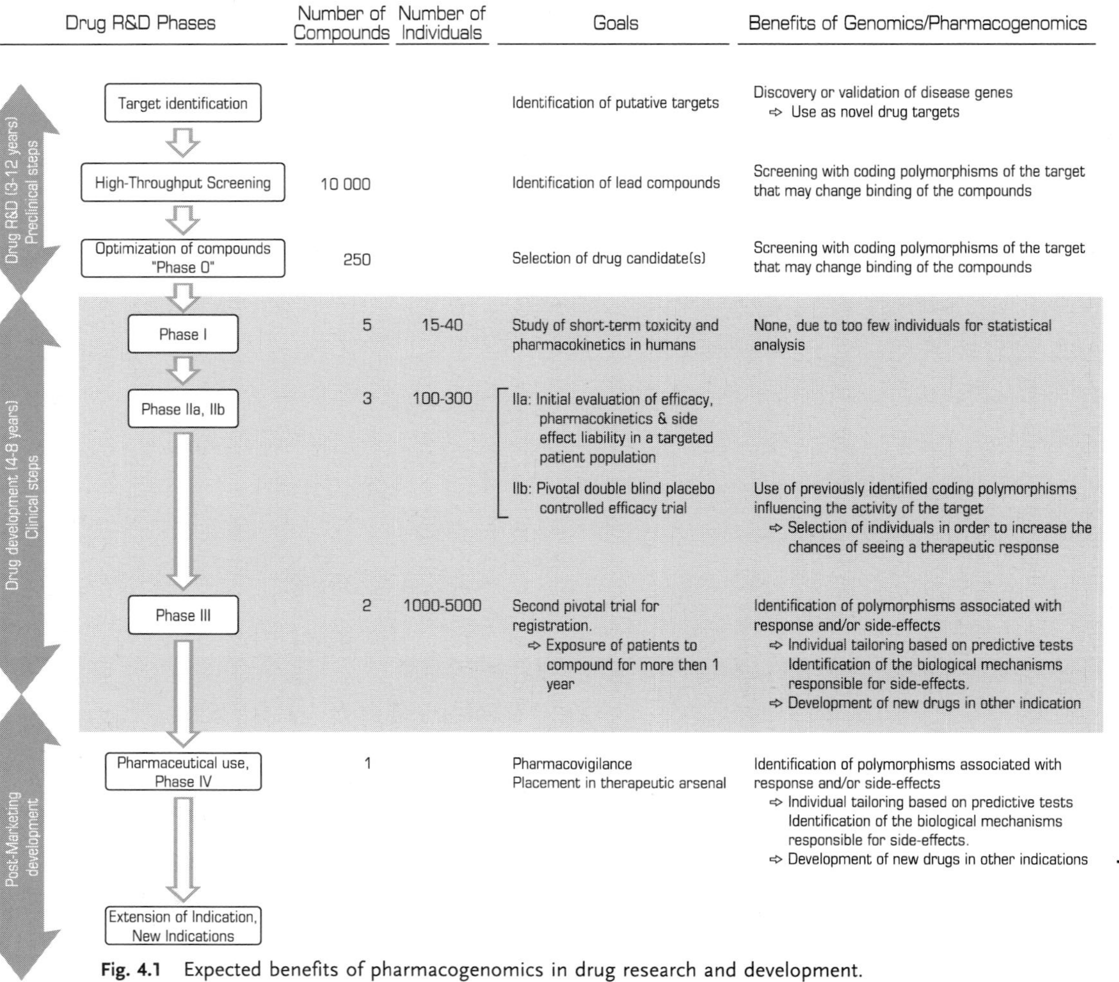

Fig. 4.1 Expected benefits of pharmacogenomics in drug research and development.

tively [80]. As a whole, current drug therapy is based on the targeting of approximately 500 proteins [80], a number that should increase as there are at least 30000–40000 protein-coding genes in the human genome [62, 74].

Pharmacogenomics offers interesting opportunities to discover novel targets responsible for drug response. Such an approach is particularly useful for drugs with unclear mechanisms of action, such as drugs having multiple binding sites. In theory, such information can be further used to design novel compounds that are specific for a targeted gene product. However, the validity of such an approach, which has never been used on a large scale, remains to be determined.

4.7.3 Pre-Clinical Development Phase

Optimization of lead compounds is a main step in the development of new drug compounds. This involves an extensive study of the pharmacology (especially pharmacokinetics) and toxicology of the compounds. These determine if the compound should enter clinical development. Here, pharmacogenomics offers the possibility of eliminating candidates that interact with frequently polymorphic proteins, such as drug – metabolizing enzymes responsible for variable drug response and toxicity. Another application is predictive toxicology based on gene expression profiling. The objective of this line of research is to correlate specific profiles with toxicity in animals and to develop predictive models and databases from this information.

4.7.4 Pre-Marketing Clinical Trials

Phase I: The preliminary studies in humans are designed to evaluate the pharmacokinetics and the acute toxicity/tolerability of the potential drug. Usually, phase I studies are carried out in a very small population [15–40] of healthy male volunteers. Thus, the nature of the population tested and the small number of individuals involved limits the scope of pharmacogenomics to the study of genetic variability in drug metabolism.

Phase II: The first part of phase II clinical trials (phase IIa) typically involves the initial assessment of compound efficacy in a limited number of patients [2–3 concentrations with 50 or more patients per group or arm of the study) in comparison with controls, and is used to determine the *m*ean *e*ffective *d*ose (MED) of the compound in the disease, the frequency of administration and the therapeutic efficacy and index (dose-producing efficacy versus dose-eliciting side effects) of a potential new drug. Genotyping of individuals from this portion of the clinical trial process for key proteins having an influence on the pharmacokinetics of candidate drugs that may be altered by the disease, can be very helpful in the design and the selection of patients for subsequent stages of the trial. In many instances, patients – in contrast to healthy volunteers – have better tolerance to new medications. The second part of phase II (phase IIb) is a confirmatory efficacy trial, usually double blind, placebo-controlled, which, depending on the power (number of patients and controls included in the study), can be used as a pivotal trial for registration purposes. Frequently, the effects of a new compound are compared with those of a reference (if available) already on the market to assess the efficacy of the new compound on a dose/dose comparative basis.

Phase III: The entry of a new compound into phase III clinical trials is a key step in the drug development process since such trials involve cohorts of a thousand patients or more in each arm and thus are extremely expensive. This phase is used to further demonstrate the efficacy and safety of the new compound through double-blind randomized studies. Phase II trials also involve assessment of potential drug–drug interactions, use in the elderly population and continuous exposure of a cohort of at least 1000 patients to the compound for a year. Pharmacogenomics may be key in the selection of individuals included in a phase III trial. Patients may be chosen not only according to their clinical phenotype, but also according to their genotype, for instance by genotyping the genes coding for the proteins with which the drug interacts. Another approach is to take into account the contributing genetic factors, as co-variables, in the analysis of the clinical trials. Pharmacogenomics can also be used during the trial to identify key genes involved in the efficacy and/or ADR profile of the drug, thus making it possible to enhance the safety and efficacy of the drug. However, information regarding individual prescriptions will have to be validated prospectively.

4.7.5
Post-Marketing Phase IV

Taking place after marketing authorization and continuing as long as the product is marketed, this phase places considerable emphasis on the cost-to-benefit ratio (pharmacoepidemiology) of new drugs so that they can be included in state and federal formularies. The use of pharmacogenomics in this stage of drug assessment will provide a better understanding of the exact relationship between drug activity/toxicity and polymorphisms. Drug interactions and the effect of the drug on pathologies other than those tested in the phase III trial are also studied. Here pharmacogenomics can also be used to increase the safety and/or efficacy of the treatment by helping to identify of important genetic variations responsible for the ADRs. However, the problem with rare or long-term ADRs involves the recruitment of a sufficient number of patients possessing the trait to design an association study. Phase IV studies also involve the assessment of alternative uses of a drug that may only become apparent when the drug is used in patients. For instance, valproate, originally used as an anticonvulsant, was found post approval to be effective in the treatment of bipolar affective disorder and in the prophylaxis of migraine.

4.7.6
Targeting Drugs to the Individual

In addition to facilitating the selection of individuals for clinical trails to reduce side effect risk, pharmacogenomics is anticipated to be useful in helping to select drugs for the individuals that will have the most benefit and least side effects [81]. Drugs likely to be tested are those that are metabolized by specific enzymatic pathways and/or that have a narrow therapeutic window and/or are used to treat disorders for which several other lines of treatment are available. The concept of using pharmacogenomics to tailor drug treatment to the individual has become an ethic issue for public health [82] and raises a profit maker question for the pharmaceutical industry. If one drug is replaced by 2, 3 or 10 new ones that are tailored to individual patient subsets, unless the time and cost of development are reduced, the cost of drug development will increase by the same factors of 2, 3 or 10 placing additional cost burdens on the health care system.

In some rare instances, pharmacogenomics has clearly been useful in medical practice. For example, testing for mutations in the dihydropyrimidine dehydrogenase gene, which codes for a rate-limiting enzyme in 5-FU catabolism, is now recommended before initiation of patient treatment in order to prevent severe toxicity in individuals having a mutation of this gene [14]. Furthermore, inactivating mutations of TPMT are frequently screened for before initiation of treatment with azathioprine, an immunosuppressant drug used in organ transplantation, to avoid fatal hematologic toxicity [83]. However, in most cases, multiple variable factors are thought to participate to a phenotype. Thus, an individual genetic factor revealed by pharmacogenomics will only have a limited influence on the overall variability, resulting in the fact that pharmacogenomics would be probabilistic. For instance, Drazen et al. have shown that a 5-lipooxygenase (ALOX5] promoter genotype is associated with a diminished clinical response to treatment with the novel antiasthmatic compound, ABT-761 (an analog of Zileuton) [48]. The genotype has a 100% positive predictive value for non-response to the drug. However, because the susceptibility genotype is uncommon, less than 10% of non-responders can be attributed to this genotype, while other variable factors are thought to determine non-response to the drug.

4.8
Conclusion

Pharmacogenomics has opened a new area in the field of pharmacology by pinpointing genetic factors involved in drug response. Association studies appear to be well suited for the study of such complex traits. In the past years, some significant results have been obtained either by direct association studies or more rarely by indirect study design. However, the wide use of such studies remained restricted by the unavailability of relevant genetic information. The near completion of the human genome sequencing and the availability of high-density SNP maps now provide the required tools. We can, therefore, anticipate that within the next two years association studies will be widely used in order to elucidate drug response factors. Whether genome-wide analysis or candidate gene association studies will be most successful remains a subject of debate. However, it is clear that high-throughput genotyping methods, massive calculation capacity, and high-quality patient collections are and will be key factors for the success of these studies.

The place of pharmacogenomics in the drug development process remains a pivotal question. We anticipate that pharmacogenomics will have a major impact by reducing both the resources and the time required for drug development. Firstly, pharmacogenomics appears to be a valuable tool in the identification of new therapeutic targets. Secondly, it should improve the design of clinical trials through patient stratification and re-definition of disease state. Thirdly, understanding the genetic basis for drug ADME, therapeutic response and ADRs should lead to better medicine prescription and therapeutic strategies. Although pharmacogenomics has not had the success initially planned, it has now the potential to improve health care delivery in the near future.

Acknowledgements

The authors would like to thank Michael Williams, Said Meguenni and Sandrine Macé for discussions and input.

4.9 References

1 Lu AY. Drug-metabolism research challenges in the new millennium: individual variability in drug therapy and drug safety. Drug Metab Dispos **1998**; 26[12]:1217–1222.

2 Emsley RA. Partial response to antipsychotic treatment: the patient with enduring symptoms. J Clin Psychiatry **1999**; 60(Suppl 23]:10–13.

3 Lazarou J, Pomeranz BH, Corey PN. Incidence of adverse drug reactions in hospitalized patients: a meta-analysis of prospective studies. JAMA **1998**; 279[15]:1200–1205.

4 Evans WE, Relling MV. Pharmacogenomics: translating functional genomics into rational therapeutics. Science **1999**; 286[5439]:487–491.

5 Nagao T, Mauer AM. Concordance for drug-induced aplastic anemia in identical twins. N Engl J Med **1969**; 281[1]:7–11.

6 Alda M, Grof P, Grof E, Zvolsky P, Walsh M. Mode of inheritance in families of patients with lithium-responsive affective disorders. Acta Psychiatr Scand **1994**; 90[4]:304–310.

7 Gonzalez FJ, Skoda RC, Kimura S, Umeno M, Zanger UM, Nebert DW et al. Characterization of the common genetic defect in humans deficient in debrisoquine metabolism. Nature **1988**; 331[6155]:442–446.

8 Nelson DR, Koymans L, Kamataki T, Stegeman JJ, Feyereisen R, Waxman DJ et al. P450 superfamily: update on new sequences, gene mapping, accession numbers and nomenclature. Pharmacogenetics **1996**; 6[1]:1–42.

9 Linder MW, Prough RA, Valdes R. Pharmacogenetics: a laboratory tool for optimizing therapeutic efficiency. Clin Chem **1997**; 43[2]:254–266.

10 GUENGERICH FP. Mechanism-based inactivation of human liver microsomal cytochrome P-450 IIIA4 by gestodene. Chem Res Toxicol **1990**; 3[4]:363–371.

11 FELIX CA, WALKER AH, LANGE BJ, WILLIAMS TM, WINICK NJ, CHEUNG NK et al. Association of CYP3A4 genotype with treatment-related leukemia. Proc Natl Acad Sci USA **1998**; 95[22]:13176–13181.

12 ANDERSSON T, REGARDH CG, DAHL-PUUSTINEN ML, BERTILSSON L. Slow omeprazole metabolizers are also poor S-mephenytoin hydroxylators. Ther Drug Monit **1990**; 12[4]:415–416.

13 MACKENZIE PI, OWENS IS, BURCHELL B, BOCK KW, BAIROCH A, BELANGER A et al. The UDP glycosyltransferase gene superfamily: recommended nomenclature update based on evolutionary divergence. Pharmacogenetics **1997**;7[4]:255–269.

14 RELLING MV, HANCOCK ML, RIVERA GK, SANDLUND JT, RIBEIRO RC, KRYNETSKI EY et al. Mercaptopurine therapy intolerance and heterozygosity at the thiopurine S-methyltransferase gene locus. J Natl Cancer Inst **1999**; 91[23]:2001–2008.

15 SCHUETZ EG, SCHINKEL AH. Drug disposition as determined by the interplay between drug-transporting and drug-metabolizing systems. J Biochem Mol Toxicol **1999**; 13[3–4]:219–222.

16 SCHINKEL AH. The physiological function of drug-transporting P-glycoproteins. Semin Cancer Biol **1997**; 8[3]:161–170.

17 HOFFMEYER S, BURK O, VON RICHTER O, ARNOLD HP, BROCKMOLLER J, JOHNE A et al. Functional polymorphisms of the human multidrug-resistance gene: multiple sequence variations and correlation of one allele with P-glycoprotein expression and activity *in vivo*. Proc Natl Acad Sci USA **2000**; 97[7]:3473–3478.

18 KRAGH-HANSEN U, BRENNAN SO, GALLIANO M, SUGITA O. Binding of warfarin, salicylate, and diazepam to genetic variants of human serum albumin with known mutations. Mol Pharmacol **1990**; 37[2]:238–242.

19 HERVE F, GOMAS E, DUCHE JC, TILLEMENT JP. Evidence for differences in the binding of drugs to the two main genetic variants of human alpha 1-acid glycoprotein. Br J Clin Pharmacol **1993**; 36[3]:241–249.

20 ARRANZ MJ, MUNRO J, BIRKETT J, BOLONNA A, MANCAMA D, SODHI M et al. Pharmacogenetic prediction of clozapine response. Lancet **2000**; 355[9215]:1615–1616.

21 SCHUMACHER J, SCHULZE TG, WIENKER TF, RIETSCHEL M, NOTHEN MM. Pharmacogenetics of the clozapine response. Lancet **2000**; 356[9228]:506–507.

22 KUIVENHOVEN JA, JUKEMA JW, ZWINDERMAN AH, DE KNIJFF P, MCPHERSON R, BRUSCHKE AV et al. The role of a common variant of the cholesteryl ester transfer protein gene in the progression of coronary atherosclerosis. The Regression Growth Evaluation Statin Study Group. N Engl J Med **1998**; 338[2]:86–93.

23 HOLLOWAY JW, DUNBAR PR, RILEY GA, SAWYER GM, FITZHARRIS PF, PEARCE N et al. Association of beta2-adrenergic receptor polymorphisms with severe asthma. Clin Exp Allergy **2000**; 30[8]:1097–1103.

24 LIPWORTH BJ, DEMPSEY OJ, AZIZ I. Functional antagonism with formoterol and salmeterol in asthmatic patients expressing the homozygous glycine-16 beta[2]-adrenoceptor polymorphism. Chest **2000**; 118[2]:321–328.

25 JEUNEMAITRE X, INOUE I, WILLIAMS C, CHARRU A, TICHET J, POWERS M et al. Haplotypes of angiotensinogen in essential hypertension. Am J Hum Genet **1997**; 60[6]:1448–1460.

26 MARTINELLI I, SACCHI E, LANDI G, TAIOLI E, DUCA F, MANNUCCI PM. High risk of cerebral-vein thrombosis in carriers of a prothrombin-gene mutation and in users of oral contraceptives. N Engl J Med **1998**; 338[25]:1793–1797.

27 BROWN RL, POLLOCK AN, COUCHMAN KG, HODGES M, HUTCHINSON DO, WAAKA R et al. A novel ryanodine receptor mutation and genotype-phenotype correlation in a large malignant hyperthermia New Zealand Maori pedigree. Hum Mol Genet **2000**; 9[10]:1515–1524.

28 STEEN VM, LOVLIE R, MACEWAN T, MCCREADIE RG. Dopamine D3-receptor gene variant and susceptibility to tardive

dyskinesia in schizophrenic patients. Mol Psychiatry **1997**; 2[2]:139–145.

29 Ishiguro H, Okuyama Y, Toru M, Arinami T. Mutation and association analysis of the 5′ region of the dopamine D3 receptor gene in schizophrenia patients: identification of the Ala38Thr polymorphism and suggested association between DRD3 haplotypes and schizophrenia. Mol Psychiatry **2000**; 5[4]:433–438.

30 Kuzushita N, Hayashi N, Kanto T, Takehara T, Tatsumi T, Katayama K et al. Involvement of transporter associated with antigen processing 2 (TAP2] gene polymorphisms in hepatitis C virus infection. Gastroenterology **1999**; 116[5]:1149–1154.

31 Yee LJ, Tang J, Herrera J, Kaslow RA, van Leeuwen DJ. Tumor necrosis factor gene polymorphisms in patients with cirrhosis from chronic hepatitis C virus infection. Genes Immun **2000**; 1[6]:386–390.

32 Powell EE, Edwards-Smith CJ, Hay JL, Clouston AD, Crawford DH, Shorthouse C et al. Host genetic factors influence disease progression in chronic hepatitis C. Hepatology **2000**; 31[4]:828–833.

33 Lechmann M, Schneider EM, Giers G, Kaiser R, Dumoulin FL, Sauerbruch T et al. Increased frequency of the HLA-DR15 (B1*15011] allele in German patients with self-limited hepatitis C virus infection. Eur J Clin Invest **1999**; 29[4]:337–343.

34 Yee LJ, Tang J, Gibson AW, Kimberly R, Van Leeuwen DJ, Kaslow RA. Interleukin 10 polymorphisms as predictors of sustained response in antiviral therapy for chronic hepatitis C infection. Hepatology **2001**; 33[3]:708–712.

35 Hijikata M, Ohta Y, Mishiro S. Identification of a single nucleotide polymorphism in the MxA gene promoter (G/T at nt -88] correlated with the response of hepatitis C patients to interferon. Intervirology **2000**; 43[2]:124–127.

36 Matsushita M, Hijikata M, Ohta Y, Iwata K, Matsumoto M, Nakao K et al. Hepatitis C virus infection and mutations of mannose-binding lectin gene MBL. Arch Virol **1998**; 143[4]:645–651.

37 Enomoto N, Sakuma I, Asahina Y, Kurosaki M, Murakami T, Yamamoto C et al. Comparison of full-length sequences of interferon-sensitive and resistant hepatitis C virus 1b. Sensitivity to interferon is conferred by amino acid substitutions in the NS5A region. J Clin Invest **1995**; 96[1]:224–230.

38 Pawlotsky JM, Pellerin M, Bouvier M, Roudot-Thoraval F, Germanidis G, Bastie A et al. Genetic complexity of the hypervariable region 1 (HVR1] of hepatitis C virus (HCV): influence on the characteristics of the infection and responses to interferon alfa therapy in patients with chronic hepatitis C. J Med Virol **1998**; 54[4]:256–264.

39 Pawlotsky JM, Germanidis G, Frainais PO, Bouvier M, Soulier A, Pellerin M et al. Evolution of the hepatitis C virus second envelope protein hypervariable region in chronically infected patients receiving alpha interferon therapy. J Virol **1999**; 73[8]:6490–6499.

40 Cambien F, Poirier O, Nicaud V, Herrmann SM, Mallet C, Ricard S et al. Sequence diversity in 36 candidate genes for cardiovascular disorders. Am J Hum Genet **1999**; 65[1]:183–191.

41 Weiss KM. Is there a paradigm shift in genetics? Lessons from the study of human diseases. Mol Phylogenet Evol **1996**; 5[1]:259–265.

42 Lander ES, Schork NJ. Genetic dissection of complex traits. Science **1994**; 265[5181]:2037–2048.

43 Spielman RS, McGinnis RE, Ewens WJ. Transmission test for linkage disequilibrium: the insulin gene region and insulin-dependent diabetes mellitus (IDDM). Am J Hum Genet **1993**; 52[3]:506–516.

44 Devlin B, Roeder K. Genomic control for association studies. Biometrics **1999**; 55[4]:997–1004.

45 Pritchard JK, Rosenberg NA. Use of unlinked genetic markers to detect population stratification in association studies. Am J Hum Genet **1999**; 65[1]:220–228.

46 Rannala B, Mountain JL. Detecting immigration by using multilocus genotypes. Proc Natl Acad Sci USA **1997**; 94[17]:9197–9201.

47 Fallin D, Schork NJ. Accuracy of haplotype frequency estimation for biallelic loci, via the expectation-maximization algorithm for unphased diploid genotype data. Am J Hum Genet **2000**; 67[4]:947–959.

48 Drazen JM, Yandava CN, Dube L, Szczerback N, Hippensteel R, Pillari A et al. Pharmacogenetic association between ALOX5 promoter genotype and the response to anti-asthma treatment. Nature Genet **1999**; 22[2]:168–170.

49 Rieder MJ, Taylor SL, Clark AG, Nickerson DA. Sequence variation in the human angiotensin converting enzyme. Nature Genet **1999**; 22[1]:59–62.

50 Peterson LE, Barnholtz JS, Page GP, King TM, de Andrade M, Amos CI. A genome-wide search for susceptibility genes linked to alcohol dependence. Genet Epidemiol **1999**; 17(Suppl 1]:S295–300.

51 Bonnen PE, Story MD, Ashorn CL, Buchholz TA, Weil MM, Nelson DL. Haplotypes at ATM identify coding-sequence variation and indicate a region of extensive linkage disequilibrium. Am J Hum Genet **2000**; 67[6]:1437–1451.

52 Lonjou C, Collins A, Morton NE. Allelic association between marker loci. Proc Natl Acad Sci USA **1999**; 96[4]:1621–1626.

53 Lam JC, Roeder K, Devlin B. Haplotype fine mapping by evolutionary trees. Am J Hum Genet **2000**; 66[2]:659–673.

54 Muller-Myhsok B, Abel L. Genetic analysis of complex diseases. Science **1997**; 275[5304]:1328–1329; discussion 1329–1330.

55 Zhao JH, Curtis D, Sham PC. Model-free analysis and permutation tests for allelic associations. Hum Hered **2000**; 50[2]:133–139.

56 Xiong M, Akey J, Jin L. The haplotype linkage disequilibrium test for genome-wide screens: its power and study design. Pac Symp Biocomput **2000**; 26[3]:675–686.

57 Fallin D, Cohen A, Essioux L, Chumakov I, Blumenfeld M, Cohen D et al. Genetic analysis of case/control data using estimated haplotype frequencies: application to APOE locus variation and Alzheimer's disease. Genome Res **2001**; 11[1]:143–151.

58 El-Omar EM, Carrington M, Chow WH, McColl KE, Bream JH, Young HA et al. Interleukin-1 polymorphisms associated with increased risk of gastric cancer. Nature **2000**; 404[6776]:398–402.

59 Bickeboller H, Campion D, Brice A, Amouyel P, Hannequin D, Didierjean O et al. Apolipoprotein E and Alzheimer disease: genotype-specific risks by age and sex. Am J Hum Genet **1997**; 60[2]:439–446.

60 Sachidanandam R, Weissman D, Schmidt SC, Kakol JM, Stein LD, Marth G et al. A map of human genome sequence variation containing 1.42 million single nucleotide polymorphisms. Nature **2001**; 409[6822]:928–933.

61 Jorde LB. Linkage disequilibrium and the search for complex disease genes. Genome Res **2000**; 10[10]:1435–1444.

62 Venter JC, Adams MD, Myers EW, Li PW, Mural RJ, Sutton GG et al. The sequence of the human genome. Science **2001**; 291[5507]:1304–1351.

63 Ruano G, Kidd KK. Direct haplotyping of chromosomal segments from multiple heterozygotes via allele-specific PCR amplification. Nucleic Acids Res **1989**; 17[20]:8392.

64 Ruano G, Kidd KK, Stephens JC. Haplotype of multiple polymorphisms resolved by enzymatic amplification of single DNA molecules. Proc Natl Acad Sci USA **1990**; 87[16]:6296–6300.

65 Woolley AT, Guillemette C, Li Cheung C, Housman DE, Lieber CM. Direct haplotyping of kilobase-size DNA using carbon nanotube probes. Nature Biotechnol **2000**; 18[7]:760–763.

66 Breen G, Harold D, Ralston S, Shaw D, St Clair D. Determining SNP allele frequencies in DNA pools. Biotechniques **2000**; 28[3]:464–466, 468, 470.

67 Roses AD. Pharmacogenetics and the practice of medicine. Nature **2000**; 405[6788]:857–865.

68 Poirier J, Delisle MC, Quirion R, Aubert I, Farlow M, Lahiri D et al. Apolipoprotein E4 allele as a predictor of cholinergic deficits and treatment outcome in Alzheimer disease. Proc Natl Acad Sci USA **1995**; 92[26]:12260–12264.

69 Young RA. Biomedical discovery with DNA arrays. Cell **2000**; 102[1]:9–15.

70 Wang X, Feuerstein GZ. Suppression subtractive hybridisation: application in the discovery of novel pharmacological targets. Pharmacogenomics **2000**; 1[1]:101–108.

71 Rininger JA, DiPippo VA, Gould-Rothberg BE. Differential gene expression technologies for identifying surrogate markers of drug efficacy and toxicity. Drug Discov Today **2000**; 5[12]:560–568.

72 Kruglyak L. Prospects for whole-genome linkage disequilibrium mapping of common disease genes. Nature Genet **1999**; 22[2]:139–144.

73 McCarthy JJ, Hilfiker R. The use of single-nucleotide polymorphism maps in pharmacogenomics. Nature Biotechnol **2000**; 18[5]:505–508.

74 Lander ES, Linton LM, Birren B, Nusbaum C, Zody MC, Baldwin J et al. Initial sequencing and analysis of the human genome. Nature **2001**; 409[6822]:860–921.

75 Wheeler DL, Church DM, Lash AE, Leipe DD, Madden TL, Pontius JU et al. Database resources of the National Center for Biotechnology Information. Nucleic Acids Res **2001**; 29[1]:11–16.

76 Liang F, Holt I, Pertea G, Karamycheva S, Salzberg SL, Quackenbush J. Gene index analysis of the human genome estimates approximately 120000 genes. Nature Genet **2000**; 25[2]:239–240.

77 Claverie JM. Gene number. What if there are only 30,000 human genes? Science **2001**;291[5507]:1255–1257.

78 Kuhlmann J. Alternative strategies in drug development: clinical pharmacological aspects. Int J Clin Pharmacol Ther **1999**; 37[12]:575–583.

79 Craig AM, Malek M. Market structure and conduct in the pharmaceutical industry. Pharmacol Ther 1995;66[2]:301–337.

80 Drews J. Drug discovery: A historical perspective. Science **2000**; 287[5460]:1960–1964.

81 Berry S. Drug discovery in the wake of genomics. Trends Biotechnol **2001**; 19[7]:239–240.

82 Margolin J. From comparative and functional genomics to practical decisions in the clinic: a view from the trenches. Genome Res **2001**; 11[6]:923–925.

83 Krynetski EY, Evans WE. Pharmacogenetics as a molecular basis for individualized drug therapy: the thiopurine S-methyltransferase paradigm. Pharm Res **1999**; 16[3]:342–349.

84 Taube J, Halsall D, Baglin T. Influence of cytochrome P-450 CYP2C9 polymorphisms on warfarin sensitivity and risk of over-anticoagulation in patients on long-term treatment. Blood **2000**; 96[5]:1816–1819.

85 Ibeanu GC, Blaisdell J, Ferguson RJ, Ghanayem BI, Brosen K, Benhamou S et al. A novel transversion in the intron 5 donor splice junction of CYP2C19 and a sequence polymorphism in exon 3 contribute to the poor metabolizer phenotype for the anticonvulsant drug S-mephenytoin. J Pharmacol Exp Ther **1999**; 290[2]:635–640.

86 Joober R, Benkelfat C, Brisebois K, Toulouse A, Turecki G, Lal S et al. T102C polymorphism in the 5HT2A gene and schizophrenia: relation to phenotype and drug response variability. J Psychiatry Neurosci **1999**; 24[2]:141–146.

87 Inayama Y, Yoneda H, Sakai T, Ishida T, Nonomura Y, Kono Y et al. Positive association between a DNA sequence variant in the serotonin 2A receptor gene and schizophrenia. Am J Med Genet **1996**; 67[1]:103–105.

88 Ohmichi N, Iwai N, Uchida Y, Shichiri G, Nakamura Y, Kinoshita M. Relationship between the response to the angiotensin converting enzyme inhibitor imidapril and the angiotensin converting enzyme genotype. Am J Hypertens **1997**; 10[8]:951–955.

89 Vasku A, Soucek M, Znojil V, Rihacek I, Tschoplova S, Strelcova L et al. Angiotensin I-converting enzyme and angiotensinogen gene interaction and prediction of essential hypertension. Kidney Int **1998**; 53[6]:1479–1482.

90 Hunt SC, Geleijnse JM, Wu LL, Witteman JC, Williams RR, Grobbee DE. Enhanced blood pressure response to mild

sodium reduction in subjects with the 235T variant of the angiotensinogen gene. Am J Hypertens **1999**; 12[5]:460–466.

91 Amouyel P, Vidal O, Launay JM, Laplanche JL. The apolipoprotein E alleles as major susceptibility factors for Creutzfeldt-Jakob disease. The French Research Group on Epidemiology of Human Spongiform Encephalopathies. Lancet **1994**; 344[8933]:1315–1318.

92 Taylor DR, Drazen JM, Herbison GP, Yandava CN, Hancox RJ, Town GI. Asthma exacerbations during long term beta agonist use: influence of beta[2] adrenoceptor polymorphism. Thorax **2000**; 55[9]:762–767.

93 Deeb SS, Fajas L, Nemoto M, Pihlajamaki J, Mykkanen L, Kuusisto J et al. A Pro12Ala substitution in PPARgamma2 associated with decreased receptor activity, lower body mass index and improved insulin sensitivity. Nature Genet **1998**; 20[3]:284–287.

94 Hansen T, Echwald SM, Hansen L, Moller AM, Almind K, Clausen JO et al. Decreased tolbutamide-stimulated insulin secretion in healthy subjects with sequence variants in the high-affinity sulfonylurea receptor gene. Diabetes **1998**; 47[4]:598–605.

95 Reis AF, Ye WZ, Dubois-Laforgue D, Bellanne-Chantelot C, Timsit J, Velho G. Association of a variant in exon 31 of the sulfonylurea receptor 1 (SUR1] gene with type 2 diabetes mellitus in French Caucasians. Hum Genet **2000**; 107[2]:138–144.

96 Malloy PJ, Eccleshall TR, Gross C, Van Maldergem L, Bouillon R, Feldman D. Hereditary vitamin D resistant rickets caused by a novel mutation in the vitamin D receptor that results in decreased affinity for hormone and cellular hyporesponsiveness. J Clin Invest **1997**; 99[2]:297–304.

97 Lucotte G, Mercier G, Burckel A. The vitamin D receptor FokI start codon polymorphism and bone mineral density in osteoporotic postmenopausal French women. Clin Genet **1999**; 56[3]:221–224.

5
Genomics Applications that Facilitate the Understanding of Drug Action and Toxicity

L. Mike Furness

Abstract

As we advance in the 21st century, genomics has passed a number of major landmarks, including the completion of a number of eukaryotic genome sequences. During the last 10 years, genomics has also been rapidly integrated into the processes involved in drug discovery and development, and has begun to make significant in roads into plant science, agrochemical production, food science, and even into cosmetics and personal health care. In this chapter, we will limit our discussions to the areas relating to drugs and toxins, and primarily, the role of genomics in the modern drug discovery and development process in the pharmaceutical industry. There will be brief coverage of the platform technologies that are now routinely being implemented into the pharmaceutical industry as well as a discussion of other areas where these technologies will have impact in the future.

5.1
Platform Technologies

In this section we will aim to review some of the key technologies used in the field of genomics and those related areas that have either had a major impact in the field of genomic technologies. Examples of the applications of these technologies will be provided in the next section.

5.1.1
Genomics

In recent years, many new technologies have been introduced into the life sciences to speed up and automate routine tasks, to the extent that what were once considered to be complex, specialized tasks 10 years ago, are now routine tasks. One obvious example is DNA sequencing. In the 20–30 years since the initial publication of Sanger's method for dideoxynucleotide sequencing [1], we have progressed from months of dedicated work to sequence individual genes to the stage where the task of sequencing whole genomes has become a routine, industrialized process. As a result of this we now have 2 drafts of the Human Genome published [2, 3], as well as many other eukaryotic, prokaryotic and plant species (a good review of these can be found at *http://www-fp.mcs.anl.gov/*~gaasterland/genomes.html).

Sequencing is one key example of genomic technology that has had a far broader effect on the world as a whole. There are two key foci in the sequencing world – genome sequencing and cDNA or expressed sequence tag (EST) sequencing. In the former, we gain information on gene structure and distribution [4–6], as well as a more accurate map on which to locate disease-linked genes and genetic changes [7, 8]. In the latter case, mRNA is isolated from different cells or tissues and used to generate a cDNA library, representing the pool of genes transcribed within that cell or tissue. By picking a few thousand clones from each of these and sequencing short regions to generate ESTs, we gain information on the genes transcribed in that cell or tissue and a crude measure of their relative abundance. Collation of cDNA sequence data from large numbers of libraries also gives us a rough measure of the tissue distribution of these cDNAs [9–11].

More recently, the progress in automation in molecular biology has enabled these genes, or gene transcripts, to be arrayed of solid matrices in an ordered array, which is then used to identify accurate expression levels of large numbers of genes in parallel. Initially these arrays were on nylon or nitroceulluose membranes at relatively low density, but more recently a number of alternatives have been generated. These fall into two main categories of microarrays: arrays of short [20–60 bases long) synthetic oligonuleotides, with several overlapping oligonucleotides representing each gene [12–15], or arrays of cDNA or genome fragments [16–18]. The scalability of both of these methods has been drastically increased by the introduction of robotics and silicon chip technologies, allowing thousands of oligonucleotides to be synthesized on a solid surface in an area a few millimeters across, or thousands of DNA fragments to be spotted on to a similar area of nylon, glass or plastic. The populations of expressed genes from different cells or tissues are then labeled with fluorescent tags and hybridized to the array of DNA fragments. Any of the expressed gene transcripts in this fluorescent probe that are complementary to the DNA fragments on the array will hybridize to the corresponding spot on the array. The array is then scanned with a laser at the appropriate wavelength, and fluorescence measurements are made at each location on the array, giving a measure of how many gene transcripts are seen in that sample. By comparing the levels of all these genes between many samples, we can begin to understand what molecular changes are occurring at the transcription level during biological changes (see below). The key advantage of array technology is the ability to scale the process of expression analysis to several thousand genes per sample, compared to a few dozen for traditional methods such as Northern blot analysis. Oligonucleotide technology has some advantages, such as the ability to better distinguish between very closely related homologs, as long as the homologs were known and this information was included in the chip design. However, spotted arrays of cDNAs have an advantage in that the sequence of each clone need not be known before it is spotted, so it is potentially faster to generate customized arrays using this format. Once the full genome sequence for any organism has been published and detailed, then it should be possible to generate a single, genome-wide array using either format, and at this point, the costs for arrays will be significantly reduced.

Reverse transcription-polymerase chain reaction (RT-PCR) can also be used to characterize the levels of gene expression in samples [19–21]. In this case we must have information about the sequence being studied to generate sequence-specific PCR primers. One of these primers is labeled with a specific fluorescent tag, and by monitoring the fluorescence increase during PCR, it is possible to gain an accurate measure of how many copies of the sequence were present in the sample initially. While this method gives a potentially more accurate measurement of copy number than arrays [22, 23], it cannot readily be scaled to run over such a large number of genes in parallel.

A more recent area of growth has been that of pharmacogenomics and pharmacogenetics. For the sake of this review, we will use the definitions previously published [24], i.e., pharmacogenomics describes the effect of drugs on the genome, whereas pharmacogenetics describes the effect of the genetics of a protein on how it interacts with a drug. The former utilizes a range of technologies including microarrays [12–18], but also including differential display PCR (DD-PCR) [25], serial analysis of gene expression (SAGE) [26], GeneCalling [27], MPSS [28] and TOGA [29] technologies, to name but a few. In the case of microarrays, the technology is considered to be a "closed" technology, in that you can only measure the changes in genes that are on the array. The latter are "open" technologies, in that they allow identification of the differentially expressed genes between two or more cell populations. The difference becomes moot once whole genome and coding sequence within a genome has been identified. In the case of pharmacogenetics, a plethora of methods have been reported for the analysis of polymorphic changes between individuals in populations. These are well covered in a number of reviews [30–34], and will be treated in other chapters of this book, so minimal coverage will be addressed within this chapter.

5.1.2
Proteomics

Up until the last few years, proteomic analysis (the analysis of the spectrum of proteins and protein isoforms in a particular cell or tissue type) has been carried out almost exclusively by 2D gel electrophoresis, based on the paper by O'Farrell in 1977 [35]. Most technology changes to date have been refinements of the separation of gels, but major advances have taken place in the analysis of the proteins after separation. The improvements in sensitivity and resolution of mass spectroscopy, combined with more sensitive peptide sequencing methods [36–38], have allowed more of the components of a proteome to be defined. However, the field of protein array technology has now begun to grow significantly, with techniques ranging from analyzing proteins expressed from spotted bacterial cultures (that can also be used for DNA/RNA hybridization studies) [39, 40], through "affibody" technology [41], protein–protein interaction maps [42], to protein chips [43, 44] Several good overviews of the current state of proteomic technologies and companies can be found elsewhere [40, 45, 46]

5.1.3 Bioinformatics

Perhaps one of the most important, and still rapidly growing areas in genomics, is that of bioinformatics. The ability of "wet lab" scientists to generate huge amounts of data using high-throughput technologies, miniturization and automation, has led to a need for larger and more complex data management tools, and the development of new ways of analyzing and viewing the data sets. The field can currently be divided into a few key areas:

- *Data Management*
  This is fundamental to the progress of genomics (and many other areas) of science. Generating data in a common exchangeable format, with a common lexicon of terms [47] in a single non-redundant location is a major goal. A number of examples exist, such as the DNA and protein sequence data in GenBank, EMBL or SwissProt [48–50].

- *Primary Sequence Assembly and Comparison (DNA and RNA)*
  Initially compiling short DNA sequence fragments from the high-throughput sequencing technologies, to generate contiguous regions of genomic sequence, and complete gene sequences. To this end algorithms such as BLAST [51], FastA [52] and Phred [53] have been developed to help analyze sequences, showing their similarity to known sequences, and aiding in identification of functional roles for novel genes. Before the sequence of the human genome was released, predictions of gene numbers had been made based on other species, which led to estimates of 28,000–34,000 genes [54]. With the advent of the first complete draft human genome sequence and maps available to researchers, predictions of gene numbers have been quoted as 35,000–120,000, based on comparison of EST sequences [55, 56]. This discrepancy could be resolved by the fact that each gene in the genome gives rise to multiple gene transcripts by differential splicing – from these figures it would suggest an average of ~ 3 gene transcripts per gene. When comparing protein sequences, we can compare the whole sequence, or for novel proteins with less obvious homology, a number of tools are available for "feature" searching, i.e., small pieces of protein sequence known to be involved in specific roles, such as calcium or ATP binding. These include algorithms such as BLOCKS, PRINTS, PFAM and PROSITE [50].

- *SNP and Other Polymorphism Data*
  SNPs, or single nucleotide polymorphims, are the subtle changes in the genome sequence that can lead to alterations either within single proteins by altering the protein coding sequence, or whole pathways, if the mutation affects

the function of promoters for instance. The genome sequencing efforts have identified a number of these mutations as a "by-product" of the redundancy within the sequence, but this area will be extensively covered elsewhere in this book.

- *Secondary Structure Prediction and Alignment*
  A number of tools have been around for several years that aim to predict likely areas of secondary structure from a primary protein sequence or based on an RNA sequence [50]. These models have varying degrees of accuracy of prediction, with some of the best reaching up to 70% accuracy [57].

- *Tertiary structure prediction and comparison*
  One "Holy Grail" in protein structure is to develop tools that accurately predict three-dimensional structures of proteins from their primary sequence information [58, 59]. Many of the best tools to date only go part of the way by using known three-dimensional structures from proteins which share similar primary sequence to model the possible structures of new proteins. This technology still has a long way to go, but the potential rewards would be enormous, allowing a genome sequence to be translated into targets for therapeutic intervention *in silico*, in relatively short periods of time.

- *Protein-Ligand Interactions*
  At this point we are at the crossroads where bioinformatics meets chemoinformatics. Many companies now use X-ray crystallography or NMR structures to generate models of proteins/targets with which known drugs and other ligands interact, to enable a better understanding of the interaction, with the ultimate goals of deriving more potent and selective drugs for those targets [60].

- *Expression Databases*
  With the ability to rapidly generate both DNA, RNA and protein expression data, more resources are now being applied to addressing many of the issues seen in the "sequence era", such as data management and stadardized formats, but a new level of complexity has been added. As well as the increase in potential data generated per unit time, new and more complex tools are needed to analyze how these gene and protein interactions are collated, to represent the pathways and interactions in normal cell metabolism, and the changes associated with disease and xenobiotic challenge. These databases are just beginning to emerge [50], and tools previously used for applications in other fields, such as clinical research, astronomy and physics, are now being adapted to handle biological data (e.g., [61]).

- *Protein-Protein Interactions*
  As well as protein expression analysis, proteomics in its broadest sense also includes protein–protein interaction mapping. This can range from small studies using a single protein or peptide to "fish out" any proteins that interact with it

from a complex pool, such as a cell lysate, to looking at how all proteins in an organism interact with each other [62–64].

- *Metabolic Pathway Analysis and Prediction*
  We have grown used to seeing large charts displaying known metabolic pathways lining the corridors of many research labs. These represent only a small proportion of the interactions that are likely to take place, and tools are being developed to allow inference of function and position of novel proteins in the known pathways, or to display how changes in the metabolic flux occur in response to changing environmental parameters, and the net effect this has on the pathways (e.g., [65]).

- *Virtual Cells, Organs and Organisms*
  A number of research programs are underway to model electronically the functions of single cells or organs, which aim ultimately at generating a virtual body, in which properties, such as response to drugs, could be modeled before moving to real biology. While this work is at an early stage, some impressive work has been done in a number of areas, including modeling cardiac function (e.g., [66]).

- *Informatics*
  To some extent, the use of statistical tools such as k-means clustering, self-organizing maps and principal component analyses [67, 68], only address the first stage of the analysis of expression data by reducing the sets of gene and/or protein sequences that we need to characterize in more depth. The really labor-intensive part of the bioinformatics process associated with any form of expression data is to try and make biological sense of these "cropped" gene or protein lists. Some genes will be readily identifiable by the experimenter as having involvement in, e.g., specific metabolic pathways, but a large number of the genes will either be known, but not instantly recognized by the experimenter, or more and more often, these may be novel, uncharacterized gene sequences. In all these cases, an enormous amount of reading and literature searching is normally the next stage. To address this, there has been a recent rapid growth in the number and diversity of search tools that will provide ways to highlight common features in the literature associated with large numbers of genes simultaneously. This includes tools like EDGAR [69], MedMiner [70], ThemeMap [71,72], and many others [73, 74].

Analysis of biological data has now become far more complex, and there is a drive to develop software to allow disparate data sets, such as sequence, literature, clinical data and expression analyses, to all be accessible and interlinked. This allows movement between information systems and provides more complex meta-analyses of these data sets, allowing a holistic view of biological research, in place of the current fragmented view we have available to us. This will ultimately lead to the blurring of boundaries between different disciplines, such as the areas of chemoinformatics and bioinformatics within pharmaceutical and biotechnology companies, to link chemistry data with biological and clinical data. Hopefully the cross-fertilization of disparate disciplines will fuel new developments in analysis methods, as can be seen by the migration of physicists, mathematicians and statisticians to work on these new problems in the life sciences.

5.2 The Pharmaceutical Process

Initially a range of disease areas will be the focus of a pharmaceutical company. Within each area, biological molecules, usually proteins, which will potentially influence the onset or course of a disease are identified by a variety of methods, and could, therefore, be targets for pharmacological intervention. These range from identifying targets reported in the literature to studies, in cell lines or animal models of the disease, through to identifying novel DNA sequences with similarity to known drug targets.

Once a potential therapeutic target, such as a G protein coupled receptor (GPCR), protease or ion channel, has been identified, an assay system will be set up to test compounds for activity against the target. This can range from a small number of assays where the target is well characterized and the chemistry understood, to a full blown high-throughput screen (HTS), in which several hundred thousand compounds are screened against the assay to identify which, if any, show activity.

Following the HTS, a number of "hits" (compounds showing some activity against the target) will be identified. These "hits" will then be analyzed in lower-throughput assays to gain a better understanding of the potency and selectivity of each compound against the target. Those which show sufficient potency and selectivity are then progressed to further medicinal chemistry to identify compounds

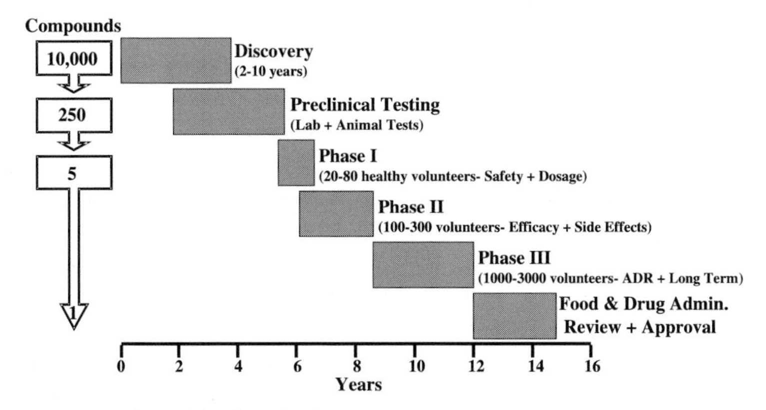

Fig. 5.1 Outline of the drug development process.

that are available with similar structure and/or synthesize related chemical structures for assaying to identify which show improved characteristics.

Once a "lead series" has been identified, the range of tests expands significantly. From this point on, there may also be involvement of computational chemists to model the possible sites of compound interactions on the target. There will be involvement of toxicologists, pharmacokineticists and drug metabolism groups to try and understand if the drug is likely to have any adverse side effects, its ability to be absorbed by the body and reach target organs, and any effects that the normal human metabolic processes may have on the structure and function of the compound.

After completing preclinical testing, a company files an Investigational New Drug (IND) application with the regulators (the FDA in the U.S.), so that clinical studies in man can begin. The IND shows results of all experiments to this point, a detailed proposal for the clinical study, the expected mode of action for the drug, and any side effects observed. All clinical trials will also be reviewed and approved by the Institutional Review Board (IRB) at the clinic where the trials will be run.

Assuming that all the appropriate approvals have been met at this point, then the phase I clinical study can begin. This normally involves 20–80 healthy volunteers who will take the drug for a limited time to help determine the drug's safety and pharmacokinetic properties in humans. If the drug passes phase I trials, it will enter phase II, which uses a larger group [100–300] including patients suffering from the disease against which the drug is targeted, to allow the clinicians to determine how efficacious the drug is in treating the disease. Success at this stage leads to phase III studies, during which an even larger group of patients, 1000–3000, are treated with the drug in a clinical setting, and the patients are closely monitored by the clinician to identify efficacy and any adverse events associated with the drug treatment.

On completion of phase III trials, the data will be checked to see that it fulfils all the criteria required to generate a viable, marketable drug. The company will then file a New Drug Application (NDA), with the intention of proving the efficacy and safety of the drug in this therapeutic application. The NDA will contain all the clinical data and all relevant preclinical data for review by the FDA. Application reviews were 16.2 months on average in 1997 [75].

Even when the drug is approved, the work is still not complete. Although the drug is available for doctors to prescribe, the drug company must continue to provide regular reports relating to any adverse effects seen in patients treated with the drug and any QC data. In some cases, the FDA may also require that the drug company follow up with phase IV trials to identify any affects associated with taking the drug over prolonged periods.

From the detail shown above, it is easy to see why drug discovery and development is such a costly process. One drug in every 5000–10 000 entering preclinical studies makes it to the clinic. A recent estimate put the cost of bringing one drug to market as US$ 500 million [75], and up to 70% of that cost is associated with compound failures in these studies. DiMasi et al. have suggested that reductions

Tab. 5.1 Drugs withdrawn due to safety issues identified after marketing (79)

Drug	*Approval date*	*Date withdrawn*	*Indication*	*Sales at withdrawal*
Duract	Jul-97	Jun-98	Short-term pain management	$50 million
Pondimin	–	Sep-97	Obesity	$106 million
Posicor	Jun-97	Jun-98	Hypertension and chronic stable angina pectoris	$41 million
Propulsid	Jun-93	Mar-00	Gastroesophageal reflux disease	$975 million
Raxar	Nov-97	Oct-99	Community-acquired infections	$17 million
Redux	Apr-96	Sep-97	Obesity management	$114 million
Rezulin	Jan-97	Mar-00	Type 2 diabetes	$600 million
Rotashield	Aug-98	Oct-99	Rotavirus vaccine	$43 million
Seldane	May-85	Dec-97	Seasonal allergic rhinitis	$90 million
Trovan	Dec-97	Jun-99	Infections	$160 million

in drug development times by 1–2 years could generate a 20–23% reduction in costs for drug development [76], and that a 2% increase of drugs reaching the clinic from phase I trials would reduce costs by up to 10% [77].

The potential commercial rewards for companies that successfully deliver new medicines are substantial – a recent survey shows that in the U.S. alone an average prescription drug generated US$1.3 million per day in revenue in 2000, and in the case of Prilosec® (an anti-ulcer drug), this figures leapt up to US $ 11.2 million per day [78]. Given the scale of these rewards, there is an enormous drive within the pharmaceutical and biotechnology industries to identify new drug targets and to better understand the actions of the drugs currently on the market. This is poignantly borne out by the fact that in the last 3 years, at least 10 drugs that have been successfully licensed for the clinic were withdrawn due to adverse effects identified later (Table 5.1).

5.3 Application to the Pharmaceutical Industry

Havng discussed the platform technologies and the general principles of pharmaceutical drug discovery and development in previous sections, this section will give some examples of how these technologies have been applied within a pharmaceutical arena. There will be little coverage on the literature associated with pharmacogenetics, as this will be covered extensively in other chapters of this book. Instead, we will concentrate primarily on the applications of expression technologies (EST sequencing, expression profiling and proteomics) to the area of pharmaceutical drug discovery and development, and medical research.

5.3.1
Understanding Biology and Disease

In Table 5.2, some examples of the application of expression technologies are listed, including EST library sequencing, which has primarily been used to identify novel gene transcripts present in specific tissues of interest [9, 82–92], or from specific chromosomes [93]. There is a limit in resolution of this approach, mainly due to the limited sizes of most datasets generated. As a cDNA library generated from a single mRNA population may contain as many as 10,000,000 clones, sequencing 5,000 clones only allows random selection of 0.5% of the genes, so the tendency is to identify more abundant genes, unless we pre-screen or select these clones, by methods such as differential display or suppression subtraction hybridization. This means that an enormous amount of sequencing must be carried out to generate significant sets of data for comparison. In the public arena, the Bodymap project from Kyoto University [50] and the efforts of TIGR (The Institute for Genome Research) [50], e.g., have focussed on sequencing ESTs from a wide range of normal tissues, allowing the production of an "electronic Northern" analysis of gene distribution. Similar efforts have also been happening within private companies, such as Human Genome Sciences and Incyte Genomics. The greatest value of the databases lies in generating very large data sets from a broad and diverse set of tissues, both normal and diseased, as well as from model animals and pathogenic organisms.

Once data from thousands of different sequences derived from thousands of different tissues has been compiled, bioinformatic analysis of the data allows us to identify genes that show restricted patterns of expression, assisting in both annotation of function, and association with phenotype. This analysis can be extended to look for novel markers or targets, by analyzing those novel sequences that are co-expressed with known markers or targets. This has been applied to a number of diseases already, such as prostate cancer and Parkinson's disease [120, 164].

One additional application that has been driven by both genomic and cDNA sequencing is the discovery of SNPs, or single nucleotide polymorphisms. These have a number of applications, including "traditional" genetic linkage. For example, Roses et al. [336] have shown the relationship between certain ApoE4 polymorphisms and the incidence of Alzheimer's disease, Barrosso et al. [337] showed that polymorphic changes in the PPAR receptor, the target for the glitazone antidiabetic drugs, has a strong linkage with diabetes and obesity, and Martinez et al. [338] showed a correlation between bronchodilator efficiency and the sequence polymorphisms in the β-adrenoreceptor. All of these genes had already been identified, but the SNP identification associated with the genes is allowing us a greater depth of understanding of the function of the expressed proteins, and their association with disease phenotypes. Equally, the identification of polymorphic variants in drug metabolizing enzymes, such as the cytochrome P450s, is helping us to understand how different metabolic activities can be genetically regulated, leading to variations in the activity of these enzymes [339]. The identification of many

Tab. 5.2 Examples of the applications of genomic expression technologies. N. B. while this may not be an exhaustive list, the content is intended to allow researchers from many different areas to find starting points in the literature related to their own areas of interest

Biological system studied	***References***
EST databases	
Brain anatomy	82, 91
Cardiovascular anatomy	85, 86
Cochlear anatomy	90
Embryo anatomy	83
Granulocyte anatomy	87
Inner renal medulla	92
Multiple tissue anatomy	88
Skeletal muscle anatomy	84, 89
Chromosome 7 specific genes	93
Gene Expression	
Cancer	
B cell lymphoma typing	94
Breast cancer	95–100
Cancer cell lines	101–103
Colon tumor	104
Cutaneous malignant melanoma typing	105
Doxorubicin	106
Estrogen receptor drug profiling	107
Head and neck squamous cell carcinoma	108
Head and neck tumors	109
Leukemia	110
Lung cancer	111
Lung squamous cell carcinoma	112
Oligodendroglioma	113
Ovarian cancer	114–116
Pancreatic cancer	117
Prostate cancer	118–120
Tamoxifen	121
Thyroid tumors	122
Infectious Disease/Pathogens	
Bordtella pertussis infection changes	123
C. albicans	124
C. elegans germline	125
CMV infection	126, 127
Drug effects on *H. influenzae*	128
E. coli	129
HBV infection	130
HDL-deficiency	131
HPV 31 induction	132
HSV 1 activation	133
Listeria monocytogenes infection	134
Mycobacterium	135, 136
P. aeruginosa infection	137
S. cerevisiae	138–143

Tab. 5.2 (continued)

Biological system studied	***References***
CNS disease	
Aging	144–146
Alcoholism	147
Alzheimer's neurons	148
Amphetamine	149
Brain anatomy	150, 151
Huntington's Disease	152, 153
Hypoxic induction	154
Mania/psychosis	155
Microglial activation	156
Multiple sclerosis	157
Neuronal anatomy	158
Neuronal cell death	159
Nutritional regulation	160
Photoreceptor	161
Schizophrenia	162–164
Sleep induction	165
THC treatment	166
Cardiovascular disease	
Atherosclerosis	167, 168
Myocardial infarction	169, 174
Ischemia	170
Cardiac growth and development	171
Cholesterol loading	172, 173
Cardia hypertrophy	175
Inflammation	
Haematopoeitic differentiation	176
Interferon gene regulation	177
Inflammatory Bowel Disease	178
Immune system	179–182
Pharmacology/Toxicology	
Dimethylnitrosamine hepatotoxicity	183
Carbon tetrachloride hepatotoxicity	184
Phenobarbital hepatotoxicity	185
Mitogen-induced versus regerative growth in liver	186
Arsenic-induced liver disease	187
Toxicogenomic databases	188–194

Tab. 5.2 (continued)

Biological system studied	***References***
Other areas	
Adipogenesis	195, 196
Apoptosis	197, 198
Cell cycle	188, 200
Diabetes	201
Genotoxic stress	202, 203
Gravity	204
Pancreatic β cells	205
PBMC anatomy	206
Peroxisome proliferation	207
Prostate anatomy	208
Renal function	209, 210
Secreted and membrane associated gene products	211
Testis-specific	212
Thyroid hormone	213, 214
Protein Expression	
Cancer	
Tumor classification	215, 216
Bladder cancer	217–220
Colonic crypts versus polyps	221
Small intestine and colon in APC mutant mice	222
Prostate cancer and cell line comparisons	223
Hepatocellular carcinoma cell line HCC-M pathology	224
Colon carcinoma cells line LIM 1215 pathology	225
Normal colon mucosa versus colorectal carcinoma	226
Effects of FGF2 on MCF-7 breast cancer cells	227
B-cell chronic lymphocytic leukemia	228
Neuroblastoma cell pathology	229
CNS	
Pathology in p53 knockout mice	230
Schizophrenia	231–233
Hippocampal pathology	235–237
Alzheimer's disease	237
Mouse brain	238
Myelin formation in c-myc knockout mic	239
Bipolar disorder	240
Stroke	241
Rat cerebellar development	242
CSF analysis in cerebral autosomal dominant arteriopathy	243
Cardiovascular disease	
Cardiac hypertrophy	244, 245
Atrophy of rat sloeus	246
PKCe compleses in cardioprotection	247
Hypoxia	248
Bovine dilated cardiomyopathy	249

Tab. 5.2 (continued)

Biological system studied	***References***
Inflammation	
Acute inflammation versus arthritis	250
Stress-induced fibroblast senescence	251
Lymphoblastoid cell pathology	234
Monocyteyic lysosomal pathology	252
PDGF receptor pathway	253
Human Jurkat T-cell pathology	254
Human braonch lavage alveolar fluid	255
TNF-stimulated fibroblast lysates	256
Infectious disease/pathogens	
C. elegans	257, 258
H. pylori	259–261
M. genitalium	262, 263
S. cerevisiae	63, 64, 264–268
Vaccinia virus	269
C. elegans	270, 271
S. pneumoniae	272
S. melliferum	273
H. influenzae	274–277
E. coli	278–281
B. subtilis	282
B. garinii	283
P. aeruginosa	284, 285
L. monocytogenes	286
S. typhimurium	287
Synechocystis PCC6803	288
Epstein-Barr virus	289
M. tuberculosis	290
Other areas	
Liver pathology	291
Gender differences	292
Mammary epithelial cells	293–295
Luteinizing hormone changes on mouse ovary	296
Na^+ transporter distribution in NHE3 and NCC mice	297
Degradtion of alphaB crystallin	298
Rat tooth enamel development	299
Hypothermia on CHO cell pathology	300
Rat dermal papilloma pathology	301
Identification of MAP kinase signalling pathways	302
Mitochondrial pathology	303–305
Human relfelx tear	306
Human epidermal transit amplifying cell markers	307
Sperm pathology	308
Golgi pathology	309, 310
Proteasomal interactions	311
Rat kidney cortex versus medulla	312

Tab. 5.2 (continued)

Biological system studied	***References***
Pharmacology/Toxicology	
Cyclosporine A	313
Multiple drug comparisons	314–317
Oltipraz	318
SDX PGU 693	319
Peroxisome proliferators	320–323
2 acetyl aminofluorene	324
Methapyrilene	325
Naenopin versis TNFalpha	326
Glomerular nephrotoxicity	327
Acetaminophen versus 3-acetamidophenol	328
Alpha1 protease inhibitor	329
Drug-induced hepatomegaly	330
Lovastatin	331
JP8 jet fuel toxicity	332–334
Lead toxicity	335

new genes and gene transcripts will undoubtedly give rise to many more genes with polymorphic changes in activity or function.

With the advent of the gene expression technologies, especially microarraying, we have moved from simultaneously studying the detailed expression analysis of a few genes with Northern blots to hundreds of genes on filters or using RT-PCR, and up to thousands using microarrays. By virtue of this increase in throughput, whole genomes have been studied on single arrays in a number of pathogens such as *Candida albicans* [124], *Saccharomyces cerevisiae* [138–143], CMV [126, 127], and *Haemophilus influenzae* [128], to name but a few examples. If we look at some of the examples using *S. cerevisiae,* we have seen cell cycle- [138] or signal-dependent pathways analyzed [140], and in one example, by co-ordinating deletion mutants with pharmacology and expression analysis, functional assignment of novel genes and even new pharmacological targets were identified [139].

When we start to look at higher organisms, we can approach the generation of microarrays based on genome sequence [340], cDNA sequence [18, 341, 342], or a combination of both. Most array strategies aim to generate minimal redundancy to allow maximum numbers of different genes to be present on each microarray. The bioinformatics associated with this becomes several orders of magnitude more complex than the sequence analyses. To start with, even a single microarray can generate thousands of data points; if we extend this to an example with an array of 10,000 genes hybridized with probes from tumor and matched non-tumor tissue, each from 5 patients at 2 times, we suddenly have 200,000 expression datapoints, or $\sim$5 Mb of data. Obviously if this is extended to a larger clinical study of say 50 patients samples weekly over 2 years, we suddenly need to store, let alone analyze, 25 Gb of data. The data collection is by no means trivial, but in comparative terms is relatively straightforward. A number of commercial and academic tools are appearing to analyze such data sets. A recent review testing some of these showed that in many cases, a "large" dataset (cited as 13,000 genes in 17 experimental conditions) was problematic for almost all the tools tested [67, 68]. While this could be partly related to issues with hardware, there is still a long way to go before we have tools to analyze and visualize significant sets of data, and render the results in a comprehensible format. That said, even with the limited tools we currently possess, some extraordinary data analysis has been done. The examples included are here to try and give some idea of the breadth of application of the technologies in use today, and more examples are listed in Table 5.2.

In cancer research there have been some seminal papers in typing clinical samples to discriminate between pathological types of tumors. Golub et al. [110] measured expression of 6,817 genes in parallel on an oligonucleotide array. Initially this was probed with a "test" set of 38 samples (27 from ALL lymphomas, and 11 from AML lymphomas) and a subset of 1,100 genes were identified as showing significant correlation with one tumor type over the other. Further analysis of the 1,100 genes revealed a subset of 50 genes that were used as "class predictors". When these were applied to a further 34 leukemic samples, 100% prediction accuracy was obtained, and 29 samples of these gave strong predictions. In a similar way, Bertucci et al. [95], showed that using a set of 176 selected genes arrayed on a filter, 34 primary breast carcinomas could be readily characterized as belonging to two molecularly distinct subgroups, characterized by different clinical outcomes after chemotherapy. Furthermore, differential gene expression patterns could distinguish tumors with lymph node metastases and by estrogen receptor status. Bittner et al. [105] profiled expression from malignant melanoma samples from 42 samples (19 primary cluster melanoma, 12 non-clustered melanoma, 4 uveal melanoma and 7 controls) across 8,150 genes. Analysis of the data allowed the identification of a transcriptionally discrete subset of genes which chracterized melanomas that form primitive tubular networks *in vitro*, also seen in some highly aggressive metastatic melanomas. Perou et al. [98] examined 65 surgical breast tumor samples from 42 patients on a microarray of 8,102 genes. 20 of these tumors had been sampled twice, pre-dosing and after 16 weeks doxorubicin treatment. Gene expression patterns were closer between different tumor samples taken from the same patient than from other patients, but 8 subsets of co-expressed genes were identified that could be related to variations in physiology, including endothelial cells, stromal cells, normal cells, B lymphocytes, T lymphocytes, macrophages, basal epithelial cells, and luminal epithelial cells. Watson et al. [102] took the approach of trying to correlate WHO classifications of oligodendriogliomas with gene expression patterns, using oligonucleotide arrays containing 1,879 genes. Using expression data from 1,100 of these genes, the tumors could be divided into two distinct groups that corresponded with their histological grades. By gene clustering analyses, 196 of these transcripts were shown to discriminate between the different tumor grades.

In pathogen research, yeast has been particularly well studied, in part because the genome sequence is available, and all the predicted coding regions can be distilled on to a single microarray. Spellman et al. [138] used *S. cerevisiae* cultures that had been synchronized in three different ways to identify a set of 800 genes that were associated with the cell cycle. When the cells were treated with the G1 cyclin Cln3p or the B-type cyclin Clb2p, ~500 of these genes responded, to one, or both, treatments. *C. albicans* has also been studied, e.g., by Staib et al. [124]. They showed that activation of the virulence-associated gene family of secreted aspartic proteases (SAPs) was differentially regulated, adapting to the host niche being infected. In a similar way, changes in gene expression in *Listeria monocytogenes* have been studied by Cohen et al. [134], both on oligonucleotide arrays representing 6,800 genes, and on cDNA filter arrays containing either 588 or 18,376 genes. These studies revealed that 74 genes were consistently upregulated, and 23 downregulated, in THP-1 cells upon infection with *L. monocytogenes. Pseudomonas aeruginosa* infection of A549 human lung pneumocytes has been studied in a similar way, using microarrays of 1,506 genes. Ichikawa et al. [137] identified 680 usuable signals, of which 24 genes showed at least 2-fold change in expression, 3 h after infection. In addition, 16 genes showed a greater than 2-fold increase in expression when comparing wild-type PAK strains versus PAK-NP. 13 of these genes were common to the time course study run with the wild-type PAK.

In CNS research, a number of neurological pathologies have been studied. Individual neurons from early- and late-stage Alzheimer's disease have been studied by Chow et al. [148] using antisense RNA profiling. When checked by typing with *in situ* hybridization, they were able show that the same characterization was found by individual neuron profiling. However, when compared with earlier neocortex studies using Northern blot analyses, the single cell profiling showed discrepancies. Models of Huntigdon's disease have been characterized by Luthi-Carter et al. [152], in which changes in striatal signaling genes were seen. Transcriptional differences have also been identified in multiple sclerosis (MS) brains, when Whitney et al. [157] compared MS acute lesion samples with normal white matter on microarrays of 5,000 genes. 29 genes showed significant upregulation in at least two plaque samples compared to normal white matter. The results were also confirmed for several of the genes using immunohistochemistry. Schizophrenia has also been studied, with two recent papers showing analysis of clinical samples. In the first [162], Hakak et al. screened 24 prefrontal cortex samples [12 schizophrenic – 9 male and 3 female, and 12 normal – 4 male and 8 female), against 6,000 genes on oligonucleotide arrays. 72 genes up-regulated and 17 down-regulated, covering a range of biological processes, but most notably, a down-regulation in all the myelination genes studied, suggesting a potential disruption in oligodendrocyte function. In the second paper [163], Mirnics et al. carried out a similar study, screening 22 clinical samples [18 male and 4 female) derived from area 9 of the brains, against cDNA microarrays with over 9,800 genes. Of these, 3,735 were detected in all six array comparisons, and 4.8% of these showed significant changes in schizophrenic patients [2.6% up-regulated and 2.2% down-regulated in schizophrenics). These genes fell into several classes, but most notable was a decrease in genes coding for proteins involved in regulation of presynaptic function. Neuronal apoptosis has been another area of high interest within the neurosciences. Chiang et al. [159] showed how the changes in gene expression during neuronal apoptosis are co-regulated and synchronized. Microarrays were generated with ~7,000 genes from rat brain cDNA libraries and screened with probes derived from rat cerebellar granule neurons. These were challenged by potassium withdrawal, combined potassium and serum withdrawal, and addition of kainic acid. In all cases, numbers of genes were seen to change over 3-fold (83 genes, 790 genes and 86 genes, respectively), including well-characterized apoptosis-associated genes. However, they also identified a subset of 26 of these genes that were regulated in all the challenges, suggesting a central "core" mechanism.

Cardiovascular disease has also been well studied by expression profiling. Friddle et al. [175] studied the changes in gene expression associated with induction and regression of cardiac hypertrophy in mouse models. Arrays of 4,000 genes were screened with probes derived from the left ventricles of mice that had been treated with either vehicle, angiotensin II, or isoproterenol. 55 genes were shown to be regulated during induction and regression, of which 32 were only changed in induction, and 8 were only changed in regression. Lyn et al. [170] also showed changes in cardiac anatomy in disease, by looking at the myocardium of mice 24 h after ischemia was induced by surgical restriction of the left coronary artery. Of the 588 genes screened on the filter, only 6 showed a greater than 5-fold increase in expression in the ischemic heart ventricles. Stanton et al. [174] looked at changes in myocardial infarction, using a rat model. In this case, 7,000 genes were monitored from 2–16 weeks after the infarction, in both the left ventricle free wall and the interventricular septum. 731 differentially expressed genes were identified, many associated with the cytoskeleton, extracellular matrix, contractility and metabolism. Sehl et al. [171] studied expression changes in rat associated with cadiac growth, development and in response to injury. The microarrays contained 86 known and 989 anonymous cDNAs obtained from sham-operated and 6 week post-myocardial infarction samples. 58 genes were identified as associated with myocardial development (12 known, 36 not associated with cardiac development, and 12 novel ESTs). After myocardial infarction, the genes associated with stress and wound healing were changed, as were 14 genes not previously associated with the disease.

Moving on to look at the application of proteomics to expression analysis for target discovery, we add an extra layer of complexity, and at the same time, resolution. We add complexity from the additional variants that can exist with regard to post-translational modification, such as glycosylation, methylation and phosporylation. As such, each functional protein may retain the same primary protein sequence, but the modifications can affect the protein's localization, activity, and proteolyic processing to generate peptidic signals. To this extent, we gain extra resolution of how a smaller subset of proteins change, and effect change, within the biological system. As two-dimensional electrophoresis tends to be limited to resolving a few thousand spots on a gel, the protein applications currently do not offer the breadth of genome coverage of gene expression data, but for focussed questions about specific protein populations, the information can be far more detailed.

Overall, the technology has been applied to a number of human diseases [215–257] and the pathology and physiology of several different microorganisms [258–291].

In one example, Steiner et al. [294] looked at differences in the proteome of male and female Wistar rats, identifying 6 proteins that were male-specific, one female-specific, and seven others that were statistically different between the sexes. They also identified four sets of protein spots that appeared to represent polymorphic proteins. The technology has also been used to examine changes in proteome between normal and diseased tissues for heart disease and breast cell pathology. In the former, Dunn et al. [244] reviewed the literature to date relating to heart disease, citing examples ranging from dilated cardiomyopathy and cardiac antigens in disease, through to animal models of heart disease. In the former, Page et al. [293] compared normal human luminal and myoepithelial breast cells from 10 reduction mammoplasties. By comparing 43,302 proteins detected, 1,738 non-redundant proteins were identified. 170 proteins showed greater than 2-fold change between cell types, and 51 were annotated by mass spectroscopy. A further 134 proteins were also annotated that did not show differential expression between cell types. In a further example demonstrating disease studies, Doherty et al. [249] studied the changes in plasma proteins from 3 populations – before versus after inflammation induced by typhoid vaccination, patients suffering from rheumatoid arthritis (RA) versus controls, and RA patients treated with tenidap versus piroxicam. By comparing 19 plasma proteins, they could see similarities between acute inflammation from vaccination and chronic inflammation from rheumatoid arthritis. The pattern of changes with tenidap was also distinguishable from that of piroxicam treatment.

5.3.2
Target Identification and Validation

One of the big driving forces in understanding disease pathology and physiology better, has been to identify new points of intervention for therapy, either using small molecules or biological approaches. The initial focus of much of the EST and genome sequencing was to identify novel homologs of known drug targets, as this may help identify adverse effects caused by a lack of target protein isoform, or may provide new targets for intervention. Historically, there have been approximately 500 proteins against which therapeutics have been targeted [344]. It has been estimated that this may only represent 10% of all possible targets for therapeutic intervention [345], meaning there are plenty of new targets available, offering advantages in novelty, and potentially exclusivity, of targets moving forward. That said, most of the known targets to date fall into a small number of protein families, with the majority being G-protein coupled receptors (GCPRs), proteases or ion channels. The EST approach has some initial advantages over the genome approach, as the expressed RNA is already spliced into actual coding sequence, so the clones used to generate the EST sequence data will contain partial, or sometimes full length, transcripts as reagents available for follow-up work.

Some examples include the identification of 3 novel GPCRs by Marchese et al. [346]. These were found by EST database searching, combined with "wet" biology, identifying three receptors, GPR27 and GPR30 (both show broad brain distribution), and GPR35 (shows localization to the intestine). Bertilsson et al. [347] identified a protein, hPAR, which appears to be a human ortholog of the mouse PXR1 receptor, involved in signaling for CYP3A induction. Two papers have shown examples of cell death-associated novel proteins – PARP2 [348], related to PARP, is involved in DNA repair, and RIP3 [349], that appears to modulate RIP and TNF receptor-1 effects on the NFkB signaling pathway. A number of other therapeutically relevant homologs have also been identified from EST databases, including a homolog of an eosinophil granule protein [350], corin (novel transmembrane serine protease from human heart) [351], and PDE9A, a novel human cGMP-specific phosphodiesterase [352].

If we look at how array technology has had an impact on target validation, there are many reports, using gene knockouts to identify gene's involved in key biological pathways in disease, and the effect that nullifying the gene's action has on physiology. For yeast there are examples from Marton et al. [353] and from Hughes et al. [354]. In the first paper [353], calcineurin, the proposed biological target for FK506 and cyclosporine A, was knocked out in yeast. Despite the absence of the proposed drug target, other biological effects were seen when the yeast were treated with the drugs, and each drug showed different "non-specific" effects. In the latter paper [354], 300 different combinations of yeast gene knockouts, and/or chemical treatments, were run on microarrays. In this case the authors started to identify functional roles for novel genes by the effects knocking them out had on genome-wide gene expression. They could also see how treatment of the yeast with different compounds effected global transcription changes. In one example, treatment of yeast with dyclonine (a topical anaesthetic), was profiled on microarrays, and showed close similarity to the effects of knockouts of erg2 in the ergosterol biosynthesis pathway. A number of further studies then confirmed that erg2 did appear to have a role in the effects of dyclonine. That information, combined with erg2 sharing sequence similarity with the human sigma receptor, led to a possible explanation of the mechanism of action of the drug.

In proteomics, we can select a number of examples. In the study of bladder cancer, several papers have been published showing application of 2D polyacrylamide gel electrophoresis [2D-PAGE). Celis et al. [219] analyzed 150 bladder tumors, of which 6 showed protein expression patterns corresponding to squamous cell carcinomas (SCCs), which had not been clear from histopathology alone. A number of proteins were shown to be regulated, with a limited subset being common to multiple samples, including psoriasin that was passed in urine and may be able to act as a clinically accessible surrogate marker. Psoriasis-associated fatty acid-binding protein (PA-FABP) was another common marker, which in a later paper [307] was also identified as a novel marker of human transit amplifying cells, using both 2D-PAGE and microarrays. Human bladder transitional cell carcinomas (TCCs) have also been examined by 2D-PAGE [217] and a number of regulated proteins identified, including 5 that were upregulated in at least 3 of the 4 tissues (tryptophanyl-tRNA synthetase, the interferon gamma-inducible protein gamma 3, manganese superoxide dismutase, and two unknowns) and 1 downregulated (aldose reductase). Many of these changes were also seen when primary cultures from TCCs were examined. Finally in the area of cancer, Voss et al. [228], looked for correlation between clinical data and 2D-PAGE data in human B-cell chronic lymphocytic leukemia (B-CLL). Using statistical methods to analyze the 2D-PAGE data from 24 patients, they were able to discriminate between patient groups with defined chromosomal characteristics or those who had expression levels which did correlate with clinical parameters. The patients with shortest survival times showed changes in hsp27, protein disulfide isomerase, and a number of redox enzymes.

Johnston-Wilson et al. [233] used 2D-PAGE analysis to study a number of neurological diseases, including bipolar disorder, major depressive disorder, and schizophrenia. *Post mortem* tissues from the frontal cortices of 89 individuals, including some with each of the three disorders, and some non-psychiatric controls, were examined. 6 proteins were identified that decrease when compared to the controls, including four forms of glial fibrillary acidic protein (GFAP) and one each of dihydropyrimidinase-related protein 2 and ubiquinone cytochrome c reductase core protein 1. Two additional spots showed increases in level in one or more diseases when compared to control (carbonic anhydrase 1 and fructose biphosphate aldolase C). In a final example, Lewis et al. [373] selectively activated and inhibited MKK1/2 to study the downstream signaling pathways. 25 targets of the signaling pathway were identified, of which only 5 had previously been identified as effectors of this pathway. The newly identified effectors suggest that this pathway may also be involved in nuclear transport, nucleotide excision repair, nucleosome assembly, membrane trafficking, and cytoskeletal regulation.

5.3.3 Drug Candidate Identification and Optimization

In one recent example of gene expression for pharmacological analysis, Scherf et al. [102] treated 60 different human cancer cell lines with 60,000 compounds. These same 60 cell lines were analyzed by expression analysis [103], and the cell lines were clustered based on expression profiles. The clusters were significantly different from those based on their response to drugs.

In a separate study on the effects of captopril on rats after myocardial infarction, Davis et al. [169] identified 37 genes that were seen to change between myocardial infarct and sham operated rats. 10 of these gene changes were inhibited by captopril treatment. The 37 genes clustered into 11 functional groups, of which 6 included at least one of the 10 genes inhibited by captopril.

Zajchowski et al. [107] showed comparison of expression profiles between 38 different estrogen receptor-modulating compounds. An intial study was run based on the results of profiling of 24 combinations of cells and genes with estrogen, tamoxifen, raloxifene and ICI 164384 (a pure ER antagonist). Using the optimized assay panel derived from these studies, the 38 compounds were then profiled and classified into 8 groups. The compound gene expression profiles predicted uterine-stimulatory activity, and one of the groups of compounds was assessed for activity in attenuating bone loss in ovarectomized rats. In a similar multi-compound study [343], our analysis of data from dose and time studies from 29 compound treatments in a single human cell line generated well over 2,300 arrays, each containing nearly 10,000 gene transcripts, generating over 20,000,000 data points. Even after cropping the data and collapsing repeats to averaged data sets, we were left with well over 300 columns of data for over 7,000 genes remained (see Figure 5.2).

Several papers have also been published showing effects of drugs on microrganisms. One by Gmuender et al. [128] showed comparison of modes of action of two antibiotics, novobiocin (an ATPase inhibitor) and ciproflaxin (a DNA supercoiling inhibitor). The expression patterns appeared to reflect the expected changes in cellular metabolism in *H. influenzae*, with novobiocin-changed expression rates of several genes, probably as a result of reduction in transcription by inhibition of supercoiling, whereas ciproflaxin stimulated DNA repair systems. A

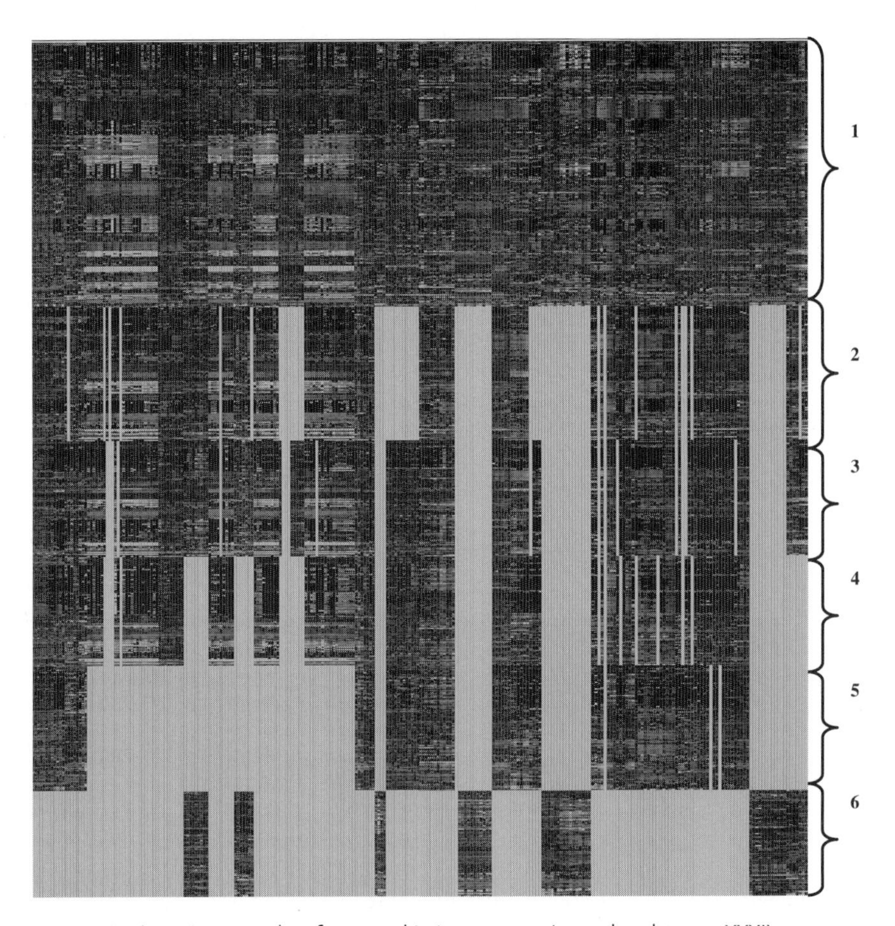

Fig. 5.2 *In vitro* microarray data from xenobiotic treatment (see color plates, p. XXXI).

paper from Bammert et al. [140] looked at comparisons between the effects of ergosterol biosynthesis inhibitors compared to specific gene deletions in the ergosterol pathway. Using 8 inhibitors (clotrimazole, fluconazole, itraconazole, ketoconazole, voriconazole, terbinafine, amorolifine, and PNU-144248E) and deleting 3 yeast genes (ERG2, ERG5, or ERG6], they identified 234 responsive genes, which included most of the ergosterol pathway genes. All the drugs appeared to show the same mode of action, and the transcriptional changes also included 36 mitochondrial genes and several other genes related to the ergosterol pathway and oxidative stress. Chambers et al. [127] described a study looking at cytomegalovirus (CMV) genome changes seen when CMV-infected fibroblasts were treated with cycloheximide or ganciclovir. The microarrays were made using oligonucleotides to represent every ORF in the CMV genome. The expression of 4 immediate early genes was identified by treating the cells with cycloheximide, and the early and late classes of genes were identified using gangciclovir-treated cells. Only 13 ORFs showed changes over 10-fold with ganciclovir treatment, one of these, UL130, showing over 900-fold increase in expression. In a final example, Wilson et al. [136] studied the effect of isoniazid treatment on the genome expression of *Mycobacterium tuberculosis*. A number of genes were induced that would have been predicted by the drug's mode of action, including several genes encoding type II fatty acid synthase enzymes and trehalose dimycolyl transferase. Several other apparently unrelated genes were activated, including efpA, fadE23, fadE24 and ahpC, which may mediate some of the toxic responses to the drug.

In a proteomic example monitoring changes associated with single drugs, Steiner et al. [331] examined alterations in protein expression in liver when F344 rats were treated with lovastatin, an HMG-CoA reductase inhibitor. The proteins identified showed that lovastatin not only evoked changes in the cholesterol biosynthesis pathway, as may be expected from a drug used to lower lipid levels, but also changes in carbohydrate metabolism. A number of cellular stress response proteins were also seen to change, which may reflect some form of toxic response to the drug. Edvardsson et al. [321] studied the effects of WY 14,643, a PPARα (peroxisome proliferator-activated receptor alpha) agonist in the livers of ob/ob mice (a disease model for insulin resistance and diabetes). At least 16 spots were seen to be upregulated, of which 14 were components of the peroxisomal fatty acid metabolism pathway. A number of papers have also been published by Anderson et al., including one looking at a 2D-PAGE database of changes applied to gene regulation and drug effect studies [315], and a paper specifically on the study of xenobiotic toxicity (using aroclor 1254, phenobarbital, cycloheximide and carbon tetrachloride) in mouse livers [316]. Myers et al. [216] have published a study that looks at the correlation of protein spots with both protein expression patterns and response to 3,989 compounds in 60 cancer cell types. The latter appeared to show much stronger correlation than the former, suggesting the pharmacology comparisons may be more robust that biological comparisons.

5.3.4
Safety and Toxicology Studies

The use of EST databases would not be an obvious point to start from for traditional toxicology studies. However, some published examples have shown how these databases can help to identify new pathways and enzymes within drug metabolism and pharmacological response. Bertilsson et al. [355] identified a new nuclear receptor, hPAR, which was activated by pregnanes and some clinical drugs, including rifampicin that has been shown to induce CYP3A4. The paper suggests that hPAR could be the human ortholog of mouse PXR1, and that they regulate overlapping target genes in response to distinct CYP3A4 activators, as pregnenolone 16α-carbonitrile activates PXR1, but not hPAR. Within Incyte, detailed sequence analysis has led to the identification of numerous new members of the drug metabolism families, including, e.g., 200 potentially novel enzymes and 20 novel transporters [356].

A number of recent reviews have discussed the overall concept of gene expression analysis applied to toxicology and drug metabolism, in the new area of toxicogenomics [357–372]. The technology has primarily been focussed on trying to identify gene expression patterns that associate with specific phenotypic outcomes, such as hepatotoxicity, nephrotoxicity, or a general toxic effect. The screening of development compounds in a similar system could potentially allow the rapid identification of toxicity, enabling the prioritization of development compounds in the lead optimization process.

Amundsen et al. [203] used microarrays to identify gene changes associated with genotoxic response. In the study, ML-1 cells were γ-irradiated and significant changes were seen in 48 genes, 30 of which had not previously been associated with ionizing radiation exposure. 13 of these genes, including BAX, CIP1, MDM2 and WAF1 and 9 others, were studied in 12 irradiated human cell lines. The responses varied in different cell types, but FRA-1 and ATF3 did appear to show some p53-associated component to their radiation induction. Bhattacharje et al. [183] looked at the changes in gene expression in dimethylnitrosamine (DMN)-induced hepatotoxicity using differential display. 48 cDNAs were identified, 23 were induced (including C3 and SAA), and 25 repressed (including CYP7, MUP and a myeloid differentiation protein gene). In a further example, Waring et al. [194] compared gene expression changes in rat hepatocytes treated with 15 known hepatotoxins. Using single dose (20 μM) and time (24 h) points for each, 179 of the 973 genes studied showed at least a 2-fold change in expression. Hierarchical clustering gave rise to 3 clusters – one included araclor 1254 and 3-methylcholanthrene, a second included carbon tetrachloride, methotrexate and monocrotaline, and the third included allyl alcohol, carbamazepine, and indomethacin. Dimethylnirosamine showed the greatest difference in expression from any of the other compounds. Compound effects have also been studied in yeast. Jelinsky et al. [142] looked at the effect of treating yeast with an alkylating agent (methyl methanesulfonate) using oligonulceotide arrays of 6,200 yeast transcripts. 403 transcripts showed changes, of which 325 were increased and 76 decreased. 18 of the

21 genes known to be inducible by DNA-damaging agents were identified, but most of the newly identified genes were more strongly induced. 48 of the 50 transcripts showed similar responses on Northern blots. In a second yeast example, Jia et al. [143] compared changes in expression of 1,529 genes during the treatment of yeast with sulfameturon methyl. 191 genes were induced and 131 repressed by 15 min treatment, and these numbers changed to 251 and 129 genes, respectively, after 4 h treatment. At a higher dose, initial expression changes were similar, but with time, more stress response and DNA damage genes were upregulated.

The study of the behavior of drugs and toxins in biological systems is probably one of the best reported areas in proteomics, so a few representative examples will be discussed here. Aicher et al. [313] looked at the effects cyclosporine A has on the kidneys of different species. In rats, decreases in calbindin-D, urinary calcium wasting and intratubular corticomedullary calcifications are seen. However, in dogs and monkeys that showed very limited nephrotoxic effects with cyclosporine A treatment, renal calbindin-D levels remained unchanged at the protein level. When this was compared to human kidneys (from cyclosporine A-treated kidney transplant patients), a decrease was seen in calbindin-D, indicating that it may be marker for cyclosporine A toxicity. Founoulakis et al. [328] generated a mouse protein database by comparing protein expression profiles of livers from acetaminophen-treated mice and 3-acetamidophenol-treated mice (3-acetamidophenol is a non-toxic regioisomer of acetaminophen). 256 spots were identified, and 35 of the identified proteins were differentially regulated by treatment with acetaminophen or 3-acetamidophenol. Most of the changes seen with 3-acetamidophenol occurred in a subset of the proteins changed by acetaminophen, and many of the changed proteins had been previously identified as having a role in acetaminophen hepatotoxicity. Newsholme et al. [330] looked at protein changes involved in increase in hepatocellular rough endoplasmic reticulum in Sprague-Dawley rats, induced by treatment with a substituted pyrimidine derivative. Livers increased in weight by 37%, and the 2D-PAGE analysis showed 5 down-regulated and 12 up-regulated proteins, including contrapsin-like protein inhibitor-6, the most up-regulated, and methionone adenosyltransferase (a catalyst in methionine/ATP metabolism), and mitochondrial HMG-CoA synthase (involved in cholesterol synthesis). Cutler et al. [327] looked at glomerular nephrotoxicity induced by puromycin aminonucleoside in rats by analyzing proteins excreted in urine. In a final example, Moller et al. [317] identified 12 proteins that showed a response to daunorubicin-induced stress of EPP85-181P human pancreas adenocarcinoma cells. Of these, 3 showed dose-dependent changes (cytokeratin 19, keratin K7 and Drg-1) whereas the other 9 showed changes with all doses (TCP1α, TCP1β, Trp-synthetase, EF1α, hnRNP H, ECP51, Grp78/Bip, and two forms of Hsp60).

5.4 Application to the Medical Research Community

While most of the technologies described so far have been applied to academic research and the drug discovery processes, some of the applications described in the previous sections can equally well apply to the medical research community. Here the focus will be on some examples of the applications that are more exclusively relevant to this community, and will include aspects relevant to clinical research, both inside and outside the pharmaceutical development process.

As there are already complete chapters relating to the drug metabolizing proteins (Chapters 9, 10, 11) [375], these will not be discussed here, but there will be reference to some recent reviews of clinical applications [376–382]. However, in the area of expression profiling, clinical applications are beginning to emerge [383], not just from the basic research to understand disease biology at the transcription level, as discussed in earlier sections, but also for clinical screening, both of patients and pathogens.

One example, published by Johnson et al. [384], examined changes in gene expression in multidrug-resistant tuberculosis patients undergoing human interleukin 2 immunotherapy, using RT-PCR, and differential display. Decreases in γ-interferon were seen in peripheral blood mononuclear cells, but at the site of delayed-type hypersensitivity, levels of both γ-interferon and interleukin 2 (IL2] were elevated during IL2 therapy given after exposure to purified protein derivative of tuberculin. The differential display analysis showed several other genes that changed expression during IL2 therapy, including cytochrome oxidase I, heterogeneous nuclear ribonuclear protein G, CD63, clathrin heavy chain, and adaptin. These may prove to be useful surrogate markers for leukocyte activation at a mycobacterial antigen-specific response site as similar changes were observed in PBMCs that were challenged with *M. tuberculosis* and IL2. In a pathogen-related example, Li et al. [385] typed and subtyped 7 human influenza A and B viruses, using cDNA arrays generated from 26 different portions of the hemagglutanin, neuraminidase and matrix protein genes.

If we now look at how proteomics has been applied, there is potentially more short-term clinical utility, as many of the readily accessible biological materials, such as urine, serum, and blood, contain little or no RNA, somewhat limiting the application of expression technologies. Earlier there were clinically relevant examples described around bladder cancers [219, 220] and in leukemia [228]. Unlu et al. [243] examined cerebrospinal fluid from CADASIL (cerebral autosomal dominant arteriopathy with subcortical infarcts and leukoencephalopathy) patients with Notch-3 mutations using 2D-PAGE. When compared to 6 control samples, a single spot was deleted in the CADASIL patients, later identified as human complement factor B, suggesting a role for the alternative complement pathway in the disease pathogenesis. Molloy et al. [306] looked at the protein components of human reflex tears, and by generating sequence tags for 30 spots, 6 proteins were identified in the SWISS-PROT database (lipocalin, lysozyme, lactotransferrin, zinc-alpha-2 glycoprotein, cystatin S, cystatin SN). The majority of the most abundant proteins showed heterogeneity on the gel, probably due to post-translational modifications. One novel protein isolated, named lacryglobin, showed strong homology to mammoglobin, which is overexpressed in breast cancer, indicating a possibility that proteomic screening of tears may be a new route for diagnosis of disease. Two further recent publications have shown application of 2D-PAGE to broncheoalveolar lavage (BAL) samples, one looking at the changes in protein expression in interstitial lung diseases [386] and the other looking for changes associated with alpha 1-protease inhibitor therapy in cystic fibrosis patients [329]. In the former paper, Wattiez et al. [386] looked at BAL samples from individual patients with diseases such as sarcidosis, idiopathic pulmonary fibrosis, and hypersensitivity pneumonitis. In the latter example, Griese et al. [329] looked at proteome changes in 8 young adult cystic fibrosis patients after dosing daily by inhalation of alpha 1-protease inhibitor over an 8 week period. Total protein and high molecular-weight proteins in general were reduced after dosing, and there was a reduction in the proteolysis of surfactant protein A.

5.5 Conclusions

An enormous amount of interest in pharmacogenomics has developed over the last few years, especially in the areas of drug metabolism and efficacy. It has been reported that 30% of patients taking statins, 35% of patients taking beta blockers, and up to 50% of patients taking tricyclic antidepressants show no response to pharmacological intervention [387]. In addition to drugs with variable efficacy based on the genetics of the target protein, safety testing in animals appears to be 70% successful, large numbers of drugs fail to reach the market, and those reaching the market that are then later withdrawn making drug development both time-consuming and expensive, both for the pharmaceutical industry and the subsequent effects on drug costs for patients. Application of genomic technologies to develop better understanding of the biology of disease and the physiological effects of drugs and toxins will hopefully lead to the generation of new tools more rapidly, and successfully assist in the discovery and development of new drugs.

Acknowledgements
I would like to thank my grandfather, Dr. Gosta Oberg, as my initial inspiration for moving into the medical biosciences. I would also like to thank Kate, Elizabeth and Charlotte for putting up with my time at home working on this chapter.

5.6 References

1 Sanger F et al. DNA sequencing with chain-terminating inhibitors. Proc Natl Acad Sci USA **1977**; 74:5463–5467.
2 International Human Genome Sequencing Consortium. Initial sequencing and analysis of the human genome. Nature **2001**; 409:860–921.
3 Venter JC et al. The sequence of the human genome. Science **2001**; 29:1304–1351.
4 de Souza SJ et al. Identification of human chromosome 22 transcribed sequences with ORF expressed sequence tags. Proc Natl Acad Sci USA **2000**; 97:12690–12693.
5 Sklonick J et al. From genes to protein structure and function: novel applications of computational approaches in the genomic era. TIBTECH **2000**; 18:34–39.
6 Fischer D et al. Predicting structures for genome proteins. Curr Opin Struct Biol **1999**; 9:208–211.
7 Gray IC et al. Single nucleotide polymorphisms as tools in human genetics. Hum Mol Genet **2000**; 9:2403–2408.
8 Saunders AM et al. The role of apolipoprotein E in Alzheimer's disease: pharmacogenomics target selection. Biochim Biophys Acta **2000**; 1502:85–94.
9 Adams MD et al. 3,400 new expressed sequence tags identify diversity of transcripts in human brain. Nature Genet **1993**; 4:256–267.
10 Ewing B et al. Analysis of expressed sequence tags indicates 35,000 human genes. Nature Genet **2000**; 25:232–234.
11 Okubo K et al. Large scale cDNA sequencing for analysis of quantitative and qualitative aspects of gene expression. Nature Genet **1992;** 2:173–179.
12 Kane MD et al. Assessment of the sensitivity and specificity of oligonucleotide [50mer) microarrays. Nucleic Acids Res **2000**; 28:4552–4557.
13 Lipshulz RJ et al. High density synthetic oligonucleotide arrays. Nature Genet **1999**; 21 (Suppl):20–24.
14 Freeman WM et al. Fundamentals of DNA hybridization arrays for gene expression analysis. BioTechniques **2000**; 29:1042–1055.
15 Hughes TR et al. Expression profiling using microarrays fabricated by an ink-jet oligonucleotide synthesiser. Nature Biotechnol **2001**; 19:342–347.
16 Schena M et al. Quantitative monitoring of gene expression patterns with a complementary DNA microarray. Science **1995**; 270:467–470.
17 Bowtell DDL Options available – from start to finish – for obtaining expression data by microarray. Nature Genet **1999**; 21 (Suppl):25–32.
18 Schena M et al. Microarrays: biotechnology's discovery platform for functional genomics. TIBTECH **1998**; 7:301–306.
19 Medhurst AD et al. The use of TaqMan RT-PCR assays for semiquantitative analysis of gene expression in CNS tissues and disease models. J Neurosci Methods **2000**; 98:9–20.
20 Harrison DC et al. The use of quantitative RT-PCR to measure mRNA expression in a rat model of focal ischemia – caspase-3 as a case study. Brain Res Mol Brain Res **2000**; 75:143–149.
21 Kreuzer KA et al. LightCycler technology for the quantitation of bcr/abl fusion transcripts. Cancer Res **1999**; 59:3171–3174.
22 Yoshikawa T et al. Isolation of novel mouse genes differentially expressed in brain using cDNA microarray. Biochem Biophys Res Commun **2000**; 275:532–537.
23 Elek J et al. Microarray-based expression profiling in prostate tumors. In vivo **2000**; 14:173–182.
24 Bailey DS et al. Pharmacogenomics – it's not just pharmacogenetics. Curr Opin Biotechnol **1999**; 9:595–601.
25 Liang P et al. Analysis of altered gene expression by differential display. Methods Enzymol **1995**; 254:304–321.
26 Bartlett J. Technology evaluation: SAGE, Genzyme Molecular Oncology. Curr Opin Mol Ther **2001**; 3:85–96.
27 Gould Rothberg BE. The use of animal models in expression pharmacogenomics analyses. Pharmacogenomics J **2001**; 1:48–58.

28 Brenner S et al. Gene expression analysis by massively parallel signature sequencing (MPSS) on microbead arrays. Nature Biotechnol **2000**; 18:630–634.

29 Sutcliffe JG et al. TOGA: an automated parsing technology for analysing expression of nearly all genes. Proc Natl Acad Sci USA **2000**; 97:1976–1981.

30 Meyer UA et al. Molecular mechanisms of genetic polymorphisms of drug metabolism. Ann Rev Pharmacol Toxicol **1997**; 37:269–296.

31 Walow W Pharmacogenetics in biological perspective. Pharmacol Rev **1997**; 49:369–379.

32 Linder MW et al. Pharmacogenetics: a laborartory tool for optimising therapeutic efficiency. Clin Chem **1997**; 43:254–266.

33 Weaver T High throughput SNP discovery and typing for genome-wide genetic analysis. New Technologies for Life Sciences: a Trends Guide **2000**; 6:36–42.

34 Bullingham R Pharmacogenomics: how gene variants can ruin a good drug. Curr Drug Discovery **2001**; March:17–20.

35 O'Farrell PH et al. Two-dimensional polyacrylamide gel electrophoretic fractionation. Methods Cell Biol **1977**; 16:407–420.

36 Chalmers MJ et al. Advances in mass spectrometry for proteome analysis. Curr Opin Biotechnol **2000**; 11:384–390.

37 Gevaert K et al. Protein identification methods in proteomics. Electrophoresis **2000**; 21:1145–1154.

38 Celis JE et al. 2D protein electrophoresis: can it be perfected? Curr Opin Biotechnol **1999**; 10:16–21.

39 Bussow K et al. A human cDNA library for high-throughput protein expression screening. Genomics **2000**; 65:1–8.

40 Haab BB Advances in protein microarray technology for protein expression and interaction profiling. Curr Opin Drug Discovery Dev **2001**; 4:116–123.

41 Larsson M et al. High-throughput protein expression of cDNA products as a tool in functional genomics. J Biotechnol **2000**; 80:143–157.

42 Legrain P et al. Genome-wide protein interaction maps using two-hybrid systems. FEBS Lett **2000**; 480:32–36.

43 Emili AQ et al. Large-scale functional analysis using peptide or protein arrays. Nature Biotechnol **2000**; 18:393–397.

44 Fung ET et al. Protein biochips for differential profiling. Curr Opin Biotechnol **2001**; 12:65–69.

45 Schneider J The next wave of molecular medicine. R&D Directions **2000**; March:65–70.

46 Pandey A et al. Proteomics to study genes and genomces. Nature **2000**; 405:837–846.

47 Gene Ontology Consortium. Gene Ontology: tool for the unification of biology. Nature Genet **2000**; 25:25–29.

48 O'Donovan C et al. The human proteome initiative (HPI). TIBTECH **2001**; 19:178–181.

49 Gelbart WM. Databases in genomic research. Science **1998**; 282:659–661.

50 Multiple authors The 2001 Database Issue. Nucleic Acids Res **2001**:29

51 Altschul SF et al. Basic local alignment search tool. J Mol Biol **1990**; 215:403–410.

52 Smith TF et al. Comparison of biosequences. Adv Appl Math **1981**; 2:482–489.

53 Ewing B et al. Base-calling of automated sequencer traces using Phred I: Accuracy assessment. Genome Res **1998**; 8:175–185.

54 Roest Croellius H et al. Estimate of human gene number provided by genome-wide analysis using *Tetradon nigroviridis* DNA sequence. Nature Genet **2000**; 25:235–238.

55 Ewing B et al. Analysis of expressed sequence tags indicates 35,000 human genes. Nature Genet **2000**; 25:232–234.

56 Liang F et al. Gene index analysis of the human genome estimates approximately 120,000 genes. Nature Genet **2000**; 25:239–240.

57 Rost B et al. Combining evolutionary information and neural networks to predict protein secondary structure. Proteins **1994**; 19:55–72.

58 Skonick J et al. From genes to protein structure and function: novel applications of computational approaches in the genomic era. TIBTECH **2000**; 18:34–39.

59 Westhead DA et al. Protein structure prediction. Curr Opin Biotechnol **1998**; 9:383-389.

60 Bailey DS et al. New tools for quantifying molecular diversity. Pharmainformatics: A Trends Guide **1999**; 4:6–9.

61 Gilbert D et al. Interactive visualization and exploration of relationships between biological objects. TIBETCH **2000**; 18:487–494.

62 Rain J-C et al. The protein-protein interaction map of *Helicobacter pylori*. Nature **2001**; 409:211–215.

63 Ito T et al. A comprehensive two-hybrid analysis to explore the yeast protein interactome. Proc Natl Acad Sci USA **2001**; 98:4569–4574.

64 Uetz P et al. A comprehensive analysis of protein-protein interactios in *Saccharomyces cerevisiae*. Nature **2000**; 403:623–627.

65 Schuster S et al. A general definition of metabolic pathways useful for systematic organization and analysis of complex metabolic networks. Nature Biotechnol **2000**; 18:326–332.

66 http://www.esc.auckland.ac.nz/Groups/Bioengineering/Movies/index.html

67 Goodman N Sampling the menu of microarray software. Genome Technol **2001**; March:40–50.

68 Sherlock G Analysis of large-scale gene expression data. Curr Opin Immunol **2000**; 12:201–205.

69 Rindflesch TC et al. EDGAR: Extraction of drugs, genes and relations from biomedical literature. Pac Symp Biocomput **2000**; 517–528.

70 Tanabe L et al. MedMiner: an internet text-mining tool for biomedical information, with application to gene expression profiling. BioTechniques **1999**; 27:1210–1217.

71 http://www.omniviz.com

72 Studt T Datamining tools keep you ahead of the flood. Drug Discovery Dev **2000**; August:30–36.

73 Adrade MA et al. Automated extraction of information in molecular biology. FEBS Lett **2000**; 47612–47617.

74 Hull R et al. Text-based information systems for drug discovery. Curr Opin Drug Discovery Dev **1999**; 2:186–196.

75 http://www.phrma.org

76 DiMasi JA et al. Cost of innovation in the pharmaceutical industry. J Health Econ **1991**; 10:107–142.

77 DiMasi JA et al. Research and development costs for new drugs by therapeutic category: a study of the US pharmaceutical industry. Pharmacoeconomics **1995**; 7:152–169.

78 Getz KA et al. Breaking the development speed barrier: assessing successful practices of the fastest drug developing companies. DIJ **2000**; 34:725–736.

79 MedAdNews **2000**; May:104.

80 Attwood TK et al. Progress with the PRINTS protein fingerprint database. Nucleic Acids Res **1996**; 24:182–188.

81 http://www.sanger.ac.uk/Pfam/

82 Adjaye J et al. cDNA libraries from single human preimplantation embryos. Genomics **1997**; 46:337–344.

83 Bortoluzzi S et al. The human adult skeletal muscle transcriptional profile reconstructed by a novel computational approach. Genome Res **2000**; 10:344–349.

84 Dempsey AA et al. A cardiovascular EST repertoire: progress and promise for understanding cardiovascular disease. Mol Med Today **2000**; 6:231–237.

85 Ewing RM et al. EST databases as multiconditional gene expression datasets. Pac Symp Biocomput **2000**; 5:430–432.

86 Hwang DM et al. A genome-based resource for molecular cardiovascular medicine. Circulation **1997**; 96:4146–4203.

87 Itoh K et al. Expression profile of active genes in granulocytes. Blood **1998**; 92:1432–1441.

88 Kawamoto S et al. BodyMap: a collection of 3' ESTs for analysis of human gene expression information. Genome Res **2000**; 10:1817–1827.

89 Pietu G et al. The Genexpress IMAGE Knowledge Base of the human muscle transcriptome: A resource of structural, functional, and positional candidate gene for muscle physiology and pathologies. Genome Res **1999**; 9:1313–1320.

90 Soto-Prior A et al. Identification of preferentially expressed cochlear genes by systematic sequencing of a rat cochlea cDNA library. Mol Brain Res **1997**; 47:1–10.

91 Stewart GJ et al. Sequence analysis of 497 mouse brain ESTs expressed in the substantia nigra. Genomics **1997**; 39:147–153.

92 Takenaka M et al. Gene expression profiling of the collecting duct in the mouse renal inner medulla. Kidney Int **2000**; 57:19–24.

93 Touchman JW et al. 2006 expressed-sequence tags derived from human chromosome 7-enriched cDNA libraries. Genome Res **1997**; 7:281–292.

94 Alizedah AA et al. Distinct types of diffuse large B-cell lymphoma identified by gene expression profiling. Nature **2000**; 403:503–511.

95 Bertucci F et al. Gene expression profiling of primary breast carcinomas using arrays of candidate genes. Hum Mol Genet **2000**; 9:2981–2991.

96 Hedenfalk I et al. Gene expression profiles in hereditary breast cancer. N Engl J Med **2001**; 344:539–548.

97 Kuang WW et al. Differential screening and suppression subtractive hybridisation identified genes differentially expressed in an estrogen receptor-positive breast carcinoma cell line. Nucleic Acids Res **1998**; 26:1116–1123.

98 Perou CM et al. Molecular portraits of human breast tumours. Nature **2000**; 406:747–752.

99 Perou CM et al. Distinctive gene expression patterns in human mammary epithelial cells and breast cancers. Proc Natl Acad Sci USA **1999**; 96:9212–9217.

100 Sgroi DC et al. *In vivo* gene expression profile analysis of human breast cancer progression. Cancer Res **1999**; 59:5656–5661.

101 Ross DT et al. Systematic variation in gene expression patterns in human cancer cell lines. Nature Genet **2000**; 24:227–235.

102 Scherf U et al. A gene expression database for the molecular pharmacology of cancer. Nature Genet **2000**; 24:236–244.

103 Weinstein JN et al. An information-intensive approach to the molecular pharmacology of cancer. Science **1997**; 275:343–349.

104 Alon U et al. Broad patterns of gene expression revealed by clustering analysis of tumor and normal colon tissues probed by oligonucleotide arrays. Proc Natl Acad Sci USA **1999**; 96:6745–6750.

105 Bittner M et al. Molecular classification of cutaneous malignant melanoma by gene expression profiling. Nature **2000**; 406:536–540.

106 Kudoh K et al. Monitoring the expression profiles of doxorubicin-induced and doxorubicin-resistant cancer cells by cDNA microarray. Cancer Res **2000**; 60:4161–4166.

107 Zajchowski DA et al. Identification of selective estrogen receptor modulators by their gene expression fingerprints. J Biol Chem **2000**; 275:15885–15894.

108 Villaret DB et al. Identification of genes overexpressed in head and neck squamous cell carcinoma using a combination of complementary DNA subtraction and microarray analysis. Laryngoscope **2000**; 11:374–381.

109 Gottschlich S et al 1. Differentially expressed genes in head and neck cancer. Laryngoscope **1998**; 108:639–644.

110 Golub TR et al. Molecular classification of cancer: class discovery and class prediction by gene expression monitoring. Science **1999**; 286:531–537.

111 Hellman GM et al. Gene expression profiling of cultured human bronchial epithelial and lung carcinoma cells. Toxicol Sci **2001**; 61:154–163.

112 WangT et al. Identification of genes differentially over-expressed in lung squamous cell carcinoma using combination of cDNA subtraction and microarray analysis. Oncogene **2000**; 19:1519–1528.

113 Watson MA et al. Gene expression profiling with oligonucleotide microarrays distinguishes World Health Organisation grade of oligodendrogliomas. Cancer Res **2001**; 61:1825–1829.

114 Ono K et al. Identification by cDNA microarray of genes involved in ovarian carcinogenesis. Cancer Res **2000**; 60:5007–5011.

115 Wang K et al. Monitoring gene expression profile changes in ovarian carcinomas using cDNA microarray. Gene **1999**; 229:101–108; 269–273.

116 Welsh JB et al. Analysis of gene expression profiles in normal and neoplastic

ovarian tissue samples identifies candidate molecular markers of epithelial ovarian cancer. Proc Natl Acad Sci USA **2001**; 98:1176–1181.

117 Gress TM et al. Identification of genes with specific expression in pancreatic cancer by cDNA representational difference analysis. Genes Chromosom Cancer **1997**; 17:1–7.

118 Bubendorf L et al. Hormone therapy failure in human prostate cancer: analysis by complementary DNA and tissue microarrays. J Natl Cancer Inst **1999**; 91:1758–1764.

119 Elek J et al. Microarray-based expression profiling in prostate tumors. In vivo **2000**; 14:173–182.

120 Walker MG et al. Prediction of gene function by genome-scale expression analysis: prostate cancer-associated genes. Genome Res **1999**; 9:1198–1203.

121 Hilsenbeck SG et al. Statistical analysis of array expression data as applied to the problem of tamoxifen resistance. J Natl Cancer Inst **1999**; 91:453–459.

122 Gonsky R et al. Identification of rapid turnover transcripts overexpressed in thyroid tumors and thyroid cancer cell lines: use of a targeted differential RNA display method to select for mRNA subsets. Nucleic Acids Res **1997**;25:3823–3831.

123 Belcher CE et al. The transcriptional responses of respiratory epithelial cells to *Bordetella pertussis* reveal host defensive and pathogen counter-defensive strategies. Proc Natl Acad Sci USA **2000**; 97:13847–13852.

124 Staib P et al. Differential activation of a *Candida albicans* virulence gene family during activation. Proc Natl Acad Sci USA **2000**; 97:6102–6107.

125 Reinke V et al. A global profile of germline gene expression in *C. elegans*. Mol Cell **2000**; 6:605–618.

126 Kenzelmann M et al. Transcriptome analysis of fibroblast cells immediate-early after human cytomegalovirus infection. J Mol Biol **2000**; 304:741–751.

127 Chamber J et al. DNA microarrays of the complex human cytomegalovirus genome: profiling kinetic class with drug sensitivity of viral gene expression. J Virol 1999; 73:5757–5766

128 Gmuender H et al. Gene expression changes triggered by exposure of *Haemophilus influenzae* to novobiocin or ciproflaxin: combined transcription and translation analysis. Genome Res **2001**; 11:28–42.

129 Selinger DW et al. RNA expression analysis using a 30 base pair resolution *Escherichia coli* genome array. Nature Biotechnol **2000**; 18:1262–1268.

130 Oh S-W et al. Identification of differentially expressed genes in human hepatoblastoma cell line (HepG2] and HBV-X transfected hepatobalstoma cell line (HepG2-4X). Col Cells **1998**; 8:212–218.

131 Callow MJ et al. Microarray expression profiling identifies genes with altered expression in HDL-deficient mice. Genome Res **2000**; 10:2022–2029.

132 Chang YE et al. Microarray analysis identifies interferon-inducible genes and Stat-1 as major transcriptional targets of human papillomavirus type 31. J Virol **2000**; 74:4174–4182.

133 Tal-Singer R et al. Use of differential display reverse transcription-PCR to reveal cellular changes during stimuli that result in Herpes simplex virus type 1 reactivation from latency: upregulation of immediate-early cellular response genes TIS7, Interferon, and Interferon regulatory actor-1. J Virol **1998**; 72:1252–1261.

134 Cohen P et al. Monitoring cellular responses to *Listeria monocytogenes* with oligonucleotide arrays. J Biol Chem **2000**; 275:11181–11190.

135 Troesch A et al. *Mycobacterium* species identification and rifampin resistance testing with high-density DNA probe arrays. J Clin Microbiol **1999**; 37:49–55.

136 Wilson M et al. Exploring drug-induced alterations in gene expression in *Mycobacterium tuberculosis* by microarray hybridisation. Proc Natl Acad Sci USA **1999**; 96:12833–12838.

137 Ichikawa JK et al. Interaction of *Pseudomonas aeruginosa* with epithelial cells: identification of differentially regulated genes by expression microarray analysis of humans. Proc Natl Acad Sci USA **2000**; 97:9659–9664.

138 Spellman PT et al. Comprehensive identification of cell cycle-regulated genes of the yeast *Saccharomyces cerevisiae* by microarray hybridisation. Mol Biol Cell **1998**; 9:3273–3297.

139 Hughes TR Functional discovery via a compendium of expression profiles. Cell **2000**; 102:109–126.

140 Bammert GF et al. Genome-wide expression patterns in *Saccharomyces cerevisiae*: comparison of drug treatments and genetic alterations affecting biosynthesis of ergosterol. Antimicrob Agents Chemother **2000**; 44:1255–1265.

141 Nau ME et al. Technical assessment of the Affymetrix yeast expression genechip YE6100 platform in a heterologous model of genes that confer resistance to antimalarial drugs in yeast. J Clin Microbiol **2000**; 38:1901–1908.

142 Jelinsky SA et al. Global response of *Saccharomyces cerevisiae* to an alkylating agent. Proc Natl Acad Sci USA **1999**; 96:1486–1491.

143 Jia MH et al. Global expression profiling of yeast treated with an inhibitor of amino acid biosynthesis, sulfometron methyl. Physiol Genomics **2000**; 3:83–92.

144 Goyns MH et al. Differential display analysis of gene expression indicates that age-related changes are restricted to a small cohort of genes. Mech Ageing Dev **1998**; 101:73–90.

145 Jiang CH et al. The effects of aging on gene expression in the hypothalamus and cortex of mice. Proc Natl Acad Sci USA **2001**; 98:1930–1934.

146 Lanahan A et al. Selective alteration of long-term potentiation-induced transcriptional response in hippocampus of aged, memory-impaired rats. J Neurosci **1997**; 17:2875–2885.

147 Leowohl JM et al. Application of DNA microarrays to study human alcoholism. J Biomed Sci **2001**; 8:28–36.

148 Chow N et al. Expression profiles of multiple genes in single neurons of Alzheimer's disease. Proc Natl Acad Sci USA **1998**; 95:9620–9625.

149 Wang XB et al. Subtracted differential display: genes with amphetamine-altered expression patterns include calcineurin. Mol Brain Res **1998**; 53:344–347.

150 Sandberg R et al. Regional and strain-specific gene expression mapping in the adult mouse brain. Proc Natl Acad Sci USA **2000**; 97:11038–11043.

151 Wen X et al. Large-scale temporal gene expression mapping of central nervous system development. Proc Natl Acad Sci USA **1998**; 95:334–339.

152 Luthi-Carter R et al. Decreased expression of striatal signalling genes in a mouse model of Huntington's disease. Hum Mol Genet **2000**; 9:1259–1271.

153 Richardson PJ et al. Correlating physiology with gene expression in striatal cholinergic neurones. J Neurochem **2000**; 74:839–846.

154 Bae M-K et al. Identification of genes differentially expressed by hypoxia in hepatocellular carcinoma cells. Biochem Biophys Res Comm **1998**; 243:158–162.

155 Niculescu AB et al. Identifying a series of candidate genes for mania and psychosis: a convergent functional genomics approach. Physiol Genomics **2000**; 4:83–91.

156 Thakker-Varia S et al. Gene expression in activated brain microglia: identification of a proteinase inhibitor that increases microglial cell number. Mol Brain Res **1998**; 56:99–107.

157 Whitney LW et al. Analysis of gene expression in multiple sclerosis lesions using cDNA microarrays. Ann Neurol **1999**; 46:425–428.

158 Luo L et al. Gene expression profiles of laser-captured adjacent neuronal subtypes. Nature Med **1999**; 5:117–122.

159 Chiang LW et al. An orchestrated gene expression component of neuronal programmed cell death revealed by cDNA array analysis. Proc Nat Acad Sci USA **2001**; 98:2814–2819.

160 Cousins RJ Nutritional regulation of gene expression. Am J Med **1999**; 106:20S–23S.

161 Swanson DA et al. A differential hybridisation scheme to identify photoreceptor-specific genes. Genome Res **1997**; 7:513–521.

162 Hakak Y et al. Genome-wide expression analysis reveals dysregulation of myelination-related genes in chronic schizophrenia. Proc Natl Acad Sci USA **2001**; 98:4766–4751.

163 Mirnics K et al. Molecular characterisation of schizophrenia viewed by microarray analysis of gene expression in prefrontal cortex. Neuron **2000**; 28:53–67.

164 Walker MG et al. Pharmaceutical target discovery using Guilt-by-Association: schizophrenia and Parkinson's disease genes. ISMB **1999**: 282–286.

165 Cirelli C et al. Differences in brain gene expression between sleep and waking as revealed by mRNA differential display and cDNA microarray technology. J Sleep Res **1999**; 8(Suppl 1]:44–52.

166 Kittler JF et al. Large-scale analysis of gene expression changes during acute and chronic delta9-THC in rats. Physiol Genomics **2000**; 3:175–185.

167 Davies PF et al. A spatial approach to transcriptional profiling: mechanotransduction and the focal origin of atherosclerosis. TIBTECH **1999**; 17:347–350.

168 De Waard B et al. Serial analysis of gene expression to assess the endothelial cell response to an atherogenic stimulus. Gene **1999**; 226:1–8.

169 Jin HJ et al. Effects of early angiotensin-converting enzyme inhibition on cardiac gene expression after acute myocardial infarction. Circulation **2001**; 103:736–742.

170 Lyn D et al. Gene expression profile in mouse myocardium after ischemia. Physiol Genomics **2000**; 2:93–100.

171 Sehl PD et al. Application of cDNA microarays in determining molecular phenotype in cardiac growth, development, and response to injury. Circulation **2000**; 101:1990–1999.

172 Shiffman D et al. Large scale gene expression analysis of cholesterol-loaded macrophages. J Biol Chem **2000**; 275:37324–37332.

173 Shiffman D et al. Gene expression profiling of cardiovascular disease models. Curr Opin Biotechnol **2000**; 11:598–601.

174 Stanton LW et al. Altered patterns of gene expression in response to myocardial infarction. Circ Res **2000**; 86:939–945.

175 Friddle CJ et al. Expression profiling reveals distinct sets of genes altered during induction and regression of cardiac hypertrophy. Proc Natl Acad Sci USA **2000**; 97:6745–6750.

176 Bond HM et al. Identification of differential display of transcripts regulated during hematopoietic differentiation. Stem Cells **1998**; 16:136–143.

177 Der SD et al. Identification of gene differentially regulated by interferon α, β, or γ using oligonucleotide arrays. Proc Natl Acad Sci USA **1998**; 95:15623–15628.

178 Dieckgraefe BK et al. Analysis of mucosal gene expression in inflammatory bowel disease by parallel oligonucleotide arrays. Physiol Genomics **2000**; 4:1–11.

179 Ollila J et al. Stimulation of B and T cells activates expression of transcription and differentiation factors. Biochem. Biophys Res Comm **1998**; 249:475–480.

180 Rogge L et al. Transcript imaging of the development of human T helper cells using oligonucleotide arrays. Nature Genet **2000**; 25:96–101.

181 Staudt, LM et al. Genomic views of the immune system. Ann Rev Immunol **2000**; 18:829–859.

182 Marrack P et al. Genomic-scale analysis of gene expression in resting and activated T cells. Curr Opin Immunol **2000**; 12:206–209.

183 Bhattacharjee A et al. Molecular dissection of dimethylnitrosamine (DMN)-induced hepatotoxicity by mRNA differential display. Toxicol Appl Pharmacol **1998**; 150:186–195.

184 Date M et al. Differential expression of transforming growth factor-β and its receptors in hepatocytes and nonparenchymal cells of rat liver after CCl_4 administration. J Hepatol **1998**; 28:572–581.

185 Frueh FW et al. Extent and character of Phenobarbital-mediated changes in gene expression in the liver. Mol Pharmacol **1997**; 51:363–369.

186 Holden PR et al. Immediate-early gene expression during regenerative and mitogen-induced liver growth in the rat. J Biochem Toxicol **1998**; 12:79–82.

187 Lu T et al. Application of cDNA microarray to the study of arsenic-induced liver diseases in the population of Guizhou, China. Toxicol Sci **2001**; 59:185–192.

188 Bristol DW et al. The NIEHS Predictive Toxicology Evaluation Project. Environ Health Persp **1996**; 104(Suppl. 5]:1001–1010.

189 FARR S et al. Concise Review: Gene expression applied to toxicology. Toxicol Sci **1999**; 50:1–9.

190 FIELDEN MR et al. Challenges and limitations of gene expression profiling in mechanistic and predicitive toxicology. Toxicol Sci **2001**; 60:6–10.

191 GERHOLD D et al. Monitoring expression of genes involved in drug metabolism and toxicology using DNA micorarrays. Physiol Genomics **2001**; 5:161–170.

192 LOVETT RA Toxicologists brace for genomics revolution. Science **2000**; 289:536.

193 MACGREGOR JT et al. *In vitro* tissue models in risk assessment: report of a consensus-building workshop. Toxicol Sci **2001**; 59:17–36.

194 WARING JF et al. Microarray analysis of hepatotoxins *in vitro* reveals a correlation between expression profiles and mechanisms of toxicity. Toxicol Lett **2001**; 120:359–368.

195 NADLER ST et al. The expression of adipogenic genes is decreased in obesity and diabetes mellitus. Proc Natl Acad Sci USA **2000**; 97:11371–11376.

196 WU Z et al. Transcriptional activation of sdipogenesis. Curr Opin Cell Biol **1999**; 11:689–694.

197 VOEHRINGER DW et al. Gene microarray identification of redox and mitochondrial elements that control resistance or sensitivity to apoptosis. Proc Natl Acad Sci USA **1999**; 97:2680–2685.

198 WANG Y et al. Identification of the genes responsive to etoposide-induced apoptosis: application of DNA chip technology. FEBS Lett **1999**; 445:269–273.

199 CHO RJ et al. Transcriptional regulation and function during the human cell cycle. Nature Genet **2001**; 27:48–54.

200 LAUB MT et al. Global analysis of the genetic network controlling a bacterial cell cycle. Science **2000**; 291:2144–2148.

201 WILSON SB et al. Multiple differences in gene expression in regulatory V24JQ T cells from identical twins discordant for type I diabetes. Proc Natl Acad Sci USA **2000**; 97:7411–7416.

202 MENICHINI P et al. A gene trap approach to isolate mammalian gene involved in the cellular response to genotoxic stress. Nucleic Acids Res **1997**; 25:4803–4807.

203 AMUNDSON SA et al. Fluorescent cDNA microarray hybridisation reveals complexity and heterogeneity of cellular genotoxic stress responses. Oncogene **1999**; 18:3666–3672.

204 HAMMOND TG et al. Mechanical culture conditions effect gene expression: gravity-induced changes on the space shuttle. Physiol Genomics **2000**; 3:163–173.

205 WEBB GC et al. Expression profiling of pancreatic β cells: glucose regulation of secretory and metabolic pathway genes. Proc Natl Acad Sci USA **2000**; 97:5773–5778.

206 WALKER J et al. Gene expression profiling in human peripheral blood mononuclear cells using high-density filter-based cDNA microarrays. JIM **2000**; 239:167–179.

207 CORTON JC et al. Down-regulation of cytochrome P450 2C family members and positive acute-phase response gene expression by peroxisome proliferator chemicals. Mol Pharmacol **1998**; 54:463–473.

208 NELSON PS et al. Comprehensive analysis of prostate gene expression: convergence of expressed sequence tag databases, transcript profiling and proteomics. Electrophoresis **2000**; 21:1823–1831.

209 HSIAO L-L et al. Prospective use of DNA microarrays for evaluating renal function and disease. Curr Opin Nephrol Hypertens **2000**; 9:253–258.

210 TAKENAKA M et al. Isolation of genes identified in mouse renal proximal tubule by comparing different gene expression profiles. Kidney Int **1998**; 53:562–572.

211 DIEHN M et al. Large-scale identification of secreted and membrane-associated gene products using DNA microarrays. Nature Genet **2000**; 25:58–62.

212 ANDREWS J et al. Gene discovery using computational and microarray analysis of transcription in the *Drosophila melanogaster* testis. Genome Res **2000**; 10:2030–2043.

213 BUBENDORF L et al. Hormone therapy failure in human prostate cancer: analysis by complementary DNA and tissue microarrays. J Natl Cancer Inst **1999**; 91:1758–1764.

214 Feng X et al. Thyroid hormone regulation of hepatic genes *in vivo* detected by complementary DNA microarray. Mol Endocrinol **2000**; 14:947–955.

215 Alaiya AA et al. Cancer proteomics: from identification of novel markers to creation of artificial learning models for tumor classification. Electrophoresis **2000**; 21:1210–1217.

216 Myers TG et al. A protein expression database for the molecular pharmacology of cancer. Electrophoresis **1997**; 18:647–653.

217 Aboagye-Mathiesen G et al. Interferon gamma regulates a unique set of proteins in fresh human bladder transitional cell carcinomas. Electrophoresis **1999**; 20:344–348.

218 Celis A et al. Short-term culturing of low-grade superficial bladder transitional cell carcinomas leads to changes in the expression levels of several proteins involved in key cellular activities. Electrophoresis **1999**; 20:355–361.

219 Celis JE et al. Proteomics and immunohistochemistry define some of the steps involved in the squamous differentiation of the bladder transitional epithelium: a novel strategy for identifying metaplastic lesions. Cancer Res **1999**; 59:3003–3009.

220 Ostergaard M et al. Proteome profiling of bladder squamous cell carcinomas: identification of markers that define their degree of differentiation. Cancer Res **1997**; 57:4111–4117.

221 Cole AR et al. Proteomic analysis of colonic crypts from normal, multiple intestinal neoplasia and p53-null mice: a comparison with colonic polyps. Electrophoresis **2000**; 21:1772–1781.

222 Minowa T et al. Proteomic analysis of the small intestine and colon epithelia of adenomatous polyposis coli gene-mutant mice by two-dimensional gel electrophoresis. Electrophoresis **2000**; 21:1782–1786.

223 Ornstein DK et al. Proteomic analysis of laser capture microdissected human prostate cancer and *in vitro* prostate cell lines. Electrophoresis **2000**; 21:2235–2242.

224 Seow TK et al. Two-dimensional electrophoresis map of the human hepatocellular carcinoma cell line, HCC-M, and identification of the separated proteins by mass spectrometry. Electrophoresis **2000**; 21:1787–1813.

225 Simpson RJ et al. Proteomic analysis of the human colon carcinoma cell line (LIM 1215]: development of a membrane protein database. Electrophoresis **2000**; 21:1707–1732.

226 Stulik J et al. Protein abundance alterations in matched sets of macroscopically normal colon mucosa and colorectal carcinoma. Electrophoresis **1999**; 20:3638–3646.

227 Vercoutter-Edouart AS et al. Proteomic detection of changes in protein synthesis induced by fibroblast growth factor-2 in MCF-7 human breast cancer cells. Exp Cell Res **2001**; 262:59–68.

228 Voss T et al. Correlation of clinical data with proteomics profiles in 24 patients with B-cell chronic lymphocytic leukemia. Int J Cancer **2001**; 91:180–186.

229 Wimmer K et al. Two-dimensional separations of the genome and proteome of neuroblastoma cells. Electrophoresis **1996**; 17:1741–1751.

230 Araki N et al. Comparative analysis of brain proteins from p53-deficient mice by two-dimensional electrophoresis. Electrophoresis **2000**; 21:1880–1889.

231 Edgar PF et al. Comparative proteome analysis of the hippocampus implicates chromosome 6q in schizophrenia. Mol. Psychiatry **2000**; 5:85–90.

232 Edgar PF et al. A comparative proteome analysis of hippocampal tissue from schizophrenic and Alzheimer's disease individuals. Mol Psychiatry **1999**; 4:173–178.

233 Johnston-Wilson NL et al. Disease-specific alterations in frontal cortex brain proteins in schizophrenia, bipolar disorder, and major depressive disorder. The Stanley Neuropathology Consortium. Mol Psychiatry **2000**; 5:142–149.

234 Joubert-Caron R et al. Protein analysis by mass spectrometry and sequence database searching: a proteomic approach to identify human lymphoblastoid cell line proteins. Electrophoresis **2000**; 21:2566–2575.

235 Edgar PF et al. Proteome map of the human hippocampus. Hippocampus **1999**; 9:644–650.

236 Edgar PF Comparative proteome analysis. Tissue homogenate from normal human hippocampus subjected to two-dimensional gel electrophoresis and Coomassie blue protein staining. Mol Psychiatry **2000**; 5:85–90.

237 Edgar PF et al. A comparative proteome analysis of hippocampal tissue from schizophrenic and Alzheimer's disease individuals. Mol. Psychiatry **1999**; 4:173–178.

238 Gauss C et al. Analysis of the mouse proteome. (I) Brain proteins: separation by two-dimensional electrophoresis and identification by mass spectrometry and genetic variation. Electrophoresis **1999**; 20:575–600.

239 Jensen NA et al. Proteomic changes associated with degeneration of myelin-forming cells in the central nervous system of c-myc transgenic mice. Electrophoresis **1998**; 19:2014–2020.

240 Johnston-Wilson NL et al. Disease-specific alterations in frontal cortex brain proteins in schizophrenia, bipolar disorder, and major depressive disorder. The Stanley Neuropathology Consortium. Mol Psychiatry **2000**; 5:142–149.

241 Sironi L et al. Acute-phase proteins before cerebral ischemia in stroke-prone rats: identification by proteomics. Stroke **2001**; 32:753–760.

242 Taoka M et al. Protein profiling of rat cerebella during development. Electrophoresis **2000**; 21:1872–1879.

243 Unlu M et al. Detection of complement factor B in the cerebrospinal fluid of patients with cerebral autosomal dominant arteriopathy with subcortical infarcts and leukoencephalopathy disease using two-dimensional gel electrophoresis and mass spectrometry. Neurosci Lett **2000**; 282:149–152.

244 Dunn MJ Studying heart disease using the proteomic approach. DDT **2000**; 5:76–84.

245 Arnott D et al. An integrated approach to proteome analysis: identification of proteins associated with cardiac hypertrophy. Anal Biochem **1998**; 258:1–18.

246 Isfort RJ et al. Proteomic analysis of the atrophying rat soleus muscle following denervation. Electrophoresis **2000**; 21:2228–2234.

247 Ping P et al. Functional proteomic analysis of protein kinase C epsilon signaling complexes in the normal heart and during cardioprotection. Circ Res **2001**; 88:59–62.

248 Reinheckel T et al. Adaptation of protein carbonyl detection to the requirements of proteome analysis demonstrated for hypoxia/reoxygenation in isolated rat liver mitochondria. Arch Biochem Biophys **2000**; 376:59–65.

249 Weekes J et al. Bovine dilated cardiomyopathy: proteomic analysis of an animal model of human dilated cardiomyopathy. Electrophoresis **1999**; 20:898–906.

250 Doherty NS et al. Analysis of changes in acute phase plasma proteins in an acute inflammatory response and in rheumatoid arthritis using two-dimensional gel electrophoresis. Electrophoresis **1998**; 19:355–363.

251 Dierick JF et al. Transcriptome and proteome analysis in human senescent fibroblasts and fibroblasts undergoing premature senescence induced by repeated sublethal stresses. Ann NY Acad Sci **2000**; 908:302–305.

252 Journet A et al. Towards a human repertoire of monocytic lysosomal proteins. Electrophoresis **2000**; 21:3411–3419.

253 Soskic V et al. Functional proteomics analysis of signal transduction pathways of the platelet-derived growth factor beta receptor. Biochemistry **1999**; 38:1757–1764.

254 Thiede B et al. A two dimensional electrophoresis database of a human Jurkat T-cell line. Electrophoresis **2000**; 21:2713–2720.

255 Wattiez R et al. Human bronchoalveolar lavage fluid protein two-dimensional database: study of interstitial lung diseases. Electrophoresis **2000**; 21:2703–2712.

256 Yanagida M et al. Matrix assisted laser desorption/ionization-time of flight-mass spectrometry analysis of proteins detected by anti-phosphotyrosine antibody on two-dimensional-gels of fibrolast cell lysates after tumor necrosis factor-alpha stimulation. Electrophoresis **2000**; 21:1890–1898.

257 Lai C-H et al. Identification of novel human genes evolutionarily conserved in *Caenorhabditis elegans* by comparative

proteomiocs. Genome Res **2000**; 10:703–711.

258 Walhout AJM et al. Protein interaction mapping in *C.elegans* using proteins involved in vulval development. Science **2000**; 287:116–122.

259 Jungblut PR et al. Comparative proteome analysis of *Helicobacter pylori*. Mol Microbiol **2000**; 36:710–725.

260 McAtee CP et al. Characterization of a *Helicobacter pylori* vaccine candidate by proteome techniques. J Chromatogr B Biomed Sci Appl **1998**; 714:325–333.

261 Nilsson CL et al. Identification of protein vaccine candidates from *Helicobacter pylori* using a preparative two-dimensional electrophoretic procedure and mass spectrometry. Anal Chem **2000**; 72:2148–2153.

262 Balasubramanian S et al. Proteomics of *Mycoplasma genitalium*: identification and characterization of unannotated and atypical proteins in a small model genome. Nucleic Acids Res **2000**; 28:3075–3082.

263 Betts JC et al. Comparison of the proteome of *Mycobacterium tuberculosis* strain H37Rv with clinical isolate CDC 1551. Microbiology **2000**; 146:3205–3216.

264 Zhu H et al. Analysis of yeast protein kinases using protein chips. Nature Genet **2000**; 26:283–289.

265 Futcher B et al. A sampling of the yeast proteome. Mol Cell Biol **1999**; 19:7357–7368.

266 Joubert R et al. Two-dimensional gel analysis of the proteome of lager brewing yeasts. Yeast **2000**; 16:511–522.

267 Norbeck J et al. Two-dimensional electrophoretic separation of yeast proteins using a non-linear wide range (pH 3–10) immobilized pH gradient in the first dimension; reproducibility and evidence for isoelectric focusing of alkaline (pI >7) proteins. Yeast **1997**; 13:1519–1534.

268 Pardo M et al. A proteomic approach for the study of *Saccharomyces cerevisiae* cell wall biogenesis. Electrophoresis **2000**; 21:3396–3410.

269 McCraith S et al. Genome-wide analysis of vaccinia virus protein-protein interactions. Proc Natl Acad Sci USA **2000**; 97:4879–4884.

270 Lai C-H et al. Identification of novel human genes evolutionarily conserved in *Caenorhabditis elegans* by comparative proteomiocs. Genome Res **2000**; 10:703–711.

271 Walhout AJM et al. Protein interaction mapping in *C. elegans* using proteins involved in vulval development. Science **2000**; 287:116–122.

272 Cash P et al. A proteomic analysis of erythromycin resistance in *Streptococcus pneumoniae*. Electrophoresis **1999**; 20:2259–2268.

273 Cordwell SJ et al. Characterisation of basic proteins from *Spiroplasma melliferum* using novel immobilised pH gradients. Electrophoresis **1997**; 18:1393–1398.

274 Fountoulakis M et al. Reference map of the low molecular mass proteins of *Haemophilus influenzae*. Electrophoresis **1998**; 19:1819–1827.

275 Fountoulakis M et al. Two-dimensional map of basic proteins of *Haemophilus influenzae*. Electrophoresis **1998**; 1:761–766.

276 Langen H et al. Two-dimensional map of the proteome of *Haemophilus influenzae*. Electrophoresis **2000**; 21:411–429.

277 Link AJ et al. Identifying the major proteome components of *Haemophilus influenzae* type-strain NCTC 8143. Electrophoresis **1997**; 18:1314–1334.

278 Han MJ et al. Proteome analysis of metabolically engineered *Escherichia coli* producing Poly[3-hydroxybutyrate). J Bacteriol **2001**; 183:301–308.

279 Molloy MP et al. Proteomic analysis of the *Escherichia coli* outer membrane. Eur J Biochem **2000**; 267:2871–2881.

280 Slonczewski JL et al. Acid and base regulation in the proteome of *Escherichia coli*. Novartis Foundation Symp **1999**; 221:75–83.

281 Thomas GH. Completing the *E. coli* proteome: a database of gene products characterised since the completion of the genome sequence. Bioinformatics **1999**; 15:860–861.

282 Hirose I et al. Proteome analysis of *Bacillus subtilis* extracellular proteins: a two-dimensional protein electrophoretic study. Microbiology **2000**; 146:65–75.

283 Jungblut PR et al. Comprehensive detection of immunorelevant *Borrelia gari-*

nii antigens by two-dimensional electrophoresis. Electrophoresis **1999**; 20:3611–3622.

284 Quadroni M et al. Proteome mapping, mass spectrometric sequencing and reverse transcription-PCR for characterization of the sulfate starvation-induced response in *Pseudomonas aeruginosa* PAO1. Eur J Biochem **1999**; 266:986–996.

285 Malhotra S et al. Proteome analysis of the effect of mucoid conversion on global protein expression in *Pseudomonas aeruginosa* strain PAO1 shows induction of the disulfide bond isomerase, dsbA. J Bacteriol **2000**; 182:6999–7006.

286 Phan-Thanh L et al. A proteomic approach to study the acid response in *Listeria monocytogenes*. Electrophoresis **1999**; 20:2214–2224.

287 Qi SY et al. Proteome of *Salmonella typhimurium* SL1344: identification of novel abundant cell envelope proteins and assignment to a two-dimensional reference map. J Bacteriol **1996**; 178:5032–5038.

288 Sazuka T et al. Towards a proteome project of cyanobacterium *Synechocystis* sp. strain PCC6803: linking 130 protein spots with their respective genes. Electrophoresis **1997**; 18:1252–1258.

289 Toda T et al. Proteomic analysis of Epstein-Barr virus-transformed human B-lymphoblastoid cell lines before and after immortalization. Electrophoresis **2000**; 21:1814–1822.

290 Urquhart BL et al. Comparison of predicted and observed properties of proteins encoded in the genome of *Mycobacterium tuberculosis* H37Rv. Biochem Biophys Res Comm **1998**; 253:70–79.

291 Kristensen DB et al. Proteome analysis of rat hepatic stellate cells. Hepatology **2000**; 32:266–277.

292 Steiner S et al. Protein variability in male and female Wistar rat liver proteins. Electrophoresis **1995**; 16:1969–1976.

293 Page MJ et al. Proteomic definition of normal human luminal and myoepithelial breast cells purified from reduction mammoplasties. Proc Natl Acad Sci USA **1999**; 96:12589–12594.

294 Wu CC et al. Proteomics reveal a link between the endoplasmic reticulum and lipid secretory mechanisms in mammary epithelial cells. Electrophoresis **2000**; 21:3470–3482.

295 Wu CC et al. Proteomic analysis of two functional states of the Golgi complex in mammary epithelial cells. Traffic **2000**; 1:769–782.

296 Brockstedt E et al. Luteinizing hormone induces mouse vas deferens protein expression in the murine ovary. Endocrinology **2000**; 141:2574–2581.

297 Brooks HL et al. Profiling of renal tubule Na^+ transporter abundances in NHE3 and NCC null mice using targeted proteomics. J Physiol **2001**; 530:359–366.

298 Colvis CM et al. Tracking pathology with proteomics: identification of *in vivo* degradation products of alpha-B-crystallin. Electrophoresis **2000**; 21:2219–2227.

299 Hubbard MJ Proteomic analysis of enamel cells from developing rat teeth: big returns from a small tissue. Electrophoresis **1998**; 19:1891–1900.

300 Kaufmann H et al. Influence of low temperature on productivity, proteome and protein phosphorylation of CHO cells. Biotechnol Bioeng **1999**; 63:573–682.

301 Kristensen DB et al. Analysis of the rat dermal papilla cell proteome. Exp Dermatol **1999**; 8:339–340.

302 Lewis TS et al. Identification of novel MAP kinase pathway signaling targets by functional proteomics and mass spectrometry. Mol Cell **2000**; 6:1343–1354.

303 Lopez MF et al. High-throughput profiling of the mitochondrial proteome using affinity fractionation and automation. Electrophoresis **2000**; 21:3427–3440.

304 Reinheckel T et al. Adaptation of protein carbonyl detection to the requirements of proteome analysis demonstrated for hypoxia/reoxygenation in isolated rat liver mitochondria. Arch Biochem Biophys **2000**; 376:59–65.

305 Scharfe C et al. MITOP, the mitochondrial proteome database: 2000 update. Nucleic Acids Res **2000**; 28:155–158.

306 Molloy MP et al. Establishment of the human reflex tear two-dimensional polyacrylamide gel electrophoresis reference map: new proteins of potential diagnostic value. Electrophoresis **1997**; 18:2811–2815.

307 O'Shaughnessy RF et al. PA-FABP, a novel marker of human epidermal transit amplifying cells revealed by 2D protein gel electrophoresis and cDNA array hybridisation. FEBS Lett **2000**; 486:149–154.

308 Shetty J et al. Human sperm proteome: immunodominant sperm surface antigens identified with sera from infertile men and women. Biol Reprod **1999**; 61:61–69.

309 Taylor RS et al. Proteomics of rat liver Golgi complex: minor proteins are identified through sequential fractionation. Electrophoresis **2000**; 21:3441–3459.

310 Wu CC et al. Proteomic analysis of two functional states of the Golgi complex in mammary epithelial cells. Traffic **2000**; 1:769–782.

311 Verma R et al. Proteasomal proteomics: identification of nucleotide-sensitive proteasome-interacting proteins by mass spectrometric analysis of affinity-purified proteasomes. Mol Biol Cell **2000**; 11:3425–3439.

312 Witzmann FA et al. Differential expression of cytosolic proteins in the rat kidney cortex and medulla: preliminary proteomics. Electrophoresis **1998**; 19:2491–2497.

313 Aicher L et al. New insights into cyclosporine A nephrotoxicity by proteome analysis. Electrophoresis **1998**; 19:1998–2003.

314 Truffa-Bachi P et al. Proteomic analysis of T cell activation in the presence of cyclosporin A: immunosuppressor and activator removal induces *de novo* protein synthesis. Mol Immunol **2000**; 37:21–28.

315 Anderson NL et al. An updated two-dimensional gel database of rat liver proteins useful in gene regulation and drug effect studies. Electrophoresis **1995**; 16:1977–1981.

316 Anderson NL et al. Effects of toxic agents at the protein level: quantitative measurement of 213 mouse liver proteins following xenobiotic treatment. Fund Appl Toxicol **1987**; 8:39–50.

317 Moller A et al. Two-dimensional gel electrophoresis: a powerful method to elucidate cellular responses to toxic compounds. Toxicology **2001**; 160:129–138.

318 Anderson NL et al. Effects of oltipraz and related chemoprevention compounds on gene expression in rat liver. J Cell Biochem **1995**; 22(Suppl):108–116.

319 Arce A et al. Changes in the liver protein pattern of female Wistar rats treated with the hypoglycaemic agent SDX PGU 693. Life Sci **1998**; 63:2243–2250.

320 Chevalier S et al. Proteomic analysis of differential protein expression in primary hepatocytes induced by EGF, tumour necrosis factor alpha or the peroxisome proliferator nafenopin. Eur J Biochem **2000**; 267:4624–4634.

321 Edvardson U et al. A proteome analysis of livers from obese (ob/ob) mice treated with peroxisome proliferator WY14,643. Electrophoresis **1999**; 20:935–942.

322 Giometti CS et al. A comparative study of the effects of clofibrate, ciprofibrate, WY14,643, and di-[2-ethylhexyl)-phthalate on liver protein expression in mice. Appl. Theoret Electrophoresis **1991**; 2:101–107.

323 Macdonald N et al. PPARalpha-dependent alteration of GRP94 expression in mouse hepatocytes. Biochem Biophys Res Comm **2000**; 277:699–704.

324 Kaderbhai MA et al. Alteration in the enzyme activity and polypeptide composition of rat endoplasmic reticulum during acute exposure to 2-acetylaminofluorene. Chem Biol Interact **1982**; 39:279–299.

325 Richardson FC et al. Dose responses in rat hepatic protein modification and expression following exposure to the rat hepatocarcinogen methapyrilene. Carcinogenesis **1994**; 15:325–329.

326 Chevalier S et al. Proteomic analysis of differential protein expression in primary hepatocytes induced by EGF, tumour necrosis factor alpha or the peroxisome proliferator nafenopin. Eur J Biochem **2000**; 267:4624–4634.

327 Cutler P et al. An integrated proteomic approach to studying glomerular nephrotoxicity. Electrophoresis **1999**; 20:3647–3658.

328 Fountoulakis M et al. Two-dimensional database of mouse liver proteins: changes in hepatic protein levels following treatment with acetaminophen or its nontoxic regioisomer 3-acetamidophenol. Electrophoresis **2000**; 21:2148–2161.

329 Griese M et al. Reduced proteolysis of surfactant protein A and changes of the bronchoalveolar lavage fluid proteome by inhaled alpha 1-protease inhibitor in cystic fibrosis. Electrophoresis **2001**; 22:165–171.

330 Newsholme SJ et al. Two-dimensional electrophoresis of liver proteins: characterization of a drug-induced hepatomegaly in rats. Electrophoresis **2000**; 21:2122–2128.

331 Steiner S et al. Proteomics to display lovastatin-induced protein and pathway regulation in rat liver. Electrophoresis **2000**; 21:2129–2137.

332 Witzmann FA et al. Proteomic analysis of simulated occupational jet fuel exposure in the lung. Electrophoresis **1999**; 20:3659–3669.

333 Witzmann FA et al. Proteomic analysis of the renal effects of simulated occupational jet fuel exposure. Electrophoresis **2000**; 21:976–984.

334 Witzmann FA et al. Toxicity of chemical mixtures: proteomic analysis of persisting liver and kidney protein alterations induced by repeated exposure of rats to JP-8 jet fuel vapor. Electrophoresis **2000**; 21:2138–2147.

335 Witzmann FA et al. Regional protein alterations in rat kidneys induced by lead exposure. Electrophoresis **1999**; 20:943–951.

336 Roses AD. Pharmacogenetics and the practice of medicine. Nature **2000**; 405:857–865.

337 Barroso I et al. Dominant negative mutations in human PPAR-γ associated with severe insulin resistance, diabetes mellitus and hypertension. Nature **1999**; 402:880–883.

338 Martinez FD et al. Association between genetic polymorphisms of the 2-adrenoceptor and response to albuterol in children with and without a history of wheezing. J Clin Invest **1997**; 100:3184–3188.

339 Meyer UA et al. Molecular mechanisms of genetic polymorphisms of drug metabolism. Ann Rev Pharmacol Toxicol **1997**; 37:269–296.

340 Shoemaker DD et al. Experimental annotation of the human genome using microarray technology. Nature **2001**; 409:922–927.

341 Braxton S et al. The integration of microarray information in the drug development process. Curr Opin Biotechnol **1998**; 9:643–649.

342 Heller RA et al. Discovery and analysis of inflammatory disease-related genes using cDNA microarrays. Proc Natl Acad Sci USA **1997**;94:2150–2155.

343 Furness,L.M. Expression Databases for Pharmaceutical Lead Optimisation. Proceedings of the 13th Noordwijkerhout-Camerino Symposium [2001]. Trends in Drug Research III. Pharmacochemistry Library, volume 32 Ed. H. van der Goot (Elsevier).

344 Goodman and Gilman's The Pharmacological Basis of Therapeutics (McGraw Hill).

345 Drews JJ Drug Discovery: A Historical Perspective. Science **2000**; 287:1960–1964.

346 Marchese AM et al. Discovery of three novel orphan G-protein-coupled receptors. Genomics **1999**; 56:12–21.

347 Bertilsson G et al. Identification of a human nuclear receptor defines a new signaling pathway for CYP3A induction. Proc Natl Acad Sci USA **1998**; 95:12208–12213.

348 Ame J-C et al. PARP-2, A novel mammalian DNA damage-dependent poly(ADP-ribose) polymerase. J Biol Chem **1999**; 274:17860–17868.

349 Sun X et al. RIP3, a novel apoptosis-inducing kinase. J Biol Chem **1999**; 274:16871–16875.

350 Plager DA et al. A novel and highly divergent homolog of human eosinophil granule major basic protein. J Biol Chem **1999**; 274:14464–14473.

351 Yan W et al. Corin, a mosaic transmembrane serine protease encoded by a novel cDNA from human heart. J Biol Chem **1999**; 274:14926–14935.

352 Fisher DA et al. Isolation and characterization of PDE9A, a novel human cGMP-specific phosphodiesterase. J Biol Chem **1998**; 273:15559–15564.

353 Marton MJ et al. Drug target validation and identification of secondary drug tar-

get effects using DNA microarrays. Nature Med **1998**; 4:1293–1301.
354 Hughes TR et al. Functional discovery via a compendium of expression profiles. Cell **2000**; 102:109–126.
355 Bertilsson G et al. Identification of a human nuclear receptor defines a new signaling pathway for CYP3A induction. Proc Natl Acad Sci USA **1998**; 95:12208–12213.
356 H.Ring, personal communication.
357 Afshari CA et al. Application of complementary DNA microarray technology to carcinogen identification, toxicology, and drug safety evaluation. Cancer Res **1999**; 59:4759–4760.
358 Bartosiewicz M et al. Development of a toxicological gene array and quantitative assessment of this technology. Arch Biochem Biophys **2000**; 376:66–73.
359 Bulera SJ et al. RNA expression in the early characterisation of hepatotoxicants in Wistar rats by high-density DNA microarrays. Hepatology **2001**; 33:1239–1258.
360 Gerhold D et al. Monitoring expression of genes involved in drug metabolism and toxicology using microarrays. Physiol. Genomics **2001**; 5:161–170.
361 Nuwaysir EF et al. The advent of toxicogenomics. Mol Carcinog **1999**; 24:153–159.
362 Pennie WD. Use of cDNA microarrays to probe and understand the toxicological consequences of altered gene expression. Toxicol Lett **2000**; 112–113; 473–477.
363 Pennie WD et al. Application of genomics to the definition of the molecular basis for toxicity. Toxicol Lett **2001**; 120:353–358.
364 Pennie WD et al. The principles and practice of toxicogenomics: Applications and opportunities. Toxicol Sci **2000**; 54:277–283.
365 Rininger JA et al. Differential gene expression technologies for identifying surrogate markers of drug efficacy and toxicity. DDT **2000**; 5:560–568.
366 Rockett JC et al. Application of DNA arrays to toxicology. Environ Health Persp **1999**; 107:681–685.
367 Rockett JC et al. Differential gene expression in drug metabolism and toxicology: practicalities, problems and potential. Xenobiotica **1999**; 29:655–691.
368 Rodi CR et al. Revolution through genomics in investigative and discovery toxicology. Toxicol Pathol **1999**; 27:107–110.
369 Steiner S et al. Expression profiling in toxicology – potentials and limitations. Toxicol Lett **2000**; 112–113; 467–471.
370 Stevens JL et al. Linking gene expression to mechanisms of toxicity. Toxicol Lett **2000**; 112–113; 479–486.
371 Todd MD et al. Emerging technologies for accelerated toxicity evaluation of potential drug candidates. Curr Opin Drug Discovery Dev **1999**; 2:58–68.
372 Waring JF et al. The impact of genomics-based technologies on drug safety evaluation. Ann Rev Pharmacol Toxicol **2000**; 40:335–352.
373 Lewis TS et al. Identification of novel MAP kinase pathway signaling targets by functional proteomics and mass spectrometry. Mol Cell **2000**; 6:1343–1354.
374 Wattiez R et al. Human bronchoalveolar lavage fluid protein two-dimensional database: study of interstitial lung diseases. Electrophoresis **2000**; 21:2703–2712.
375 Meyer U Introduction to pharmacogenomics: Promises, opportunities, and limitations. (Chapter 1).
376 Bell J The new genetics in clinical practice. BMJ **1998**; 316:618–620.
377 Issa AM Ethical considerations in clinical pharmacogenomics research. TIPS **2000**; 21:247–249.
378 Lin K-M et al. The evolving science of pharmacogenetics. Psychopharmacol Bull **1997**; 32:205–217.
379 Shork NJ et al. The use of genetic information in large-scale clinical trials: applications to Alzheimer research. Alzheimer Dis Ass Disord **1996**; 10(Suppl 1]:22–26.
380 Vermes A et al. Individualisation of cancer therapy based on cytochrome P450 polymorphism: a pharmacogenetic approach. Cancer Treat Rev **1997**; 23:321–339.
381 West WL et al. Interpatient variability: genetic predisposition and other genetic factors. J Clin Pharmacol **1997**; 37:635–648.
382 Wolf CR et al. Pharmacogenetics. BMJ **2000**; 320:987–990.

383 Friend SH How DNA microarrays and expression profiling will affect clinical practice. BMJ **1999**; 319:1–2.

384 Johnson BJ et al. Differential gene expression in response to adjunctive recombinant human interleukin-2 immunotherapy in multidrug-resistant tuberculosis patients. Infect Immunity **1998**; 66:2426–2433.

385 Li J et al. Typing and subtyping influenza virus using DNA microarrays and multiplex reverse transcriptase PCR. J Clin Microbiol **2001**; 39:696–704.

386 Wattiez R et al. Human bronchoalveolar lavage fluid protein two-dimensional database: study of interstitial lung diseases. Electrophoresis **2000**; 21:2703–2712.

387 Tanne JH The new world in designer drugs. BMJ **1998**; 316:1930.

6
The Role of Pharmacogenomics in Drug Discovery and Therapeutics

Klaus Lindpaintner

6.1
Introduction

There can be no doubt that the advances of molecular biology and molecular genetics and genomics, and of the associated methods and technologies has had major impact on our understanding of biology and drug action, and these tools are quintessential and indispensable for future progress in biomedicine and health care. The interface between these methods and concepts, and the discovery, development, and use of new medicines is being recognized as a new "discipline", or facet of biomedical science, termed pharmacogenetics and pharmacogenomics.

6.2
Definition of Terms

There is widespread indiscriminate use of, and thus confusion about the terms "pharmacogenetics" and "pharmacogenomics". While no universally accepted definition exists, there is emerging consensus on the differential connotation of the two terms (see Tab. 6.1).

6.2.1
Pharmacogenomics

Pharmacogenomics, and its close relative toxicogenomics are etymologically linked to "-genomics", the study of the genome and of the entirety of expressed and non-expressed genes. These two fields of study are concerned with a comprehensive, genome-wide assessment of the effects of certain interventions, mainly drugs or toxicants. Pharmacogenomics is concerned with the systematic assessment of how chemical compounds modify the overall expression pattern in certain tissues of interest. In contrast to pharmacogenetics, pharmacogenomics does not focus on differences from one person to the next with regard to the drug's effects, but rather focuses on differences among several drugs or compounds with regard to a "generic" set of expressed or non-expressed genes (most commonly

Tab. 6.1 Terminology

- **Pharmacogenetics**
  - Differential effects of a drug – *in vivo* – in different patients, dependent on the presence of inherited gene variants
  - Assessed primarily genetic (SNP) and genomic (expression) approaches
  - A concept to provide more patient/disease-specific health care
  - *One drug – many genomes (i.e., different patients)*
  - *Focus: patient variability*
- **Pharmacogenomics**
  - Differential effects of compounds – *in vivo* or *in vitro* – on gene expression, among the entirety of expressed genes
  - Assessed by expression profiling
  - A tool for compound selection/drug discovery
  - *Many "drugs" (i.e., early-stage compounds) – one genome [i.e., "normative" genome (database, technology platform)]*
  - *Focus: compound variability*

using quantitative measures of expression) and their (possible) association with phenotype characteristics.

6.2.2 Pharmacogenetics

In contrast, the term "-genetics" relates etymologically to the presence of individual properties as a consequence of having inherited them. Thus, the term pharmacogenetics describes the interactions between drug and individuals' characteristics (which may be related to inborn traits to a larger or lesser extent). Pharmacogenetics, therefore, is based on observations of clinical efficacy and/or the safety and tolerability profile of a drug in individuals – the phenotype – and tests the hypothesis that inter-individual differences in the observed response may be associated with the presence or absence of individual-specific biological markers that may allow prediction of individual drug response. Such markers are most commonly polymorphisms at the level of the nuclear DNA, but conceivably also other types of nucleic acid-derived data, such as quantitative gene expression measurements, which serve as surrogates for the presence of underlying variants in the DNA.

Thus, although both pharmacogenetics and pharmacogenomics refer to the evaluation of drug effects using nucleic acid technology, the directionalities of their approaches are distinctly different: pharmacogenetics represents the study of *differences among a number of individuals with regard to clinical response to a particular drug*, whereas pharmacogenomics represents the study of *differences among a number of compounds with regard to gene expression response in a single (normative) genome/expressome*. Accordingly, the fields of intended use are distinct: the former will help in the clinical setting to find the best medicine for a patient, the latter in the setting of pharmaceutical research and development to find the best drug candidate from a given series of compounds under evaluation.

6.3 Pharmacogenomics: Finding New Medicines Quicker and More Efficiently

Once a screen (assay) has been set up in a drug discovery project, and lead compounds are identified, the major task becomes the identification of an optimized clinical candidate molecule among the many compounds synthesized by clinical chemists. Conventionally, such compounds are screened in a number of animal or cell models for efficacy and toxicity, experiments that – while having the advantage of being conducted in the *in vivo* setting – commonly take significant amounts of time and depend entirely on the similarity between the experimental animal condition/setting and its human counterpart, i.e., the validity of the model.

Although such experiments will never be entirely replaced by expression profiling on either the nucleic acid (genomics) or the protein (proteomics) level, these technique offers powerful advantages and complimentary information. First, efficacy and profile of induced changes can be assessed in a comprehensive fashion (within the limitations – primarily sensitivity and completeness of transcript representation) of the technology platform used. Second, these assessments of differential efficacy can be carried out much more expeditiously than in conventionally used, physiology-based animal models. Third, the complex pattern of expression changes revealed by such experiments may provide new insights into possible biological interactions between the actual drug target and other biomolecules, and thus reveal new elements, or branch points of a biological pathway. Fourth, increasingly important, these tools serve to determine specificity of action among members of gene families that may be highly important for both efficacy and safety of a new drug. It must be borne in mind that any and all such experiments are limited by the coefficient of correlation with which the surrogate markers examined are linked to the desired *in vivo* physiological action of the compound.

As a subcategory of this approach, toxicogenomics is increasingly evolving as a powerful adjuvant to classic toxicological testing. As pertinent databases are being created from experiments with known toxicants, revealing expression patterns that may potentially be predictive of longer-term toxic liabilities of compounds, future drug discovery efforts should benefit by insights allowing earlier "killing" of compounds likely to cause such complications.

It is imperative, however, to understand the probabilistic nature of such experiments: a promising profile on pharmacogenomic and toxicogenomic screens will enhance the likelihood of having selected an ultimately successful compound, and will achieve this goal quicker than conventional animal experimentation, but will do so only with a certain likelihood of success. The less reductionist approach of the animal experiment will still be needed. It is to be anticipated, however, that such approaches will constitute an important, time- and resource-saving first evaluation or screening step that will help to focus and reduce the number of animal experiments that will ultimately need to be conducted.

6.4 Pharmacogenetics: More Targeted, More Effective Medicines for our Patients

6.4.1 Genes and Environment

It is common knowledge that today's pharmacopea – inasmuch as it represents enormous progress compared with what our physicians had only 15 or 20 years ago – is far from perfect. Many patients respond only partially, or fail to respond altogether, to the drugs they are given, and others suffer serious adverse events. If we accept, reasonably, that all common complex diseases – i.e., the health problems that are the main contributors to public and private health spending – are the results of complex, multifactorial interactions between inborn predispositions and susceptibilities on the one hand, and external, environmental factors on the other, then the problem of inter-individual variance of response to medication is but one of the aspects of this complexity, and may, likewise, be assumed to have as much to do with external influences (e.g., non-compliance, wrong dose) as with inherent (i.e., inherited, genetically determined) ones.

Clearly, a better, more fundamental understanding of the nature of genetic predispositions to disease, and of pathology and of drug action on the molecular level, is essential for future progress in health care. Current progress in molecular biology and genetics has indeed provided us with some of the prerequisite tools to reach this more refined understanding.

Drugs, among all the "environmental factors" that we are exposed to, may be particularly likely to "interact" specifically and selectively with the genetic properties of a given individual, as their potency pitches them into a narrow "therapeutic window", precariously balanced between potent potions and perilous poisons. We would predict that, based on a patient's innate, individual biological makeup – as it affects the interaction with a drug – one or the other of these properties may manifest itself; this phenomenon is covered by the term pharmacogenetics.

6.4.2 An Attempt at a Systematic Classification

Several conceptually very different scenarios of such individual-specific drug response may be distinguished (see Tab. 6.2). They include, on the one hand, differential *pharmacokinetics*, due to inter-individual differences in absorption, distribution, metabolism (with regard to both activation of prodrugs, inactivation of the active molecule, and generation of derivative molecules with biological activity), or excretion of the drug. In any of these cases, differential effects are observed due to the presence at the intended site of action either of inappropriate concentrations of the pharmaceutical agent, or of inappropriate metabolites, or of both. Pharmacogenetics, as it relates to pharmacokinetics, has of course been recognized as an entity ever since Archibald Garrod's seminal observations and his visionary interpretation as inter-individual differences in detoxification of drugs.

Tab. 6.2 Pharmacogenetics systematic classification

- **Pharmacokinetics**
  - Absorption
  - Metabolism
    - Activation of prodrugs
    - De-activation
    - Generation of biologically active metabolites
  - Distribution
  - Elimination
- **Pharmacodynamics**
  - Causative drug action: related to molecular pathology
  - Palliative drug action: related to molecular physiology

We have since come to understand the underlying genetic causes for many of the previously known differences in enzymatic activity, most prominently with regard to the P450 enzyme family, and these have been the subject of recent reviews [1, 2].

On the other hand, inter-individual differences in a drug's effects may also be observed in the presence of appropriate concentrations of the intended compound at the intended site of action, i.e., be due to differential *pharmacodynamics*. Here, two conceptually quite different conceptual scenarios may be distinguished that relate to the two principal mechanisms by which drugs act: etiology-specific and palliative.

The former relates to drugs that work by targeting, and mitigating or correcting the actual cause of the disease or one of its etiologically contributing elements. In contrast, palliative drugs modulate disease phenotype-relevant (but not disease cause-relevant) pathways that are not dysfunctional but can be used to counterbalance the effect of a disease-causing, dysfunctional pathway. These drugs do not directly address the underlying cause or etiological contribution.

There is general agreement today that any of the major clinical diagnoses, such as diabetes or cancer, are comprised of a number of etiologically (i.e., at the molecular level) more or less distinct subcategories. In the case of an etiologically acting drug this implies that it will only be appropriate in a fraction of the patients that carry the clinical diagnosis; namely in those in whom the dominant molecular etiology, or at least one of the contributing etiological factors matches the mechanism of the drug given. A schematic (Figure 6.1) is enclosed to help clarify these somewhat complex concepts, in which a hypothetical case of a complex disease is depicted where excessive function of one of the trait-controlling pathways causes symptomatic disease – assume, e.g., the trait is blood pressure, and the associated disorder is hypertension (for the case of a defective function of a pathway, an analogous schematic could be constructed, and again for a deviant function). Since a causative treatment will only work if the mechanism it addresses is indeed contributing to the patient's disease (Figure 6.1 A, B, C), such a treatment may be ineffective if that mechanism is not operative (Figure 6.1 D, E). Thus, unrecognized and undiagnosed disease heterogeneity at the molecular level provides

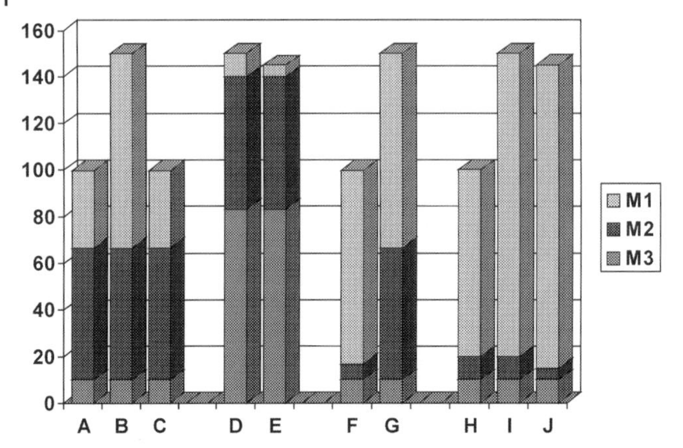

Fig. 6.1 A: Normal physiology: three molecular mechanisms (M1, M2, M3) contribute to a trait. B: Diseased physiology D1: derailment (cause/contribution) of molecular mechanism 1 (M1). **C:** Diseased physiology D1: causal treatment T1 (aimed at M1). D: Diseased physiology D3: derailment (cause/contribution) of molecular mechanism 3 (M3). E: Diseased physiology D3, treatment T1: treatment does not address cause. F: Diseased physiology D1, palliative treatment T2 (aimed at M2). G: Diseased physiology D1, palliative treatment T2; T2-refractory gene variant in M2. H: Normal physiology variant: differential contribution of M1 and M2 to normal trait. I: Diseased physiology D1-variant: derailment of mechanism M1. J: Diseased physiology D1-variant: treatment with T2.

an important explanation for differential drug response and likely represents a substantial fraction of what we today somewhat indiscriminately subsume under the term "pharmacogenetics".

On the other hand, in the case of a drug that works palliatively, molecular variations in the structure of the drug's biological target that affect the target's interaction with a drug, as well as inter-individual differences in the activity of the targeted pathways (and thus in the relative disease-counterbalancing effect of inhibiting or enhancing them) provide a second, conceptually different explanation for differential drug response based on pharmacodynamics. Thus, a palliative treatment (Figure 6.1 F) may not be effective either if the target molecule represents a variant that does not respond to the treatment (Figure 6.1 G), or if the particular mechanism targeted by the palliative drug is not phenotype-relevant in the patient in question, due to a genetic variant or other reasons (Figure 6.1 H, I, J). Here we are faced with disease etiology-unrelated, inter-individual variability as the root cause for differential drug response.

6.4.2.1 **Pharmacogenetics as a Consequence of "Subclinical" Differential Diagnosis**

An increasingly sophisticated and precise diagnosis of disease, arising from a deeper, more differentiated understanding of pathology at the molecular level, that will subdivide today's clinical diagnoses into molecular subtypes, will foster medical advances which, if considered from the viewpoint of today's clinical diagnosis, will appear as "pharmacogenetic" phenomena. However, the sequence of events commonly expected as characteristic for a "pharmacogenetic scenario" – namely, exposing patients to the drug, recognizing a differential [i.e., (quasi-)bimodal-] response pattern, discovering a marker predicting this response, and creating a diagnostic product to be co-marketed with the drug henceforth – is likely to be reversed. Rather, we will search for a new drug specifically, and *a priori*, based on a new diagnosis (i.e., a newly found ability to diagnose a molecular sub-entity of a previously more encompassing, broader, and less precise clinical disease definition). Thus, pharmacogenetics will not be so much about finding the "right medicine for the right patient", but about finding the "right medicine for the disease (-subtype)", as we have aspired to do all along throughout the history of medical progress. This is, in fact, good news: the conventional "pharmacogenetic scenario" would invariably present major challenges from both a regulatory and a business development and marketing standpoint, as it will confront development teams with a critical change in the drug's profile at a very late point during the development process. In addition, the timely development of an approvable diagnostic in this situation is difficult at best, and its marketing as an "add-on" to the drug a less than attractive proposition to diagnostics business. Thus, the "practice" of pharmacogenetics will, in many instances, be marked by progress along the very same path that has been one of the main avenues of medical progress for the last several hundred years: differential diagnosis.

Rather, the sequence of events in this case would likely involve, first, the development of an *in vitro* diagnostic test as a stand-alone product that may even be marketed on its own merits, allowing the physician to establish an accurate, state-of-the-art diagnosis of the molecular subtype of the patient's disease. Sometimes such a diagnostic may prove helpful even in the absence of specific therapy by guiding the choice of existing medicines and/or of non-drug treatment modalities such as specific changes in diet or lifestyle. Availability of such a diagnostic – as part of the more sophisticated understanding of disease – will undoubtedly foster and stimulate the search for new, more specific drugs; and once such drugs are found, availability of the specific diagnostic will be important for carrying out the appropriate clinical trials. This will allow a prospectively planned, much more systematic approach towards clinical and business development, with a commensurate greater chance of actual realization and success.

In practice, some extent of guesswork will remain, due to the nature of common complex disease. First, all diagnostic approaches will ultimately only provide a measure of probability, not of certainty: thus, although the variances of patient response among patients who do or do not carry the drug-specific sub-diagnosis will be smaller, there will still be a distribution of differential responses; thus, although by-and-large the drug will work better in the "responder" group, there will be some who respond less ore not at all in that group, and conversely, not everyone belonging to the non-responder group will completely fail to respond, depending ultimately on the relative magnitude with which the particular mechanism contributes to the disease. Thus, it is important to bear in mind that even in the case of fairly obvious bimodality individual patients will still fall into a distribution pattern of responses, and all predictions as to responder- or non-responder status will be of a probabilistic nature (Figure 6.2). In addition, based on our current understanding of the polygenic and heterogeneous nature of these disorders, we will – even in an ideal world where we would know about all possible susceptibility gene variants for a given disease and have treatments for them – only be able to exclude, in any one patient, those that do not appear to contribute to the disease, and, therefore, rule out certain treatments. We will, however, most likely find ourselves left with a small number – two to four, perhaps – of potentially disease-contributing gene variants whose relative contribution to the disease will be very difficult, if not impossible, to rank in an individual patient. Likely then, trial and error, and this great intangible quantity of "physician experience" will still play an important role, albeit on a more limited and sub-selective basis.

Today, the most frequently cited example for this category of "pharmacogenetics" is trastuzamab (HERCEPTIN®), a humanized monoclonal antibody directed against the her-2-oncogene. This breast cancer treatment is prescribed based on the level of her-2-oncogene expression in the patient's tumor tissue. Differential diagnosis at the molecular level not only provides an added level of diagnostic sophistication, but also actually represents the prerequisite for choosing the appropriate therapy. Because trastuzamab specifically inhibits a "gain-of-function" variant of the oncogene, it is ineffective in the 2/3 of patients who do not "overexpress" the drug's target, whereas it significantly improves survival in the 1/3 of patients that constitute the "sub-entity" of the broader diagnosis "breast cancer" in whom the gene is expressed [3]. [Some have argued against this being an example of "pharmacogenetics", because the parameter for patient stratification (i.e., for differential diagnosis) is the somatic gene expression level rather than a particular "genotype" data [4]. This is a difficult argument to follow, since in the case of a treatment effect-modifying germ line mu-

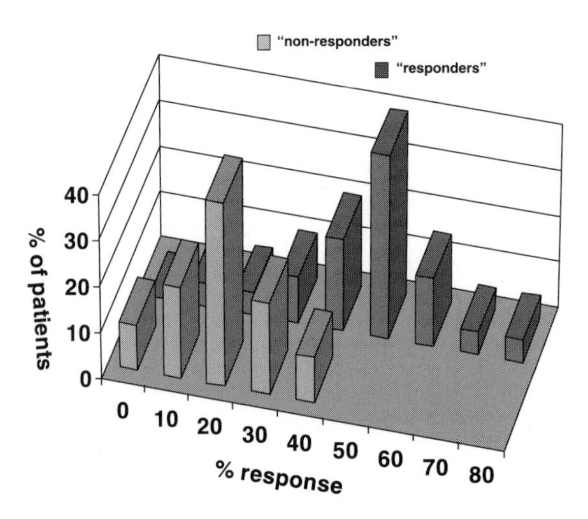

Fig. 6.2 Hypothetical example of bimodal distribution according to 'a' marker that indicates "non-responder" or "responder" status. Note that in both cases a distribution is present, with overlaps, thus, the categorization into "responders" or "non-responders" based on the marker must be understood to convey only the probability to belong to one or the other group.

tation it would obviously not be the nuclear gene variant *per se*, but also its specific impact on either structure/function or on expression of the respective gene/gene product that would represent the actual physiological corollary underlying the differential drug action. Conversely, an *a priori* observed expression difference is highly likely to reflect a – potentially as yet undiscovered – sequence variant. Indeed, as pointed out below, there are a number of examples where the connection between genotypic variant and altered expression has already been demonstrated [5, 6].]

Another example, although still hypothetical, of how proper molecular diagnosis of relevant pathomechanisms will significantly influence drug efficacy, is in the evolving class of anti-AIDS/HIV drugs that target the CCR5 cell surface receptor [7, 8, 9]. These drugs are predicted to be ineffective in the rare patients who carry the delta-32 variant, but who nevertheless have contracted AIDS or test HIV-positive (most likely due to infection with an SI-virus phenotype that utilizes CXCR4) [10, 11].

It should be noted that the pharmacogenetically relevant molecular variant need not affect the primary drug target, but may equally well be located in another molecule belonging to the system or pathway in question, both up- or downstream in the biological cascade with respect to the primary drug target.

6.4.2.2 Pharmacogenetic Effects of Palliative Drugs due to Structural Target Diversity

The alternative scenario, where differential drug response and/or safety occurs with regard to a "palliative" drug is likely to pose, as discussed, considerably greater difficulty in planning and executing a clinical development program because, presumably, it will be more difficult to anticipate or predict differential responses *a priori*. When such a differential response occurs, it will also potentially be more difficult to find the relevant marker(s), unless it happens to be among the "obvious" candidate genes implicated in the disease physiopathology or the treatment's mode of action. Although screening for molecular variants of these genes, and testing for their possible associations with differential drug response is a logical first step, if unsuccessful, it may be necessary to embark on an unbiased genome-wide screen, using single nucleotide polymorphisms (SNPs) as molecular flagpoles. Despite recent progress in high-throughput genotyping, the obstacles that will have to be overcome on the technical, data analysis, and cost levels are formidable. They will limit the deployment of such programs, at least for the foreseeable future, to select cases in which there are very solid indications for doing so, based on clinical data showing a near-categorical (e.g., bimodal) distribution of treatment outcomes. Even then, we may expect to encounter for every success – that will be owed to a favorably strong linkage disequilibrium across considerable genomic distance in the relevant chromosomal region – as many or more failures, in cases where the culpable gene variant cannot be found due to the higher recombination rate or other characteristics of the stretch of genome that it is located on.

Several of the more persuasive examples we have accumulated to date for such palliative drug-related pharmacogenetic effects have been observed in the field of asthma. The treatment of asthma relies on an array of drugs aimed at modulating different "generic" pathways, thus mediating bronchodilation or anti-inflammatory effects. Pharmacogenetic effects have been demonstrated in situations where these pathways do not respond as expected. Thus, molecular variants of the β-2-adrenoceptor have been shown associated with differential treatment response to β-2-agonists [13, 14]. Individuals carrying one or two copies of a variant allele that contains a glycine in place of arginine in position 16 were found to have a 3- and 5-fold reduced response to the agonist, respectively. This was shown in both *in vitro* [15, 16] and *in vivo* [16] studies to correlate with an enhanced rate of agonist-induced receptor downregulation, but no difference in gene transcriptional or translational activity, or agonist binding. In contrast, a second polymorphism affecting position 19 of the beta upstream peptide has been shown to affect translation (but not transcription) of the receptor itself, with a 50% decrease in receptor numbers associated with the variant allele – which happens to be in strong linkage disequilibrium with a variant allele position 16 in the receptor. The simultaneous presence of both mutations would thus be predicted to result in low expression and enhanced downregulation of an otherwise functionally normal receptor, depriving patients carrying such alleles of the benefits of effective bronchodilation as a "palliative" (i.e., non-causal) countermeasure to their pathological airway hyper-reactivity. (In the schematic depicted in Figure 6.1, the common type of beta-receptor response would be represented by situation F, the variant by situation G.) Importantly, there is no evidence that any of the allelic variants encountered are associated with the prevalence or incidence, and thus potentially the etiology of the underlying disease [17, 18].

Similarly, inhibition of leukotriene synthesis proved clinically ineffective in a small fraction of patients who carried only non-wild-type alleles of the 5-lipoxygenase promoter region [12]. These allelic variants had previously been shown to be associated with decreased transcriptional activity of the gene [5]. It stands to reason – consistent with the clinical observations – that in the presence of already reduced 5-lipoxygenase activity pharmacological inhibition may be less effective (corresponding to situations H–J in Figure 6.1). Of note, again, there is no evidence for a primary, disease-causing or -contributing role of 5-lipoxygenase variants; all of them were observed at equal frequencies in affected and non-affected individuals [5].

Pharmacogenetic stratification allows not only recognition of responders and non-responders with regard to the intended treatment effect, but also with regard to undesirable responses, i.e., the occurrence of adverse effects. An example for this scenario is provided by the well-documented "pharmacogenetic" association between molecular sequence variants of the 12S rRNA, a mitochondrion-encoded gene, and aminoglycoside-induced ototoxicity [19]. Intriguingly, the mutation that is associated with susceptibility to ototoxicity renders the sequence of the human 12S rRNA similar to that of the bacterial 12S rRNA gene, and thus effectively turns the human 12S rRNA into the (bacterial) target for aminoglycoside drug action – presumably mimicking the structure of the bacterial binding site of the drug [20]. As in the other examples, presence of the 12S rRNA mutation *per se* has no primary, drug treatment-independent effect.

Analogously, within one species such "molecular mimicry" may occur: adverse events may arise if the selectivity of a drug is lost because a gene that belongs to the same gene family as the primary target, loses its "identity" *vis-à-vis* the drug and attains, based on its structural similarity with the principal target, similar affinity to the drug. Depending on the biological role of the "imposter" molecule, adverse events may occur. Although we currently have no clear actual examples for this, it is certainly imaginable for classes of receptors and enzymes.

6.4.2.3 **Different Classes of Markers**

Pharmacogenetic phenomena, as pointed out previously, need not be restricted to the observation of a direct association between allelic sequence variation and phenotype, but may extend to a broad variety of indirect manifestations of underlying, but often unrecognized, sequence variation. Thus, differential methylation of the promoter region of O6-methylguanine-DNA-methylase has recently been reported to be associated with differential efficacy of chemotherapy with alkylating agents. If methylation is present, expression of the enzyme that rapidly reverses alkylation and induces drug resistance is inhibited, and therapeutic efficacy is greatly enhanced [21].

6.4.2.4 **Complexity is to be Expected**

In the real world, it is likely that not only one of the scenarios depicted, but a combination of several ones may affect how well a patient responds to a given treatment, or how likely it is that he or she will suffer an adverse event. Thus, a fast-metabolizing patient with poor-responder pharmacodynamics may be particularly unlikely to gain any benefit from taking the drug in question, while a slow-metabolizing status may counterbalance in another patient the same pharmacodynamics, whereas a third patient, being a slow metabolizer and displaying normal pharmacodynamics, may be more likely to suffer adverse events. In all of them, both the pharmacokinetic and pharmacodynamics properties may result from the interaction of several of the mechanisms described above. In addition, we know of course that co-administration of other drugs, or even the consumption of certain foods, may affect and further complicate the picture for any given treatment.

6.5
Pharmacogenetic Testing for Drug Efficacy vs. Safety

In principle, pharmacogenetic approaches may be useful both to raise efficacy and to avoid adverse events, by stratifying patient eligibility for a drug according to appropriate markers. In both cases, clinical decisions and recommendations must be supported by data that have undergone rigorous biostatistical scrutiny (Tab. 6.3). Based on the substantially different prerequisites for and opportunities to acquiring such data, and to applying them to clinical decision making, we ex-

Tab. 6.3 Required minimal criteria for acceptability of (pharmaco-)genetic–epidemiological studies [21]

Requirement	*Assessment*
Reproducibility	Validation of molecular methods
Objectivity/Interpretation	Blinding of investigators as to clinical data
Delineation of case group	Detailed description of in- and exclusion criteria
Adequacy of case spectrum	To support claims of the study
Delineation of comparison group	Appropriate to support claims regarding disease/indication
Adequacy of comparison group	Negative and positive controls required
Quantitative summary of results	Specify magnitude of difference as well as statistical support

pect the use of pharmacogenetics for enhanced efficacy to be considerably more common than for the avoidance of adverse events.

The likelihood that adequate data on efficacy in a subgroup may be generated is reasonably high, given the fact that unless the drug is viable in a sizeable number of patients, it will probably not be developed for lack of a viable business case, or at least only in the protected environment of orphan regulations. Implementation of pharmacogenetic testing to stratify for efficacy, provided that safety in the non-responder group is not an issue, will primarily be a matter of physician preference and sophistication, and potentially of third-party payer directives, but would appear less likely to become a matter of regulatory mandate. Indeed, an argument can be made against depriving those who carry the "non-responder" genotype of eligibility for the drug, but who individually, of course, may respond to the drug with a certain, albeit lower probability. From a regulatory aspect, use of pharmacogenetics for efficacy, if adequate safety data exist, appears largely unproblematic – the worst-case scenario (a genotypically inappropriate patient receiving the drug) resulting in treatment without expected beneficial effect, but with no increased odds to suffer adverse consequences, i.e., much of what one would expect under conventional paradigms.

The utility and clinical application of pharmacogenetic approaches towards improving safety, in particular with regard to serious adverse events, will meet with considerably greater hurdles and is, therefore, less likely expected to become reality. A number of reasons are cited for this: first, in the event of serious adverse events associated with the use of a widely prescribed medicine, withdrawal of the drug from the market is usually based almost entirely on anecdotal evidence from a rather small number of cases – in accordance with the Hippocratic mandate "*primum non nocere*". If the sample size is insufficient to statistically demonstrate a significant association between drug exposure and event, it will most certainly be insufficient to allow meaningful testing for genotype–phenotype correlations; this becomes progressively more difficult as many markers are tested and the number of degrees of freedom applicable to any analysis continues to rise. Therefore, the fraction of attributable risk shown to be associated with a given at-risk (combination of) genotype(s) would have to be very substantial for regulators to accept such data. Indeed, the low prior probability of the event will, by definition, result in an expected equally low positive (or negative) predictive value. Second, the very nature of safety issues raises the hurdles substantially because in this situation the worst case scenario – administration of the drug to the "wrong" patient – will result in higher odds to harm to the patient. Therefore, it is likely that the practical application of pharmacogenetics towards limiting adverse events will be restricted to diseases with dire prognosis, where a high medical need exists, where the drug in question offers unique potential advantages (usually bearing the characteristics of a "life saving" drug), and where the tolerance even for relatively severe side effects is *a priori* substantial, and accepted in favor of the drug's beneficial effects. This applies primarily to areas like oncology or HIV/AIDS. In most other indications, the sobering biostatistical and regulatory considerations discussed represent barriers that are unlikely to be overcome easily; and the proposed, conceptually highly attractive, routine deployment of pharmacogenetics as a generalized drug surveillance practice following the introduction of a new pharmaceutical agent [22] faces these as well as formidable economic hurdles.

6.6 Ethical – Societal Aspects of Pharmacogenetics

No discussion about the use of genetic/genomic approaches to health care can be complete without considering their impact on the ethical, societal, and legal level. Arguments have been advanced that genotype determinations for pharmacogenetic characterization, in contrast to "genetic" testing for primary disease risk assessment, are less likely to raise potentially sensitive issues with regard to patient confidentiality, the misuse of genotyping data or other nucleic acid-derived information, and the possibility of stigmatization. While this is certainly true when pharmacogenetic testing is compared to predictive genotyping for highly penetrant Mendelian disorders, it is not apparent why in common complex disorders issues surrounding predictors of primary disease risk would be any more or less sensitive than those pertaining to predictors of likely treatment success/failure. Indeed, two lines of reasoning may actually indicate an increased potential for ethical issues and complex confrontations among the various stakeholders to arise from pharmacogenetic data.

First, while access to genotyping and other nucleic acid-derived data related to disease susceptibility can be strictly limited, the very nature of pharmacogenetic data calls for a rather more liberal position regarding use: if this information is to serve its intended purpose, i.e., improving the patients chance for successful treatment, then it is essential that it is shared among at least a somewhat wider circle of participants in the health care process. Thus, the prescription for a drug that is limited to a group of patients with a particular genotype will inevitably disclose the receiving patient's genotype to anyone of a large number of individuals involved in the patients care at the medical and administrative level. The only way to limit this quasi-public disclosure of this patient's genotype data would be if he or she were to sacrifice the benefits of the indicated treatment for the sake of data confidentiality.

Second, patients profiled to carry a high disease probability along with a high likelihood for treatment response may be viewed, from the standpoint of, e.g., insurance risk, as quite comparable to patients displaying the opposite profile, i.e., a low risk to develop the disease, but a high likelihood not to respond to medical treatment, if the disease indeed occurs. For any given disease risk, then, patients less likely to respond to treatment would be seen as a more unfavorable insurance risk, particularly if non-responder status is associated with chronic, costly illness rather than with early mortality, the first case having much more far-reaching economic consequences. The pharmacogenetic profile may thus, under certain circumstances, even become a more important (financial) risk assessment parameter than primary disease susceptibility, and would be expected – inasmuch as it represents but one stone in the complex-disease mosaic – to be treated with similar weight, or lack thereof, as other genetic and environmental risk factors.

Practically speaking, the critical issue is not only, and perhaps not even predominantly, the sensitive nature of the information, and how it is, if at all, disseminated and disclosed, but how and to what end it is used. Obviously, generation and acquisition of personal medical information must always be contingent on the individual's free choice and consent, as must be all application of such data for specific purposes. Beyond this, however, there is today an urgent need for the requisite dialog and discourse among all stakeholders within society to develop and endorse a set of criteria by which the use of genetic, indeed of all personal medical information should occur. It will be critically important that society as a whole endorses, in an act of solidarity with those destined to develop a certain disease, guidelines that support the beneficial and legitimate use of the data in the patient's interest while at the same time prohibiting their use in ways that may harm the individual, personally, financially, or otherwise. As long as we trust our political decision processes to reflect societal consensus, and as long as such consensus reflects the principles of justice and equality, the resulting set of principles should assert such proper use of medical information. Indeed, both aspects – data protection and patient/subject protection, are seminal components of the mandates included in the WHO's "Proposed International Guidelines on Ethical Issues in Medical Genetics and Genetic Services" [23] which mandate autonomy, beneficence, no maleficence, and justice.

6.7 Summary

Pharmacogenetics, in the different scenarios included in this term, will represent an important new avenue towards understanding disease pathology and drug action, and will offer new opportunities of stratifying patients to achieve optimal treatment success. As such, it represents a logical, consequent step in the history of medicine – evolution, rather than revolution. Its implementation will take time, and will not apply to all diseases and all treatments equally. If society finds ways to sanction the proper use of this information, thus allowing and protecting its unencumbered use for the patient's benefit, important progress in health care will be made.

6.8 References

1 Dickins M, Tucker G. Drug disposition: To phenotype or genotype. Int J Pharm Med **2001**; 15:70–73; also see: *http://www.imm.ki.se/CYPalleles/.*

2 Evans WE, Relling MV. Pharmacogenomics: Translating functional genomics into rational therapies. Science **1999**; 206:487–491; also see: *http://www.sciencemag.org/feature/data/1044449.shl/.*

3 Baselga J, Tripathy D, Mendelsohn J, Baughman S, Benz CC, Dantis L, Sklarin NT, Seidman AD, Hudis C, Moore J, Rosen PP, Twaddell T, Henderson IC, Norton L. Phase II study of weekly intravenous recombinant humanized anti-p185(HER2) monoclonal antibody in patients with HER2/neu-overexpressing metastatic breast cancer. J Clin Oncol **1996**; 14:737–744.

4 Haseltine WA. Not quite pharmacogenomics (letter; comment). Nature Biotechnol **1998**; 16:1295.

5 In KH, Asano K, Beier D, Grobholz J, Finn PW, Silverman EK, Silverman ES, Collins T, Fischer AR, Keith TP, Serino K, Kim SW, De Sanctis GT, Yandava C, Pillari A, Rubin P, Kemp J, Israel E, Busse W, Ledford D, Murray JJ, Segal A, Tinkleman D, Drazen JM. Naturally occurring mutations in the human 5-lipoxygenase gene promoter that modify transcription factor binding and reporter gene transcription. J Clin Invest **1997** Mar 1;99(5):1130–1137.

6 McGraw DW, Forbes SL, Kramer LA, Liggett SB. Polymorphisms of the 5′ leader cistron of the human beta2-adrenergic receptor regulate receptor expression. J Clin Invest **1998**; 102:1927–1932.

7 Huang Y, Paxton WA, Wolinsky SM, Neumann AU, Zhang L, He T et al. The role of a mutant CCR5 allele in HIV-1 transmission and disease progression. Nature Med **1996**; 2:1240–1243.

8 Dean M, Carrington M, Winkler C, Huttley GA, Smith MW, Allikmets R et al. Genetic restriction of HIV-1 infection and progression to AIDS by a deletion of the CKR5 structural gene. Science **1996**; 273:1856–1862.

9 Samson M, Libert F, Doranz BJ, Rucker J, Liesnard C, Farber CM et al. Resistance to HIV-1 infection in Caucasian individuals bearing mutant alleles of the CCR-5 chemokine receptor gene. Nature **1996**; 382:722–725.

10 O'Brien TR, Winkler C, Dean M, Nelson JAE, Carrington M, Michael NL et al. HIV-1 infection in a man homozygous for CCR5 32. Lancet **1997**; 349:1219.

11 Theodorou I, Meyer L, Magierowska M, Katlama C, Rouzious C. Seroco Study Group. HIV-1 infection in an individual homozygous for CCR5 32. Lancet **1997**; 349:1219–1220.

12 Drazen JM, Yandava CN, Dube L, Szczerback N, Hippensteel R, Pillari A, Israel E, Schork N, Silverman ES, Katz DA, Drajesk J. Pharmacogenetic association between ALOX5 promoter genotype and the response to anti-asthma treatment. Nature Genet **1999**; 22:168–170

13 Martinez FD, Graves PE, Baldini M, Solomon S, Erickson R. Association between genetic polymorphisms of the beta2-adrenoceptor and response to albuterol in children with and without a history of wheezing. J Clin Invest **1997**; 100:3184–3148.

14 Tan S, Hall IP, Dewar J, Dow E, Lipworth B. Association between beta 2-adrenoceptor polymorphism and susceptibility to bronchodilator desensitisation in moderately severe stable asthmatics. Lancet **1997**; 350:995–999.

15 Green SA, Turki J, Innis M, Liggett SB. Amino-terminal polymorphisms of the human beta 2-adrenergic receptor impart distinct agonist-promoted regulatory properties. Biochemistry **1994**; 33:9414–9419.

16 Green SA, Turki J, Bejarano P, Hall IP, Liggett SB. Influence of beta 2-adrenergic receptor genotypes on signal transduction in human airway smooth muscle cells. Am J Respir Cell Mol Biol **1995**; 13(1):25–33.

17 Reihsaus E, Innis M, MacIntyre N, Liggett SB. Mutations in the gene encoding for the beta 2-adrenergic receptor in normal and asthmatic subjects. Am J Respir Cell Mol Biol **1993**; 8:334–349.

18 Dewar JC, Wheatley AP, Venn A, Morrison JFJ, Britton J, Hall IP. 2 adrenoceptor polymorphisms are in linkage disequilibrium, but are not associated with asthma in an adult population. Clin Exp All **1998**; 28:442–448.

19 Fischel-Ghodsian N. Genetic factors in aminoglycoside toxicity. Ann NY Acad Sci **1999**; 884:99–109.

20 Hutchin T, Cortopassi G. Proposed molecular and cellular mechanism for aminoglycoside ototoxicity. Antimicrob Agents Chemother **1994**; 38:2517–2520.

21 Esteller M, Garcia-Foncillas J, Andion E, Goodman, SN, OF Hidalgo, Vanaclocha V, Baylin SB, Herman JG. Inactivation of the DNA-repair gene mgmt and the clinical response of gliomas to alkylating agents. N Engl J Med **2000**; 343:1350–1354.

22 Roses A. Pharmacogenetics and future drug development and delivery. Lancet **2000**; 355:1358–1361.

23 Proposed international guidelines on ethical issues in medical genetics and genetic services *http://www.who.int/ncd/hgn/hgnethic.htm.*

7
Pharmacogenomics and Drug Design

Philip Dean, Paul Gane and Edward Zanders

Abstract

The Human Genome Project has reached an important stage with the recent declaration that the first draft is now complete. In the new post-genomic world, the problem facing scientists is how to maximize the use of the newly acquired data for improving health care. It is estimated that the number of therapeutic targets available for drug discovery will increase from the current number of 600–1,000 to perhaps as many as 5,000–10,000. In addition to the challenges that this number provides to the pharmaceutical industry, there is the issue of sequence variation through single nucleotide polymorphisms (SNPs), some of which may impact upon the way that the body handles drug treatment. This may be due to a direct effect on the binding site of the protein target through non-conservative alterations of the amino acid sequence, or else through indirect effects on drug metabolizing enzymes. In order to meet these challenges, the marriage of structural proteomics and computer-aided small molecule design will provide opportunities for creating new molecules *in silico*; these may be designed to bind to selected pharmacogenetic variants of a protein in order to overcome the non-responsiveness of certain patient groups to a particular medicine. The basic aspects of these technologies, and their applicability to selected targets showing structural variation, form the basis of this chapter.

7.1
Introduction

Much has been written concerning the expense and problems encountered by the modern pharmaceutical industry in developing novel drugs. As a further difficulty, the patient population (market) is becoming fragmented due to genetic variation in the response to medicines resulting from alterations in the drug target or in the metabolism of the compound once ingested. The relatively new discipline of pharmacogenetics is concerned with the inheritance of these variations, measured using single nucleotide polymorphism (SNP) analysis at the level of genomic DNA. Pharmacogenomics, on the other hand, is relevant when the genetic variation leads to changes in protein expression or conformation of key ligand binding sites. This may result in loss of drug efficacy due to lack of binding; when the loss of binding occurs in drug metabolizing enzymes, this may result in levels of compound in the blood that exceed safety thresholds.

While these aspects of human genetics are problematical for both patients and pharmaceutical companies alike (for obviously different reasons), there are ways forward; these are emerging from new drug discovery technologies, including *in silico* approaches to drug design. The relevant technologies will be described in this chapter, but first we shall discuss the background to lead discovery and pharmacogenomics by concentrating on the structural aspects of the protein targets for small molecule drugs.

7.2
The Need for Protein Structure Information

Traditional drug discovery has relied upon the chemical modification of biologically active natural products or high-throughput screening of compound libraries to obtain "hits" that may be converted to "leads" and ultimately drugs. This process is generally inefficient, since the nature of compound collections used for screening is often a reflection of the historical activity of the company in question. This means that most compounds will be limited in coverage of the variety of targets encountered in drug research, namely enzymes, receptors and ion channels [1]. Even the advent of combinatorial chemistry in the mid 1990s has failed to deliver a noticeable increase in good drug leads (these issues are discussed in [2]).

The problems highlighted above have been compounded by the increase in targets afforded by the genome sequencing projects, culminating in the recent publications of the human sequence [3, 4]. The realization that it will be impossible to find suitable small molecule candidates for every potential drug target using high-throughput screening alone, has driven the search for alternatives based on an understanding of the three-dimensional structure of the protein. The structural information on the ligand-binding site may then be used for the *in silico* design of compounds that make strong interactions with appropriate residues within these sites; alternatively, existing small molecule structures may be docked into the sites and optimized using medicinal chemistry techniques.

The availability of three-dimensional protein structures is clearly one of the rate-limiting steps in this process. Publicly available structures, derived by X-ray diffraction or NMR techniques, are deposited in the Protein Data Bank (PDB), and currently number over 15,000 entries. There is considerable redundancy in this, however, with the number of single entries for human proteins being approximately 500. Due to the technical difficulties associated with certain classes of protein there is a strong bias towards enzymes in the database. This excludes, therefore, the G protein-coupled receptors (GPCR) and ion channel classes that make up a large proportion of drug targets. A number of public and private structural proteomics initiatives have been established in order to increase the number and variety of structures that may be used for studies of protein folding or drug discovery [5]. Where structures are not available, it is possible in many instances to create a homology model using the structure of a related protein as a guide [6]. The success of drug design corresponds to the % identity between the sequence and the known structure. Despite these advances (including the structural determination of the GPCR rhodopsin [7]), it has to be accepted that some proteins will not yield to current structural techniques. This does not mean, however, that *in silico* drug design techniques cannot be used with these proteins, since ligand-based design can be employed (see Sect. 7.9).

7.3
Protein Structure and Variation in Drug Targets – the Scale of the Problem

Variation in the response of individual patients to medicines is an important issue that is currently being addressed by the pharmaceutical industry [8]. It would be useful to have some idea of the scale of the problem through determining the number of SNPs in the human genome and their relative effects on both the level of expression and functional activity of the protein target. From the perspective of rational drug design, it is necessary to consider the alteration of the tertiary structure of the translated protein in which function is retained, but not the binding of an existing drug. This is directly analogous to the situation with the microbial targets (*e.g.*, HIV reverse transcriptase) in which sequence variations affect the structure of the target, resulting in the problem of drug resistance [9].

An analysis of 1.42 million non-redundant human SNPs has been recently published by an international collaborative group [10] (plus additional commentary [11]). The majority of these are in repetitive regions, whereas 60,000 lie within coding and untranslated regions of exons. Bearing in mind that many more SNPs are being identified in the public and private domains, the total number that are potentially able to disrupt protein structure, while small in comparison with the total genome complement, is still likely to be significant in terms of pharmaceutical research opportunities. This study gave an overview of the total number of SNPs available within the public domain as of the first quarter of 2001.

An analysis of polymorphisms in coding regions was published by Lander's group in 1999 [12]. They identified 560 SNPs (392 in coding regions) in 106 genes of relevance to cardiovascular disease, neuropsychiatry and endocrinology. Only a minority of polymorphic changes gave rise to non-conservative amino acid substitutions, most likely due to evolutionary selection against deleterious mutations in the human genome.

Nevertheless, there is considerable activity in identifying SNPs within protein targets for current marketed drugs. The whole purpose of the study of genetic variations in drug responses is to identify patients who may benefit from a particular medicine and avoid wasteful (or dangerous) prescription to others [8]. Examples of drug targets and detoxification systems that have been studied using SNP or other mutational analyses are listed in Table 7.1 (see [13] for a more comprehensive list).

Tab. 7.1 Examples of SNPs that influence drug metabolism or disease state

Drug target	*Disease*	*Comments*	*Reference*
NMDAR1 receptor	Schizophrenia	SNPs in coding region, but no functional significance	14
Quinone oxidoreductase, Sulfotransferase	Drug metabolism	6 out of 22 coding region SNPs gave amino acid substitutions	15
75 candidate genes for blood pressure homeostasis	Hypertension	874 candidate SNPs with 387 within coding sequence; 54% of these predicted to change protein sequence	16
Thiopurine S-methyltransferase	Drug metabolism	SNPs in promoter region, introns and 3'UTR, but not coding region	17
41 candidate genes	Ischemic heart disease	SNPs restricted to ethnic group (Japanese) used in study	18
CYP3A	Drug metabolism	SNPs cause alternative splicing and protein truncation	19
β_2-adrenergic receptor	Asthma	SNPs cause alteration of ligand binding	20
BCR-ABL tyrosine kinase	Cancer	Point mutation causes resistance to STI-571 compound	21

These comparatively early studies are beginning to highlight a number of points, including the fact that some SNPs are specific for particular ethnic populations. In addition, some drug targets appear not to have amino acid sequence variation as a result of SNP polymorphism, and therefore will not require a number of different ligands to accommodate this. However, since variations in tertiary structure resulting from mutations in the coding region offer opportunities for structure-based design, some relevant examples will be discussed in detail below.

7.4 Mutations in Drug Targets Leading to Changes in the Ligand Binding Pocket

7.4.1 β_2-Adrenergic Receptor

This well characterized drug target for anti-asthma medications represents one of the earliest examples of a natural mutation leading to an alteration in ligand binding [20]. Mutation of Thr^{164} to Ile in the fourth transmembrane-spanning domain

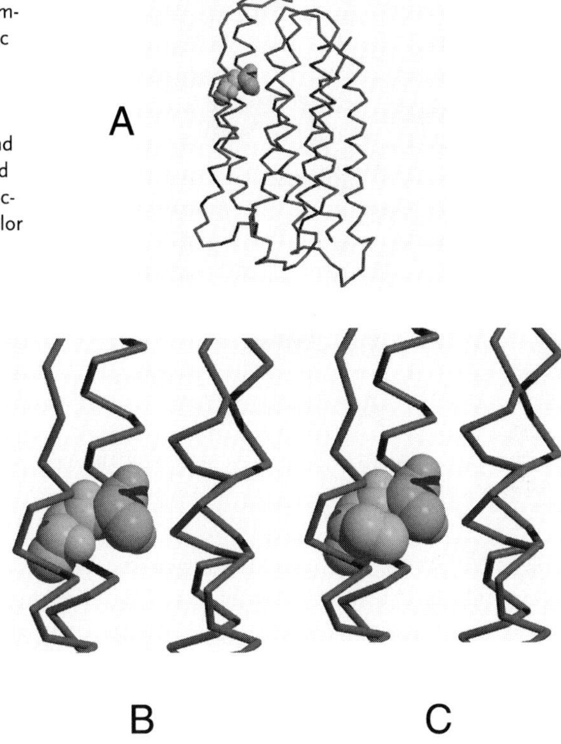

Fig. 7.1 Mutation of Thr[164] to Ile (cyan) in the fourth transmembrane region of the β_2-adrenergic receptor leads to steric interference with Ser[165] (green) within the active site of the receptor. (**A**) general domain structure and position of the residues, (**B**) wild type, (**C**) mutant showing interaction of Ile[164] with Ser[165] (see color plates, p. XXXII).

of this G protein-coupled receptor occurs at low frequency in the population tested, but gives rise to altered ligand binding through interaction with an adjacent Ser[165] as illustrated in Figure 7.1.

This example is one where the accurate three-dimensional structure of the protein is unknown; under these circumstances, it is necessary to create a computer model. The development of inhibitors that are designed to overcome the effects of this mutation could not be based on the accurate structure of a ligand-binding site; here, ligand-based design would be appropriate (see Sect. 7.9).

7.4.2
STI-571 and BCR-ABL

STI-571 is a 2-phenylamino pyrimidine chemotherapeutic agent that targets the tyrosine kinase domain of the BCR-ABL oncogene present in chronic myelogenous leukemia (CML) [22]. After many years of research into compounds that are selective for specific oncogenes, this compound is showing real clinical benefit against CML [23]. Unfortunately, due to the selective pressure on tumor cells to evolve resistance, a specific mutation in BCR-ABL inhibits the action of STI-571 [21]. Since crystal structures of ligand bound into the wild-type protein are avail-

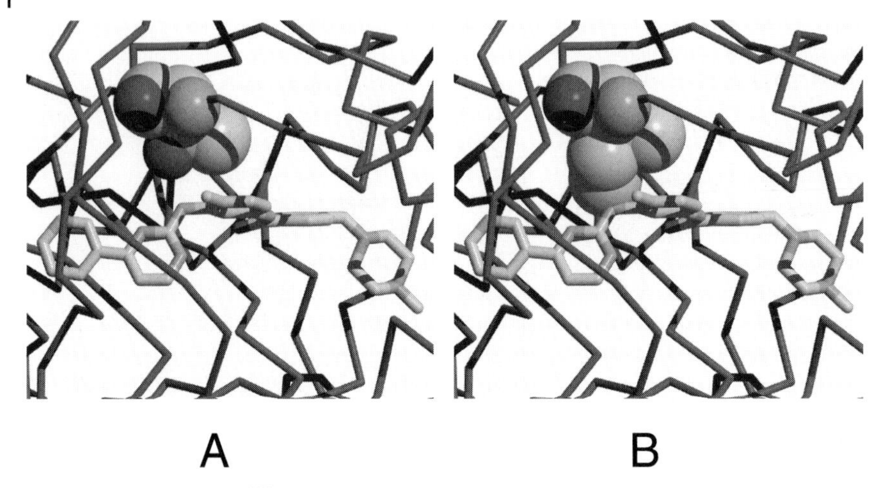

Fig. 7.2 Mutation of Thr^{315} (**A**) to Ile (**B**) in BCR-ABL interferes with the binding of STI-571 (see color plates, p. XXXIII).

able (PDB: 1IEP), it is possible to visualize the effects of a C-T nucleotide mutation resulting in a substitution of Thr^{315} by Ile (Figure 7.2). The change of amino acid results in disruption of a key hydrogen bond between oxygen on the threonine and a secondary amine nitrogen on STI-571.

In this example, structure-based design of inhibitors to the mutant enzyme may be undertaken using available X-ray structures and homology models.

7.5
Resistance to Human Immunodeficiency Virus (HIV)

The human retrovirus HIV can be controlled using chemotherapy directed at the reverse transcriptase and aspartyl protease encoded by the viral genome; as with other microbial pathogens, however, resistance to drug therapy becomes a major problem. Figure 7.3 shows a crystal structure (PDB: 1HXW) of the HIV protease, where mutated amino acids (shown in cyan) lead to disrupted binding of the clinically effective inhibitor ritonavir [24].

7.6
In silico Design of Small Molecules

Changes induced by SNPs (or other mutations) in the protein targets of drugs may result in the medicine being ineffective. In this case, the relevant disease may remain untreated, or else an expensive drug discovery effort will have to be undertaken to identify new molecules that are able to affect the mutant protein. Given that most current small molecule discovery strategies rely on expensive

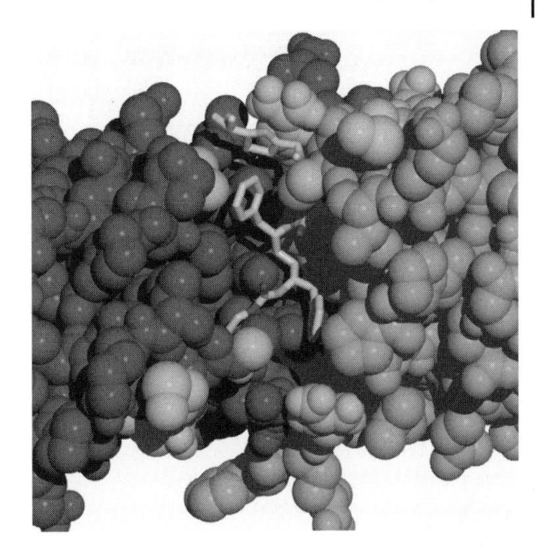

Fig. 7.3 Binding of the HIV protease inhibitor ritonavir. Amino acids highlighted in cyan are mutated in resistant strains of the virus and tend to occur at the extremities of the inhibitor (see color plates, p. XXXIII).

high-throughput screening and lead optimization programs, this is an unattractive prospect. Advances in computer-aided design, however, may make this process more efficient and less costly through creation of virtual small molecules that can by optimized for interaction with the mutant protein before any chemical synthesis is undertaken. The remainder of this chapter will be concerned with an overview of this exciting and rapidly evolving area of research.

7.7 Automated Drug Design Methods

The pharmaceutical industry faces a large increase in the number of therapeutic targets available for exploration. New methods must be developed for drug discovery that can be automated, are cheap and that optimally explore the chemical space available for drug design. The number of different possible chemical structures has been estimated to be of the order 10^{200}, with perhaps 10^{60} structures with drug-like properties [25]; these numbers can be compared with the estimated mass of the visible universe at 10^{56} g. Thus it would not be possible to make a single gram of each potential drug molecule. In contrast, the number of different chemical structures synthesized is small by comparison, and has not yet reached 10^{8}. Clearly, *in silico* methods will have to be developed to cope with searches through the vastness of chemical space.

Design paradigms fall into two types:

- structure-based methods that use three-dimensional structural information about the site,
- ligand-based methods that can be used for drug discovery in the absence of structural information.

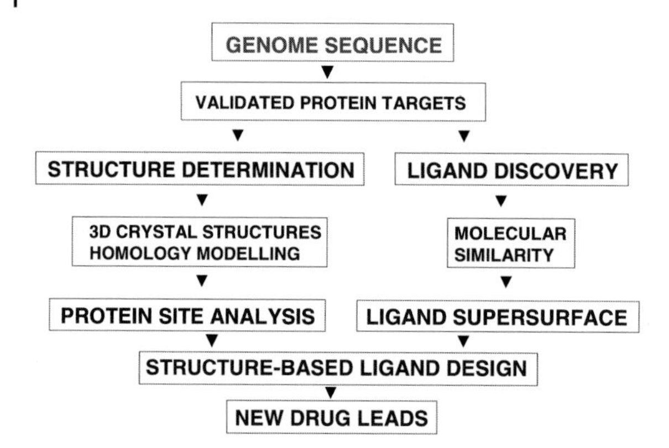

Fig. 7.4 Scheme for drug design utilizing genomic data. The left track illustrates the process of drug design based on structural determination of binding sites. The right track outlines design from known ligands.

The applicability of structure-based methods is critically dependent on the availability of relevant structural data. The Structural Genomics Initiative [5] is an international project that seeks to provide high-quality crystallographic or NMR data about new proteins. While promising for future drug discovery efforts, the rate of introduction of new structures into the public domain is still too slow. To speed this up, alternative methods for the generation of three-dimensional data using homology modeling techniques have been introduced. These are less precise, but show increasingly better performances as elucidated by the Computer Aided-Structure Prediction (CASP) challenges [26]. The logical next step would be to automate protein structure prediction in order to provide extensive coverage of the human proteome. One of the main drivers for this effort is the need for three-dimensional structural models of whole gene families for use in drug design.

Ligand-based design methods are built upon identified molecular similarities in known active ligands. The similarity methods have to cope with finding unspecified partial molecular similarities within the ligand data sets for freely flexible ligands. There are often many solutions to molecular similarity and choices have to be made as to which one to use. Figure 7.4 summarizes the two approaches to drug design.

7.8
Structure-Based Drug Design

The automation of structure-based design can be divided into key areas for development:
- identification of binding sites,
- automated analysis of binding sites,
- prioritization of strategies for design within sites,
- *de novo* design of scaffolds (active templates),
- elaboration of scaffolds into combinatorial libraries.

The identification of binding sites can be initiated from protein sequence information. Many sites show conservation of structural motifs; for example, many enzymatic catalytic sites have highly conserved local sequences within a gene family. Many of these motifs can be identified at the annotation stage of ascribing a function to a novel gene. Furthermore, the identification of co-factor binding sites can also provide additional clues to function. At the moment, the prediction of the molecular structure of the substrate has not been solved precisely, although docking algorithms could provide clues for potential substrate fitting into binding sites.

Ligands bind to their sites through specific molecular interactions, including hydrogen bonds, electrostatic interactions, steric interactions, interactions through hydration and hydrophobic interactions. This pattern lies at the surfaces of both ligand and binding site, so that complementary interactions are optimized for efficient binding. If the target proteins have 3D coordinate data, computer algorithms can be used to automatically analyze the binding features and categorize them on a grid map of the protein surface. The features mapped onto grid points are known as site points. Structurally and functionally related proteins may then be compared by superposing the binding sites and aligning the pattern of chemical features. This facility is important in pharmacogenomics where we may wish either to design compounds that bind to a class of sites within related proteins, or to design compounds specific for a particular member of the protein family. In the latter case, the aim is to acquire selectivity. Figure 7.5 illustrates the family of caspase enzymes; caspases 1, 4 and 5 form a separate group – the ICE sub-family, the rest are in the CED-3 subfamily.

The CED-3 sub-family shows very similar three-dimensional structures. Figure 7.6 shows the superpositions of the Cα atoms. The structural backbones of the proteins within the family are closely superposed.

The concept of dividing sites into a collection of site points for design provides a method for directing design to specific regions. In some respects this process has parallels to drug design from pharmacophores where a small number of points are used to search compound collections by virtual screening. 4-point pharmacophores provide three-dimensionality to the search. However, the number of site points encountered in a site, often about 30, creates a combinatorial problem for selection [27]. Suppose that there are n site points; if r site point interactions are needed for a molecule to have measurable affinity, the number of combinations of r site points, $C(n,r)$ is given by

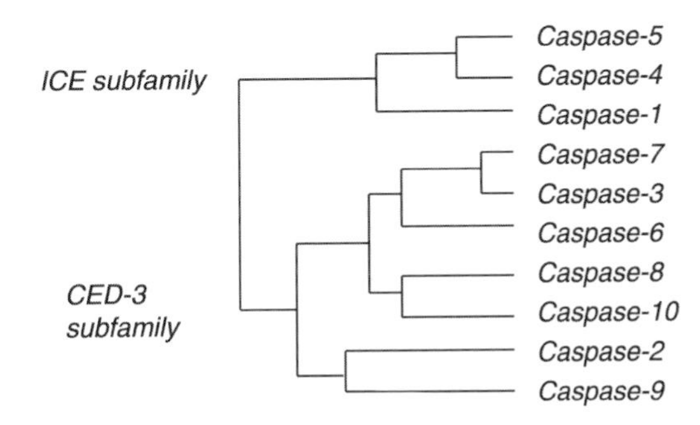

Fig. 7.5 Sequence similarities of the caspase family.

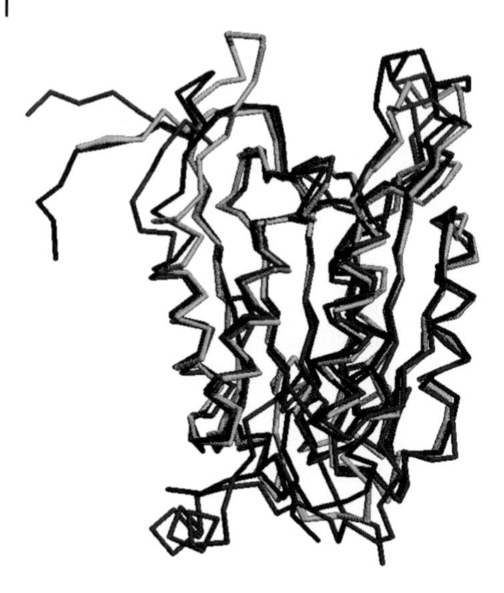

Fig. 7.6 Superposed Cα atoms from a selection of proteins in the CED-3 family of caspases.

$$C(n, r) = n!/(r!(n - r)!)$$

$C(n, r)$ is a maximum when $r=n/2$. For the case where $n=30$ and $r=5$, there are 140000 subsets of 5 site points. Each subset would be a strategy for design. However, some strategies would lead to preferred designs, e.g., a strategy that included interactions with the catalytic residues would be expected to consistently perform well in designing inhibitors. Moreover, if there were crystallographic evidence of substrates or known inhibitors binding to certain residues, these would form a basic set for *de novo* design. Other methods for prioritizing design strategies could be built on hydrogen bonding regions, where there are overlaps in hydrogen bonding probability. In the pharmacogenomic problem of design within a family of sites, the choice of design strategy is paramount for obtaining selectivity. The selection of site points should be such that the set chosen for the target protein should be as different as possible from nearly corresponding sets within the family of proteins.

The affinity of the designed drug molecule for its site is governed by the free energy of interaction ΔG. ΔG can be determined experimentally from the equation

$$\Delta G = -RT \ln K_b$$

Where K_b is the experimental binding constant, R is the gas constant and T is the temperature. The free energy has an enthalpic and entropic component.

$$\Delta G = \Delta H - T\Delta S$$

Where ΔH is the change in enthalpy on binding and ΔS is the change in entropy.

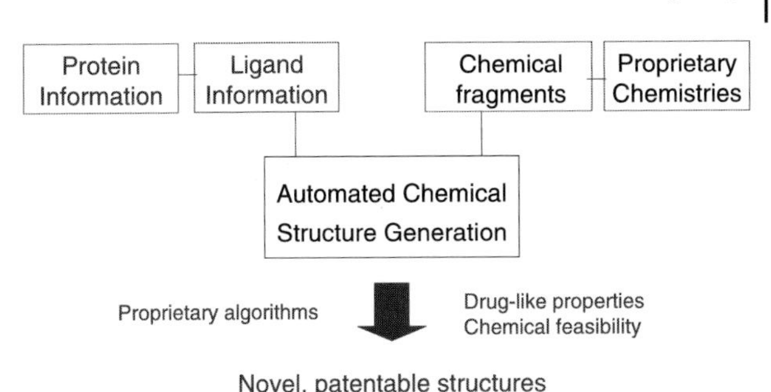

Fig. 7.7 Scheme for *de novo* design.

De novo methods for drug design attempt to build novel molecules within specified regions of the site. Given the large number of molecules that could be created, it is important to assess the relative differences in predicted binding affinity for the designed ligands and the site. This is achieved using a scoring function based on the equations described above.

A composite scheme for *de novo* design is shown in Figure 7.7.

Molecules are built from small molecular fragments, the latter being derived from databases of drug molecules such as the World Drug Index, or from proprietary compound collections. Fragments are labeled for attachment and substitution positions. Molecules are then built into the site by a stochastic assembly process from randomly selected fragments. This process is controlled by chemical rules that govern the addition, removal or exchange of a fragment on the evolving skeleton. At each modification, the assembled structure is fitted in the site by positional and conformational transformations, with the scoring function being used to monitor the prediction of the free energy of the interaction between ligand and site. An optimal assembly procedure is performed using combinatorial optimization routines, such as simulated annealing or genetic algorithms.

Simulated annealing is an ergodic optimization method, meaning that given sufficient time, the algorithm converges on the optimal solution irrespective of the initial starting position. If time constraints are placed on the algorithm run time, it behaves non-ergodically and different solutions can be obtained that are acceptable, but suboptimal. In structure assembly this is important, since a variety of molecular structures can be produced, thus giving a choice of different chemistries. These solutions can be analyzed statistically to identify different classes of structures that fit a chosen starting strategy.

Constraints can be placed on the structure assembly to limit the size of the molecule being generated so that it can be used as a template for further decoration by combinatorial chemistry. Virtual combinatorial chemical libraries can be created round the designed template to explore regions close to where the template has been designed to bind. The in-site enumeration of combinatorial libraries designed for a particular site is very important for chemical genomics, since it offers a way of designing molecules to distinguish between mutational differences in a particular site.

7.9 Ligand-Based Drug Design

In the absence of useful three-dimensional information about the site, the designer can only build novel molecular structures based on molecular similarity with existing active molecules. Ligand-based design requires the resolution of a number of technical issues. These are itemized below.

1. A set of molecules is needed to extract spatially distributed chemical features for inclusion in drug design.
2. Structurally dissimilar, active molecules provide more information about the site than a series of active close structural homologs.
3. The molecules in the initial ligand set may have numerous torsion angles making identification of the active conformation difficult.
4. The choice of molecular similarity procedure needs to be made from a wide variety of available methods, preferably including one that includes a search for partial molecular similarity. This allows a more useful set of similarities to be identified.
5. Molecules within the ligand set could have different binding modes which need to be identified before selecting a subset for a design strategy.
6. Molecules within the ligand set, with similar binding modes, need to be superposed such that the site points inferred from the corresponding ligand points show common spatial positions.
7. The constraining supersurface of the set of molecules needs to be constructed to provide limits for automated design procedures.
8. Since there is no information about the site, a scoring function based on free energy methods cannot be used.

A molecular similarity method that is applicable to this method of design has recently been published as the algorithm SLATE [28]. This algorithm takes a set of molecules and compares them in a pairwise fashion for molecular similarity. The molecules are allowed to flex. In the case of potential hydrogen bonding interactions between the ligand points and possible complementary site points, site points are projected from the ligand surfaces. Hydrophobic regions are handled as local regional centroids. The molecular description is reduced to a set of points with associated properties. These sets of points can be matched using distance geometry and the search optimized by simulated annealing through conformational space. The procedure generates a similarity matrix; partial similarity is handled by null correspondences [27].

An application of this procedure to the design of novel compounds for the histamine H_3 receptor has been published [29]. Similarity data suggested that four hydrogen bonding site points, together with two lipophilic regions could be identified with a common spatial distribution. Furthermore, the constraining supersurface showed strong similarity with each of the selected ligand conformations. From these *in silico* designs, histamine H_3 antagonists were synthesized and showed high affinity for the site. Thus computational methods could be used to design potent ligands without any prior knowledge of the sequence or structure of the receptor.

7.10
Future Directions

New developments in pharmacogenomics will impact on drug design at three main levels

- the interaction of the drug with its receptor binding site,
- the absorption and distribution of the drug,
- the elimination of the drug from the body.

If we are ever to achieve personalized drug therapies, the above issues will have to be addressed.

The bulk of this chapter has discussed SNPs in terms of mutated drug binding sites and the resulting changes that occur in response to medicines. In each of these cases, drug design strategies may need to be modified to develop drug molecules that would be selective and effective in the mutated site.

Absorption, in general, is treated as a physicochemical transport process based on computations of logP (the octanol/water partition coefficient) and solubility governed by factors such as polar surface area on the molecule. It is conceivable that SNPs in drug transporter genes will affect the pharmacokinetic properties of compounds and, therefore, these may have to be taken into consideration in the design process.

Elimination mechanisms may include active excretion of drug molecules from cells, as with multiple drug resistance that arises in cancer chemotherapy. Patients with genotypes for particular MDR pathways could be identified and alternative treatments provided. Pharmacogenetic differences in the response to drugs can often be related to population differences in drug metabolizing enzymes. Crystal structures, or homology models of these enzymes can be used to screen compounds designed *de novo* before synthesis has begun. This can be done at two levels:

- The virtual compounds can be input into metabolism prediction programs, such as Metabolexpert [30] or Meteor [31], to identify principal pathways of expected drug metabolism.
- The virtual compounds can be screened against structural models of the metabolizing enzymes, including the known SNP variants. These procedures are becoming widely adopted for the cytochrome P450 isozymes involved in oxidative drug metabolism.

Genomic pre-screening of patients for SNP mutations in drug metabolism will improve the utility of clinical trials by focusing attention on the response of specific subpopulations to drug treatment. The consequence of this approach is that it should be possible to industrialize the creation of individual therapies, although at a significant increase in cost.

7.11 Conclusions

Pharmacogenomics will have a major influence on drug discovery, through drug design, as well clinical practice. The subtle changes in biomolecular structure caused by small SNP-induced amino acid alterations will profoundly affect the search for personalized medicines. In the future, greater emphasis will be placed on very precise differences in drug structures that select between mutated proteins. As well as being scientifically challenging, this approach will clearly have significant economic effects on the pharmaceutical industry.

7.12 References

1 Drews JJ. Drug discovery: A historical perspective. Science **2000**; 287:1960–1964.
2 Bailey D, Brown D. High-throughput chemistry and structure-based design: survival of the smartest. Drug Discovery Today **2001**; 6:57–59.
3 International Human Genome Sequencing Consortium. Initial sequencing and analysis of the human genome. Nature **2001**; 409:860–921.
4 Venter JC, Adams MD, Myers EW, Li PW, Mural RJ, Sutton GG et al. The sequence of the human genome. Science **2001**; 291:1304–1351.
5 Burley SK. An overview of structural genomics. Nature Struct Biol **2000**; 7 (Suppl):932–934.
6 Maggio ET, Ramnarayan K. Recent developments in computational proteomics. Trends Biotechnol **2001**; 19:266–272.
7 Palczewski K, Kumasaka T, Hori T, Behnke CA, Motoshima H, Fox BA, Le Trong I, Teller DC, Okada T, Stenkamp RE, Yamamoto M, Miyano M. Crystal structure of rhodopsin: a G protein-coupled receptor. Science **2000**; 289:739–745.
8 Roses AD. Pharmacogenetics and the practice of medicine. Nature **2000**; 405:857–865.
9 Richman, DE. HIV chemotherapy. Nature **2001**; 410:995–1001.
10 The International SNP Map Working Group. A map of human genome sequence variation containing 1.42 million single nucleotide polymorphisms. Nature **2001**; 409:928–933.
11 Marth G, Yeh R, Minton M, Donaldson R, Li Q, Duan S, Davenport R, Miller RD, Kwok PY. Single-nucleotide polymorphisms in the public domain: how useful are they? Nature Genet **2001**; 4:371–372.
12 Cargill M, Altshuler D, Ireland J, Sklar P, Ardlie K, Patil N, Shaw N, Lane CR, Lim EP, Kalyanaraman N, Nemesh J, Ziaugra L, Friedland L, Rolfe A, Warrington J, Lipshutz R, Daley GQ, Lander ES. Characterization of single-nucleotide polymorphisms in coding regions of human genes. Nature Genet **1999**; 3:231–238.
13 Evans WE, Relling MV. Pharmacogenomics: translating functional genomics into rational therapeutics. Science **1999**; 286:487–491.
14 Rice SR, Niu N, Berman DB, Heston LL, Sobell JL. Identification of single nucleotide polymorphisms (SNPs) and other sequence changes and estimation of nucleotide diversity in coding and flanking regions of the NMDAR1 receptor gene in schizophrenic patients. Mol Psychiatry **2001**; 6:274–284.
15 Iida A, Sekine A, Saito S, Kitamura Y, Kitamoto T, Osawa S, Mishima C, Nakamura Y. Catalog of 320 single nucleotide polymorphisms (SNPs) in 20 quinone oxidoreductase and sulfotransferase genes. J Hum Genet **2001**; 46:225–240.

16 Halushka MK, Fan JB, Bentley K, Hsie L, Shen N, Weder A, Cooper R, Lipshutz R, Chakravarti A. Patterns of single-nucleotide polymorphisms in candidate genes for blood-pressure homeostasis. Nature Genet **1999**; 22:239–247.

17 Seki T, Tanaka T, Nakamura Y. Genomic structure and multiple single-nucleotide polymorphisms (SNPs) of the thiopurine S-methyltransferase (TPMT) gene. J Hum Genet **2000**; 45:299–302.

18 Ohnishi Y, Tanaka T, Yamada R, Suematsu K, Minami M, Fujii K, Hoki N, Kodama K, Nagata S, Hayashi T, Kinoshita N, Sato H, Sato H, Kuzuya T, Takeda H, Hori M, Nakamura Y. Identification of 187 single nucleotide polymorphisms (SNPs) among 41 candidate genes for ischemic heart disease in the Japanese population. Hum Genet **2000**; 106:288–292.

19 Kuehl P, Zhang J, Lin Y, Lamba J, Assem M, Schuetz J, Watkins PB, Daly A, Wrighton SA, Hall SD, Maurel P, Relling M, Brimer C, Yasuda K, Venkataramanan R, Strom S, Thummel K, Boguski MS, Schuetz E. Sequence diversity in CYP3A promoters and characterization of the genetic basis of polymorphic CYP3A5 expression. Nature Genet **2001**; 27:383–391.

20 Green SA, Cole G, Jacinto M, Innes M, Ligett SB. A polymorphism of the human β_2 adrenergic receptor within the fourth transmembrane domain alters ligand binding and functional properties of the receptor. J Biol Chem **1993**; 268:23116–23121.

21 Gorre ME, Mohammed M, Ellwood K, Hsu N, Paquette R, Rao PN, Sawyers CL. Clinical resistance to STI-571 cancer therapy caused by BCR-ABL gene mutation or amplification. Science **2001**; 293:876–880.

22 Druker BJ, Tamura S, Buchdunger E, Ohno S, Segal GM, Fanning S, Zimmermann J, Lydon NB. Effects of a selective inhibitor of the Abl tyrosine kinase on the growth of Bcr-Abl positive cells. Nature Med **1996**; 2:561–566.

23 Druker BJ, Sawyers CL, Kantarjian H, Resta DJ, Reese SF, Ford JM, Capdeville R, Talpaz M. Activity of a specific inhibitor of the BCR-ABL tyrosine kinase in the blast crisis of chronic myeloid leukemia and acute lymphoblastic leukemia with the Philadelphia chromosome N. Engl J Med **2001**; 344:1084–1086.

24 Boden D, Markowitz M. Resistance to human immunodeficiency virus type 1 protease inhibitors. Antimicrob Agents Chemother **1998**; 42:2775–2783.

25 Klebe G. Virtual screening: An alternative or complement to high throughput screening. Perspectives in Drug Discovery and Design **2000**; 20:9 (preface).

26 *http://PredictionCenter.llnl.gov*

27 Dean PM. Defining molecular similarity and complementarity for drug design. In: Molecular Similarity in Drug Design (Dean PM, ed.), Blackie Academic & Professional **1995**; 1–23.

28 Mills JEJ, De Esch IJP, Perkins TDJ, Dean PM. SLATE: a method for the superposition of flexible ligands. J Comput Aided Mol Design **2001**; 15:81–96.

29 De Esch IJP, Mills JEJ, Perkins TDJ, Romeo G, Hoffmann M, Wieland K, Leurs R, Menge WMPB, Nederkoorn PHJ, Dean PM, Timmerman H. Development of a pharmacophore model for histamine H_3 receptor antagonists, using the newly developed molecular modeling program SLATE. J Med Chem **2001**; 44:1666–1674.

30 Darvas F. Predicting metabolic pathways by logic programming. J Mol Graph **1988**; 6:80–86.

31 Greene N, Judson PN, Langowski JJ, Marchant CA. Knowledge-based expert systems for toxicity and metabolism prediction: DEREK, StAR and METEOR. SAR and QSAR in Environmental Research. **1999**; 10:299–314.

8
The Pharmacogenomics of Human P-Glycoprotein

Martin F. Fromm and Michel Eichelbaum

Abstract

Active transport processes are now recognized as important determinants of drug absorption and elimination. The role of the *MDR1* gene product P-glycoprotein for drug disposition in humans is now well established. It is an ATP-dependent efflux transporter which translocates its substrates out of cells. Multiple structurally unrelated drugs including HIV protease inhibitors, immunosuppressants, cardiac drugs and β-adrenoceptor antagonists are substrates of P-glycoprotein. Inhibition and induction of P-glycoprotein have been identified as mechanisms of drug interactions in humans. Recently, multiple mutations in the *MDR1* gene have been identified. The C3435T mutation was associated with reduced intestinal P-glycoprotein levels and higher plasma concentrations of the P-glycoprotein substrate digoxin. The implications of genetically determined differences in P-glycoprotein function for drug disposition, therapeutic outcome and risk for development of certain diseases will be summarized in this chapter.

8.1
Introduction: Importance of Active Transport Mechanisms for Uptake, Tissue Distribution and Elimination of Xenobiotics

The *MDR1* gene product P-glycoprotein (ABCB1) is a membrane protein which functions as an ATP-dependent exporter of xenobiotics from cells. It was first described in tumor cells where it contributed to the occurrence of *m*ulti*d*rug *r*esistance (MDR) against anti-cancer agents [1]. In addition, it is expressed in normal tissues with excretory function such as intestine, liver and kidneys, in capillary endothelial cells of brain, placenta and testis and in peripheral blood cells [2–4]. Expression of P-glycoprotein in these normal tissues is believed to be a protective mechanism against xenobiotics. Intestinal P-glycoprotein has been shown to limit absorption of the immunosuppressant cyclosporine, the cardiac glycoside digoxin and the β-adrenoceptor antagonist talinolol in humans [5–7]. Moreover, induction and inhibition of P-glycoprotein are underlying mechanisms of drug interactions [6, 8].

It is well established that mutations of genes encoding xenobiotic metabolizing enzymes (e.g., members of the cytochrome P450 family such as CYP2D6) and drug targets (e.g., receptors) determine the efficacy or toxicity of a broad variety of drugs and carcinogens. Recently, multiple mutations were also found in the human *MDR1* gene [9, 10]. One of those mutations (C3435T), which is located at a wobble position in exon 26 and has a frequency of 28.6% for the homozygous genotype in Caucasians [10, 11], was associated with a lower P-glycoprotein expression in the small intestine in comparison to subjects homozygous for the wild-type allele [10]. Accordingly, after oral administration of the P-glycoprotein substrate digoxin subjects homozygous for this mutation had significantly higher plasma concentrations than the remainder of the population. Moreover, a reduced P-glycoprotein function was observed in peripheral blood cells of subjects with the TT genotype.

8.2 Structure of the Human *MDR1* Gene

P-glycoprotein is a member of the ATP-binding cassette (ABC) superfamily of membrane proteins. After its first description in cultured cells selected for MDR [1], it was cloned from mouse and human cells [12, 13]. The *MDR1* gene is located on chromosome 7q21. P-glycoprotein is a 170 kDa phosphorylated and glycosylated protein and consists of 1,280 amino acids. It has two homologous halves, which contain six hydrophobic transmembrane domains and an ATP-binding site. Both halves are separated by a flexible linker polypeptide. A hypothetical two-dimensional model of human P-glycoprotein is shown in Figure 8.1. In mice two

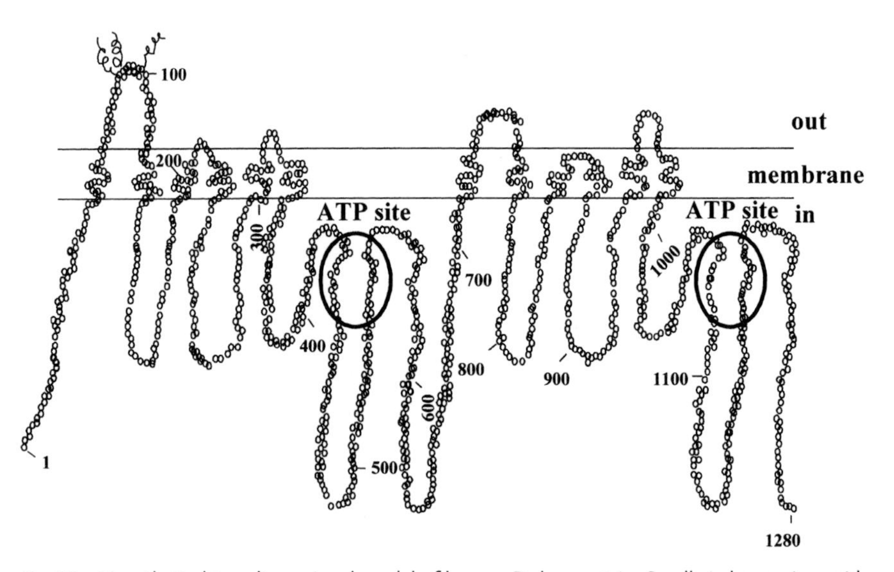

Fig. 8.1 Hypothetical two-dimensional model of human P-glycoprotein. Small circles: amino acid residues; large circles: ATP sites; squiggly lines: N-linked glycosylation sites (modified from [15]).

drug-transporting *mdr1* genes have been described, *mdr1a* and *mdr1b*. Structural analysis by electron microscopy and image analysis revealed that P-glycoprotein approximates a cylinder of about 10 nm in diameter with a maximum height of 8 nm. Since the depth of the lipid bilayer is about 4 nm, about one half of the molecule is within the membrane. P-glycoprotein has a large central pore of about 5 nm. It forms a large aqueous chamber within the membrane, which is open at the extracellular face and closed at the cytoplasmic face of the membrane [14]. Extensive mutational analysis revealed that two halves of the human P-glycoprotein interact to form a single transporter and that the major drug binding domains reside in or near transmembrane domains 5, 6 and 11, 12 [15]. Since it appears that P-glycoprotein detects and ejects its substrates before they reach the cytoplasm, it was suggested that P-glycoprotein acts as a "hydrophobic vacuum cleaner", which detects and removes its substrate from the lipid bilayer [16]. Another model suggests that P-glycoprotein acts as a flippase, carrying its substrate from the inner leaflet of the lipid bilayer to the outer leaflet [17].

8.3 P-glycoprotein Expression in Healthy Tissues

P-glycoprotein is not only expressed in tumor cells, but also in cells of several healthy tissues. In liver it was detected in the biliary canalicular surface of hepatocytes and the apical surface of small biliary ductules. In the small intestine and colon, it is localized in the apical surface of columnar epithelial cells, and in kidneys it is found in the brush border membrane of proximal tubules. Moreover, it is detectable on the apical surface of small ductules in the pancreas and on the surface of cells in the medulla and cortex of adrenals [2].

P-glycoprotein is also expressed in endothelial cells of capillaries of the central nervous system [3] and the sub-apical surface of the choroid plexus epithelium [18], which forms the blood-cerebrospinal fluid barrier. In addition to the blood-brain barrier it is expressed in capillary endothelial cells forming the blood-testis barrier [3]. The microvillus border of syncytiotrophoblasts of human placenta also expressed P-glycoprotein [19]. Finally, MDR1 mRNA was found in several leukocyte lineages with highest expression in $CD56^+$ cells followed by lower levels in $CD8^+$, $CD4^+$, $CD15^+$, $CD19^+$ and $CD14^+$ cells [4].

8.4 Function of P-Glycoprotein

The tissue distribution of P-glycoprotein yields important clues to its function. In most tissues it is localized to the apical (luminal) membrane of polarized epithelial cell layers. This location suggests that P-glycoprotein extrudes its substrates from the epithelial cells into the adjacent lumen. It is anticipated that P-glycoprotein plays an important role as a protective mechanism against naturally occurring toxins ingested with food. Intestinal P-glycoprotein will limit absorption of toxins, whereas renal (via direct secretion and possibly prevention of re-absorption from the lumen of tubules) and hepatic P-glycoprotein promotes elimination of these toxins into urine and bile. Moreover, P-glycoprotein expression in the blood-brain and blood-testis barrier will limit entry of these substances into brain and testis. P-glycoprotein also appears to be associated with hormone transport and reproduction. It is expressed in adrenal gland, uterus and placenta. Since P-glycoprotein is capable of transporting the corticosteroid hormones cortisol, corticosterone and aldosterone [20], it was suggested that P-glycoprotein contributes to secretion of steroids from these tissues. Possible roles of P-glycoprotein function have been described for the hematological compartment, for cell volume regulation, lipid transport and cell death and differentiation [21, 22].

Studies on P-glycoprotein expressing cell lines (Caco-2, L-MDR1) and P-glycoprotein knock-out mice yielded important insights into the role of this protein in drug transport [23, 24]. P-glycoprotein transports a wide range of structurally unrelated, hydrophobic and amphiphatic drugs such as anticancer agents, cardiac drugs, HIV protease inhibitors, β-adrenoceptor antagonists and antihistamines. A list of P-glycoprotein substrates is provided in Table 8.1. For example, the cardiac glycoside digoxin is a substrate of P-glycoprotein [25–28]. P-glycoprotein (*mdr1a*) knock-out mice had 35-fold and 2-fold higher concentrations than P-glycoprotein-expressing control animals in brain and plasma, respectively [27]. Moreover, control animals excreted about 16% of an intravenously administered dose within 90 min directly into the gut lumen [29]. In contrast, only 2% of a given dose is found in the gut lumen of P-glycoprotein knock-out mice. Intestinal secretion of digoxin in P-glycoprotein-expressing animals could completely be blocked by oral administration of the P-glycoprotein inhibitor PSC-833 [30].

Digoxin is a substrate of P-glycoprotein and it is not metabolized to a major extent in humans. Nevertheless, multiple drug interactions have been observed leading to increased or decreased digoxin plasma concentrations. A common and serious drug interaction occurs between digoxin and quinidine resulting in 2- to 3-fold increased digoxin plasma concentrations and digoxin toxicity [31, 32]. Pharmacokinetic studies reported enhanced absorption, reduced biliary and renal excretion of digoxin during administration of quinidine [33–36]. The following lines of evidence indicate that inhibition of P-glycoprotein-mediated digoxin transport is a major mechanism underlying the digoxin-quinidine interaction. First, polarized, P-glycoprotein-mediated digoxin transport in Caco-2 cell monolayers could be inhibited by low concentrations of quinidine [8]. Second, identical quinidine serum concentrations caused a significant increase in digoxin plasma concentrations in P-glycoprotein-expressing animals, whereas this interaction was not observed in P-glycoprotein knock-out mice [8]. Increased digoxin plasma concentrations have not only been reported during concomitant quinidine therapy, but also with patients simultaneously treated with verapamil, propafenone, cyclosporine, itraconazole, and amiodarone. Many of these drugs are also known to interact with P-glycoprotein; thus, inhibition of P-glycoprotein-mediated digoxin elimination may be a common mechanism leading to increased digoxin plasma concentrations.

Tab. 8.1 Summary of drugs, which are substrates of P-glycoprotein

Drug	*Reference*	*Drug*	*Reference*
Anticancer agents		Antibiotics	
– actinomycin D	[73]	– erythromycin	[82]
– etoposide	[41]	– levofloxacin	[83]
– docetaxel	[74]	– sparfloxacin	[84]
– doxorubicin	[75]		
– daunorubicin	[75]	Steroids	
– irinotecan	[76]	– dexamethasone	[27]
– mitomycin C	[77]		
– mitoxantrone	[77]	Lipid-lowering agents	
– paclitaxel	[78]	– atorvastatin	[85]
– teniposide	[77]	– lovastatin	[86]
– topotecan	[77]		
– vinblastine	[79]	Cacium channel blocker	
– vincristine	[24]	and metabolites	
		– diltiazem	[87]
Cardiac drugs		– mibefradil	[88]
– β-acetyldigoxin	[80]	– verapamil	[89]
– α-methyldigoxin	[80]	– D-617	[89]
– digitoxin	[80]	– D-620	[89]
– digoxin	[27]		
– quinidine	[8]	β-Adrenoceptor antagonists	
		– bunitrolol	[90]
HIV protease inhibitors		– celiprolol	[91]
– amprenavir	[56]	– talinolol	[92]
– indinavir	[38]		
– nelfinavir	[38]	H_1 antihistamines	
– saquinavir	[38]	– fexofenadine	[47]
– ritonavir	[40]	– terfenadine	[86]
Immunosuppressants		H_2 antihistamines	
– cyclosporine A	[81]	– cimetidine	[93]
– tacrolimus	[81]	– ranitidine	[93]
Antiemetic drugs		Others	
– domperidon	[41]	– debrisoquine	[86]
– ondansetron	[41]	– losartan	[94]
		– morphine	[27]
Antidiarrheal agents		– phenytoin	[41]
– loperamide	[41]	– rifampin	[95]

8.4.1 Intestinal P-Glycoprotein

In addition to the observations of increased digoxin plasma concentrations due to comedications, there have been reports of reduced digoxin plasma concentrations during treatment with the antibiotic rifampin, which has been shown to induce P-glycoprotein in human colon carcinoma cell lines [37]. Since intestinal P-glycoprotein appears to have a major impact on drug absorption, the hypothesis was tested, whether induction of intestinal P-glycoprotein could contribute to reduced digoxin plasma concentrations during treatment with rifampin in humans [6]. Indeed, rifampin significantly reduced AUC (area under the plasma concentration-time curve) of orally administered digoxin and caused a 3.5-fold increase in intestinal P-glycoprotein obtained from duodenal biopsies. Moreover, AUC of oral digoxin was negatively correlated with intestinal P-glycoprotein content, indicating that P-glycoprotein in the duodenum is a determinant of digoxin plasma concentrations [6].

The introduction of HIV protease inhibitors was a considerable step forward in the treatment of HIV infection. However, bioavailability of HIV protease inhibitors is low or variable. Recently indinavir, nelfinavir, saquinavir and ritonavir were identified as substrates of P-glycoprotein [38–40]. Transepithelial translocation of these drugs across a polarized, P-glycoprotein-expressing monolayer of Caco-2 cells, which is a well established system for studying intestinal drug transport, was considerably higher in the basal to apical direction (corresponding to drug secretion into the gut lumen) than to the opposite direction. The influence of intestinal P-glycoprotein on bioavailability of HIV protease inhibitors was further highlighted by intravenous and oral administration of these drugs to P-glycoprotein knock-out mice and control animals [38]. After intravenous administration of indinavir, nelfinavir and saquinavir, plasma concentrations were not different between both groups of animals indicating that mechanisms other than P-glycoprotein (such as CYP3A-mediated drug metabolism) are more important for elimination after intravenous drug administration. However, after oral administration of these drugs, plasma concentrations were 2- to 5-fold higher in P-glycoprotein knock-out mice as compared to control animals. Taken together these data point to a major role of intestinal P-glycoprotein for limiting bioavailability of HIV protease inhibitors.

8.4.2 P-Glycoprotein and the Blood-Brain Barrier

The blood-brain barrier is formed by capillary endothelial cells in the brain. Similar to gut wall mucosa, the blood-brain barrier was considered a passive anatomical structure determining brain entry of molecules, e.g., by lipophilicity and protein binding. It is now well established that active efflux by P-glycoprotein contributes to brain permeability of drugs in addition to the factors mentioned above. Insights into the role of P-glycoprotein in the blood-brain barrier were again obtained by P-glycoprotein knock-out mice. The particular importance of P-glycoprotein to the brain in comparison to most other tissues is shown by the considerably higher accumulation of several drugs in the brain of P-glycoprotein knock-out mice than in the plasma. Brain entry of the HIV protease inhibitors indinavir, nelfinavir and saquinavir is reduced due to P-glycoprotein expression in the endothelial cells of the blood-brain barrier. Thus, it was speculated that the ability of these drugs to achieve therapeutic concentrations in the brain is limited, thereby creating a potential sanctuary for viral replication [38].

Therapeutic effects of centrally active drugs also depend on adequate brain concentrations. Accordingly, centrally acting drugs such as haloperidol, clozapine, midazolam and flunitrazepam easily enter the CNS and are not substrates of P-glycoprotein [41]. The peripherally acting opioid loperamide, which penetrates the CNS poorly and is used as an antidiarrheal agent, is a substrate of P-glycoprotein. P-glycoprotein knock-out animals have 14-fold higher brain concentrations after oral administration of loperamide in comparison to control animals. Moreover, P-glycoprotein-deficient mice show typical, morphine-like effects after administration of loperamide [41]. The relevance of these data was confirmed by a study in healthy volunteers. Administration of loperamide alone did not cause any respiratory depression, but respiratory depression occurred during co-administration of loperamide with the P-glycoprotein inhibitor quinidine [42]. Taken together, these data indicate that P-glycoprotein in the blood-brain barrier is the major cause of the selective peripheral effects of loperamide in humans.

8.4.3 P-Glycoprotein in Other Tissues

The importance of P-glycoprotein expression in the blood-testis barrier, peripheral leukocytes and the kidneys for drug disposition and disease risk is discussed in Sections 8.7.2 and 8.7.3.

8.5 Identification of *MDR1* Mutations and Their Consequences for Function

In addition to the environmental factors described above, which modify expression and function of human P-glycoprotein, there is also increasing knowledge on naturally occurring mutations in the *MDR1* gene and their potential relevance to drug disposition. A list of *MDR1* gene mutations and their allelic frequencies is provided in Table 8.2. The first mutations in normal cells were described by Mickley et al. (G2677T, G2995A, [9]). A first systematic screen of the *MDR1* gene was conducted by Hoffmeyer et al. [10]. They analyzed all 28 exons of the *MDR1* gene including the core promoter region and the intron-exon boundaries from healthy Caucasian individuals and detected 15 SNPs (single nucleotide polymorphisms) with 6 SNPs located in the coding region. Subsequently, Cascorbi et al. [11] analyzed 461 German volunteers for the frequencies of the previously described mutations and the presence of additional mutations. Taken together, 20 SNPs have

Tab. 8.2 *MDR1* genetic variants in Caucasian individuals

#	*Location*	*Position*	*Allele*	*Effect*	*Allelic frequency [%]*	*Genotype*	*Genotype frequency [%]*	*Reference*
1	exon 1b	exon 1b/12	T	noncoding		T/T	88.2	[10]
			C	(?)		T/C	11.8	
						C/C	0.0	
2	intron 1	exon 2 −1	G	initiation	91.0 G	G/G	82.0	[10]
			A	of	9.0 A	G/A	18.0	
				translation		A/A	0.0	
3	exon 2	cDNA 61	A	21 Asn	88.8	A/A	78.5	[10, 11]
			G	21 Asp	11.2	A/G	20.6	
						G/G	0.9	
4	intron 4	exon 5 −35	G			G/G	98.8	[10]
			C			G/C	1.2	
						C/C	0.0	
5	intron 4	exon 5 −25	G			G/G	70.5	[10]
			T			G/T	26.0	
						T/T	3.5	
6	intron 6	exon 6+139	C		62.8	C/C	39.0	[10]
			T		37.2	C/T	47.5	
						T/T	13.4	
7	intron 6	exon 6+145	C			C/C	97.6	[10]
			T			C/T	2.4	
						T/T	0.0	
8	exon 11	cDNA 1199	G	400 Ser	94.5	G/G	88.9	[10]
			A	400 Asn	5.5	G/A	11.1	
						A/A	0.0	
9	exon 12	cDNA 1236	C	wobble	59.0	C/C	34.4	[10, 11]
			T		41.0	C/T	49.2	
						T/T	16.4	
10	intron 12	exon 12+44	C		95.1	C/C	90.2	[10]
			T		4.9	C/T	9.8	
						T/T	0.0	
11	intron 16	exon 17–76	T		53.8	T/T	28.4	[10]
			A		46.2	T/A	50.8	
						A/A	20.8	
12	intron 17	exon 17+137	A			A/A	98.8	[10]
			G			A/G	1.2	
						G/G	0.0	

Tab. 8.2 (continued)

#	*Location*	*Position*	*Allele*	*Effect*	*Allelic frequency [%]*	*Genotype*	*Genotype frequency [%]*	*Reference*
13	exon 21	cDNA 2677	G	893 Ala	56.5	G/G	30.9	[11]
			T	893 Ser	41.6	G/T	49.2	
			A	893 Thr	1.9	T/T	16.1	
						G/A	2.0	
						T/A	1.8	
						A/A	0.0	
15	exon 24	cDNA 2995	G	999 Ala		G/G	89.0	[9]
			A	999 Thr		G/A	11.0	
						A/A	0.0	
16	exon 26	cDNA 3320	A	1107 Gln	99.8	A/A	99.6	[11]
			C	1107 Pro	0.2	A/C	0.4	
						C/C	0.0	
17	exon 26	cDNA 3396	C	wobble		C/C	99.5	[10]
			T			C/T	0.5	
						T/T	0.0	
18	exon 26	cDNA 3421	T	1141 Ser		T/T	99.5	[49]
			A	1141 Thr		T/A	0.5	
						A/A	0.0	
19	exon 26	cDNA 3435	C	wobble	46.1	C/C	20.8	[10]
			T		53.9	C/T	50.5	
						T/T	28.6	

The positions of the polymorphisms correspond to positions of *MDR1* cDNA with the first base of the ATG start codon set to 1 (GenBank accession number M14758). Mutations located in introns are given as position downstream (–) or upstream (+) of the respective exon according to the genomic organization of *MDR1* as described by Chen et al. [96].

been described so far in Caucasian individuals. These mutations including allelic and genotype frequencies are summarized in Table 8.2. Seven of those mutations alter the amino acid sequence of P-glycoprotein. The A61G SNP, which is located close to the N-terminus of P-glycoprotein, leads to an amino acid exchange from Asn to Asp. The G1199A mutation (Ser400Asn) is located in the cytoplasmic loop close to the first ATP-binding domain. The most frequent SNP (G2677T/A) leading to amino acid exchanges from Ala to Ser or Thr is located in the second transmembrane domain. The G2995A mutation is also located in the second transmembrane domain, but closer to the second ATP-binding domain. Finally, two rare SNPs located in exon 26 (A3320C and T3421A) leading to amino acid changes were reported. The T307C (Phe-Leu) mutation described in the initial paper by Hoffmeyer [10] was not detectable in the larger population investigated by Cascorbi et al. [11]. The location of these mutations in the *MDR1* gene and within the structure of P-glycoprotein are shown in Figure 8.2.

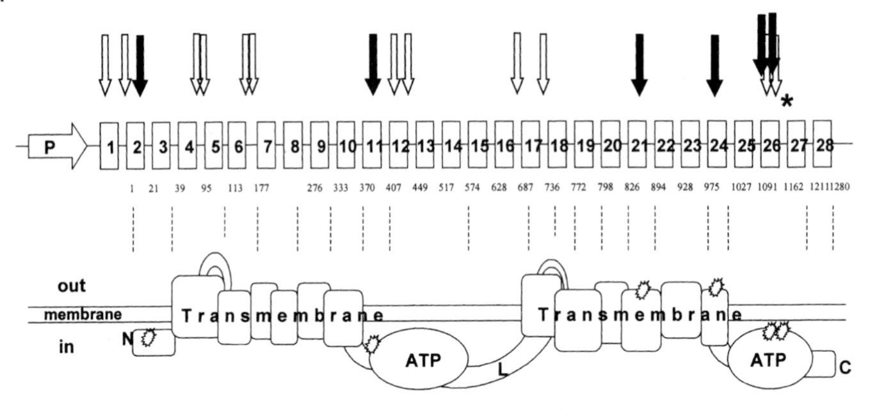

Fig. 8.2 Single nucleotide polymorphisms of the human *MDR1* gene (compiled from [10, 11, 97]). The locations of the identified polymorphisms (white arrow: non-coding polymorphisms; black arrow: polymorphisms leading to amino acid exchanges) in relation to the exon-intron structure of the *MDR1* gene and the corresponding P-glycoprotein structure. * Mutation with described functional consequences (modified from [97]; with permission from Dr. U. Brinkmann).

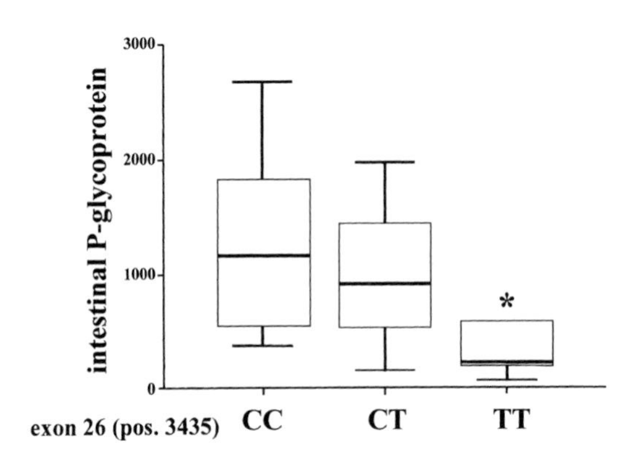

Fig. 8.3 Expression of P-glycoprotein in human small intestine according to genotype at position 3435 of the *MDR1* gene (modified from [10]; with permission from Dr. U. Brinkmann).

The only polymorphism described so far that is associated with altered P-glycoprotein function is the silent mutation in exon 26 at position 3435 (C3435T). Several lines of evidence indicate an association of this polymorphism with altered transporter function. First, individuals homozygous for the mutation at position 3435 (TT) had significantly lower P-glycoprotein levels in the small intestine in comparison to the remainder of the population (Figure 8.3, [10]). Second, the same genotype-dependent differences were found in human kidneys (H. Brauch and U. Brinkmann, personal communication). Third, in accordance with low intestinal P-glycoprotein content, subjects with the TT genotype had higher plasma concentrations of the P-glycoprotein substrate digoxin as compared to subjects with the wild-type genotype [10]. Finally, P-glycoprotein function determined by efflux of the P-glycoprotein substrate rhodamine 123 from $CD56^+$ natural killer cells

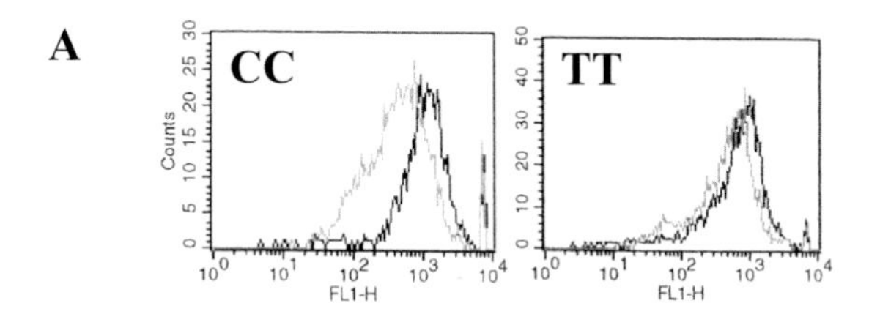

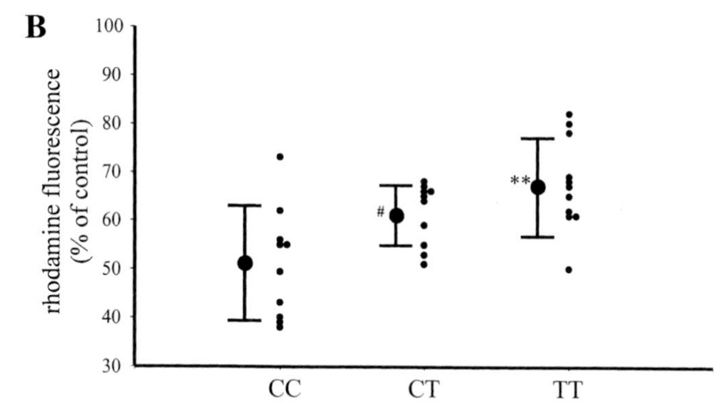

Fig. 8.4 P-glycoprotein function in $CD56^+$ natural killer cells of healthy volunteers according to *MDR1* genotype at position 3435. Panel A: Rhodamine fluorescence of $CD56^+$ natural killer cells in two healthy individuals with different *MDR1* genotypes in exon 26 (position 3435) after an efflux period of 10 min (gray line) compared to control cells (i.e., fluorescence of cells from the same individuals incubated for 10 min in medium containing 2.5 μM of the P-glycoprotein inhibitor PSC-833; black line). Panel B: Individual results (and mean ± SD) of rhodamine fluorescence of $CD56^+$ natural killer cells in 31 healthy individuals with different *MDR1* genotypes in exon 26 (position 3435) after an efflux period of 10 min as % of control values (** CC vs. TT: $p<0.01$, # CC vs. CT: $p<0.05$; modified from [43]).

was impaired in healthy volunteers with the TT genotype, unlike in individuals with the CC genotype (Figure 8.4, [43]). Moreover, MDR1 mRNA showed a trend towards lower values in the TT as compared to subjects with the CT or CC genotype [43]. The C3435T polymorphism has a high frequency in the Caucasian population with genotype frequencies of 20.8, 50.5 and 28.6% for CC, CT and TT, respectively [11]. The underlying mechanism for the association of the silent C3435T mutation with altered P-glycoprotein expression and function has not yet been established. It could be speculated that the functional consequences of this polymorphism arise from its linkage to the G2677T mutation (Ala893Ser) or to other unidentified changes in the promoter/enhancer region or in sequences that are important for mRNA processing [10].

Additional studies on the impact of *MDR1* polymorphisms on the disposition of the H_1 antagonist fexofenadine, which is a P-glycoprotein substrate and that is not metabolized in humans, revealed conflicting results. It was reported in an abstract that genetic variations in the *MDR1* gene (wild-type 1236C and 2677G and 3435C vs. mutations 1236T and 2677T and 3435T) impact on fexofenadine disposition in U.S. Caucasians. Those individuals who carry wild-type alleles at the respective positions have higher AUC [44]. These data stand in contrast to the results of a fexofenadine study in Caucasians living in Germany (no differences in fexofenadine disposition in subjects with 3435 TT vs. CC or 2677GG vs. TT; own data) and to the increased digoxin plasma concentrations in subjects carrying mutations at position 3435 [10]. The reasons for those discrepancies are unknown. One could speculate that differences in European and North American diet (e.g., salt content) account for differential regulation and expression of drug transporters determining fexofenadine disposition [45, 46]. Moreover, fexofenadine disposition might not only depend on P-glycoprotein function, since OATP-A has been shown to mediate cellular uptake of this antihistaminic drug [47].

8.6 Racial Differences in Frequency of *MDR1* Mutations

Recent studies indicate that the 3435C high-expression allele is considerably more frequent in African populations in comparison to Caucasian and Asian populations [48, 49]. The interethnic differences observed for the C3435T polymorphism are summarized in Table 8.3. One could speculate that the higher frequency of the CC genotype observed in Africans as compared to the Caucasian and Asian populations results from a selective advantage offered by this genotype against gastrointestinal tract infections, which are endemic in tropical countries. This hy-

Tab. 8.3 Interethnic differences in allele and genotype frequencies of the *MDR1* exon 26 C3435T polymorphism

Population		*Allele frequency*		*Genotype frequency*			*Reference*
	n	*C*	*T*	*CC*	*CT*	*TT*	
Caucasian, Germany	188	0.52	0.48	0.28	0.48	0.24	[10]
Caucasian, Germany	461	0.46	0.54	0.21	0.51	0.29	[11]
Caucasian, UK	190	0.48	0.52	0.24	0.48	0.28	[48]
Portuguese	100	0.43	0.57	0.22	0.42	0.36	[48]
Japanese	50	0.57	0.43	0.34	0.46	0.20	[49]
South-west Asians	89	0.34	0.66	0.15	0.38	0.47	[48]
Chinese	132	0.53	0.47	0.32	0.42	0.26	[48]
Filipino	60	0.59	0.41	0.38	0.42	0.20	[48]
Saudi	96	0.55	0.45	0.37	0.38	0.26	[48]
Ghanaian	206	*0.83*	0.17	*0.67*	0.34	0.00	[48]
Ghanaian	172	*0.90*	0.10	*0.83*	0.16	0.02	[49]
Kenyan	80	*0.83*	0.17	*0.70*	0.26	0.04	[48]
Sudanese	51	*0.73*	0.27	*0.52*	0.43	0.06	[48]
African American	88	*0.84*	0.16	*0.68*	0.31	0.01	[48]
African American	41	*0.78*	0.22	*0.61*	0.34	0.05	[49]

pothesis is supported by findings that P-glycoprotein plays a role in the defense against both bacterial and viral infections (see below). It was recently reported that clearance of the P-glycoprotein substrates cyclosporine and tacrolimus is reduced in Caucasians compared to African Americans [50, 51]. Since the *MDR1* genotype was not reported in these studies, it remains to be determined whether the lower frequency of the low expression T-allele in the African American than in the Caucasian population contributed to interethnic differences in disposition of these P-glycoprotein substrates.

In accordance with the allele frequencies mentioned above, it was reported that the 2677T mutation was present in 42% of Caucasians and only 13% of African Americans with the 2677T mutation being frequently linked to the 3435T mutation [44]. In addition, a new mutation in exon 26 (T3421A) leading to an amino acid exchange was identified in Ghanaians and African Americans (AA genotype Ghanaians vs. African Americans vs. Caucasians: 1.2% vs. 2.4% vs. 0%; [49]). Another report indicates that the CC genotype at position 1236 is considerably more frequent among Caucasians than in Japanese subjects (34.4% vs. 14.6%, [52]). Whether this difference is of any functional relevance remains to be determined.

8.7 *MDR1* Mutations and the Potential Risk for Idiopathic or Spontaneous Diseases

8.7.1 P-Glycoprotein and Ulcerative Colitis

The etiology of inflammatory bowel diseases such as ulcerative colitis is not completely understood. One of the current hypotheses suggests that either an enhanced or aberrant immunologic responsiveness to constituents of the GI lumen is involved in chronic inflammation. There is evidence indicating that the intestinal microflora is one factor in the pathogenesis of intestinal inflammation and that disruption of the protective epithelial cell barrier promotes development of inflammatory bowel disease [53, 54]. The physiological role of intestinal P-glycoprotein might be to prevent entry of bacterial toxins into the gut wall mucosa. This hypothesis is supported by the work of Panwala et al. [55], who showed that *mdr1a* P-glycoprotein knock-out mice are susceptible to developing a severe, spontaneous intestinal inflammation when maintained under specific pathogen-free conditions. Moreover, treatment of the P-glycoprotein knock-out mice with oral antibiotics prevented development of disease and resolved active intestinal inflammation. Studies are currently underway to determine whether individuals carrying the C3435T low-expression mutation in the human *MDR1* gene are more susceptible to developing ulcerative colitis.

8.7.2
P-Glycoprotein and HIV Infection

There is increasing evidence that disease risk and therapeutic outcome of the HIV infection are linked to P-glycoprotein function in several ways. Highly active anti-retroviral therapy (HAART), which includes treatment with HIV protease inhibitors, has considerably improved the clinical management of HIV infection. The HIV protease inhibitors indinavir, nelfinavir, saquinavir, ritonavir and amprenavir are transported by P-glycoprotein [38, 40, 56]. Some of them have a low bioavailability (e.g., saquinavir) and highly variable plasma concentrations, which might lead to subtherapeutic plasma concentrations in some patients. Although intestinal CYP3A4 certainly contributes to that variability, experiments with P-glycoprotein knock-out mice indicate that P-glycoprotein localized in the brush border membrane of enterocytes limits absorption by pumping the drug back into the gut lumen [38]. Whether genetically determined differences in P-glycoprotein function affect plasma concentrations of HIV protease inhibitors in humans has to be determined. Therapeutic efficacy of HIV protease inhibitors might further be affected in humans by P-glycoprotein expression at the blood-brain barrier and the blood-testis barrier, thereby limiting drug entry in these compartments and creating sanctuary sites for the virus. Animal experiments have shown considerably reduced brain and testis concentrations of HIV protease inhibitors compared to plasma [38, 56, 57]. Pharmacological inhibition of P-glycoprotein resulted in increased tissue concentrations of the HIV protease inhibitors in these tissues. Studies are currently underway in order to determine the impact of mutations in the *MDR1* gene (e.g., C3435T) for P-glycoprotein expression in the blood-brain barrier.

P-glycoprotein is also expressed in peripheral blood subpopulations, with the highest expression in $CD56^+$ natural killer cells followed by lower expression in $CD8^+$, $CD4^+$, $CD15^+$, $CD19^+$ and $CD14^+$ cells [4]. P-glycoprotein expressing $CD56^+$ natural killer cells from individuals homozygous for the C3435T mutation of the human *MDR1* gene have reduced P-glycoprotein function (determined by inhibition of rhodamine 123 efflux by PSC-833) as compared to the rest of the population [43]. Lee et al. [39] found a reduced inhibition of HIV replication by ritonavir, saquinavir and indinavir in cells with a high P-glycoprotein content, which could be restored by inhibitors of P-glycoprotein function. Although P-glycoprotein function is highly heterogeneous within $CD4^+$ cells [58], it will be important to determine possible genetically determined differences in P-glycoprotein function for this major cellular target of HIV protease inhibitors. The data described above indicate that individuals with a low P-glycoprotein expression might benefit more from HIV protease inhibitor therapy. However, using a human $CD4^+$ T-leukemic cell line, Lee et al. [59] showed that HIV virus production was greatly reduced when P-glycoprotein was overexpressed. This could indicate that cells with high P-glycoprotein expression may be relatively resistant to HIV infection, but once they are infected, it might be more difficult to eradicate the virus using HIV protease inhibitors [59].

8.7.3
P-Glycoprotein and Renal Cell Carcinoma

P-glycoprotein is expressed in the brush border membrane of proximal tubular cells [2, 3]. It mediates active secretion of its substrates (e.g., digoxin) into urine. Similar to its role as a protective barrier in the gut wall mucosa and at the blood-brain barrier, renal P-glycoprotein is likely to function as a protective mechanism against toxic substances in the glomerulum filtrate by prevention of re-absorption. Thus, individuals with a low renal P-glycoprotein expression would potentially be exposed to higher concentrations of toxic agents and should be more susceptible to their damaging effects. Indeed, there is data that indicates that the C3435T polymorphism is associated with reduced renal P-glycoprotein expression (H. Brauch, personal communication). In normal renal tissues quantitative immunohistochemistry revealed a lower P-glycoprotein expression in individuals homozygous for the mutation (TT) than in patients homozygous for the wild-type allele. Moreover, the TT genotype appears to be a risk factor for the development of renal cell carcinoma. In patients with non-clear cell renal cell carcinoma, the frequency of homozygous carriers of the T allele was significantly increased in comparison to healthy controls. Thus, the results of this study are in accordance with the hypothesis that subjects with low renal P-glycoprotein expression are less protected from intracellular accumulation of carcinogenic compounds. However, it must be emphasized that the quantity of P-glycoprotein expression is not the only factor determining the risk of renal cell carcinoma. For example, Longuemaux et al. [60] provided evidence that the presence of mutations in several xenobiotic-metabolizing enzymes (CYP1A1, GSTs, NAT2) was also associated with an increased risk for the development of renal cell carcinoma.

8.8
Aspects of P-Glycoprotein Regulation

The factors controlling basal expression of P-glycoprotein in human tissues are poorly understood. The transcriptional control of mouse *mdr1a* and *mdr1b* and of human *MDR1* has recently been summarized by M. Müller [61]. In brief, the human *MDR1* promoter contains an inverted CCAAT box, which is known to be a core sequence of the Y-box, a GC element and several putative recognition sites for transcription factors including AP1, NF-Y, Y-box binding protein (YB) and pregnane X receptor (PXR). Recently, a role for both NF-Y and Sp1 in the transcriptional activation of the *MDR1* gene by genotoxic stress has been reported [62]. Moreover, p53 inactivation that is seen in cancers most likely leads to selective resistance to chemotherapeutic agents because of up-regulation of P-glycoprotein expression [63]. Its expression is also increased by reactive oxygen species due to a process involving NF-κB activation [64]. Finally, it is now well established that induction of intestinal P-glycoprotein is the mechanism involved in the reduction of plasma digoxin concentrations during concomitant treatment with rifampin (see above). Similarly, the induction of intestinal P-glycoprotein by rifampin was the underlying mechanism for reduced plasma concentrations of the P-glycoprotein substrate talinolol [7]. Due to the fact that the nuclear receptor PXR is involved in the induction of CYP3A4 by xenobiotics [65–67] and that CYP3A4 and P-glycoprotein are often co-induced, it was speculated whether a similar mechanism is involved in the induction of P-glycoprotein. Indeed, using the human colon carcinoma cell line LS174T it could be shown that induction of P-glycoprotein by rifampin is mediated by a DR4 motif in the upstream enhancer at about 8 kb, to which PXR binds [68].

Similar to rifampin, the herbal medicine St. John's wort frequently used as a treatment for mild depression, induces CYP3A4 via activation of PXR [69, 70], which can cause severe drug interactions with cyclosporine and HIV protease inhibitors [71, 72]. It has recently been shown that during treatment with St. John's wort reduced plasma concentrations of the P-glycoprotein substrate digoxin are due to the induction of intestinal P-glycoprotein by constituents of St. John's wort [70]. It should be noted that induction of P-glycoprotein (e.g., by rifampin) shows considerable interindividual variability, with some subjects having almost no increase in P-glycoprotein expression. Whether this variability is genetically determined or not remains to be clarified.

8.9
Conclusions

P-glycoprotein is expressed in organs with excretory function such as small and large intestine, liver and kidneys. Moreover, it limits tissue penetration of drugs due to its expression in the blood-brain and blood-testis barriers and in the placenta. It determines absorption, tissue distribution and elimination of a wide range of structurally unrelated drugs such as anticancer agents, HIV protease inhibitors, cardiac drugs and β-adrenoceptor antagonists. Inhibition of P-glycoprotein function results in increased drug concentrations and is now a well recognized mechanism of drug interactions in humans. On the other hand, P-glycoprotein is induced by rifampin resulting in reduced drug concentrations of concomitantly administered P-glycoprotein substrates. In addition to these exogenous factors affecting P-glycoprotein function, we are now beginning to understand genetic factors determining inter-individual variability in P-glycoprotein expression and function. It appears that mutations in the *MDR1* gene have an impact on drug disposition within and among ethnic populations. Moreover, an individual's P-glycoprotein function might determine the risk for certain diseases and the therapeutic outcome from treatment with P-glycoprotein substrates. However, the relative importance of variability in P-glycoprotein function due to exogenous and genetic factors for drug disposition, therapeutic outcome and disease risk needs to be clarified in future studies.

Acknowledgements
Our work cited is supported by grants of Deutsche Forschungsgemeinschaft (FR 1298/2-1, Bonn, Germany), Bundesministerium für Bildung und Forschung (01 GG 9846 and 31 0311782, Bonn, Germany), and Robert Bosch Foundation (Stuttgart, Germany).

8.10 References

1 Juliano RL, Ling V. Biochim Biophys Acta **1976**; 455:152–162.
2 Thiebaut F, Tsuruo T, Hamada H, Gottesman MM, Pastan I, Willingham MC. Proc Natl Acad Sci USA **1987**; 84, 7735–7738.
3 Cordon-Cardo C, O'Brien JP, Casals D, Rittman-Grauer L, Biedler JL, Melamed MR, Bertino JR. Proc Natl Acad Sci USA **1989**; 86:695–698.
4 Klimecki WT, Futscher BW, Grogan TM, Dalton WS. Blood **1994**; 83:2451–2458.
5 Lown KS, Mayo RR, Leichtman AB, Hsiao HL, Turgeon DK, Schmiedlin R, Brown MB, Guo W, Rossi SJ, Benet LZ, Watkins PB. Clin Pharmacol Ther **1997**; 62:248–260.
6 Greiner B, Eichelbaum M, Fritz P, Kreichgauer H-P, von Richter O, Zundler J, Kroemer HK. J Clin Invest **1999**; 104:147–153.
7 Westphal K, Weinbrenner A, Zschiesche M, Franke G, Knoke M, Oertel R, Fritz P, von Richter O, Warzok R, Hachenberg T, Kauffmann HM, Schrenk D, Terhaag B, Kroemer HK, Siegmund W. Clin Pharmacol Ther **2000**, 68:345–355.
8 Fromm MF, Kim RB, Stein CM, Wilkinson GR, Roden DM. Circulation **1999**; 99:552–557.
9 Mickley LA, Lee JS, Weng Z, Zhan Z, Alvarez M, Wilson W, Bates SE, Fojo T. Blood **1998**; 91:1749–1756.
10 Hoffmeyer S, Burk O, von Richter O, Arnold HP, Brockmöller J, Johne A, Cascorbi I, Gerloff T, Roots I, Eichelbaum M, Brinkmann U. Proc Natl Acad Sci USA **2000**; 97:3473–3478.
11 Cascorbi I, Gerloff T, Johne A, Meisel C, Hoffmeyer S, Schwab M, Schäffeler E, Eichelbaum M, Brinkmann U, Roots I. Clin Pharmacol Ther **2001**; 69:169–174.
12 Chen CJ, Chin JE, Ueda K, Clark DP, Pastan I, Gottesman MM, Roninson IB. Cell **1986**; 47:381–389.
13 Gros P, Ben NY, Croop JM, Housman DE. Nature **1986**; 323:728–731.
14 Rosenberg MF, Callaghan R, Ford RC, Higgins CF. J Biol Chem **1997**; 272:10685–10694.
15 Ambudkar SV, Dey S, Hrycyna CA, Ramachandra M, Pastan I, Gottesman MM. Annu Rev Pharmacol Toxicol **1999**; 39:361–398.
16 Gottesman MM, Pastan I. Annu Rev Biochem **1993**; 62:385–427.
17 Higgins CF, Gottesman MM. Trends Biochem Sci **1992**; 17:18–21.
18 Rao VV, Dahlheimer JL, Bardgett ME, Snyder AZ, Finch RA, Sartorelli AC, Piwnica-Worms D. Proc Natl Acad Sci USA **1999**; 96:3900–3905.
19 MacFarland A, Abramovich DR, Ewen SW, Pearson CK. Histochem J **1994**; 26:417–423.
20 Ueda K, Okamura N, Hirai M, Tanigawara Y, Saeki T, Kioka N, Komano T, Hori R. J Biol Chem **1992**; 267:24248–24252.
21 Schinkel AH. Sem Cancer Biol **1997**; 8:161–170.
22 Johnstone RW, Ruefli AA, Smyth MJ. Trends Biochem Sci **2000**; 25:1–6.
23 Schinkel AH, Mayer U, Wagenaar E, Mol CA, van Deemter L, Smit JJ, van der Valk MA, Voordouw AC, Spits H, van Tellingen O, Zijlmans JM, Fibbe

WE, Borst P. Proc Natl Acad Sci USA **1997**; 94:4028–4033.

24 Schinkel AH, Smit JJ, van Tellingen O, Beijnen JH, Wagenaar E, van Deemter L, Mol CA, van der Valk MA, Robanus-Maandag EC, te Riele HP, Berns AJM, Borst P. Cell **1994**; 77:491–502.

25 Tanigawara Y, Okamura N, Hirai M, Yasuhara M, Ueda K, Kioka N, Komano T, Hori R. J Pharmacol Exp Ther **1992**; 263:840–845.

26 de Lannoy IA, Silverman M. Biochem Biophys Res Commun **1992**; 189:551–557.

27 Schinkel AH, Wagenaar E, van Deemter L, Mol CA, Borst P. J Clin Invest **1995**; 96:1698–1705.

28 Cavet ME, West M, Simmons NL. Br J Pharmacol **1996**; 118:1389–1396.

29 Mayer U, Wagenaar E, Beijnen JH, Smit JW, Meijer DK, van Asperen J, Borst P, Schinkel AH. Br J Pharmacol **1996**; 119:1038–1044.

30 Mayer U, Wagenaar E, Dorobek B, Beijnen JH, Borst P, Schinkel AH. J Clin Invest **1997**; 100:2430–2436.

31 Leahey EB, Jr., Reiffel JA, Drusin RE, Heissenbuttel RH., Lovejoy WP, Bigger JT, Jr. JAMA **1978**; 240:533–534.

32 Doering W, König E. Med Klinik **1978**; 73:1085–1088.

33 Hager WD, Fenster P, Mayersohn M, Perrier D, Graves P, Marcus FI, Goldman S. N Engl J Med **1979**; 300:1238–1241.

34 Pedersen KE, Christiansen BD, Klitgaard NA, Nielsen-Kudsk F. Eur J Clin Pharmacol **1983**; 24:41–47.

35 Angelin B, Arvidsson A, Dahlqvist R, Hedman A, Schenck-Gustafsson K. Eur J Clin Invest **1987**; 17:262–265.

36 de Lannoy IA, Koren G, Klein J, Charuk J, Silverman M. Am J Physiol **1992**; 263:F613–622.

37 Schuetz EG, Beck WT, Schuetz JD. Mol Pharmacol **1996**; 49:311–318.

38 Kim RB, Fromm MF, Wandel C, Leake B, Wood AJJ, Roden DM, Wilkinson GR. J Clin Invest **1998**; 101:289–294.

39 Lee CG, Gottesman MM, Cardarelli CO, Ramachandra M, Jeang KT, Ambudkar SV, Pastan I, Dey S. Biochemistry **1998**; 37:3594–3601.

40 Alsenz J, Steffen H, Alex R. Pharm Res **1998**; 15:423–428.

41 Schinkel AH, Wagenaar E, Mol CA, van Deemter L. J Clin Invest **1996**; 97:2517–2524.

42 Sadeque AJ, Wandel C, He H, Shah S, Wood AJ. Clin Pharmacol Ther **2000**; 68:231–237.

43 Hitzl M, Drescher S, van der Kuip H, Schäffeler E, Fischer J, Schwab M, Eichelbaum M, Fromm MF. Pharmacogenetics **2001**; 11:1–6.

44 Kim RB, Leake B, Choo E, Dresser GK, Kubba SV, Schwarz UI, Taylor A, Xie H-G, Stein CM, Wood AJJ, McKinsey J, Schuetz EG, Schuetz JD, Wilkinson GR. Drug Metab Rev **2000**; 32:199 (Abstract).

45 Wilkinson GR. Adv Drug Devel Rev **1997**; 27:129–159.

46 Darbar D, Fromm MF, Dell'Orto S, Kim RB, Kroemer HK, Eichelbaum M, Roden DM. Circulation **1998**; 98:2702–2708.

47 Cvetkovic M, Leake B, Fromm MF, Wilkinson GR, Kim RB. Drug Metab Dispos **1999**; 27:866–871.

48 Ameyaw M-M, Regateiro F, Li T, Liu X, Tariq M, Mobarek A, Thornton N, Folayan GO, Githang'a J, Indalo A, Ofori-Adjei D, Price-Evans D, McLeod HL. Pharmacogenetics **2001**; 11:217–221.

49 Schaeffeler E, Eichelbaum M, Brinkmann U, Penger A, Asante-Poku S, Zanger UM, Schwab M. Lancet **2001**; 358:383–384.

50 Min DI, Lee M, Ku YM, Flanigan M. Clin Pharmacol Ther **2000**; 68:478–486.

51 Mancinelli LM, Frassetto L, Floren LC, Dressler D, Carrier S, Bekersky I, Benet LZ, Christians U. Clin Pharmacol Ther **2001**; 69:24–31.

52 Ito S, Ieiri I, Tanabe M, Suzuki A, Higuchi S, Otsubo K. Pharmacogenetics **2001**; 11:175–184.

53 Hermiston ML, Gordon JI. Science **1995**; 270:1203–1207.

54 Dianda L, Hanby AM, Wright NA, Sebesteny A, Hayday AC, Owen MJ. Am J Pathol **1997**; 150:91–97.

55 Panwala CM, Jones JC, Viney JL. J Immunol **1998**; 161:5733–5744.

56 Polli JW, Jarrett JL, Studenberg SD, Humphreys JE, Dennis SW, Brouwer KR, Woolley JL. Pharm Res **1999**; 16:1206–1212.

57 Choo EF, Leake B, Wandel C, Imamura H, Wood AJ, Wilkinson GR, Kim RB. Drug Metab Dispos **2000**; 28:655–660.

58 Huisman MT, Smit JW, Schinkel AH. AIDS **2000**; 14:237–242.

59 Lee CG, Ramachandra M, Jeang KT, Martin MA, Pastan I, Gottesman MM. FASEB J **2000**; 14:516–522.

60 Longuemaux S, Delomenie C, Gallou C, Mejean A, Vincent-Viry M, Bouvier R, Droz D, Krishnamoorthy R, Galteau MM, Junien C, Beroud C, Dupret JM. Cancer Res **1999**; 59:2903–2908.

61 Müller M. Semin Liver Dis **2000**; 20:323–337.

62 Hu Z, Jin S, Scotto KW. J Biol Chem **2000**; 275:2979–2985.

63 Thottassery JV, Zambetti GP, Arimori K, Schuetz EG, Schuetz JD. Proc Natl Acad Sci USA **1997**; 94:11037–11042.

64 Thevenod F, Friedmann JM, Katsen AD, Hauser IA. J Biol Chem **2000**; 275:1887–1896.

65 Bertilsson G, Heidrich J, Svensson K, Asman M, Jendeberg L, Sydow-Backman M, Ohlsson R, Postlind H, Blomquist P, Berkenstam A. Proc Natl Acad Sci USA **1998**; 95:12208–12213.

66 Blumberg B, Evans RM. Genes Dev **1998**; 12:3149–3155.

67 Lehmann JM, McKee DD, Watson MA, Willson TM, Moore JT, Kliewer SA. J Clin Invest **1998**; 102:1016–1023.

68 Geick A, Eichelbaum M, Burk O. J Biol Chem **2001**; 276:14581–14587.

69 Moore LB, Goodwin B, Jones SA, Wisely GB, Serabjit-Singh CJ, Willson TM, Collins JL, Kliewer SA. Proc Natl Acad Sci USA **2000**; 97:7500–7502.

70 Dürr D, Stieger B, Kullak-Ublick GA, Rentsch KM, Steinert HC, Meier PJ, Fattinger K. Clin Pharmacol Ther **2000**; 68:598–604.

71 Fugh-Berman A. Lancet **2000**; 355:134–138.

72 Piscitelli SC, Burstein AH, Chaitt D, Alfaro RM, Falloon J. Lancet **2000**; 355:547–548.

73 Jette L, Murphy GF, Leclerc JM, Beliveau R. Biochem Pharmacol **1995**; 50:1701–1709.

74 Wils P, Phung-Ba V, Warnery A, Lechardeur D, Raeissi S, Hidalgo IJ, Scherman D. Biochem Pharmacol **1994**; 48:1528–1530.

75 Bart J, Groen HJ, Hendrikse NH, van der Graaf WT, Vaalburg W, de Vries EG. Cancer Treat Rev **2000**; 26:449–462.

76 Sugiyama Y, Kato Y, Chu X. Cancer Chemother Pharmacol **1998**; 42(Suppl): S44–S49

77 Relling MV. Ther Drug Monit **1996**; 18:350–356.

78 Sparreboom A, van Asperen J, Mayer U, Schinkel AH, Smit JW, Meijer DK, Borst P, Nooijen WJ, Beijnen JH, van Tellingen O. Proc Natl Acad Sci USA **1997**; 94:2031–2035.

79 Hoki Y, Fujimori A, Pommier Y. Cancer Chemother Pharmacol **1997**; 40:433–438.

80 Pauli-Magnus C, Mürdter T, Godel A, Mettang T, Eichelbaum M, Klotz U, Fromm MF. Naunyn Schmiedebergs Arch Pharmacol **2001**; 363:337–343.

81 Saeki T, Ueda K, Tanigawara Y, Hori R, Komano T. J Biol Chem **1993**; 268:6077–6080.

82 Schuetz EG, Yasuda K, Arimori K, Schuetz JD. Arch Biochem Biophys **1998**; 350:340–347.

83 Ito T, Yano I, Tanaka K, Inui KI. J Pharmacol Exp Ther **1997**; 282:955–960.

84 de Lange E, Marchand S, van den Berg D, van der Sandt IC, de Boer AG, Delon A, Bouquet S, Couet W. Eur J Pharm Sci **2000**; 12:85–93.

85 Wu X, Whitfield LR, Stewart BH. Pharm Res **2000**; 17:209–215.

86 Kim RB, Wandel C, Leake B, Cvetkovic M, Fromm MF, Dempsey PJ, Roden MM, Belas F, Chaudhary AK, Roden DM, Wood AJJ, Wilkinson GR. Pharm Res **1999**; 16:408–414.

87 Saeki T, Ueda K, Tanigawara Y, Hori R, Komano T. FEBS Lett **1993**; 324:99–102.

88 Wandel C, Kim RB, Guengerich FP, Wood AJ. Drug Metab Dispos **2000**; 28:895–898.

89 Pauli-Magnus C, von Richter O, Burk O, Ziegler A, Mettang T, Eichelbaum

M, Fromm MF. J Pharmacol Exp Ther **2000**; 293:376–382.

90 Matsuzaki J, Yamamoto C, Miyama T, Takanaga H, Matsuo H, Ishizuka H, Kawahara Y, Kuwano M, Naito M, Tsuruo T, Sawada Y. Biopharm Drug Dispos **1999**; 20:85–90.

91 Karlsson J, Kuo SM, Ziemniak J, Artursson P. Br J Pharmacol **1993**; 110:1009–1016.

92 Wetterich U, Spahn-Langguth H, Mutschler E, Terhaag B, Rosch W, Langguth P. Pharm Res **1996**; 13:514–522.

93 Collett A, Higgs NB, Sims E, Rowland M, Warhurst G. J Pharmacol Exp Ther **1999**; 288:171–178.

94 Soldner A, Benet LZ, Mutschler E, Christians U. Br J Pharmacol **2000**; 129:1235–1243.

95 Schuetz EG, Schinkel AH, Relling MV, Schuetz JD. Proc Natl Acad Sci USA **1996**; 93:4001–4005.

96 Chen CJ, Clark D, Ueda K, Pastan I, Gottesman MM, Roninson IB. J Biol Chem **1990**; 265:506–514.

97 Kerb R, Hoffmeyer S, Brinkmann U. Pharmacogenomics **2001**; 2:51–64.

9
Pharmacogenomics of Drug Transporters

Rommel G. Tirona and Richard B. Kim

Abstract

It is now widely accepted that the drug disposition process represents an interplay between the expressed function and activity of drug metabolizing enzymes and membrane bound transporters. Recent advances relating to the cloning and expression of individual transporters have revealed the presence of a wide variety of transporters capable of drug uptake as well as efflux transport. Moreover, these transporters appear to be importantly involved in the absorption, distribution and elimination of a large number of drugs in clinical use. Accordingly, genetic polymorphisms in drug transporters may be one of the key determinants of interindividual and interethnic variability in drug disposition. In this chapter, we summarize pertinent molecular, biochemical and physiological aspects of transporters associated with the cellular uptake and efflux of drugs. In addition, we present a compilation of currently known genetic polymorphisms in drug transporters. The impact of these polymorphisms to drug disposition and disease is also discussed.

9.1
Introduction

Membrane transporters represent an important class of proteins responsible for regulating cellular and physiological solute and fluid balance. With the sequencing of the human genome, it has been estimated that approximately 500–1,200 genes code for transport proteins [1, 2]. From a drug disposition perspective, only a small fraction of these transporters are currently known to significantly interact with drugs. In particular, transporters that are localized at key gateway tissues within the body such as the intestine [3, 4], liver [5, 6], kidney [7, 8], placenta [9], and brain [10, 11] are critical modulators of drug absorption, tissue distribution and elimination and have been the subject of recent reviews.

It has now been established that genetic polymorphisms in drug metabolizing enzymes such as the cytochrome P450s (CYP) and the phase II enzyme, thiopurine methyltransferase, are responsible for inter-individual variability in response and adverse reactions [12, 13]. However, at the present time, the impact of polymorphisms in drug transporters on variability in drug disposition has not been clearly demonstrated, although this area of research is still in its infancy. There has only been progress in the identification and characterization of a number of proteins expected to be importantly associated with drug transport. However, determination of the *in vivo* relevance of particular transporters in pharmacokinetics continues to be elusive probably because of the presence of multiple transporters within a given tissue with overlapping substrate specificities.

In this chapter, we review the current state of knowledge in the area of pharmacogenomics of drug transporters. Focus is directed towards transporters that are known to or are implicated in regulating drugs disposition (Table 9.1). For more information on a comprehensive list of transporters, readers are directed to visit the Internet web page *www.med.rug.nl/mdl/tab3.htm/*. Included in the present compilation are the members of the organic anion transporting polypeptides (OATPs), organic anion transporters (OATs), organic cation transporters (OCTs) and peptide transporters (PepTs) thought to be importantly involved in the cellular uptake of endogenous compounds and drugs. In addition, transporters which mediate the cellular efflux of drugs, such as multidrug resistance proteins (MDRs) and multidrug resistance-related proteins (MRPs), will also be discussed. The molecular, biochemical and physiological aspects of each transporter will be presented as well as details on the identity and functional relevance of genetic polymorphisms and mutations. A current (July 2001) and comprehensive list of drug transporter polymorphisms is detailed in Tables 9.2 and 9.3 based on published data, GenBank cDNA sequences and available information from the public single nucleotide polymorphism database, dbSNP (available at *www.ncbi.nlm.nih.gov/SNP/*).

9.2
Organic Anion Transporting Polypeptide Family (OATP)

9.2.1
OATP-A

Rat oatp1, the first member of this family, cloned by Jacquemin and colleagues [14] from liver, was shown to have the properties of a sodium-independent organic anion transporter. These initial discoveries quickly lead to the isolation of the first human OATP, OATP-A (previously named OATP), from human liver [15]. Although OATP-A has been reported to be expressed in various tissues including liver, brain, lung, kidney and testes by Northern blot analysis [15] and in the liver by immunoblotting [16], others have found a restricted distribution only in brain [17]. This result is supported by data demonstrating widespread OATP-A mRNA in the brain [18] and immunodetectable protein in brain capillary endothelial cells [19], thus indicating a role for this transporter in regulating blood-brain permeability of solutes [19]. Recently, OATP-A mRNA was detected in biliary epithelial cells [20] although confirmation of protein expression and function has not been

Tab. 9.1 Summary of drug transporters

Family	*Gene*	*Gene symbol*	*Tissues* [1)]	*Polarity* [2)]	*Chromosome*	*Ref. seq./ GenBank*
OAT	OAT1	SLC22A6	K, B	BL	11q13.1-q13.2	NM_004790
	OAT2	SLC22A7	L	BL	6p21.2-p21.1	NM_006672
	OAT3	SLC22A8	K, B	BL	11q11	AB042595
	OAT4	–	K, Pl	?	11	AB026116
OCT	OCT1	SLC22A1	K, L, B, I	BL	6q26	NM_003057
	OCT2	SLC22A2	K, B, I	BL	6q26	NM_003058
	OCT3	SLC22A3	Pl, K, B, I	BL?	6q27	NM_021977
	OCTN1	SLC22A4	K	?	5	NM_003059
	OCTN2	SLC22A5	K, L, B, I	AP	5q31	NM_003060
OATP	OATP-A	SLC21A1	B	BL	12	U21943
	OATP-B	SLC21A9	K, L, B, I	BL	11q13	NM_007256
	OATP-C	SLC21A6	L	BL	12p	NM_006446
	OATP-D	SLC21A11	ubiquitous	?	15q26	NM_013272
	OATP-E	SLC21A12	ubiquitous	BL?	20	NM_016354
	OATP-F	SLC21A14	B	?	12p12.3-p14.3	NM_017435
	OATP8	SLC21A8	L	BL	12p12	NM_019844
PepT	PeptT1	SLC15A1	I, K	AP	13q33-q34	NM_005073
	PeptT2	SLC15A2	K	AP	13p13-q26.1	NM_021082
MDR	MDR1	ABCB1	K, L, B, I	AP	7q21.1	NM_000927
	MDR3	ABCB4	L	AP	7q21.1	NM_000443
	SPGP	ABCB11	L	AP	2q24	NM_003742
MRP	MRP1	ABCC1	ubiquitous	BL	16p13.1	NM_004996
	MRP2	ABCC2	K, L, B, I	AP	10q24	NM_000392
	MRP3	ABCC3	L, A, Pl, K, I	BL	17q22	NM_003786
	MRP4	ABCC4	ubiquitous	BL	13q32	NM_005845
	MRP5	ABCC5	ubiquitous	BL	3q27	NM_005688
	MRP6	ABCC6	L, K	BL, AP	16p13.1	NM_001171
	MRP7	ABCC10	K, B	?	6p12-21	–
BCRP	BCRP	ABCG2	L, I, Pl	?	4q22	NM_004827

1) L, liver; K, kidney; B, brain; I, small intestine; Pl, placenta; A, adrenal gland; Lu, lung.
2) BL, basolateral; AP, apical.

determined. In addition, the combination of amino acid similarities as well as syntenic chromosomal localizations has led to the suggestion that OATP-A may be orthologous to rat Oatp and, therefore, expressed at the apical membrane of intestinal epithelia [21]. Whether this is the case for OATP-A requires further investigation. What is clearer, however, is that OATP-A is truly multispecific in that it is capable of transporting diverse compounds including bromosulfophthalein (BSP) [15], bile acids [15], steroid sulfates [18, 22], bulky organic cations [22–24], fexofenadine [25], thyroid hormones [26] and opioid peptides [19]. As with all human OATPs, the driving force for solute transport by OATP-A has not been eluci-

Tab. 9.2 Summary of genetic polymorphisms in drug uptake transporters

Gene	*Polymorphism*	*Position*	*Effect*	*Function* [1)]	*References*
OAT1	G40A	Exon	G14S	?	[45]
	G51A	Exon	Synonymous	↔	[42]
	G252T	Exon	Synonymous	↔	[43]
	C480G	Exon	Synonymous	↔	[44]
	C852A	Exon	Synonymous	↔	
OAT2	?	?	?	?	[50]
OAT3	?	?	?	?	[53]
OAT4	?	?	?	?	[54]
OCT1	*T156C*	Exon	Synonymous	↔	[56]
	G480C	Exon	*L160F*	?	[57]
	A12226	Exon	*M408V*	?	[198]
OCT2	*G390T*	Exon	Synonymous	↔	[57]
OCT3	*C360T*	Exon	Synonymous	↔	[65, 199]
OCTN1	*C555T*	Exon	Synonymous	↔	[69]
	T917C	Exon	*I306T*	?	
	C1182G	Exon	*T394T*	?	
	C1183G	Exon	*L395V*	?	
	C1507T	Exon	*L503F*	?	
OCTN2	113 bp del	Exon	Loss of 2TMD	ø	[72]
	5 ins C	Exon	Frameshift	ø	[71]
	C285T	Exon	Synonymous	↔	[73]
	G396A	Exon	W132X	ø	[74]
	GIVS8 -1A	Intron	Splice	ø	[76]
	C505T	Exon	R169W	ø	[78]
	G506A	Exon	R169Q	ø	[77]
	A632G	Exon	Y211C	ø	[80]
	G725T	Exon	G242V	ø	[79]
	G807A	Exon	Synonymous	↔	[79]
	C844T	Exon	R282X	ø	[200]
	T847C	Exon	W283R	ø	
	C902A	Exon	A301D	ø	
	G978A	Exon	Synonymous	↔	
	T1051C	Exon	W351R	ø	
	1203 ins A	Exon	Y401X	ø	
	G1336T	Exon	V446F	ø	
	1345 del G	Exon	458X	ø	
	G1354A	Exon	E452K	ø	
	C1433T	Exon	P478L	ø	
OATP-A	?	?	?	?	[15]
OATP-B	A644T	Exon	D215V	?	[30]
	C663T	Exon	Synonymous	↔	[26]
	T1175C	Exon	*I392T*	?	
	C1457T	Exon	S486F	?	

Tab. 9.2 (continued)

Gene	Polymorphism	Position	Effect	Function 1)	References
OATP-C	T217C	Exon	F73L	↓	[17]
	T245C	Exon	V82A	↓	[31]
	A388G	Exon	N130D	↔	[32]
	G411A	Exon	Synonymous	↔	[30]
	G455A	Exon	R152K	↔	[36]
	C463A	Exon	P155T	↔	
	A467G	Exon	E156G	↓	
	T521C	Exon	V174A	↓	
	C571T	Exon	Synonymous	↔	
	G721A	Exon	D241N	↔	
	T1058C	Exon	I1353T	↓	
	A1294G	Exon	N432D	↔	
	A1385G	Exon	D462G	↔	
	G1463C	Exon	G488A	↓	
	A1964G	Exon	D655G	↓	
	A2000G	Exon	E667G	↓	
OATP-D	*C222G*	Exon	*D74E*	?	[30]
	TA604-605AT	Exon	Y202I	?	
OATP-E	*G232A*	Exon	*V78I*	?	[30, 37]
OATP-F	C1927T	Exon	Synonymous	↔	[39]
OATP8	*A1557G*	Exon	Synonymous	↔	[34,38
PeptT1	*T1347C*	Exon	Synonymous	↔	[85]
PeptT2	*C141T*	Exon	Synonymous	↔	[95]
	G171A	Exon	*R57H*	?	
	G1162A	Exon	Synonymous	↔	
	T1225C	Exon	*S409P*	?	
	A1527G	Exon	*K509R*	?	
	A2110C	Exon	*M704L*	?	

1) ↔, does not alter transport activity; ↑, increased transport activity; ↓, decreased transport activity; ø, null function; ?, unknown effect on transport function.
Polymorphisms in *italics* were obtained from dbSNP (*www.ncbi.nlm.nih.gov/SNP/*) on July 2001.

dated. Studies involving rat OATPs suggest possible bicarbonate or glutathione exchange mechanisms [27–29]. Little is known about the existence of genetic polymorphisms in OATP-A and their possible functional consequences.

9.2.2 OATP-B

Tamai et al. [30] cloned OATP-B from human brain and transporter mRNA has been detected in a number of other tissues including liver, lung, kidney, placenta, brain, heart and small intestine [26, 30]. Within the liver, OATP-B protein is local-

Tab. 9.3 Summary of genetic polymorphisms in drug efflux transporters

Gene	*Polymorphism*	*Position*	*Effect*	*Function* [1)]	*References*
BCRP	G71T	Exon	A24V	?	[188]
	C496G	Exon	Q166E	?	[189]
	T623C	Exon	F208S	?	
	G1086A	Exon	Synonymous	↔	
	A1444G	Exon	T482G	?	
	C1445G	Exon	T482R	?	
MDR1	A61G	Exon	N21D	?	[201]
	T307C	Exon	F103L	?	[202]
	A548G	Exon	N183S	?	[203]
		Exon	G185V	↑	[117]
	G1199A	Exon	S400N	?	[204]
	C1236T	Exon	Synonymous	↔	[205]
	C1474T	Exon	R4892C	?	[206]
	C2650T	Exon	Synonymous	↔	[207]
	G2677T	Exon	A893S	↑	[208]
	G2677A	Exon	A893T	?	[209]
	A2956G	Exon	M986V	?	[210]
	G2995A	Exon	A999T	?	
	A3320C	Exon	Q1107P	?	
	C3396T	Exon	Synonymous	?	
	T3421A	Exon	S1141T	?	
	C3435T	Exon	Synonymous	↓	
MDR3	A79G	Exon	Splice	ø	[165]
	287-1005 del	Exon	Frameshift	ø	[172]
	394-400 del	Exon	Frameshift	ø	[177]
	T412C	Exon	W138R	↓	[173]
	A523G	Exon	T175V	↓	[176]
	C959T	Exon	S320F	↓	[175]
	G1037T	Exon	S346I	↓	[174]
	A1184G	Exon	E395G	↓	
	A1270G	Exon	T424A	↓	
	G1275A	Exon	V425M	↓	
	1327 ins A	Exon	Frameshift	ø	
	A1621T	Exon	I541F	ø	
	C1637A	Exon	A546D	↓	
	T1667T	Exon	L556R	↓	
	A1691G	Exon	D564G	↓	
	1712 del T	Exon	Frameshift	ø	
	C1906T	Exon	Q636X	ø	
	A1954G	Exon	R652G	?	
	T2132C	Exon	F711S	↓	
	C2869T	Exon	C957X	ø	
	2943-52 del	Exon	Frameshift	↓	
	G2947A	Exon	G983S	↓	
	C3481T	Exon	P1161S	↓	

Tab. 9.3 (continued)

Gene	*Polymorphism*	*Position*	*Effect*	*Function* [1)]	*References*
MRP1	Gene Deletion	Exon	No Protein	ø	[101]
	G128C	Exon	C43S	?	[117]
	C218T	Exon	T73I	?	[116]
	C350T	Exon	T117M	?	[115]
	T825C	Exon	Synonymous	↔	
	T1062C	Exon	Synonymous	↔	
	T1684C	Exon	Synonymous	?	
	G1898A	Exon	R633Q	↔	
	C2001T	Exon	Synonymous	↔	
	C2007T	Exon	Synonymous	?	
	G2012T	Exon	G671V	?	
	G2168A	Exon	R723Q	↔	
	C2665T	Exon	Synonymous	↔	
	T2694C	Exon	Synonymous	?	
	G3173A	Exon	R1058Q	↔	
	G4002A	Exon	Synonymous	↔	
	C4524T	Exon	Synonymous	?	
	C4535T	Exon	S1512L		
MRP2	G1249A	Exon	V417I	?	[124]
	T1815+2A	Intron	Splice	ø	[211]
	C1967+2T	Intron	Splice	ø	[212]
	C2302T	Exon	R768W	ø	[132]
	C2366T	Exon	S789F	ø	[213]
	T2439+2C	Intron	Splice	ø	[133]
	C3196T	Exon	R1066X	ø	
	C3972T	Exon	Synonymous	↔	
	A4145G	Exon	Q1382R	ø	
	Del 4170-5	Exon	Del R,M	ø	
	G4348A	Exon	A1450T	?	
MRP3	T124C	Exon	C42R	?	[141]
	G258A	Exon	Synonymous	↔	[139]
	C1031G	Exon	A344G	?	[140]
	C1580G	*Exon*	*T527R*	?	[138]
	C1583G	*Exon*	*A528G*	?	[137]
	C1633T	Exon	Synonymous	↔	[142]
	T1706A	Exon	F569Y	?	
	C3932T	Exon	Synonymous	↔	
	C4048G	Exon	L1362V	?	
	A4509G	Exon	Synonymous	↔	
MRP4	*T669C*	Exon	Synonymous	↔	[214]
	C1497T	Exon	Synonymous	↔	

Tab. 9.3 (continued)

Gene	*Polymorphism*	*Position*	*Effect*	*Function* [1]	*References*
MRP5	GC527-8CG	Exon	R176P	?	[155]
	A723G	Exon	Synonymous	↔	[153]
	G1146A	Exon	Synonymous	↔	[154]
	C1185T	Exon	Synonymous	↔	[142]
	AGC1198-200GGT	Exon	S400G	?	
	C1200T	Exon	Synonymous	↔	
	A1741G	Exon	I581V	?	
	T1782C	Exon	Synonymous	↔	
	T3624C	Exon	Synonymous	↔	
	C4148A	Exon	T1383N	?	
MRP6	Gene Deletion	Exon	No protein	ø	[160]
	G189C	Exon	Synonymous	↔	[157]
	T190C	Exon	W64R	?	[158]
	G549A	Exon	Synonymous	↔	[161]
	C681G	Exon	Y227X	ø	[115]
	960 del C	Exon	Frameshift	ø	[111]
	T1233C	Exon	Synonymous	↔	[163]
	G1245A	Exon	Synonymous	↔	[164]
	C1552T	Exon	R518X	ø	[162]
	T1841C	Exon	V614A	↔	[112]
	C1890G	Exon	Synonymous	↔	
	Del 1967-89	Exon	Frameshift	ø	
	C2490T	Exon	Synonymous	↔	
	GIVS21+1T	Intron	Splice	ø	
	G3341C	Exon	R1114P	ø	
	C3412T	Exon	R1138W	ø	
	G3413A	Exon	R1138Q	ø	
	C3421T	Exon	R1141X	ø	
	C3490T	Exon	R1164X	ø	
	G3736-1A	Splice	Exon 27	ø	
	3775 del T	Exon	Frameshift	ø	
	3798 del T	Exon	Frameshift	ø	
	G3803A	Exon	R1268Q	↔	
	C3940T	Exon	R1314W	ø	
	4243 ins AGAA	Exon	Frameshift	ø	
MRP7	?	?	?	?	[215]

Tab. 9.3 (continued)

Gene	*Polymorphism*	*Position*	*Effect*	*Function* [1)]	*References*
SPGP		Exon	S114R	ø	[182]
	695 del 1bp	Exon	Frameshift	ø	[183]
	G713T	Exon	G238V	ø	
	A890G	Exon	E297G	ø	
	908 del G	Exon	303X	ø	
		Exon	C336S	ø	
	A1381G	Exon	K461E	ø	
	A1445G	Exon	D482G	ø	
	C1723T	Exon	R575X	ø	
		Exon	S593R	ø	
	G2944A	Exon	G982R	ø	
	C3169T	Exon	R1057X	ø	
	3213 del 1bp	Exon	Frameshift	ø	
	C3268T	Exon	R1090X	ø	
	C3457T	Exon	R1153C	ø	
	3767-8 in C	Exon	Frameshift	ø	
	G3803A	Exon	R1268Q	ø	

1) ↔, does not alter transport activity; ↑, increased transport activity; ↓, decreased transport activity; ø, null function; ?, unknown effect on transport function.
Polymorphisms in *italics* were obtained from dbSNP (*www.ncbi.nlm.nih.gov/SNP/*) on July 2001.

ized to the basolateral membrane of hepatocytes [26], indicating a blood-to-liver uptake role for this transporter. Kullak-Ublick and colleagues [26] revealed that the substrate specificity of OATP-B appears quite limited, with only BSP and steroid sulfates being transported from a large group of compounds tested. Given that OATP-B is one of several OATPs (OATP-C and OATP8) expressed on the basolateral membrane of the hepatocyte and due to its restricted substrate specificity, the importance of this transporter in the hepatic elimination of drugs is currently unclear. However, its broad tissue expression suggests OATP-B could regulate drug distribution. Initial reports detailed the presence of a SNP in OATP-B at a single codon (486) that alters an amino acid (Ser/Phe) [26, 30] (Table 9.2). Furthermore, inspection of OATP-B cDNA sequences deposited in GenBank combined with data from the public SNP database, further add to the list of non-synonymous polymorphisms (Table 9.2). Functional characterization of these OATP-B variants will be required to assess their possible implications on drug disposition.

9.2.3 OATP-C

OATP-C, a liver-specific member of this transporter family, was cloned by number of groups [17, 30–32] and often referred by aliases such as liver-specific transporter 1 (LST-1) and OATP2. Immunohistochemical analysis proved OATP-C expression on the basolateral membrane of hepatocytes [32]. As with OATP-A, OATP-C

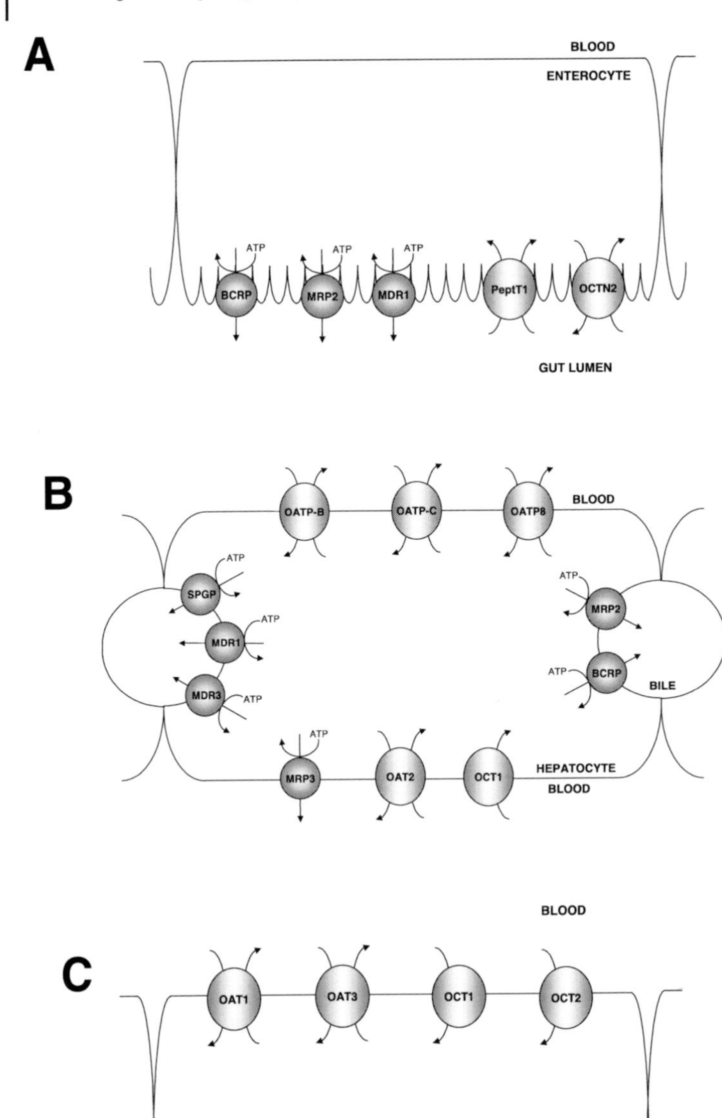

Fig. 9.1 Schematic diagram depicting drug transporters and their subcellular localization in the human small intestinal enterocyte (**A**), hepatocyte (**B**), and renal tubular cell (**C**).

transports a broad range of compounds such as bile acids [17], sulfate and glucuronide conjugates [32], thyroid hormones [17], peptides [26] and drugs such as pravastatin [31, 33] and methotrexate [34, 36]. Importantly, bilirubin and its glucuronides were identified as physiological substrates for OATP-C [35]. Given its liver-specific tissue distribution pattern and the capacity for transporting a multiplicity of chemical structures, it is likely that OATP-C plays an important role in the hepatocellular elimination of drugs. On this basis, the *OATP-C* gene was screened in a population of African- and European-Americans and a series of functional polymorphisms were characterized [36]. Non-synonymous polymorphisms affecting highly conserved amino acids within the putative transmembrane regions and extracellular loop 5 were associated with severely compromised transport function *in vitro*. Cell surface trafficking defects proved to be responsible for altered transport function for many of these OATP-C variants. A few functional polymorphisms were found at a relatively high frequency in African- and European-Americans raising the possibility that these OATP-C variants may be responsible for inter-individual variation in drug disposition. However, the *in vivo* relevance of OATP-C polymorphisms remains to be determined.

9.2.4 OATP-E

Current information indicates that OATP-E [30] is ubiquitously expressed in tissues [30, 37]. Some substrates transported by OATP-E include estrone-sulfate [30], prostaglandin E_2 [30] and taurocholate [37]. The capacity for T_3 and T_4 transport and the wide tissue distribution suggests that OATP-E is largely responsible for the peripheral uptake of thyroid hormone [37]. Further studies are required to assess whether OATP-E is an important determinant of drug distribution.

9.2.5 OATP8

OATP8 is similar to OATP-C at both the amino acid level (80% amino acid identity) and liver-specific tissue distribution [38]. In addition, OATP8 shares transport substrates with OATP-C such as BSP, steroid sulfates and glucuronides, thyroid hormone, bile acids, peptide compounds and methotrexate [26, 34, 38], albeit with some differences in affinity. In contrast to OATP-C, unconjugated bilirubin is not transported by OATP8 [35]. Interestingly, OATP8 is the only member of the human OATP family known to transport digoxin [26]. Recently, this transporter has been found to be expressed in certain gastric, colon and pancreatic cancers. It has been suggested that OATP8 expression may alter tumor sensitivity to methotrexate treatment [34]. Currently, the contribution of OATP8 in comparison to other hepatic OATPs in the overall hepatocellular uptake of drugs is unclear.

9.2.6
Other OATPs

There is evidence for the presence of other human OATPs including OATP-D [30], OATP-F (NM_017435] [39] and OATPRP4 (NM_030958). Further studies are required to determine the impact of these transporters on the drug distribution and elimination.

9.3
Organic Anion Transporter Family (OAT)

9.3.1
OAT1

The classical basolateral organic anion transporter in kidney was first cloned from the rat (rOAT1) [40, 41], leading to the subsequent identification of a human ortholog by several groups [42, 43–45]. OAT1, expressed mainly in the proximal tubule, mediates the uptake of the prototypical organic anion, *p*-aminohippurate (PAH) [42–44] from blood through exchange with intracellular dicarboxylates [43, 44]. The rat ortholog of OAT1 has been shown to directly transport or interact with at least 100 known drugs/compounds including salicylate, methotrexate, indomethacin, penicillin and captopril (see review by Sekine et al. [46]). This is in contrast to human OAT1, which appears to have a much narrower substrate specificity [43] and to date, has only been reported to transport PAH and the antiviral nucleotide analogs, cidofovir and adefovir [47]. Therefore, the contribution of OAT1 compared to other OATs in the renal elimination of drugs remains to be clarified. Little information is available with regard to the identification and evaluation of polymorphisms in OAT1. Nevertheless, an assessment of four submitted OAT1 cDNA sequences suggests the presence of several synonymous and one non-synonymous SNPs (Table 9.2).

9.3.2
OAT2

Initially identified and presumed a liver-specific transport protein named NLT [48], rat OAT2 was functionally characterized by Sekine and colleagues [49]. These investigators found rOAT2 transported several anions such as PAH, salicylate and PGE_2. In contrast with earlier findings with rOAT1 [49], rOAT2 did not appear to be energized by dicarboxylate exchange. Immunohistochemical analysis revealed a basolateral expression of rOAT2 [48] suggesting a role in the blood to hepatocyte uptake of organic anions. Although, the human ortholog, OAT2/hNLT, has been identified [50] and the chromosomal location determined [51], little is known about its function.

9.3.3 OAT3

The first OAT3 transporter identified from human kidney, designated hOAT3* [44], showed low homology with the rat ortholog, rOAT3 [52] and was incapable of transporting any of the anionic or cationic substances tested. However, rOAT3 transports PAH as well as the cationic drug cimetidine [52]. Recently, another OAT3 cDNA has been reported [53] bearing 85% similarity with hOAT3* with expression on the basolateral membrane of the proximal tubule. Functional studies demonstrated transport of several compounds including PAH, methotrexate, cimetidine and estrone sulfate, indicating that OAT3 has a broader substrate specificity than OAT1 [53]. These findings indicate that OAT3 may be a key regulator of renal anionic drug clearance.

9.3.4 OAT4

Within the OAT family, OAT4 is the only transporter expressed at appreciable levels in both the placenta and in the kidney [54]. The membrane localization of OAT4 within these tissues has not been examined. Steroid sulfates, and ochratoxin A are efficient transport substrates of OAT4, whereas PAH is weakly transported [54]. The functional importance of OAT4 in regulating placental permeability and renal drug elimination is currently unknown.

9.4 Organic Cation Transporter Family (OCT)

9.4.1 OCT1

Following the initial cloning of an organic cation transporter OCT1 from rat kidney [55], a human OCT1 ortholog was identified [56, 57]. OCT1 appears to be expressed predominantly in liver [56, 57] which differs from the rat where it is found also in the kidney, intestine and colon [55]. Although not demonstrated, it is likely that OCT1 is localized to the basolateral membrane of human hepatocytes in a position similar to the rat ortholog [58]. This would suggest that OCT1 may be responsible for the initial uptake of organic cations into liver. Membrane potential has been shown to drive the OCT1-mediated uptake of small protonated molecules such as 1-methyl 1,4-phenylpyridinium (MPP^+) [56], tetraethylammonium (TEA) [57] and N-1-methylnicotinamide (NMN) [57] as well as larger, bulkier (type II) cations including N-methyl-quinine, N-methyl-quinidine and quinidine (at pH 6) [24]. Rat OCT1 transports the catecholamines dopamine, 5-HT, noradrenaline, adrenaline and tyramine [59] but whether hOCT behaves similarly has not been reported. Recently, Jonker and colleagues demonstrated that mice with targeted disruption of OCT1 had much reduced liver levels of TEA and MPP^+ following intravenous administration [60] indicating that mOCT1 was important for the basolateral hepatic uptake of these substrates *in vivo*. Since TEA was efficiently excreted into urine in the knock-out mice, other renal organic cation transporters such as OCT2 apparently compensate for the loss of mOCT1 [60]. Since a role for OCT1 in governing the pharmacokinetics of drugs *in vivo* is becoming better understood, characterizing genetic polymorphisms in this transporter would be of interest. At the time of writing, the public SNP database describes several polymorphisms including two non-synonymous variations (Table 9.2) within its records.

9.4.2 **OCT2**

Okuda et al. [61] isolated rat OCT2 from kidney which was subsequently followed by the identification of human OCT2 [57]. OCT2 mRNA has been detected in kidney and brain, and immunohistochemical methods indicated that OCT2 is localized to the apical membrane of the distal tubule [57]. This is in striking disagreement with available information regarding rat OCT2, which is clearly localized at the basolateral membrane of the proximal tubule [62, 63]. Whether or not species differences exist in the localization of OCT2 in kidney needs further clarification. For now, the importance of OCT2 in the renal elimination of organic cations and drugs in humans is questionable. In the brain, OCT2 expression has been found in neurons by *in situ* hybridization and immunohistochemical methods [64]. This finding is of particular pharmacological importance since OCT2 is known to mediate the sodium-independent transport of monoamine neurotransmitters and the anti-Parkinsonian drug amantadine [64].

9.4.3 **OCT3**

OCT3, cloned from a human kidney carcinoma cell line, was termed the extraneuronal catecholamine transporter (EMT) as a result of demonstration that the protein mediated the transport of the neurotransmitters adrenaline, noradrenaline and the neurotoxin 1-methyl-4-phenylpyridinium (MPP^+) [65]. Among the tissues examined, the highest expression of OCT3 was seen in placenta [66–68], intestine and heart, with lower levels observed in kidney and brain. Given that mouse OCT3 mRNA is detected in the renal proximal tubule [67], it is possible that human OCT3 is expressed similarly and may contribute to the renal secretion of organic cations.

9.5 Novel Organic Cation Transporter Family (OCTN)

9.5.1 OCTN1

OCTN1 was initially cloned from human fetal liver and Northern analysis showed transporter expression in a variety of tissues including kidney and skeletal muscle but not in liver [69]. The exact subcellular localization of OCTN1 in kidney has not been described and, therefore, it is not known whether the transporter acts in an influx or efflux mode. Studies have shown OCTN1-mediated uptake of cationic compounds including TEA, quinidine, pyrilamine and verapamil [70]. Only modest transport function was detected for the zwitterionic compound carnitine [70]. Membrane potential-independent and pH-dependent transport properties of OCTN1 indicate a proton/antiport mechanism [70]. Taken together, the tissue distribution and substrate specificity suggest a potential role for OCTN1 in regulating the distribution and excretion of drugs. SNPs in OCTN1 are catalogued in the public database (Table 9.2) and several include non-synonymous amino acid changes. The functional consequences of these polymorphisms are currently unknown.

9.5.2 OCTN2

Shortly after the identification of OCTN1, another member of this subfamily, OCTN2, was reported [71, 72]. OCTN2 was characterized as the sodium-dependent, high-affinity carnitine transporter having tissue distribution in kidney, muscle, heart, placenta and pancreas [71]. This important finding quickly led to the discovery that mutations in OCTN2 cause primary systemic carnitine deficiency [73–75]. To date, over 15 mutations have been identified in OCTN2 which are associated with this disease (Table 9.2) [76–80]. Furthermore, experiments assessing the function of several mutant OCTN2 variants have unequivocally demonstrated a loss of carnitine transport activity. In addition to carnitine, other substrates for OCTN2 include TEA [72], quinidine [81], verapamil [81] and cephalosporins containing quaternary nitrogen groups [82]. In contrast to the zwitterion, carnitine, OCTN2-mediated transport of cationic substrates does not appear to require sodium. Expression of rat OCTN2 to the apical membrane of renal proximal tubule cells [83] supports the notion that this transporter may be essential for the reabsorption of filtered carnitine and plays a role in the lumenal efflux of cationic drugs. The latter is supported by the finding that mice lacking functional OCTN2 have greater plasma level and decreased renal clearance of TEA after intravenous administration [84]. Therefore, known and unidentified mutations and polymorphisms in OCTN2 are expected to affect drug disposition in humans.

9.6 Peptide Transporter Family (PepT)

9.6.1 PepT1

The human PepT1 transporter was cloned from intestine [85] after initial identification and characterization of the rabbit ortholog [86]. PepT1 is localized to the brush border of intestinal epithelial cells [87, 88] and the S1 segment of apical proximal tubules [89], and facilitates substrate absorption from the gut as well as renal tubular solute reabsorption. Proton co-transport [86] drives the PepT1-mediated cellular uptake of endogenous substrates/nutrients including di- and tripeptides [86], and δ-aminolevulinic acid [90]. Drugs such as β-lactam antibiotics [86, 91], the antiviral drugs valaciclovir [92] and valganciclovir [93] and the angiotensin-converting enzyme inhibitor captopril [94] are known to be transported by PepT1. Mutations or functional polymorphisms in PepT1 have not yet been reported but if present, they would be expected to have pathophysiological and pharmacological consequences.

9.6.2 PepT2

PepT2, first identified by Liu et al. [95], was found to be highly expressed in kidney but not the intestine. PepT2 is localized to the apical proximal tubule membranes of the S3 segment [89] and participates in the renal reabsorption of filtered substrates. Recent evidence expands the tissue expression of PepT2 to the choroid plexus [96, 97], the peripheral nervous system [98] and lung epithelia [99]. Substrates for PepT2 are generally similar to PepT1 although differences in affinity exist [100]. Upon screening the public SNP database, several non-synonymous polymorphisms were encountered (Table 9.2).

9.7 Multidrug Resistance-Associated Proteins (MRP)

9.7.1 MRP1

Cole and colleagues identified MRP1 from a lung cancer cell line selected with doxorubicin [101]. Expression of MRP1 has been associated with cellular resistance to chemotherapeutic agents including vinca alkaloids, taxol, methotrexate and etoposide and may play a role in cancer chemotherapy [102]. Compounds such as *p*-aminohippurate and glucuronide, sulfate and glutathione conjugates have also been shown to be transported by MRP1. The ubiquitous distribution of MRP1 on the basolateral membrane of various normal epithelial tissues [103] suggests that MRP1 may alter tissue levels of administered drugs. Particularly important is the role of MRP1 in maintaining the blood-cerebrospinal fluid (CSF) barrier as result of its strategic localization in choroid plexus epithelium [104, 105]. However, low or absent expression in brain capillary endothelial cells suggests MRP1 does not contribute to the maintenance of the blood-brain barrier [106–108]. Expression of MRP1 in Sertoli cells has been shown to protect the testes against the damaging and antifertility effects of an anticancer drug [109]. A role in regulating intestinal drug absorption has not been established but murine MRP1 appears to be expressed in crypt cells but not enterocytes in the small intestine and in colonic epithelial cells [110]. As for the kidney, MRP1 is expressed only in the distal tubules and collecting ducts [110].

Polymorphisms in MRP1 have been described in studies linking a connective tissue disease, pseudoxanthoma elasticum, with mutations in MRP6. Two groups have described a large chromosomal deletion encompassing the entire MRP1 and MRP6 genes on chromosome 16 [111, 112]. Aside from the manifestations of pseudoxanthoma elasticum, those affected by this MRP1/6 deletion are otherwise phenotypically normal [112], consistent with findings from MRP1 knockout mice [113, 114]. SNPs in MRP1 have also been described in studies examining the genetic basis for pseudoxanthoma elasticum [111, 112, 115, 116] as well as another study that screened for MRP1 polymorphisms in a population of Japanese [117]. Further studies are required to examine the impact of MRP1 polymorphisms on the disposition of drugs.

9.7.2 MRP2

First cloned from rat as the canalicular multispecific organic anion transporter (cMoat) [118–120], MRP2 was characterized as a major apical efflux pump of organic anions in liver. By demonstration of its absence in the jaundiced TR^- and Eisai rat strains [118–120] as well as in humans affected by a specific form of hyperbilirubinemia known as Dubin-Johnson syndrome [121], it was determined that bilirubin glucuronide conjugates were important substrates of MRP2 [122]. In addition to glucuronide and glutathione conjugates, non-conjugated compounds such as *p*-aminohippurate [123], vinblastine [124], and telmisartan [125] are also transported by MRP2, indicating broad substrate specificity. In addition to its expression in liver, MRP2 is localized to the apical domain of the proximal tubule [126]. In the small intestine, MRP2 is expressed on the apical membrane of enterocytes [127] where highest expression is found in the duodenum [128]. Studies in rat indicate that intestinal MRP2 is involved in the secretion of organic anions into the gut lumen [129] and its absence in MRP2-deficient rats is associated with the increased bioavailability of the carcinogen PhIP [130]. Finally, the drug permeation barrier in the placenta is partially attributable to the expression of MRP2 on the apical membrane of the syncytiotrophoblast [55].

Known genetic lesions in MRP2 causing Dubin-Johnson syndrome are varied and range from point mutations to base pair deletions leading to missense mutations, premature stop codons and aberrant RNA splicing (Table 9.3). These mutations are associated with the complete absence of immunochemically detectable MRP2 in affected individuals [121, 131, 132]. For one mutation involving a two amino acid change located between the Walker A motif and the ATP binding cassette signature, the lack of canalicular expression of MRP2 was attributable to impaired protein maturation and sorting [133]. Although the altered disposition of many compounds has been well characterized in TR^- [134] and Eisai [135] rats, little is known regarding drug disposition among patients with Dubin-Johnson syndrome aside from the characteristic changes in serum BSP pharmacokinetics and the biliary excretion of the BSP-GSH conjugate [136]. Whether intestinal, renal and placental membrane permeation by drugs is affected by MRP2 mutations also remains to be determined. Moreover, the functional consequences of MRP2 polymorphisms not associated with Dubin-Johnson syndrome [117] in terms of drug disposition, need to be examined.

9.7.3 MRP3

Cloned by several groups [137–142], MRP3 is expressed mainly in liver, colon, small intestine, adrenal glands and to a lesser extent in kidney, pancreas and prostate. In liver, MRP3 is localized to the basolateral membrane of hepatocytes and cholangiocytes [140, 141]. Studies *in vitro* have demonstrated MRP3 can mediate the transport of glucuronide and GSH conjugates [141, 143], MTX [143], bile acid conjugates [143, 144] as well as conferring chemotherapeutic resistance to the cytotoxic effects of etoposide [141, 145] and vincristine [145]. Upregulation of MRP3 is thought to protect hepatocytes during MRP2 deficiency [140, 146] and during obstructive cholestasis [147, 148] or chemical challenge (e.g., phenobarbital) [147]. Recently St. Pierre and colleagues [55] have shown MRP3 expression in fetal blood endothelia and in syncytiotrophoblasts of term placentas, suggesting a role for this transporter in the export of fetal bile salts.

The combination of GenBank cDNA sequence comparison and data from the public SNP database reveal the presence of several SNPs in MRP3, many of which involve non-synonymous amino acid changes (Table 9.3). These polymorphisms have not been functionally characterized and, therefore, the clinical pharmacological impact of such MRP3 variants remains unknown.

9.7.4 MRP4

MRP4 is an ubiquitously expressed transporter with highest expression in the prostate where it can be localized to the basolateral membrane of tubuloacinar cells [149]. Schuetz and colleagues [150] demonstrated that high copy number and overexpression of MRP4 in a human T-lymphoid cell line was associated with cytotoxic resistance and increased cell efflux of an acyclic nucleoside phosphonate drug PMEA. Furthermore, these cells were more resistant to nucleoside analog drugs such as AZT, 3TC and d4T in comparison with non-MRP4 overexpressing cells. Other recent studies have shown MRP4-mediated resistance to methotrexate [149], and the purine analogs 6-mercaptopurine and 6-thioguanine [151]. Similar to previous findings with MRP5 (see below), MRP4 was found to transport the cyclic nucleotides cAMP and cGMP [151]. Due to its unique substrate specificity and structural divergence [152], the physiologic role of MRP4 may differ from other members of this transporter family. Whether genetic polymorphisms in MRP4 impact on drug disposition or pharmacodynamics *in vivo* remains to be determined.

9.7.5 MRP5

Also named MOAT-C and pABC11, MRP5 appears to be expressed in every tissue examined [142, 153, 154]. Heterologous expression of MRP5 in polarized cells revealed sorting of the transporter to the basolateral membrane [155]. Functional studies *in vitro* demonstrate that MRP5 directly transports a variety of fluorochromes [153], and the GSH conjugate of chlorodintrobenzene [155]. Similar to MRP4, MRP5 overexpression confers drug resistance towards PMEA, 6-mercaptopurine and thioguanine [155] as well as to cadmium chloride and antimonyl tartrate [153]. Jedlitschky et al. [156] showed that MRP5 was capable of transporting the natural cyclic nucleotides cGMP and cAMP. Furthermore, cGMP transport could be blocked with the phosphodiesterase inhibitor sildenafil, suggesting a dual mode of action for this class of drugs. With its ubiquitous distribution in the body, it is possible that MRP5 plays a role in the tissue disposition and elimination of drugs. In this regard, the MRP5-deficient mouse model [155] would be a valuable tool to test this hypothesis. In comparing available MRP5 cDNA sequences in GenBank, variations were found suggesting the existence of SNPs (Table 9.3).

9.7.6 MRP6

MRP6 (also known as MOAT-E, ARA and MLP-1) is expressed predominantly in the liver and kidney [157, 158] and using immunochemical analysis, Madon et al. [159] have demonstrated a basolateral and canalicular localization of rat MRP6 in liver. So far, only a limited substrate specificity for rMRP6 has been defined, including the peptidomimetic compound BQ-123 but not other typical MRP substrates (e.g., LTC_4, E-17β-G, GSSG, DNP-SG, TLCAS, TCA) [159]. Anthracycline resistance has been associated with overexpression of MRP6 in tumor cells [160] but this occurred through co-amplification of MRP1 [157]. Recently, mutations in MRP6 have been found to cause pseudoxanthoma elasticum, a condition involving elastic fiber calcification in skin, retina and arteries [111, 115, 116]. A variety of missense, splice, insertion, and deletion mutations are associated with this condition (Table 9.3). Other polymorphisms detected during genetic screens were not causative for pseudoxanthoma elasticum [112, 115, 162–164] and their functional relevance remains to be determined. Interestingly, a number of investigators [111, 115] have demonstrated MRP6 mRNA in fibroblasts, retinal and vascular epithelia, while others [161] were not able to detect MRP6 mRNA or protein in fibroblasts. Thus determining the endogenous substrate for this predominantly hepatic and renally expressed transporter will be key to understanding its role in the pathogenesis of pseudoxanthoma elasticum.

9.8 Multidrug Resistance (P-Glycoprotein) Family (MDR)

9.8.1 MDR1

The MDR1 gene, coding the apical drug efflux pump P-gp, has been established as an important regulator of cancer chemotherapeutic drug resistance as well as drug bioavailability and tissue distribution. Not surprisingly, there has been much interest in the study of functional polymorphisms in MDR1. Readers are directed to refer to Chapter 8 in this volume regarding a comprehensive review on this topic. For reasons of completeness we provide a partial list of MDR1 polymorphisms in Table 9.3.

9.8.2 MDR3

MDR3 [165] is expressed almost exclusively in liver on the bile canalicular membrane [166]. Mice lacking the MDR3 ortholog (mdr2) develop liver disease as a result of an inability to secrete phospholipids and cholesterol into bile [167], a severe phenotype that can be reversed by introduction of the MDR3 [168, 169]. MDR3 specifically translocates phosphatidylcholine from the inner to the outer plasma membrane leaflet [170, 171]. Mutations in MDR3 cause a spectrum of liver diseases including progressive familial intrahepatic cholestasis type 3 (PFIC3) [172–174], intrahepatic cholestasis of pregnancy [175, 176] and a form of cholesterol gallstone disease [177]. The genetic lesions in MDR3 (Table 9.3) responsible for these phenotypes include nucleotide deletions, insertions, and missense mutations. Often these mutations are associated with the lack of detectable or low-level MDR3 protein expression in liver [172, 173]. At least one variant (R652G) is considered a polymorphism since it occurs at a frequency similar to unaffected individuals. In terms of drug disposition, Smith and colleagues [178] have demonstrated that several MDR1 substrates such as digoxin, paclitaxel and vinblastine were transported by MDR3, albeit at a low rate. Furthermore, Huang and Vore [179] have shown that the mouse ortholog of MDR3 (MDR2) is essential for the biliary excretion of indocyanine green even though it is not a substrate of MDR2. Therefore, polymorphisms in MDR3 could influence the rate hepatic drug elimination and deserves further investigation.

9.8.3
SPGP (BSEP)

Sister of P-glycoprotein (Spgp) was originally cloned from pig liver as a closely related member of MDR1/P-gp [180]. Later, rat Spgp was isolated and was shown to be localized on the canalicular membrane of hepatocytes [181] and determined to be the bile salt export pump of the liver (bsep) [181]. Shortly thereafter, the human *SPGP* gene was isolated and mutations in the transporter were found to cause progressive familial intrahepatic cholestasis type 2 (PFIC2) [182]. A series of mutations including frameshift, missense and premature termination codons were found in PFIC2 patients (Table 9.3) [182, 183]. Many of those patients harboring these genetic lesions displayed an absence of immunochemically detectable SPGP in liver [183]. The severe SPGP null phenotype found in humans is different from that in spgp knockout mice which develop non-progressive but persistent intrahepatic cholestasis [184]. This is attributed to alternate metabolic and transport pathways for bile acids in mice that are different to humans. Other than bile acids, little is known about drug substrates for SPGP, although some data indicates that vinblastine [185] and the non-steroidal anti-inflammatory drug sulindac [186] may be substrates of the transporter. Clearly, a cautious approach is required for drug selection and dosing in patients with SPGP null alleles due to the severity of liver disease. However, polymorphisms in SPGP that decrease but maintain a level of transport activity may also exist. Given the recent data demonstrating the importance of bile acid transport and homeostasis to the activation of drug metabolizing enzymes [187], polymorphisms in SPGP may significantly alter the pharmacokinetics of many drugs.

9.9
"White" ABC Transporter Family

9.9.1
BCRP

Breast Cancer Resistance Protein (BCRP, also known as MXR or ABCP), first cloned from mitoxantrone and anthracycline-resistant breast and colon cancer cells [188, 189] is a half-transporter efflux pump believed to function as a homo- or hetero-dimer. Following its identification, BCRP-mediated drug resistance was observed for topoisomerase inhibitors including camptothecins [190, 191] and indolocarbazoles [192]. In normal tissues, BCRP was detected in placental syncytiotrophoblasts, hepatocyte canalicular membrane, apical intestinal epithelia and vascular endothelial cells [193]. These findings support the important role BCRP plays in modulating topotecan bioavailability, fetal exposure and hepatic elimination [194]. Considering that the substrates and tissue distributions for BCRP overlap somewhat with MDR1 and MRPs [195], additional studies will be required to define the relative contribution of each of these transporters in the overall and tissue-specific distribution of drugs *in vivo*. As with other transporters, functional polymorphisms in BCRP could explain inter-individual variability in the disposition of certain drugs. Indeed, 34 SNPs in the BCRP gene have been recently identified during large-scale sequencing of 313 genes in individuals of various ethnicity [196].

9.10
Conclusions

It was originally estimated that SNPs in the human genome occur at a rate of 1 in every 1,200 to 1,900 base pairs [1, 197]. However, a recent report in which of 313 genes in 82 individuals of diverse ethnic origins were screened, indicates a much higher rate of polymorphism that averages 1 SNP in every 185 nucleotides [196]. The rate of polymorphism within gene coding regions was estimated at 3.4 per kilobase. Therefore, it is likely that a large number of coding region SNPs exists among drug transporter genes. Many of those SNPs are expected to cause amino acid changes that directly alter protein function. Furthermore, the rate of SNPs in non-coding regions such as promoter, introns, splice junctions and 3′ untranslated regions is greater than within exons [196] suggesting that genetic polymorphisms that invoke alterations in transcriptional activation and splicing may also prove to be relevant in determining drug transporter function.

As SNP discovery technologies become mainstream research tools, the importance of genotype-phenotype relationships will continue to be a focus in pharmacogenomic research. Not only will characterization of drug transporter polymorphisms enhance our insight of the molecular mechanisms involved in transporter function, it is likely that such findings will become important components of individualized drug therapy in the future.

Acknowledgements
Supported by USPHS grants GM54724 and GM31304.

9.11
References

1 Venter JC, Adams MD, Myers EW, Li PW, Mural RJ, Sutton GG et al. The sequence of the human genome. Science **2001**; 291(5507):1304–1351.

2 Lander ES, Linton LM, Birren B, Nusbaum C, Zody MC, Baldwin J et al. Initial sequencing and analysis of the human genome. Nature **2001**; 409(6822):860–921.

3 Tsuji A, Tamai I. Carrier-mediated intestinal transport of drugs. Pharm Res **1996**; 13(7):963–977.

4 Suzuki H, Sugiyama Y. Role of metabolic enzymes and efflux transporters in the absorption of drugs from the small intestine. Eur J Pharm Sci **2000**; 12(1):3–12.

5 Suzuki H, Sugiyama Y. Transport of drugs across the hepatic sinusoidal membrane: sinusoidal drug influx and efflux in the liver. Semin Liver Dis **2000**; 20(3):251–263.

6 Keppler D, König J. Hepatic secretion of conjugated drugs and endogenous substances. Semin Liver Dis **2000**; 20(3):265–272.

7 Inui KI, Masuda S, Saito H. Cellular and molecular aspects of drug transport in the kidney. Kidney Int **2000**; 58(3):944–958.

8 Dresser MJ, Leabman MK, Giacomini KM. Transporters involved in the elimination of drugs in the kidney: organic anion transporters and organic cation transporters. J Pharm Sci **2001**; 90(4): 397–421.

9 Ganapathy V, Prasad PD, Ganapathy ME, Leibach FH. Placental transporters relevant to drug distribution across the maternal-fetal interface. J Pharmacol Exp Ther **2000**; 294(2):413–420.

10 Gao B, Meier PJ. Organic anion transport across the choroid plexus. Microsc Res Tech **2001**; 52(1):60–64.

11 Tamai I, Tsuji A. Transporter-mediated permeation of drugs across the blood-brain barrier. J Pharm Sci **2000**; 89(11):1371–1388.

12 Evans WE, Relling MV. Pharmacogenomics: translating functional genomics into rational therapeutics. Science **1999**; 286(5439):487–491.

13 Meyer UA. Pharmacogenetics and adverse drug reactions. Lancet **2000**; 356(9242):1667–1671.

14 Jacquemin E, Hagenbuch B, Stieger B, Wolkoff AW, Meier PJ. Expression cloning of a rat liver Na(+)-independent organic anion transporter. Proc Natl Acad Sci USA **1994**; 91(1):133–137.

15 Kullak-Ublick GA, Hagenbuch B, Stieger B, Schteingart CD, Hofmann AF, Wolkoff AW et al. Molecular and functional characterization of an organic anion transporting polypeptide cloned from human liver. Gastroenterology **1995**; 109(4):1274–1282.

16 Kullak-Ublick GA, Glasa J, Boker C, Oswald M, Grutzner U, Hagenbuch B et al. Chlorambucil-taurocholate is transported by bile acid carriers expressed in human hepatocellular carcinomas. Gastroenterology **1997**; 113(4):1295–1305.

17 Abe T, Kakyo M, Tokui T, Nakagomi R, Nishio T, Nakai D et al. Identification of a novel gene family encoding human liver-specific organic anion transporter LST-1. J Biol Chem **1999**; 274(24):17159–17163.

18 Kullak-Ublick GA, Fisch T, Oswald M, Hagenbuch B, Meier PJ, Beuers U et al. Dehydroepiandrosterone sulfate (DHEAS): identification of a carrier protein in human liver and brain. FEBS Lett **1998**; 424(3):173–176.

19 Gao B, Hagenbuch B, Kullak-Ublick GA, Benke D, Aguzzi A, Meier PJ. Organic anion-transporting polypeptides mediate transport of opioid peptides across blood-brain barrier. J Pharmacol Exp Ther **2000**; 294(1):73–79.

20 Chignard N, Mergey M, Veissiere D, Parc R, Capeau J, Poupon R et al. Bile acid transport and regulating functions in the human biliary epithelium. Hepatology **2001**; 33(3):496–503.

21 Walters HC, Craddock AL, Fusegawa H, Willingham MC, Dawson PA. Expression, transport properties, and chromosomal location of organic anion trans-

porter subtype 3. Am J Physiol Gastrointest Liver Physiol **2000**; 279(6):G1188–G1200.

22 Bossuyt X, Muller M, Meier PJ. Multispecific amphipathic substrate transport by an organic anion transporter of human liver. J Hepatol **1996**; 25(5):733–738.

23 van Montfoort JE, Hagenbuch B, Fattinger KE, Muller M, Groothuis GM, Meijer DK et al. Polyspecific organic anion transporting polypeptides mediate hepatic uptake of amphipathic type II organic cations. J Pharmacol Exp Ther **1999**; 291(1):147–152.

24 van Montfoort JE, Muller M, Groothuis GM, Meijer DK, Koepsell H, Meier PJ. Comparison of "type I" and "type II" organic cation transport by organic cation transporters and organic anion-transporting polypeptides. J Pharmacol Exp Ther **2001**; 298(1):110–115.

25 Cvetkovic M, Leake B, Fromm MF, Wilkinson GR, Kim RB. OATP and P-glycoprotein transporters mediate the cellular uptake and excretion of fexofenadine. Drug Metab Dispos **1999**; 27(8):866–871.

26 Kullak-Ublick GA, Ismair MG, Stieger B, Landmann L, Huber R, Pizzagalli F et al. Organic anion-transporting polypeptide B (OATP-B) and its functional comparison with three other OATPs of human liver. Gastroenterology **2001**; 120(2):525–533.

27 Satlin LM, Amin V, Wolkoff AW. Organic anion transporting polypeptide mediates organic anion/HCO_3-exchange. J Biol Chem **1997**; 272(42):26340–26345.

28 Li L, Lee TK, Meier PJ, Ballatori N. Identification of glutathione as a driving force and leukotriene C4 as a substrate for oatp1, the hepatic sinusoidal organic solute transporter. J Biol Chem **1998**; 273(26):16184–16191.

29 Li L, Meier PJ, Ballatori N. Oatp2 mediates bidirectional organic solute transport: a role for intracellular glutathione. Mol Pharmacol **2000**; 58(2):335–340.

30 Tamai I, Nezu J, Uchino H, Sai Y, Oku A, Shimane M et al. Molecular identification and characterization of novel members of the human organic anion transporter (OATP) family. Biochem Biophys Res Commun **2000**; 273(1):251–260.

31 Hsiang B, Zhu Y, Wang Z, Wu Y, Sasseville V, Yang WP et al. A novel human hepatic organic anion transporting polypeptide (OATP2). Identification of a liver-specific human organic anion transporting polypeptide and identification of rat and human hydroxymethylglutaryl-CoA reductase inhibitor transporters. J Biol Chem **1999**; 274(52):37161–37168.

32 König J, Cui Y, Nies AT, Keppler D. A novel human organic anion transporting polypeptide localized to the basolateral hepatocyte membrane. Am J Physiol Gastrointest Liver Physiol **2000**; 278(1):G156–G164.

33 Nakai D, Nakagomi R, Furuta Y, Tokui T, Abe T, Ikeda T et al. Human liver-specific organic anion transporter, LST-1, mediates uptake of pravastatin by human hepatocytes. J Pharmacol Exp Ther **2001**; 297(3):861–867.

34 Abe T, Unno M, Onogawa T, Tokui T, Kondo TN, Nakagomi R et al. Lst-2, a human liver-specific organic anion transporter, determines methotrexate sensitivity in gastrointestinal cancers. Gastroenterology **2001**; 120(7):1689–1699.

35 Cui Y, König J, Leier I, Buchholz U, Keppler D. Hepatic uptake of bilirubin and its conjugates by the human organic anion transporter SLC21A6. J Biol Chem **2001**; 276(13):9626–9630.

36 Tirona RG, Leake BF, Merino G, Kim RB. Polymorphisms in *OATP-C*: Identification of multiple allelic variants associated with altered transport activity among European- and African-Americans. J Biol Chem **2001**; 276(38):35669–35675.

37 Fujiwara K, Adachi H, Nishio T, Unno M, Tokui T, Okabe M et al. Identification of thyroid hormone transporters in humans: different molecules are involved in a tissue-specific manner. Endocrinology **2001**; 142(5):2005–2012.

38 König J, Cui Y, Nies AT, Keppler D. Localization and genomic organization of a new hepatocellular organic anion transporting polypeptide. J Biol Chem **2000**; 275(30):23161–23168.

39 Pizzagalli F, Hagenbuch B, Bottomley KM, Meier PJ. Identification of a new human organic anion transporting polypeptide OATP-F. GenBank Accession No AF260704 **2001**.

40 Sekine T, Watanabe N, Hosoyamada M, Kanai Y, Endou H. Expression cloning and characterization of a novel multispecific organic anion transporter. J Biol Chem **1997**; 272(30):18526–18529.

41 Sweet DH, Wolff NA, Pritchard JB. Expression cloning and characterization of ROAT1. The basolateral organic anion transporter in rat kidney. J Biol Chem **1997**; 272(48):30088–30095.

42 Hosoyamada M, Sekine T, Kanai Y, Endou H. Molecular cloning and functional expression of a multispecific organic anion transporter from human kidney. Am J Physiol **1999**; 276(1 Pt 2):F122–F128.

43 Lu R, Chan BS, Schuster VL. Cloning of the human kidney PAH transporter: narrow substrate specificity and regulation by protein kinase C. Am J Physiol **1999**; 276(2 Pt 2):F295–F303.

44 Race JE, Grassl SM, Williams WJ, Holtzman EJ. Molecular cloning and characterization of two novel human renal organic anion transporters (hOAT1 and hOAT3). Biochem Biophys Res Commun **1999**; 255(2):508–514.

45 Bahn A, Prawitt D, Buttler D, Reid G, Enklaar T, Wolff NA et al. Genomic structure and *in vivo* expression of the human organic anion transporter 1 (hOAT1) gene. Biochem Biophys Res Commun **2000**; 275(2):623–630.

46 Sekine T, Cha SH, Endou H. The multispecific organic anion transporter (OAT) family. Pflügers Arch **2000**; 440(3):337–350.

47 Cihlar T, Lin DC, Pritchard JB, Fuller MD, Mendel DB, Sweet DH. The antiviral nucleotide analogs cidofovir and adefovir are novel substrates for human and rat renal organic anion transporter 1. Mol Pharmacol **1999**; 56(3):570–580.

48 Simonson GD, Vincent AC, Roberg KJ, Huang Y, Iwanij V. Molecular cloning and characterization of a novel liver-specific transport protein. J Cell Sci **1994**; 107:1065–1072.

49 Sekine T, Cha SH, Tsuda M, Apiwattanakul N, Nakajima N, Kanai Y et al. Identification of multispecific organic anion transporter 2 expressed predominantly in the liver. FEBS Lett **1998**; 429(2):179–182.

50 Kim RB, Leake BF, Cvetkovic M. Molecular cloning of a human liver-specific transporter, NLT. GenBank Accession No AF097518 **1998**.

51 Kok LD, Siu SS, Fung KP, Tsui SK, Lee CY, Waye MM. Assignment of liver-specific organic anion transporter (SLC22A7) to human chromosome 6 bands p21.2 → p21.1 using radiation hybrids. Cytogenet Cell Genet **2000**; 88(1–2):76–77.

52 Kusuhara H, Sekine T, Utsunomiya-Tate N, Tsuda M, Kojima R, Cha SH et al. Molecular cloning and characterization of a new multispecific organic anion transporter from rat brain. J Biol Chem **1999**; 274(19):13675–13680.

53 Cha SH, Sekine T, Fukushima JI, Kanai Y, Kobayashi Y, Goya T et al. Identification and characterization of human organic anion transporter 3 expressing predominantly in the kidney. Mol Pharmacol **2001**; 59(5):1277–1286.

54 Cha SH, Sekine T, Kusuhara H, Yu E, Kim JY, Kim DK et al. Molecular cloning and characterization of multispecific organic anion transporter 4 expressed in the placenta. J Biol Chem **2000**; 275(6):4507–4512.

55 St. Pierre MV, Serrano MA, Macias RI, Dubs U, Hoechli M, Lauper U et al. Expression of members of the multidrug resistance protein family in human term placenta. Am J Physiol Regul Integr Comp Physiol **2000**; 279(4):R1495–R1503.

56 Zhang L, Dresser MJ, Gray AT, Yost SC, Terashita S, Giacomini KM. Cloning and functional expression of a human liver organic cation transporter. Mol Pharmacol **1997**; 51(6):913–921.

57 Gorboulev V, Ulzheimer JC, Akhoundova A, Ulzheimer-Teuber I, Karbach U, Quester S et al. Cloning and characterization of two human polyspecific organic cation transporters. DNA Cell Biol **1997**; 16(7):871–881.

58 Meyer-Wentrup F, Karbach U, Gorboulev V, Arndt P, Koepsell H. Membrane localization of the electrogenic cation transporter rOCT1 in rat liver. Biochem Biophys Res Commun **1998**; 248(3):673–678.
59 Breidert T, Spitzenberger F, Grundemann D, Schomig E. Catecholamine transport by the organic cation transporter type 1 (OCT1). Br J Pharmacol **1998**; 125(1):218–224.
60 Jonker JW, Wagenaar E, Mol CA, Buitelaar M, Koepsell H, Smit JW et al. Reduced hepatic uptake and intestinal excretion of organic cations in mice with a targeted disruption of the organic cation transporter 1 (oct1 [Slc22a1]) gene. Mol Cell Biol **2001**; 21(16):5471–5477.
61 Okuda M, Saito H, Urakami Y, Takano M, Inui K. cDNA cloning and functional expression of a novel rat kidney organic cation transporter, OCT2. Biochem Biophys Res Commun **1996**; 224(2):500–507.
62 Sweet DH, Miller DS, Pritchard JB. Basolateral localization of organic cation transporter 2 in intact renal proximal tubules. Am J Physiol Renal Physiol **2000**; 279(5):F826–F834.
63 Karbach U, Kricke J, Meyer-Wentrup F, Gorboulev V, Volk C, Loffing-Cueni D et al. Localization of organic cation transporters OCT1 and OCT2 in rat kidney. Am J Physiol Renal Physiol **2000**; 279(4):F679–F687.
64 Busch AE, Karbach U, Miska D, Gorboulev V, Akhoundova A, Volk C et al. Human neurons express the polyspecific cation transporter hOCT2, which translocates monoamine neurotransmitters, amantadine, and memantine. Mol Pharmacol **1998**; 54(2):342–352.
65 Grundemann D, Schechinger B, Rappold GA, Schomig E. Molecular identification of the corticosterone-sensitive extraneuronal catecholamine transporter. Nature Neurosci **1998**; 1(5):349–351.
66 Verhaagh S, Schweifer N, Barlow DP, Zwart R. Cloning of the mouse and human solute carrier 22a3 (Slc22a3/SLC22A3) identifies a conserved cluster of three organic cation transporters on mouse chromosome 17 and human 6q26–q27. Genomics **1999**; 55(2):209–218.
67 Wu X, Huang W, Ganapathy ME, Wang H, Kekuda R, Conway SJ et al. Structure, function, and regional distribution of the organic cation transporter OCT3 in the kidney. Am J Physiol Renal Physiol **2000**; 279(3):F449–F458.
68 Kekuda R, Prasad PD, Wu X, Wang H, Fei YJ, Leibach FH et al. Cloning and functional characterization of a potential-sensitive, polyspecific organic cation transporter (OCT3) most abundantly expressed in placenta. J Biol Chem **1998**; 273(26):15971–15979.
69 Tamai I, Yabuuchi H, Nezu J, Sai Y, Oku A, Shimane M et al. Cloning and characterization of a novel human pH-dependent organic cation transporter, OCTN1. FEBS Lett **1997**; 419(1):107–111.
70 Yabuuchi H, Tamai I, Nezu J, Sakamoto K, Oku A, Shimane M et al. Novel membrane transporter OCTN1 mediates multispecific, bidirectional, and pH-dependent transport of organic cations. J Pharmacol Exp Ther **1999**; 289(2):768–773.
71 Tamai I, Ohashi R, Nezu J, Yabuuchi H, Oku A, Shimane M et al. Molecular and functional identification of sodium ion-dependent, high affinity human carnitine transporter OCTN2. J Biol Chem **1998**; 273(32):20378–20382.
72 Wu X, Prasad PD, Leibach FH, Ganapathy V. cDNA sequence, transport function, and genomic organization of human OCTN2, a new member of the organic cation transporter family. Biochem Biophys Res Commun **1998**; 246(3):589–595.
73 Nezu J, Tamai I, Oku A, Ohashi R, Yabuuchi H, Hashimoto N et al. Primary systemic carnitine deficiency is caused by mutations in a gene encoding sodium ion-dependent carnitine transporter. Nature Genet **1999**; 21(1):91–94.
74 Wang Y, Ye J, Ganapathy V, Longo N. Mutations in the organic cation/carnitine transporter OCTN2 in primary carnitine deficiency. Proc Natl Acad Sci USA **1999**; 96(5):2356–2360.
75 Lamhonwah AM, Tein I. Carnitine uptake defect: frameshift mutations in the human plasmalemmal carnitine transporter gene. Biochem Biophys Res Commun **1998**; 252(2):396–401.

76 Mayatepek E, Nezu J, Tamai I, Oku A, Katsura M, Shimane M et al. Two novel missense mutations of the OCTN2 gene (W283R and V446F) in a patient with primary systemic carnitine deficiency. Hum Mutat **2000**; 15(1):118.

77 Wang Y, Taroni F, Garavaglia B, Longo N. Functional analysis of mutations in the OCTN2 transporter causing primary carnitine deficiency: lack of genotype-phenotype correlation. Hum Mutat **2000**; 16(5):401–407.

78 Wang Y, Kelly MA, Cowan TM, Longo N. A missense mutation in the OCTN2 gene associated with residual carnitine transport activity. Hum Mutat **2000**; 15(3):238–245.

79 Vaz FM, Scholte HR, Ruiter J, Hussaarts-Odijk LM, Pereira RR, Schweitzer S et al. Identification of two novel mutations in OCTN2 of three patients with systemic carnitine deficiency. Hum Genet **1999**; 105(1/2):157–161.

80 Burwinkel B, Kreuder J, Schweitzer S, Vorgerd M, Gempel K, Gerbitz KD et al. Carnitine transporter OCTN2 mutations in systemic primary carnitine deficiency: a novel Arg169Gln mutation and a recurrent Arg282ter mutation associated with an unconventional splicing abnormality. Biochem Biophys Res Commun **1999**; 261(2):484–487.

81 Ohashi R, Tamai I, Yabuuchi H, Nezu JI, Oku A, Sai Y et al. Na(+)-dependent carnitine transport by organic cation transporter (OCTN2): its pharmacological and toxicological relevance. J Pharmacol Exp Ther **1999**; 291(2):778–784.

82 Ganapathy ME, Huang W, Rajan DP, Carter AL, Sugawara M, Iseki K et al. β-lactam antibiotics as substrates for OCTN2, an organic cation/carnitine transporter. J Biol Chem **2000**; 275(3):1699–1707.

83 Tamai I, China K, Sai Y, Kobayashi D, Nezu J, Kawahara E et al. Na(+)-coupled transport of L-carnitine via high-affinity carnitine transporter OCTN2 and its subcellular localization in kidney. Biochim Biophys Acta **2001**; 1512(2):273–284.

84 Ohashi R, Tamai I, Nezu JI, Nikaido H, Hashimoto N, Oku A et al. Molecular and physiological evidence for multifunctionality of carnitine/organic cation transporter OCTN2. Mol Pharmacol **2001**; 59(2):358–366.

85 Liang R, Fei YJ, Prasad PD, Ramamoorthy S, Han H, Yang-Feng TL et al. Human intestinal H^+/peptide cotransporter. Cloning, functional expression, and chromosomal localization. J Biol Chem **1995**; 270(12):6456–6463.

86 Fei YJ, Kanai Y, Nussberger S, Ganapathy V, Leibach FH, Romero MF et al. Expression cloning of a mammalian proton-coupled oligopeptide transporter. Nature **1994**; 368(6471):563–566.

87 Ogihara H, Saito H, Shin BC, Terado T, Takenoshita S, Nagamachi Y et al. Immuno-localization of H^+/peptide cotransporter in rat digestive tract. Biochem Biophys Res Commun **1996**; 220(3):848–852.

88 Sai Y, Tamai I, Sumikawa H, Hayashi K, Nakanishi T, Amano O et al. Immunolocalization and pharmacological relevance of oligopeptide transporter PepT1 in intestinal absorption of beta-lactam antibiotics. FEBS Lett **1996**; 392(1):25–29.

89 Shen H, Smith DE, Yang T, Huang YG, Schnermann JB, Brosius FC, III. Localization of PEPT1 and PEPT2 proton-coupled oligopeptide transporter mRNA and protein in rat kidney. Am J Physiol **1999**; 276(5 Pt 2):F658–F665.

90 Doring F, Walter J, Will J, Focking M, Boll M, Amasheh S et al. Delta-aminolevulinic acid transport by intestinal and renal peptide transporters and its physiological and clinical implications. J Clin Invest **1998**; 101(12):2761–2767.

91 Ganapathy ME, Brandsch M, Prasad PD, Ganapathy V, Leibach FH. Differential recognition of beta-lactam antibiotics by intestinal and renal peptide transporters, PEPT 1 and PEPT 2. J Biol Chem **1995**; 270(43):25672–25677.

92 Ganapathy ME, Huang W, Wang H, Ganapathy V, Leibach FH. Valacyclovir: a substrate for the intestinal and renal peptide transporters PEPT1 and PEPT2. Biochem Biophys Res Commun **1998**; 246(2):470–475.

93 Sugawara M, Huang W, Fei YJ, Leibach FH, Ganapathy V, Ganapathy ME. Transport of valganciclovir, a ganci-

clovir prodrug, via peptide transporters PEPT1 and PEPT2. J Pharm Sci **2000**; 89(6):781–789.

94 Zhu T, Chen XZ, Steel A, Hediger MA, Smith DE. Differential recognition of ACE inhibitors in *Xenopus laevis* oocytes expressing rat PEPT1 and PEPT2. Pharm Res **2000**; 17(5):526–532.

95 Liu W, Liang R, Ramamoorthy S, Fei YJ, Ganapathy ME, Hediger MA et al. Molecular cloning of PEPT 2, a new member of the H^+/peptide cotransporter family, from human kidney. Biochim Biophys Acta **1995**; 1235(2):461–466.

96 Novotny A, Xiang J, Stummer W, Teuscher NS, Smith DE, Keep RF. Mechanisms of 5-aminolevulinic acid uptake at the choroid plexus. J Neurochem **2000**; 75(1):321–328.

97 Teuscher NS, Novotny A, Keep RF, Smith DE. Functional evidence for presence of PEPT2 in rat choroid plexus: studies with glycylsarcosine. J Pharmacol Exp Ther **2000**; 294(2):494–499.

98 Groneberg DA, Doring F, Nickolaus M, Daniel H, Fischer A. Expression of PEPT2 peptide transporter mRNA and protein in glial cells of rat dorsal root ganglia. Neurosci Lett **2001**; 304(3):181–184.

99 Groneberg DA, Nickolaus M, Springer J, Doring F, Daniel H, Fischer A. Localization of the peptide transporter PEPT2 in the lung: implications for pulmonary oligopeptide uptake. Am J Pathol **2001**; 158(2):707–714.

100 Inui K, Terada T, Masuda S, Saito H. Physiological and pharmacological implications of peptide transporters, PEPT1 and PEPT2. Nephrol Dial Transplant **2000**; 15 (Suppl 6):11–13.

101 Cole SPC, Bhardwaj G, Gerlach J, Mackie J, Grant C, Almquist K et al. Overexpression of a transporter gene in a multidrug-resistant human lung cancer cell line. Science **1992**; 258:1650–1654.

102 Allen JD, Brinkhuis RF, van Deemter L, Wijnholds J, Schinkel AH. Extensive contribution of the multidrug transporters P-glycoprotein and Mrp1 to basal drug resistance. Cancer Res **2000**; 60(20):5761–5766.

103 Flens MJ, Zaman GJ, van d, V, Izquierdo MA, Schroeijers AB, Scheffer GL et al. Tissue distribution of the multidrug resistance protein. Am J Pathol **1996**; 148(4):1237–1247.

104 Rao VV, Dahlheimer JL, Bardgett ME, Snyder AZ, Finch RA, Sartorelli AC et al. Choroid plexus epithelial expression of MDR1 P glycoprotein and multidrug resistance-associated protein contribute to the blood-cerebrospinalfluid drug-permeability barrier. Proc Natl Acad Sci USA **1999**; 96(7):3900–3905.

105 Wijnholds J, deLange EC, Scheffer GL, van den Berg DJ, Mol CA, van d, V et al. Multidrug resistance protein 1 protects the choroid plexus epithelium and contributes to the blood-cerebrospinal fluid barrier. J Clin Invest **2000**; 105(3):279–285.

106 Sun H, Johnson DR, Finch RA, Sartorelli AC, Miller DW, Elmquist WF. Transport of fluorescein in MDCKII-MRP1 transfected cells and mrp1-knockout mice. Biochem Biophys Res Commun **2001**; 284(4):863–869.

107 Seetharaman S, Barrand MA, Maskell L, Scheper RJ. Multidrug resistance-related transport proteins in isolated human brain microvessels and in cells cultured from these isolates. J Neurochem **1998**; 70(3):1151–1159.

108 Regina A, Koman A, Piciotti M, El Hafny B, Center MS, Bergmann R et al. Mrp1 multidrug resistance-associated protein and P-glycoprotein expression in rat brain microvessel endothelial cells. J Neurochem **1998**; 71(2):705–715.

109 Wijnholds J, Scheffer GL, van d, V, van d, V, Beijnen JH, Scheper RJ et al. Multidrug resistance protein 1 protects the oropharyngeal mucosal layer and the testicular tubules against drug-induced damage. J Exp Med **1998**; 188(5):797–808.

110 Peng KC, Cluzeaud F, Bens M, van Huyen JP, Wioland MA, Lacave R et al. Tissue and cell distribution of the multidrug resistance-associated protein (MRP) in mouse intestine and kidney. J Histochem Cytochem **1999**; 47(6):757–768.

111 Bergen AA, Plomp AS, Schuurman EJ, Terry S, Breuning M, Dauwerse H et

al. Mutations in ABCC6 cause pseudoxanthoma elasticum. Nature Genet **2000**; 25(2):228–231.

112 Meloni I, Rubegni P, De Aloe G, Bruttini M, Pianigiani E, Cusano R et al. Pseudoxanthoma elasticum: Point mutations in the ABCC6 gene and a large deletion including also ABCC1 and MYH11. Hum Mutat **2001**; 18(1):85.

113 Lorico A, Rappa G, Finch RA, Yang D, Flavell RA, Sartorelli AC. Disruption of the murine MRP (multidrug resistance protein) gene leads to increased sensitivity to etoposide (VP-16) and increased levels of glutathione. Cancer Res **1997**; 57(23):5238–5242.

114 Wijnholds J, Evers R, van Leusden MR, Mol CA, Zaman GJ, Mayer U et al. Increased sensitivity to anticancer drugs and decreased inflammatory response in mice lacking the multidrug resistance-associated protein. Nature Med **1997**; 3(11):1275–1279.

115 Le Saux O, Urban Z, Tschuch C, Csiszar K, Bacchelli B, Quaglino D et al. Mutations in a gene encoding an ABC transporter cause pseudoxanthoma elasticum. Nature Genet **2000**; 25(2):223–227.

116 Perdu J, Germain DP. Identification of novel polymorphisms in the pM5 and MRP1 (ABCC1) genes at locus 16p13.1 and exclusion of both genes as responsible for pseudoxanthoma elasticum. Hum Mutat **2001**; 17(1):74–75.

117 Ito S, Ieiri I, Tanabe M, Suzuki A, Higuchi S, Otsubo K. Polymorphism of the ABC transporter genes, MDR1, MRP1 and MRP2/cMOAT, in healthy Japanese subjects. Pharmacogenetics **2001**; 11(2):175–184.

118 Paulusma CC, Bosma PJ, Zaman GJR, Bakker CTM, Otter M, Scheffer GL et al. Congenital jaundice in rats with a mutation in a multidrug resistance-associated protein gene. Science **1996**; 271(5252):1126–1128.

119 Buchler M, König J, Brom M, Kartenbeck J, Spring H, Horie T et al. cDNA cloning of the hepatocyte canalicular isoform of the multidrug resistance protein, cMrp, reveals a novel conjugate export pump deficient in hyperbilirubinemic mutant rats. J Biol Chem **1996**; 271(25):15091–15098.

120 Ito K, Suzuki H, Hirohashi T, Kume K, Shimizu T, Sugiyama Y. Molecular cloning of canalicular multispecific organic anion transporter defective in EHBR. Am J Physiol **1997**; 272(1 Pt 1): G16–G22.

121 Kartenbeck J, Leuschner U, Mayer R, Keppler D. Absence of the canalicular isoform of the MRP gene-encoded conjugate export pump from the hepatocytes in Dubin-Johnson syndrome. Hepatology **1996**; 23(5):1061–1066.

122 Kamisako T, Leier I, Cui Y, König J, Buchholz U, Hummel-Eisenbeiss J et al. Transport of monoglucuronosyl and bisglucuronosyl bilirubin by recombinant human and rat multidrug resistance protein 2. Hepatology **1999**; 30(2):485–490.

123 Leier I, Hummel-Eisenbeiss J, Cui Y, Keppler D. ATP-dependent para-aminohippurate transport by apical multidrug resistance protein MRP2. Kidney Int **2000**; 57(4):1636–1642.

124 Evers R, Kool M, van Deemter L, Janssen H, Calafat J, Oomen LC et al. Drug export activity of the human canalicular multispecific organic anion transporter in polarized kidney MDCK cells expressing cMOAT (MRP2) cDNA. J Clin Invest **1998**; 101(7):1310–1319.

125 Nishino A, Kato Y, Igarashi T, Sugiyama Y. Both cMOAT/MRP2 and another unknown transporter(s) are responsible for the biliary excretion of glucuronide conjugate of the nonpeptide angiotensin II antagonist, telmisaltan. Drug Metab Dispos **2000**; 28(10):1146–1148.

126 Schaub TP, Kartenbeck J, König J, Spring H, Dorsam J, Staehler G et al. Expression of the MRP2 gene-encoded conjugate export pump in human kidney proximal tubules and in renal cell carcinoma. J Am Soc Nephrol **1999**; 10(6):1159–1169.

127 Fromm MF, Kauffmann HM, Fritz P, Burk O, Kroemer HK, Warzok RW et al. The effect of rifampin treatment on intestinal expression of human MRP transporters. Am J Pathol **2000**; 157(5):1575–1580.

128 Mottino AD, Hoffman T, Jennes L, Vore M. Expression and localization of multidrug resistant protein mrp2 in rat small intestine. J Pharmacol Exp Ther **2000**; 293(3):717–723.

129 Gotoh Y, Suzuki H, Kinoshita S, Hirohashi T, Kato Y, Sugiyama Y. Involvement of an organic anion transporter (canalicular multispecific organic anion transporter/multidrug resistance-associated protein 2) in gastrointestinal secretion of glutathione conjugates in rats. J Pharmacol Exp Ther **2000**; 292(1):433–439.

130 Dietrich CG, de Waart DR, Ottenhoff R, Schoots IG, Elferink RP. Increased bioavailability of the food-derived carcinogen 2-amino-1-methyl-6-phenylimidazo(4,5-b)-pyridine in MRP2-deficient rats. Mol Pharmacol **2001**; 59(5):974–980.

131 Paulusma CC, Kool M, Bosma PJ, Scheffer GL, ter Borg F, Scheper RJ et al. A mutation in the human canalicular multispecific organic anion transporter gene causes the Dubin-Johnson syndrome. Hepatology **1997**; 25(6):1539–1542.

132 Tsujii H, König J, Rost D, Stockel B, Leuschner U, Keppler D. Exon-intron organization of the human multidrug-resistance protein 2 (MRP2) gene mutated in Dubin-Johnson syndrome. Gastroenterology **1999**; 117(3):653–660.

133 Keitel V, Kartenbeck J, Nies AT, Spring H, Brom M, Keppler D. Impaired protein maturation of the conjugate export pump multidrug resistance protein 2 as a consequence of a deletion mutation in Dubin-Johnson syndrome. Hepatology **2000**; 32(6):1317–1328.

134 Jansen PL, Peters WH, Lamers WH. Hereditary chronic conjugated hyperbilirubinemia in mutant rats caused by defective hepatic anion transport. Hepatology **1985**; 5(4):573–579.

135 Kurisu H, Kamisaka K, Koyo T, Yamasuge S, Igarashi H, Maezawa H et al. Organic anion transport study in mutant rats with autosomal recessive conjugated hyperbilirubinemia. Life Sci **1991**; 49(14):1003–1011.

136 Abe H, Okuda K. Biliary excretion of conjugated sulfobromophthalein (BSP) in constitutional conjugated hyperbilirubinemias. Digestion **1975**; 13(5):272–283.

137 Kiuchi Y, Suzuki H, Hirohashi T, Tyson CA, Sugiyama Y. cDNA cloning and inducible expression of human multidrug resistance associated protein 3 (MRP3). FEBS Lett **1998**; 433(1–2):149–152.

138 Uchiumi T, Hinoshita E, Haga S, Nakamura T, Tanaka T, Toh S et al. Isolation of a novel human canalicular multispecific organic anion transporter, cMOAT2/MRP3, and its expression in cisplatin-resistant cancer cells with decreased ATP-dependent drug transport. Biochem Biophys Res Commun **1998**; 252(1):103–110.

139 Fromm MF, Leake B, Roden DM, Wilkinson GR, Kim RB. Human MRP3 transporter: identification of the 5′-flanking region, genomic organization and alternative splice variants. Biochim Biophys Acta **1999**; 1415(2):369–374.

140 König J, Rost D, Cui Y, Keppler D. Characterization of the human multidrug resistance protein isoform MRP3 localized to the basolateral hepatocyte membrane. Hepatology **1999**; 29(4):1156–1163.

141 Kool M, van der LM, de Haas M, Scheffer GL, de Vree JM, Smith AJ et al. MRP3, an organic anion transporter able to transport anti-cancer drugs. Proc Natl Acad Sci USA **1999**; 96(12):6914–6919.

142 Belinsky MG, Bain LJ, Balsara BB, Testa JR, Kruh GD. Characterization of MOAT-C and MOAT-D, new members of the MRP/cMOAT subfamily of transporter proteins. J Natl Cancer Inst **1998**; 90(22):1735–1741.

143 Zeng H, Liu G, Rea PA, Kruh GD. Transport of amphipathic anions by human multidrug resistance protein 3. Cancer Res **2000**; 60(17):4779–4784.

144 Hirohashi T, Suzuki H, Takikawa H, Sugiyama Y. ATP-dependent transport of bile salts by rat multidrug resistance-associated protein 3 (Mrp3). J Biol Chem **2000**; 275(4):2905–2910.

145 Zeng H, Bain LJ, Belinsky MG, Kruh GD. Expression of multidrug resistance protein-3 (multispecific organic anion transporter-D) in human embryonic kid-

ney 293 cells confers resistance to anticancer agents. Cancer Res **1999**; 59(23):5964–5967.

146 Hirohashi T, Suzuki H, Ito K, Ogawa K, Kume K, Shimizu T et al. Hepatic expression of multidrug resistance-associated protein-like proteins maintained in eisai hyperbilirubinemic rats. Mol Pharmacol **1998**; 53(6):1068–1075.

147 Ogawa K, Suzuki H, Hirohashi T, Ishikawa T, Meier PJ, Hirose K et al. Characterization of inducible nature of MRP3 in rat liver. Am J Physiol Gastrointest Liver Physiol **2000**; 278(3):G438–G446.

148 Soroka CJ, Lee JM, Azzaroli F, Boyer JL. Cellular localization and up-regulation of multidrug resistance-associated protein 3 in hepatocytes and cholangiocytes during obstructive cholestasis in rat liver. Hepatology **2001**; 33(4):783–791.

149 Lee K, Klein-Szanto AJ, Kruh GD. Analysis of the MRP4 drug resistance profile in transfected NIH3T3 cells. J Natl Cancer Inst **2000**; 92(23):1934–1940.

150 Schuetz JD, Connelly MC, Sun D, Paibir SG, Flynn PM, Srinivas RV et al. MRP4: A previously unidentified factor in resistance to nucleoside-based antiviral drugs. Nature Med **1999**; 5(9):1048–1051.

151 Chen ZS, Lee K, Kruh GD. Transport of cyclic nucleotides and estradiol 17-beta-D-glucuronide by multidrug resistance protein 4: Resistance to 6-mercaptopurine and 6-thioguanine. J Biol Chem **2001**; 276(36):33747–33754.

152 Gerloff T, Stieger B, Hagenbuch B, Madon J, Landmann L, Roth J et al. The sister of P-glycoprotein represents the canalicular bile salt export pump of mammalian liver. J Biol Chem **1998**; 273(16):10046–10050.

153 McAleer MA, Breen MA, White NL, Matthews N. pABC11 (also known as MOAT-C and MRP5), a member of the ABC family of proteins, has anion transporter activity but does not confer multidrug resistance when overexpressed in human embryonic kidney 293 cells. J Biol Chem **1999**; 274(33):23541–23548.

154 Suzuki T, Sasaki H, Kuh HJ, Agui M, Tatsumi Y, Tanabe S et al. Detailed structural analysis on both human MRP5 and mouse mrp5 transcripts. Gene **2000**; 242(1/2):167–173.

155 Wijnholds J, Mol CA, van Deemter L, de Haas M, Scheffer GL, Baas F et al. Multidrug-resistance protein 5 is a multispecific organic anion transporter able to transport nucleotide analogs. Proc Natl Acad Sci USA **2000**; 97(13):7476–7481.

156 Jedlitschky G, Burchell B, Keppler D. The multidrug resistance protein 5 functions as an ATP-dependent export pump for cyclic nucleotides. J Biol Chem **2000**; 275(39):30069–30074.

157 Kool M, van der LM, de Haas M, Baas F, Borst P. Expression of human MRP6, a homologue of the multidrug resistance protein gene MRP1, in tissues and cancer cells. Cancer Res **1999**; 59(1):175–182.

158 Belinsky MG, Kruh GD. MOAT-E (ARA) is a full-length MRP/cMOAT subfamily transporter expressed in kidney and liver. Br J Cancer **1999**; 80(9):1342–1349.

159 Madon J, Hagenbuch B, Landmann L, Meier PJ, Stieger B. Transport function and hepatocellular localization of mrp6 in rat liver. Mol Pharmacol **2000**; 57(3):634–641.

160 Longhurst TJ, O'Neill GM, Harvie RM, Davey RA. The anthracycline resistance-associated (ara) gene, a novel gene associated with multidrug resistance in a human leukaemia cell line. Br J Cancer **1996**; 74(9):1331–1335.

161 Ringpfeil F, Lebwohl MG, Christiano AM, Uitto J. Pseudoxanthoma elasticum: mutations in the MRP6 gene encoding a transmembrane ATP-binding cassette (ABC) transporter. Proc Natl Acad Sci USA **2000**; 97(11):6001–6006.

162 Germain DP, Remones V, Perdu J, Jeunemaitre X. Identification of two polymorphisms (c189G>C; c190T>C) in exon 2 of the human MRP6 gene (ABCC6) by screening of Pseudoxanthoma elasticum patients: possible sequence correction? Hum Mutat **2000**; 16(5):449.

163 Germain DP, Perdu J, Remones V, Jeunemaitre X. Homozygosity for the R1268Q mutation in MRP6, the pseudoxanthoma elasticum gene, is not disease-causing. Biochem Biophys Res Commun **2000**; 274(2):297–301.

164 Ringpfeil F, Nakano A, Uitto J, Pulkkinen L. Compound heterozygosity for a recurrent 16.5-kb Alu-mediated deletion mutation and single-base-pair substitutions in the ABCC6 gene results in pseudoxanthoma elasticum. Am J Hum Genet **2001**; 68(3):642–652.
165 Van der Bliek AM, Baas F, Ten Houte DL, Kooiman PM, van d, V, Borst P. The human mdr3 gene encodes a novel P-glycoprotein homologue and gives rise to alternatively spliced mRNAs in liver. EMBO J **1987**; 6(11):3325–3331.
166 Smit JJ, Schinkel AH, Mol CA, Majoor D, Mooi WJ, Jongsma AP et al. Tissue distribution of the human MDR3 P-glycoprotein. Lab Invest **1994**; 71(5):638–649.
167 Smit JJ, Schinkel AH, Oude Elferink RP, Groen AK, Wagenaar E, van Deemter L et al. Homozygous disruption of the murine mdr2 P-glycoprotein gene leads to a complete absence of phospholipid from bile and to liver disease. Cell **1993**; 75(3):451–462.
168 Smith AJ, de Vree JM, Ottenhoff R, Oude Elferink RP, Schinkel AH, Borst P. Hepatocyte-specific expression of the human MDR3 P-glycoprotein gene restores the biliary phosphatidylcholine excretion absent in Mdr2 (–/–) mice. Hepatology **1998**; 28(2):530–536.
169 Crawford AR, Smith AJ, Hatch VC, Oude Elferink RP, Borst P, Crawford JM. Hepatic secretion of phospholipid vesicles in the mouse critically depends on mdr2 or MDR3 P-glycoprotein expression. Visualization by electron microscopy. J Clin Invest **1997**; 100(10):2562–2567.
170 Smith AJ, Timmermans-Hereijgers JL, Roelofsen B, Wirtz KW, van Blitterswijk WJ, Smit JJ et al. The human MDR3 P-glycoprotein promotes translocation of phosphatidylcholine through the plasma membrane of fibroblasts from transgenic mice. FEBS Lett **1994**; 354(3):263–266.
171 van Helvoort A, Smith AJ, Spring H, Fritzsche I, Schinkel AH, Borst P et al. MDR1 P-glycoprotein is a lipid translocase of broad specificity, while MDR3 P-glycoprotein specifically translocates phosphatidylcholine. Cell **1996**; 87(3):507–517.
172 De Vree JM, Jacquemin E, Sturm E, Cresteil D, Bosma PJ, Aten J et al. Mutations in the MDR3 gene cause progressive familial intrahepatic cholestasis. Proc Natl Acad Sci USA **1998**; 95(1):282–287.
173 Jacquemin E, de Vree JM, Cresteil D, Sokal EM, Sturm E, Dumont M et al. The wide spectrum of multidrug resistance 3 deficiency: from neonatal cholestasis to cirrhosis of adulthood. Gastroenterology **2001**; 120(6):1448–1458.
174 Chen HL, Chang PS, Hsu HC, Lee JH, Ni YH, Hsu HY et al. Progressive familial intrahepatic cholestasis with high gamma-glutamyltranspeptidase levels in taiwanese infants: role of MDR3 gene defect? Pediatr Res **2001**; 50(1):50–55.
175 Jacquemin E, Cresteil D, Manouvrier S, Boute O, Hadchouel M. Heterozygous non-sense mutation of the MDR3 gene in familial intrahepatic cholestasis of pregnancy. Lancet **1999**; 353(9148):210–211.
176 Dixon PH, Weerasekera N, Linton KJ, Donaldson O, Chambers J, Egginton E et al. Heterozygous MDR3 missense mutation associated with intrahepatic cholestasis of pregnancy: evidence for a defect in protein trafficking. Hum Mol Genet **2000**; 9(8):1209–1217.
177 Rosmorduc O, Hermelin B, Poupon R. MDR3 gene defect in adults with symptomatic intrahepatic and gallbladder cholesterol cholelithiasis. Gastroenterology **2001**; 120(6):1459–1467.
178 Smith AJ, van Helvoort A, van Meer G, Szabo K, Welker E, Szakacs G et al. MDR3 P-glycoprotein, a phosphatidylcholine translocase, transports several cytotoxic drugs and directly interacts with drugs as judged by interference with nucleotide trapping. J Biol Chem **2000**; 275(31):23530–23539.
179 Huang L, Vore M. Multidrug resistance p-glycoprotein 2 is essential for the biliary excretion of indocyanine green. Drug Metab Dispos **2001**; 29(5):634–637.
180 Childs S, Yeh RL, Georges E, Ling V. Identification of a sister gene to P-glyco-

protein. Cancer Res **1995**; 55(10):2029–2034.

181 Childs S, Yeh RL, Hui D, Ling V. Taxol resistance mediated by transfection of the liver-specific sister gene of P-glycoprotein. Cancer Res **1998**; 58(18):4160–4167.

182 Strautnieks SS, Bull LN, Knisely AS, Kocoshis SA, Dahl N, Arnell H et al. A gene encoding a liver-specific ABC transporter is mutated in progressive familial intrahepatic cholestasis. Nature Genet **1998**; 20(3):233–238.

183 Jansen PL, Strautnieks SS, Jacquemin E, Hadchouel M, Sokal EM, Hooiveld GJ et al. Hepatocanalicular bile salt export pump deficiency in patients with progressive familial intrahepatic cholestasis. Gastroenterology **1999**; 117(6):1370–1379.

184 Wang R, Salem M, Yousef IM, Tuchweber B, Lam P, Childs SJ et al. Targeted inactivation of sister of P-glycoprotein gene (spgp) in mice results in nonprogressive but persistent intrahepatic cholestasis. Proc Natl Acad Sci USA **2001**; 98(4):2011–2016.

185 Lecureur V, Sun D, Hargrove P, Schuetz EG, Kim RB, Lan LB et al. Cloning and expression of murine sister of P-glycoprotein reveals a more discriminating transporter than MDR1/P-glycoprotein. Mol Pharmacol **2000**; 57(1):24–35.

186 Bolder U, Trang NV, Hagey LR, Schteingart CD, Ton-Nu HT, Cerre C et al. Sulindac is excreted into bile by a canalicular bile salt pump and undergoes a cholehepatic circulation in rats. Gastroenterology **1999**; 117(4):962–971.

187 Staudinger JL, Goodwin B, Jones SA, Hawkins-Brown D, MacKenzie KI, LaTour A et al. The nuclear receptor PXR is a lithocholic acid sensor that protects against liver toxicity. Proc Natl Acad Sci USA **2001**; 98(6):3369–3374.

188 Doyle LA, Yang W, Abruzzo LV, Krogmann T, Gao Y, Rishi AK et al. A multidrug resistance transporter from human MCF-7 breast cancer cells. Proc Natl Acad Sci USA **1998**; 95(26):15665–15670.

189 Miyake K, Mickley L, Litman T, Zhan Z, Robey R, Cristensen B et al. Molecular cloning of cDNAs which are highly overexpressed in mitoxantrone-resistant cells: demonstration of homology to ABC transport genes. Cancer Res **1999**; 59(1):8–13.

190 Maliepaard M, van Gastelen MA, de Jong LA, Pluim D, van Waardenburg RC, Ruevekamp-Helmers MC et al. Overexpression of the BCRP/MXR/ABCP gene in a topotecan-selected ovarian tumor cell line. Cancer Res **1999**; 59(18):4559–4563.

191 Brangi M, Litman T, Ciotti M, Nishiyama K, Kohlhagen G, Takimoto C et al. Camptothecin resistance: role of the ATP-binding cassette (ABC), mitoxantrone-resistance half-transporter (MXR), and potential for glucuronidation in MXR-expressing cells. Cancer Res **1999**; 59(23):5938–5946.

192 Komatani H, Kotani H, Hara Y, Nakagawa R, Matsumoto M, Arakawa H et al. Identification of breast cancer resistant protein/mitoxantrone resistance/placenta-specific, ATP-binding cassette transporter as a transporter of NB-506 and J-107088, topoisomerase I inhibitors with an indolocarbazole structure. Cancer Res **2001**; 61(7):2827–2832.

193 Maliepaard M, Scheffer GL, Faneyte IF, van Gastelen MA, Pijnenborg AC, Schinkel AH et al. Subcellular localization and distribution of the breast cancer resistance protein transporter in normal human tissues. Cancer Res **2001**; 61(8):3458–3464.

194 Jonker JW, Smit JW, Brinkhuis RF, Maliepaard M, Beijnen JH, Schellens JH et al. Role of breast cancer resistance protein in the bioavailability and fetal penetration of topotecan. J Natl Cancer Inst **2000**; 92(20):1651–1656.

195 Litman T, Brangi M, Hudson E, Fetsch P, Abati A, Ross DD et al. The multidrug-resistant phenotype associated with overexpression of the new ABC half-transporter, MXR (ABCG2). J Cell Sci **2000**; 113(Pt 11):2011–2021.

196 Stephens JC, Schneider JA, Tanguay DA, Choi J, Acharya T, Stanley SE et al. Haplotype variation and linkage disequilibrium in 313 human genes. Science **2001**; 293(5529):489–493.

197 Sachidanandam R, Weissman D, Schmidt SC, Kakol JM, Stein LD, Marth G et al. A map of human genome sequence variation containing 1.42 million single nucleotide polymorphisms. Nature **2001**; 409(6822):928–933.

198 Grundemann D, Schomig E. Gene structures of the human non-neuronal monoamine transporters EMT and OCT2. Hum Genet **2000**; 106(6):627–635.

199 Wieland A, Hayer-Zillgen M, Bonisch H, Bruss M. Analysis of the gene structure of the human (SLC22A3) and murine (Slc22a3) extraneuronal monoamine transporter. J Neural Transm **2000**; 107(10):1149–1157.

200 Tang NL, Ganapathy V, Wu X, Hui J, Seth P, Yuen PM et al. Mutations of OCTN2, an organic cation/carnitine transporter, lead to deficient cellular carnitine uptake in primary carnitine deficiency. Hum Mol Genet **1999**; 8(4):655–660.

201 Hoffmeyer S, Burk O, von Richter O, Arnold HP, Brockmöller J, Johne A et al. Functional polymorphisms of the human multidrug-resistance gene: multiple sequence variations and correlation of one allele with P-glycoprotein expression and activity *in vivo*. Proc Natl Acad Sci USA **2000**; 97(7):3473–3478.

202 Cascorbi I, Gerloff T, Johne A, Meisel C, Hoffmeyer S, Schwab M et al. Frequency of single nucleotide polymorphisms in the P-glycoprotein drug transporter MDR1 gene in white subjects. Clin Pharmacol Ther **2001**; 69(3):169–174.

203 Tanabe M, Ieiri I, Nagata N, Inoue K, Ito S, Kanamori Y et al. Expression of P-glycoprotein in human placenta: relation to genetic polymorphism of the multidrug resistance (MDR)-1 gene. J Pharmacol Exp Ther **2001**; 297(3):1137–1143.

204 Ameyaw MM, Regateiro F, Li T, Liu X, Tariq M, Mobarek A et al. MDR1 pharmacogenetics: frequency of the C3435T mutation in exon 26 is significantly influenced by ethnicity. Pharmacogenetics **2001**; 11(3):217–221.

205 Kim RB, Leake BF, Choo EF, Dresser GD, Kubba SV, Schwarz UI et al. Identification of functionally important MDR1 variant alleles among European- and African-Americans. Clin Pharmacol Ther **2001**; 70:189–199.

206 Decleves X, Chevillard S, Charpentier C, Vielh P, Laplanche JL. A new polymorphism (N21D) in the exon 2 of the human MDR1 gene encoding the P-glycoprotein. Hum Mutat **2000**; 15(5):486.

207 Hitzl M, Drescher S, van der KH, Schaffeler E, Fischer J, Schwab M et al. The C3435T mutation in the human MDR1 gene is associated with altered efflux of the P-glycoprotein substrate rhodamine 123 from CD56+ natural killer cells. Pharmacogenetics **2001**; 11(4):293–298.

208 Rund D, Azar I, Shperling O. A mutation in the promoter of the multidrug resistance gene (MDR1) in human hematological malignancies may contribute to the pathogenesis of resistant disease. Adv Exp Med Biol **1999**; 457:71–75.

209 Mickley LA, Lee JS, Weng Z, Zhan Z, Alvarez M, Wilson W et al. Genetic polymorphism in MDR-1: a tool for examining allelic expression in normal cells, unselected and drug-selected cell lines, and human tumors. Blood **1998**; 91(5):1749–1756.

210 Kioka N, Tsubota J, Kakehi Y, Komano T, Gottesman MM, Pastan I et al. P-glycoprotein gene (MDR1) cDNA from human adrenal: normal P-glycoprotein carries Gly185 with an altered pattern of multidrug resistance. Biochem Biophys Res Commun **1989**; 162(1):224–231.

211 Toh S, Wada M, Uchiumi T, Inokuchi A, Makino Y, Horie Y et al. Genomic structure of the canalicular multispecific organic anion-transporter gene (MRP2/cMOAT) and mutations in the ATP-binding-cassette region in Dubin-Johnson syndrome. Am J Hum Genet **1999**; 64(3):739–746.

212 Wada M, Toh S, Taniguchi K, Nakamura T, Uchiumi T, Kohno K et al. Mutations in the canalicular multispecific organic anion transporter (cMOAT) gene, a novel ABC transporter, in patients with hyperbilirubinemia II/Dubin-Johnson syndrome. Hum Mol Genet **1998**; 7(2):203–207.

213 Kajihara S, Hisatomi A, Mizuta T, Hara T, Ozaki I, Wada I et al. A splice mutation in the human canalicular multispecific organic anion transporter gene causes Dubin-Johnson syndrome. Biochem Biophys Res Commun **1998**; 253(2):454–457.

214 Lee K, Belinsky MG, Bell DW, Testa JR, Kruh GD. Isolation of MOAT-B, a widely expressed multidrug resistance-associated protein/canalicular multispecific organic anion transporter-related transporter. Cancer Res **1998**; 58(13):2741–2747.

215 Hopper E, Belinsky MG, Zeng H, Tosolini A, Testa JR, Kruh GD. Analysis of the structure and expression pattern of MRP7 (ABCC10), a new member of the MRP subfamily. Cancer Lett **2001**; 162(2):181–191.

10
Pharmacogenomics of Asthma Treatment

Lyle J. Palmer, Eric S. Silverman, Jeffrey M. Drazen and Scott T. Weiss

Abstract

Asthma is the most common chronic childhood disease in the developed nations, and is a complex disorder that has high pharmacoeconomic costs. Studies of the genetic etiology of asthma offer a means of better understanding its pathogenesis, with the goal of improving preventive strategies, diagnostic tools and therapies. Even though we do not understand the cause of most asthma, we have pharmacological treatments that are effective, although not uniformly so. For a given treatment, there are some patients who respond, while there are others for whom the response is minimal. Since it is reasonable to hypothesize that some of the variance in treatment response may be due to DNA sequence variants that change the encoded proteins and modify a given patient's response to drugs, attempts to detect specific polymorphisms in genetic loci contributing to variation in individual response to therapy have been undertaken. Concomitantly, the technology for detecting single nucleotide polymorphisms (SNPs) has undergone rapid development, accompanied by equally rapid developments in functional genomics, genetic statistics and bioinformatics. This chapter reviews both current and potential future contributions of pharmacogenomics to our understanding and treatment of asthma.

10.1
Introduction

Asthma is the most serious of the atopic diseases and has become epidemic, affecting more than 155 million individuals in the developed world. It is the most common chronic childhood disease in developed nations [1], and carries a very substantial direct and indirect economic cost worldwide [2]. A number of pharmacological treatments have been developed for asthma. These treatments have a modest efficacy overall, due in part to widely variable individual responses to asthma drugs. Because of such variability, it is clear that some of the substantial resources expended on asthma medication, estimated to exceed U.S. $ 3 billion per annum in the U.S. alone [3], would be better spent targeting those patients who would benefit the most. At present there are no proven methods of effectively predicting response and prospectively targeting asthma treatment.

Asthma is characterized by variable symptoms such as wheeze, shortness of breath and coughing and is usually associated with airway inflammation, with variably reduced spirometric indices [4, 5], with increased non-specific airway responsiveness (AR) to spasmogens [6, 7] and increased levels of serum immunoglobulin E (IgE) and eosinophils [8–10]. The symptoms of asthma are primarily due to excessive airway narrowing, which leads to an increased resistance to airflow, especially during forced expiration, and produces characteristic spirometric findings. A cardinal feature of asthma is that airway narrowing is *reversible*; either spontaneously or as the result of therapy.

The prevalence of asthma and other allergic diseases has risen over the past two decades in developed nations [11, 12]. During the same period, the genetic etiology of asthma has been increasingly emphasized as a means of better understanding its pathogenesis, with the ultimate goal of improving preventive strategies, diagnostic tools and therapies [13, 14]. Concomitant technical developments in molecular genetics and in the use of polymorphisms directly derived from DNA sequence have occurred, and extensive catalogs of DNA sequence variants across the human genome have begun to be constructed [15, 16].

As for most complex human diseases, asthma is a syndrome rather than a distinct disease, and probably has multiple environmental and genetic determinants [13]. A component of this complexity is a highly variable response to pharmacological therapy among individual patients with asthma [17, 18]. Pharmacogenomics is the study of the genomic basis of individual variable response to therapy. Ideally, we would be able to stratify a population needing treatment into those likely, or unlikely, to respond to treatment as well as those likely, or unlikely, to experience adverse side effects [19]. This chapter summarizes the state of the art in asthma pharmacogenomics and suggests future directions and contributions of pharmacogenomics to our understanding and treatment of asthma.

10.2
Pharmacogenomic Pathways and Phenotypes

At the cellular level, eosinophils, mast cells, alveolar macrophages, lymphocytes and neutrophils recruited to the airways of asthmatics produce a variety of inflammatory mediators, such as histamine, kinins, neuropeptides, and leukotrienes, which lead to airway smooth muscle constriction and obstruction of airflow, and the perpetuation of airway inflammation [20, 21]. An understanding of the inflammatory processes and the molecular pathways of these mediators has led to the development and widespread use of several pharmacologic agents that mitigate airway inflammation and bronchoconstriction.

There are four major classes of asthma pharmacotherapy currently in widespread use [22, 23]:

1) beta$_2$-agonists (β-agonists) used by inhalation for the relief of airway obstruction (e.g., albuterol, salmeterol, fenoterol);

2) glucocorticosteroids for both inhaled and systemic use (e.g., beclomethasone, triamcinolone, prednisone);
3) theophylline and its derivatives, used for both the relief of bronchospasm and the control of inflammation; and
4) inhibitors of the cysteinyl-leukotriene pathway (e.g., montelukast, pranlukast, zafirlukast, zileuton).

The pharmacodynamics and molecular biology of the possible pathways involved in the action of these pharmacological therapies have been extensively reviewed [24–28].

When asthmatics are treated according to established guidelines [29], most compliant patients can be effectively managed with minimal morbidity. However, there is substantial heterogeneity in therapeutic response to each asthma drug. Although all four classes of asthma drugs are effective on average when examined in large clinical drug trials, many studies indicate that there is significant variation in inter-individual response [18, 30]. Individuals metabolize pharmaceutical agents differently, show differences in dose-response relationships to commonly used asthma drugs, and have a range of susceptibilities to adverse side effects from pharmacologic agents. Variability in individual asthma treatment response may be due to many factors, including the severity and type of disease, treatment compliance, intercurrent illness, other medication taken (drug-drug interaction), environmental exposures, and age [17]. However, there is reason to believe that genetic factors may comprise a substantial component of the observed treatment variance. Comparison of the inter- and intra-person variance in treatment response has suggested that up to 80% of such variance in Caucasians may have a genetic basis [18].

Although there are data on the variability of the treatment response for each of these classes of agents, there are no systematic studies on the reasons for variance in the treatment response to steroids or theophylline. Therefore, this chapter will focus on the specific pharmacogenomics of β-agonists and inhibitors of the cysteinyl-leukotriene pathway [18, 30] and on general considerations related to pharmacogenomic mechanisms.

While many pharmacogenomic mechanisms are possible, genetic variants may alter response to drugs in three main ways (Table 10.1) [31]:

Tab. 10.1 Pharmacogenetic mechanisms with implications for asthma treatment

1. Genetic variants associated with altered uptake, distribution or metabolism of the agent administered (Pharmacokinetic)
2. Genetic variants resulting in an unintended action of a drug outside of its therapeutic indication (Idiosyncratic)
3. Genetic variation in the drug target or a component of the drug pathway leading to altered drug efficacy, and genetic variants leading to differences in the expression of a physiological phenotype such that a given target may not be disease-associated in a given patient (Pharmacodynamic)

1) Variation in individual metabolism of a drug, especially in enzymes involved in the catabolism or excretion of drug. The best known example is the highly genetically diverse cytochrome P450 system, recognized to have many pharmacogenomic effects [32, 33]. No genes implicated in the modulation of this kind of pharmacogenomic mechanism in asthma have yet been discovered. Although the variations observed in theophylline response among asthma patients may result from variance in the catabolism of theophylline, there have been no genetic studies of such a mechanism.
2) Individual variation of adverse effects of a drug that are not based on the drug's action. A striking pulmonary example is the variation in the metabolism of isoniazid and its side effects [34, 35]; however, a genetic basis to the side effects of asthma treatments has not yet been established. It is possible that the variation in incidence of adverse effects of inhaled glucocorticoids (e.g., glaucoma [36], cataracts [37] or the rate of bone loss [38]) may be genetically determined, but there have been no data establishing a specific gene or locus associated with these adverse effects.
3) Genetic variance in the drug treatment target or target pathways. All of the currently available data on asthma pharmacogenomics fall into this single mechanistic category, in which a population is conceptually divided into responders and non-responders, and analysis of specific DNA variants is used in an attempt to determine which patients would belong in each of these groups [19].

Before proceeding further, we will briefly define and explain genetic methods and terminology used in this field.

10.3 Genetic Association Analysis Using Single Nucleotide Polymorphisms (SNPs)

The growing recognitions of the limitations of linkage analysis in complex human diseases [39, 40] has seen a shift in emphasis away from linkage analysis and microsatellite markers towards SNP genotyping and different analytical strategies based on association and haplotype analysis [41–44]. Association analyses are now recognized as essential for localizing susceptibility loci, and they are intrinsically more powerful than linkage analyses in detecting weak genetic effects [45]. Linkage analysis is also inherently unsuitable for pharmacogenomic studies, as drug response data cannot generally be obtained from multiple family members [46].

Genetic polymorphism arises from mutation. The simplest class of polymorphism derives from a single-base mutation that substitutes one nucleotide for another, termed a "single nucleotide polymorphism" or SNP (pronounced "*snip*") [47]. SNPs are recognized through a variety of techniques that exploit the known DNA sequence variant and use restriction enzymes, variable amplification by PCR, oligonucleotide probing, or single-base extension sequencing reactions with dideoxynucleotides [47]. SNPs may be found in coding or regulatory regions of a gene and thus can directly affect gene function or expression. However, most SNPs do not alter gene structure or function in any way and, therefore, may not be directly associated with any change in phenotype [48]. Thus, it is important to ascertain whether the DNA sequence variant under consideration is potentially functional (i.e., could lead to the observed biology) or is a marker in linkage disequilibrium with another DNA sequence variant that is the actual cause of the variable treatment response [49].

The last decade has seen dramatic increases in molecular genetic technologies that can potentially be used to understand the biological basis of asthma pharmacogenomics [15]. Because of their potential biological importance, the common SNPs in the human genome increasingly have been the subject of large-scale cataloging projects funded by both government and industry groups [43, 50, 51]. The process of SNP discovery in the human genome is increasing exponentially [42, 43, 52–55], and the generation of SNP maps from such high-throughput sequencing projects [47, 56–58] may add to the process of gene discovery in asthma research. There are now many large projects sponsored by government or industry devoted to large-scale SNP discovery [16]. The current focus in asthma genetics and pharmacogenetics is thus on SNP discovery leading to the creation of SNP catalogs and on improving technologies for SNP genotyping. The Pharmacogenetics Knowledge Base [*http://pharmgkb.stanford.edu/*] includes data of relevance to asthma pharmacogenomics. Limitations related to cost and the current incomplete status of SNP databases [16] has meant that the association analysis of SNPs in asthma pharmacogenomics has so far been limited to polymorphisms within biologically plausible candidate loci. Many investigators interested in a specific pathway have independently sought to identify sequence variants.

There are a number of potential advantages to using SNPs to investigate the pharmacogenomics of complex human diseases compared to other types of genetic polymorphism [50, 59]. First, the frequency of SNPs across the human genome is higher than for any other type of polymorphism (such as repeat sequences or insertion/deletion polymorphisms) [58]. SNPs are found in exons, introns, promoters, enhancers and intergenic regions, allowing them to be used as markers in dense positional cloning investigations using both randomly distributed markers and markers clustered within genes [59, 60]. Second, groups of adjacent SNPs may exhibit patterns of linkage disequilibrium and haplotypic diversity that could be used to enhance gene mapping [61] and which may highlight recombination "hot-spots" [62]. Finally, there is good evidence that SNPs are less mutable than other types of polymorphism [63, 64]. The resultant greater stability may allow more consistent estimates of linkage disequilibrium and gene-phenotype associations.

Although less common than SNPs, another type of genetic polymorphism is the variable nucleotide tandem repeat (VNTR), also called "microsatellite" [65]. This is a group of DNA bases, ranging from a dinucleotide to a heptanucleotide (or larger structure), that are repeated at a particular locus. VNTRs have a large number of different alleles (i.e., repeat lengths) at a given genetic locus, and the mutation rate is generally higher than that found in SNPs. If the VNTR is functional, e.g., a triplet repeat leading to the insertion or deletion of a repeated amino acid in a protein, it could have a pharmacogenomic effect. Both SNPs and VNTRs that have pharmacogenomic effects in asthma have been identified.

Problems in genetic association analyses arise from the now well-described general limitations of investigating gene-phenotype associations in complex human diseases involving multiple interacting genetic and environmental factors [16, 40, 66]. Some of these relate to technical sequencing and genotyping issues. More fundamentally, the growing focus on SNP genotyping has made it clear that concomitant statistical advances in the linkage disequilibrium mapping of complex traits will also be required [67–69] (see Section 10.5).

10.4 Previous Studies of Asthma Pharmacogenomics

The number of biologically plausible candidate genes that might be involved in the pharmacodynamic and pharmacokinetic determination of response to asthma therapy is very large [16–18]. Each pathway is complex and modulated by numerous interacting cytokines, chemokines, growth factors, and cofactors. All of the receptors, signal transduction components and transcriptional factors for effector genes must also be considered pharmacogenomic candidates. Although genetic information has only recently begun to be integrated into the asthma clinical trial setting, there is now a growing list of candidate genes investigated for association with treatment response in asthma. Genetic polymorphisms have begun to be described that directly or indirectly alter an asthmatic's response to therapy and can, therefore, be used to predict the response to certain asthma drugs, thereby maximizing efficacy and avoiding adverse effects. All asthma pharmacogenomic studies published to date have essentially been *post hoc* genetic studies undertaken using DNA and phenotypic data from subjects who had been enrolled in a conventional clinical trial.

10.4.1 Pharmacogenomics of β-Agonists in Asthma

The pharmacodynamics and molecular biology of the possible pathways involved in the action of β-agonists have been extensively reviewed [24, 25]. β-agonists act via binding to β_2-adrenergic receptors (β_2AR), a cell surface G protein-coupled receptor. Responses to this drug currently represent the most investigated pharmacogenomic pathway in asthma [70–72].

Efforts to explain individual differences in response to β-agonist and sporadic reports of tachyphylaxis have focused on the β_2AR gene, an intronless gene localized to chromosome 5q31-32, because of its direct interaction with β-agonist and its central role in the β-agonist pathway [73]. The primary reasons for the focus on the β_2AR gene and its relationship to treatment response are

1) β-agonists are the most commonly prescribed asthma medications, and

2) there is great controversy among clinicians as to the toxicity and appropriate usage of these drugs.

A total of 13 polymorphisms in the gene and its transcriptional regulator beta-upstream peptide (BUP) have been identified [73]. Three closely linked polymorphisms, two coding block SNPs at amino acid positions 16 and 27, and a SNP in the BUP were found to be common (i.e., allele frequency >0.15) in the general Caucasian population [74, 75].

Although initial studies suggested a relationship between the Gly16 polymorphism and increased risk of severe asthma [76] and increased airways responsiveness [77], subsequent associations with asthma, allergy and airways responsiveness have been inconsistent [74, 76–81].

The common coding variants (β16 and β27) within the β_2AR gene have been shown *in vitro* to be functionally important [70, 82]. The Gly16 receptor exhibits enhanced down-regulation *in vitro* following agonist exposure [82]. In contrast, Arg16 receptors are more resistant to down-regulation. Owing to the strong linkage disequilibrium between these two genes, individuals who are Arg/Arg at position 16 are much more likely to be Glu/Glu at position β27; conversely, individuals who are Gly/Gly at position 16 are much more likely to be Gln/Gln at position β27. The position β27 genotype reduces but does not fully negate the effect of the position β16 polymorphisms with regard to down-regulation of phenotypes *in vitro* [70, 82].

Previous studies of treatment response phenotypes have either been negative [83] or limited by a small sample size [84]. The largest study to date is based on a multicenter, placebo-controlled, double-blind trial involving 255 mild asthmatics who were randomized to either receive two puffs of albuterol 4 times a day on a regularly scheduled basis or treatment only as needed. There was an approximate 6 puffs daily difference in inhaled albuterol between the two treatment groups. The initial data on all patients in the primary trial suggested that there was no difference in AM and PM peak expiratory flow. The investigators concluded that regular use of albuterol was not more associated with adverse events than was the use of albuterol as needed [85]. However, when the results from 190 of the 255 randomized subjects were stratified by genotype at the β16 and β27 polymorphisms, a decrease in morning peak expiratory flow was noted among the patients who were Arg/Arg homozygotes at position 16 and who regularly used albuterol [71, 86]. During the 4-week run-out period, in which all patients used albuterol only as needed, the patients with the Arg/Arg genotypes who had regularly used albuterol during the trial had a morning peak expiratory flow of 30.5 ± 10.1 L min^{-1} lower than Arg/Arg patients who had used albuterol on an as-needed basis during the entire trial. The difference in morning peak expiratory flow between the Arg/Arg regular albuterol users and the Gly/Gly regular albuterol users was roughly 20 L min^{-1} (Figure 10.1).

The *in vitro* data and the results from this clinical trial were synthesized in a so-called "dynamic model" of receptor kinetics by Liggett [72]. According to this theory, Gly/Gly homozygous individuals are already down-regulated as a result of exposure to endogenous catecholamines. Thus, the recurring exogenous exposure to

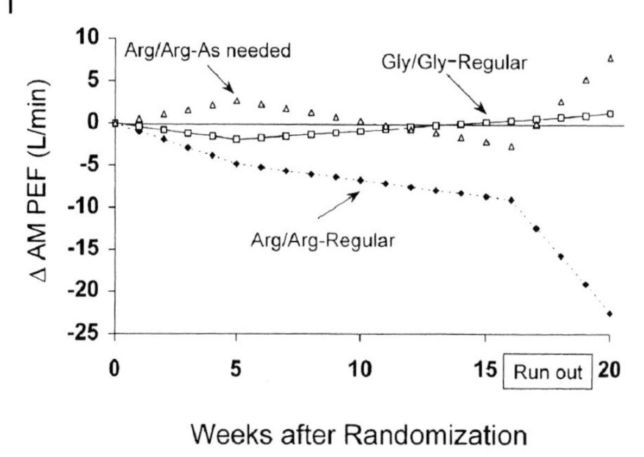

Fig. 10.1 β_2AR genotype predicts response to β-agonist treatment. The figure shows the time course of the change in morning peak expiratory flow (*A.M.* PEF) among different Arg16Gly genotypes in response to β-agonist treatment. Over the treatment and run-out period, Arg16Gly-*Arg/Arg* patients who received regularly scheduled β-agonist treatment (*Arg/Arg*-Regular) experienced a 30.5 ± 10.1 $\mathrm{L\,min^{-1}}$ decline in *A.M.* peak expiratory flow relative to those who received as-needed treatment (*Arg/Arg*-As needed) (p>5 0.012). Arg16Gly/*Gly* patients were not affected by regular treatment (*Gly/Gly*-Regular). Thus, regular treatment was associated with a 23.8 ± 9.5 $\mathrm{L\,min^{-1}}$ decline in peak expiratory flow in response to β-agonist treatment Arg16Gly-Arg/Arg patients relative to Arg16Gly-Gly/Gly (p>5 0.012). Run-out 5 predetermined 4-wk period when regular β-agonist use had been discontinued. (Reprinted with permission from Israel et al. 2000 [71]).

β-agonist would lead to tachyphylaxis, which would be more apparent in the Arg/Arg patients because their receptors had not yet been down-regulated. In this model, the initial response to albuterol in β-agonist naïve patients would be depressed in those who are Gly/Gly homozygous, because their receptors had been endogenously down-regulated to a greater extent than the receptors of patients who are Arg/Arg.

This idea is supported by the results of a study of the bronchodilator response following administration of a single dose of albuterol in children [80]. The study group consisted of 191 normal children and 78 children with a history of wheezing (37 of whom had a diagnosis of asthma). Both the asthmatic and normal children showed a significantly greater percentage of bronchodilator responses in the homozygous Arg16 group. When the groups were compared, the homozygous Arg16 children were at approximately 5-fold higher risk of a positive bronchodilator response to albuterol than were Gly16/Gly16 children.

While the dynamic model of receptor kinetics may explain the results of both clinical trials and epidemiologic studies, other explanations remain possible. It could be that other genes acting in concert with the β_2AR are important in determining pharmacologic response in this pathway. Alternatively, Israel and co-workers did not genotype polymorphisms in the 5 leader cistron [71, 86], which is known to be in linkage disequilibrium with the β16 polymorphism and may influence expression of the β_2 adrenergic receptor.

The issue of the use of haplotypes in asthma pharmacogenomics remains unresolved. Haplotypes are linear combinations of SNPs along a chromosome; use of haplotypes among SNPs within a gene may enhance the detection of phenotype-genotype associations. Drysdale and colleagues investigated molecular haplotypes of the 13 SNPs in the promoter and coding regions of the β_2AR gene [87]. A study of the common haplotypes of these SNPs in 121 Caucasian patients with asthma found that certain haplotypes appeared to affect receptor function differentially and also appeared to correlate with clinical phenotypes [87]. Although this approach is likely to be more powerful than focusing on a single SNP locus, it divides the population into multiple small groups, thus requiring large sample sizes to identify a biological effect.

10.4.2
Pharmacogenomics of Leukotrienes in Asthma

The leukotrienes are a family of polyunsaturated lipoxygenated eicosatetraenoic acids that are derived from arachidonic acid and exhibit a wide range of pharmacological and physiological actions [88]. Three enzymes are involved in the formation of the leukotrienes: 5-lipoxygenase (ALOX5, also known as 5-LO), LTC_4 synthase and LTA_4 epoxide hydrolase. ALOX5 is the enzyme required for the production of both the cysteinyl leukotrienes (LTC_4, LTD_4 and LTE_4) and LTB_4. ALOX5 activity partially determines the level of bronchoconstrictor leukotrienes present in the airways, and pharmacological inhibition of the action of ALOX5 or antagonism of the action of the cysteinyl leukotrienes at their receptor is associated with an amelioration of asthma [89–91].

The ALOX5 gene promoter contains numerous consensus binding sites for many known transcription factors. ALOX5 transcriptional activity has been shown to be dependent in part on transcription factor binding to an Sp-1 binding motif (–GGGCGG–) located about 100 bp upstream from the ATG start site. This *cis* element is of particular interest because it is highly polymorphic, with 3–6 tandem repeats, 5 being the most common, identified in Caucasians and African Americans [92]. Moreover, these VNTR mutations were found to have significant functional consequences in the context of promoter-reporter constructs [93], such that constructs containing more or fewer copies of the VNTR compared to the most commonly occurring 5-repeat constructs were found to have diminished activity. This has resulted in the hypothesis that patients with VNTRs other than the wild type, i.e., 5 repeats of the sequence –GGGCGG– in the core promoter, will exhibit diminished transcription of the ALOX5 gene and hence produce fewer leukotrienes. If this were true, patients harboring the mutant genotype would likely not respond to anti-leukotriene treatment because their disease would be mediated by other factors.

This hypothesis was investigated in a retrospective analysis of the response to an ALOX5 inhibitor, ABT-761, which is clinically similar to zileuton. In 221 patients with asthma who received either high-dose ABT-761 ($n=114$) or placebo ($n=107$] treatment, approximately 6% of asthmatic patients had no wild-type allele

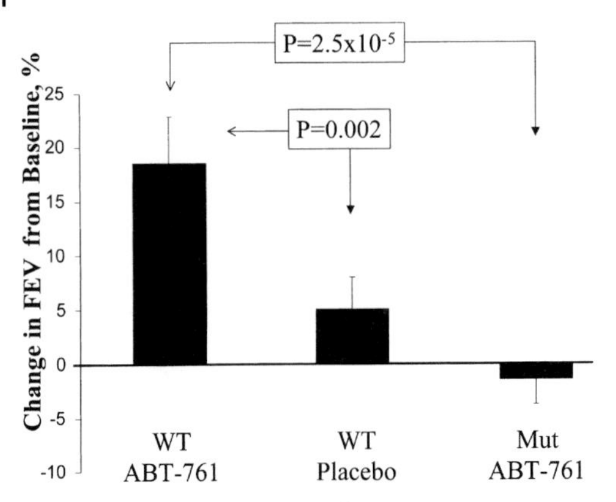

Fig. 10.2 ALOX5 genotype predicts anti-leukotriene response. Outcome of clinical trial of ABT-761, an ALOX5 inhibitor similar to Zileuton, stratified by genotype. Improvement in FEV_1 from pre-treatment baseline at 84 days of treatment was significantly greater for subjects possessing the wild-type (WT) genotype treated with ABT-761 (300 mg/day) compared with subjects possessing any ALOX5 mutant (Mut) allele. (Modified from Drazen et al. 1999 [94]).

at the ALOX5 promoter locus and a diminished response to ABT-761 treatment [94] (Figure 10.2). These findings were consistent with the hypothesis that repeats of the –GGGCGG– sequence other than the wild type are associated with decreased gene transcription and ALOX5 production.

A study of the effects of this polymorphism on lung function response observed following treatment with the cysteine leukotriene antagonist zafirlukast [95] found that patients without wild-type alleles at the ALOX5 promoter locus had a decrease in the FEV_1 in response to treatment with zafirlukast, while patients harboring at least one copy of the wild-type allele at this locus showed improvement in their FEV_1. These data confirm the findings with ABT-761 and lend weight to the yet unproven suggestion that this VNTR modifies treatment responses through modification of the synthesis of the leukotrienes.

Another enzyme of the ALOX5 pathway, LTC_4 synthase, is responsible for the adduction of glutathione at the C-6 position of the arachidonic acid backbone [96]. There is a known SNP in the LTC_4-synthase promoter, A–444C, with a C allele frequency of about 20% in normal subjects and 30% in patients with severe asthma [97]. The -444C allele creates an activator protein-2 binding sequence that appears to be functional. For example, the introduction of a H4TF-2 decoy oligonucleotide into LTC(4)S-positive, differentiated HL-60 cells that were –444C genotype decreased accumulation of LTC_4, and transfection of COS-7 cells with the –444C variant of the promoter increased expression of a β-galactosidase reporter construct [98]. These data suggested that the –444C variant is associated with enhanced cysteinyl leukotriene production.

Sampson and colleagues found that, among asthmatic subjects treated with zafirlukast (20 mg bid), those homozygous for the A allele ($n=10$ subjects) at the –444 locus had a lower FEV_1 response than those with the C/C or C/A ($n=13$) genotype. These findings provide evidence of a second pharmacogenomic locus in the ALOX5 pathway. It will be of interest to determine whether individuals possessing multiple variant alleles at loci in the ALOX5 pathway will have an additive response to treatment aimed at this pathway. Since any loss of function will influence leukotriene production, we predict that any one DNA sequence variant associated with a decrease in cysteinyl leukotriene synthesis will also be associated with decreased response to therapy.

10.5 Statistical Issues

The SNP genotyping effort has caused a broad re-examination of mapping methodologies and study designs in complex human disease [40, 99, 100]. The testing of large numbers of SNPs for association with one or more traits raises important statistical issues regarding the appropriate false positive rate of the tests and the level of statistical significance to be adopted given the multiple testing involved [40]. The required methodological development in genetic statistics is non-trivial given the complexity of common diseases like asthma [16].

The current trend in genetic analysis of complex human diseases is away from family-based strategies using microsatellite markers towards SNP genotyping and different analytical strategies based on association and haplotype analysis [16, 41, 42]. Since response to asthma treatment varies with age, and the number and type of asthma medications is changing rapidly, it is unlikely that family-based asthma treatment data will be available in the foreseeable future. In the absence of these data, case-control association studies are the approach of choice. Case-control association analyses are now recognized as being well suited for localizing susceptibility loci [101], and they are intrinsically more powerful than linkage analyses in detecting weak genetic effects [45].

The two major statistical issues in asthma pharmacogenomics relate to population stratification and statistical power.

10.5.1 Genetic Heterogeneity and Population Stratification

In addition to variation in allele frequencies, there is also a high degree of variation in the strength of linkage disequilibrium in a given chromosomal region among populations of different origins [102] and also between different genomic regions [103, 104]. Such genetic heterogeneity is a major challenge to gene discovery in asthma [13]. Among the limitations of case control association studies is the potential that undetected population stratification will produce misleading evidence of association.

Nearly all pharmacogenomic studies conducted to date in asthma have been case control studies. A major criticism of genetic case control studies of human diseases has been potentially undetected population stratification. Population stratification may cause spurious associations in a case control study when allelic frequencies vary across subpopulations in a study cohort. For example, if there is an imbalance in ethnic group representation between the case and control cohorts, one could detect a spurious association [105]. Such population stratification may result from recent admixture or from poorly matched cases and controls. Genotyping of panels of commonly occurring, unlinked SNPs, chosen without regard to the phenotype of interest, can be used to assure that case and control populations are genetically homogenous. Methods have recently been developed to assess population stratification and, if necessary, to test correctly for association in the presence of such stratification [106–108]. However, neither systematic testing for population stratification nor application of these new statistical methods has yet been incorporated into any pharmacogenomic studies, including studies of asthma.

10.5.2 Statistical Power

Growing experience with complex disease genetics has made clear the need to minimize type I error in genetic studies [41, 109]. Power is especially an issue for SNP-based association studies of susceptibility loci for phenomenon such as response to pharmacological therapy, which are extremely heterogeneous and which are likely to involve genes of small individual effect. Table 10.2 shows some simple estimation of required sample sizes of cases needed to detect a true odds ratio (OR) of 1.5 with 80% power and type I error probability (α) of either 0.05 or 0.005.

Tab. 10.2 Sample size requirements for case control analyses of SNPs (2 controls per case; detectable difference of OR ≥1.5; power=80%) (reprinted with permission from Palmer and Cookson 2001 [16])

Allele frequency [1]	*Dominant model* [3]			*Recessive model* [4]		
	Exposure [2]	*No. Cases required*		*Exposure* [2]	*No. Cases required*	
		α=0.05	α=0.005		α=0.05	α=0.005
10%	19%	430	711	1%	6113	10070
20%	36%	311	516	4%	1600	2637
30%	51%	308	512	9%	769	1269
40%	64%	354	590	16%	485	802
50%	75%	456	762	25%	363	602
60%	84%	661	1107	36%	311	516

1) Allele frequency in controls.
2) Exposure (=prevalence) in controls assuming a diallelic locus with a dominant or recessive allele at Hardy-Weinberg equilibrium.
3) OR of 1.5 between cases and controls for possession of at least one copy of disease-associated SNP by case.
4) OR of 1.5 between cases and controls for possession of two copies of disease-associated SNP by case.

Power calculations assumed that there are two controls for each case and a SNP that operates as if it was a simple binary factor to which a proportion of the population are exposed in a manner directly related to the genotypic frequency (e.g., for 19% exposure, equivalent to a dominant allele at Hardy-Weinberg equilibrium with a prevalence of 10%).

Both mode of inheritance (dominant, recessive) and allele frequency can have dramatic effects on required sample sizes (Table 10.2). Even for the "best case scenario" – a common SNP acting in a dominant fashion – a relatively large sample size of more than 300 cases (a total sample size of >900 subjects) is required at an α of 0.05 (Table 10.2).

Multiple testing issues are likely in many genetic association studies of candidate loci where either multiple SNPs in one gene, multiple SNPs in several loci, or both [110] are tested, suggesting that an α of 0.005 is probably more realistic than an α of 0.05. Using the α of 0.005 or assuming an uncommon SNP (allele frequency ≤ 0.10) that acts in a recessive fashion points to the need for very large sample sizes, i.e., more than 10,000 cases (outside of the scope of even most phase III trials). Table 10.2 assumes an effect size (OR=1.5) which, in the context of a common multifactorial disease such as asthma, may be quite large. Assuming a smaller effect may be more realistic for many genes, and would lead to concomitantly higher required sample sizes. Simulation studies have also suggested that genes of small effect are not likely to be detectable by association studies in sample sizes of less than 500 [69].

While these power calculations are simple and fairly conservative, they clearly demonstrate that the sample sizes used in many of the small case control pharmacogenomic association studies conducted to date had insufficient power to detect even a large effect associated with a SNP. This suggests that larger-scale studies than most of those currently being performed will be needed. As other researchers have suggested [111], the integration of genetic information into clinical trials will likely require a paradigm shift in the conduct of clinical trials with regard to size and cost. A central problem has been that the parameters of the mutation(s) affecting drug response (mode of inheritance, allele frequency, effect size) are not generally known at the start of an asthma clinical trial. Study design remains one of the areas most in need of attention in asthma pharmacogenomics.

10.6 Future Directions and Issues

The ultimate goals of pharmacogenomics are to understand the role that sequence variation among individuals and populations plays in the variability of responses to pharmaceuticals, and to use this information both to tailor an individual's therapeutic regimen in order to maximize efficacy and minimize side effects and to expedite targeted drug discovery and development. The sequencing of the human genome, pursued both by government and industry, is rapidly informing us as to genetic structure and diversity [112]. The availability of a complete reference sequence for the human genome together with new technical advances in high-throughput drug metabolism and pharmacokinetic screening [113], combinatorial chemistry [114] and in bioinformatics and genomics will likely accelerate the discovery process in asthma pharmacogenomics.

The frequency and penetrance of a sequence variant affecting responsiveness to a particular drug and potential interactions with other genetic and environmental factors must ultimately be assessed in multiple population-based samples. A SNP must be relatively common and have a significant impact upon phenotype to be important at the population level in determining treatment response. These criteria become particularly important when extrapolating from specific clinical trials to general clinical use in the highly heterogeneous populations where asthma is most common and which are the current major markets for asthma therapeutics [17, 115]. As SNP-associated pharmacogenomic, diagnostic and gene-environment effects are discovered and utilized to further our understanding of asthma pathophysiology, the study of genetic heterogeneity will become increasingly important. The issue of ethnic diversity has not been seriously addressed in asthma pharmacogenomic studies; most studies have taken place in Caucasian samples. The availability of case control and general population samples from other major population groups will be critical for pharmacogenomic studies in non-Caucasian populations. It is clear that large well-characterized cohort studies of population-based and ethnically diverse samples will be critical to the future success of any diagnostic SNP-based pharmacogenomic tests and for cost effectiveness studies. Indeed, we contend that failure to archive DNA for pharmacogenomic analysis in a large asthma treatment trial would be a significant waste of resources.

The ability to determine response to therapy and potential disease severity prior to the onset of treatment using diagnostic technologies would be potentially of great benefit in asthma. The understanding of asthma pathophysiology may then enter into the realm of clinical and population genetics. As for all diagnostic genetic tests, the utility and ultimate success of diagnostic testing for pharmacogenomic susceptibilities using SNPs in a particular population would depend upon the extent and nature of disease heterogeneity, the frequency of the high-risk allele and the concomitant attributable risk, the penetrance of a specific allele, the ability to define a useful risk model including other genetic factors, important environmental risk factors and interactions between the SNP and factors such as age and gender [42, 116]. In addition, there are both technical problems with routine genetic testing, largely related to false negatives, and important ethical and psychosocial concerns that remain unresolved [116–118].

Thus far, asthma pharmacogenomics has been limited to the candidate gene model. A new direction that is technically feasible at present, yet remains unexplored in asthma pharmacogenomics, are SNP-based whole genome screens for variants associated with variation in drug response [111]. Other future directions for pharmacogenomics research in asthma include the use of pharmacogenomic data for the study of gene-environment interactions in determining response to pharmacological therapy and for homogeneity testing and improving study design [16].

10.7 Conclusions

Pharmacogenomic approaches to asthma offer great potential to improve our understanding and treatment of this disorder, but they also offer significant challenges. However, significant progress has been made and it is now possible to genotype asthmatics at a few loci and to use this information to make therapeutic decisions that improve drug efficacy and mitigate complications. Despite this progress, accompanied by rapid technical progress in SNP genotyping technologies, further research is required. A large number of groups are currently active in addressing methodological problems in SNP genotyping, genetic statistics, and study design, and technological advances in disequilibrium mapping using SNPs and functional genomics will likely accelerate our understanding of the pharmacogenomics of asthma.

Current research in asthma pharmacogenomics has highlighted associations between SNPs in the β-adrenergic receptors and modified response to regular inhaled β-agonist treatments (e.g., albuterol). Variants within the 5-lipoxygenase gene has been suggested to predict the response to the anti-leukotrienes in asthmatic subjects. Confirmation of these findings together with the current rapid creation of new knowledge may mark the beginning of the clinical use of genotyping at an individual level as an adjunct to pharmacotherapy for asthma.

Acknowledgements
This work was supported in part by grant 5 U01 HL65899-02 from the National Heart, Lung and Blood Institute of the NIH.

10.8 References

1 Asher MI, Keil U, Anderson HR et al. International Study of asthma and allergies in childhood (ISAAC): rationale and methods. Eur Respir J **1995**; 8:483–491.
2 Lenney W. The burden of pediatric asthma. Pediatr Pulmonol Suppl **1997**; 15:13–16.
3 Weiss KB, Sullivan SD. The health economics of asthma and rhinitis. I. Assessing the economic impact. J Allergy Clin Immunol **2001**; 107:3–8.
4 Enright PL, Lebowitz MD, Cockroft DW. Physiologic measures: pulmonary function tests. Asthma outcome. Am J Respir Crit Care Med **1994**; 149:S9–18; discussion S19–20.
5 Sherrill DL, Lebowitz MD, Halonen M, Barbee RA, Burrows B. Longitudinal evaluation of the association between pulmonary function and total serum IgE. Am J Respir Crit Care Med **1995**; 152:98–102.
6 Burrows B, Sears MR, Flannery EM, Herbison GP, Holdaway MD, Silva PA. Relation of the course of bronchial responsiveness from age 9 to age 15 to allergy. Am J Respir Crit Care Med 1995; 152:1302–1308.
7 Marsh D, Neely J, Breazeale D et al. Linkage analysis of IL4 and other chromosome 5q31.1 markers and total serum immunoglobin E concentrations. Science **1994**; 264:1152–1156.
8 Burrows B, Martinez F, Halonen M, Barbee R, Cline M. Association of asthma with serum IgE levels and skin-test

reactivity to allergens. N Eng J Med **1989**; 320:271–277.

9 Zimmerman B, Enander I, Zimmerman R, Ahlstedt S. Asthma in children less than 5 years of age: eosinophils and serum levels of the eosinophil proteins ECP and EPX in relation to atopy and symptoms. Clin Exp Allergy **1994**; 24:149–155.

10 Bousquet J, Chanez P, Vignola AM, Lacoste JY, Michel FB. Eosinophil inflammation in asthma. Am J Respir Crit Care Med **1994**; 150:S33–38.

11 Woolcock AJ. Worldwide trends in asthma morbidity and mortality. Explanation of trends. Bull Int Union Against Tuberculosis Lung Dis **1991**; 66:85–89.

12 McNally NJ, Phillips DR, Williams HC. The problem of atopic eczema: aetiological clues from the environment and lifestyles. Soc Sci Med **1998**; 46:729–741.

13 Palmer LJ, Cookson WOCM. Genomic approaches to understanding asthma. Genome Res **2000**; 10:1280–1287.

14 Ober C, Moffatt MF. Contributing factors to the pathobiology. The genetics of asthma. Clin Chest Med **2000**; 21:245–261.

15 Shi MM. Enabling large-scale pharmacogenetic studies by high-throughput mutation detection and genotyping technologies. Clin Chem **2001**; 47:164–172.

16 Palmer LJ, Cookson WOCM. Using single nucleotide polymorphisms (SNPs) as a means to understanding the pathophysiology of asthma. Respir Res **2001**; 2:102–112.

17 Hall IP. Pharmacogenetics of asthma. Eur Respir J **2000**; 15:449–451.

18 Drazen JM, Silverman EK, Lee TH. Heterogeneity of therapeutic responses in asthma. Br Med Bull **2000**; 56:1054–1070.

19 Stephens JC. Single-nucleotide polymorphisms, haplotypes, and their relevance to pharmacogenetics. Mol Diagn **1999**; 4:309–317.

20 Holgate ST. Asthma: a dynamic disease of inflammation and repair. Ciba Foundation Symposium **1997**; 206:5–28.

21 Hogg JC. The pathology of asthma. Apmis **1997**; 105:735–745.

22 National Asthma Education Program. Guidelines for the diagnosis and treatment of asthma II. Bethesda: National Institutes of Health, **1997**.

23 National Heart LaBI. NHLBI/WHO workshop report: global strategy for asthma management and prevention. Global initiative for asthma. Bethesda: National Heart, Lung and Blood Institute, **1995**.

24 Silverman R. Treatment of acute asthma. A new look at the old and at the new. Clin Chest Med **2000**; 21:361–379.

25 Howarth PH, Beckett P, Dahl R. The effect of long-acting beta2-agonists on airway inflammation in asthmatic patients. Respir Med **2000**; 94 (Suppl F):22–25.

26 Gibbs MA, Camargo CA, Jr., Rowe BH, Silverman RA. State of the art: therapeutic controversies in severe acute asthma. Acad Emerg Med **2000**; 7:800–815.

27 Sorkness CA. Leukotriene receptor antagonists in the treatment of asthma. Pharmacotherapy **2001**; 21:34S–37S.

28 Blake KV. Drug treatment of airway inflammation in asthma. Pharmacotherapy **2001**; 21:3S–20S.

29 NIH. Global initiative for asthma. Global strategy for asthma management and prevention. NHLBI/WHO Workshop Report. Washington: National Institutes of Health, National Heart, Lung and Blood Institute, **1995**.

31 Malmstrom K, Rodriguez-Gomez G, Guerra J et al. Oral montelukast, inhaled beclomethasone, and placebo for chronic asthma. A randomized, controlled trial. Montelukast/Beclomethasone Study Group. Ann Intern Med **1999**; 130:487–495.

31 Nebert DW, Weber WW. Pharmacogenetics. In: Pratt WB, Taylor P eds. Principles of Drug Action: The Basis of Pharmacology. New York: Churchill Livingston, **1990**.

32 Meyer UA. Pharmacogenetics: the slow, the rapid, and the ultrarapid. Proc Natl Acad Sci USA **1994**; 91:1983–1984.

33 Weber WW. Acetylation pharmacogenetics: experimental models for human toxicity. Fed Proc **1984**; 43:2332–2337.

34 Evans DAP, Manley KA, McKusick VA. Genetic control of isoniazid metabolism

in man. British Medical Journal **1960**; 2:485–491.

35 Hughes HB. Metabolism of isoniazid in man as related to the occurrence of peripheral neuritis. American Review of Tuberculosis **1954**; 70:266–273.

36 Macris N. Glucocorticoid use and risks of ocular hypertension and glaucoma. JAMA **1997**; 277:1929; discussion 1930.

37 Cumming RG, Mitchell P, Leeder SR. Use of inhaled corticosteroids and the risk of cataracts. N Engl J Med **1997**; 337:8–14.

38 Wong CA, Walsh LJ, Smith CJ et al. Inhaled corticosteroid use and bone-mineral density in patients with asthma. Lancet **2000**; 355:1399–1403.

39 Kruglyak L, Lander E. High-resolution genetic mapping of complex traits. Am J Hum Genet **1995**; 56:1212–1223.

40 Risch NJ, Merikangas K. The future of genetic studies of complex human diseases. Science **1996**; 273:1516–1517.

41 Risch NJ. Searching for genetic determinants in the new millennium. Nature **2000**; 405:847–856.

42 Schork NJ, Fallin D, Lanchbury JS. Single nucleotide polymorphisms and the future of genetic epidemiology. Clin Genet **2000**; 58:250–264.

43 Gray IC, Campbell DA, Spurr NK. Single nucleotide polymorphisms as tools in human genetics. Hum Mol Genet **2000**; 9:2403–2408.

44 Keavney B. Genetic association studies in complex diseases. J Hum Hypertens **2000**; 14:361–367.

45 Elston R. The genetic dissection of multifactorial traits. Clinical and Experimental Allergy **1995**; 2:103–106.

46 Kleyn PW, Vesell ES. Genetic variation as a guide to drug development. Science **1998**; 281:1820–1821.

47 Marth GT, Korf I, Yandell MD et al. A general approach to single-nucleotide polymorphism discovery. Nat Genet **1999**; 23:452–456.

48 Collins A, Lonjou C, Morton NE. Genetic epidemiology of single nucleotide polymorphisms. Proc Natl Acad Sci USA **1999**; 96:15173–15177.

49 Rosenthal N, Schwartz RS. In search of perverse polymorphisms. N Engl J Med **1998**; 338:122–124.

50 Collins FS, Patrinos A, Jordan E, Chakravarti A, Gesteland R, Walters L. New goals for the U.S. Human Genome Project: 1998–2003. Science **1998**; 282:682–689.

51 Masood E. As consortium plans free SNP map of human genome [news]. Nature **1999**; 398:545–546.

52 Martin ER, Lai EH, Gilbert JR et al. SNPing away at complex diseases: analysis of single-nucleotide polymorphisms around APOE in Alzheimer disease. Am J Hum Genet **2000**; 67:383–394.

53 Eberle MA, Kruglyak L. An analysis of strategies for discovery of single-nucleotide polymorphisms. Genet Epidemiol **2000**; 19:29–35.

54 Bentley DR. The Human Genome Project – an overview. Med Res Rev **2000**; 20:189–196.

55 Roberts L. Human genome research. SNP mappers confront reality and find it daunting [news]. Science **2000**; 287:1898–1899.

56 Velculescu VE, Zhang L, Vogelstein B, Kinzler KW. Serial analysis of gene expression. Science **1995**; 270:484–487.

57 Schena M, Shalon D, Davis RW, Brown PO. Quantitative monitoring of gene expression patterns with a complementary DNA microarray. Science **1995**; 270:467–470.

58 Wang DG, Fan JB, Siao CJ et al. Large-scale identification, mapping, and genotyping of single-nucleotide polymorphisms in the human genome. Science **1998**; 280:1077–1082.

59 Collins FS, Guyer MS, Chakravarti A. Variations on a theme: cataloging human DNA sequence variation. Science **1997**; 278:1580–1581.

60 Kruglyak L. The use of a genetic map of biallelic markers in linkage studies. Nature Genet **1997**; 17:21–24.

61 Nickerson DA, Whitehurst C, Boysen C, Charmley P, Kaiser R, Hood L. Identification of clusters of biallelic polymorphic sequence-tagged sites (pSTSs) that generate highly informative and automatable markers for genetic linkage mapping. Genomics **1992**; 12:377–387.

62 Chakravarti A. It's raining SNPs, hallelujah? [news]. Nature Genet **1998**; 19:216–217.

63 Stallings RL, Ford AF, Nelson D, Torney DC, Hildebrand CE, Moyzis RK. Evolution and distribution of (GT)n repetitive sequences in mammalian genomes. Genomics **1991**; 10:807–815.

64 Brookes AJ. The essence of SNPs. Gene **1999**; 8:177–186.

65 Beier DR, Dushkin H, Sussman DJ. Mapping genes in the mouse using single-strand conformation polymorphism analysis of recombinant inbred strains and interspecific crosses. Proc Natl Acad Sci USA **1992**; 89:9102–9106.

66 Palmer LJ, Cookson WOCM. Atopy and asthma. In: Genetic Analysis of Multifactorial Diseases (Sham PC, Bishop T, Eds). London: BIOS Scientific Publishers, **2000**:215–237.

67 Terwilliger JD, Weiss KM. Linkage disequilibrium mapping of complex disease: fantasy or reality? Curr Opin Biotechnol **1998**; 9:578–594.

68 Zhao LP, Aragaki C, Hsu L, Quiaoit F. Mapping of complex traits by single-nucleotide polymorphisms. Am J Hum Genet **1998**; 63:225–240.

69 Long AD, Langley CH. The power of association studies to detect the contribution of candidate genetic loci to variation in complex traits. Genome Res **1999**; 9:720–731.

70 McGraw DW, Forbes SL, Kramer LA et al. Transgenic overexpression of beta[2]-adrenergic receptors in airway smooth muscle alters myocyte function and ablates bronchial hyperreactivity. J Biol Chem **1999**; 274:32241–32247.

71 Israel E, Drazen JM, Liggett SB et al. The effect of polymorphisms of the beta[2]-adrenergic receptor on the response to regular use of albuterol in asthma. Am J Respir Crit Care Med **2000**; 162:75–80.

72 Liggett SB. Pharmacogenetics of beta-1- and beta-2-adrenergic receptors. Pharmacology **2000**; 61:167–173.

73 Liggett SB. The pharmacogenetics of beta2-adrenergic receptors: relevance to asthma. J Allergy Clin Immunol **2000**; 105:487–492.

74 Dewar J, Wheatley A, Venn A, Morrison J, Britton J, Hall I. β_2-adrenoreceptor polymorphisms are in linkage disequilibrium but are not associated with asthma in an adult population. Clin Exp Allergy **1998**; 28:442–448.

75 McGraw DW, Forbes SL, Kramer LA, Liggett SB. Polymorphisms of the 5′ leader cistron of the human beta2-adrenergic receptor regulate receptor expression. J Clin Invest **1998**; 102:1927–1932.

76 Reihsaus E, Innis M, MacIntyre N, Liggett SB. Mutations in the gene encoding for the β_2-adrenergic receptor in normal and asthmatic subjects. Am J Respir Cell Mol Biol **1993**; 8:334–339.

77 D'Amato M, Vitiani LR, Petrelli G et al. Association of persistent bronchial hyperresponsiveness with beta2-adrenoceptor (ADRB2) haplotypes. A population study. Am J Respir Crit Care Med **1998**; 158:1968–1973.

78 Liggett S. Genetics of β2-adrenergic receptor variants in asthma. Clin Exp Allergy **1995**; 25:89–94.

79 Weir TD, Mallek N, Sandford AJ et al. β2-Adrenergic receptor haplotypes in mild, moderate and fatal/near fatal asthma. Am J Respir Crit Care Med **1998**; 158:787–791.

80 Martinez FD, Graves PE, Baldini M, Solomon S, Erickson R. Association between genetic polymorphisms of the beta2-adrenoceptor and response to albuterol in children with and without a history of wheezing. J Clin Invest **1997**; 100:3184–3188.

81 Summerhill E, Leavitt SA, Gidley H, Parry R, Solway J, Ober C. β(2)-adrenergic receptor Arg16/Arg16 genotype is associated with reduced lung function, but not with asthma, in the Hutterites. Am J Respir Crit Care Med **2000**; 162:599–602.

82 Green S, Turki J, Innis M, Liggett S. Amino-terminal polymorphisms of the human β2-adrenergic receptor impart distinct agonist-promoted regulatory properties. Biochemistry **1994**; 33:9414–9419.

83 Hancox RJ, Sears MR, Taylor DR. Polymorphism of the β2-adrenoceptor and

the response to long-term beta2-agonist therapy in asthma. Eur Respir J **1998**; 11:589–593.

84 Tan S, Hall IP, Dewar J, Dow E, Lipworth B. Association between beta 2-adrenoceptor polymorphism and susceptibility to bronchodilator desensitisation in moderately severe stable asthmatics. Lancet **1997**; 350:995–999.

85 Drazen JM, Israel E, Boushey HA et al. Comparison of regularly scheduled with as-needed use of albuterol in mild asthma. Asthma Clinical Research Network. N Engl J Med **1996**; 335:841–847.

86 Israel E, Drazen JM, Liggett SB et al. Effect of polymorphism of the beta(2)-adrenergic receptor on response to regular use of albuterol in asthma. Int Arch Allergy Immunol **2001**; 124:183–186.

87 Drysdale CM, McGraw DW, Stack CB et al. Complex promoter and coding region beta 2-adrenergic receptor haplotypes alter receptor expression and predict *in vivo* responsiveness. Proc Natl Acad Sci USA **2000**; 97:10483–10488.

88 Samuelsson B, Dahlen SE, Lindgren JA, Rouzer CA, Serhan CN. Leukotrienes and lipoxins: structures, biosynthesis, and biological effects. Science **1987**; 237:1171–1176.

89 Israel E, Rubin P, Kemp JP et al. The effect of inhibition of 5-lipoxygenase by zileuton in mild-to-moderate asthma. Ann Intern Med **1993**; 119:1059–1066.

90 Holgate ST, Bradding P, Sampson AP. Leukotriene antagonists and synthesis inhibitors: new directions in asthma therapy. J Allergy Clin Immunol **1996**; 98:1–13.

91 Drazen JM, Israel E, O'Byrne PM. Treatment of asthma with drugs modifying the leukotriene pathway. N Engl J Med **1999**; 340:197–206.

92 In KH, Asano K, Beier D et al. Naturally occurring mutations in the human 5-lipoxygenase gene promoter that modify transcription factor binding and reporter gene transcription. J Clin Invest **1997**; 99:1130–1137.

93 Silverman ES, Du J, De Sanctis GT et al. Egr-1 and Sp1 interact functionally with the 5-lipoxygenase promoter and its naturally occurring mutants. Am J Respir Cell Mol Biol **1998**; 19:316–323.

94 Drazen JM, Yandava CN, Dube L et al. Pharmacogenetic association between ALOX5 promoter genotype and the response to anti-asthma treatment. Nature Genet **1999**; 22:168–170.

95 Anderson W, Kalberg C, Edwards L et al. Effects of polymorphisms in the promoter region of 5-lipoxygenase and LTC4 synthase on the clinical response to Zafirlukast and Fluticasone. Eur Respir J **2000**; 16:183s.

96 Lam BK, Austen KF. Leukotriene C-4 synthase – a pivotal enzyme in the biosynthesis of the cysteinyl leukotrienes. Am J Respir Crit Care Med **2000**; 161:16–19.

97 Sampson AP, Siddiqui S, Buchanan D et al. Variant LTC[4] synthase allele modifies cysteinyl leukotriene synthesis in eosinophils and predicts clinical response to zafirlukast. Thorax **2000**; 55(Suppl 2):28–31.

98 Sanak M, Pierzchalska M, Bazan-Socha S, Szczeklik A. Enhanced expression of the leukotriene C(4) synthase due to overactive transcription of an allelic variant associated with aspirin-intolerant asthma. Am J Respir Cell Mol Biol **2000**; 23:290–296.

99 Lander E, Schork N. Genetic dissection of complex traits. Science **1994**; 265:2037–2048.

100 Weeks D, Lathrop G. Polygenic disease: methods for mapping complex disease traits. Trends Genet **1995**; 11:513–519.

101 Silverman EK, Palmer LJ. Case-control association studies for the genetics of complex respiratory diseases. Am J Respir Cell Mol Biol **2000**; 22:645–648.

102 Zavattari P, Deidda E, Whalen M et al. Major factors influencing linkage disequilibrium by analysis of different chromosome regions in distinct populations: demography, chromosome recombination frequency and selection. Hum Mol Genet **2000**; 9:2947–2957.

103 Watkins WS, Zenger R, O'Brien E et al. Linkage disequilibrium patterns vary with chromosomal location: a case study from the von Willebrand factor region. Am J Hum Genet **1994**; 55:348–355.

104 Jorde LB, Watkins WS, Carlson M et al. Linkage disequilibrium predicts physical distance in the adenomatous polyposis coli region. Am J Hum Genet **1994**; 54:884–898.

105 Ewens W, Spielman R. The transmission/disequilibrium test: history, subdivision, and admixture. Am J Hum Genet **1995**; 57:455–464.

106 Pritchard JK, Rosenberg NA. Use of unlinked genetic markers to detect population stratification in association studies. Am J Hum Genet **1999**; 65:220–228.

107 Pritchard JK, Stephens M, Rosenberg NA, Donnelly P. Association mapping in structured populations. Am J Hum Genet **2000**; 67:170–181.

108 Reich DE, Goldstein DB. Detecting association in a case-control study while correcting for population stratification. Genet Epidemiol **2001**; 20:4–16.

109 Lander E, Kruglyak L. Genetic dissection of complex traits: guidelines for interpreting and reporting linkage results. Nature Genetics **1995**; 11:241–247.

110 Witte JS, Elston RC, Cardon LR. On the relative sample size required for multiple comparisons. Stat Med **2000**; 19:369–372.

111 Cardon LR, Idury RM, Harris TJ, Witte JS, Elston RC. Testing drug response in the presence of genetic information: sampling issues for clinical trials. Pharmacogenetics **2000**; 10:503–510.

112 Broder S, Venter JC. Sequencing the entire genomes of free-living organisms: the foundation of pharmacology in the new millennium. Annu Rev Pharmacol Toxicol **2000**; 40:97–132.

113 White RE. High-throughput screening in drug metabolism and pharmacokinetic support of drug discovery. Annu Rev Pharmacol Toxicol **2000**; 40:133–157.

114 Houghten RA. Parallel array and mixture-based synthetic combinatorial chemistry: tools for the next millennium. Annu Rev Pharmacol Toxicol **2000**; 40:273–282.

115 The International Study of Asthma and Allergies in Childhood (ISAAC) Steering Committee. Worldwide variation in prevalence of symptoms of asthma, allergic rhinoconjunctivitis, and atopic eczema: ISAAC. Lancet **1998**; 351:1225–1232.

116 Yan H, Kinzler KW, Vogelstein B. Tech.sight. Genetic testing – present and future. Science **2000**; 289:1890–1892.

117 Van Ommen GJ, Bakker E, den Dunnen JT. The human genome project and the future of diagnostics, treatment, and prevention. Lancet **1999**; 354(Suppl 1): 15–10.

118 Ross LF, Moon MR. Ethical issues in genetic testing of children. Arch Pediatr Adolesc Med **2000**; 154:873–879.

11
Endothelial Cells are Targets for Hydroxy Urea: Relevance to the Current Therapeutic Strategy in Sickle Cell Disease

Claudine Lapouméroulie, Manuel Brun, Marie Hélène Odièvre and Rajagopal Krishnamoorthy

11.1
Hydroxy Urea

Hydroxy urea (HU), a hydroxylated derivative of urea, has long been used in the treatment of various forms of neoplastic disorders such as polycythemia vera, head and neck cancer, chronic granulocytic leukemia, and more recently in sickle cell disease, β thalassemia and HIV infection [1].

HU is an inhibitor of ribonucleotide reductase, a rate-limiting enzyme which catalyzes the conversion of ribonucleotides into deoxyribonucleotides. HU is thus a cytotoxic agent as it has the ability to inhibit DNA synthesis. Consequently, HU can affect only cells that are actively synthesizing DNA and, therefore, a drug of S-phase cell-cycle specific. Moreover, HU-mediated inhibition of ribonucleotide reductase is reversible, implying that the action of HU will exhibit a relatively straight forward concentration–time course dependence [2–4].

HU, a freely water-soluble molecule, crosses the intestinal wall and other cells by passive diffusion [5, 6], and tissue concentration of HU rapidly matches its blood concentration [7]. The oral bioavailability of HU is nearly complete and hence therapeutically simple to administrate. HU undergoes biotransformation and is converted into urea by a yet-to-be identified hepatic P_{450} monooxygenase (CYP) enzyme [8, 9]. Elimination of HU and its metabolites involves both renal and non-renal mechanisms.

Proposed alternative mechanisms for the cytotoxic effect of HU include direct DNA damage and DNA repair inhibition. Nevertheless there is a large consensus that the main pathway of HU-induced cytotoxicity is mediated by its ability to inhibit DNA synthesis, which can be achieved by maintaining a concentration of 0.5 mmol L^{-1} of HU [10]. The expression of a variety of genes has been shown to be upregulated such as human globin genes, pro-inflammatory cytokines, interferon α-receptor, c-jun and a few such as c-myc have been shown to be downregulated by HU. Since most of these genes exhibit cell-cycle stage-specific expression, the direct effect of HU on transcriptional up- or downregulation is difficult to demonstrate.

11.2
Sickle Cell Anemia

Sickle cell anemia (SCA), a chronic hereditary hemolytic anemia, is one of the most prevalent monogenic disorders in the world and the first human disease to be defined at the molecular level [11]. This autosomal recessive disorder is characterized by a single amino acid substitution (glutamic acid to valine) in the β-subunit of the hemoglobin (Hb) tetramer. This abnormal Hb (HbS) alters the quaternary structure of Hb and thereby its physicochemical characteristics. For example, the solubility of HbS in the deoxygenated state is only 10% that of deoxygenated normal HbA. Indeed, upon deoxygenation, HbS forms a viscous solution within the red blood cells (RBC) and generates polymers. Such intracellular polymerization of HbS distorts the erythrocyte membrane resulting in an abnormal red cell shape named "sickle cell" and renders them susceptible to lysis. In sickle erythrocyte, HbS polymerizes as the cells traverse the microvasculature. The pathophysiological hallmark of SCA is an unpredictable episodic occurrence of painful crisis believed to be the result of vascular occlusion by the non-deformable rigid sickle cells in the microvasculature. This blockage leads to tissue ischemia, bony infarcts and associated bone pain and collectively termed vasoocclusive crisis (VOC) (Figure 11.1). One of the most important features of all hemoglobinopathies, in particular SCA, is the remarkable diversity of their clinical spectrum. An extremely mild clinical course of the disease observed among SCA patients from India and Eastern provinces of Saudi Arabia contrasts with the very severe forms prevalent in many parts of Africa, where an afflicted child fails to thrive above five years of age. Along with such population-specific diversity in clinical and hematological phenotype, interindividual variability within a population is another characteristic of SCA.

Given that the primary pathogenic defect in SCA is the abnormal tendency of HbS to polymerize under hypoxic conditions, over the last two decades therapeutic attempts were essentially oriented towards identifying factors that could inhibit or abolish the tendency of HbS to polymerize [12]. However, this approach was unsuccessful. Skepticism arose in recent years regarding this dogma (ascribing sickle cell vasoocclusion solely to deoxygenation-induced polymerization of HbS and the resultant sickling) as additional data emerged that pointed out that the time required for the development of cell sickling is, for most RBCs, actually longer than the microvascular transit time. Now several lines of evidence support the hypothesis that the key participant in vasoocclusion events is the predilection of erythrocytes from SCA patients for enhanced adhesiveness to vascular endothelium. These cells are perhaps an order of magnitude more adhesive than normal cells [13]. Interestingly, activated endothelial cells (activated by a variety of mediators of inflammation) seem to adhere more strongly to young sickle erythrocytes than the quiescent endothelial cells [13]. Indeed inflammatory cytokines such as tumor necrosis factor-α (TNFα) and Interleukin-I (IL-I) are elevated during steady state and during VOC in SCA patients. Further studies revealed that the sickle–endothelial cell adherence is a receptor-mediated process and different receptor-mediated adherence pathways have been described to date [13].

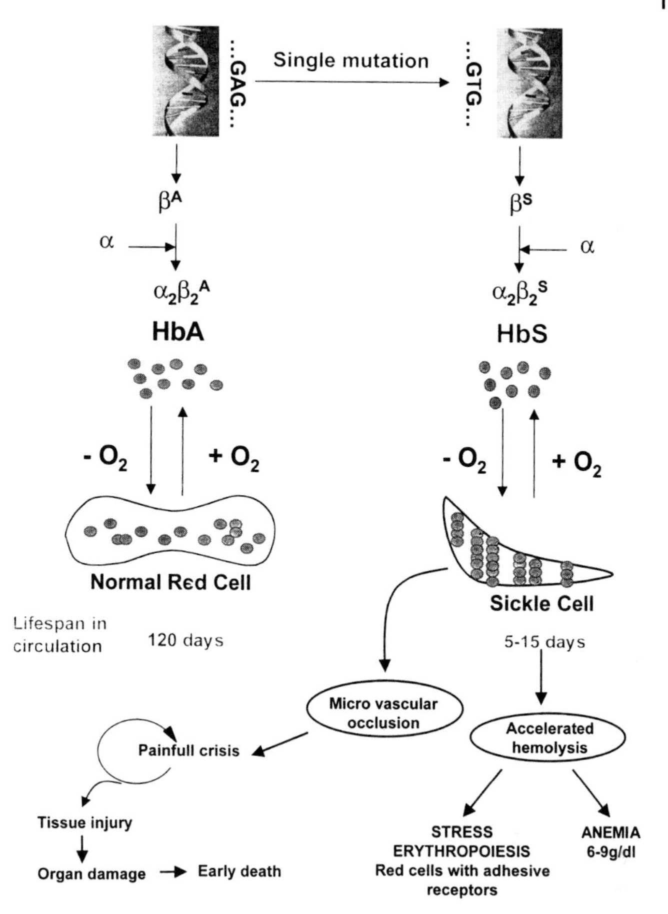

Fig. 11.1 Pathophysiologic scheme of sickle cell anemia (see color plates, p. XXXIV).

(1) The integrin adhesion receptor $\alpha 4\beta 1$, also called very late antigen (VLA-4), is found in greater numbers on sickle reticulocytes than normal reticulocytes. These cells adhere to endothelial cells by binding specifically to endothelial vascular cell adhesion molecule I (VCAM-1), whose expression is upregulated by inflammatory cytokines. This provides a mechanistic link between vasoocclusion and infection, one of the recognized triggers for VOC in SCA [14].

(2) Sickle erythrocytes have increased surface expression of CD36. Microvascular endothelial cells do express CD36. Adhesion between these CD36 molecules of the two cells types can be mediated via the bridging ligand "thrombospondin". Indeed plasma thrombospondin levels are increased in SCA patients with pain crisis [15].

(3) Sickle reticulocytes express the $\beta 3$ integrin GPIIb/IIIa, which binds to the high molecular weight von Willebrand factor (vWF). This complex can then

bind to the endothelium through the vitronectin receptor (VNR). In this context, it is of interest to note that inflammatory cytokines can enhance the endothelial expression of VNR and release of vWF multimers [16].

Endothelial damage is believed to be the consequence of sickle erythrocyte adhesion to endothelial cells. Indeed, histological sections have shown the existence of damaged endothelium in direct contact with sickle erythrocytes [17]. Much more recently, circulating activated microvascular endothelial cells have been found in SCA patients. There is a tendency towards higher levels of circulating endothelial cells at the onset of acute VOC [18]. Together, the data suggest that endothelial cells play a key role in the vascular pathology of SCA. However, further studies are necessary to fully appreciate the role of receptors, ligands and other known and unknown adhesogenic and vasoactive factors in the episodic and temporal initiation/propagation of infarctive events associated with SCA. Overall, both molecular and cellular studies have revealed new mechanisms involved in sickle cell vasoocclusion, which plays a causal role in vasoendothelial injury. Thus, although the central abnormality in SCA is confined to the red cells, the resulting vasculopathy due to endothelial damage occurs virtually in all organs and ultimately leads to organ dysfunction.

11.3
Hydroxy Urea Therapy in Sickle Cell Anemia

A number of clinical therapeutic trials have been carried out in the last decade that attempt to modify the clinical outcome of SCA by decreasing the relative concentration of HbS within the red cells. Based on the finding that the rate of polymerization of HbS is considerably reduced by the presence of fetal Hb (HbF) in the RBC, therapeutic attempts were made to increase the HbF level in SCA patients. Studies both in animals and humans have suggested that cytostatic drugs could be efficient in accomplishing this task. One among them, notably HU, became the choice of several clinical trials for the following reasons: this drug has long been used in treating some neoplastic conditions as well as congenital congestive heart disease as a myeloablative agent, and it has been shown to enhance HbF expression. Given the limited available therapeutic options for SCA, this drug was allowed to be used in clinical trials (in both adults and pediatric patients) for treating SCA patients. A trial of HU was so successful in decreasing symptom severity that it was prematurely ended to allow placebo patients to receive the medication [19]. It is useful to recall again that the major objective of using HU is to enhance the expression of HbF in SCA patients.

Several independent trials have shown that:

- HU is capable of enhancing HbF expression in a substantial number of patients.
- Increment is quite variable from patient to patient (0–30% increment in HbF).

- HU significantly reduced the number of vasoocclusive crisis, hospitalization, rate of transfusion and, most importantly, the incidence of one of the deadly complications of this disease, called "Acute chest syndrome" [19].

However, quite unexpectedly, the following observations were also made:

- Degree of increment in HbF does not correlate consistently with the reduction in VOC and in overall clinical response.
- Clinically the best responders for VOC had the largest decrements in white cell counts, especially neutrophils [20].
- Increments in HbF were much higher among pediatric patients, and more children were "responders" with sustained increase in HbF than adults [21].
- Clinical benefit of HU therapy vanishes rapidly if the therapy is discontinued (frequently due to poor compliance) even if the patients still maintain a high level of HbF (unpublished data).

Although HU may mediate some of its clinical benefits through its positive effect on HbF expression, these data also suggest that it may, by a yet-to-be defined mechanism of action, modulate the clinical severity of SCA. One possible pathway is that among the two incriminated interacting cell partners (sickle erythrocyte and endothelial cell) involved in vaso-adhesion and occlusion, HU may also affect the phenotype status of endothelial cells so that its adhesogenic (structural) and/or vasoregulatory (functional) properties are modified in a favorable manner.

11.4
Issues

Although the clinical benefits of HU therapy are unequivocal for a large number of patients, there are still some serious limitations:

(1) Since HU needs to be taken lifelong in the treatment of non-neoplastic conditions such as SCA (since the inhibition of ribonucleotide reductase is reversible) a major concern is its long term secondary effects. Several studies have shown the potential leukomegenic effect of HU in myeloproliferative disorders [22], [23]. Such concern becomes quite legitimate in sickle cell patients with permanently expanded erythropoiesis in whom the use of HU is at the limit of marrow toxicity. Hence alternative therapies must be sought for.

(2) Both interindividual variations (responders and non-responders) and variable intraindividual temporal efficacy (short-term responders) for HbF expression may reflect the inherited differences in the metabolic pathways of the drug, in their targets, or in their adaptative mechanisms. These aspects remain to be clarified.

In order to design alternative therapeutic strategies and to identify variances that specify the SCA patient population who will benefit from a safe and effective therapy, the present strategies of pharmacogenomics will involve either a targeted candidate gene approach or a global genomic search. We have opted for the former because the knowledge accumulated so far (although indirect) provides a leading thread concerning the mechanism of action of HU (other than its inhibitory effect on ribonucleotide reductase) and its potential targets in SCA. While the hematopoietic system was the intended target in early clinical trials, it is clear now that the significant clinical benefit of HU therapy may be mediated by its effect on other cell types, too. We postulated that the endothelial cells lining the vascular wall, which are in the forefront of the vasoocclusive events, could be the additional target for HU.

11.5 Experimental Study Design

In order to explore the effect of HU on endothelial cells, two different human microvascular endothelial cells were studied: one transformed bone marrow endothelial cell (TrHBMEC), a gift from B. Weskler, and the other, a human umbilical vein endothelial-pulmonary epithelial hybrid cell line (EA-hy 926), kindly provided by C.S. Edgell. Human vascular endothelial cells from blood vessels, although having many functions in common, do differ in some features depending upon their origin. Hence we examined the effect of HU on human endothelial cells of micro- and macrovasculature. The experimental culture conditions were as described before [24] [25]. The cells at confluence (after cell attachment) were exposed for 48 h to varying concentrations of HU (62.5 to 500 μM) both in the presence and absence of the pro-inflammatory cytokines TNFα and Interferon γ, each at 100 U mL^{-1} unless otherwise indicated to simulate the therapeutic dose range used in patients and the intrinsic inflammatory status of SCA patients, respectively.

Culture supernatants were collected under various experimental conditions for the endothelin-1 (ET-1), soluble VCAM-1 (sVCAM-1), and soluble ICAM-1 (sICAM-1) assay. The results are expressed in pg mL^{-1} 5×10^{-5} plated cells for ET-1 and in ng mL^{-1} 5×10^{-5} plated cells for both sVCAM-1 and sICAM-1. Cells were harvested for FACS analysis for the membrane-bound adhesive receptors (mVCAM-1, mICAM-1 and mPECAM-1) using specific monoclonal antibodies. Similarly, RNA transcripts of various candidate targets mentioned above were extracted from these cultures by real-time quantitative using a ABI Prism 7700 sequence detector (PE Applied Biosystem, Foster City, CA, USA). After retrotranscription of the total RNA using reverse transcriptase from Moloney Murine Leukemia Virus and a mixture of random primers, real-time quantitative PCR was performed on total RNA from HU-treated and untreated endothelial cells, and the differential expression of target transcripts were then assessed.

The PCR primers and the fluorogenic probes for the studied targets were designed using the Primer Express Software 1.7 (PE Applied Biosystem, Foster City, CA, USA). To normalize the quantitative data, specific probes for the TATA-binding protein mRNA were used as an internal control.

11.6
Major Effect of HU

The study on the differential expression of various candidate targets by two different endothelial cells in culture revealed that the major effect of HU was restricted to ET-1 and ICAM-1 and to a limited extent to VCAM-1 in both cell types [26]. Hence only these factors will be discussed in detail.

The amount of released ET-1 peptide was different for the two cell types at basal conditions, and was not altered by the presence of pro-inflammatory cytokines. The presence of HU significantly decreased the ET-1 peptide release from these two cell types (52% and 64% reduction for TrHBMEC and EA-hy 926 cells, respectively) under basal culture conditions (Figure 11.2). The magnitude of reduction

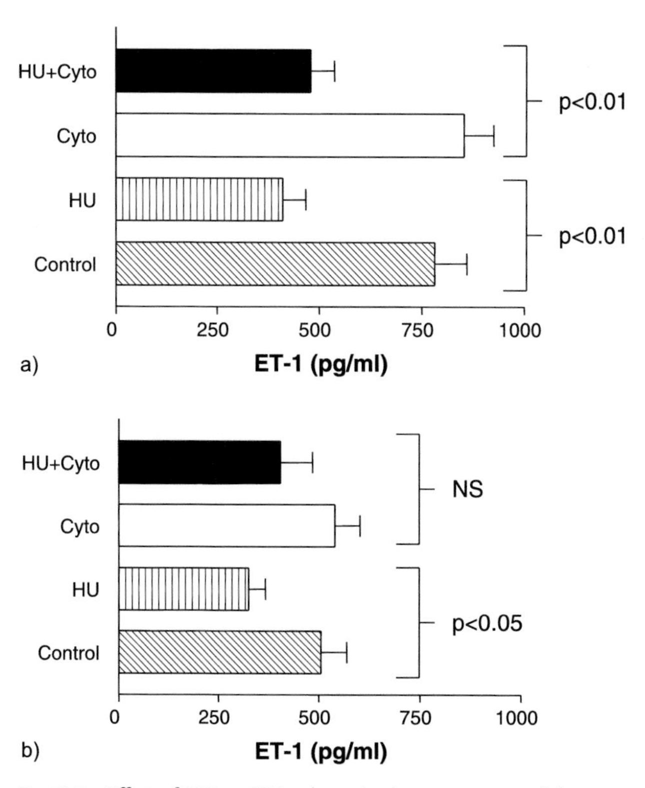

Fig. 11.2 Effect of HU on ET-1 release in the supernatant of the TrHBMEC (**a**) and EA hy 926 (**b**) cell cultures. Results of the quantitative assessment of ET-1 by Elisa, from at least four independent experiments in duplicates, are expressed in pg of ET-1 mL^{-1} supernatant per 5×10^5 plated cells. Control: basal culture conditions, HU: cells treated with HU 250 mM during 48 h, Cyto: cells treated with cytokines TNFα and IFNγ at 100 U/mL^{-1} during 48 h, HU+cyto: combination of HU and cytokines.

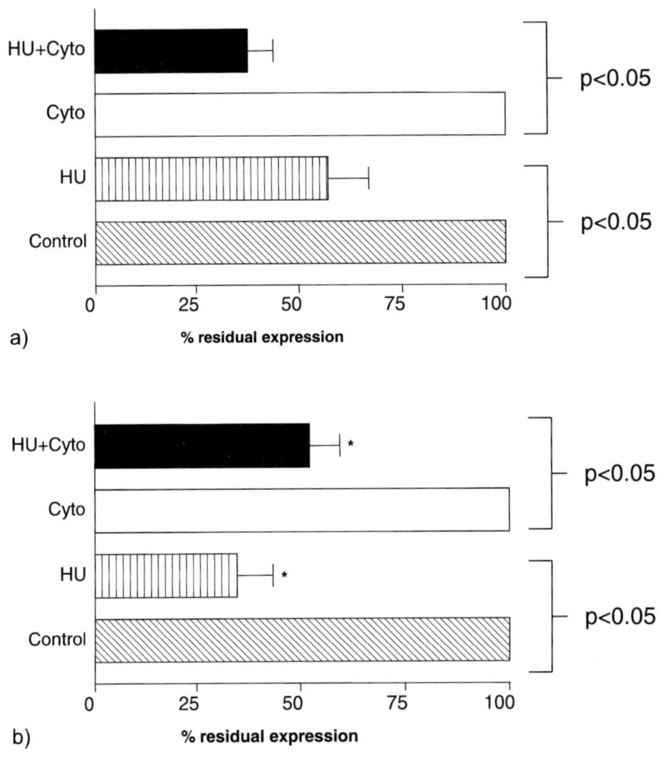

Fig. 11.3 Effect of HU on ET-1 mRNA expression in the TrHBMEC (**a**) and EA-hy 926 (**b**) endothelial cells in culture. Quantitative real-time PCR was used to assess the level of ET-1 mRNA in at least four independent experiments in duplicate. Results are expressed in percentage of residual ET-1 mRNA expression for HU-treated cells as compared to the control (culture with or without cytokines). The TATA-binding protein mRNA was used as an internal control. The abbreviations are the same as in the legend for Figure 11.2.

for both cells types was similar in the presence of inflammatory cytokines. In the presence of HU, there was also a concomitant decrease in ET-1 mRNA (63% and 48% reduction for TrHBMEC and EA-hy 926 cells, respectively) as revealed by real-time quantitative PCR analysis. The results obtained are shown in Figure 11.3. When the TrHBMEC cells were incubated with HU concentrations ranging from 62.5 μM to 500 μM, the decrease in ET-1 peptide release and mRNA levels were found to be dose-dependent (Figure 11.4). We also show that this phenomenon is reversible. Indeed, after exposure of cells to HU (without cytokines) and its removal after 48 h, the expression of ET-1 mRNA was found to be 39% of the basal level but rose to 77% (Figure 11.5). In the presence of HU and cytokines, ET-1 mRNA expression was only 35% of the basal level, but after removal of HU, there was a marked recovery (89%). A similar trend, although less marked

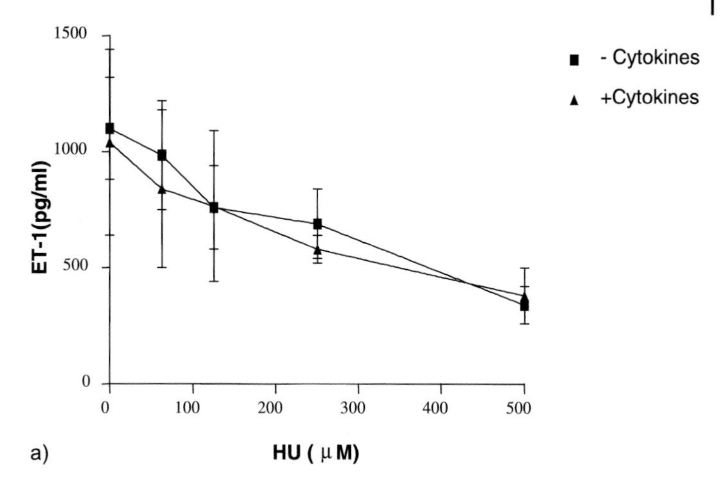

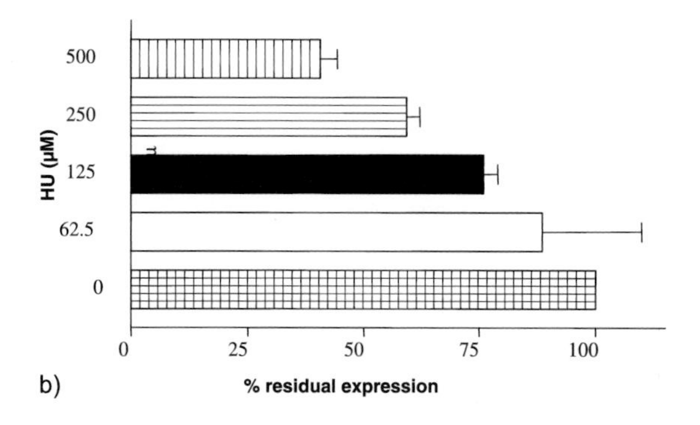

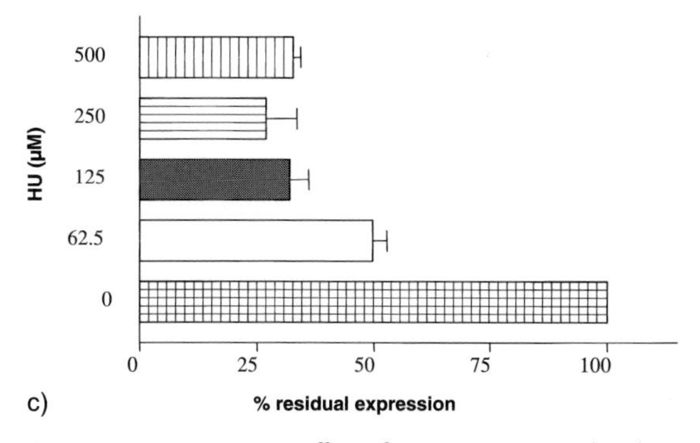

Fig. 11.4 Dose–response effect of HU on ET-1 peptide release (**a**) from TrHBMEC cells incubated with various concentrations of HU during 48 h in the presence (▲) and absence (■) of cytokines (TNFα and IFNγ at 100 U mL^{-1}). Under the same conditions, quantitative mRNA analysis was also performed and the residual percentage of expression, in the presence (**b**) and absence (**c**) of pro-inflammatory cytokines is given.

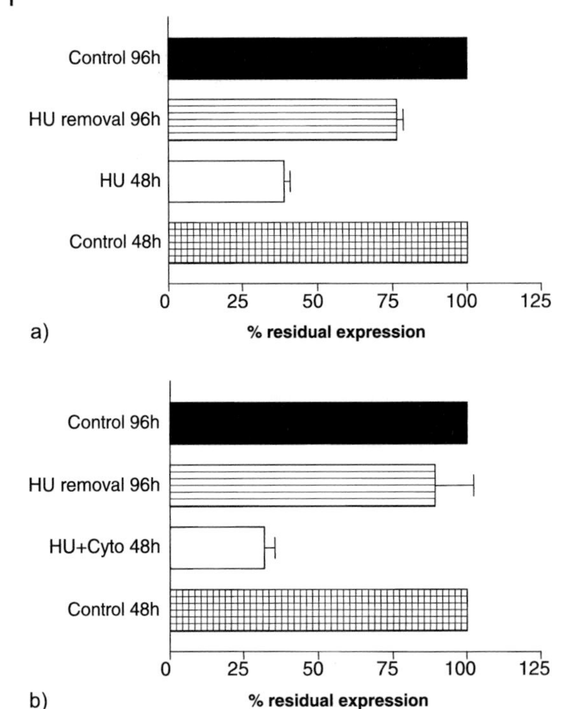

Fig. 11.5 Restoration of ET-1 mRNA expression after the removal of HU. TrHBMEC cells were treated with HU at 250 μM for 48 h and the culture medium was then replaced by a fresh one free of HU (HU removal) and incubated for an additional 48 h. The percentage of residual expression of ET-1 mRNA from three independent experiments is given both in the presence (**a**) and absence (**b**) of cytokines.

(needs extended culture duration), was noted for ET-1 peptide release (Figure 11.6). Differences in the response of endothelial cells from the micro- and macrovasculature were observed in these experiments.

Flow cytometry was used to investigate the effect of HU on the expression of mICAM-1 by TrHBMEC and EA-hy 926 cells. The results are based on the analysis of 5000 events. In the presence of HU alone, TrHBMEC cells (Figure 11.7 a) exhibited a slight increase in mICAM-1 expression as compared to the basal values. In the presence of pro-inflammatory cytokines a significant increase in mICAM-1 expression was observed. When these cells were incubated both with HU and cytokines, a synergistic effect on the expression of mICAM-1 was observed. As shown in Figure 11.7 b, the overall trend was quite similar for EA-hy 926 cells. HU, in presence of pro-inflammatory cytokines, induces a two-fold increase in the release of sICAM-1 in the culture supernatant (19.84 ng $mL^{-1} \pm 10.16$) as compared to that of cytokines alone (9.19 ng $mL^{-1} \pm 5.75$). Under basal conditions, (without cytokines), no sICAM-1 was detectable in the supernatant of TrHBMEC cells (Figure 11.7 c).

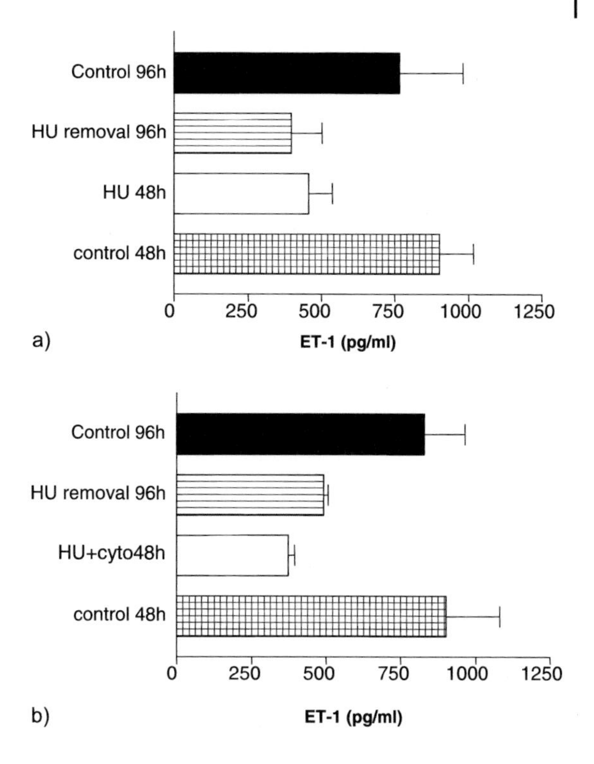

Fig. 11.6 Restoration of ET-1 peptide release (pg mL^{-1}) after the removal of HU in the presence (**a**) and in the absence of (**b**) cytokines. The other experimental conditions were identical to those described in the legend to Figure 11.5.

11.7 ICAM-1

Intercellular adhesion molecule-1 (ICAM-1) is a membrane-bound glycoprotein belonging to the immunoglobulin superfamily. It is constitutively expressed in vascular endothelium and upregulated by cytokines during inflammation. It plays a key role in leukocyte adhesion to endothelial cells as well as in immune response. It has been shown that sickle erythrocytes, under flow conditions upregulate the expression of ICAM-1 mRNA and membrane-bound ICAM-1 protein production (up to 6-fold). The release of the soluble form of ICAM-1 (sICAM-1) was also enhanced. Although normal RBCs elicit a qualitatively similar response, the overall magnitude of increase is much smaller [27].

Although ICAM-1 seems not to be involved in the sickle cell adhesion to vascular endothelium, it may exacerbate vasoocclusion by promoting leukocyte adhesion. In this context, it is remarkable to note that in HU-treated SCA patients, the strongest correlation was found between total white cell count and severity of crisis rather than with erythrocyte-related parameters [28]. The current consensus is that leukocyte endothelium adhesion may initiate vasoocclusion followed by RBC sequestration and entrapment in the microvascular lumen with ensuing painful crisis. Thus overexpression of ICAM-1 is expected to promote VOC. The data

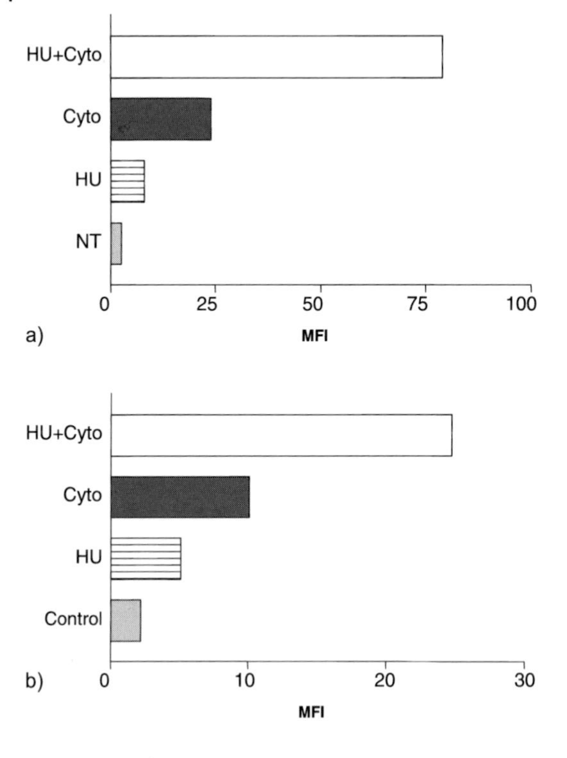

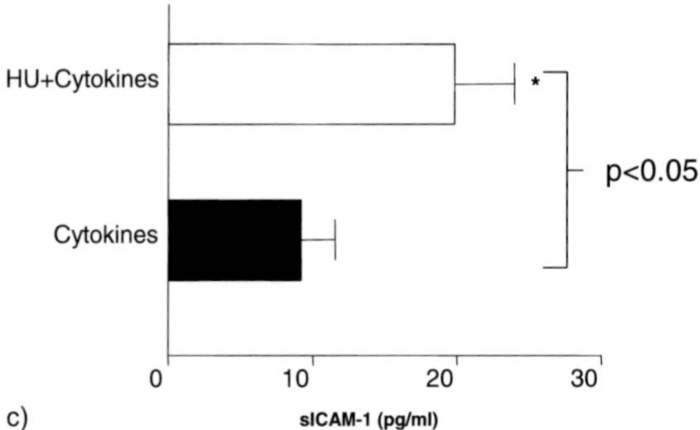

Fig. 11.7 Effect of HU on mICAM-1 expression in the TrHBMEC (**a**) and EA-hy 926 (**b**). These cells were incubated with HU 250 μM for 48 h with or without 100 U mL^{-1} of TNFα and IFNγ. mICAM-1 cellular expression was analyzed by flow cytometry. Results are the Mean Fluorescent Index (MFI) of one representative experiment, with overall trend in three other independent experiments being comparable. Parallel estimation of sICAM-1 release in the culture supernatant of TrHBMEC cells (6 independent experiments) revealed that without cytokines sICAM-1 was not detectable in the supernatant for the basal conditions. The results of HU-treated cells (**c**) in the presence of cytokines showed a significant increase in release of sICAM-1 ($p<0.05$).

showing that HU enhances the expression of mICAM-1 by endothelial cells *in vitro* is quite disparaging in the face of remarkable clinical benefits conferred by HU *in vivo*. One of the explanations to this paradox may reside with the HU-induced marked increase in sICAM-1 parallel to that of mICAM-1. Indeed, it has been shown that sICAM-1 and soluble selectins significantly decrease the neutrophil-endothelial adhesion *in vitro*. Hence the beneficial effect of HU may be mediated through enhanced release of soluble forms of adhesive receptors [29, 30], which may outweigh the effect of mICAM-1. These data suggest that anti-adhesion therapies may become a useful alternative in the treatment of SCA.

11.8
ET-1

ET-1 is a physiologically long-acting mediator of vasoconstriction and inflammation, secreted by vascular endothelial cells and vascular smooth muscle cells. Plasma levels of ET-1 are significantly elevated in SCA patients in the steady state as compared to normal individuals and during painful crisis [31]. Elevated ET-1 levels have been shown to contribute to the development of ACS [32]. *In vitro* studies have demonstrated that sickle erythrocytes interact with vascular endothelial cells to stimulate the expression of ET-1 mRNA and ET-1 peptide release from these cells in a specific manner. Both ET-1 production and release seem to be mediated by a yet-to-be characterized soluble diffusible factor(s) from sickle erythrocytes, since the direct contact between sickle erythrocytes and endothelial cells is not required [33]. Elevated plasma ET-1 in SCA patients is considered as a marker for endothelial damage. By its vasoconstrictor property, ET-1 could decrease the vascular diameter of blood vessels locally and cause the transit time of sickle erythrocytes to be longer. Thus ET-1 could contribute to the vasoocclusion through exacerbated sickling and entrapment of sickled erythrocytes.

Elevated ET-1 in SCA patients, even in the steady state, may play an important role in the dehydration of sickle erythrocytes and the resulting enhanced intraerythrocytic HbS polymerization. Indeed, it has been shown that ET-1 activates Ca^{2+}- gated K^{+} channels in mouse erythrocytes [34]. ET-1, as a pro-inflammatory agonist, has been shown to induce the production of inflammatory cytokines by monocytes. One of the cytokines, namely TNFα enhances the adherence of sickle erythrocytes to vascular endothelium [35]. In addition, endothelins upregulate the expression of endothelial adhesion molecules such as ICAM-1, VCAM-1 and E-selectin, which participate in the recruitment of white cells to the site of inflammation. The overall conclusions that can be drawn from these data is that ET-1 plays a critical role in the vasospasm and inflammation that result in VOC. The major effect of HU in ameliorating the clinical symptoms of SCA likely results from its ability to inhibit the chronically activated ET-1 expression in SCA patients.

In conclusion, the use of specific endothelin antagonists must be considered in the treatment of sickle cell disease as an alternative to HU, whose potential long-term mutagenic/carcinogenic effect is still under debate [36], especially in children.

11.9
Conclusion

As in any field, extrapolation of *in vitro* data to the *in vivo* context warrants caution. Nevertheless, as revealed in this study, experimentally demonstrated down-regulation of ET-1 gene expression by HU corroborates the *in vivo* findings where ET-1 levels correlate with the clinical stage of SCA. Similarly, the balance between soluble and membrane-bound adhesive receptors of endothelial and other cells may dictate the triggering of acute clinical events. Individualized therapies for

SCA may be envisaged based on biological parameters such as ET-1 and sICAM-1, which are the presently known inter- and intra-individual endothelial cell-related variables in SCA. Ongoing research will hopefully decipher the transcriptional signature of HU in endothelial cells and lead to other treatment options for this simple monogenic yet a complex disorder.

11.10 References

1 Gwilt PR, Tracewell WG. Pharmacokinetics and pharmacodynamics of hydroxyurea. Clin Pharmacokinet **1998**, 34:347–358.

2 Yarbro JW, Kennedy BJ, Barnum CP. Hydroxyurea inhibition of DNA synthesis in ascites tumor. Proc Natl Acad Sci USA **1965**; 53:1033–1035.

3 Krakoff IH, Brown NC, Reichard P. Inhibition of ribonucleoside diphosphate reductase by hydroxyurea. Cancer Res **1968**; 28:1559–1565.

4 Graslund A, Ehrenberg A, Thelander L. Characterization of the free radical of mammalian ribonucleotide reductase. J Biol Chem **1982**; 257:5711–5715.

5 Evered DF, Selhi HS. Transport characteristics of two carcinostatic compounds, hydroxyurea and hadacidin, with rat small intestine. Biochem J **1972**; 126:26P.

6 Morgan JS, Creasey DC, Wright JA. Evidence that the antitumor agent hydroxyurea enters mammalian cells by a diffusion mechanism. Biochem Biophys Res Commun **1986**; 134:1254–1259.

7 Fabricius E, Rajewsky F. Determination of hydroxyurea in mammalian tissues and blood. Rev Eur Etud Clin Biol **1971**; 16:679–683.

8 Colvin M, Bono VH Jr. The enzymatic reduction of hydroxyurea to urea by mouse liver. Cancer Res. **1970**; 30:1516–1519.

9 Andrae U. Evidence for the involvement of cytochrome P-450-dependent monooxygenase(s) in the formation of genotoxic metabolites from N-hydroxyurea. Biochem Biophys Res Commun **1984**; 118:409–415.

10 Yarbro JW. Mechanism of action of hydroxyurea. Semin Oncol **1992**; 19:1–10.

11 Pauling L, Singer SJ, Wells JC. Sickle cell anemia: a molecular disease. Science. **1949**; 110:543–548.

12 Bunn HF. Pathogenesis and treatment of sickle cell disease. N Engl J Med 1997; 337:762–769.

13 Hebbel R. Adhesive interactions of sickle erythrocytes with endothelium. J Clin Invest **1997**; 100:S83–86.

14 Gee BE, Platt OS. Sickle reticulocytes adhere to VCAM-1. Blood **1995**; 85:268–274.

15 Browne PV, Hebbel RP. CD36-positive stress reticulocytosis in sickle cell anemia. J Lab Clin Med **1996**; 127:340–347.

16 Wick T, Kaye N, Jensen W. Unusually large von Willebrand factor multimers increase adhesion of sickle erythrocytes to human endothelial cells under controlled flow. N Engl J Med **1982**; 337:1584–1590.

17 Klug PP, Kaye N, Jensen WN. Endothelial cell and vascular damage in the sickle cell disorders. Blood Cells **1982**; 8:175–184.

18 Solovey A, Lin Y, Browne P, Choong S, Wayner E, Hebbel RP. Circulating activated endothelial cells in sickle cell anemia. N Engl J Med **1997**; 337:1584–1590.

19 Charache S, Terrin ML, Moore RD, Dover GJ, Barton FB, Eckert SV, McMahon RP, Bonds DR. Effect of hydroxyurea on the frequency of painful crises in sickle cell anemia. Investigators of the Multicenter Study of Hydroxyurea in Sickle Cell Anemia [see comments]. N Engl J Med **1995**; 332:1317–1322.

20 Charache S, Barton FB, Moore RD, Terrin ML, Steinberg MH, Dover GJ, Ballas SK, McMahon RP, Castro O, Orringer EP. Hydroxyurea and sickle cell anemia. Clinical utility of a myelosuppressive "switching"

agent. The Multicenter Study of Hydroxyurea in Sickle Cell Anemia. Medicine (Baltimore) **1996**; 75:300–326.

21 Maier-Redelsperger M, Labie D, Elion J. Long-term hydroxyurea treatment in young sickle cell patients. Curr Opin Hematol **1999**; 6:115–120.

22 Sterkers Y, Preudhomme C, Lai JL, Demory JL, Caulier MT, Wattel E, Bordessoule D, Bauters F, Fenaux P. Acute myeloid leukemia and myelodysplastic syndromes following essential thrombocythemia treated with hydroxyurea: high proportion of cases with 17p deletion. Blood **1998**; 91:616–622.

23 Weinfeld A, Swolin B, Westin J. Acute leukaemia after hydroxyurea therapy in polycythaemia vera and allied disorders: prospective study of efficacy and leukaemogenicity with therapeutic implications. Eur J Haematol **1994**; 52:134–139.

24 Schweitzer KM, Vicart P, Delouis C, Paulin D, Drager AM, Langenhuijsen MM, Weksler BB. Characterization of a newly established human bone marrow endothelial cell line: distinct adhesive properties for hematopoietic progenitors compared with human umbilical vein endothelial cells. Lab Invest **1997**; 76:25–36.

25 Edgell CJ, McDonald CC, Graham JB. Permanent cell line expressing human factor VIII-related antigen established by hybridization. Proc Natl Acad Sci USA. **1983**; 80:3734–3737.

26 Brun M, Bourdoulous S, Couraud PO, Elion J, Krishnamoorthy R, Lapouméroulie C. Effect of hydroxyurea on endothelial cells in culture [abstract] 25th annual meeting of the National Sickle Cell Disease Program. New York, NY, USA **2001**; p 42a.

27 Shiu YT, Udden MM, McIntire LV. Perfusion with sickle erythrocytes up-regulates ICAM-1 and VCAM-1 gene expression in cultured human endothelial cells. Blood **2000**; 95:3232–3241.

28 Platt OS, Brambilla DJ, Rosse WF, Milner PF, Castro O, Steinberg MH, Klug PP. Mortality in sickle cell disease. Life expectancy and risk factors for early death. N Engl J Med **1994**; 330:1639–1644.

29 Ohno N, Ichikawa H, Coe L, Kvietys PR, Granger DN, Alexander JS. Soluble selectins and ICAM-1 modulate neutrophil-endothelial adhesion and diapedesis in vitro. Inflammation. **1997**; 21:313–324.

30 Kusterer K, Bojunga J, Enghofer M, Heidenthal E, Usadel KH, Kolb H, Martin S. Soluble ICAM-1 reduces leukocyte adhesion to vascular endothelium in ischemia-reperfusion injury in mice. Am J Physiol **1998**; 275:G377–380.

31 Rybicki AC, Benjamin LJ. Increased levels of endothelin-1 in plasma of sickle cell anemia patients [letter]. Blood **1998**; 92:2594–2596.

32 Hammerman SI, Kourembanas S, Conca TJ, Tucci M, Brauer M, Farber H.W. Endothelin-1 production during the acute chest syndrome in sickle cell disease. Am J Respir Crit Care Med **1997**; 156:280–285.

33 Phelan M, Perrine SP, Brauer M, Faller DV. Sickle erythrocytes, after sickling, regulate the expression of the endothelin-1 gene and protein in human endothelial cells in culture. J Clin Invest **1995**; 96:1145–1151.

34 Rivera A, Rotter MA, Brugnara C. Endothelins activate Ca(2+)-gated K(+) channels via endothelin B receptors in CD-1 mouse erythrocytes. Am J Physiol **1999**; 277:C746–754.

35 Vordermeier S, Singh S, Biggerstaff J, Harrison P, Grech H, Pearson TC, Dumonde DC, Brown KA. Red blood cells from patients with sickle cell disease exhibit an increased adherence to cultured endothelium pretreated with tumour necrosis factor (TNF). Br J Haematol **1992**; 81:591–597.

36 Hanft VN, Fruchtman SR, Pickens CV, Rosse WF, Howard TA, Ware RE. Acquired DNA mutations associated with *in vivo* hydroxyurea exposure. Blood **2000**; 95:3589–3593.

12
Pharmacogenomics and Complex Cardiovascular Diseases – Clinical Studies in Candidate Genes

Bernhard R. Winkelmann, Markus Nauck, Michael M. Hoffmann
and Winfried März

Abstract

Pharmacogenetics of candidate genes is the focus in clinical research, while pharmacogenomics with array-based transcript profiling using high-density oligonucleotide arrays ("RNA/DNA chips") is still the domain of pharmaceutical drug discovery. A major limitation to pharmacogenetic/-genomic research is our incomplete knowledge in defining the complex clinical phenotypes and the complexity of physiologic pathways interacting with environmental stimuli in individuals susceptible to diseases such as coronary artery disease, myocardial infarction, hypertension or type 2 diabetes mellitus and related traits.

Although the renin–angiotensin–aldosterone system (RAAS) is clearly implicated in cardiovascular disease, the impact of genetic variation in the RAAS on disease susceptibility and the response to cardiovascular drugs remains to be elucidated. Another prominent example are G-protein-coupled β-adrenergic receptors, which mediate positive inotropic, chronotropic and dromotropic effects of endogenous catecholamines, and play a major role in thermogenesis and lipid mobilization in adipose tissue. Genetic variants of these receptors with different functionalities have been identified. However, initial results need to be confirmed in adequately powered studies. Future research strategies should focus on many mutations (i. e., haplotypes) in many genes and evaluate entire pathways in a cluster of conditions, instead of investigating just one genetic variant in a complex phenotype misclassified as one disease.

12.1
Introduction

Pharmacogenomics is an emerging field in drug development and clinical medicine and may be defined operationally as the use of genomics for the identification of new drug targets as well as the study of the association between genetic variation and individual differences in drug response. Thus pharmacogenomics makes use of the technological advances in high-throughput DNA and mRNA analysis to elucidate genetic determinants of drug effects and toxicity, and to study the effects of therapeutic agents on the pattern of gene expression in tissues. Pharmacogenomics focuses on genes, RNA transcripts, and their encoded proteins (i.e., the genome, transcriptome and proteome) and seeks to define the effects of drugs on gene expression patterns and protein synthesis in cells, tissues, and organ systems. It is not enough to only investigate nucleotide sequences, because even if cells in a human population were based on essentially the same genome (i.e., 100% identity in nucleotide sequence), operationally they would still differ considerably due to epigenetic modification and exposure of cells to environmental factors. Therefore, in order to assess the variability of responses to therapeutic interventions from the disease, population or individual human point of view, the heterogeneity among cells and tissues in their genome, transcriptome and proteome profiles needs to be assessed in parallel at all three levels. Operome is a new concept unifying those three levels, and operomics is defined as the molecular research of the operome [1]. Thus, strictly speaking, pharmacogenomics is technically "pharmaco-operomics".

Pharmacogenetics, in contrast, may be viewed as the subset of pharmacogenomics focusing on the genome, particularly the variation in nucleotide sequence of candidate genes with respect to drug action. The pharmacogenetic field is moving rapidly to assemble a large collection of polymorphisms in order to relate genetic diversity to drug response. In this respect, it is important to define a quantifiable clinical drug response, that is rather a quantitative phenotype (i.e., blood pressure response) as opposed to the qualitative phenotype (i.e., hypertension) of classical genetics [2].

Phenotypic consequences of polymorphisms with respect to drug response are presumably associated with changes at the RNA and protein level. Thus, array-based transcript profiling may be used in clinical trials to analyze patient tissues in response to drug therapy. However, in the past, expression-based *in vivo* studies were mostly a domain in cancer research where RNA can be obtained from biopsies and surgical specimen [3, 4]. In the cardiovascular field, high-density oligonucleotide arrays have been used in an *ex vivo* investigation of cardiac tissue from explanted hearts [5]. The progress in this new field of research is dramatic. Today, at least 50,000 RNAs contained on a single microarray can be investigated and the results stored in a database within 48 hours – a quantity of data that takes 20 years to generate by one researcher using Northern blot analysis.

Proteins remain the markers of choice since they are more closely related to the clinical phenotype. However, not the study of one or a few biochemical marker proteins (used as surrogates in clinical studies long ago before the new field of pharmacogenomics emerged), but the study of all proteins in blood or tissues (i.e., proteomics) by methods such as two-dimensional gel electrophoresis coupled with mass spectrometry distinguishes pharmacogenomic research on proteins [6]. While the impact of large-scale gene expression studies on drug discovery and drug development is a reality today, this is not yet the case with clinical cardiovascular medicine, where the focus has been on the effects of genetic variation in candidate genes on drug action. It is, therefore, pharmacogenetics, as a "subspecialty" of pharmacogenomics, that will be primarily discussed here. However, mRNA expression profiling (i. e., transcriptomics) and proteomics may emerge as powerful approaches for directly identifying predictive markers (or protein expression patterns) in blood in the future [6].

12.2 Complexity of Clinical Phenotypes

Complex genetic diseases are also labeled as multifactorial diseases and are defined as clinical entities where risk conditions have been identified, but the actual causes of disease remain largely unknown [7]. Examples are atherosclerosis-related phenotypes, i. e., coronary artery disease, myocardial infarction, stroke, or peripheral arterial occlusive disease. The metabolic disorders type 2 diabetes mellitus and related traits such as obesity, metabolic syndrome, or primary arterial hypertension are further examples of complex genetic diseases that often coexist as risk factors in individuals with atherosclerotic phenotypes. For all these phenotypes, rare monogenetic causes exist, but these are of minor importance from a population point of view and will not be discussed here. As a heterogeneous group, individuals with atherosclerosis generally have one or more risk factors including dyslipidemia and environmental factors (i. e., smoking, nutrition). An isolated atherosclerotic phenotype without the additional clinical risk factors mentioned above is practically non-existent.

It is important to realize, that diseases such as myocardial infarction or type 2 diabetes represent a heterogeneous group of several distinct subphenotypes definable by clinical or biochemical characteristics. Thus, contradictory findings in pharmacogenomic studies may only not be the consequence of a lack of a major isolated gene effect (of the gene variant studied) and chance findings, but also be caused by the variability in the mix of distinct clinical phenotypes hidden beneath a common characterization such as type 2 diabetes and modulated by differences in the environment between studies.

12.3 Limitations

This overview concentrates on exemplary candidate genes and discusses the impact of select genetic variants on drug action in cardiovascular phenotypes. It is impossible to be exhaustive, neither with regard to the number of candidate genes, to the genetic variants within those genes, nor to the distinct cardiovascular phenotypes investigated in pharmacogenomic and pharmacogenetic studies in humans. Although pharmacogenetic studies have been performed as early as in the 1960s, it is only lately with the availability of new genomic tools and technologies that the number of publications has tripled. A Medline search using the MESH indexing term pharmacogenetic as keyword identified 1,165 articles, while 22 were found for the keyword pharmacogenomic (not a MESH term), and a

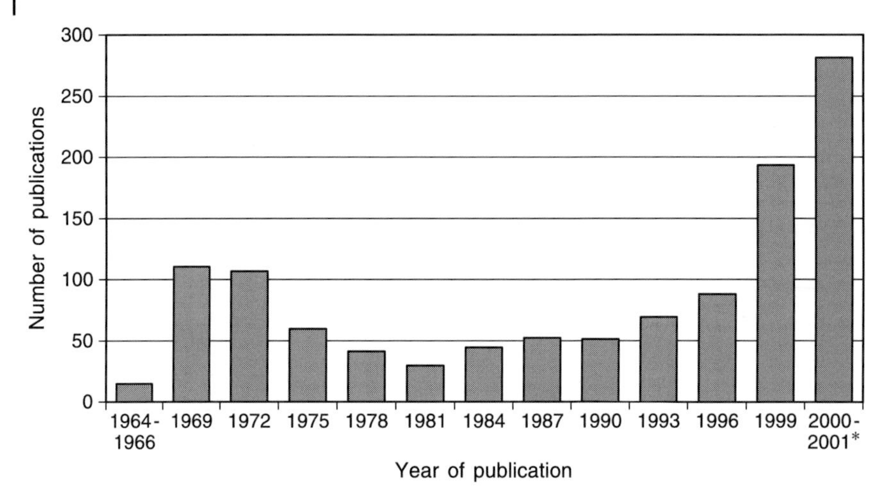

Fig. 12.1 Number of publications identified in a Medline search using pharmacogenetics as keyword (* Search for the entire year 2000 and the first half of 2001).

mere 4 articles for the combination pharmacogenetic/-genomic and cardiovascular. There was a peak in publications indexed under the keyword pharmacogenetics in the early seventies, and an exponential rise in numbers lately. As of July 2001, the number of articles indexed in the first half of 2001 under the heading pharmacogenetics rose to 164, more than in any entire previous year (Figure 12.1).

12.4 Studies of the Impact of Nucleotide Sequence Variation on Drug Effects

12.4.1 Angiotensin I-Converting Enzyme Gene

Angiotensin I-converting enzyme (ACE) is an ubiquitous carbopeptidase (EC 3.4.15.1), which is expressed in the membrane of the endothelial lining of the vasculature. It is a Zn^{2+}-metallopeptidase which acts on many substrates, particularly on two important cardiovascular hormonal regulatory systems. It cleaves angiotensin I to generate the active hormone angiotensin II, and, in parallel, it degrades bradykinin to an inactive metabolite. Since the enzyme has two active sites exposed at the luminal surface of the endothelium, gene duplication is suggested during its evolution [8]. A genetic variant explaining half of the interindividual variability in plasma ACE activity has been described. Although probably not the causative genetic defect, the intronic variation seems to be in tight linkage with the causative variant, which is up to now unknown. This ACE insertion/deletion (ACE ID) gene polymorphism results from the presence or absence of a 287-base pair DNA fragment in intron 16 of the ACE gene. Plasma ACE activity is almost twice as high in carriers homozygous for the D allele as compared to individuals with the II genotype [9, 10].

ACE is not rate-limiting for the production of angiotensin II, and angiotensin II in plasma and tissues appears independent from ACE activity or ACE genotype [11, 12]. However, ACE genotype-dependent differences of endothelial function have been reported not only in plasma [13], but also in human arteries [14]. Prasad et al. reported in a study of 56 patients with endothelial dysfunction that enalaprilat potentiated the coronary microvascular and epicardial response to acetylcholine in carriers of the DD genotype, despite similar responses to acetylcholine at baseline [15]. The authors speculated that the conversion of Ang II from angiotensin I is higher in individuals with a DD genotype, which, however, stands in contrast to published data [11, 12].

Although initial positive findings on a relationship between clinical phenotypes and ACEID polymorphism, such as myocardial infarction [16] and hypertension [17], were not confirmed subsequently [10, 11, 18–20], this does not rule out a minor effect of ACEID genotype. It is almost to be expected that a single genetic variant can contribute but a minor part to a complex genetic disease [21]. Although the genetic variant might still be associated with an altered antihypertensive response, especially using ACE inhibitors, this was not the case, neither in controlled studies in healthy subjects [22] nor in clinical studies in patients [23].

The chronic blood pressure lowering effects of ACE inhibitors cannot be explained by a reduction in angiotensin II. Once a critical level of ACE suppression has been reached, the ACE activity itself cannot become rate-limiting for Ang II production because non-ACE enzyme-dependent pathways are recruited [24]. Due to the fact that Ang II plasma concentration returns to normal, or is even slightly increased after feedback stimulation of renin during ACE blockade, the long term benefit of ACE inhibitors needs to be explained by other mechanisms such as increased bioavailability of NO via the bradykinin pathway or vasodilatory effects of Ang II metabolites such as Ang-(1-7) [25, 26]. The role of receptor downregulation in this context is unclear. Buikema et al. suggested that the Ang II receptor density may be diminished in DD subjects [14] and Hopkins et al. suggested, that long-term exposure to elevated Ang II levels results in a decrease in angiotensin receptor density [24]. The importance of bradykinin metabolism is supported indirectly by the lack of effect of chronic ACE inhibition on Ang II.

ACE as kininase II promotes the degradation of bradykinin and seems rate-limiting in the bradykinin pathway, which may explain the effects linked to ACE genotype and differences in ACE activity. The results of several studies investigating endothelial function in different vascular beds support such reasoning. The enhancement of serotonin-induced vasodilation in forearm blood flow by enalapril, but not by valsartan, an angiotensin AT_1 receptor blocker, has been explained by a decreased degradation of bradykinin with subsequent increased nitric oxide release via bradykinin-induced stimulation of nitric oxide synthase [27]. Furthermore, infusion of bradykinin into the brachial artery of 27 volunteers resulted in a stepwise decrease in bradykinin concentrations depending on ACE ID polymorphism [28]. In another study of intra-arterial administration of bradykinin in 28 men under salt-controlled conditions, forearm blood flow improved most in those subjects who carried the ACE D allele [29]. It was reasoned that in carriers of an ACE D allele endogenous degradation of bradykinin is accelerated and the response to exogenous bradykinin would thus be most pronounced [30]. However, the opposite has been reported for the coronary response to bradykinin: epicardial bradykinin response was depressed in patients with high ACE levels or with the ACE DD genotype [31]. In another study on the coronary vasomotion in 177 patients with coronary atherosclerosis, the response to acetylcholine was not modulated by ACE genotype, but the vasodilator response to exogenous nitric oxide was depressed in DD patients. Thus patients who had the D allele appeared to have increased vascular smooth muscle tone that was counterbalanced by higher basal nitric oxide activity in carriers of a D allele [32]. A link between ACEID genotype and a blunted endothelium-dependent vasodilation (i.e., endothelial function assessed by forearm blood flow) has been reported in healthy volunteers and in hypertensive patients by other research groups, as well [33, 34].

Taken together, ACEID genotype seems to modulate endothelial vasodilator response as demonstrated by investigation of several physiological mediators of the endothelium and after acute [15] and chronic treatment with an ACE inhibitor [35]. High ACE activity associated with the ACE DD genotype has been linked to an excess in left ventricular growth and decreased endurance capability during structured physical training [36, 37]. The AT_1 receptor blocker losartan did not prevent the excess in left ventricular hypertrophy as seen in DD subjects. The authors speculated that effects on left ventricular growth might either be mediated by the effects of angiotensin II on other (non-AT_1) receptors or by a lower degradation of growth-inhibitory kinins (i.e., bradykinin) [37].

Finally, a pharmacokinetic interaction between betablocker therapy and ACEID genotype has been observed in patients with congestive heart failure. Almost 90% of patients were treated with an ACE inhibitor. In the entire cohort of 328 patients transplant-free survival was significantly poorer in patients with a D allele. This adverse outcome was prevented by concomitant treatment with betablocker and enhanced in patients without betablockers [38].

In conclusion, the impact of the ACEID genotype on clinical (disease) phenotypes such as myocardial infarction and hypertension is weak to non-existent. However, the ACE gene polymorphism seems to interact with physiological pathways such as endothelial function and exercise-induced left ventricular growth. A variety of functional effects modulated by ACEID genotype were observed in human studies on endothelial function after infusion of vasodilator and vasopressor substances. Interestingly, such results fit well with experimental findings on ACE knockout mice. Although basal blood pressure remained unaffected in mice with only one active ACE gene (blood pressure was lower though in mutants without any active ACE gene), the response to angiotensin I and bradykinin was altered from wild types indicating compensatory adaptive responses [39].

12.4.2
Angiotensinogen Gene

A polymorphism in the angiotensinogen gene, M235T, where the T allele is associated with higher plasma angiotensinogen has been linked to elevated blood pressure [40, 41] and myocardial infarction [41, 42]. However, such findings were not reproduced in later studies [43, 44]. Even in our own study that reported a positive association of the M235T angiotensinogen gene polymorphism with diastolic blood pressure the gene variant explained only 2.5% of the variance in diastolic pressure [41]. In the Copenhagen City Heart study where 9,100 men and women were genotyped for the M235T and T174M polymorphisms, no significant differences across genotypes, alone or in combination, were observed in systolic or diastolic blood pressures. However, elevated blood pressure (defined as a blood pressure ≥140 mmHg systolic, and/or ≥90 mmHg diastolic or treatment with antihypertensive medication) was significantly more frequent in women homozygous for the 235 T allele or for both the 235 T and 174 T allele, but not so in men [45].

There is evidence that angiotensinogen gene variants modestly affect blood pressure [46, 47]. However, neither the M235T, nor the T174M polymorphisms, both located in exon 2 of the AGT gene, seem to directly affect function, secretion, or metabolism of angiotensinogen. Furthermore, the frequency of the 235T (or 174 M) allele varies substantially from one ethnic group to another. The 235T gene variant, e.g., is seen in ~ 35% of Caucasians, ~ 75% of Asians and >90% in Africans [46]. These dramatic differences between populations clearly suggest that blood pressure in these populations does not depend to a large extent on such gene variants alone. Finally, plasma angiotensinogen, which is produced to a large extent in adipose tissues, and whose synthesis is regulated by estrogen, glucocorticoids, angiotensin II and thyroid hormones, is about 30% increased in Caucasian women compared to men, despite similar blood pressures between genders [48].

The blood pressure response to ACE inhibitor monotherapy in 125 previously untreated hypertensives varied according to M235T angiotensinogen genotype in a retrospective study. Carriers of at least one T allele had a significantly larger fall in blood pressure after administration of the ACE inhibitor compared to subjects carrying the MM wild type, both for systolic and diastolic pressures. The average fall in systolic blood pressure was 14 mmHg for MM subjects, compared to 22 mmHg and 21 mmHg for MT and TT subjects, respectively [49]. In that study no effect on blood pressure response was observed for the ACEID and AT1 receptor A1166C polymorphisms, however, the latter two polymorphisms were significantly associated with pretreatment systolic and diastolic pressures, while the angiotensinogen 235T gene variant was not [49]. In contrast, no effect of the M235T angiotensinogen gene variant or the ACEID gene polymorphism on the blood pressure response to several antihypertensive agents was observed in a study of 107 patients, who underwent serial ambulatory blood pressure monitoring [23]. Interestingly, Schunkert et al. observed in a population-based cohort of the German MONICA study that those who carried at least one 235 T allele were 1.6 times more likely to use antihypertensive medication, and twice as likely to be treated with two or more antihypertensive drugs [50].

The investigators of the Trials of Hypertension Prevention Program genotyped 1,509 subjects participating in phase II of the program for the angiotensinogen gene–6 promoter polymorphism, in which an A for G base pair substitution is associated with higher angiotensinogen levels [51]. This gene variant at position – 6, upstream of the initial transcription site, is in almost complete linkage disequilibrium with the 235 T allele [46]. Therefore, a TT genotype at position 235 is comparable to an AA genotype at –6. Moderately obese subjects with a mean diastolic blood pressure between 83 and 89 mmHg (averaged over 3 baseline visits) were randomized to 1 of 4 treatment groups: usual care, sodium reduction, weight loss, or combined reduction. There was no difference in pretreatment blood pressures in any of the four treatment groups with an GG, AG, or AA genotype. As an example of the genotype-specific results, the main study found net 4.0/2.8 mmHg and 1.1/0.6 mmHg decreases in systolic/diastolic blood pressure in the combined reduction group after 6 and 36 months (the effect in the sodium or weight reduction alone treatment groups was somewhat less). There was no significant difference in blood pressure response across genotypes after 6 months in any of the four treatment groups. At 36 months, a significantly higher reduction in diastolic blood pressure was observed in subjects with an AA genotype in the sodium retention (–2.2 ± 1.1 mmHg) and weight reduction treatment groups (–2.4 ± 1.2 mmHg) compared to AG or GG genotypes, despite the fact that neither change in sodium nor in weight varied across genotypes. A similar non-significant trend was observed for systolic blood pressure. However, no such effects were seen in the usual care or combined reduction treatment groups. Although differences in blood pressure of 2 to 3 mmHg can have a significant public health impact, the overall impact of the angiotensinogen promoter polymorphism on blood pressure response seems to be small and rather inconsistent (i.e., findings were not reproduced in the combined treatment group). Although the authors provide a lengthy discussion concerning these findings [51], results might well be chance findings. Lack of sufficient sample size could explain these discrepant results. Although the overall study sample was > 1,500 individuals, the sample size was < 400 in each of the four different treatment groups. According to one calculation at least 728 patients need to be studied in order to detect a mean difference in the fall of blood pressure of 5 mmHg with 80% power at the 0.05 level [23]. However the actual difference in the fall of blood pressure between genotypes was even smaller in the Trials of Hypertension Prevention Program. Therefore, it will be difficult to detect any significant single gene effect on blood pressure lowering response unless large sample sizes are studied (i.e., > 1,000 individuals).

In a mechanistic study of renovascular response to angiotensin II infusion, Hopkins et al. showed that the expected decrease in renal blood flow after a mild pressor dose of angiotensin II was least in TT subjects [52]. Thus, the M235T angiotensinogen gene polymorphism was associated with a blunted renovascular response.

Taken together it seems, that the angiotensinogen gene plays a significant but modest role in human blood pressure variance. However, genetic findings may vary greatly depending on study design and the populations studied. There is also a need for better phenotyping of the hypertensive population. With regard to the contribution of the angiotensinogen and epithelial amiloride-sensitive sodium channel genes it is too early to propose any dietary recommendations and specific drug treatment according to patients' genotype [53].

12.4.3 Drug Metabolizing Enzymes and Drug Transporters

There is extensive knowledge of genetic variation in drug metabolizing enzymes, e.g., polymorphisms of cytochrome P450 enzymes (2A6, 2C9, 2C19, 2D6) or drug transporters such as the MDR-1 gene and its protein product P-glycoprotein. Genetically determined variation in such genes may determine plasma levels of a drug, if the molecule is metabolized to a large extent by the corresponding pathway. This has been demonstrated in phase I studies in healthy volunteers. From such data, a genotype-based dose adjustment has been suggested for the CYP2D6 polymorphism: poor metabolizers should only receive 30% of the standard 100 mg dose of metoprolol, while ultrafast metabolizers would need a dose increase to 140% [54]. However, whether such dose adjustment has any relevant clinical impact remains to be determined in adequately powered patient studies, especially since other mechanisms contribute to fluctuations in plasma levels of a drug. For example, components in the diet (i. e., grapefruit juice) may interfere with drug absorption.

However, in case of concomitant medication, treatment with another agent competing for the same metabolizing enzyme may result in dramatic increases in plasma levels of the drug with the poorer affinity. The risk of adverse events may increase dramatically in the case of drugs with a high intestinal absorption rate, but a poor bioavailability, since plasma levels may rise 10 to 20-fold. Examples are the increased incidence of rhabdomyolysis if statins metabolized via CYP3A4 (i. e., lovastatin, simvastatin, cerivastatin) are taken together with other CYP3A4 substrates (i. e., ciclosporin, erythromycin, gemfibrozil) [55], as well as the increased incidence of myopathies with the combination of simvastatin and mibefradil, which led to withdrawal of this promising new T-channel blocker agent in 1998 one year after its introduction, and lately, to the withdrawal of cerivastatin in August 2001.

12.4.4 Genetic Polymorphisms of the β-Adrenoreceptors

Beta-1, beta-2, and beta-3 adrenergic receptors are G-protein-coupled receptors. Beta-1 and beta-2 receptors mediate the positive inotropic, chronotropic, and dromotropic effects of the endogenous catecholamines epinephrine and norepinephrine. The beta-3 subtype seems to play a role in regulating thermogenesis and lipid mobilization in brown and white adipose tissue. Several coding and promoter polymorphisms of these receptors have been identified. Clinical studies in asthma revealed that polymorphisms of the β_2-adrenoreceptor have disease modifying effects and alter the response to treatment [56, 57]. Furthermore, the Arg16Gly polymorphism in the human β_2-adrenergic receptor, substituting the amino acid glycine for arginine, produced functional *in vitro* differences: the Gly allele was associated with enhanced β_2-downregulation after prolonged agonist stimulation *in vitro* [58]. The Arg16Gly polymorphism was also associated with hypertension in African Carribeans and in Caucasians [59, 60] and with altered physiological function *in vivo* when vasodilation to a β_2-agonist was significantly reduced [61].

Polymorphisms of the beta adrenergic receptors have also been studied in patients with heart failure and cardiomyopathy, or other complex and rather ill-defined phenotypes. In patients with heart failure due to ischemic or idiopathic dilated cardiomyopathy, the Thr164Ile polymorphism in the β_2-adrenoreceptor was significantly associated with survival rate at one year [62]. Similarly, the Ser49Gly polymorphism of the β_1-adrenoreceptor gene has been linked to the improved survival of patients with idiopathic cardiomyopathy [63]. However, sample size was limited in those studies and results need to be confirmed in adequately powered studies.

The Arg389Gly polymorphism of the β_1-adrenoreceptor gene displayed functional differences in *in vitro* studies [64]. However, this did not translate into a functional difference when studied in humans [65]. Whether this is due to the lack of adequately representing the genetic variation of a gene by restricting the analysis to a single nucleotide variant, instead of analyzing the entire genetic variability by determining haplotypes remains to be seen. An argument for such an approach comes from a study of the β_2-adrenoreceptor in which 13 single nucleotide polymorphisms, when analyzed separately, could not predict the *in vivo* responsiveness to the β_2-agonists, whereas a haplotype did predict functional difference in humans [57]. The Trp64Arg polymorphism of the β_3-adrenoreceptor has been investigated in several metabolic phenotypes including obesity, insulin resistance and diabetes [66–68]. However, the overall significance of these rather inconsistent findings is doubtful [69].

It is intriguing to note that the 13 SNPs reported for the β_2-adrenoreceptor gene could theoretically lead to 2^{13} different combinations. However only 12 haplotypes were observed [57]. To date, no such a study has been reported for the β_1-adrenoreceptor gene, in which 18 polymorphisms giving rise to only 11 haplotypes are known [70].

Apart from the evidence that investigating a single nucleotide variant in order to disentangle a putative functional variation of a gene product might be the wrong approach (i.e., haplotypes being better), it is obvious that the adrenoreceptor represents only part of the entire signaling cascade and pathways within cells and tissues. For example, downstream signaling components, such as the G-proteins may modulate receptor function. Indeed, such functional variants have been described [71, 72]. Thus, the heterogeneity in phenotypes and genetic variation between studies, may explain inconsistent findings among apparently similar studies. Without investigating entire pathways the gene-to-phenotype approach is doomed to fail. Feedback mechanisms may lead to downregulation, and upstream or downstream genetic variation may counterbalance the functional effects of a mutation.

12.5 Conclusions

Although the mathematics of pathway analysis has been developed [73–75], the integration of genomics and proteomics is in its infancy. We still need to understand how components of the cell interact in healthy cellular physiology and in disease. How does drug treatment, nutrition and other environmental stimuli alter cellular function? The common forms of hypertension, diabetes mellitus or dilated cardiomyopathy are not monogenic, but complex genetic diseases. Furthermore, the clinical labeling as diabetic or hypertensive suggests we are dealing with only one phenotype. This, however, is certainly not the case from a causality point of view. One should consider describing diabetic or hypertensives syndromes, similar to the metabolic syndrome, since such terminology immediately implies the complexity of causes and the paucity of our knowledge. Thus, there is a need to revise scientific thinking when dealing with complex phenotypes and to stress the fact that such complex phenotypes are caused by a cluster of similar, but genetically and environmentally distinct complex interactions. While the search for candidate genes in monogenic disorders were a recipe for success in the past, this strategy does not necessarily lead to success in disentangling complex genetic disorders in the future. It is thus time to move from the reductionist paradigm to an integrative approach in the study of complex biological systems [76].

12.6 References

1 Hanash SM. Operomics: Molecular analysis of tissues from DNA to RNA to protein. Clin Chem Lab Med **2000**; 38:805–813.

2 Nebert DW. Pharmacogenetics and pharmacogenomics: why is this relevant to the clinical geneticist? Clin Genet **1999**; 56:247–258.

3 Clark EA, Golub TR, Lander ES, Hynes RO. Genomic analysis of metastasis reveals an essential role for RhoC. Nature **2000**; 403:532–535.

4 Alizadeh AA, Eisen MB, Davis RE, Ma C, Lossos IS, Rosenwald A, Boldrick JC, Sabet H, Tran T, Yu X, Powell JI, Yang L, Marti GE, Moore T, Husdson J Jr, Lu L, Lewis DB, Tibshirani R, Sherlock G, Chan WC, Greiner TC, Weisenburger DD, Armitage JO, Warnke R, Staudt LM et al. Distinct types of diffuse large B-cell lymphoma identified by gene expression profiling. Nature **2000**; 406:503–511.

5 Yang J, Moravec CS, Sussman MA, DiPaola NR, Fu D, Hawthorn L, Mitchell CA, Young JB, Francis GS, McCarthy PM, Bond M. Decreased SLIM1 expression and increased gelsolin expression in failing human hearts measured by high-density oligonucleotide arrays. Circulation **2000**; 102:3046–3052.

6 Kleyn PW, Vesell ES. Genetic variation as a guide to drug development. Science **1998**; 281:820–821.

7 Sing CF, Haviland MB, Reilly SL. Genetic architecture of common multifactorial diseases. In. Variation in the human genome. Ciba Foundation Symposium 197. Wiley: Chichester. **1996**: 211–232.

8 Soubrier F, Alhenc-Gelas F, Hubert C, Allegrini J, John M, Tregear et al. Two putative active centers in human angiotensin I-converting enzyme revealed by molecular cloning. Proc Natl Acad Sci USA **1988**; 85:9386–9390.

9 Rigat B, Alhenc-Gelas F, Cambien F, Soubrier F. An insertion/deletion polymorphism in the angiotensin I-converting gene accounting for half the variance of serum levels. J Clin Invest **1990**; 86:1343–1346.

10 Winkelmann BR, Nauck M, Klein B, Russ AP, Böhm BO, Siekmeier R, Ihnken K, Verho M, Gross W, März W. Deletion polymorphism of the angiotensin I-converting enzyme gene Is associated with increased plasma angiotensin-converting enzyme activity but not with increased risk for myocardial infarction and coronary artery disease. Ann Intern Med **1996**; 125:19–25.

11 Harrap SB, Davidson R, Connor JM, Soubrier F, Corvol P, Fraser R, Foy CJW, Watt GCM. The angiotensin I-converting enzyme gene and presdisposition to high blood pressure. Hypertension **1993**; 21:455–460.

12 Perry GJ, Tatsuhiko M, Wei C-C, Xu XY, Chen Y-F, Oparil S, Lucchesi P, Dell'Italia LJ. Genetic variation in angiotensin-converting enzyme does not prevent development of cardiac hypertrophy or upregulation of angiotensin II in response to aortocaval fistula. Circulation **2001**; 103:1012–1016.

13 Ueda S, Elliott H, Morton J, Connell J. Enhanced pressor response to angiotensin I in normotensive men with the deletion genotype (DD) for angiotensin-converting enzyme. Hypertension **1995**; 25:1266–1269.

14 Buikema H, Pinto YM, Rooks G, Grandjean JG, Schunkert H, van Gilst WH. The deletion polymorphism of the angiotensin-converting enzyme gene is related to phenotypic differences in human arteries. Eur Heart J **1996**; 17:787–794.

15 Prasad A, Narayanan S, Husain S, Padder F, Waclawiw M, Epstein N, Quyyumi AA. Insertion-deletion polymorphism of the ACE gene modulates reversibility of endothelial dysfunction with ACE inhibition. Circulation **2000**; 102:35–41.

16 Cambien F, Poirier O, Lecerf L, Alun E, Jean-Pierre C, Dominique A, Gerald L, Jean-Marie B, Lucienne B, Sylvain R, Laurence T, Philippe A, Francois AG, Florent S. Deletion polymorphism in the angiotensin I-converting enzyme is a potent risk factor for myocardial infarction. Nature **1992**; 359:641–644.

17 Duru K, Farrow S, Wang J-M, Lockette W, Kurz T. Frequency of a deletion polymorphism in the gene for angiotensin converting enzyme is increased in African-Americans with hypertension. Am J Hypertens **1994**; 7:759–762.

18 Agerholm-Larsen B, Nordestgaard BG, Tybjaerg-Hansen A. ACE Gene Polymorphism in Cardiovascular Disease. Meta-Analyses of Small and Large Studies in Whites. Arterioscler Thromb Vasc Biol **2000**; 20:484–492.

19 Lachurié ML, Azizi M, Guyene TT, Alhenc-Gelas F, Menard J. Angiotensin-converting enzyme gene polymorphism has no influence on the circulating renin-angiotensin-aldosterone system or blood pressure in normotensive subjects. Circulation **1995**; 91:2933–2942.

20 Beige J, Zilch O, Hohenbleicher H, Ringel J, Kunz R, Distler A, Sharma AM. Genetic variants of the renin-angiotensin system and ambulatory blood pressure in essential hypertension. J Hypertens **1997**; 15:503–508.

21 Winkelmann BR, Hager J, Kraus WE, Merlini P, Keavney B, Grant PJ, Muhlestein JB, Granger CB. Genetics in coronary heart disease: current knowledge and research principles. Am Heart J **2000**; 140[Suppl.]:S11–S26.

22 Todd GP, Chadwick IG, Higgins KS, Yeo WW, Jackson PR, Ramsay LE. Relation between changes in blood pressure and serum ACE activity after a single

dose of enalapril and ACE genotype in healthy subjects. Br J Clin Pharmacol **1995**; 39:131–134.

23 Dudley C, Keavney B, Casadei B, Conway J, Bird R, Ratcliffe P. Prediction of patient responses to antihypertensive drugs using genetic polymorphisms: investigation of renin-angiotensin system genes. J Hypertens **1996**; 14:259–262.

24 Petrie MC, Padmanaban N, McDonald JE, Hillier C, Connell JMC, McMurray JJV. Angiotensin converting enzyme (ACE) and non-ACE dependent angiotensin II generation in resistance arteres from patients with heart failure and coronary heart disease. J Am Coll Cardiol **2001**; 37:1056–1061.

25 Hopkins PN, Lifton RP, Hollenberg NK, Jeunematre X, Hallouin MC, Skuppin J, Williams CS, Dluhy RG, Lalouel JM, Williams RR, Williams GH. Blunted renal vascular response to angiotensin II is associated with a common variant of the angiotensinogen gene and obesity. J Hypertens **1996**; 14:199–207.

26 Brosnihan KB, Li P, Ferrario CM. Angiotensin-(1-7) dilates canine coronary arteries through kinins and nitric oxide. Hypertension **1996**; 27[part 2]:523–528.

27 van Ampting JMA, Hijmering ML, Beutler JJ, van Etten RE, Koomans HA, Rabelink TJ, Stroes ESG. Vascular effects of ACE inhibition independent of the renin-angiotensin system in hypertensive renovascular disease. Hypertension **2001**; 37:40–45.

28 Murphey LJ, Gainer JV, Vaughan DE, Brown NJ. Angiotensin-converting enzyme insertion/deletion polymorphism modulates the human *in vivo* metabolism of bradykinin. Circulation. **2000**; 102:829–832.

29 Gainer JV, Stein CM, Neal T, Vaughan DE, Brown NJ. Interactive effect of ethnicity and ACE insertion/deletion polymorphism on vascular reactivity. Hypertension **2001**; 37:46–51.

30 Brown NJ, Blais C Jr, Gandhi SK, Aam A. ACE insertion/deletion genotype affects bradykinin metabolism. J Cardiovasc Pharmacol **1998**; 32:373–377.

31 Prasad A, Husain S, Schenke W, Mincemoyer R, Epstein N, Quyyumi AA. Contribution of bradykinin B2 receptor dysfunction to abnormal coronary vasomotion in humans. J Am Coll Cardiol **2000**; 36:1467–1473.

32 Prasad A, Narayanan S, Waclawiw MA, Epstein N, Quyyumi AA. The insertion/deletion polymorphism of the angiotensin-converting enzyme gene determines coronary vascular tone and nitric oxide activity. J Am Coll Cardiol **2000**; 36:1579–1586.

33 Butler R, Morris AD, Burchell B, Struthers AD. DD angiotensin-converting enzyme gene polymorphism is associated with endothelial dysfunction in normal humans. Hypertension **1999**; 33:1164–1168.

34 Perticone F, Ceravolo R, Maio R, Ventura G, Zingone A, Perrotti N, Mattioli PL. Angiotensin-converting enzyme gene polymorphism is associated with endothelium-dependent vasodilation in never treated hypertensive patients. Hypertension **1998**; 31:900–905.

35 Anderson TJ, Elstein E, Haber H, Charbonneau F. Comparative study of ACE-inhibition, angiotensin II antagonism, and calcium channel blockade on flow-mediated vasodilation in patients with coronary disease (BANFF study). J Am Coll Cardiol **2000**; 35:60–66.

36 Montgomery H, Clarkson P, Barnard M, Bell J, Brynes A, Dollery C, Hajnal J, Hemingway H, Mercer D, Jarman P, Marshall R, Prasad K, Rayson M, Saeed N, Talmud P, Thomas L, Jubb M, World M, Humphries S. Angiotensin-converting-enzyme gene insertion/deletion polymorphism and response to physical training. Lancet **1999**; 353:541–45.

37 Myerson SG, Montgomery HE, Whittingham M, Jubb M, World MJ, Humphries SE, Pennelll DJ. Left ventricular hypertrophy with exercise and ACE gene insertion/deletion polymorphism – a randomized controlled trial with losartan. Circulation **2001**; 103;226–230.

38 McNamara DM, Holubkov R, Janosko K, Palmer A, Wang JJ, MacGowan GA, Murali S, Rosenblum WD, London B, Feldman AM. Pharmacogenetic interactions between β-blocker therapy and the

angiotensin-converting enzyme deletion polymorphism in patients with congestive heart failure. Circulation **2001**; 103:1644–1648.

39 Tian B, Meng QC, Chen Y-F, Krege JH, Smithies O, Oparil S. Blood pressures and cardiovascular homeostasis in mice having reduced or absent angiotensin-converting enzyme gene function. Hypertension **1997**; 30[part I]:128–133.

40 Jeunemaitre X, Soubrier F, Kotelevtsev YV, Lifton RP, Williams CS, Charru A, Hunt SC, Hopkins N, Williams RR, Lajouel J-M, Corvol P. Molecular basis of human hypertension: role of angiotensinogen. Cell **1992**; 71:169–180.

41 Winkelmann BR, Russ AP, Nauck M, Klein B, Böhm BO, Maier V, Zotz R, Matheis G, Wolf A, Wieland H, Gross W, Galton DJ, März W. Angiotensinogen M235T polymorphism is associated with plasma angiotensinogen and cardiovascular disease. Am Heart J **1999**; 137:698–705.

42 Katsuya T, Koike G, Yee TW, Sharpe N, Jackson R, Norton R, Horiuchi M, Pratt RE, Dzau VJ, MacMahon S. Association of angiotensinogen gene T235 variant with increased risk of coronary heart disease. Lancet **1995**; 345:1600–1603.

43 Sethi AA, Tybjaerg-Hansen A, Gronholdt M-LM, Steffensen R, Schnohr P, Nordestgaard BG. Angiotensinogen mutations and risk for ischemic heart disease, myocardial infarction, and ischemic cerebrovascular disease. Ann Intern Med **2001**; 134:941–954.

44 Tiret L, Ricard S, Poirier O, Arveiler D, Cambou J-P, Luc G, Evans A, Nicaud V, Cambien F. Genetic variation at the angiotensinogen locus in relation to blood pressure and myocardial infarction: the ECTIM study. J Hypertens **1995**; 13:311–317.

45 Sethi AA, Nordestgaard BG, Agerholm-Larsen B, Frandsen E, Jensen G, Tybjaerg-Hansen A. Angiotensinogen polymorphisms and elevated blood pressure in the general population – the Copenhagen City Heart study. Hypertension **2001**; 37:875–881.

46 Corvol P, Jeunemaitre X. Molecular genetics of human hypertension: Role of angiotensinogen. Endocrine Rev **1997**; 18:662–677.

47 Kunz R, Kreutz R, Beige J, Distler A, Sharma AM. Association between the angiotensinogen 235T-variant and essential hypertension in whites – a systematic review and methodological appraisal. Hypertension **1997**; 30:1331–1337.

48 Cooper RS, Guo X, Rotimi CN, Luke A, Ward R, Adeyemo A, Danilov SM. Heritability of angiotensin – converting enzyme and angiotensinogen. Hypertension **2000**; 35:1141–1147.

49 Hingorani AD, Stevens PA, Hopper R, Dickerson JEC, Brown MJ. Renin-angiotensin system gene polymorphisms influence blood pressure and the response to angiotensin converting enzyme inhibition. J Hypertens **1995**; 13[pt. II]:1602–1609.

50 Schunkert H, Hense H-W, Gimenez-Roqueplo AP. The angiotensinogen T235 variant and the use of antihypertensive drugs in a population-based cohort. Hypertension **1997**; 29:628–633.

51 Hunt SC, Cook NR, Oberman A, Cutler JA, Hennekens CH, Allender PS, Walker WG, Whelton PK, Williams RR. Angiotensinogen genotype, sodium reduction, weight loss, and prevention of hypertension – trials of Hypertension Prevention, Phase II. Hypertension **1998**; 32:393–401.

52 Hopkins PN, Lifton RP, Hollenberg NK. Blunted renal vascular response to angiotensin II is associated with a common variant of the angiotensinogen gene and obesity. J Hypertens **1996**; 14:199–207.

53 Corvol P, Persu A, Gimenez-Roqueplo A-P, Jeunemaitre X. Seven lessons from two candidate genes in human essential hypertension. Hypertension **1999**; 33:1324–1331.

54 Meisel C, Roots I, Cascorbi I, Brinkmann U, Brockmöller J. How to manage individualized drug therapy: application of pharmacogenetic knowledge of drug metabolism and transport. Clin Chem Lab Med **2000**; 38:869–876.

55 Pierce LR, Wysowski DK, Gross TP. Myopathy and rhabdomyolysis associated with lovastatin-gemfibrozil combination therapy. JAMA **1990**; 264:71.
56 Ligett SB. Pharmacogentics of beta-1 and beta-2 adrenergic receptors. Pharmacology **2000**; 61:167–173.
57 Drysdale CM, McGraw DW, Stack CB, Stephens JC, Judson RS, Nandabalan K, Arnold K, Ruano G, Liggett SB. Complex promoter and coding region β_2-adrenergic receptor haplotypes alter receptor expression and predict *in vivo* responsiveness. Proc Natl Acad Sci USA **2000**; 97:10483–10488.
58 Green SA, Turki J, Innis M, Liggett SB. Amino-terminal polymorphisms of the human β_2-adrenergic receptor impart distinct agonist-promoted regulatory properties. Biochemistry **1994**; 33:9414–9419.
59 Kotanko P, Binder A, Tasker J, DeFreitas P, kamdar S, Clark AJ, Skrabal F, Caulfield M. Essential hypertension in African Carribeans associates with a variant of the beta2-adrenoreceptor. Hypertension **1997**; 30:773–776.
60 Timmermann B, Mo R, Lift FC, Gerdts E, Busjahn A, Omvik P, Li GH, Schuster H, Wienker TF, Hoehe MR, Lund-Johansen P. Beta-2-adrenoreptor genetic variation is associated with genetic predisposition to essential hypertension: the Bergen blood pressure study. Kidney Int **1998**; 53:1455–1460.
61 Gratze G, fortin J, Labugger R, Binder A, Kotanko P, Timmermann B, Luft F, Hoehe M, Skrabal F. β_2-adrenergic receptor variants affect resting blood pressure and agonist-induced vasodialtion in young adult caucasians. Hypertension **1999**; 33:1425–1430.
62 Liggett SB, Wagoner LE, Craft LL, Hornun RW, Hoit BD, McIntosh TC, Walsh RA. The Ile164 beta 2-adrenergic recepor polymorphism adversely affects the outcome of congestive heart failure. J Clin Invest **1998**; 102:1534–1539.
63 Börjesson M, Magnusson Y, Hjalmarson A, Andersson B. A novel polymoprhism in the gene encoding for the beta1-adrenergic receptor is associated with survival in patients with heatrt failure. Eur Heart J **2000**; 21:1853–1858.
64 Mason DA, Moore JD, Green SA, Ligett SB. A gain-of-function polymorphism in a G-protein coupling domain of the human β_1-adrenergic receptor. J Biol Chem **1999**; 274:12670–12674.
65 O'Shaughnessy KM, Fu B, Dickerson C, Thurston D, Brown MJ. The gain-of-function G389R variant of the β_1-adrenoreceptor does not influence blood pressure or heart rate response to β-blockade in hypertensive subjects. Clin Science **2000**; 99:233–238.
66 Azuma N, Yoshimasa Y, Nishimura H, Yamamoto Y, Masuzaki H, Suga J, Shigemoto M, Matsuoka N, Tanaka T, Satoh N, Igaki T, Miyamoto Y, Itoh H, Yoshimasa T, Hosoda K, Nishi S, Nakao K. The significance of the Trp64Arg mutation of the β_3-adrenergic receptor gene in impaired glucose tolerance, non-insulin-dependent diabetes mellitus, and insulin resistance in Japanese subjects. Metabolism **1998**; 47:456–460.
67 Gagnon J, Mauriège P, Roy S, Sjöström D, Chagnon YC, Dionne FT, Oppert J-M, Pérusse L, Sjöström L, Bouchard C. The Trp64Arg mutation of the β_3-adrenergic receptor gene has no effect on obesity phenotypes in the Québec family study and swedish obese subjects cohorts. J Clin Invest **1996**; 98:2086–2089.
68 Sipiläinen R, Uusitupa M, Heikkinen S, Rissanen A, Laakso M. Polymorphism of the β_3-adrenergic receptor gene affects basal metabolic rate in obese Finns. Diabetes **1997**; 46:77–80.
69 Mauriège P, Bouchard C. The Trp64Arg mutation in the β_3-adrenoreceptor gene of doubtful significance for obesity and insulin resistance. Lancet **1996**; 348:698–699.
70 Podlowski S, Wenzel K, Luther HP, Müller J, Bramlage P, Baumann G, Felix SB, Speer A, Hetzer R, Köpke K, Hoehe MR, Wallukat G. β_1-Adrenoreceptor gene variations: a role in idiopathic cardiomyopathy? J Mol Med **2000**; 78:87–93.
71 Rosskopf D, Busch S, manthey I, Suffert W. G protein β3 gene: structure, promoter, and additonal polymorphisms. Hypertension **2000**; 36:33–41.

72 Siffert W, Rosskopf S, Siffert G, Busch S, Moritz A, Erbel R et al. Association of a human G-protien $\beta 3$ subunit variant with hypertension. Nature Genet **1998**; 18:45–48.

73 Fell DA. Metabolic control analysis: A survey of its theoretical and experimental development. Biochem J **1992**; 286:313–330.

74 Fell DA, Thomas S. Physiologic control of metabolic flux: the requirement for multisite modulation. Biochem J **1995**; 311:35–39.

75 Thomas S, Fell DA. The role of multiple enzyme activation in metabolic flux control. Advan Enzyme Regul **1998**; 38:65–85.

76 Palsson B. The challenges of in silico biology – moving from a reductionist paradigm to one that views cells as systems will necessitate changes in both the culture and the practice of research. Nature Biotechnol **2000**; 18:1147–1150.

13
Pharmacogenomics of Lipid-Lowering Agents

Michael M. Hoffmann, Bernhard R. Winkelmann, Heinrich Wieland
and Winfried März

Abstract

Cardiovascular disease is the major cause of death in North America and Europe. As shown by several large placebo-controlled intervention studies, the correction of dyslipoproteinemias with bile acid sequestrants, fibrates, niacin or 3-hydroxy-3-methylglutaryl coenzyme A reductase inhibitors substantially reduces the risk of future coronary events. The response to these lipid-lowering drugs is modified by a number of factors like gender, age, concomitant disease, additional medication, and genetic determinants. Even among carefully selected patients of clinical trials, individual responses vary greatly. At the time being there are no established biochemical or clinical parameters to distinguish between responders, non-responders, and patients who will develop adverse, potentially life-threatening events. Monogenetic disorders of the lipid metabolism such as familial hypercholesterolemia or type III hyperlipoproteinemia can produce severe premature atherosclerosis. Due to their low frequency, however, their contribution to the overall burden of cardiovascular disease is small. On the other hand, polymorphisms in genes of the lipoprotein metabolism (e.g., apolipoprotein E) are associated with plasma lipoprotein concentrations, explaining a substantial fraction of the variance of low density lipoprotein (LDL) or high density lipoprotein (HDL) in the general population. The recent advances in pharmacogenomics, e.g., the characterization of new variants and haplotypes of lipoprotein-related genes, will deepen our understanding of lipid and lipoprotein metabolism and of the individual response to lipid-lowering drugs.

13.1
Introduction

Changes in the concentrations of lipoproteins, in particular increases in low density lipoproteins (LDL), triglyceride-rich lipoproteins, and decreases in high density lipoproteins (HDL), are among the most important causes of atherosclerosis. Dyslipidemias result from the interaction of environmental risk factors and multiple predisposing genes. Among the genetic factors affecting lipoprotein metabolism are monogenetic disorders producing severe clinical phenotypes, e.g., familial hypercholesterolemia (due to mutations in the LDL receptor gene) or Tangier disease (due to mutations in the ABC-A1 gene). Although some of these disorders belong to the most frequent inborn errors of metabolism in humans, they are too rare to make significant contributions to the variance of LDL or HDL cholesterol concentrations observed in the general population. In the last decade, population studies showed that polymorphisms of genes involved in lipoprotein metabolism determine a substantial fraction of the variance of concentrations of circulating lipoproteins. Although the benefits of lipid-lowering therapy have been shown in many patient populations, the individual variation in response is large. In the case of LDL lowering by statins, responses may vary from decreases by 10–70% [1]. It is reasonable to assume that these differences, at least in part, relate to the genetic diversity.

13.2 The Metabolism of Plasma Lipoproteins

There are three major routes of lipid transport in plasma: the exogenous, the endogenous and the reverse cholesterol transport pathway. The exogenous pathway is fed by dietary lipids, which are incorporated into chylomicrons and subsequently released into the intestinal lymph. These particles enter the bloodstream via the thoracic duct, thus by-passing the liver. Lipoprotein lipase (LPL), which resides on the lumenal surface of the capillary endothelium, hydrolyzes the triglyceride moiety of the chylomicrons. The liberated free fatty acids are taken up by tissues such as adipose, for storage, and muscle, for oxidation. As a result of the hydrolysis process, chylomicrons are converted to smaller remnant particles, excess surface components (phospholipids and apolipoproteins) being transferred to HDL. The remnant particles become enriched in cholesterol and acquire apo E from HDL. apo E is needed as ligand for lipoprotein receptors in the liver, because apo B-48, the major apolipoprotein of chylomicrons, lacks the receptor binding domain of apo B-100. In summary thus, there are two major steps in the catabolism of chylomicrons: hydrolysis of triglycerides in the circulation and receptor-mediated catabolism of cholesterol in the liver.

In the endogenous pathway the liver is the source of triglycerides and cholesterol, which are secreted as components of VLDL. Like chylomicrons, VLDL undergo lipolysis in the circulation to give rise to IDL. A significant portion of the IDL is rapidly taken up by the liver. The remainder undergoes further lipolysis by LPL and another lipolytic enzyme, hepatic triglyceride lipase (HTGL), leading to the formation of LDL. LDL particles contain most of the cholesterol in blood. Their only protein constituent is apo B-100. In the periphery, LDL is taken up via the LDL-r and provides cholesterol for the synthesis of cell membranes and steroid hormones. However, roughly two thirds of the LDL are catabolized by the liver.

The reverse cholesterol pathway is mediated by HDL. HDL is formed from precursor particles originating from the intestine and the liver. In addition, surface material derived from the catabolism of chylomicrons is a source of HDL particles. Nascent HDL particles mobilize free cholesterol from peripheral cells. This process is mediated by several proteins, e.g., apo AI and transmembrane ATP-binding cassette molecules like ABC-A1. The HDL-associated enzyme lecithin:cholesterol acyltransferase (LCAT) immediately esterifies the free cholesterol. The esterified cholesterol is then transferred to the pool of apo B-100 containing lipoproteins or delivered directly to the liver, a process which is mediated by the action of cholesterol ester transfer protein (CETP).

13.3 Pharmacogenomics of Lipid-Lowering Agents

13.3.1 Bile Acid Sequestrants (Resins)

Resins, like cholestyramine and colestipol, impede the recycling of bile acids by trapping them in the lumen of the intestine [2]. As a consequence the hepatic conversion of cholesterol to bile acid is increased by up-regulation of the *cholesterol-7-α*-hydroxylase (CYP7) [3], the rate-limiting enzyme of bile acid synthesis. CYP7 activity seems to be inversely correlated with plasma cholesterol levels [4]. There exists at least one common polymorphism within the regulatory region of the CYP7 gene (C-278A [5] or A-204C [6]). Depending on the population studied, the C-278A polymorphism accounted for 1–15% of the variation of LDL cholesterol [5, 6]. The effect of this SNP on the regulation of CYP7 has not been evaluated in detail. It is, therefore, difficult to predict whether it will influence the lipid-lowering effect of bile acid sequestrants or HMG-CoA reductase inhibitors. Empirical evidence for this possibility has also not been provided so far. The *apolipoprotein E* genotype, which has shown to influence the plasma cholesterol level (see statins), had no effect on the hypolipidaemic efficacy of colestipol [7].

13.3.2 Fibrates

The most pronounced effects of fibrates are to decrease plasma triglyceride-rich lipoproteins. In addition, fibrates slightly reduce LDL cholesterol and substantially raise HDL cholesterol. Further, they reduce small dense LDL, a highly atherogenic subfraction of LDL. Fibrates act on the transcription of genes involved in lipoprotein metabolism by activating transcription factors belonging to the nuclear hormone receptor family, the peroxisome proliferator-activated receptors (PPARs) (for review, see [8, 9]). PPARα is predominantly expressed in tissues that metabolize high amounts of fatty acids [10], like liver, kidney, heart, and muscle.

The hypotriglyceridemic action of fibrates involves combined effects on LPL and apolipoprotein CIII (apo CIII). LPL is up-regulated [11], whereas apo CIII, an inhibitor of LPL, is down-regulated [12], leading to enhanced hydrolysis of triglyceride-rich lipoproteins. Moreover, fibrates decrease apo B and VLDL production [13]. In rodents, fibrates enhance intracellular fatty acid metabolism. Whether this mechanism plays a role in humans is still under investigation. In humans the expression of apo A-I, apo AII and of ABC-A1 [14] is stimulated, providing an explanation for the raise in HDL cholesterol during fibrate therapy. Kinetic analyses have revealed that fibrates increase the receptor-mediated clearance of LDL. This is, however, most likely due to changes in the composition of LDL towards more receptor-active particles rather than to up-regulation of the LDL receptor itself [15].

No polymorphisms have been described within the PPAR responsive elements (PPRE) of the promoters of LPL, apo CIII, apo AI and apo AII, which might influence directly the binding of these transcription factors. On the other hand there are several possible polymorphisms in the target genes of PPARα, which might interact with the action of fibrates action, e.g., *LPL* D9N, N291S and S447X. Carriers of the truncated LPL variant S447X, which is associated with higher plasma LPL activity, might have greater benefit, whereas carriers of LPL9N and 291S, who have lower plasma LPL activity, might have a smaller benefit from fibrate therapy, but this has not been proven experimentally so far.

Several SNPs in the *PPARα* gene have been published recently: a G/A transversion in intron 3, R131Q, and L162V [16–18]. In all studies the frequency of the minor allele was lower than 10%. There was no evidence that the mutations within the coding region of PPAR have a major role in type 2 diabetes, although they might have a borderline impact on LDL cholesterol levels [16, 17]. In the SENDCAP study, bezafibrate-treated V162 allele carriers (13 patients) showed a 2-fold greater lowering of total cholesterol (–0.90 vs. –0.42 mmol L^{-1}, $p=0.04$] and non-HDL-C (–1.01 vs. –0.50 mmol L^{-1}, $p=0.04$) than L162 allele homozygotes (109 patients) [17]. As bezafibrate is not PPARα-specific, but also interacts with PPARγ and PPARβ/δ, the effects of the V162 variation might even be greater in the case of other, more specific fibrates. In view of the small number of V162 carriers, these results obviously need to be reproduced in other studies.

There are a few studies investigating the role of polymorphisms at the *apolipoprotein B* gene in modulating the response to fibrates. Although the apo B *Xba*I and signal peptide insertion/deletion polymorphisms might influence the baseline level of LDL cholesterol, they do not influence the response to fibrate therapy [19, 20]. The reports concerning the *apolipoprotein E* locus are conflicting [21–25]. It is important to note that all of these studies are very small ($n=63-230$) and, therefore, their power to detect an effect of the apo E genotype is low.

13.3.3
Niacin (Nicotinic Acid)

Niacin reduces plasma LDL cholesterol, lipoprotein (a), triglycerides and raises HDL cholesterol in all types of hyperlipoproteinemia [26]. Although available on the market for more than 40 years, the mechanisms of action of niacin are poorly understood. Putative mechanisms are the activation of adipose tissue LPL, diminished HTGL activity, a reduced hepatic production and release of VLDL, and compositional changes in LDL leading to higher affinity to the LDL receptor. Potential candidates for modulators of the action of niacin are LPL, HTGL, the apo AI-CIII-AIV gene cluster, LCAT, CETP, and MTP, but empirical data is lacking up to now.

13.3.4 Probucol

Primarily sold as antioxidant probucol is serving as an efficient cholesterol-lowering agent, reducing both LDL and HDL cholesterol without affecting plasma triglyceride levels. The decrease in HDL, mainly a reduction in HDL_2, is a direct consequence of increased CETP activity in plasma [27]. It is still a matter of debate whether this indicates an enhanced reverse cholesterol transport. Unfortunately, there is no data relating polymorphisms in the CETP gene to the efficacy of probucol therapy. Probucol lowers LDL cholesterol in homozygous LDL receptor deficiency [28], providing evidence that probucol may increase LDL receptor-independent catabolism of LDL.

One of the first pharmacogenomic studies investigating lipid-lowering drugs was published by Nestruck et al. in 1987 [29], describing that carriers of at least one apo E4 allele who received probucol showed the greatest cholesterol reduction in comparison to those without an apo E4 allele. These data were confirmed in a second study by Eto et al. [30].

13.3.5 HMG-CoA Reductase Inhibitors (Statins)

The most successful strategies to reduce the concentration of LDL in the circulation involve the up-regulation of the LDL receptor activity by depleting the regulatory pool of cholesterol in the liver. Inhibitors of 3-hydroxy-3-methylglutaryl coenzyme A (HMG-CoA) reductase constitute the most powerful single class of hypolipidemic drugs currently available. Their efficacy in reducing coronary morbidity and mortality has been established by large secondary and primary intervention trials. The expression of the LDL-r gene is regulated by the intracellular cholesterol pool through sterol-responsive element binding proteins (SREBPs) 1 and 2 (for review, see [31, 32]). The precursors of SREBPs are anchored in the membrane of the endoplasmic reticulum. When the sterol content of a cell decreases, SREBP processing proteins including SREBP cleavage activating protein (SCAP), site 1 protease (S1P), and site 1 protease (S2P) act synergistically to release the amino-terminal domain of SREBP by proteolysis. These active domains are subsequently transferred into the nucleus where they activate the transcription of the genes of the LDL-r and of enzymes involved in the biosynthesis of cholesterol. Beyond this, the SREBPs up-regulate genes involved in the production of free fatty acids, including acetyl-CoA carboxylase and fatty acid synthase [33, 34].

Inter-individual variability of the cholesterol-lowering efficacy may relate to the metabolic processing of the drugs themselves. *Genetic polymorphisms of drug metabolizing enzymes* give rise to three categories of biochemical phenotypes:

- extensive metabolism of a drug is characteristic of the normal population;
- ultraextensive metabolism results in increased drug metabolism and is an autosomal dominant trait arising from gene duplication; and
- poor metabolism is associated with the accumulation of specific drug substrates and is typically an autosomal recessive trait requiring mutation/deletion of two alleles.

Atorvastatin, cerivastatin, lovastatin, and simvastatin are all substrates of cytochrome P450 (CYP) 3A4 [35]. Cerivastatin is in addition metabolized by CYP2C8, while pravastatin is not significantly metabolized by any of the CYPs. Fluvastatin is metabolized by CYP2C9, which to a minor degree also contributes to the metabolism of lovastatin and simvastatin [35]. CYP2D6, a monooxygenase displaying several genetic variants [36, 37] has a minor role only in the metabolism of statins. Current knowledge of the relationship between genetic variants of the cytochrome P450s and the clinical efficacy of statins is rather limited. Clinically, however, such information will be of value, in particular in the identification of patients susceptible to rare, but potentially life-threatening adverse events of statin therapy such as myositis and rhabdomyolysis.

Mutations within the SREBPs and the SREBP processing proteins (SCAP, S1P, S2P) have intensively been searched, especially in patients with familial hypercholesterolemia. So far, however, only four polymorphic sites within SCAP [38, 39], one within the promoter of SREBP-1a [40], and five mutations in SREBP-2 [41] have been published. Yet, the impact of these polymorphisms and mutations on the response to statins has not been evaluated.

Familial hypercholesterolemia (FH) is an autosomal dominantly inherited disease caused by mutations in the gene for the LDL receptor. Up to now more than 680 distinct mutations, distributed over the entire gene, have been described [42]. Heterozygous FH individuals express only half the number of functional LDL-r and, therefore, have a markedly raised plasma cholesterol and usually present with premature coronary artery disease. Homozygous FH individuals are more severely affected and may succumb before the age of maturity. The prevalence of heterozygous FH is approximately 1 in 500 in Caucasians.

Heterozygous FH subjects have successfully been treated with statins [43–45], and cholesterol lowering has also been observed in LDL-r negative, homozygous carriers [46]. The type of the mutation has been shown to impact on the cholesterol-lowering effect of statins [45]. Thus, although characterization of the molecular defect in FH patients may not be relevant to their immediate clinical management, those with a particular mutation may need more aggressive lipid-lowering treatment to reach LDL cholesterol levels recommended to reduce the risk of coronary heart disease.

Apolipoprotein AI (apo AI) is the major apolipoprotein of HDL and plays an important role in the formation of mature HDL and the reverse cholesterol transport. HDL concentrations are largely determined by the rate of synthesis of apo AI in the liver. As a consequence deficiency of apo AI results in an almost complete absence of HDL and in accelerated atherosclerosis. In the promoter of the apo AI gene, a GA substitution at position –75 is common in the general population. A recent meta-analysis has shown that the minor allele A is associated with mildly elevated apo A-I levels in healthy individuals [47]. In a small study, 58 male subjects were treated with atorvastatin ($40\,mg\,d^{-1}$) or placebo in a cross-over design. Carriers of the apo AI –75 A allele ($n=15$) showed a smaller response to atorvastatin treatment in the fasting as well as in the postprandial state [48]. Larger studies are needed to confirm these results. However, the overall effect of this polymorphism appears to be small.

Apolipoprotein AIV (apo AIV) is produced in the intestine and is found in chylomicrons, VLDL and HDL. It may modulate enzymes involved in lipoprotein metabolism and may serve as a saturation signal [49]. In a study with 144 participants the apo AIV His360Glu polymorphism showed no significant effect on cholesterol lowering in response to statin therapy [50].

Apolipoprotein B is the only apolipoprotein of LDL particles and responsible for the receptor-mediated uptake of LDL. Therefore, it is obvious that mutations and polymorphisms of the apo B gene may modulate the lipid response to statins. Familial defective apo B-100 (FDB) is a group of autosomal dominantly inherited disorders, in which the cellular uptake of LDL from the blood is diminished due to mutations within the apo B-100 receptor binding domain [51]. A number of point mutations of the putative receptor binding domain of apo B-100 have been identified. Only three of these mutations have so far been proven to produce binding-defective apo B-100. Apparently the most frequent one is apo B-100 ($arg^{3500} \rightarrow gln$) [52]. We and others identified homozygous FDB patients [53, 54]. Hypercholesterolemia was less severe in these subjects as compared to patients homozygous for FH in whom the LDL receptor is defective. Using a stable isotope labeling technique, we studied the turnover *in vivo* of lipoproteins in the fasting state in our FDB homozygous patient [55]. As expected, the residence time of LDL apo B-100 was prolonged 3.6-fold in homozygous FDB, but the production rate of LDL apo B-100 was approximately half of normal. This resulted from an enhanced removal of apo E containing LDL precursors by LDL receptors, which may be up-regulated as a consequence of the decreased flux of LDL-derived cholesterol into hepatocytes. The availability of apo E for the receptor-mediated removal of remnant particles may also explain why FDB patients, homozygous or heterozygous, similarly respond to statins compared to individuals with other types of hypercholesterolemia. Numerous frequent polymorphisms have been identified at the apo B locus. Among these, a (silent) polymorphic *Xba*I site has extensively been examined. In most studies, presence of the *Xba*I cutting site was associated with moderately increased LDL cholesterol. One study addressing the impact of this polymorphism on response to lovastatin treatment (20 or $40\,mg\,d^{-1}$; $n=211$) was negative [56].

Among known genetic variants of genes related to lipoprotein metabolism, the *apolipoprotein E polymorphism* determines the greatest fraction (around 5%) of the population variance of LDL cholesterol [6]. In humans, there are three common alleles designated ε_2, ε_3, ε_4, giving rise to three homozygous and three heterozygous genotypes (for review, see [57]). The polymorphism of apo E affects the concentration of LDL by modifying the expression of hepatic LDL-r. By virtue of its preferential association with triglyceride-rich lipoproteins and due to stronger binding to lipoprotein receptors, apo E4 enhances the catabolism of remnants. Consequently, hepatic LDL-r are down-regulated and LDL plasma levels increase. For this reason, apo E4 is associated with increased LDL cholesterol and atherosclerosis. The ε_2 allele exerts an opposite effect on lipoprotein levels. apo E2 is defective in binding to lipoprotein receptors. This decreases the flux of remnant-derived cholesterol into the liver, up-regulates hepatic LDL-r and lowers LDL cholesterol. Ultimately, apo E2 may thus confer protection against the development of vascular disease. For yet unknown reasons, however, one out of twenty apo E2/2 homozygotes develops type III hyperlipoproteinemia, a disorder characterized by accumulation of excessive amounts of cholesterol-rich remnant lipoproteins derived from the partial catabolism of chylomicrons and very low-density lipoproteins.

Reports on the effects of the apo E polymorphism on the efficacy of hypolipidemic drugs are conflicting. There are several negative reports [25, 56, 58, 59] and a few publications, describing a lower cholesterol reduction in apo E4 carriers [7, 50, 60, 61] (for review, see [58]). In view of the fact that the apo E polymorphism is a strong predictor of baseline LDL cholesterol, it is surprising that there is a weak interaction only, if any at all, between the apo E genotype and the change in the LDL cholesterol concentration on statin treatment. On the other hand, most of the studies addressing this issue included patients with severe forms of hyperlipoproteinemia in which the influence of apo E might be less than in polygenic hypercholesterolemia.

In an elegant paper, Gerdes et al. [59] examined whether the risk of death or a major coronary event in survivors of myocardial infarction (MI) was related to the apo E genotype and whether risk reduction brought about by simvastatin was different between genotypes. They analyzed 5.5 years of follow-up data of 966 Danish and Finish myocardial infarction survivors enrolled in the Scandinavian Simvastatin Survival Study and found that MI survivors with the apo E4 allele have a nearly 2-fold increased risk of death, and that treatment with simvastatin abolished excess mortality. They concluded that the effect of apo E4 may involve mechanisms unrelated to serum lipoproteins because

- baseline lipid levels did not differ between apo E genotypes,
- E4 carriers and patients with other genotypes were equally responsive to simvastatin treatment in terms of LDL cholesterol lowering [59].

It would be very interesting to go back into the other cohorts, in which no difference in cholesterol reduction between the genotypes has been seen and to examine, whether the statin treatment also abolished excess mortality of apo E4 carriers.

In 1998 the REGRESS group published data, which showed that the *Taq*IB polymorphism in intron 1 of the *cholesterol-ester transfer protein* (CETP) gene predicts whether men with coronary artery disease would benefit from treatment with pravastatin or not [60]. Pravastatin therapy slowed the progression of coronary atherosclerosis in B1B1 carriers but not in B2B2 carriers who represented 16% of the patients. In the meantime the effect of this polymorphic site on HDL cholesterol and CETP plasma levels was confirmed by other investigators [61], but at least in the WOSCOPS (West of Scotland Coronary Prevention Study) study it was not possible to confirm the interaction between *Taq*IB genotype and pravastatin treatment [62]. REGRESS was an angiography-based trial in men with pre-existing coronary disease, whereas WOSCOPS was a primary prevention study in men with elevated LDL cholesterol. Possibly the different populations and primary endpoints of these studies are the reason for the inconsistent results.

Hepatic triglyceride lipase (HTGL) catalyzes the hydrolysis of triglycerides of HDL and remnant lipoproteins like IDL and LDL. Further it is involved in their uptake in the liver. Whether HTGL is pro- or antiatherogenic is still a matter of debate [63]. Recently, a CT polymorphism at position –514 (–480) in the promoter of the HTGL gene has been described which is in complete linkage disequilibrium with three other polymorphic sites within the promoter (G-250A, T-710C, A-763G) [64]. The common C allele is associated with higher HTGL activity and an atherogenic lipid profile, characterized by lower levels of HDL_2-cholesterol and dense LDL particles [65]. In a small study of 25 men with dyslipoproteinemia and established CAD, undergoing lipid-lowering therapy with 40 mg daily of lovastatin and colestipol, subjects with the CC genotype had the greatest decrease in HTGL activity, the greatest improvement in LDL density, and the greatest increase in HDL_2 cholesterol [66]. Consistently, the CC homozygous subjects had the greatest angiographic improvements. The authors concluded that the HTGL gene –514 CT polymorphism predicts 16% of the change in coronary stenosis produced by lipid-lowering therapy. There are several other polymorphisms within the coding region of the HTGL gene, which influence the activity of the lipase [67]. It would be interesting to see whether the effects of the C-514 allele could be reproduced for other variants.

Lipoprotein (a) is an independent risk factor for coronary artery disease [68]. It consists of two components: an LDL particle and apolipoprotein (a) which are linked by a disulfide bridge. Apo(a) reveals a genetically determined size polymorphism, resulting from a variable number of plasminogen kringle IV-type repeats [69]. Statins either do not affect Lp(a) or may even increase Lp(a) [70, 71]. In a study of 51 FH patients, treated with 40 $mg\,d^{-1}$ pravastatin, it has been shown that the increase in Lp(a) was greatest in patients with the low molecular-weight apo(a) phenotypes [70].

Recently, within the *stromelysin-1* promoter a functional 5A/6A polymorphism has been described [72]. Stromelysin-1 is a member of metalloproteinases that degrade extracellular matrix. *In situ* hybridization and histopathological studies suggest that stromelysin-1 activity is important in connective tissue remodeling associated with atherogenesis and plaque rupture. Patients homozygous for the 6A allele showed greater progression of angiographic disease than those with other genotypes [72]. In the REGRESS study (Regression Growth Evaluation Study) patients within the placebo group with the 5A6A or 6A6A genotype had more clinical events than patients with the 5A5A genotype. In the pravastatin group, the risk of clinical events in patients with 5A6A or 6A6A genotypes was lower, compared with placebo [73]. Similar data were obtained for the incidence of repeat angioplasty. These beneficial changes were independent of the effects of pravastatin on the lipid level, raising the possibility that pravastatin exerts pleiotropic effects on stromelysin-1 expression or activity. Up to now there are two studies, one with gemfibrozil (LOCAT) [74] and the REGRESS study conducted with pravastatin [73], suggesting that the stromelysin-1 promoter polymorphism confers a genotype-specific response to medication.

13.4 Conclusion

Lipid-lowering pharmacotherapy is one of the most recent advances in the treatment of heart disease and atherosclerosis. Genetic variants of genes involved in drug metabolism and genes involved in the lipoprotein metabolism can modify the response of plasma lipoproteins to these drugs. Research of the interaction of genetic factors and the efficacy of lipid-lowering agents, however, is at its very beginning. Publications on interactions between genotypes and the effects of lipid-lowering drugs on plasma lipoproteins and clinical outcomes are sporadic. Many studies have methodical limitations because the influence of genetic determinants has not been a pre-specified objective. The majority of the studies is not sufficiently powered to detect the effects of less frequent variants. In several cases initial positive results have not been confirmed in other studies. One reason might be the different genetic background in these cohorts. The concept that one polymorphism would provide sufficient information is probably too simplistic and deterministic. Haplotypes, describing at the same time variations in both regulatory elements and coding regions of a gene on the individual chromosome, might have the advantage of providing more information on genotype–phenotype relationships than individual SNPs. To use this information in daily practice, novel analytical tools are needed to make the results even of complex genetic profiles immediately available to clinicians. Genetic information will then probably help to set up indications for lipid-lowering drug therapy and to choose between an expanding number of treatment options.

13.5 References

1 Aguilar-Salinas SA, Barnett H, Schonfeld G. Metabolic modes of action of statins in the hyperlipoproteinemias. Atherosclerosis **1998**; 141:203-207.

2 Grundy SM, Ahrens EH, Jr., Salen G. Interruption of the enterohepatic circulation of bile acids in man: comparative effects of cholestyramine and ileal exclusion on cholesterol metabolism. J Lab Clin Med **1971**; 78:94–121.

3 Reihner E, Bjorkhem I, Angelin B, Ewerth S, Einarsson K. Bile acid synthesis in humans: regulation of hepatic microsomal cholesterol 7 alpha-hydroxy-

lase activity. Gastroenterology **1989**; 97:1498–1505.

4 COHEN JC. Contribution of cholesterol 7alpha-hydroxylase to the regulation of lipoprotein metabolism. Curr Opin Lipidol **1999**; 10:303–307.

5 WANG J, FREEMAN DJ, GRUNDY SM, LEVINE DM, GUERRA R, COHEN JC. Linkage between cholesterol 7alpha-hydroxylase and high plasma low-density lipoprotein cholesterol concentrations. J Clin Invest **1998**; 101:1283–1291.

6 COUTURE P, OTVOS JD, CUPPLES LA, WILSON PW, SCHAEFER EJ, ORDOVAS JM. Association of the A-204C polymorphism in the cholesterol 7alpha-hydroxylase gene with variations in plasma low density lipoprotein cholesterol levels in the Framingham Offspring Study. J Lipid Res **1999**; 40:1883–1889.

7 KORHONEN T, HANNUKSELA ML, SEPPANEN S, KERVINEN K, KESANIEMI YA, SAVOLAINEN MJ. The effect of the apolipoprotein E phenotype on cholesteryl ester transfer protein activity, plasma lipids and apolipoprotein A I levels in hypercholesterolaemic patients on colestipol and lovastatin treatment. Eur J Clin Pharmacol **1999**; 54:903–910.

8 STAELS S. DALLONGEVILLE J, AUWERX J, SCHOONJANS K, LEITERSDORF E, FRUCHART J. Mechanism of action of fibrates on lipid and lipoprotein metabolism. Circulation **1998**; 98:2088–2093.

9 FRUCHART JC, DURIEZ P, STAELS B. Peroxisome proliferator-activated receptor-alpha activators regulate genes governing lipoprotein metabolism, vascular inflammation and atherosclerosis. Curr Opin Lipidol **1999**; 10:245–257.

10 AUBOEUF D, RIEUSSET J, FAJAS L, VALLIER P, FRERING V, RIOU JP, et al. Tissue distribution and quantification of the expression of mRNAs of peroxisome proliferator-activated receptors and liver X receptor-alpha in humans: no alteration in adipose tissue of obese and NIDDM patients. Diabetes **1997**; 46:1319–1327.

11 SCHOONJANS K, PEINADO-ONSURBE J, LEFEBVRE AM, HEYMAN RA, BRIGGS M, DEEB S, et al. PPARalpha and PPARgamma activators direct a distinct tissue-specific transcriptional response via a PPRE in the lipoprotein lipase gene. EMBO J **1996**; 15:5336–5348.

12 STAELS B, VU-DAC N, KOSYKH V, SALADIN R, FRUCHART J-C, DALLONGEVILLE J. Fibrates downregulate apolipoprotein C-III expression independent of induction of peroxisomal acyl coenzyme A oxidase. J Clin Invest **1995**; 95:705–712.

13 LAMB RG, KOCH JC, BUSH SR. An enzymatic explanation of the differential effects of oleate and gemfibrozil on cultured hepatocyte triacylglycerol and phosphatidylcholine biosynthesis and secretion. Biochim Biophys Acta **1993**; 1165:299–305.

14 CHINETTI G, LESTAVEL S, BOCHER V, REMALEY AT, NEVE B, TORRA IP, et al. PPAR-alpha and PPAR-gamma activators induce cholesterol removal from human macrophage foam cells through stimulation of the ABCA1 pathway. Nature Med **2001**; 7:53–58.

15 CASLAKE MJ, PACKARD CJ, GAW A, MURRAY E, GRIFFIN BA, VALLANCE BD, et al. Fenofibrate and LDL metabolic heterogeneity in hypercholesterolemia. Arterioscler Thromb **1993**; 13:702–711.

16 VOHL MC, LEPAGE P, GAUDET D, BREWER CG, BETARD C, PERRON P, et al. Molecular scanning of the human PPARa gene. Association of the l162v mutation with hyperapobetalipoproteinemia. J Lipid Res **2000**; 41:945–952.

17 FLAVELL DM, PINEDA TORRA I, JAMSHIDI Y, EVANS D, DIAMOND JR, ELKELES RS, et al. Variation in the PPARalpha gene is associated with altered function in vitro and plasma lipid concentrations in Type 2 diabetic subjects. Diabetologia **2000**; 43:673–680.

18 SAPONE A, PETERS JM, SAKAI S, TOMITA S, PAPIHA SS, DAI R, et al. The human peroxisome proliferator-activated receptor alpha gene: identification and functional characterization of two natural allelic variants. Pharmacogenetics **2000**; 10:321–333.

19 AALTO-SETALA K, KONTULA K, MANTTARI M, HUTTUNEN J, MANNINEN V, KOSKINEN P, et al. DNA polymorphisms of apolipoprotein B and AI/CIII genes and response to gemfibrozil treatment. Clin Pharmacol Ther **1991**; 50:208–214.

20. Hayashi K, Kurushima H, Kuga Y, Shingu T, Tanaka K, Yasunobu Y, et al. Comparison of the effect of bezafibrate on improvement of atherogenic lipoproteins in Japanese familial combined hyperlipidemic patients with or without impaired glucose tolerance. Cardiovasc Drugs Ther **1998**; 12:3–12.

21 Manttari M, Koskinen P, Ehnholm C, Huttunen JK, Manninen V. Apolipoprotein E polymorphism influences the serum cholesterol response to dietary intervene blood is diminished due to mutations within the apo B-100 receptor binding domain [51]. A number of point mutations of the putativd its relation to E polymorphism. Orv Hetil **1994**; 135:735–741.

23 Nemeth A, Szakmary K, Kramer J, Dinya E, Pados G, Fust G, et al. Apolipoprotein E and complement C3 polymorphism and their role in the response to gemfibrozil and low fat low cholesterol therapy. Eur J Clin Chem Clin Biochem **1995**; 33:799–804.

24 Yamada M. Influence of apolipoprotein E polymorphism on bezafibrate treatment response in dyslipidemic patients. J Atheroscler Thromb **1997**; 4:40–44.

25 Sanllehy C, Casals E, Rodriguez-Villar C, Zambon D, Ojuel J, Ballesta AM, et al. Lack of interaction of apolipoprotein E phenotype with the lipoprotein response to lovastatin or gemfibrozil in patients with primary hypercholesterolemia. Metabolism **1998**; 47:560–565.

26 Walldius G, Wahlber G. Nicotinic acid and its derivates. In: Lipoproteins in Health and Disease (Betteridge DJ, Illingworth DR, Shepherd J, Eds.). London: Arnold; **1999**, 1181–1197.

27 Chiesa G, Michelagnoli S, Cassinotti M, Gianfranceschi G, Werba JP, Pazzucconi F, et al. Mechanisms of high-density lipoprotein reduction after probucol treatment: changes in plasma cholesterol esterification/transfer and lipase activities. Metabolism **1993**; 42:229–235.

28 Yamamoto A, Matsuzawa Y, Kishino B, Hayashi R, Hirobe K, Kikkawa T. Effects of probucol on homozygous cases of familial hypercholesterolemia. Atherosclerosis **1983**; 48:157–166.

29 Nestruck AC, Bouthillier D, Sing CF, Davignon J. Apolipoprotein E polymorphism and plasma cholesterol response to probucol. Metabolism **1987**; 36:743–747.

30 Eto M, Sato T, Watanabe K, Iwashima Y, Makino I. Effects of probucol on plasma lipids and lipoproteins in familial hypercholesterolemic patients with and without apolipoprotein E4. Atherosclerosis **1990**; 84:49–53.

31 Brown MS, Goldstein JL. The SREBP pathway: regulation of cholesterol metabolism by proteolysis of a membrane-bound transcription factor. Cell **1997**; 89:331–340.

32 Osborne TF. Sterol regulatory element binding protein (SREBPs): Key regulators of nutritional homeostasis and insulin action. J Biol Chem **2000**; 275:32379–32382.

33 Bennett MK, Lopez JM, Sanchez HB, Osborne TF. Sterol regulation of fatty acid synthase promoter. Coordinate feedback regulation of two major lipid pathways. J Biol Chem **1995**; 270:25578–25583.

34 Lopez JM, Bennett MK, Sanchez HB, Rosenfeld JM, Osborne TE. Sterol regulation of acetyl coenzyme A carboxylase: a mechanism for coordinate control of cellular lipid. Proc Natl Acad Sci USA **1996**; 93:1049–1053.

35 Herman RJ. Drug interactions and the statins. Cmaj 1999;161:1281-1286.

36 Linder MW, Prough RA, Valdes R, Jr. Pharmacogenetics: a laboratory tool for optimizing therapeutic efficiency. Clin Chem **1997**; 43:254–266.

37 Tanaka E. Update: genetic polymorphism of drug metabolizing enzymes in humans. J Clin Pharm Ther **1999**; 24:323–329.

38 Nakajima T, Ota N, Kodama T, Emi M. Isolation and radiation hybrid mapping of a highly polymorphic CA repeat sequence at the SREBP cleavage-activating protein (SCAP) locus. J Hum Genet **1999**; 44:350–351.

39 Iwaki K, Nakajima T, Ota N, Emi M. A common Ile796Val polymorphism of the human SREBP cleavage-activating pro-

tein (SCAP) gene. J Hum Genet **1999**; 44:421–422.

40 Vedie B, Jeunemaitre X, Megnien JL, Atger V, Simon A, Moatti N. A new DNA polymorphism in the 5' untranslated region of the human SREBP-1a is related to development of atherosclerosis in high cardiovascular risk population. Atherosclerosis **2001**; 154:589–597.

41 Muller PJ, Miserez AR. Mutations in the gene encoding sterol-regulatory element-binding protein-2 in hypercholesterolaemic subjects. Atherosclerosis Supplements **2001**; 2:69.

42 Heath KE, Gahan M, Whittall RA, Humphries SE. Low-density lipoprotein receptor gene (LDLR) world-wide website in familial hypercholesterolaemia: update, new features and mutation analysis. Atherosclerosis **2001**; 154:243–246.

43 Karayan L, Qiu S, Betard C, Dufour R, Roederer G, Minnich A, et al. Response to HMG CoA reductase inhibitors in heterozygous familial hypercholesterolemia due to the 10-kb deletion ("French Canadian mutation") of the LDL receptor gene. Arterioscler Thromb **1994**; 14:1258–1263.

44 Vuorio AF, Ojala JP, Sarna S, Turtola H, Tikkanen MJ, Kontula K. Heterozygous familial hypercholesterolaemia: the influence of the mutation type of the low-density-lipoprotein receptor gene and PvuII polymorphism of the normal allele on serum lipid levels and response to lovastatin treatment. J Intern Med **1995**; 237:43–48.

45 Heath KE, Gudnason V, Humphries SE, Seed M. The type of mutation in the low density lipoprotein receptor gene influences the cholesterol-lowering response of the HMG-CoA reductase inhibitor simvastatin in patients with heterozygous familial hypercholesterolaemia. Atherosclerosis **1999**; 143:41–54.

46 Feher MD, Webb JC, Patel DD, Lant AF, Mayne PD, Knight BL, et al. Cholesterol-lowering drug therapy in a patient with receptor-negative homozygous familial hypercholesterolaemia. Atherosclerosis **1993**; 103:171–180.

47 Juo SH, Wyszynski DF, Beaty TH, Huang HY, Bailey-Wilson JE. Mild association between the A/G polymorphism in the promoter of the apolipoprotein A-I gene and apolipoprotein A-I levels: a meta-analysis. Am J Med Genet **1999**; 82:235–241.

48 Ordovas JM, Vargas C, Santos A, Tayler TD, Daly JA, Augustine J, et al. The G/A promoter polymorphism at the apoAI gene locus predicts individual variability in fasting and postprandial responses to the HMG CoA reductase inhibitor atorvastatin. Circulation **1999**; 100(Suppl I):I–239.

49 Tso P, Liu M, Kalogeris TJ. The role of apolipoprotein A-IV in food intake regulation. J Nutr **1999**; 129:1503–1506.

50 Ordovas JM, Lopez-Miranda J, Perez-Jimenez F, Rodriguez C, Park JS, Cole T, et al. Effect of apolipoprotein E and A-IV phenotypes on the low density lipoprotein response to HMG CoA reductase inhibitor therapy. Atherosclerosis **1995**; 113:157–166.

51 Fisher E, Scharnagl H, Hoffmann MM, Kusterer K, Wittmann D, Wieland H, et al. Mutations in the apolipoprotein (apo) B-100 receptor-binding region: Detection of apo B-100 (Arg3500–>Trp) associated with two new haplotypes and evidence that apo B-100 (Glu3405–>Gln) diminishes receptor-mediated uptake of LDL. Clin Chem **1999**; 45:1026–1038.

52 Soria LF, Ludwig EH, Clarke HR, Vega GL, Grundy SM, McCarthy BJ. Association between a specific apolipoprotein B mutation and familial defective apolipoprotein B-100. Proc Natl Acad Sci USA **1989**; 86:587–591.

53 März W, Baumstark M, Scharnagl H, Ruzicka V, Buxbaum S, Herwig J, et al. Accumulation of "small dense" low density lipoproteins in a homozygous patient with familial defective apolipoprotein B-100 results from heterogenous interaction of LDL subfractions with the LDL receptor. J Clin Invest **1993**; 92:2922–2933.

54 Myant NB. Familial defective apolipoprotein B-100: a review, including some comparisons with familial hypercholesterolaemia [published erratum appears in Atherosclerosis 1994 Feb;105[2]:253]. Atherosclerosis **1993**; 104:1–18.

55 Schaefer JR, Scharnagl H, Baumstark MW, Schweer H, Zech LA, Seyberth H, et al. Homozygous familial defective apolipoprotein B-100. Enhanced removal of apolipoprotein E-containing VLDLs and decreased production of LDLs. Arterioscler Thromb Vasc Biol **1997**; 17:348–353.

56 Ojala JP, Helve E, Ehnholm C, Aalto-Setala K, Kontula KK, Tikkanen MJ. Effect of apolipoprotein E polymorphism and XbaI polymorphism of apolipoprotein B on response to lovastatin treatment in familial and non-familial hypercholesterolaemia. J Intern Med **1991**; 230:397–405.

57 Mahley RW, Huang Y. Apolipoprotein E: from atherosclerosis to Alzheimer's disease and beyond. Curr Opin Lipidol **1999**; 10:207–217.

58 Hoffmann MM, Winkelmann BR, Wieland H, Marz W. The significance of genetic polymorphisms in modulating the response to lipid-lowering drugs. Pharmacogenomics **2001**; 2:107–121.

59 Gerdes LU, Gerdes C, Kervinen K, Savolainen M, Klausen IC, Hansen PS, et al. The apolipoprotein epsilon4 allele determines prognosis and the effect on prognosis of simvastatin in survivors of myocardial infarction: a substudy of the Scandinavian simvastatin survival study. Circulation **2000**; 101:1366–1371.

60 Kuivenhoven JA, Jukema JW, Zwinderman AH, De Knijff P, McPherson R, Bruschke AVG, et al. The role of a common variant of the cholesteryl ester transfer protein gene in the progression of coronary atherosclerosis. N Engl J Med **1998**; 338:86–93.

61 Ordovas JM, Cupples LA, Corella D, Otvos JD, Osgood D, Martinez A, et al. Association of cholesteryl ester transfer protein-TaqIB polymorphism with variations in lipoprotein subclasses and coronary heart disease risk: the Framingham study. Arterioscler Thromb Vasc Biol **2000**; 20:1323–1329.

62 Freeman DJ, Wilson V, McMahon AD, Packard CJ, Gaffney D. A polymorphism of the Cholesteryl Ester Transfer Protein (CETP) gene predicts cardiovascular events in the West of Scotland Coronary Prevention Study (WOSCOPS). Atherosclerosis **2000**; 151:91.

63 Santamarina-Fojo S, Haudenschild C, Amar M. The role of hepatic lipase in lipoprotein metabolism and atherosclerosis. Curr Opin Lipidol **1998**; 9:211–219.

64 Guerra R, Wang J, Grundy SM, Cohen JC. A hepatic lipase (LIPC) allele associated with high plasma concentrations of high density lipoprotein cholesterol. Proc Natl Acad Sci USA **1997**; 94:4532–4537.

65 Zambon A, Hokanson JE, Brown BG, Brunzell JD. Evidence for a new pathophysiological mechanism for coronary artery disease regression : hepatic lipase-mediated changes in LDL density. Circulation **1999**; 99:1959–1964.

66 Zambon A, Deeb SS, Brown BG, Hokanson JE, Bertocco S, Brunzell JD. A common hepatic lipase gene promoter variant determines clinical response to intensive lipid lowering treatment. Atherosclerosis **2000**; 151:266.

67 Nie L, Niu S, Vega GL, Clark LT, Tang A, Grundy SM, et al. Three polymorphisms associated with low hepatic lipase activity are common in African Americans. J Lipid Res **1998**; 39:1900–1903.

68 Marcovina SM, Koschinsky ML. Lipoprotein(a) as a risk factor for coronary artery disease. Am J Cardiol **1998**; 82:57U–66U.

69 Utermann G. Genetic architecture and evolution of the lipoprotein(a) trait. Curr Opin Lipidol **1999**; 10:133–141.

70 Klausen IC, Gerdes LU, Meinertz H, Hansen FA, Faergeman O. Apolipoprotein(a) polymorphism predicts the increase of Lp(a) by pravastatin in patients with familial hypercholesterolaemia treated with bile acid sequestration. Eur J Clin Invest **1993**; 23:240–245.

71 Marz W, Grutzmacher P, Paul D, Siekmeier R, Schoeppe W, Gross W. Effects of lovastatin [20-80 mg daily) on lipoprotein fractions in patients with severe primary hypercholesterolemia. Int J Clin Pharmacol Ther **1994**; 32:92–97.

72 Ye S, Watts GF, Mandalia S, Humphries SE, Henney AM. Preliminary report: genetic variation in the human stromelysin promotor is associated with progression of coronary atherosclerosis. Br Heart J **1995**; 73:209–215.

73 de Maat MP, Jukema JW, Ye S, Zwinderman AH, Moghaddam PH, Beekman M, et al. Effect of the stromelysin-1 promoter on efficacy of pravastatin in coronary atherosclerosis and restenosis. Am J Cardiol **1999**; 83:852–856.

74 Humphries SE, Luong LA, Talmud PJ, Frick MH, Kesaniemi YA, Pasternack A, et al. The 5A/6A polymorphism in the promoter of the stromelysin-1 (MMP-3] gene predicts progression of angiographically determined coronary artery disease in men in the LOCAT gemfibrozil study. Lopid Coronary Angiography Trial. Atherosclerosis **1998**; 139:49–56.

14
Pharmacogenomics of Chemotherapeutic Agents in Cancer Treatment

Federico Innocenti, Lalitha Iyer and Mark J. Ratain

Abstract

This chapter describes how genetic differences among patients may change therapeutic outcome in cancer chemotherapy. The therapeutic window of anticancer agents is narrow and, in most cases, patient are treated at dose levels that are close to those maximally tolerated. Inter-patient genetic differences altering pharmacokinetics and/or pharmacodynamics might result in unpredictable outcome. Severe toxicity in genetically predisposed patients is predominantly associated with mutations in drug metabolism enzyme genes. Intolerance to chemotherapy is clearly demonstrated in subsets of patients receiving 6-mercaptopurine (inactivated by thiopurine methyltransferase, TPMT), 5-fluorouracil (inactivated by dihydropyrimidine dehydrogenase, DPD), irinotecan [the active metabolite 7-ethyl-10-hydroxycamptothecin (SN-38) is inactivated by UDP-glucuronosyltransferase 1A1, UGT1A1], amonafide (metabolized by N-acetyltransferase 2, NAT2). Moreover, cancer patients carrying a mutation in the methylenetetrahydrofolate reductase (MTHFR) gene are highly susceptible to myelosuppressive effects of CMF regimen (cyclophosphamide+methotrexate+5-fluorouracil). It is emerging that not only toxicity, but also response to chemotherapy could be influenced by pharmacogenetic determinants. As a matter of fact, recent studies highlighted the correlation between mutations in glutathione-S-transferase (GST) and thymidylate synthase (TS) genes and patients' response to chemotherapy.

14.1
Pharmacological Treatment of Cancer and Importance of Pharmacogenomics

Chemotherapy of cancer is part of a multimodal treatment including surgery and radiation therapy. In drug-sensitive tumors, the major obstacle to successful chemotherapy is the occurrence of resistance, and is related to the impossibility to administer curative doses of drugs due to the occurrence of toxicity. Reduced intensity of treatment allows the emergence of cell clones with a resistant phenotype as a result of somatic mutations occurring in surviving cells. Maximizing tumor exposure while reducing the risk of intolerable toxicity is a mandatory task, in particular because chemotherapy is the only alternative available in cases of inoperable and metastatic disease.

Cancer patients are treated at doses close to those maximally tolerated, making anticancer agents a class of drugs with a very narrow therapeutic window, defined as the interval between the dose required to produce a therapeutic effect and that responsible for toxicity. Current modes to administer anticancer drugs do not take into account differences among individuals. Cancer patients receive fixed doses "normalized" by body surface area, a very imprecise approach of dose individualization. Existing differences between individuals in pharmacokinetics and pharmacodynamics imply that some patients might benefit from chemotherapy, but others might experience adverse reactions without any therapeutic advantage. The explosion of genetic investigation of human molecular biology is increasingly demonstrating that unpredictability in patient outcome could be due to genetic differences in the way drugs are handled and react with targets of action.

Multiple steps occur from drug administration to pharmacological effect on normal and neoplastic tissues. Processes mediated by membrane transporters and metabolizing enzymes are critical for achieving effective intracellular drug concentrations. At intracellular level, killing action of cytotoxic drugs is dependent upon drug activation/inactivation pathways, levels of the molecular target of action, mechanisms of DNA repair, and balance between pro- and anti-apoptotic pathways. Theoretically, genetic mutations can lead to reduced or increased efficiency in each of these processes. So far, germ line mutations in drug metabolizing enzymes have been demonstrated to be major determinants of severe toxicity in genetically predisposed patients. The field of cancer pharmacogenomics is extraordinarily expanding, and recent findings also pointed out that the detection of both germ line and somatic genetic polymorphisms in detoxifying enzymes and in molecular targets of action can be used as predictors of response and outcome of chemotherapy. Somatic mutations in the tumor can be also used to select appropriate chemotherapy treatment after surgery in order to overcome drug resistance.

In cancer patients with normal liver/kidney function who receive single agent chemotherapy, the possible presence of genetic determinants of toxicity/response could be argued by the following observations:

(1) high interpatient variability in pharmacokinetic parameters of active drug,
(2) bimodal distribution of area under the concentration versus time curve (AUC) metabolic ratios of inactive metabolite to active drug,
(3) occurrence of severe toxicity after the first cycle of treatment, and re-occurrence of toxicity in following cycles, even at reduced doses.

Once a candidate gene has been identified, anticancer drug therapy can be rationalized by means of genetic principles. The current application of pharmacogenetics in cancer chemotherapy suggests that pharmacogenetic differences among patients can be identified, and the therapeutic window of new and old anticancer agents can be enlarged. Patient predisposition to severe toxicity and genetic markers of response should be characterized prospectively, in order to eliminate toxicity of ineffective therapy and allow more rational search for new therapies in patients who cannot benefit from conventional chemotherapy.

14.2
Pharmacogenetic Determinants of Toxicity after Cancer Chemotherapy

Among the cancer patient population, a subgroup of patients is genetically predisposed to develop more prolonged and severe toxicity. Mutations in genes coding drug inactivating/activating enzymes, as well as enzymes involved in reduced folate metabolism, can be responsible for intolerance to standard doses of several drugs currently used in cancer treatment.

14.2.1
6-Mercaptopurine and TPMT Pharmacogenetics

14.2.1.1 Clinical Use and Toxicity of 6-MP in Childhood Acute Lymphoblastic Leukemia

6-mercaptopurine (6-MP) is a purine analog used in the cure of childhood acute lymphoblastic leukemia (ALL), the most common malignancy in children. Leukemic clones are eradicated from the bone marrow after an intense poly-chemotherapeutic regimen. To maintain clinical remission, patients receive daily oral 6-MP in combination with methotrexate (MTX) for a period of two-three years, during which children are under a permanent condition of intentionally "controlled" myelotoxicity, a surrogate endpoint to monitor treatment efficacy. Most ALL protocols include individual tailoring of 6-MP dose depending on white blood cell count. Hematological toxicity is not related to 6-MP dosage but to the conversion of 6-MP into active metabolites [1, 2]. At least 2/3 of children with ALL are disease free for five years and appear cured after the termination of chemotherapy. In the past two decades, it became evident that 6-MP is the cornerstone of the maintenance chemotherapy and patient outcome could be improved by understanding the complex pharmacology of 6-MP. Childhood ALL can be regarded as a curable disease, but about 1/3 of children will not be cured. About 80% of first relapses in ALL occur in the hematopoietic tissues, and adequate bone marrow exposure is of considerable importance.

14.2.1.2 Metabolism of 6-MP – Activating and Inactivating Pathways and Their Clinical Relevance

At cellular level, 6-MP is transformed in a number of active and inactive metabolites, and in the bone marrow, the balance between activation and inactivation of 6-MP is the main determinant of its antiproliferative effect. Similar to other antimetabolites, 6-MP is a prodrug lacking any cytotoxic activity and needs to be activated [3] (Figure 14.1). The first step is 6-MP transformation into 6-thioinosine monophosphate (6-TIMP), which is subsequently converted to 6-thioguanine tri-

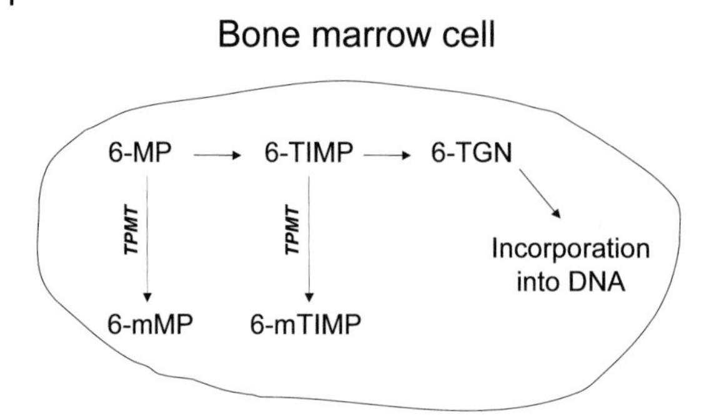

Fig. 14.1 6-MP metabolism in bone marrow cells. 6-TIMP, 6-thioinosine monophosphate; 6-mMP, 6-methylmercaptopurine; 6-mTIMP, 6-methylthioinosine monophosphate.

phosphate nucleotides (6-TGN). DNA incorporation of 6-TGN mediates 6-MP antileukemic activity, interfering with DNA ligase, endonuclease, and polymerase functions [4]. The amount of 6-MP that can be activated in the bone marrow depends upon the extent of 6-MP methylation by thiopurine methyltransferase (TPMT) [5, 6]. Although methylated 6-MP and 6-TIMP are inhibitors of the purine salvage pathway, their contribution to the overall cytotoxicity of 6-MP is not as relevant as 6-TGN production. Adequate activation of 6-MP to 6-TGN at bone marrow level is required for increasing the probability of better outcome [7, 8]. By indirectly regulating the size of 6-TGN production, TPMT is the prime determinant of 6-MP antileukemic effect. TPMT is genetically polymorphic, introducing a major factor of variability in outcome of ALL patients.

14.2.1.3 Reduced Tolerance to 6-MP in Patients with Genetic Impairment of TPMT Activity

Several case reports evidenced that ALL patients with reduced TPMT activity are intolerant to standard doses of 6-MP. Similar findings were reported also in patients with skin or autoimmune disorders, as well as in transplantation patients receiving the 6-MP analog azathioprine. ALL patients with genetic deficiency in TPMT accumulate 6-TGN to toxic concentrations, leading to severe and prolonged myelosuppression associated with bone marrow hypoplasia. Due to the latency of 6-MP cytotoxicity, these effects are not evident after the first daily administrations of 6-MP, becoming manifest generally after two-three weeks. In the presence of excessive hematological toxicity, TPMT-deficient patients can be exposed to life-threatening opportunistic infections, and drug treatment is discontinued until the bone marrow has recovered. 6-MP total dose delivered at the end of the maintenance is a critical factor for outcome of ALL patients [9, 10]. As a consequence, when ALL patients are not able to complete the scheduled treatment, they are more susceptible to drug resistance and disease recurrence. In one patient with TPMT deficit, no therapy could be administered for half of the maintenance period [11]. In another case report, 6-MP standard dosage was reduced by 15-folds to avoid intolerable myelosuppression persisting for 3 weeks [12]. Along with myelosuppression, these patients experienced severe gastrointestinal toxicity, permanent alopecia, and less severe mucositis [11, 12], and support therapy including erythrocytes and platelet transfusion and antibiotics is required.

The impact of TPMT genetic make up on patient outcome has been established in two clinical trials. In one study, TPMT-deficient heterozygotes and homozygotes received full doses of 6-MP only for 65% and 7% of the maintenance period, respectively. On the contrary, wild-type patients tolerated full doses of 6-MP for 84% of the maintenance period [13]. In another study, no clear distinction in 6-MP tolerance was noted between wild-type and heterozygous patients, probably due to differences in the intensity of previous treatment protocols affecting bone marrow sensitivity. However, the only patient who was mutant homozygous could not receive 6-MP for half of the maintenance period [11]. Finally, the first case of life-threatening myelosuppression in a TPMT-deficient patient receiving 6-thioguanine (a 6-MP analog and TPMT substrate) was recently observed during the consolidation phase of ALL [14].

14.2.1.4 Prediction of TPMT Deficiency

In order to identify patients at higher risk of toxicity after 6-MP, either phenotyping or genotyping procedures are successfully used. The choice between phenotyping or genotyping is dependent upon the availability of such techniques in the laboratory. TPMT activity is already routinely measured in some centers (sometimes coupled to the measurement of 6-TGN production in erythrocytes) and about 90–95% of the TPMT-deficient phenotypes are concordant with their genotype. ALL patients receive blood transfusions when they experience severe anemia, and the use of TPMT activity measurement in erythrocytes is not reliable. In these cases, genotyping should be indicated. Ideally, genotyping associated with phenotyping should be used, since there are still unknown mutations accounting for at least 10% of cases of TPMT phenotypic deficiency.

TPMT Phenotyping

TMPT activity in human erythrocytes is transmitted as an autosomic codominant trait [15] and is trimodally distributed, with 89–94% of the individuals having high, 6–11% intermediate, and 0.3% low activity [7, 15–17] (Figure 14.2). The measurement of TPMT activity in erythrocytes closely reflects the ability of bone marrow to inactivate 6-MP. TPMT activity is inversely related to erythrocyte 6-TGN levels [7, 13, 18, 19], and children with low TPMT activity and very high 6-TGN levels experienced profound myelotoxicity [20, 21]. Moreover, TPMT phenotype in erythrocyte reflects that in leukemic blasts [22]. Patients with intermediate TPMT activity had a 5-fold greater cumulative incidence of dose reductions than subjects with high activity [13], and TPMT activity has been inversely related to the time of treatment withdrawal due to cytopenia [21].

TPMT Genotyping

Ten *TPMT* variants associated with low enzyme activity have been described, and *TPMT*2*, *TPMT*3A*, and *TPMT*3C* account for about 80–95% of the TPMT-defi-

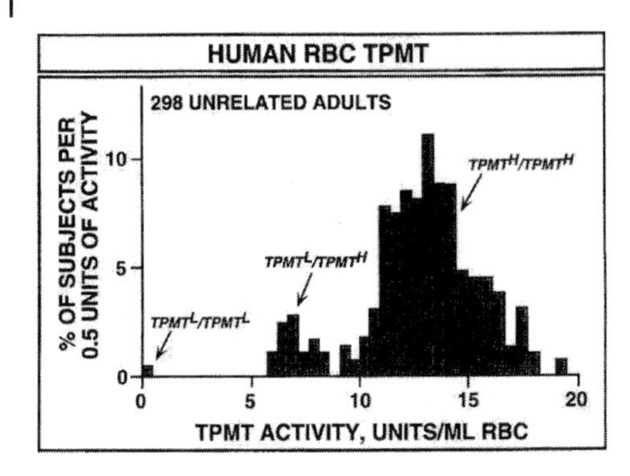

Fig. 14.2 Trimodal distribution of human TPMT activity in erythrocytes. High (*TPMT^H/TPMT^H*), intermediate (*TPMT^L/TPMT^H*), and low (*TPMT^L/TPMT^L*) metabolizer genotypes are indicated.

cient phenotype [16, 23–27]. *TPMT* genotype and phenotype are highly concordant. Wild-type individuals have high TPMT activity, while heterozygotes and homozygotes for one variant allele have intermediate and low activity, respectively [24] (Figure 14.2). *TPMT*3A* allele comprises two non-synonymous single nucleotide polymorphisms in exons 7 (G460A) and 10 (A719G) and is the most common variant (frequency of 3.2–5.7% in Caucasians). *TPMT*3A* represents 55–86% of all defective variants, and has been found in about 55% of deficient phenotypes. The frequency of *TPMT*2* (G238C in exon 5) and *TPMT*3C* (A719G in exon 10) is about 0.2–0.8% in Caucasians [16, 24, 28, 29].

14.2.1.5 **6-MP Dose Adjustment in ALL Patients**

Clinical experience in TPMT-deficient patients suggests that they should receive 5-10% of the planned 6-MP dose. With regard to wild-type patients with high TPMT activity (about 90% of ALL children), the molecular basis of the interpatient differences in TPMT activity is still unclear. Up to 5-fold variability in TPMT activity has been found in wild-type patients [24, 30], suggesting the possible contribution of differences in *TPMT* gene expression. A polymorphism in the variable tandem repeat region of the *TPMT* promoter has been proposed to modulate TPMT activity, however, the magnitude of this modulation is probably not relevant enough to explain such differences in TPMT activity in wild-type individuals [31–34]. It would be scientifically reasonable to treat this subgroup with higher 6-MP doses to avoid underdosing, in particular because patients with high TPMT activity and low 6-TGN are more at risk for relapse [7]. However, 6-MP dose escalation in the absence of toxicity paradoxically reduces dose intensity (i.e., the amount of drug delivered per unit of time) because of increased toxicity, and full doses of 6-MP are still recommended in wild-type patients [9, 35]. The future challenge in 6-MP pharmacogenetics is to identify the genetic basis of TPMT variability in wild-type patients.

14.2.2
5-Fluorouracil and DPD Pharmacogenetics

14.2.2.1 Clinical Use and Toxicity of 5-Fluorouracil

5-Fluorouracil (5-FU) is a pyrimidine analog widely used in the treatment of colorectal, breast, and head and neck cancers. In combination with leucovorin (folinic acid), 5-FU represents the standard adjuvant treatment of non-metastatic colon cancer, one of the most frequent tumors in developed countries. In metastatic disease, combinations of 5-FU/leucovorin with either irinotecan or oxaliplatin showed better efficacy compared to 5-FU/leucovorin alone [36, 37]. The main schedules for 5-FU administration are intravenous bolus given daily for 5 days every 3–4 weeks or weekly bolus. 5-FU is generally well tolerated, and highly replicating epithelial tissues are targets of its toxic action. Dose-limiting toxicity of bolus 5-FU includes nausea/vomiting, myelotoxicity, oral mucositis, diarrhea, desquamation of the palms and soles, and rarely, cardiac and neurological toxic effects.

14.2.2.2 Metabolism of 5-FU – Activating and Inactivating Pathways and Their Clinical Relevance

After intravenous administration, about 80–90% of the dose is catabolized in the liver by dihydropyrimidine dehydrogenase (DPD) [38] (Figure 14.3). The formation of the inactive 5-fluoro-5,6-dihydrouracil (5-FUH_2) by DPD is the rate-limiting step of 5-FU catabolism [39]. DPD is widely distributed among tissues, with the highest levels found in the liver. Once 5-FU entered tumor cells, its antitumor effect is mainly dependent on the extent of 5-FU anabolism. After two sequential anabolic steps involving thymidine phosphorylase (TP) and thymidine kinase

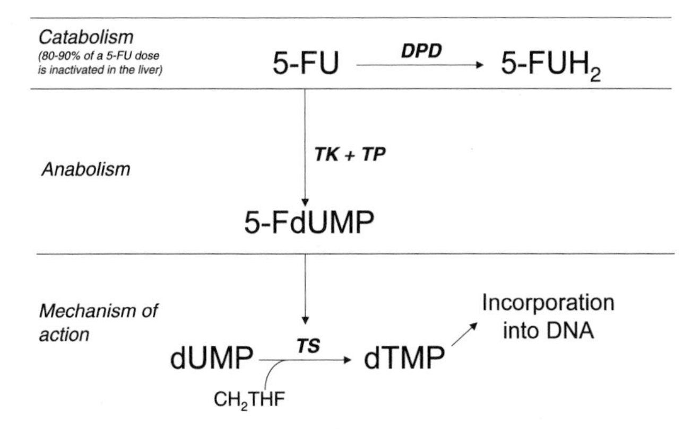

Fig. 14.3 5-FU catabolism, anabolism and mechanism of action. 5-FUH_2, 5-fluoro-5,6-dihydrouracil; 5-FdUMP, 5-fluorodeoxyuridine monophosphate; TP, thymidine phosphorylase; TK, thymidine kinase; TS, thymidylate synthase; CH_2THF, 5,10-methylenetetrahydrofolate.

(TK), 5-FU is activated to 5-fluorodeoxyuridine monophosphate (5-FdUMP). Potent inhibition of thymidylate synthase (TS) by 5-FdUMP is considered critical for 5-FU cytotoxicity. TS catalyzes the rate-limiting step of DNA synthesis, such as the conversion of dUMP into dTMP. Optimal TS function requires the formation of a covalent ternary complex consisting of TS, the folate cofactor 5,10-methylenetetrahydrofolate (CH_2THF), and 5-FdUMP. Inadequate cellular levels of 5,10-methylenetetrahydrofolate reduce the stability of the ternary complex and consequently the inhibition of TS by 5-FdUMP. For this reason, 5-FU is administered in association with folinic acid, a precursor of 5,10-methylenetetrahydrofolate [40].

14.2.2.3 Life-Threatening Toxicity in DPD-Deficient Patients

Since 1985, it was clear that reduced ability to inactivate 5-FU was heritable and could expose patients to intolerable toxicities [41]. Further observations confirmed this finding and clarified the biochemical determinant of this genetic defect (Figure 14.4) [42]. DPD activity is completely or partially deficient in about 0.1% and 3–5% of individuals [43], with at least 150 cases reported so far [40]. A neurological syndrome with thymine-uraciluria occurs in pediatric patients due to complete deficiency of DPD. In cancer patients with defective DPD, a pharmacogenetic syndrome occurs after 5-FU dosing, and 5-FU-related toxicities are severe and life-threatening. DPD-deficient patients experience grade 4 myelosuppression, along with grade 3–4 neurological and gastrointestinal toxicities. The occurrence of severe toxicity usually requires 5-FU discontinuation and empiric dose reduction in the following cycles of therapy, hospitalization, and, sometimes, evaluation of alternative chemotherapy. A few cases of toxic deaths with documented DPD defects were also reported [44–46].

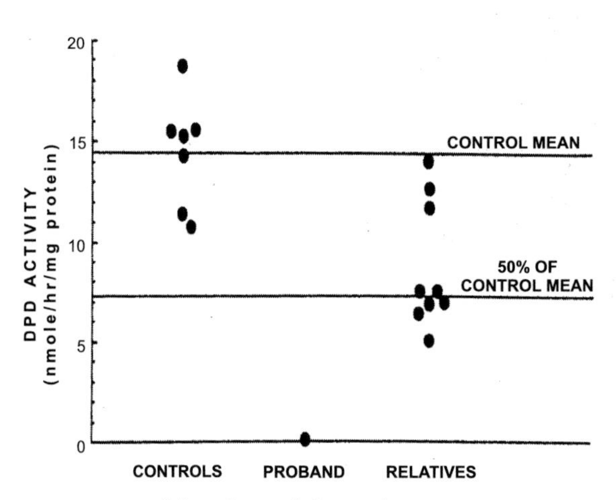

Fig. 14.4 Heritability of DPD deficient phenotype. DPD activity was measured in peripheral blood mononuclear cells from a proband, her family members, and healthy volunteers (controls).

14.2.2.4 DPD Genotype and Molecular Basis of DPD Deficiency

DPD genotype has an autosomal recessive pattern of inheritance [47]. The inactivation of one allele leading to a 50% reduction in the normal DPD activity is sufficient to trigger the development of toxicities after 5-FU treatment [48]. At least 20 mutations in *DPD* gene (*DPYD*) coding and promoter region have been reported. Among *DPYD* variant alleles, eight of them are rare polymorphisms not affecting DPD activity [49, 50]. Correlative studies between 5-FU toxicity and mutated genotypes in cancer patients were not able to clearly identify the *DPYD* mutations explaining this pharmacogenetic syndrome. Potential candidates with clinical relevance are *DPYD**2A* and *DPYD**9A*. *DPYD**2A* is a splice site mutation (intron 14 G1A) resulting in the production of a truncated mRNA. It was associated with 5-FU-related toxicity and low DPD activity in three cancer patients from different studies [48, 51, 52], and its allele frequency is low (1.3%). Among 14 cancer patients selected on the basis of low DPD activity and severe toxicity, this mutation was found in six of them [53]. However, in another study, discordance was demonstrated between *DPYD**2A* and DPD activity [50]. *DPYD**9A* is a common missense T85C mutation in exon 2 resulting in a C29R amino acid change, but its association with reduced DPD activity is still controversial. However, heterozygosity for *DPYD**9A* was found in four and eight of 14 cancer patients with severe 5-FU toxicity in two different studies [50, 53]. *DPYD**2A* and *DPYD**9A* mutations seems to have good concordance with clinical phenotype (i.e., 5-FU toxicity), but a low concordance with biochemical phenotype (i.e., DPD activity). Other mutations in the *DPYD* promoter and coding region occurred in DPD-deficient patients experiencing severe toxicity but their frequency is unknown [50, 54, 55].

14.2.2.5 Measures to Predict DPD Deficiency in Patients Receiving 5-FU

The complexity of the genetic basis of DPD deficiency implies that the identification of patients at high risk of 5-FU toxicity is mostly based on phenotypic procedures. These methods are not suitable for general use and concomitant drugs, dietary intake and other environmental factors could reduce their predictive power in cases of partial DPD deficit.

DPD Biochemical Phenotype Measured in Peripheral Blood Mononuclear Cells (PBMC)
DPD activity measured in PBMC is used as a surrogate for systemic DPD activity. DPD activity is normally distributed and highly variable among individuals (coefficient of variation of 33.9–46.6%) [43, 56–59]. DPD activity is undetectable in totally deficient patients. The majority of partially deficient patients had a DPD value $\leq$30% of the mean in the normal population, and this value is considered the cut-off for patients at higher risk of toxicity. Among patients experiencing severe toxicity after 5-FU, 36–59% of them were deficient in DPD activity [43, 53, 60]. This suggests the involvement of other determinants in the susceptibility to 5-FU toxicity. The concordance between liver and PBMC DPD activity is modest [61], and normal DPD activity in PBMC was found in one patient with very depressed liver DPD activity who died because of 5-FU toxicities [44].

Measurement of Natural Pyrimidines in Biological Fluids
In the majority of DPD defective patients experiencing severe 5-FU toxicity, abnormally high levels of natural pyrimidines are present in plasma and/or urine [62]. Moreover, endogenous dihydrouracil/uracil ratio in plasma has been proposed as a measure of 5-FU catabolic deficiency in cancer patients [63], and screening of cancer patients for these simple markers should be prospectively evaluated.

14.2.3
Irinotecan and UGT1A1 Pharmacogenetics

14.2.3.1 Clinical Use and Toxicity of Irinotecan

Irinotecan (CPT-11) is a semi-synthetic analog of the natural alkaloid camptothecin with considerable activity in colorectal cancer patients with poor prognosis due to 5-FU resistance. Moreover, the utility of irinotecan as a component of initial therapy in association with 5-FU of metastatic colorectal cancer has been recently demonstrated [37]. The most common administration schedule of irinotecan is a short (30–90 min) intravenous infusion, either once every three weeks or weekly for four weeks [64]. Common and dose limiting toxicities of irinotecan are neutropenia and delayed diarrhea. Both grade 3–4 neutropenia and diarrhea may occur in about 1/3 of patients, with variable frequency depending on the schedule of administration. Severe nausea/vomiting is reported in less than 10% of patients [65]. Increasing evidences support the correlation between toxicity and irinotecan pharmacology.

14.2.3.2 Metabolism of Irinotecan – Activating and Inactivating Pathways and Their Clinical Relevance

Although irinotecan metabolism generates at least 20 metabolites, many of them are found at trace levels in patients. Clinically relevant metabolites of irinotecan are the active metabolite 7-ethyl-10-hydroxycamptothecin (SN-38), inactive glucuronide SN-38G, and inactive aminopentane carboxylic acid (7-ethyl-10[4-N-(5-aminopentanoic acid)-1-piperidino]carbonyloxycamptothecin, APC) (Figure 14.5).

Activating Pathway
Irinotecan is a prodrug, and hydrolysis of irinotecan by the high-affinity carboxylesterase-2 enzyme in many normal tissues and tumors is responsible for activa-

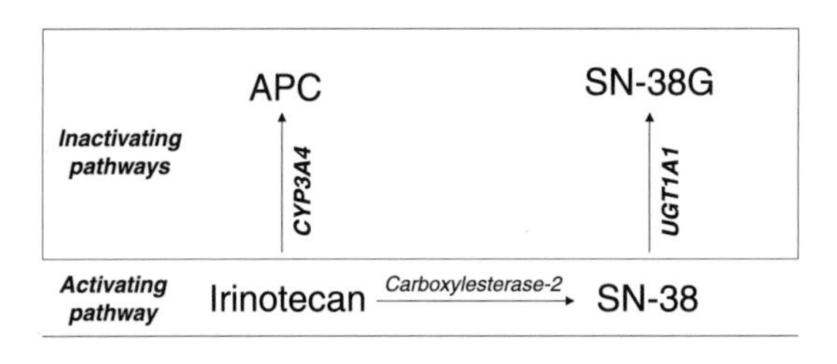

Fig. 14.5 Inactivating and activating pathways of irinotecan metabolism.

tion of irinotecan to SN-38, a potent topoisomerase I inhibitor [66–68]. Although SN-38 concentrations in plasma and urine are the lowest among all irinotecan metabolites, SN-38 formation within the tumor is critical for irinotecan antitumor activity.

Inactivating Pathways

Inactivation pathways involve oxidation of irinotecan and glucuronidation of SN-38. Oxidation of irinotecan accounts for about 15% of irinotecan dose [69], and the formation of inactive APC by cytochrome P450 (CYP) 3A4 reduces the availability of irinotecan for its activation to SN-38 [70]. The final step of sequential irinotecan metabolism is the inactivation of SN-38 by glucuronidation to SN-38G. Glucuronidation of SN-38 is the major elimination pathway of SN-38 and protects patients from irinotecan toxicity. The severity of diarrhea is dependent upon the extent of inactivation of SN-38 by glucuronidation. From preclinical experiments in nude mice, accumulation of SN-38 in the intestine is responsible for the diarrhea after irinotecan [71]. When biliary excretion of SN-38 was measured by the "biliary index" [which takes into account SN-38 glucuronidation rates normalized by irinotecan AUC (SN-38 AUC/SN-38G AUC×CPT-11 AUC)], patients with high biliary index are more likely exposed to the occurrence of severe diarrhea than patients with low biliary index (Figure 14.6). This suggests that higher glucuronidation of SN-38 in the liver may protect against irinotecan-induced intestinal toxicity as a result of reduced elimination of SN-38 in the bile [72].

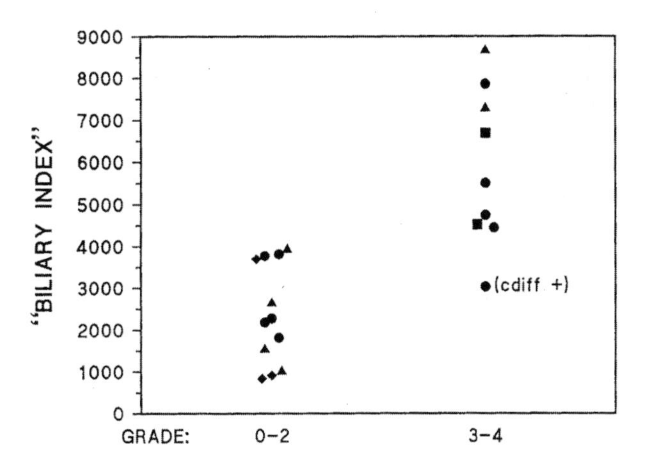

Fig. 14.6 Biliary indexes and severity of diarrhea in cancer patients after four different dose levels (◆, ▲, ●, ■) of irinotecan. Statistically significant correlation of biliary index to severity of diarrhea was shown. Cdiff+, one patient found positive for *C. difficile* toxin.

14.2.3.3 Increased Risk of Toxicity in Cancer Patients with Gilbert's Syndrome

SN-38 glucuronidation is catalyzed by the polymorphic UDP-glucuronosyltransferase 1A1 (UGT1A1) enzyme, which is responsible for bilirubin glucuronidation [73]. Among the hyperbilirubinemic syndromes caused by genetic defects in *UGT1A1* gene, a promoter polymorphism induces Gilbert's syndrome. This is an inherited disorder, characterized by mild, chronic unconjugated hyperbilirubinemia (serum bilirubin levels usually <3 mg dL^{-1}). In two cancer patients with Gilbert's syndrome, grade 4 neutropenia and/or diarrhea occurred after irinotecan. Both of them had a familiar history of Gilbert's syndrome and periodic asymptomatic increases in unconjugated bilirubin. Exaggerated toxic response to standard doses of irinotecan was associated with abnormally elevated values of biliary index [74]. These results suggested that genetically reduced inactivation of SN-38 could result in a higher risk of developing irinotecan-induced toxicity.

14.2.3.4 Gilbert's Syndrome Genotype

The genetic defect in Gilbert's syndrome is a TA insertion in the promoter region of *UGT1A1* gene, resulting in the variant allele $(TA)_7TAA$ (*UGT1A1*28*) instead of the wild-type allele $(TA)_6TAA$ [75, 76]. The presence of an additional TA repeat results in reduced UGT1A1 expression levels and activity, since transcriptional activity of the promoter decreases with the progressive increase in the number of TA repeats [77]. A wide variation in the incidences of this syndrome has been reported, ranging from 0.5% to 23% in various groups [76–80]. In addition to the $(TA)_7$ polymorphism, $(TA)_8$ and $(TA)_5$ alleles have been found in individuals from different ethnic backgrounds [77, 81, 82] and a subject with Gilbert's syndrome was found to be heterozygous for $(TA)_8$ [83].

While the majority of Gilbert's syndrome patients are $(TA)_7$ homozygotes, some patients do not have mutations at the promoter level but are heterozygotes for G211A, T1456G and C686A missense mutations in the *UGT1A1* coding region [84, 85]. G211A (G71R, *UGT1A1*6*) mutation results in a 30% (heterozygotes) and 60% (homozygotes) decrease in bilirubin glucuronidating activity, and is highly prevalent in individuals of Asian origin [86, 87], being responsible for about 60% of the Gilbert's syndrome cases among Japanese individuals [85]. G211A allele frequency of 11–13% has been reported in Asians [86, 88]. T1456G (Y486D, *UGT1A1*7*) mutation was found in two patients with Gilbert's syndrome [88, 89], but its frequency in the general population is not known. Interestingly, mutations in the *UGT1A1* coding region seem to be more frequent in Asian than Caucasian populations [88].

14.2.3.5 Gilbert's Syndrome Phenotype and SN-38 Glucuronidation

Interpatient variability in SN-38 glucuronidation is considerably high in cancer patients [72]. A 17-fold difference in SN-38 glucuronidation was found in human livers [90], and significant variability of UGT1A1 phenotype might account for differences in SN-38 inactivation. SN-38 glucuronidation in human livers was highly

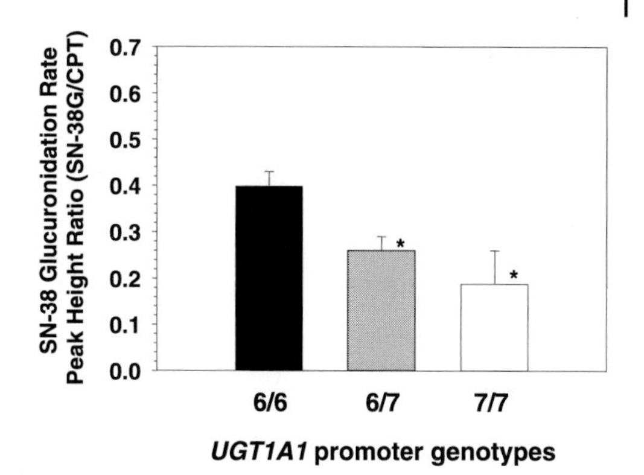

Fig. 14.7 *In vitro* glucuronidation of SN-38 in human liver microsomes genotyped for *UGT1A1* promoter polymorphism. Each bar represents the mean (±standard error) SN-38G production in livers with 6/6 (n=19), 6/7 (n=21), and 7/7 (n=4) genotype. *Significantly less than 6/6, $p<0.05$.

concordant with the *UGT1A1* promoter genotype, since glucuronidation rates of SN-38 were significantly lower in homozygotes and heterozygotes for $(TA)_7$ when compared to wild type (Figure 14.7). Patients homozygous and heterozygous for $(TA)_7$ might be expected to have at least a 50% and 25% decrease in SN-38 glucuronidation, respectively [90].

14.2.3.6 Possible Measures to Predict Patients at High Risk of Toxicity after Irinotecan

Predictive measures to classify patients as low and high SN-38 glucuronidators and consequently identify those at higher risk of toxicity are required. Gilbert's syndrome remains often undiagnosed, and ratio of conjugated to unconjugated bilirubinemia can not be considered a predictive parameter. Recent results from two clinical trials propose *UGT1A1* genotyping as a more reliable test to predict the risk of severe toxicity after irinotecan. Preliminary findings from a phase I study of irinotecan at two dose levels show that *UGT1A1* promoter genotype correlates with irinotecan pharmacokinetics and toxicity [91]. With irinotecan 300 mg m^{-2}, $(TA)_6$ wild-type patients developed grade ≤1 toxicity, while about 50% of $(TA)_7$ carriers experienced grade ≥2 diarrhea and neutropenia associated with reduced SN-38 glucuronidation. No significant differences were observed between homozygotes and heterozygotes for $(TA)_7$ and the irinotecan dose of 300 mg m^{-2} has been increased to 350 mg m^{-2}, the dose level approved by the Food and Drug Administration. Out of five 350 mg m^{-2} patients, the two of them with $(TA)_7$ allele developed grade 4 neutropenia. A recent retrospective study in Japanese patients confirmed these results [92]. Among patients with severe toxicity after irinotecan, 46% of them where $(TA)_7$ carriers. Among patients who did not experience severe toxicity, only 14% of them were $(TA)_7$ carriers. The presence of $(TA)_7$ allele was a significant risk factor for irinotecan severe toxicity. Interestingly, all three patients with a missense C686A mutation in the coding region (P229Q, *UGT1A1*27*) experienced severe toxicity, and no statistical association was found between severe toxicity and the G211A mutation.

14.2.4 Amonafide and NAT2 Pharmacogenetics

Amonafide is a DNA intercalating agent and topoisomerase II inhibitor which showed activity in breast cancer and leukemia. Highly variable and unpredictable toxicity partly caused by interindividual differences in N-acetylation have hampered its clinical development. Although a dose individualization scheme was validated for low and high metabolizers, amonafide is no longer in clinical development. The experience with amonafide remains an example of population pharmacogenetics and successful phenotyping strategy in cancer chemotherapy.

14.2.4.1 Metabolism of Amonafide and NAT2 Polymorphism

Amonafide is extensively metabolized, including N-acetylation by N-acetyltransferase 2 (NAT2) to N-acetyl-amonafide (Figure 14.8), a metabolite approximately equipotent *in vitro* with the parent drug [93]. Mutated alleles *NAT2*5A, B, C, NAT2*6A, NAT2*7, NAT2*13* and *NAT2*14* account for more than 99% of slow acetylators in Caucasian populations [94, 95]. Homozygosity for *NAT2* mutated alleles is required for the slow acetylator phenotype, and rapid acetylators include both mutant heterozygotes and wild-type individuals, the latter having significantly higher acetylation rates [96].

14.2.4.2 Dose Individualization of Amonafide Based on N-Acetylator Phenotype

Due to the polymorphic acetylation of amonafide, a phenotyping procedure for amonafide acetylation using caffeine as a probe was evaluated in cancer patients. Slow and fast acetylators of both caffeine and amonafide were identified. Fast ace-

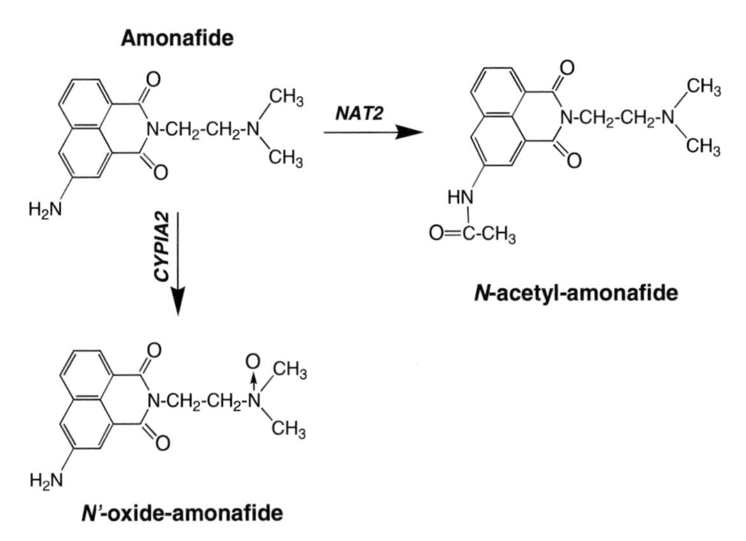

Fig. 14.8 Amonafide metabolism. Acetylation and oxidation pathways

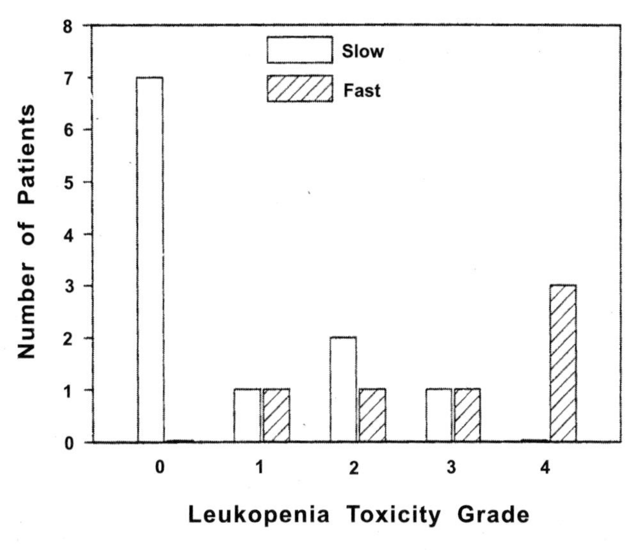

Fig. 14.9 Degree of leukopenia in cancer patients receiving amonafide. Incidence and degree of leukopenia was higher in fast acetylators compared to slow acetylators.

tylators had significantly greater myelosuppression than slow acetylators and amonafide exposure was significantly greater in fast acetylators, who would be expected to have a higher clearance of amonafide (Figure 14.9) [97]. This appeared to be unusual compared with most drugs metabolized by N-acetylation, where slow acetylators are more likely to experience adverse reactions. The unexpected behavior of amonafide was due to the inhibition of amonafide oxidation by N-acetyl-amonafide, since amonafide is a substrate for CYP1A2 and amonafide oxidation is inhibited by its acetylated metabolite [98]. Based on acetylator phenotype, a pharmacogenetic phase I study of amonafide recommended doses of 250 and 375 mg m^{-2} for fast and slow acetylators, respectively [99]. Further investigation in phase II of studies of 300 mg m^{-2} amonafide demonstrated that fixed dosing was considered inappropriate for all patients, as fast phenotypes would be expected to experience severe toxicity and slow phenotypes may be significantly underdosed. Since there was still significant interpatient variability in toxicity at these dose levels, a subsequent study attempted to develop pharmacodynamic models to individualize amonafide dosing, and the optimal model was defined by acetylator phenotype, pretreatment white blood cell count and gender [100].

14.2.5
MTHFR Gene Polymorphism in Breast Cancer Patients Receiving CMF Regimen

CMF regimen is a combination of cyclophosphamide, MTX, and 5-FU, and represents one of the treatments of choice for women with non-metastatic breast cancer, significantly increasing disease-free and overall survival. A recent report in a small series of patients described an interesting association between the occurrence of severe myelotoxicity after CMF and a single nucleotide polymorphism in the methylenetetrahydrofolate reductase (MTHFR) gene [101]. One breast cancer patient experienced grade 4 leukopenia after the first cycle of CMF. After a similar treatment regimen based on 5-FU and MTX, her mother affected by gastric cancer experienced life-threatening toxicity as well. Both patients were homozygous carriers for a single nucleotide polymorphism in the *MTHFR* gene. *MTHFR* genotyping was extended to additional five consecutive breast cancer patients experiencing severe toxicity after CMF, and four of them were found to be homozygous for the same mutation.

14.2.5.1 **MTHFR Function and Polymorphism**

Human *MTHFR* gene consists of eleven exons, and the coded enzyme converts 5,10-methylenetetrahydrofolate (CH_2THF) to 5-methyltetrahydrofolate (CH_3THF), a methyl donor in the conversion of homocysteine to methionine during protein synthesis [102] (Figure 14.10). In cell folate metabolism, MTHFR regulates the pool of folates for nucleic acid synthesis. A C677T mutation in the *MTHFR* gene codes an enzyme variant with *in vitro* thermolability and reduced catalytic activity (35% compared to wild type), leading to accumulation of plasma homocysteine in homozygous individuals [103]. In addition to this, homozygous subjects accumulate CH_2THF polyglutamates in erythrocytes at the expense of CH_3THF species, the only folate form found in erythrocytes of wild-type individuals [104]. C677T polymorphism creates a shift in the distribution of intracellular folates, creating retention of folates committed for purine and pyrimidine synthesis (i.e., CH_2THF). This polymorphism is common, with about 10% of homozygous individuals (TT) in Caucasian population [105]. Taking into account different frequencies due to ethnicity, incidence of T allele ranges from 5% to 54%.

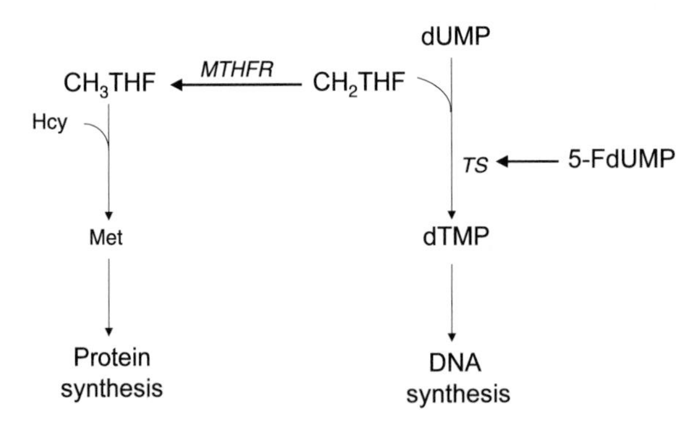

Fig. 14.10 Folate metabolism and role of MTHFR. Genetically reduced MTHFR activity affects the distribution between folate species required for protein and DNA synthesis. Higher availability of 5,10-methylenetetrahydrofolate (CH_2THF) potentiates the TS inhibition by 5-FdUMP, the active metabolite of 5-FU. Hcy, homocysteine; Met, methionine; CH_3HF, 5-methyltetrahydrofolate; TS, thymidylate synthase; 5-FdUMP, fluorodeoxyuridine monophosphate.

14.2.5.2 MTHFR Polymorphism as a Determinant of CMF Toxicity

Five of six patients with grade 4 leukopenia after CMF were TT homozygotes [101]. Qualitatively altered distribution of intracellular folates in breast cancer patients with TT genotype could have increased bone marrow sensitivity to CMF chemotherapy. When thymidylate synthase converts dUMP into dTMP, CH_2THF is required as a donor of monocarbon groups. MTHFR deficiency induced by TT genotype increases the availability of CH_2THF, potentiating 5-FU inhibition of thymidylate synthase mediated by 5-FdUMP, leading to severe myelosuppression. This genotype/phenotype association needs to be confirmed in a larger trial and the postulated biochemical mechanism further investigated. These findings highlight a possible role of MTHFR polymorphism in selecting cancer patients at higher risk of toxicity after receiving the CMF regimen.

14.3 Pharmacogenetic Determinants of Response after Cancer Chemotherapy

Recent studies focused on the importance of pharmacogenetic determinants of response in cancer patients. Screening of patients for polymorphic mutants of glutathione-S-transferase and thymidylate synthase has the potentiality to predict response and hence outcome of chemotherapy.

14.3.1 Glutathione-S-Transferase Mutations in Cancer Chemotherapy

Xenobiotic detoxification in mammalian cells is efficiently mediated by conjugation of the nucleophilic center of the compound with reduced glutathione (GSH) by glutathione-S-transferase (GST). *GST* gene mutations can lead to high phenotypic variability. Increased GST function might arise from gene duplications (ultrarapid phenotype), increased protein level due to promoter mutations, and coding mutations associated with increased enzyme efficiency. Reduced detoxification is generally related to gene deletions (null genotypes), as well as to conformational changes induced by single amino acid changes in the coding region [106].

A broad literature exists on genetically reduced GST activity as a risk factor in carcinogenesis. Less information is available on the clinical implications of polymorphic GSTs in cancer patients receiving chemotherapy. Conjugation with GST has been reported for cisplatin and alkylating agents, and germ line mutations altering GST activity could change drug pharmacokinetics. Moreover, the GSH/GST system is involved in the development of cellular resistance to cancer chemotherapy. Cancer cells protect themselves from the toxic action of chemotherapy by overexpressing GST, and the relevance of *GST* mutations for patient outcome has been investigated in both solid tumors and hematological malignancies.

14.3.1.1 GST Pharmacogenetics and Outcome in Solid Tumor Patients

In breast cancer patients, inherited mutations in *GSTP1* gene have been shown to influence treatment outcome [107]. In this study, the most commonly used chemotherapeutic agents were cyclophosphamide, 5-fluorouracil, and doxorubicin. Reactive metabolites of cyclophosphamide are conjugated with GSH by GSTP1 [108], and increased GSTP expression was reported in a doxorubicin-resistant cancer cell line [109]. Single nucleotide substitutions in the *GSTP1* coding region result in amino acid changes $Ile^{105}Val$ and $Ala^{114}Val$, and Val^{105} variant was 2-fold lower efficient than Ile^{105} in conjugating thiotepa [110]. In tumor biopsies, homozygosity for the less active Val^{105} variant improved the survival of breast cancer patients compared to those carrying Ile^{105}, and the hazard of death conferred by tumor Val^{105}/Val^{105} genotype was 30% of that of Ile^{105} patients.

Genotyping of ovarian cancer patients for null *GSTM1* and *GSTT1* revealed a poorer survival in patients with null genotype compared to wild type [111]. None of the patients with both null genotypes survived 3.5 years after diagnosis, while 43% of wild-type patients survived beyond this time. In this study, 70% of patients received single-agent carboplatin, the remaining of them being treated with alkylating agents. Reduced systemic detoxification of these compounds in patients with null genotype should have led to better response rate and survival compared to wild type, but opposite results have been observed. The most plausible explanation is the effect of reduced GST-mediated detoxification on the biology of ovarian epithelial cells. Ovarian tumors with loss of p53 function are characterized by lack of response to chemotherapy and poor outcome [112]. GST activity protects ovarian cells from chronic oxidative damage to genomic DNA potentially leading to loss of p53 function. Patients with null *GST* and loss of p53 function in ovarian cancer cells might have experienced shorter survival caused by reduced protection from oxidative damage.

14.3.1.2 GST Pharmacogenetics and Outcome in Childhood Leukemias

The observation of 3-fold increased risk of relapse in ALL patients expressing GSTM compared to non-expressors [113] prompted to investigate the association between frequency of *GST* variants and outcome of leukemic patients.

In ALL, null *GSTT1* genotype was a major determinant of initial response to prednisone therapy in patients treated according to the Berlin–Frankfurt–Munster (BFM) protocols [114]. GSTs have been implicated in cell resistance to glucocorticoid treatment, and initial response to prednisone is considered a strong predictor of outcome. A 6.7-fold reduced risk of poor response to prednisone was found in null *GSTT1* genotype patients compared to heterozygous and wild-type patients. In another trial from the BFM study group, mutated *GST* genotypes were selected for their impact on ALL relapse [115]. Null *GSTM1* and *GSTT1* conferred 2-fold and 2.8-fold reduction in risk of relapse compared to wild type, respectively. Among polymorphisms in the *GSTP1* gene, Val^{105}/Val^{105} genotype showed 3-fold decreased risk of relapse compared to other variants at codons 105 and 114. Based on these observations, null *GSTM1*, null *GSTT1* and *GSTP1* Val^{105}/Val^{105} genotypes were designated as "low risk" genotypes, and patients having at least two "low risk" genotypes had 3.5-fold reduced risk of relapse compared to patients with no "low risk" genotype. A previous study did not demonstrate any impact of null *GSTM1* and *GSTT1* genotypes for event-free survival in ALL, with only a tendency of higher central nervous system relapse-free survival in null *GSTM1* patients [116]. Compared to the BFM study, these findings are more applicable to the overall ALL population, since, in the BFM study, matching criteria led to the selection of a particular patient subgroup of the entire ALL population.

In acute myeloid leukemia (AML), intensification of both induction and post-remission chemotherapy improves overall survival but is associated with significant drug-related morbidity and mortality. When AML patients receiving standard and intensive induction chemotherapy were genotyped for null *GSTT1*, interesting results were observed [117]. In the intensive treatment arm, null *GSTT1* genotype was associated with reduced survival and increased risk of toxic death in remission compared to wild-type patients.

14.3.2 Thymidylate Synthase Gene Promoter Polymorphism and Response to 5-Flurouracil-Based Chemotherapy

Thymidylate synthase (TS) is the rate-limiting enzyme in the DNA synthetic pathway and the target for 5-FU and folate analogs (Figure 14.3). Compared to normal tissues, TS is often overexpressed in tumor cells, probably as a result of tumor suppression loss of function, gene amplification or other mechanisms. Acute induction of TS protein as well as stable amplification of TS-specific genes may be associated with resistance to fluoropyrimidine derivatives [118, 119], and an inverse correlation between tumor TS expression and clinical response was found [120–122].

14.3.2.1 Regulation of TS Gene Expression

Mechanisms regulating *TS* gene expression are not very well understood. Tumor suppressor elements modulate TS gene transcription. Translation of TS mRNA is negatively regulated by direct binding of TS protein to promoter elements on its cognate mRNA [123]. A translational regulatory element within the coding region has also been found [124], and a 6-bp deletion in the 3′-untranslated region of TS mRNA could affect mRNA stability and translation [125]. In addition to this, two, three, four and nine copies of 28-bp tandem repeated sequences have been described in the enhancer region of the *TS* promoter [126–128]. Presence of a triple repeat increased TS expression by 2.6-folds compared to double repeat in transient expression assays [127]. In patients with gastrointestinal malignancies, TS levels were significantly higher in tumor specimens homozygous for triple repeats compared to those with double repeats [129].

14.3.2.2 Genotyping of the *TS* Promoter Variable Tandem Repeat and Clinical Outcome

The chance of downstaging after radiation and 5-FU-based therapy in rectal cancer patients was related to polymorphisms in the enhancer region of the TS gene promoter in tumor specimens [130]. The presence of downstaging is an important prognostic factor. Based on the number of tandem repeats, 20% of patients had 2/2, 38.5% 2/3, and 42.5% 3/3 genotype. The relative probability of achieving downstaging for 2/2 and 2/3 patients was 3.7-fold higher than for 3/3 patients. Genotyping for tandem repeat promoter polymorphisms could be used to select the most appropriate chemotherapy in patients with 3/3 genotype, since they might respond to irinotecan or oxaliplatin that have different mechanisms of action.

14.4 Conclusion

The main focus of genetic investigation in chemotherapy over the past two decades has been to predict the occurrence of severe toxicity. The examples of the clinical pharmacogenetics of 6-MP, 5-FU, and CPT-11 demonstrated that about 10% of cancer patient population is at high risk of severe toxicity. The challenge for the future is to use pharmacogenomics as a tool for dose individualization, since, for the vast majority of patients, it is still unclear how dosage should be selected on the basis of patient genotype.

Genetic investigation in cancer patients should start as early as possible during drug development. Candidate genes playing a significant role in the pharmacology of new chemotherapeutic agents are often unknown before clinical trials. Genetic investigation in chemotherapy is complicated by the fact that multi-drug therapy is often the standard of care, confounding the results of phenotype–genotype correlation. For this reason, phase I–II trials of new single agents should include the search of genetic determinants of toxicity and response.

There is an increasing need of understanding the reasons why some cancer patients respond to chemotherapy while others experience toxicity without any therapeutic benefit. Prospective large studies with multivariate analysis will be required to confirm the results of retrospective studies that, so far, have demonstrated associations between genetic defects and response.

Acknowledgements

We wish to thank Mss. Debby Stoit and Marla Scofield for their assistance in the preparation of figures. This Pharmacogenetics of Anticancer Agents Research Group (*www.pharmacogenetics.org*) chapter was supported by grant GM61393 from National Institute of Health, Bethesda, MD.

14.5 References

1 Lennard L, Rees CA, Lilleyman JS et al. Childhood leukaemia: A relationship between intracellular 6-mercaptopurine metabolites and neutropenia. Br J Clin Pharmacol **1983**; 16:359–363.

2 Schmiegelow K, Bruunshuus I. 6-Thioguanine nucleotide accumulation in red blood cells during maintenance chemotherapy for childhood acute lymphoblastic leukemia, and its relation to leukopenia. Cancer Chemother Pharmacol **1990**; 26:288–292.

3 Tidd DM and Paterson AR. A biochemical mechanism for the delayed cytotoxic reaction of 6-mercaptopurine. Cancer Res **1974**;34:738–746.

4 Ling YH, Chan JY, Beattie KL et al. Consequences of 6-thioguanine incorporation into DNA on polymerase, ligase, and endonuclease reactions. Mol Pharmacol **1992**; 42:802–807.

5 Deininger M, Szumlanski CL, Otterness DM et al. Purine substrates for human thiopurine methyltransferase. Biochem Pharmacol **1994**; 48:2135–2138.

6 Woodson LC, Ames MM, Selassie CD et al. Thiopurine methyltransferase. Aromatic thiol substrates and inhibition by benzoic acid derivatives. Mol Pharmacol **1983**; 24:471–478.

7 Lennard L, Lilleyman JS, Van Loon J et al. Genetic variation in response to 6-mercaptopurine for childhood acute lymphoblastic leukaemia. Lancet **1990**; 336:225–229.

8 Lilleyman JS, Lennard L. Mercaptopurine metabolism and risk of relapse in childhood lymphoblastic leukaemia. Lancet **1994**; 343:1188–1190.

9 Relling MV, Hancock ML, Boyett JM et al. Prognostic importance of 6-mercaptopurine dose intensity in acute lymphoblastic leukemia. Blood **1999**; 93:2817–2823.

10 Dibenedetto SP, Guardabasso V, Ragusa R et al. 6-Mercaptopurine cumulative dose: a critical factor of maintenance therapy in average risk childhood acute lymphoblastic leukemia. Pediatr Hematol Oncol **1994**; 11:251–258.

11 McLeod HL, Coulthard S, Thomas AE et al. Analysis of thiopurine methyltransferase variant alleles in childhood acute lymphoblastic leukaemia. Br J Haematol **1999**; 105:696–700.

12 Evans WE, Horner M, Chu YQ et al. Altered mercaptopurine metabolism, toxic effects, and dosage requirement in a thiopurine methyltransferase-deficient child with acute lymphocytic leukemia. J Pediatr **1991**; 119:985–989.

13 Relling MV, Hancock ML, Rivera GK et al. Mercaptopurine therapy intolerance and heterozygosity at the thiopurine S-methyltransferase gene locus. J Natl Cancer Inst **1999**; 91:2001–2008.

14 McBride KL, Gilchrist GS, Smithson WA et al. Severe 6-thioguanine-induced marrow aplasia in a child with acute lymphoblastic leukemia and inhibited thiopurine methyltransferase deficiency. J Pediatr Hematol Oncol **2000**; 22:441-445.

15 Weinshilboum RM, Sladek SL. Mercaptopurine pharmacogenetics: monogenic inheritance of erythrocyte thiopurine methyltransferase activity. Am J Hum Genet **1980**; 32:651–662.

16 Otterness D, Szumlanski C, Lennard L et al. Human thiopurine methyltransferase pharmacogenetics: gene sequence polymorphisms. Clin Pharmacol Ther **1997**; 62:60–73.

17 McLeod HL, Lin JS, Scott EP et al. Thiopurine methyltransferase activity in American white subjects and black subjects. Clin Pharmacol Ther **1994**; 55:15–20.

18 Lennard L, Van Loon JA, Lilleyman JS et al. Thiopurine pharmacogenetics in leukemia: correlation of erythrocyte thiopurine methyltransferase activity and 6-thioguanine nucleotide concentrations. Clin Pharmacol Ther **1987**; 41:18–25.

19 Lennard L, Lilleyman JS. Variable mercaptopurine metabolism and treatment outcome in childhood lymphoblastic leukemia. J Clin Oncol **1989**; 7:1816–1823.

20 Lennard L, Gibson BE, Nicole T et al. Congenital thiopurine methyltransferase deficiency and 6-mercaptopurine toxicity

during treatment for acute lymphoblastic leukaemia. Arch Dis Child **1993**; 69:577–579.

21 LENNARD L, WELCH JC, LILLEYMAN JS. Thiopurine drugs in the treatment of childhood leukaemia: the influence of inherited thiopurine methyltransferase activity on drug metabolism and cytotoxicity. Br J Clin Pharmacol **1997**; 44:455–461.

22 MCLEOD HL, RELLING MV, LIU Q et al. Polymorphic thiopurine methyltransferase in erythrocytes is indicative of activity in leukemic blasts from children with acute lymphoblastic leukemia. Blood **1995**; 85:1897–1902.

23 SZUMLANSKI C, OTTERNESS D, HER C et al. Thiopurine methyltransferase pharmacogenetics: human gene cloning and characterization of a common polymorphism. DNA Cell Biol **1996**; 15:17–30.

24 YATES CR, KRYNETSKI EY, LOENNECHEN T et al. Molecular diagnosis of thiopurine S-methyltransferase deficiency: genetic basis for azathioprine and mercaptopurine intolerance. Ann Intern Med **1997**; 126:608–614.

25 TAI HL, KRYNETSKI EY, YATES CR et al. Thiopurine S-methyltransferase deficiency: two nucleotide transitions define the most prevalent mutant allele associated with loss of catalytic activity in Caucasians. Am J Hum Genet **1996**; 58:694–702.

26 OTTERNESS DM, SZUMLANSKI CL, WOOD TC et al. Human thiopurine methyltransferase pharmacogenetics. Kindred with a terminal exon splice junction mutation that results in loss of activity. J Clin Invest **1998**; 101:1036–1044.

27 HON YY, FESSING MY, PUI CH et al. Polymorphism of the thiopurine S-methyltransferase gene in African-Americans. Hum Mol Genet **1999**; 8:371–376.

28 COLLIE-DUGUID ES, PRITCHARD SC, POWRIE RH et al. The frequency and distribution of thiopurine methyltransferase alleles in Caucasian and Asian populations. Pharmacogenetics **1999**; 9:37–42.

29 MCLEOD HL, PRITCHARD SC, GITHANG'A J et al. Ethnic differences in thiopurine methyltransferase pharmacogenetics: evidence for allele specificity in Caucasian and Kenyan individuals. Pharmacogenetics **1999**; 9:773–776.

30 COULTHARD SA, HOWELL C, ROBSON J et al. The relationship between thiopurine methyltransferase activity and genotype in blasts from patients with acute leukemia. Blood **1998**; 92:2856–2862.

31 KRYNETSKI EY, FESSING MY, YATES CR et al. Promoter and intronic sequences of the human thiopurine S-methyltransferase (TPMT) gene isolated from a human PAC1 genomic library. Pharm Res **1997**; 14:1672–1678.

32 SPIRE-VAYRON DE LA MOUREYRE C, DEBUYSERE H, MASTAIN B et al. Genotypic and phenotypic analysis of the polymorphic thiopurine S-methyltransferase gene (TPMT) in a European population. Br J Pharmacol **1998**; 125:879–887.

33 SPIRE-VAYRON DE LA MOUREYRE C, DEBUYSERE H, FAZIO F et al. Characterization of a variable number tandem repeat region in the thiopurine S-methyltransferase gene promoter. Pharmacogenetics **1999**; 9:189–198.

34 YAN L, ZHANG S, EIFF B et al. Thiopurine methyltransferase polymorphic tandem repeat: genotype-phenotype correlation analysis. Clin Pharmacol Ther **2000**; 68:210–219.

35 WELCH JC, LILLEYMAN JS. 6-Mercaptopurine dose escalation and its effect on drug tolerance in childhood lymphoblastic leukaemia. Cancer Chemother Pharmacol **1996**; 38:113–116.

36 DE GRAMONT A, FIGER A, SEYMOUR M et al. Leucovorin and fluorouracil with or without oxaliplatin as first-line treatment in advanced colorectal cancer. J Clin Oncol **2000**; 18:2938–2947.

37 SALTZ LB, COX JV, BLANKE C et al. Irinotecan plus fluorouracil and leucovorin for metastatic colorectal cancer. Irinotecan Study Group. N Engl J Med **2000**; 343:905–914.

38 HO DH, TOWNSEND L, LUNA MA et al. Distribution and inhibition of dihydrouracil dehydrogenase activities in human tissues using 5-fluorouracil as a substrate. Anticancer Res **1986**; 6:781–784.

39 HEGGIE GD, SOMMADOSSI JP, CROSS DS et al. Clinical pharmacokinetics of 5-

fluorouracil and its metabolites in plasma, urine, and bile. Cancer Res **1987**; 47:2203–2206.

40 Diasio RB, Johnson MR. The role of pharmacogenetics and pharmacogenomics in cancer chemotherapy with 5-fluorouracil. Pharmacology **2000**; 61:199–203.

41 Tuchman M, Stoeckeler JS, Kiang DT et al. Familial pyrimidinemia and pyrimidinuria associated with severe fluorouracil toxicity. N Engl J Med **1985**; 313:245–249.

42 Harris BE, Carpenter JT, Diasio RB. Severe 5-fluorouracil toxicity secondary to dihydropyrimidine dehydrogenase deficiency. A potentially more common pharmacogenetic syndrome. Cancer **1991**; 68:499–501.

43 Lu Z, Zhang R, Diasio RB. Dihydropyrimidine dehydrogenase activity in human peripheral blood mononuclear cells and liver: population characteristics, newly identified deficient patients, and clinical implication in 5-fluorouracil chemotherapy. Cancer Res **1993**; 53:5433–5438.

44 Stephan F, Etienne MC, Wallays C et al. Depressed hepatic dihydropyrimidine dehydrogenase activity and fluorouracil-related toxicities. Am J Med **1995**; 99:685–688.

45 Milano G, Etienne MC, Pierrefite V et al. Dihydropyrimidine dehydrogenase deficiency and fluorouracil-related toxicity. Br J Cancer **1999**; 79:627–630.

46 Fleming RA, Milano GA, Gaspard MH, et al. Dihydropyrimidine dehydrogenase activity in cancer patients. Eur J Cancer **1993**; 29A:740–744.

47 Meinsma R, Fernandez-Salguero P, Van Kuilenburg AB et al. Human polymorphism in drug metabolism: mutation in the dihydropyrimidine dehydrogenase gene results in exon skipping and thymine uracilurea. DNA Cell Biol **1995**; 14:1–6.

48 Wei X, McLeod HL, McMurrough J et al. Molecular basis of the human dihydropyrimidine dehydrogenase deficiency and 5-fluorouracil toxicity. J Clin Invest **1996**; 98:610–615.

49 McLeod HL, Collie-Duguid ES, Vreken P et al. Nomenclature for human DPYD alleles. Pharmacogenetics **1998**; 8:455–459.

50 Collie-Duguid ES, Etienne MC, Milano G et al. Known variant DPYD alleles do not explain DPD deficiency in cancer patients. Pharmacogenetics **2000**; 10:217–223.

51 Van Kuilenburg AB, Vreken P, Beex LV et al. Severe 5-fluorouracil toxicity caused by reduced dihydropyrimidine dehydrogenase activity due to heterozygosity for a G → A point mutation. J Inherit Metab Dis **1998**; 21:280–284.

52 Johnson MR, Hageboutros A, Wang K et al. Life-threatening toxicity in a dihydropyrimidine dehydrogenase-deficient patient after treatment with topical 5-fluorouracil. Clin Cancer Res **1999**; 5:2006–2011.

53 van Kuilenburg AB, Haasjes J, Richel DJ et al. Clinical implications of dihydropyrimidine dehydrogenase (DPD) deficiency in patients with severe 5-fluorouracil-associated toxicity: identification of new mutations in the DPD gene. Clin Cancer Res **2000**; 6:4705–4712.

54 Collie-Duguid ES, Johnston SJ, Powrie RH et al. Cloning and initial characterization of the human DPYD gene promoter. Biochem Biophys Res Commun **2000**; 271:28–35.

55 Kouwaki M, Hamajima N, Sumi S et al. Identification of novel mutations in the dihydropyrimidine dehydrogenase gene in a Japanese patient with 5-fluorouracil toxicity. Clin Cancer Res **1998**; 4:2999–3004.

56 Etienne MC, Lagrange JL, Dassonville O et al. Population study of dihydropyrimidine dehydrogenase in cancer patients. J Clin Oncol **1994**; 12:2248–2253.

57 Ridge SA, Sludden J, Wei X et al. Dihydropyrimidine dehydrogenase pharmacogenetics in patients with colorectal cancer. Br J Cancer **1998**; 77:497–500.

58 McMurrough J, McLeod HL. Analysis of the dihydropyrimidine dehydrogenase polymorphism in a British population. Br J Clin Pharmacol **1996**; 41:425–427.

59 Grem JL, Yee LK, Venzon DJ et al. Inter- and intraindividual variation in dihydropyrimidine dehydrogenase activity in peripheral blood mononuclear cells. Cancer Chemother Pharmacol **1997**; 40:117–125.

60 Milano G, Etienne MC, Pierrefite V et al. Dihydropyrimidine dehydrogenase deficiency and fluorouracil-related toxicity. Br J Cancer **1999**; 79:627–630.

61 Chazal M, Etienne MC, Renee N et al. Link between dihydropyrimidine dehydrogenase activity in peripheral blood mononuclear cells and liver. Clin Cancer Res **1996**; 2:507–510.

62 Innocenti F, Iyer L, Ratain MJ. Pharmacogenetics: a tool for individualizing antineoplastic therapy. Clin Pharmacokinet **2000**; 39:315–325.

63 Gamelin E, Boisdron-Celle M, Guerin-Meyer V et al. Correlation between uracil and dihydrouracil plasma ratio, fluorouracil (5-FU) pharmacokinetic parameters, and tolerance in patients with advanced colorectal cancer: a potential interest for predicting 5-FU toxicity and determining optimal 5-FU dosage. J Clin Oncol **1999**; 17:1105–1110.

64 Iyer L, Ratain MJ. Clinical pharmacology of camptothecins. Cancer Chemother Pharmacol **1998**; 42(Suppl):S31–43

65 Vanhoefer U, Harstrick A, Achterrath W et al. Irinotecan in the treatment of colorectal cancer: clinical overview. J Clin Oncol **2001**; 19:1501–1518.

66 Kawato Y, Furuta T, Aonuma M et al. Antitumor activity of a camptothecin derivative, CPT-11, against human tumor xenografts in nude mice. Cancer Chemother Pharmacol **1991**; 28:192–198.

67 Humerickhouse R, Lohrbach K, Li L et al. Characterization of CPT-11 hydrolysis by human liver carboxylesterase isoforms hCE-1 and hCE-2. Cancer Res **2000**; 60:1189–1192.

68 Rivory LP, Bowles MR, Robert J et al. Conversion of irinotecan (CPT-11) to its active metabolite, 7-ethyl-10-hydroxycamptothecin (SN-38), by human liver carboxylesterase. Biochem Pharmacol **1996**; 52:1103–1111.

69 Slatter JG, Schaaf LJ, Sams JP et al. Pharmacokinetics, metabolism, and excretion of irinotecan (CPT-11) following I.V. infusion of [(14)C]CPT-11 in cancer patients. Drug Metab Dispos **2000**; 28:423–433.

70 Rivory LP, Riou JF, Haaz MC et al. Identification and properties of a major plasma metabolite of irinotecan (CPT-11) isolated from the plasma of patients. Cancer Res **1996**; 56:3689–3694.

71 Araki E, Ishikawa M, Iigo M et al. Relationship between development of diarrhea and the concentration of SN-38, an active metabolite of CPT-11, in the intestine and the blood plasma of athymic mice following intraperitoneal administration of CPT-11. Jpn J Cancer Res **1993**; 84:697–702.

72 Gupta E, Lestingi TM, Mick R et al. Metabolic fate of irinotecan in humans: correlation of glucuronidation with diarrhea. Cancer Res **1994**; 54:3723–3725.

73 Iyer L, King CD, Whitington PF et al. Genetic predisposition to the metabolism of irinotecan (CPT-11). Role of uridine diphosphate glucuronosyltransferase isoform 1A1 in the glucuronidation of its active metabolite (SN-38) in human liver microsomes. J Clin Invest **1998**; 101:847–854.

74 Wasserman E, Myara A, Lokiec F et al. Severe CPT-11 toxicity in patients with Gilbert's syndrome: two case report. Ann Oncol **1997**; 8:1049–1051.

75 Bosma PJ, Chowdhury JR, Bakker C et al. The genetic basis of the reduced expression of bilirubin UDP-glucuronosyltransferase 1 in Gilbert's syndrome. N Engl J Med **1995**; 333:1171–1175.

76 Monaghan G, Ryan M, Seddon R et al. Genetic variation in bilirubin UDP-glycuronyltransferase gene promoter and Gilbert's syndrome. Lancet **1996**; 347:578–581.

77 Beutler E, Gelbart T, Demina A. Racial variability in the UDP-glucuronosyltransferase 1 (UGT1A1) promoter: a balanced polymorphism for regulation of bilirubin metabolism? Proc Natl Acad Sci USA **1998**; 95:8170–8174.

78 Monaghan G, Foster B, Jurima-Romet M et al. UGT1*1 genotyping in a Canadian Inuit population. Pharmacogenetics **1997**; 7:153–156.

79 Owens D, Evans L. Population studies on Gilbert's syndrome. J Med Genet **1975**; 12:152–156.

80 Ando Y, Chida M, Nakayama K et al. The UGT1A1*28 allele is relatively rare in a Japanese population. Pharmacogenetics **1998**; 8:357–360.

81 Hall D, Ybazeta G, Destro-Bisol G et al. Variability at the uridine diphosphate glucuronosyltransferase 1A1 promoter in human populations and primates. Pharmacogenetics **1999**; 9:591–599.

82 Lampe JW, Bigler J, Horner NK et al. UDP-glucuronosyltransferase (UGT1A1*28 and UGT1A6*2) polymorphisms in Caucasians and Asians: relationships to serum bilirubin concentrations. Pharmacogenetics **1999**; 9:341–349.

83 Iolascon A, Faienza MF, Centra M et al. $(TA)_8$ allele in the UGT1A1 gene promoter of a Caucasian with Gilbert's syndrome. Haematologica **1999**; 84:106–109.

84 Koiwai O, Yasui Y, Hasada K et al. Three Japanese patients with Crigler-Najjar syndrome type I carry an identical nonsense mutation in the gene for UDP-glucuronosyltransferase. Jpn J Hum Genet **1995**; 40:253–257.

85 Sato H, Adachi Y, Koiwai O. The genetic basis of Gilbert's syndrome. Lancet **1996**; 347:557–558.

86 Akaba K, Kimura T, Sasaki A et al. Neonatal hyperbilirubinemia and mutation of the bilirubin uridine diphosphate-glucuronosyltransferase gene: a common missense mutation among Japanese, Koreans and Chinese. Biochem Mol Biol Int **1998**; 46:21–26.

87 Akaba K, Kimura T, Sasaki A et al. Neonatal hyperbilirubinemia and a common mutation of the bilirubin uridine diphosphate-glucuronosyltransferase gene in Japanese. J Hum Genet **1999**; 44:22–25.

88 Huang C-S, Luo G-A, Huang M-J et al. Variations of the bilirubin uridine-diphosphoglucuronosyl transferase 1A1 gene in healthy Taiwanese. Pharmacogenetics **2000**; 10:539–544.

89 Maruo Y, Sato H, Yamano T et al. Gilbert syndrome caused by a homozygous missense mutation (Tyr486Asp) of bilirubin UDP-glucuronosyltransferase gene. J Pediatr **1998**; 132:1045–1047.

90 Iyer L, Hall D, Das S et al. Phenotype-genotype correlation of *in vitro* SN-38 (active metabolite of irinotecan) and bilirubin glucuronidation in human liver tissue with UGT1A1 promoter polymorphism. Clin Pharmacol Ther **1999**; 65:576–582.

91 Iyer L, Janisch L, Das S et al. UGT1A1 promoter genotype correlates with pharmacokinetics of irinotecan (CPT-11). Proc Am Soc Clin Oncol **2000**; 19:178a (abstract).

92 Ando Y, Saka H, Ando M et al. Polymorphisms of UDP-glucuronosyltransferase gene and irinotecan toxicity: a pharmacogenetic analysis. Cancer Res **2000**; 60:6921–6926.

93 Felder TB, McLean MA, Vestal ML et al. Pharmacokinetics and metabolism of the antitumor drug amonafide (NSC-308847) in humans. Drug Metab Dispos **1987**; 15:773–777.

94 Meyer UA, Zanger UM. Molecular mechanisms of genetic polymorphisms of drug metabolism. Ann Rev Pharmacol Toxicol **1997**; 37:269–296.

95 Grant DM, Goodfellow GH, Sugamori K et al. Pharmacogenetics of the human arylamine N-acetyltransferases. Pharmacology **2000**; 61:204–211.

96 Cascorbi I, Drakoulis N, Brockmoller J et al. Arylamine N-acetyltransferase (NAT2) mutations and their allelic linkage in unrelated Caucasian individuals: correlation with phenotypic activity. Am J Human Genet **1995**; 57:581–592.

97 Ratain MJ, Mick R, Berezin F et al. Paradoxical relationship between acetylator phenotype and amonafide toxicity. Clin Pharmacol Ther **1991**; 50:573–579.

98 Ratain MJ, Rosner G, Allen SL et al. Population pharmacodynamic study of amonafide: a Cancer and Leukemia Group B study. J Clin Oncol **1995**; 13:741–747.

99 Ratain MJ, Mick R, Berezin F et al. Phase I study of amonafide dosing based on acetylator phenotype. Cancer Res **1993**; 53:2304–2308.

100 Ratain MJ, Mick R, Janish L et al. Individualized dosing of amonafide based on a pharmacodynamic model incorporating acetylator phenotype and gender. Pharmacogenetics **1996**; 6:93–101.

101 Toffoli G, Veronesi A, Boiocchi M et al. MTHFR gene polymorphism and se-

vere toxicity during adjuvant treatment of early breast cancer with cyclophosphamide, methotrexate, and fluorouracil (CMF). Ann Oncol **2000**; 11:373–374.

102 Ueland PM, Hustad S, Schneede J et al. Biological and clinical implications of the MTHFR C677T polymorphism. Trends Pharmacol Sci **2001**; 22:195–201.

103 Frosst P, Blom HJ, Milos R et al. A candidate genetic risk factor for vascular disease: a common mutation in methylenetetrahydrofolate reductase. Nature Genet **1995**;10:111–113.

104 Bagley PJ, Selhub J. A common mutation in the methylenetetrahydrofolate reductase gene is associated with an accumulation of formylated tetrahydrofolates in red blood cells. Proc Natl Acad Sci USA **1998**; 95:13217–13220.

105 Brattstrom L, Zhang Y, Hurtig M et al. A common methylenetetrahydrofolate reductase gene mutation and longevity. Atherosclerosis **1998**; 141:315–319.

106 Hayes JD, Strange RC. Glutathione S-transferase polymorphisms and their biological consequences. Pharmacology **2000**; 61:154–166.

107 Sweeney C, McClure GY, Fares MY et al. Association between survival after treatment for breast cancer and glutathione S-transferase P1 Ile105Val polymorphism. Cancer Res **2000**; 60:5621–5624.

108 Dirven HA, van Ommen B, van Bladeren PJ. Involvement of human glutathione S-transferase isoenzymes in the conjugation of cyclophosphamide metabolites with glutathione. Cancer Res **1994**; 54:6215–6220.

109 Wang K, Ramji S, Bhathena A et al. Glutathione S-transferases in wild-type and doxorubicin-resistant MCF-7 human breast cancer cell lines. Xenobiotica **1999**; 29:155–170.

110 Srivastava SK, Singhal SS, Hu X et al. Differential catalytic efficiency of allelic variants of human glutathione S-transferase Pi in catalyzing the glutathione conjugation of thiotepa. Arch Biochem Biophys **1999**; 366:89–94.

111 Howells RE, Redman CW, Dhar KK et al. Association of glutathione S-transferase GSTM1 and GSTT1 null genotypes with clinical outcome in epithelial ovarian cancer. Clin Cancer Res **1998**; 4:2439–2445.

112 Buttitta F, Marchetti A, Gadducci A et al. p53 alterations are predictive of chemoresistance and aggressiveness in ovarian carcinomas: a molecular and immunohistochemical study. Br J Cancer **1997**; 75:230–235.

113 Hall AG, Autzen P, Cattan AR et al. Expression of mu class glutathione S-transferase correlates with event-free survival in childhood acute lymphoblastic leukemia. Cancer Res **1994**; 54:5251–5254.

114 Anderer G, Schrappe M, Brechlin AM et al. Polymorphisms within glutathione S-transferase genes and initial response to glucocorticoids in childhood acute lymphoblastic leukaemia. Pharmacogenetics **2000**; 10:715–726.

115 Stanulla M, Schrappe M, Brechlin AM et al. Polymorphisms within glutathione S-transferase genes (GSTM1, GSTT1, GSTP1) and risk of relapse in childhood B-cell precursor acute lymphoblastic leukemia: a case-control study. Blood **2000**; 95:1222–1228.

116 Chen CL, Liu Q, Pui CH et al. Higher frequency of glutathione S-transferase deletions in black children with acute lymphoblastic leukemia. Blood **1997**; 89:1701–1707.

117 Davies SM, Robison LL, Buckley JD et al. Glutathione S-transferase polymorphisms and outcome of chemotherapy in childhood acute myeloid leukemia. J Clin Oncol **2001**; 19:1279–1287.

118 Berger SH, Jenh CH, Johnson LF et al. Thymidylate synthase overproduction and gene amplification in fluorodeoxyuridine-resistant human cells. Mol Pharmacol **1985**; 28:461–467.

119 Johnston PG, Drake JC, Trepel J et al. Immunological quantitation of thymidylate synthase using the monoclonal antibody TS 106 in 5-fluorouracil-sensitive and -resistant human cancer cell lines. Cancer Res **1992**; 52:4306–4312.

120 Aschele C, Debernardis D, Tunesi G et al. Thymidylate synthase protein expression in primary colorectal cancer compared with the corresponding distant

metastases and relationship with the clinical response to 5-fluorouracil. Clin Cancer Res **2000**; 6:4797–4802.

121 Johnston PG, Lenz HJ, Leichman CG et al. Thymidylate synthase gene and protein expression correlate and are associated with response to 5-fluorouracil in human colorectal and gastric tumors. Cancer Res **1995**; 55:1407–1412.

122 Leichman CG, Lenz HJ, Leichman L et al. Quantitation of intratumoral thymidylate synthase expression predicts for disseminated colorectal cancer response and resistance to protracted-infusion fluorouracil and weekly leucovorin. J Clin Oncol **1997**; 15:3223–3229.

123 Chu E, Voeller D, Koeller DM et al. Identification of an RNA binding site for human thymidylate synthase. Proc Natl Acad Sci USA **1993**; 90:517–521.

124 Lin X, Parsels LA, Voeller DM et al. Characterization of a cis-acting regulatory element in the protein coding region of thymidylate synthase mRNA. Nucleic Acids Res **2000**; 28:1381–1389.

125 Ulrich CM, Bigler J, Velicer CM et al. Searching expressed sequence tag databases: discovery and confirmation of a common polymorphism in the thymidylate synthase gene. Cancer Epidemiol Biomarkers Prev **2000**; 9:1381–1385.

126 Kaneda S, Takeishi K, Ayusawa D et al. Role in translation of a triple tandemly repeated sequence in the 5′-untranslated region of human thymidylate synthase mRNA. Nucleic Acids Res **1987**; 15:1259–1270.

127 Horie N, Aiba H, Oguro K et al. Functional analysis and DNA polymorphism of the tandemly repeated sequences in the 5′-terminal regulatory region of the human gene for thymidylate synthase. Cell Struct Funct **1995**; 20:191–197.

128 Marsh S, Ameyaw MM, Githanga J et al. Novel thymidylate synthase enhancer region alleles in African populations. Hum Mutat **2000**; 16:528.

129 Kawakami K, Omura K, Kanehira E et al. Polymorphic tandem repeats in the thymidylate synthase gene is associated with its protein expression in human gastrointestinal cancers. Anticancer Res **1999**; 19:3249–3252.

130 Villafranca E, Okruzhnov Y, Dominguez MA et al. Polymorphisms of the repeated sequences in the enhancer region of the thymidylate synthase gene promoter may predict downstaging after preoperative chemoradiation in rectal cancer. J Clin Oncol **2001**; 19:1779–1786.

15
Pharmacogenomics of the Blood–Brain Barrier

Jean-Michel Scherrmann

Abstract

The endothelial cells of brain capillaries are not the only component forming the blood–brain barrier (BBB). Pericytes and astrocyte foot processes, separated from each other by the basement membrane, are also part of the BBB. The specificity of the BBB permeability, which is commonly defined by the high transcellular electrical resistance due to the tight junctions sealing adjacent endothelial cells, depends greatly on cross-talk between these cells. The BBB is not just a physical barrier, it is also a metabolic and pharmacological barrier. These new properties are due to the expression of many genes whose products are implicated in drug metabolism, carrier-mediated transport and interaction with receptors at the BBB. The genes encoding phase 1 oxidative enzymes like CYP2D6 or the efflux P-glycoprotein transporter (P-gp) in the cerebral endothelial cells are all polymorphic at the BBB. Together with the induction or repression of these proteins, they provide the basis for genomic studies of the BBB. The many cellular components of the BBB and the relatively small amount of the brain that forms the BBB, makes it difficult to identify the genes and proteins involved. Nevertheless, transgenic *in vivo* models like P-gp knockout mice have proved to be powerful experimental tools for assessing the impact of the P-gp transporter on BBB function. Polymerase chain reaction-based subtraction cloning methods have recently led to the identification of genes that are more actively expressed at the BBB than elsewhere. These new approaches should lead to the identification of new targets for specific drugs and a clearer picture of the mechanisms underlying cerebrovascular disorders.

15.1 Basic Concepts Underlying the Pharmacogenomics of the Blood–Brain Barrier

15.1.1 The Two Barriers

There are two physiological barriers separating the brain from its blood supply; they control the entry and exit of endogenous and exogenous compounds. This allows the body to maintain a constant internal milieu in the brain, protecting it from fluctuations in circulating hormones, amino acids, ions and other nutrients, so preventing uncontrolled disturbances of the central nervous system (CNS). One is the blood–brain barrier (BBB) and the other is the blood–cerebrospinal fluid barrier (BCSFB). The general concept of a restricted passage of solutes out of the blood into the brain dates from the studies of Lewandowsky (1900) and Ehrlich (1902) [1, 2]. They observed that dyes injected intravenously were not taken up by the brain. Goldmann (1909, 1913) [3, 4] carried out experiments with trypan blue injected directly into the cerebrospinal fluid. These studies helped to differentiate the two barriers, as the dye left only the blood vessels of the choroid plexuses (for a review, see [5]). Only a few small regions of the brain, collectively known as the circumventricular organs (CVO) making up less than 1% of the cerebrovascular bed, have no barrier, enabling substances in the blood to reach the brain extracellular fluid. These are specialized tissues that act as centers of homeostatic and neurohormonal control for the body. They include the median eminence, pineal gland, organum vasculosum of the lamina terminalis, subfornical organ, subcommissural organ, area postrema and the neurohypophysis in close proximity to the ventricular system, particularly the third ventricle. The relatively free exchange of solutes between the blood and the CVO indicates that their capillaries are more permeable than are those in the rest of the brain. However, these vessels do not provide indiscriminate access to the whole of the CNS because there are tight junctions between these regions and the rest of the brain [5].

The BCSFB is located at the choroid plexuses. These plexuses float freely in the brain ventricles and are formed by epithelial cells held together at their apices by tight junctions (Figure 15.1). Beneath the epithelial cells is a stroma containing the blood vessels, which lack tight junctions. Thus, the fenestrated blood vessels of the choroid plexus allow large molecules to pass, but the tight junctions at the epithelial cell surfaces restrict their passage into the cerebrospinal fluid (CSF) [6]. Since the surface area of the human BBB is estimated to be 20 m^2 or around a thousand times greater than that of the BCSFB, the BBB is considered to be the main region controlling the neuronal environment and the uptake of drugs into the brain parenchyma. It is also the main target for delivering drugs to the brain via carrier-mediated strategies [7].

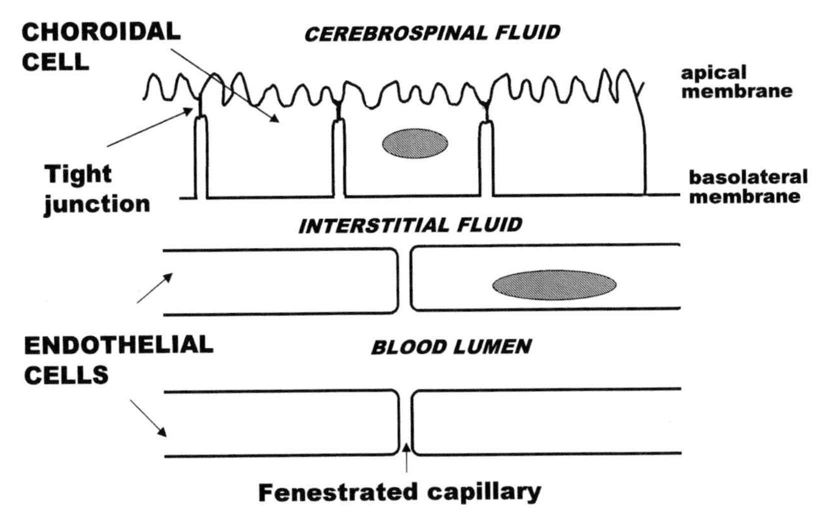

Fig. 15.1 Diagram showing a longitudinal cross-section of the blood–cerebrospinal fluid barrier at the choroid plexus. This barrier is formed by epithelial or choroid cells held together at their apices by tight junctions. The fenestrated blood vessels of the choroid plexus allow large molecules to pass, but the tight junctions between the choroid cells restrict their passage into the cerebrospinal fluid.

15.1.2 Constituents of the Blood–Brain Barrier

15.1.2.1 Endothelial Cells in the Blood–Brain Barrier

The BBB is defined by the microvasculature of the brain, which consists of a monolayer of polarized endothelial cells connected by complex tight junctions having a high electrical resistance (>1000–3000 Ω cm^{-2}). This structure prevents paracellular transport across the brain endothelium [8]. These endothelial cells are separated from the astrocyte foot processes and pericytes by a basement membrane (Figure 15.2). The astrocyte foot processes are about 20 nm from the abluminal surface of the endothelial cells and this space is mainly filled with the microvascular basement membrane and the brain extracellular fluid. The endothelial cells actively regulate vascular tone, blood flow and barrier function in the brain microvasculature. Endothelial cells are thin, about 0.1 μm thick, and thus occupy about 0.2% of the volume of the whole brain [9]. They are polarized, like epithelial cells; the luminal and abluminal endothelial membranes each segregate specific transcellular transport across the brain endothelium. This cell polarity is well illustrated by the distributions of several enzymes. Alkaline phosphatase is equally distributed between the luminal and the abluminal membranes but Na, K-ATPase and 5′-nucleotidase are present primarily on the abluminal side, and γ-glutamyl transpeptidase (γ-GTP) is found mainly on the luminal side [10]. Goldmann [4] first postulated that the brain capillaries provide the anatomical basis of a barrier in 1913, but this was not conclusively demonstrated until the 1960s, when electron microscope studies revealed that the endothelial cells of a brain capillary form electron-dense junctional contacts between two adjacent cells [11]. There are no gap junctions in brain capillaries and postcapillary venules and the tight junctions are responsible for sealing adjacent cells and maintaining cell polarity. Tight junctions are membrane microdomains made up of many specific proteins engaged in a complex membranar and intracytosolic network. Continuous tight junctions are not the only feature that makes the blood vessels of the brain different from those of other tissues. There are no detectable fenestrations or single channels between the blood and interstitial spaces. They contain fewer pinocytotic vesicles than do endothelial cells in the peripheral microvasculature and there is evidence that the majority of what appear to be independent vesicles in the endothelium cytoplasm are part of membrane invaginations that communicate with either the blood or the perivascular space [12]. The endothelial cells that form the BBB also contain many mitochondria. They occupy 8 to 11% of the cytoplasmic volume, much more than in brain regions lacking a BBB and tissues outside the CNS [13]. This indicates that the BBB also functions as a metabolic barrier, in addition to its physical barrier properties. There are highly specific transport systems carrying nutrients at the luminal or abluminal sides, or at both sides of the endothelial cell membrane. These carrier-mediated transport systems regulate the movement of nutrients between the blood and the brain. The brain capillary endothelium also bears specific receptors for circulating peptides or plasma proteins and these mediate the transcytosis of peptides or proteins through the BBB [14]. More recently, the discovery of active carrier-mediated transporters which are not involved in transporting substrate from the blood to the brain, but from the brain to blood, has greatly reinforced the barrier properties of the BBB. Most of these transmembrane proteins are located at the luminal or abluminal membranes of the endothelial cells and restrict the uptake of numerous drugs by the brain. Thus a large number of amphipathic cationic drugs are effluxed by one ABC protein, the P-glycoprotein which is present at the luminal surface of the BBB [15]. Brain vessels also have more classical types of receptors, including α- and β-adrenergic receptors and receptors for serotonin, adenosine, histamine, angiotensin and arginine vasopressin. Another specific feature of the BBB is the brain endothelium, which bears surface anionic sites differing from those found in some fenestrated and continuous endothelia. The distribution of these anionic sites on the luminal and abluminal membranes is, again, different. Experiments with proteolytic enzymes suggest that the negatively charged domains on the luminal membrane of the endothelial cells are mainly the terminal sialic acid residues of acidic glycoproteins, and the remaining few anionic domains are probably produced by heparan sulfate. An important distinction between brain endothelial cell anion distribution and that of other endothelial cells is the high density of heparan sulfate close to the tight junction domains. This could play a role in cell–cell signaling and adhesion [16]. The roles of these anionic domains in the function of the BBB remain to be established, but they actively contribute to the transcytosis of cationic proteins via adsorptive-mediated transcytosis [14]. The brain capillaries are also almost completely surrounded by other cells, like pericytes and the astrocyte foot processes.

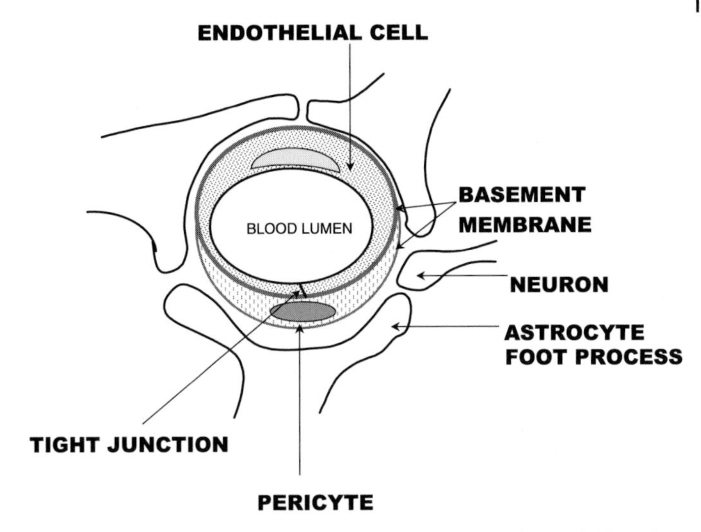

Fig. 15.2 Diagram showing a transverse cross-section of a cerebral capillary. The endothelial cells, responsible for the main barrier properties of the blood–brain barrier are separated from the astrocyte foot processes, pericytes and occasional neurons by the basement membrane. All these components make up the blood–brain barrier.

All these cellular components of the brain capillaries are joined by junctional systems. Zonal and extensive tight junctions seal the endothelial cells and gap junctions connect the endothelium to the subjacent pericyte layer, allowing their functional coupling and also weld them to the astrocyte processes. The last component of the endothelial cell network is the nerve fibers, which may be seen close to the cerebral blood vessels; these may be noradrenergic and peptidergic nerves (Figure 15.2). They influence the cerebrovascular tone and blood flow by secreting classical transmitters and a number of peptides, including substance P, neuropeptides and vasoactive intestinal peptide (VIP). This neurogenic influence could also explain the circadian variation in the permeability of the BBB under noradrenergic influence [17]. The intimate relationships between these cells make the BBB a pluricellular interface between the blood and the brain extracellular fluid.

15.1.2.2 **Pericytes in the Blood–Brain Barrier**

Pericytes lie periendothelially on the abluminal side of the microvessels (Figure 15.3). A layer of basement membrane separates the pericytes from the endothelial cells and the astrocyte foot processes. Pericytes send out cell processes which penetrate the basement membrane and cover around 20–30% of the microvascular circumference [18]. Pericyte cytoplasmic projections encircling the endothelial cells provide both a vasodynamic capacity and structural support to the microvasculature. They bear receptors for vasoactive mediators such as catecholamines, endothelin-1, VIP, vasopressin and angiotensin II. Pericytes become mark-

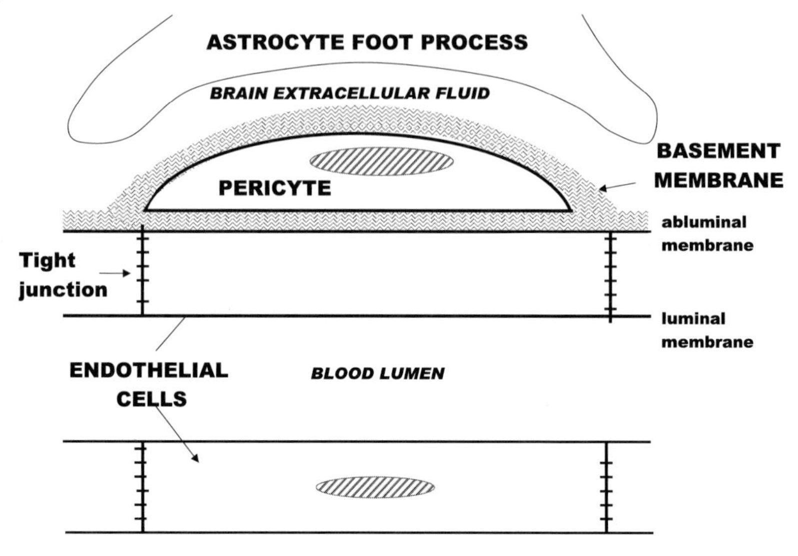

Fig. 15.3 Diagram showing a longitudinal cross-section of the blood–brain barrier, with the brain capillary endothelial cells sealed by the tight junctions and surrounded by pericytes and astrocyte foot processes. These cellular components of the BBB are separated by a basement membrane.

edly hypertrophic and hyperplasic in chronic hypertension, and bear increased amounts of marker proteins. In contrast, pericytes tend to degenerate in neurodegenerative processes, such as Alzheimer's disease, seizures and multiple sclerosis, so increasing the permeability of the BBB. Pericytes are implicated in the BBB and neuroimmune networks. They can be phagocytic in many injuries and may be actively involved in the regulation of leukocyte transmigration, antigen presentation and T-cell activation. They may provide a first line of defense against antigens infiltrating into the CNS. CNS pericytes also produce immunoregulatory cytokines such as interleukin-1β (IL-1β), IL-6 and the granulocyte–macrophage colony stimulatory factor [19].

CNS pericytes can be viewed as housekeeping scavenger cells and a second line of defense in the BBB. They are able to carry out pinocytosis and vesicular or tubular transport and also contain specific enzymes. Pericytes also play a role in hemostasis, by producing tissue factors allowing the assembly of the prothrombin complex. Thus they may be important in coagulation associated with cerebrovascular injury. Endothelial cell–pericyte bridges may also be involved in all stages of new vessel formation: pericytes guide the migrating endothelial cells, regulate their proliferation, form gap junctions and participate in the synthesis of the new basement membrane [20].

15.1.2.3 **Astrocytes in the Blood–Brain Barrier**

The blood capillaries of the CNS of vertebrates are also enveloped by a perivascular sheath of glial cells, mainly astrocytes (Figure 15.3). Immunohistochemical and morphometric studies on astrocytes and the microvasculature of the human cerebral cortex have shown that the astrocyte perivascular processes form a virtually continuous sheath around the vascular walls, with only 11% of the vessel perimeter not being covered [21]. The small portions of the vessel wall lacking an astrocyte envelope seem to be occupied by oligodendrocytes, microgliocytes and neuron bodies and processes; these adhere to the vessel endothelium-pericyte layer [22]. While the astrocytes themselves do not form the barrier, they have an important role in the development and maintenance of the BBB. Astrocytes release factors that can induce the BBB phenotype and/or the angiogenic transformation of brain endothelial cells *in vitro* and *in vivo* [23]. Astrocytes have also been co-cultured with endothelial cells in models of the BBB, and astrocyte-conditioned media can reduce the permeability of the BBB *in vitro*. Transport systems at the endothelial cells, including the glucose transporter and transport polarity, are up-regulated following exposure to astrocytes.

But the factors released by astrocytes that are responsible for these changes in permeability remain unknown, as are the cellular mechanisms underlying them. Similarly, the contribution of astroglia to BBB induction *in vivo* remains controversial [24]. BBB-competent endothelial cells may retain BBB properties following a chemically induced loss of astroglial end feet. These data suggest that astroglial factors important for the BBB phenotype may persist in the basement membrane to which the astroglial plasma membrane remains attached, or that neuronal influences are also important.

15.1.2.4 **Basement Membrane at the Blood–Brain Barrier**

The endothelial cells are separated from the pericytes and astrocytes by a basement membrane, also called the extracellular matrix (Figure 15.3). This basement membrane completely surrounds the CNS and is assembled from components of the bloodstream or secreted by the endothelial cells, pericytes and mostly by the perivascular astrocytes. Its main components are type IV collagen, fibronectin, laminins, chondroitin, and heparan sulfate from glycosaminoglycans. The most abundant component, type IV collagen, polymerizes with laminin and fibronectin proteins via protein-binding domains such as integrin and lectin receptors [25]. These components not only provide a mechanical supporting structure for the capillary wall, they are also important as a negatively charged barrier due to the chondroitin and heparan sulfate residues, in addition to the previously reported anionic properties of the luminal and abluminal membranes of the endothelial cells [26]. The basement membrane may influence the transcellular resistance of the cerebral endothelial cells and the formation of the tight junctions. The basement membrane also mediates several other processes, including cell differentiation, proliferation, migration and axon growth [27]. Astrocytes are unable to pass through the basement membrane. This is significant, since astrocytes produce all the plasminogen activators, proteases and metalloproteinases required to digest the constituents of the basement membrane [28]. Pathological changes in the permeability of the BBB are also influenced by the integrity of the basement membrane. Matrix-degrading metalloproteinases help modulate BBB permeability by breaking down the basement membrane macromolecules in such disorders as multiple sclerosis and other inflammatory processes where factors like cytokines stimulate the production of these proteolytic enzymes [29].

This overview points out the complexity of the BBB, as at least four types of cells plus the basement membrane are implicated in its structure and function. Thus many genes and the proteins they encode play a critical role in the broad pharmacological spectrum of activities carried out by the BBB. As drug responses depend on numerous proteins in the body, including metabolizing enzymes, transporters, receptors and all signaling networks mediating the response, it is very likely that there are similar gene–protein-mediated events at the BBB (Figure 15.4). The proteins involved in the formation of tight junctions and the regulatory interactions between the cerebral endothelial cells and in cross-talk between the cells of the BBB are undoubtedly targets for controlling the BBB permeability in normal and pathological conditions. The BBB permeability could be affected by

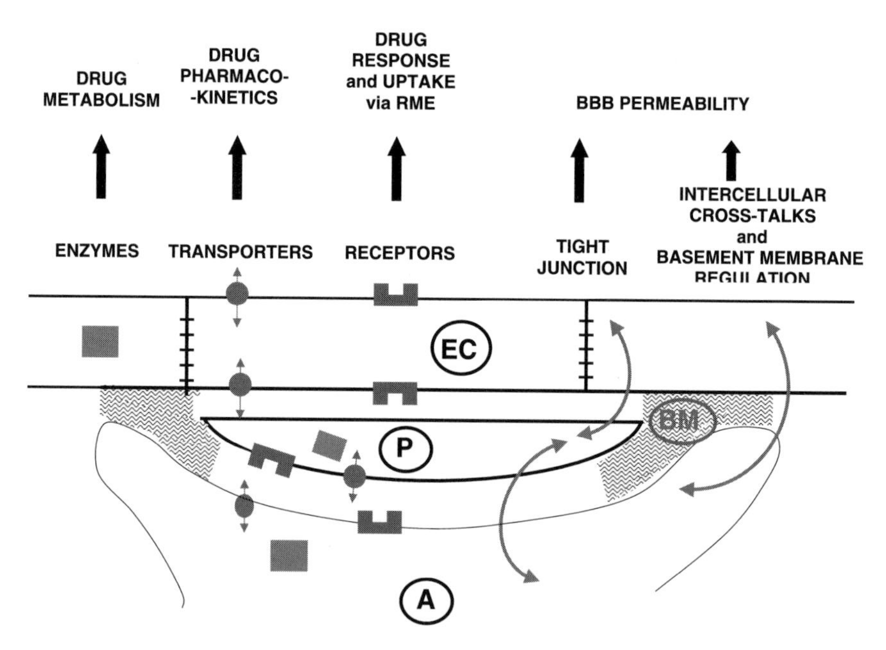

Fig. 15.4 Diagram showing the main pharmacokinetic and pharmacodynamic events mediated by the gene–protein network within the endothelial cells (EC), pericytes (P) and astrocytes (A). The intracellular cross-talk pathways that modulate the blood–brain barrier permeability are indicated by arrowheads. The basement membrane (BM) also contributes to BBB function. RME = receptor mediated endocytosis; green box = enzymes; blue circle = transporters; red box = receptors.

genetic variants under normal physiological conditions, leading to changes in the function of proteins like enzymes, transporters and receptors present at the BBB. More than 80% of all sequence variants in the human population are single nucleotide polymorphisms (SNPs), although only a few (<1%) probably change the function of the encoded proteins. Thus, identifying an SNP should help to identify the sources of differences in CNS drug responses from one patient to another, and so reduce the incidence of adverse drug reactions. For example, polymorphisms in genes encoding drug-metabolizing enzymes or transporters expressed at the BBB could affect the transport of a drug across the BBB and consequently the patient's response to it. The BBB is frequently disrupted in disease. Cerebral ischemia, bacterial meningitis, multiple sclerosis and brain injury all cause opening of the BBB, allowing solutes and water to move into the brain, leading to vasogenic brain edema. Cytokines, free radicals and proteases are the main cytotoxic factors appearing in these various unrelated diseases causing genetic defects and/or dysregulation of protein structure and function at the BBB [30]. Knowledge of their deleterious effects on the gene–protein network of the BBB architecture may help considerably our understanding of the mechanisms underlying diseases that affect the integrity of the BBB. They could also provide new therapeutic targets for restoring BBB function.

Lastly, pharmacogenomics could provide new tools for the design of more specific and active CNS pharmaceuticals. The efficacy of a broad spectrum of neuropharmaceutical drugs is often complicated by their inability to reach their site of action because of the BBB. One way to overcome this is to use carrier-mediated transport at the luminal and/or abluminal membranes of the endothelial cells of the BBB. This will provide a physiologically based drug delivery strategy for the brain by designing new chemical entities or fused proteins that can cross the BBB via these transporters.

Thus the pharmacogenomics of the BBB may lead to the discovery of the genes and proteins that are specifically produced in the BBB, and also to the discovery of the way in which they are dysregulated or defective.

15.2 The Main Gene and Protein Targets for Pharmacogenomics of the Blood–Brain Barrier

15.2.1 Drug-Metabolizing Enzymes at the Blood–Brain Barrier

The CNS contains much smaller amounts of drug-metabolizing enzymes than does the liver. The concentrations of the main enzymes in the brain, members of the cytochrome P450 (CYP) superfamily, are only 0.25% of concentration in the liver. But the brain enzymes are not uniformly distributed, as they are in the liver; they are concentrated in specific brain areas. Theoretical models have explained that drug metabolism in the CNS cannot influence drug distribution in the blood, but there are marked differences in brain tissue levels depending on the presence or absence of brain enzymes in these specific areas [31]. Thus, brain enzymes located very near receptor sites for drugs could have a very marked effect on receptor–drug exposure.

Although the absence of paracellular transport across the BBB impedes the entry of small hydrophilic compounds into the brain, low-molecular-weight lipophilic substances may pass through the endothelial cell membranes and cytosol by passive diffusion [7]. While this physical barrier cannot protect the brain against chemicals, the metabolic barrier formed by the enzymes from the endothelial cell cytosol may transform these chemicals. Compounds transported through the BBB by carrier-mediated systems may also be metabolized. Thus, L-DOPA is transported through the BBB and then decarboxylated to dopamine by the aromatic amino acid decarboxylase [7].

This metabolic barrier was first postulated for amino acid neurotransmitters in 1967 [32]. The presence of at least 30 cerebral enzymatic systems suggests that a battery of enzymes may modulate the entry of neuroactive molecules into the brain [33]. Several phase 1 enzymes, such as CYPs, monoamine oxidases (MAO-A and -B), flavin-containing monoxigenases, reductases and oxidases and phase 2 enzymes catalyzing conjugation, such as UDP-glucuronosyltransferases (UGTs) and glutathione- and sulfotransferases have been found in the rat and human CNS, as well as in isolated brain microvessels and cerebrovascular endothelial cells in primary culture [34].

These various experiments show that brain capillaries contain significantly higher enzyme activities than does the brain parenchyma itself [35, 36]. Rat brain microvessels contain several isoforms of CYP1A involved in the metabolism of aromatic polycyclic hydrocarbons [36]. The finding that CYP2D6 is responsible for debrisoquine hydroxylation and the formation of morphine from codeine O-demethylation may have pharmacogenomic significance [37]. Several of the main drug-metabolizing CYP450 enzymes, such as 1A2, 2A6, 2C9, 2C19 and 2D6 frequently have non-functional alleles, resulting in 1–10% poor metabolizers in various ethnic populations for CYP2D6. The prodrug-activating effect of CYPD6 does not occur in poor metabolizers on codeine therapy, and this contributes to between-individual variations in the potential sensitivity to CYP2D6 CNS pharmaceuticals, such as most of the tricyclic antidepressants (nortriptyline and desipramine) [31].

Other oxidative enzymes may also help activate pathways of neurotoxicity at the BBB or within the brain. For example, CYP2E1, which metabolizes ethanol, chlorinated solvents to toxic metabolites, and arachidonic acid to vasoactive compounds, could produce active metabolites resulting in oxidative stress and altering BBB permeability [38]. Between-individual variation in its induction by ethanol and an upstream polymorphism that is common in Orientals are functionally significant and related to the discovery of novel alleles in the CYP2E1 gene. One of these increases the transcription of the gene. The defective CYP2D6 expression and the wide differences in CYP1A1 and CYP2E1 inducibility result in variations in the uptake of their substrates by the brain. Other phase 1 enzymes in the mammalian brain capillaries, like MAO-B, are active enough to metabolize the lipophilic neurotoxin 1-methyl-4-phenyl-1,2,5,6-tetrahydropyridine (MPTP) into

1-methyl-4-phenyl-pyridinium (MPP^+), which is responsible for the appearance of severe Parkinson's disease-like symptoms [39]. This metabolic activity at the BBB can protect the brain against the neurotoxin MPTP. Similarly, MAO-B inactivates catecholamine-like substances that cross the BBB and may protect the brain against neurotoxic exogenous pyridine analogs [35].

The CYP3A subfamily, which contains the major isoforms expressed in the human liver and interacts with around 50% of pharmaceuticals, has not yet been found in brain endothelial cells and parenchyma, in contrast to other polarized cells like intestinal epithelial cells. The absence of this main CYP subfamily makes the brain and the BBB unique. The activity of the phase 2 enzyme, UGT 1A6, has been reported to be 6-fold greater in brain microvessels than in the brain parenchyma and polymorphisms which may affect transferases activity could also influence the distribution of drugs in the CNS [35]. To illustrate the complexity of the enzyme activities in the BBB, enzymes expressed in the pericytes and in the astrocyte end feet complement the BBB enzyme arsenal. Glutamyl aminopeptidase and aminopeptidase, which convert angiotensin II to angiotensin III and inactivate opioid peptides, are located in the pericytes. There may be species differences, such as butyrylcholinesterase which deacylates heroin to morphine is located in the pericytes of the dog brain capillaries, but in the endothelial cells of the rat brain capillaries [7]. Adenosine deaminase, which converts the cerebral vasodilatator adenosine to inosine, is found mainly in the astrocyte foot processes and may prevent the further distribution of adenosine to receptors on neurons [7]. The protective advantage of more concentrated enzymes in the brain microvasculature than in parenchyma could fail if there were to be a gene defect, as with CYP2D6, and may enhance the exposure of the brain to CNS-active drugs (e.g., tricyclic antidepressants, neuroleptics, opioids and serotonin uptake inhibitors). Similarly, the polymorphism affecting the inducibility of CYP1A1 and CYP2E1 may give rise to the large between-subject variations in the responses of the CNS to drugs.

15.2.2
Drug-Carrier Transporters at the Blood–Brain Barrier

15.2.2.1 Mono- or Bidirectional Transporters for Small Compounds

A second type of drug pharmacokinetic event at the BBB is mediated by proteins on the luminal and/or the abluminal membranes of the endothelial cells. These proteins can mediate symmetric and asymmetric drug transport. This type of transport was first discovered for nutrients that are not lipid-soluble and cannot cross the barrier by simple diffusion. Glucose and some amino acids that the brain cells cannot manufacture for themselves can be carried in both directions by facilitated osmotic diffusion [11]. Glucose was the first molecule whose passage into the brain was linked to a high-capacity transport system present on both sides of the endothelial cell membranes. The GLUT1 transporter was first cloned in 1985 and is present in the brain capillaries as well as the choroid plexus, the astroglial and neuronal cell membranes. Its state of glycosylation varies from one cell to another and it can transport about 1 $\mu mol\ min^{-1}\ g^{-1}$ at the BBB in man [40]. This capacity has been used to develop the transport into the brain of drugs linked to a D-pyranose sugar moiety. The other main types of glucose transporters are essentially present on neurons (GLUT3) or on both astroglia and neurons (GLUT5) [41].

Like the glucose carrier, the carriers for large neutral amino acids, the so-called L-system – now designated LAT – are present at both sides of the endothelial cell membranes and transport at least 10 essential amino acids. The L-transporter at the BBB has a much higher transport capacity than those in other tissues. Its marked preference for phenylalanine analogs explains why the anticancer drugs melphalan and D,L-NAM-7 are transported by the L-system, as is the L-Dopa used to treat Parkinson's disease [42].

Other CNS pharmaceuticals, such as baclofen and gabapentin, cross the BBB via the L-system. The LAT1 and LAT2 genes were recently detected in whole brain by Northern blotting and may be present in the brain endothelial cells, but the two transporters they encode have different substrate specificities. The concentration of LAT1 mRNA at the BBB is much higher (>100-fold) than in any other tissue, including other brain cells like neurons and glial cells. Thus LAT1 is probably the major transporter of large neutral amino acids at the BBB, rather than LAT2 [43, 44]. The cationic amino acid transporter which mediates the uptake of arginine, lysine and ornithine by the brain is also present at the BBB; CAT1 and CAT2 have both been cloned. Although they are both present in the brain, only CAT1 seems to be present at the BBB and may be the primary basic amino acid transporter of the BBB [45].

Other transporters can work by active mechanisms requiring a source of energy. These transporters are often present on the luminal or abluminal membranes of the endothelial cells (Figure 15.5). The sodium ion-independent amino acid transporter, system A, transports small, neutral amino acids across the abluminal membranes of the brain endothelial cells. Other systems (systems X^-, A, B^{o+}, ASC) are also involved in amino acid transport, but their pharmacological impact at the BBB remains to be elucidated [46]. Monocarboxylic acid compounds like lactic acid, ketone bodies, β-hydroxybutyrate and acetoacetate, which are abundant in the brain, are regulated by both uptake and efflux transporters at the BBB. The monocarboxylic acid transporter (MCT1) is present on both the luminal and abluminal membranes of the BBB and seems to transport substrates in both directions [47]. It is upregulated by increased glycolysis, cerebral lactate and seizure-related factors. MCT1 is responsible for the transport of several drugs, including the HMG-CoA reductase inhibitors simvastatin acid, lovastatin acid, and pravastatin. Interestingly, the anticonvulsant, valproic acid, is more efficiently transported from the brain to the blood than the other way round because an organic anion transport, probably mediated by OAT1 (organic anion transporter), counterbalances the uptake mediated by MCT1 at the BBB [48].

The BBB also has sodium- and pH-independent transporters of organic cations. They are important for the homeostasis of choline and thiamine in the brain and for the permeation of cationic drugs like propranolol, lidocaine, fentanyl, H1-an-

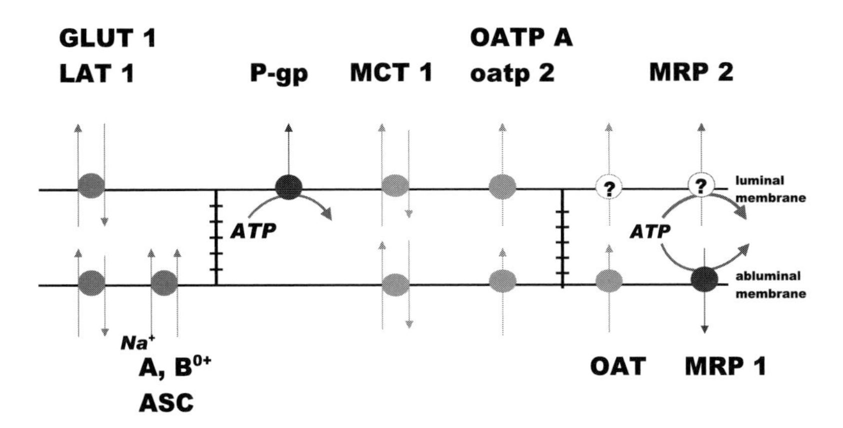

Fig. 15.5 Diagram showing some of the nutrient and drug transport processes associated with the brain capillary endothelial cells that form the BBB. Local transporters in the luminal or/and abluminal membranes are depicted as filled circles and ones whose location is more questionable or that are present at the BBB are depicted in open circles. GLUT1, LAT1, MCT1, oatp2 are present on both the luminal and abluminal membranes. This diagram shows that transport may be unidirectional or bidirectional.

tagonists and choline analogs. These organic cation transporters (OCTs) are mainly located at nerve terminals, glial cells and in the BBB. Human OCT2 is found in the brain neurons rather than at the BBB, but a new family of organic cation/carnitine transporters, the OCTNs, seems to be involved in the transport of organic cations such as mepyramine into brain capillary endothelial cells [48]. Several sodium-independent facilitative (hENT1 and hENT2) and sodium-dependent active (CNT1 and CNT2) transporters have been identified. Their presence at the BBB allows the transport of nucleoside-mimetic antivirals like azidothymidine (AZT) and dideoxycytidine (ddC) and the anticancer drug gemcitabine from the blood to the brain. But, as for valproic acid, these drugs are more efficiently excluded from the brain by active transporters sensitive to anionic compounds like probenecid.

15.2.2.2 **Peptide and Protein Transporters**

Hydrophilic peptides and proteins are frequently large molecules; they may enter the brain by carrier-mediated transport, receptor-mediated transcytosis, or by adsorptive-mediated transcytosis. Small peptides, such as di- and tripeptides are transported by the specific transporters, PepT1 and PepT2, but neither of them is present at the BBB. Nevertheless, there is saturable brain uptake of the tripeptide glutathione and of several opioid peptides, suggesting that specific transporters, as yet unidentified, are involved at the BBB. In addition to peptide transporters, adsorptive and receptor-mediated transcytosis are responsible for the transport of peptides and proteins across the BBB [49].

The human brain microvessels bear receptors for insulin, insulin-like growth factor-I, insulin-like growth factor-II, transferrin and leptin. These transporters mediate the transcytosis of these peptides and have been targeted by several strategies for delivering drugs to the brain [14]. Neither insulin nor transferrin have been coupled to the drug of interest because insulin causes hypoglycemia and the physiological concentration of transferrin in the plasma is much too high. Therefore, monoclonal murine antibodies against each of the receptors have been used to prepare high-affinity-binding vectors which still have transcytosis properties. These have been used to deliver attached peptides like VIP, nerve growth factor and brain-derived neurotrophic factors to the brain [7].

Absorption-mediated transcytosis across the BBB is mediated by mechanisms that depend on the cationic charges of peptides or proteins. The initial binding to the luminal plasma membrane is mediated by electrostatic interactions with, e.g., anionic sites on acid residues of acidic glycoproteins. These trigger absorptive endocytosis. The cationized proteins cross the BBB by adsorption-mediated transcytosis aided by the effects of the anionic charges of the abluminal membrane and those of the basement membrane. More sophisticated cationization processes have been developed using protein covalently linked to naturally occurring polyamines like putrescine, spermidine and spermine. Such covalently modified neurotrophic factors may increase the permeability of the BBB to them [50]. The recent discovery that small cationic peptide vectors can transport drugs across complex biological membranes has opened up new possibilities. Peptide vectors such as TAT peptide, SynB vectors and penetratin, have been used to deliver biologically active substances inside live cells [51]. The antineoplastic agent doxorubicin does not readily cross the BBB. Coupling it to a SynB vector significantly enhances its uptake by the brain without compromising the tight junctions. About 20-times more vectorized doxorubicin is delivered to the brain parenchyma than when free doxorubicin is used. Coupling other CNS pharmaceuticals to SynB vectors, such as penicillin, peptides like dalargin, proteins up to 67 kDa and other cytotoxic drugs, significantly improves their uptake by the brain [52].

15.2.2.3 A New Generation of Efflux Transporters

The identification of the brain-to-blood efflux transporter, P-glycoprotein (P-gp), in 1992 has added a novel property to the concept of the BBB. P-gp decreases the permeability of the BBB to lipophilic drugs by actively impeding their crossing of the luminal membranes or by transporting them out of the endothelial cells to the bloodstream [53]. P-gp was originally found as an overproduced membrane protein in multidrug resistance tumor cells, and was found to be responsible for reducing the intracellular accumulation of several anticancer drugs [48]. Transport mediated by P-gp is coupled with ATP hydrolysis and affects many substrates that have a planar structure, neutral or cationic charge and are hydrophobic. While humans have only one gene (MDR1) encoding the drug-transporter P-gp, rodents have two genes, mdr1a and mdr1b, that encode P-gps with overlapping substrate specificities [54]. The availability of mice with a disrupted mdr1a gene, the mdr1b gene, or with disrupted mdr1a and mdr1b genes, has helped to demonstrate that the P-gp in the BBB limits the entry of many drugs into the brain by actively pumping them back into the blood. Most light-microscope and electron-microscope immunochemical experiments using several specific antibodies to P-gp indicate that the luminal membrane of the brain endothelial cells normally has a high concentration of P-gp (Figure 15.5). Nevertheless, a few studies have suggested that P-gp is not present in the luminal membrane of the human and primate brain capillary endothelial cells, but is mainly on astrocyte foot processes [55]. It is quite likely that P-gp is present at both the blood luminal membrane of the brain capillary endothelial cells and in astrocytes. We have recently identified both mdr1b transcripts and a functional P-gp protein in primary cultures of rat astrocytes; we have also described greater expressions of the mdr1a gene and its resulting protein in rat brain capillaries than in astrocytes [56]. Previously, Tishler et al. found a strong immunostaining for P-gp on both blood capillaries and astrocytes in the brains of patients with intractable epilepsy [57]. While the function of P-gp in astrocytes may be questionable, its role as a barrier to the entrance of small lipophilic compounds across the luminal membranes of the brain capillary endothelial cells is now clearly established. Several functional polymorphisms of the human MDR1 gene were recently described and correlated with the synthesis and activity of P-gp *in vivo*. Analysis of the MDR1 sequence in a Caucasian population resulted in 15 SNPs, only one of which was located in exon 26, thus revealing the functional polymorphism of the MDR1 gene. About 26% of the population were homozygous for this variant and the three genotypes gave rise to poor, intermediate and high intestinal uptakes of the cardiotonic drug, digoxin, a P-gp substrate [58]. There is probably similar polymorphism at the BBB, and this might be an extremely important factor influencing between-subject variations in the uptake of a large number of pharmaceuticals by the CNS. There seem to be several other transporters that exclude drugs from the brain, in addition to P-gp. The presence at the BBB of members of the multidrug resistance-associated protein (MRPs) family, whose members preferentially transport anionic compounds, is still controversial. The seven members of the MRP family belong, like P-gp, to the ATP-binding cassette (ABC) protein superfamily. Mrp1 has been found at the BBB in isolated rat brain capillaries, primary cultures of brain capillary endothelial cells and in immortalized capillary endothelial cells, but not in human brain capillaries [59]. Another member, MRP2 has been found at the luminal membrane of the brain endothelial cells [60]. However, further studies are required to show that there are MRP transporters at the BBB (Figure 15.5). As for P-gp, a functional Mrp1 was found in primary cultured rat astrocytes [56] and it has been shown to take part in the release of glutathione disulfide from brain astrocytes under oxidative stress [61].

The efflux of organic anions seems not to be mediated by MRPs alone. Other organic anion transporters, like oatp2 and OATPA, have been detected by immunohistochemistry on both the luminal and abluminal membranes of human brain capillaries [48].

The functional characterization and identification of the proteins and genes of so many nutrient and drug transporters at the BBB over these last 25 years has led to a change in our understanding of the way drugs are transported across the BBB. This has also opened an important avenue for the development of attractive strategies for delivering drugs to the brain using some of these transporters. Therefore, the more recent discovery of efflux transporters has raised the possibility that several processes associating diffusion, influx and efflux may determine the net drug uptake into the CNS. The inducibility or functional polymorphism affecting these transporters at the BBB remain to be completely identified. This new information may shed light on the factors involved in inter-and intra-subject variations in the uptake of drugs by the CNS. It is thus possible that variations in the component molecules of the BBB may give rise to poor, intermediate and high brain drug uptake.

15.2.3
Tight Junctions, Receptors and Cell Cross-Talk at the Blood–Brain Barrier

Morphological studies on tight junctions by freeze-fracturing and ultrathin sectioning has shown that the endothelial cells of capillaries of the mammalian brain possess the most complex tight junctions in the whole vascular system. This complexity is reflected in the activities of a pair of proteins which have not yet been completely identified and whose synthesis, assembly and regulation remain to be clarified. Tight junctions are formed of pentalaminar layers that result from the fusion of the external leaflets of the partner cell membranes. Many families of integral membrane proteins, including the four-pass protein, occludin, and the products of a multigene family, the claudins (claudin-1 and claudin-5), are the main components of tight junctions [62]. The N-terminal domain of occludin regulates the transmigration of leukocytes and the C-terminal domain is involved in the control of the paracellular permeability of molecules that involves binding to a protein kinase. Corticosteroids, which decrease the paracellular uptake of sucrose by the brain, have recently been shown to increase the amounts of occludin mRNA and protein, while reducing the phosphorylation of occludin. This suggests that occludin plays a key role in the tightness of the junctions. The intracellular terminal ends of occludin also interact with other peripheral membrane proteins like zonula occludens proteins (ZO-1, ZO-2, ZO-3), cingulin and 7H6 antigen [63], leading to multiple interactions with cytoskeletal elements (actin filaments, intermediate filaments) within the endothelial cells.

Yet another family of junction adhesion molecules (JAMs) was recently located at the tight junctions of both endothelial and epithelial cells. The intracellular domain of JAM-1 also interacts with structural and signaling proteins, such as ZO-1 and cingulin. Lastly, the molecular organization of the endothelial cell junctions includes two other cell–cell contact Ca^{2+}-dependent cadherin–catenin systems. These make up the adherens junction common to all endothelial cell junctions.

The β- and γ-catenins help control the transcription of several genes. β-Catenin binds to transcription factor LEF-1 and to growth factors and probably plays a role in the translocation of ZO-1 to tight junction complexes [64]. Other proteins that mediate the cell–substrate adhesion of adherent cells may serve as signal recognition molecules. A special feature of junctions between endothelial cells is the presence of platelet–endothelial cell adhesion molecule-1 (PECAM-1), which is involved in angiogenesis and in mediating inflammatory responses such as leukocytes–endothelial interactions. The regulation and function of this unique architecture remain poorly understood. The components of intercellular junctions are rich in signaling proteins that form part of several signaling pathways, such as protein kinases and protein phosphatases. The brain capillary tight junctions are also very dynamic structures that are sensitive to environmental and cellular factors [61]. The close anatomical relationship between the end feet of astrocytes and neurons suggests that environmental neural factors may determine the characteristics of the blood–brain barrier. Cultured brain capillary endothelial cells have short, fragmented tight junctions, but these tight junctions become more dense and longer in the presence of astrocytes and their electrical resistance increases significantly [66]. Modulation of the tightness of endothelial tight junctions is an important objective of pharmacological studies. Like the unidentified soluble factors produced by astrocytes, certain agents which are released during inflammation, such as histamine, thrombin, bradykinine, cytokines and prostacyclins, may cause opening of the blood–brain barrier at the tight junctions of endothelial cells. The transient opening of the BBB allows the delivery of non-permeating drugs to the brain. The intravascular infusion of hyperosmolar solutions or the administration of agonists of the bradykinin and histamine receptors increase the permeability of the brain vascular system to anticancer agents [7, 14].

The breakdown of the BBB may also facilitate the infiltration of circulating monocytes and neutrophils into the brain and hence its invasion by bacteria or viruses [67]. Similarly, activated T-lymphocytes may cross the BBB because of the presence of adhesion molecules on the vascular endothelial cells that form the BBB. Cadherins are involved in the formation of tight junctions and integrins, including ICAM-1, VCAM-1 and PECAM-1 and selectins, which can be upregulated and overproduced following exposure to pro-inflammatory agents or activated T-cells, are all involved in the movement of immune cells across the BBB. Little is known about the second messenger systems that may be activated to facilitate the margination and diapedesis of immune cells across the BBB. Several signaling pathways, such as Rho-ICAM-1, *c-fos* and inhibitory kappaβalpha, activate the release of secondary immune response modulators by the endothelial cells like IL-6 and -8 or prostaglandins and thromboxanes. This indicates that the BBB can amplify or dampen immune responses in the brain by creating complex cell-to-cell cross talk. Further work is needed to understand how this occurs.

15.3
Pharmacogenomics of the Blood–Brain Barrier

15.3.1
Objectives for Pharmacogenomics of the Blood–Brain Barrier

The two previous paragraphs have shed light on the various cellular and non-cellular components of the BBB and on the many proteins of pharmacological interest expressed at the BBB. Combining these two observations can point to several objectives for pharmacogenomic investigations of the BBB. One would be the complete characterization of the gene–protein network of each of the component cells of the BBB. This would lead to a better understanding of the exact role of each of these components in the pharmacokinetic and pharmacodynamic events occurring at BBB and to a better definition of the BBB and all its regulation pathways. A second would be the identification of those genes that are specifically expressed at the BBB. This would clarify the specificity of the cerebral endothelium and differentiate it from the endothelia of other parts of the body. It could also open the way to the discovery of new specific targets for modulating the permeability of the BBB to drugs, e.g., via receptor-mediated transcytosis. A third would be the identification of the SNPs that may introduce polymorphisms in drug metabolism or transport. Knowledge of the regulatory mechanisms that lead to the induction or repression of enzymes or transport proteins could also be extremely valuable for predicting between-subject variations in the response of the CNS to drugs. The last objective would be the identification of the defects in genes or their encoded proteins in disease states that affect BBB permeability.

15.3.2
Current Experimental Approaches and Their Limitations

The pharmacological impact of gene and protein function or dysregulation on the BBB permeability may be investigated by *in vitro* or/and *in vivo* experiments. Animals with spontaneous gene mutations or defects, or animals that are genetically modified, may be most useful for assessing the overall influence of drug pharmacokinetics and pharmacodynamics on the response to a drug. But complementary *in vitro* studies are needed to discover the molecular mechanisms underlying the gene–protein system. Although current molecular and genomic techniques, such as DNA and protein microarrays, are more and more accessible, they cannot easily be applied to pharmacogenomic studies of the BBB. These techniques are not sensitive enough to detect endothelial cell-specific transcripts and proteins because of the extremely small amounts of specific BBB transcripts and proteins in whole brain extracts. Pardridge et al. estimated that the brain microvessels account for about 10^{-3} parts of the whole brain [68]. As the sensitivity of gene microarrays is about 10^{-4}, it is unlikely that genes selectively expressed by the brain microvessels will be detected in extracts of whole brain. This has stimulated attempts to isolate the brain capillary endothelial cells from other brain cells and to the preparation of enriched extracts for DNA and protein analysis.

15.3.2.1 *In vitro* Pharmacogenomic Studies

The analysis of gene expression in BBB cells using molecular and genomic techniques will require the preparation of pure brain endothelial cells. *In vitro* models of the BBB have evolved from isolated microvessels to primary cultures and immortalized and transfected endothelial cells over the last 25 years. As all these *in vitro* systems differ with respect to isolation procedures, cell culture conditions, mono- or co-culture and different cell types in term of origin and species, the endothelial cell phenotype and genotype may vary widely [9, 67]. The use of cultured endothelial cells has several advantages, including the availability of unlimited amounts of genetic and protein extracts and the certainty that there is no contamination by other cell components of the BBB, such as astrocyte end feet and pericytes. Nevertheless, culture conditions and the number of passages all modify the expression of BBB-specific genes like GLUT1 and γ-GTP, which can be markedly downregulated or absent. Similarly, the activities of these two specific markers of the BBB can be increased in cultured brain cells by conditioning their medium by adding, e.g., astrocyte-derived trophic factors. This great dependency of the gene and protein expression by cultured brain capillary endothelial cells on experimental factors makes this type of model inappropriate for pharmacogenomic studies on the BBB. A careful analysis of BBB-specific gene transcripts and proteins requires the isolation of brain capillaries prior to analysis of gene products. Several methods for isolation of brain microvessels under RNAase-free conditions have been published [68]. The methods that use mechanical or enzymatic tissue disruption give inhomogeneous preparations as pre-capillary arterioles, which are made up of endothelial cells and smooth muscle cells, are present. The capillaries also contain pericytes, which are 3 times more abundant than the brain capillary endothelial cells, and the abluminal membrane of isolated brain capillaries may be attached to remnants of astrocyte foot processes and nerve endings. Thus, the purity of the isolated brain capillary must be estimated by immunostaining for the glial fibrillary acidic protein (GFAP), which is specific to astrocytes, or measuring cerebrosides as possible neuronal contamination [68]. The detection of pericyte-specific proteins like smooth muscle α-actin or the surface markers CD11b may help to differentiate them from endothelial cells and astrocytes [19]. Endothelial cells may also be isolated from the brain parenchyma using magnetic microbeads cross-linked to an antibody directed against PECAM-1. This method was used to increase the P-gp content of an endothelial cell fraction 59-fold, while the negative fraction, which contained the astrocyte marker GFAP, contained no P-gp. It was recently used to demonstrate the phenotypical changes that occur in the vascular bed markers in the microvasculature of brain tumors [69].

BBB-RNA must be isolated for molecular cloning and analysis of BBB-specific transcripts, and several methods for isolating BBB-poly (A^+) mRNA in one or more steps have been described [68]). Li et al. recently reported that they obtained yields of 12 µg poly (A^+) mRNA from a single bovine cortical shell and 3.2 µg poly (A^+) mRNA from the pooled cerebral hemispheres of 21 rats [70].

The isolation of Poly (A^+) mRNA has also led to the synthesis of complementary DNA for the preparation of BBB-cDNA libraries and the construction of BBB-reporter genes. These reporter genes can be used for transfecting cultured brain endothelial cells and generating cellular tools for gene–protein mechanistic studies.

These experiments are hampered by the risk of obtaining heterogeneous isolated brain microvessels and low yields of mRNA. Analysis of BBB-proteomics may also be complicated by the polarized distributions of membrane proteins, some being found on the luminal membrane of the brain capillaries and others on the abluminal membrane. Thus the luminal membranes must be purified using density-gradient centrifugation of homogenized brain tissues following the addition of colloidal silica particles, which selectively bind to the luminal membranes, to the vascular bed. This results in a 10-fold enrichment in GLUT1 and a 17-fold enrichment in P-gp over their concentrations in isolated brain capillaries and very little contamination by basolateral membranes. This technique is suitable for detecting proteins by Western blotting [68].

Despite these limitations, quantitative polymerase chain reaction (PCR) studies on BBB-specific transcripts have been used to measure the concentration of bovine GLUT1. More recently it has been combined with subtraction cloning methods to identify the genes that are more actively expressed at the BBB than in peripheral tissues [68]. Rat brain capillary poly (A^+) RNA was used to produce tester cDNA and rat liver or kidney mRNA was used to generate driver cDNA. The two-run PCR produced cDNA and the resulting large library was used to isolate the first 50 clones. More than 80% of the genes were selectively expressed at the BBB. They included novel genes like a 2.6 kb long cDNA sequenced and named BBB-specific anion transporter type 1 (BSAT1). Several genes known to be selective for the BBB, such as the genes encoding tissue plasminogen activator, insulin-like growth factor-2, transferrin receptor, oatp2 transporter and the class I major histocompatibility complex, were also found [70]. This first study of a BBB genomics program demonstrates that numerous genes with novel sequences encoding proteins of unknown function may be selectively expressed at the BBB. This type of genomic study could also be applied to BBB-related disorders. An altered gene expression has recently been found in the cerebral capillaries of stroke-prone spontaneously hypertensive rats (SHRSP), which are a model of hypertension-induced cerebrovascular lesions. The abnormal gene results in a disturbance of the structure and function of the tight junctions of the BBB endothelial cells prior to stroke. Here, too, three cDNA fragments were found to be up-regulated by suppression subtractive hybridization between the SHRSP and stroke-resistant rat brain capillaries plus a cDNA filter screening step. These changes were associated with the pathogenesis of stroke [71].

15.3.2.2 *In vivo* Pharmacogenomic Studies

The generation of mice with disrupted genes should allow the evaluation of the pharmacokinetic and pharmacodynamic consequences of the complete, specific inhibition of particular drug enzymes, transporters or receptors.

The first gene-knockout mouse to become available was a mouse lacking detectable P-gp in the brain capillary endothelial cells. This has been used to elegantly demonstrate the real impact of P-gp in the BBB. The amounts of anticancer agents, immunosuppressants and other P-gp drug substrates in the brains of these mice were greater than in the brains of normal mice. These increased brain concentrations may dramatically modify the pharmacological activity of certain P-gp substrates like the dopamine antagonist domperidone, which only produces an anti-emetic effect in P-gp-competent mice due to its selective peripheral activity. The anti-psychotic effect of domperidone becomes its main effect when it is given to mice lacking P-gp, indicating its distribution and activity in the CNS [54]. These data suggest that a genetic deficiency in BBB P-gp may have dramatic consequences for the over-exposure of the brain parenchyma to drugs. There is also a genetic mutation affecting P-gp at the BBB in an inbred subpopulation of CF1 mice. They have a mutation in their mdr1a gene. The absence of functional P-gp at the BBB leads to effects similar to those observed in Pgp knockout mice [72]. Like the CF1 mouse, collie dogs are sensitive to the neurotoxic effects of the anti-parasitic drug ivermectin. Sequencing of their cDNA showed that they have a mutation due to the deletion of four exonic base pairs. This leads to a premature stop codon in the messenger RNA and the translation of a truncated inactive form of P-gp [73]. These data raise the question of how they can be extrapolated to humans. Humans may respond differently from mice or collie dogs to an abnormal gene and the altered gene in the knockout mice may be associated with the altered expression of other genes, perhaps with overproduction or a lack of other transporters in the BBB. Therefore, the absence of P-gp at the BBB does not decrease the tightness of the junctions, as both normal and P-gp knockout mice have similar low permeabilities to sucrose [72]. These P-gp knockout mice are undoubtedly a powerful tool for pharmacokinetic and pharmacodynamic studies, and the recent generation of mice lacking Mrp1 may help provide a better understanding of the exact role of this other ABC efflux pump at the BBB. Similarly, the generation of transgenic mice overproducing the monocarboxylic acid MCT1 transporter in the brain endothelium may be a fruitful approach to examining the function of a gene [74]. This illustrates the wide range of *in vivo* models that genetic engineering may generate. These *in vivo* transgenic models will be extremely useful for quantifying the pharmacokinetic and pharmacodynamic effects of gene modification at the BBB.

15.4 Conclusion

The recent exponential growth of genomics and proteomics in biological sciences has provided many new concepts and experimental tools for attacking the puzzling questions of the function and regulation of the BBB in normal and pathological situations. Because the BBB protects the brain from chemicals and infectious agents, a complete description of BBB-specific genes and the functions of the corresponding proteins will help provide a clearer picture of all the BBB barrier functions in the wide range of diseases (infectious, degenerative, autoimmune, metabolic, tumors) that afflict the CNS. We may well be on the threshold of a new era for the discovery of some of the mysterious properties of the BBB.

Acknowledgements
I thank Owen Parkes for editing the English text. This textbook is dedicated to two of my research colleagues at INSERM Unity 26. The first is Doctor Jeanne-Marie Lefauconnier who will retire in April 2002 after a long and fruitful career studying the amino acid transporters at the BBB. The second is Doctor Gaby Boschi, who died in October 2001. Her work has provided us with unique insights into the impact of brain microdialysis on neuropharmacokinetics.

15.5 References

1 Lewandowsky M. Zur Lehre der Cerebrospinalflüssigkeit. Z Klin Med **1900**; 40:480–494.
2 Ehrlich P. Über die Beziehungen von chemischer Konstitution, Verteilung und pharmakologischer Wirkung. In: Collected Studies in Immunity. New York: John Wiley & Sons; **1906**, 567–595.
3 Goldmann EE. Die äußere und innere Sekretion des gesunden und kranken Organismus im Lichte der „Vitalen Färbung". Beitr Klin Chir **1906**; 64:192–265.
4 Goldmann EE. Vitalfärbung am Zentralnervensystem. Abh Preuss Akad Wiss Phys Math K1 **1913**; 1:1–60.
5 Davson H, Segal MB. Blood–brain barrier. In: Physiology of the CSF and Blood–Brain Barriers (Davson H, Segal MB, Eds.), 1st Edn. Boca Raton: CRC Press; **1996**, 49–91.
6 Spector R, Johanson CE. The mammalian choroid plexus. Sci American **1989**; 61:68–74.
7 Pardridge WM. Transport of small molecules through the blood–brain barrier: biology and methodology. Adv Drug Del Rev **1995**; 17:713–731.
8 Brightman MW. Morphology of blood-brain interfaces. Exp Eye Res **1977**; 25:1–25.
9 Drewes LR. Molecular architecture of the brain microvasculature. J Mol Neurosci **2001**; 16:93–98.
10 Betz AL, Firth JA, Goldstein GW. Polarity of the blood–brain barrier: distribution of enzymes between the luminal and antiluminal membranes of brain capillary endothelial cells. Brain Res **1980**; 192:17–28.
11 Goldstein GW, Betz AL. The blood–brain barrier. Sci American **1986**; 255:74–83.
12 Broadwell RD, Balin BJ, Salcman M. Transcytotic pathway for blood-borne protein through the blood–brain barrier. Proc Natl Acad Sci USA **1998**; 85:632–636.
13 Oldendorf WH, Cornform ME, Brown WJ. The large apparent work capability of the blood–brain barrier: a study of capillary endothelial cells in brain and other tissues of the rat. Ann Neurol **1977**; 1:409–417.
14 Pardridge WM. Drug delivery to the brain. J Cereb Blood Flow Metab **1997**; 17:713–31.
15 Cordon-Cardo C, O'Brien JP, Casals D, Rittman-Grauer L, Biedler JL, Melamed MR et al. Multi-drug resistance (P-glycoprotein) is expressed by endothelial cells at blood–brain barrier sites. Proc Natl Acad Sci USA **1989**; 86:695–698.

16 Schmidley JW, Wissig SL. Anionic sites on the luminal surface of fenestrated and continuous capillaries of the CNS. Brain Res **1986**; 363:265–271.

17 Johanson BB, Isaksson O. Circadial variation of cerebral vessel vulnerability during adrenaline induced hypertension. In: Cerebral Blood Flow: Effect of Nerves and Neurotransmitters (Heistad DD, Marais ML, Eds.), 1st Edn. Amsterdam: Elsevier Biomedical; **1982**, 367–375.

18 Frank RN, Dutas S, Mancini MA. Pericyte coverage is greater in the retinal than in the cerebral capillaries of the rat. Invest Ophtalmol Visual Sci **1987**; 28:1089–1091.

19 Balabanov R, Dore-Duffy P. Role of the CNS microvascular pericyte in the blood–brain barrier. J Neurosci Res **1998**; 53:637–644.

20 Diaz-Flores L, Guttierrez R, Varela H. Angiogenesis: an update. Histol Histopathol **1994**; 9:807–843.

21 Virgintino D, Monaghan P, Robertson D, Errede M, Bertossi M, Ambrossi G et al. An immunohistochemical and morphometric study on astrocytes and microvasculature in the human cerebral cortex. Histochem J **1997**; 29:655–660.

22 Bertossi M, Virgintino D, Maiorano E, Occhiogrosso M, Roncali L. Ultrastructural and morphometric investigation of human brain capillaries in normal and peritumoral tissues. Ultrastruct Pathol **1997**; 21:41–49.

23 Stewart PA, Wiley MJ. Developing nervous tissue induces formation of blood–brain barrier characteristics in invading endothelial cells: a study using quail-chick transplantation chimeras. Dev Biol **1981**; 84:183–192.

24 Krum JM, Kenyon KL, Rosenstein JM. Expression of blood–brain barrier characteristics following neuronal loss and astroglial damage after administration of anti-thy-1 immunotoxin. Exp Neurol **1997**; 146:33–45.

25 Vorbrodt AW. Ultracytochemical characterization of anionic sites in the wall of brain capillaries. J Neurocytol **1989**; 18:359–368.

26 Washington LC, Rahman J, Klein N, Male DK. Distribution and analysis of surface charge of brain endothelium *in vitro* and *in situ*. Acta Neuropathol **1995**; 90:305–311.

27 Jucker M, Tian M, Ingram DK. Laminins in the adult and aged brain. Mol Chem Neuropathol **1996**; 28:209–218.

28 Bernstein JJ, Karps SM. Migrating fetal astrocytes do not intravasate since they are excluded from blood vessels by vital basement membrane. Int J Dev Neuroscience **1996**; 14:177–180.

29 Rosenberg GA, Kornfeld M, Estrada E, Kelley RO, Liotta LA, Stetler-Stevenson WG. TIMP-2 reduces proteolytic opening of blood–brain barrier by type IV collagenase. Brain Res **1992**; 576:203–207.

30 Banks WA. Physiology and pathology of the blood–brain barrier; implications for microbial pathogenesis, drug delivery and neurodegenerative disorders. J Neurovirol **1999**; 5:538–555.

31 Britto MR, Wedlung PJ. Cytochrome P-450 in the brain. Potential evolutionary and therapeutic relevance of localization of drug-metabolizing enzymes. Drug Met Disp **1992**; 20:446–450.

32 Van Gelder NM. A possible enzyme barrier for γ-aminobutyric acid in the central nervous system. Prog Brain Res **1967**; 29:259–268.

33 Mesnil M, Testa B, Jenner P. Xenobiotic metabolism by brain monooxygenases and other cerebral enzymes. Adv Drug Res **1984**; 13:95–207.

34 El-Bacha RS, Minn A. Drug metabolizing enzymes in cerebrovascular endothelial cells afford a metabolic protection to the brain. Cell Mol Biol **1999**; 45:15–23

35 Minn A, Ghersi-Egea JF, Perrin R, Leininger B, Siest G. Drug metabolizing enzymes in the brain and cerebral microvessels. Brain Res Rew **1991**; 16:65–82.

36 Ghersi-Egea JF, Leninger-Muller B, Suleman G, Siest G, Minn A. Localization of drug-metabolizing enzyme activities to blood–brain interfaces and circum-ventricular organs. J Neurochem **1994**; 62:1089–1096.

37 Chen ZR, Irvine RJ, Bochner F, Somogyi AA. Morphine formation in rat

brain: a possible mechanism of codeine analgesia. Life Sci **1990**; 46:1067–1074.

38 Montoliu C, Vallés S, Renau-Piqueras J, Guerri C. Ethanol-induced oxygen radical formation and lipid peroxidation in rat brain: effects of chronic alcohol consumption. J Neurochem **1994**; 63:1855–1862.

39 Chiba K, Trevor A, Castagnoli N. Metabolism of the neurotoxic tertiary amine, MPTP, by brain monoamine oxidase. Biochem Biophys Res Commun **1984**; 120:547–578.

40 Gould GW, Holman GD. The glucose transporter family: structure, function and tissue specific expression. Biochem J **1993**; 295:329–341.

41 Bauer H. Glucose transporters in mammalian brain development. In: Introduction to the Blood–Brain Barrier (Partridge WM, Ed.), 1st Edn. Cambridge: Cambridge University Press; **1998**, 175–187.

42 Cornford EM, Young D, Paxton JW. Melphalan penetration of the blood–brain barrier via the neutral amino acid transporter in tumor bearing brain. Cancer Res **1992**; 52:138–143.

43 Segawa H, Fukasawa Y, Miyamoto K, Takeda E, Endou H, Kanai Y. Identification and functional characterization of a Na^+-independent neutral amino acid transporter with broad substrate selectivity. J Biol Chem **1999**; 274:19745–19751.

44 Boado RJ, Li JY, Nagaya M, Zhang C, Partridge WM. Selective expression of the large neutral amino acid transporter at the blood–brain barrier. Proc Natl Acad Sci USA **1999**; 96:12079–12084.

45 Stoll J, Wadhwani KC, Smith QR. Identification of the cationic amino acid transporter (System y^+) of the rat blood–brain barrier. J Neurochem **1993**; 60:1956–1959.

46 Smith QR, Stoll J. Blood–brain barrier amino acid transport. In: Introduction to the Blood–Brain Barrier (Partridge WM, Ed.), 1st Edn. Cambridge: Cambridge University Press; **1998**, 188–197.

47 Gerhart DZ, Emerson BE, Zhdankina OY, Leino RL, Drewes LR. Expression of monocarboxylate transporter MCT1 by brain endothelium and glia in adult and suckling rats. Am J Physiol **1997**; 273:E 207–213.

48 Tamai T, Tsuji A. Transporter-mediated permeation of drugs across the blood–brain barrier. J Pharm Sci **2000**; 89:1371–1388.

49 Temsamani J, Scherrmann JM, Rees AR, Kaczorek M. Brain drug delivery technologies: novel approaches for transporting therapeutics. Pharm Sci Technol Today **2000**; 2:49–59.

50 Poduslo JF, Curran GL, Gill JS. Putrescine-modified nerve growth factor: bioactivity, plasma pharmacokinetics, blood–brain/nerve barrier permeability and nervous system biodistribution. J Neurochem **1998**; 71:1651–1660.

51 Derossi D, Chassaing G, Prochiantz A. Trojan peptides: the penetratin system for intracellular delivery. Trends Cell Biol **1998**; 8:84–87.

52 Rousselle C, Smirnova M, Clair P, Lefauconnier JM, Chavanieu A, Calas B et al. Enhanced delivery of doxorubicin into the brain via a peptide-vector-mediated strategy: saturation kinetics and specificity. J Pharmacol Exp Ther **2001**; 296:124–131.

53 Drion N, Lemaire M, Lefauconnier JM, Scherrmann JM. Role of P-glycoprotein in the blood–brain transport of colchicine and vinblastine. J Neurochem **1996**; 67:1688–1693.

54 Schinkel AH. P-glycoprotein, a gatekeeper in the blood–brain barrier. Adv Drug Del Rev **1999**; 36:179–194.

55 Golden PL, Pardridge WM. P-glycoprotein on astrocyte foot processes of unfixed isolated human brain capillaries. Brain Res **1999**; 819:143–146.

56 Decleves X, Regina A, Laplanche JL, Roux F, Boval B, Launay JM et al. Functional expression of P-glycoprotein and multidrug resistance-associated protein (MRP1) in primary cultures of rat astrocytes. J Neurosci Res **2000**; 60:594–601.

57 Tishler DM, Weinberg KI, Hinton DR, Barbaro N, Annett GM, Raffol C. MDR1 gene expression in brain of patients with medically intractable epilepsy. Epilepsia **1995**; 36:1–6.

58 Hoffmeyer S, Burk O, Von Richter O, Arnold HP, Brockmoller J, Johne A et al. Functional polymorphisms of the human multi-drug resistance gene: multiple sequence variations and correlation of one allele with P-glycoprotein expression and activity *in vivo*. Proc Natl Acad Sci USA **2000**; 97:3473–3478.

59 Huai-Yun H, Secrest DT, Mark KS, Carney D, Elmquist WF, Miller W. Expression of multi-drug resistance-associated protein (MRP) in brain microvessel endothelial cells. Biochem Biophys Res Commun **1998**; 243:816–820.

60 Fricker G, Miller D, Bauer B, Nobmann S, Gutmann H, Török M et al. Transport of xenobiotics across isolated brain microvessels studied by confocal microscopy. Presented at the 3rd FEBS Advanced Lecture Course – ABC**2001** Gosau, Austria.

61 Hirrlinger J, Gutterer JM, Kussmaul L, Hamprecht B, Dringen R. Microglial cells in culture express a prominent glutathione system for the defense against reactive oxygen species. Dev Neurosci **2000**; 22:384–392.

62 Kniesel U, Wolburg H. Tight junctions of the blood–brain barrier. Cell Mol Neurobiol **2000**; 20:57–76.

63 Schnittler HJ. Structural and functional aspects of intercellular junctions in vascular endothelium. Basic Res Cardiol **1998**; 93:30–39.

64 Rubin LL, Staddon JM. The cell biology of the blood–brain barrier. Annu Rev Neurosci **1999**; 22:11–28.

65 Stewart PA, Hayakawa EM. Interendothelial junctional changes underlie the developmental tightening of the blood–brain barrier. Dev Brain Res **1987**; 32:271–281.

66 Janzer RC, Raff MC. Astrocytes induce blood–brain barrier properties in endothelial cells. Nature **1987**; 325:253–257.

67 Miller DW. Immunobiology of the blood–brain barrier. J Neuro Vir **1999**; 5:570–578.

68 Boado RJ. Molecular biology of brain capillaries. In: Introduction to the Blood–Brain Barrier (Partridge WM, Ed.), 1st Edn. Cambridge: Cambridge University Press; **1998**, 151–162.

69 Demeule M, Labelle M, Regina A, Berthelet F, Beliveau R. Isolation of endothelial cells from brain, lung, and kidney: expression of the multidrug resistance P-glycoprotein isoforms. Biochem Biophys Res Commun **2001**; 281:827–834.

70 Li JY, Boado RJ, Pardridge WM. Blood–brain barrier genomics. J Cereb Blood Flow Metab **2001**; 21:61–68.

71 Kirsch T, Wellner M, Luft FC, Haller H, Lippoldt A. Altered gene expression in cerebral capillaries of stroke-prone spontaneously hypertensive rats. Brain Res **2001**; 910:106–115.

72 Dagenais C, Rousselle C, Pollack GM, Scherrmann JM. Development of an *in situ* mouse brain perfusion model and its application to mdr1a P-glycoprotein-deficient mice. J Cereb Blood Flow Metab **2000**; 20:381–386.

73 Roulet A, Puel O, Gesta S, Drag M, Soll M, Alvineri M et al. MDR Pharmacogenetics in dog. Presented at the 3rd FEBS Advanced Lecture Course. ABC **2001**. Gosau, Austria.

74 Lieno RL, Gerhart DZ, Duelli R, Enerson BE, Drewes LR. Diet-induced ketosis increases monocarboxylate transporter (MCT1) levels in rat brain. Neurochem Int **2001**; 38:519–527.

16
Pharmacogenomics and the Treatment of Neurological Disease

David B. Goldstein, Ramachandran V., Nicholas W. Wood
and Simon D. Ahorvon

Abstract

In this chapter we outline a structure for genetic association studies that can be used to search for mutations influencing the efficacy and side effects of drugs to treat neurological conditions. The basic approach is to compare the genetic make-up of populations of patients with different response profiles. We illustrate different aspects of the program with examples from epilepsy, a common neurological disease. We have chosen epilepsy for several reasons. It is a common and important condition, there is a range of drug treatments available, and yet many patients either fail to respond to most or all of the drugs or they develop side effects. Moreover the genetic basis of the epilepsies is just starting to be elucidated. Interestingly there is also a clear link between the site of action of the common anti-convulsants and the genes responsible for rare forms of the disease (see Figure 16.1). There is every reason to think that this link will also occur in the case of mutations that predispose to the common epilepsies, though examples are currently lacking. Thus, consideration of drug action, mutations underlying Mendelian epilepsies, and the biological basis of epilepsy all implicate ion channel genes as an important focus in genetic association studies for epilepsy. Another major area of interest not only for the epilepsies but much more generally is the role of the drug-metabolizing enzymes. Fortunately, given current high-throughput screening technologies and the emerging pattern of genetic variation in human populations, the list of genes directly involved in drug metabolism appears manageable in the context of systematic genetic association studies. Finally, we emphasize that identifying candidate genes for variable drug response involves a considerable amount of guesswork, and that the lists will never be definitive. Thus it remains a priority to develop genome-wide approaches which would not require assumptions about the most important genes. These approaches may take several years to be refined, however, and we argue that until these become feasible there are relatively obvious candidate genes for conditions such as response to anti-convulsants that clearly warrant detailed study.

The strategy we suggest is not the only one that could be adopted, and it is clear that it will need refinement as more is learned about patterns of variation in human populations, and especially about the genetic basis of both common diseases and variable drug reactions. We make a case, however, that the time is now ripe for using population genetic approaches to study the genetic bases of variable reaction to anti-convulsants and other drugs used to treat neurological conditions.

16.1 Introduction

The effectiveness of drugs used in neurology has considerable individual variation. Responder rates, e.g., for recently trialled antiepileptic drugs range from 28–40%, and seizures in a small number of patients are actually exacerbated. It is furthermore estimated that adverse drug reactions (ADRs) are responsible for 100 000 deaths per year in the United States, ranking ADRs between the 4th and 6th leading cause of death [1]. There is also considerable morbidity from side effects, and at least one third of patients in recent antiepileptic drug trials, e.g., report side effects. While much of this variation will be due to environmental factors, including drug–drug interactions, many lines of evidence indicate that genetic differences among people make an important contribution to efficacy and ADRs and to variable drug response more generally. Identification of the precise genetic differences underlying this variation in drug response could be used to (1) avoid drugs that have highly variable effects due to genetic differences among individuals, and (2) where appropriate to develop diagnostic tests that would permit the personalization of therapeutic treatment. In fact, many drug developers already try to avoid drugs that are metabolized largely by CYP2D6, due to considerable inter-individual differences in activity. The use of diagnostic tests, however, is currently almost absent in clinical practice. Instead, a lengthy trial and error process is often required to determine appropriate drugs and optimal doses, and this applies to most of the drugs used, e.g., in epilepsy, Parkinson's disease, neuro-psychiatric disease and neuro-oncology. There is considerable interest, therefore, in using the recent advances in human genomics to identify mutations influencing drug response, a research area now usually termed pharmacogenomics. Many aspects of modern genomics will play a part in this effort, including gene expression profiling, which is already being used, e.g., to identify "signatures" of toxicity in cell culture systems. Eventually it will also be possible to characterize genome-wide protein expression patterns for similar purposes Our focus here, however, is on how the availability of the complete genomic sequence will facilitate systematic screens for mutations that influence drug response using genetic association studies. We will use epilepsy as our prime illustrative example.

16.2 Pharmacogenomic Approaches

16.2.1 Drug Response Genes (DRG)

Beginning in the 1950s the field of pharmacogenetics has sought to identify mutations that might influence how individuals respond to drugs, and to document the frequencies of these mutations in global populations. Drug-metabolizing enzymes (DMEs) have received by far the most attention. This stems in part from their clear importance in variable drug response, and indeed many early examples of variable drug response turned out to be due to polymorphisms at drug-metabolizing enzymes such as the well known debrisoquine oxidation phenotype due to variation at the CYP2D6 gene. The focus on drug metabolism, however, is also in part methodological. The major enzymes are known for most drugs, and it is relatively easy to determine whether given mutations in the genes encoding them affect enzyme kinetics.

Of the common anti-convulsants in clinical usage many, but not all, are metabolized by the P-450 hepatic enzyme system, including carbamazepine, clobazam, clonazepam, ethosuximide, felbamate, lamotrigine, oxcarbazepine, phenobarbital, phenytoin, primidone, tiagabine and valproate [2]. This enzymatic metabolism plays a significant role in the drug serum levels attained and in drug interactions. The relation of serum level to efficacy and to side effects, and the effects of drug interactions, have duly become central issues in epilepsy therapeutics for at least two decades. *In vitro* hepatocyte cultures have elucidated the exact metabolic pathways for individual drugs, and the kinetics of the specific metabolic enzyme isoforms. It has also become clear that genetic influences play an important role in determining these kinetic properties. As yet, however, the clinical effect of genetic variation in antiepileptic drug metabolism has not been systematically studied. Other enzymatic systems are involved in antiepileptic drug metabolism also, including the hydrolyzing enzymes which metabolize levetiracetam and which are widespread in bodily tissues, but little is known about the genetic influences on their pharmacokinetics. Recently attention has expanded to include other categories of genes influencing pharmacokinetics, notably drug transporters.

The genetic factors influencing pharmacodynamics have generally received much less attention, though this is now starting to change with an increasing number of studies focused on how variants in drug targets influence both drug efficacy and adverse reactions. For example, Catalano has reviewed work on the association between polymorphisms in drug targets and the efficacy and adverse reactions of drugs for neuro-psychiatric conditions [3]. The growing attention to pharmacodynamics reflects in part our increased ability to screen large numbers of genes. Also important, however, is our rapidly improving knowledge of the biology and genetics and drug reactions and disease. For example, in the case of variable response to anti-convulsants, a very obvious starting point is the voltage gated sodium channel. Not only are a number of mutations known in genes en-

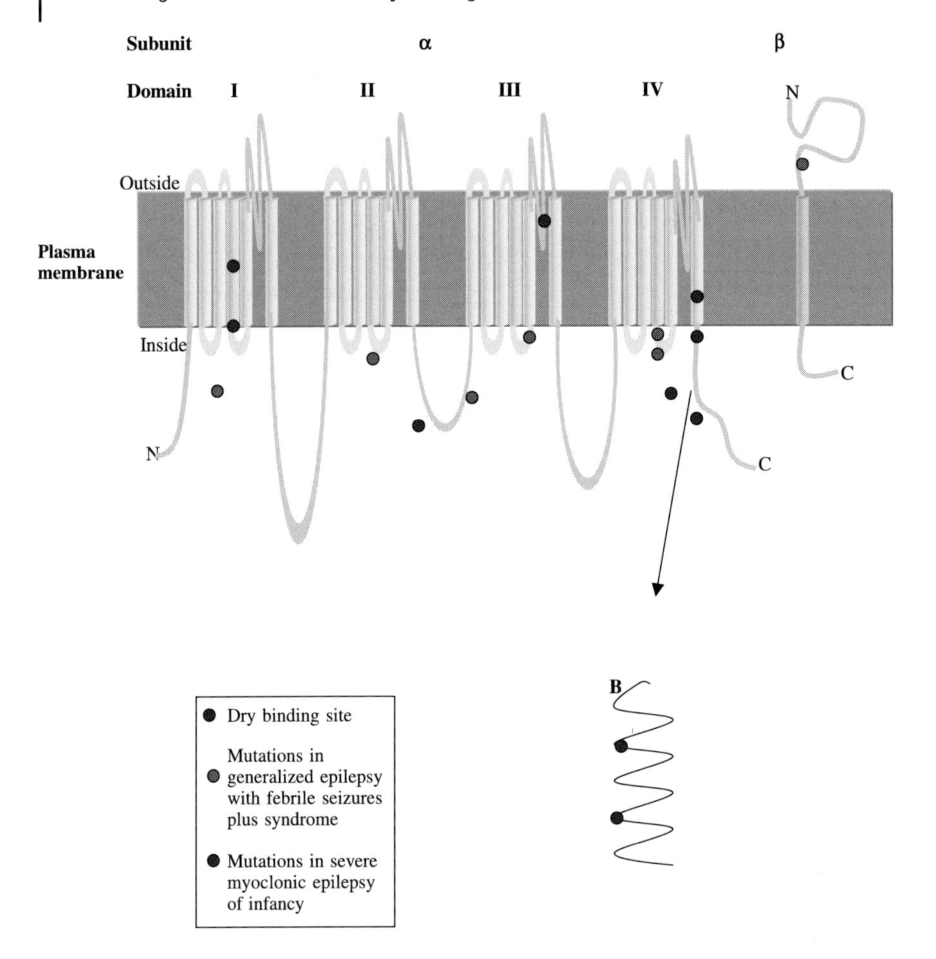

Fig. 16.1 Sodium channel structure. Schematic representation of the sodium channel subunits, α, $\beta 1$ and $\beta 2$. (A) The α-subunit consists of four homologous intracellularly linked domains (I–IV) each consisting of six connected segments (1–6). The segment 4 of each of the domains acts as the voltage sensor, physically moving out in response to depolarization resulting in activation of the sodium channel. The channel is inactivated rapidly by the linker region between III and IV docking on to the acceptor site formed by the cytoplasmic ends of S5 and S6 of domain IV. The β-subunits have a common structure, with the $\beta 1$ non-covalently bound, and $\beta 2$ linked by disulfide bonds to the α-channel (adapted from [4]). The S5/S6 and the segment linking them (P-loop) are believed to constitute the most of the pore of the channel. Specific mutations in the P-loop are associated with loss of selectivity of the channel. Mutations identified in generalized epilepsy with febrile seizures plus are denoted by red dots, while those in severe myoclonic epilepsy of infancy with black dots. The black dots denote the site of termination of the sodium channel. (B) An enlarged S6 segment of domain IV showing drug-binding site comprised of phenylalanine-1764 and tyrosine-1771 in human sodium channel $Na_v1.2$ (see color plates, p. XXXV).

coding sodium channels that cause monogenic epilepsies, but the sodium channel is thought to be the major mode of action for a number of front line anti-convulsants (Figure 16.1), and some information is available on the drug-binding domains. Given these factors it is an obvious priority to systematically screen the genes encoding the sodium channels for mutations influencing not only response to anti-convulsants, but also to test for association with epilepsy itself. Below we describe a framework for carrying out such a screen. It must be appreciated, however, that drug targets are normally components of relatively long and complex signaling pathways, and there is no reason to assume that the important genetic variants influencing response must reside in the element of this pathway that is physically bound by the drug. Thus in addition to the elements that interact directly with the drug, priority must also be given to genes encoding other steps in the relevant pathway(s). An obvious example is variable responses to anti-hypertensives targeting the renin–angiotensin pathway. It is an obvious priority to include the genes encoding this entire pathway in any effort to find mutations influencing how patients respond to these anti-hypertensives.

For convenience, we will refer to the set of genes that are relevant to response to a given drug or class of drugs as Drug Response Genes (DRG). Obviously depending on the criteria used this will be a longer or shorter list, and indeed in the limit it would probably include the entire genome. While additional candidate genes could be listed for any given drug, we suspect that in the near future candidate genes will be drawn mainly from the following categories: DMEs, transporters, and intended and unintended targets and the elements of the broader relevant pathways. The next stage after these categories are exhausted would probably be whole genome scanning, looking for variants anywhere in the genome (see below) that might influence drug response. While effective whole genome scans are currently out of reach technically and economically, there are reasons to be optimistic that some version of genome scans will be implemented in the relatively near future (see below).

16.2.2
Population Structure and Variable Drug Response

When drugs are evaluated, e.g., in Phase III study populations, individuals are usually included who have ancestry from different geographic areas. Thus during drug trials on large diverse patient populations these populations will include genetic subgroups that may respond differently, on average, to drug administration due to different frequencies of mutations at DRG.

While the importance of genetic differences among individuals in drug response is rarely questioned, the importance of average differences among populations has been the subject of considerable debate. For example, it has been claimed that enalapril, an angiotensin-converting enzyme inhibitor, is more effective in whites than in blacks [5]. Similarly, the Food and Drug Administration (FDA) has recently approved clinical trials of BiDil specifically for use in African Americans [6]. Such use of race in the context of drug response, however, has been sharply criticized on the grounds that race is a poorly defined and highly ambiguous concept in the context of human genetic variation [7]. The debate, however, is somewhat misleading. It is well known in the human population genetics community that races do not exist in the human population in any way that corresponds to the popular idea of a race as a group sharply delimited from other such groups (e.g., [8]). It is also well known that racial or ethnic labels do not provide a good guide to the genetic structure that is present in the human population. But these observations in no way imply that population structure is of no consequence to drug response. Indeed, there remains considerable concern that adverse drug reactions, efficacy, and optimal doses may show average differences in individuals with genetic ancestry from different parts of the world. There is for example a widely held, but entirely anecdotal, perception that optimal doses of many antiepileptic drugs differ in different populations, being lower in Chinese than in Caucasians and also interestingly in Europe than in the US. So these problems with racial and ethnic labels only imply that we need a more appropriate framework for dealing with population structure in the context of drug evaluation.

We have recently demonstrated that an alternative framework should be much more useful in the context of drug evaluation. Instead of using ethnicity to indicate genetic relationships, we have evaluated a framework of explicit genetic inference. Our approach is to use explicit genetic analysis to identify genetically related subclusters of individuals allowing the performance of the drug to be compared among these genetically identified clusters [9]. We must emphasize here, however, what we do and, more importantly do not, mean by related. In short, and speaking casually, we are looking for groupings of individuals that are maximally predictive of the frequencies of mutations at DRGs. We are not, however, seeking to identify the boundaries of groups that have any objective existence. In fact, the history of migrations of the human species is such that there are no such groups, but rather only graded differences among groups of people from different parts of the world. Therefore, what we are seeking is a clustering scheme that is operationally useful, as opposed to one that makes an ontological statement. We note that the complexity of human history makes this a non trivial exercise, and it is an area of continuing work.

To make a direct evaluation of this framework we assembled a set of population samples from throughout the world, including multiple populations from Europe, Asia, and Africa. In each individual we typed 39 presumably neutral microsatellite markers. These markers were used to assess the relatedness of individuals, and do not themselves have anything to do with drug response, so far as we know. We then suppressed the geographic labels indicating the origins of these individuals, and used a model-based algorithm to divide this heterogeneous group into a number of subgroups of genetically more related individuals (see [10] for a description of this method). This procedure resulted in four genetic groupings. We then looked at the frequencies of mutations at DMEs among these clusters as a simple surrogate for drug response, to assess the scope for differences among the clusters in drug response. We note that we chose DMEs because of the relatively direct connection between drug response and mutations affecting enzymatic activity. It is, e.g., much harder to determine whether and how coding changes in targets might influence drug response.

We found that for four of six of the mutations considered, there are highly significant differences among the clusters, indicating considerable scope for differences among the clusters in drug response. It is of particular interest, however, to consider the membership of these clusters. Returning to the geographic labels, we found a generally poor correspondence between casual ethnic labels and the clusters. For example, Bantu individuals from Southern Africa were largely included in one cluster, but individuals from Ethiopia were largely included in a separate cluster that included most Europeans. Finally, the two Asian populations included (from SW China and Papua New Guinea) were placed in largely separate clusters, which showed sharp differences in frequency at one of the drug-metabolizing enzymes considered. These results make clear that it is important to assess how drug response is correlated with genetic clusters. For example, as mentioned above, there is anecdotal evidence that the dosing of antiepileptic and other psychoactive drugs needs to be lower in Chinese compared with Caucasian populations, although there are few formal studies of this. Clearly, though, if this is the case, drug trials should be carried out in multiple populations and dosage recommendations from one population should not be simply generalized to all others. The need for the pharmaceutical industry to trial in different populations has been long recognized, although probably largely for economic reasons, rarely carried out. Our work provides an explicit framework for considering population-genetic structure in the context of drug evaluation.

16.2.3 Association Studies and Neurological Pharmacogenomics

The ultimate aim for pharmacogenomics is determination of the mutations that are responsible for the genetic component of inter-individual drug response. While single gene mutations with large effects are well known, most drug responses are considered to be complex traits, influenced by variation at multiple loci that show complex dependence on the environment. Unfortunately the family-based studies that proved so successful for studying monogenic diseases, are not robust to this level of complexity.

Association studies, most commonly implemented in case-control designs, are now widely viewed as the most promising alternative for studying the genetic basis of complex traits, including both common diseases and variable drug reactions. In this approach, genetic differences between cases and controls sampled from a defined population are evaluated in an effort to identify genetic differences influencing the condition of interest.

In the case of drug response data, however, the phenotypes of interest will normally be quantitative as opposed to the discrete case-control distinction that has often been used in disease studies. For example, severity of side effects and efficacy can often be assessed quantitatively.

While straightforward in principle, the optimal design of population-based approaches is complex, depending not only on the genetic basis of the disease or trait under study, but also on the genetic structure of the population from which cases and controls are selected. Although the basic approach simply tests for significant differences in allele frequencies at markers (or candidate genes) between cases and controls sampled from defined populations, genotype–phenotype correlations are notoriously difficult to interpret. Not only does the interpretation depend on the exact relationship between genetic variation and disease status [11, 12], but it must also be guided by what is known about the pattern of non-random association between alleles at different loci, usually termed linkage disequilibrium (LD). These patterns of LD in turn depend on both the demography of the population [13, 14] and on genomically localized factors such as recombination, mutation, and selection. For these reasons a description of genome-wide patterns of linkage disequilibrium is widely seen as prerequisite to effective genome-wide association studies.

Two lines of evidence recently presented indicate that this may be a more easily realized goal than has been previously thought. Studying a 500 kb stretch of 5q31, Daly et al. reported discrete blocks of high levels of LD in which haplotype diversity is very low [15]. These blocks were separated by short stretches that showed evidence of recombination. An article of Jeffries et al. published in the same issue suggests [16] that this pattern is due to the localization of recombination events to irregularly spaced hotspots. Jeffries et al. show that in a region spanning 216 kb of the Major Histocompatibility Complex (MHC) Class II region, more than 94% of the recombination events observed in sperm typings occurred in three clusters of tightly packed recombination hotspots. Most strikingly, the breakdown in LD through the region corresponded exactly with the location of these hotspots. To the extent that LD is structured into discrete blocks, it will be far easier to develop appropriate study designs for association studies. There are two important implications of this atomistic pattern of LD. First, the pattern of LD should be more easily determined than we would have expected, especially by using a hierarchical design in which a coarse set of single nucleotide polymorphisms (SNPs) are typed (or widely spaced re-sequencing carried out) in the first instance. This will identify the larger blocks, and subsequent typing would then be used to identify the smaller ones. Second, once the blocks are determined, a relatively small number of markers could be typed in case-control material and these markers, through LD, would be sufficient to represent most of the SNPs that were not directly typed [17]. For example, in one of Daly's "blocks" of linkage disequilibrium, spanning 84 kb, two haplotypes account for 96% of the chromosomes observed. Thus this entire stretch could be represented in an association study by typing a single marker, which would capture the vast majority of the variation in this genomic region. We should emphasize that this approach will succeed in identifying the mutations responsible for drug response only if they are relatively common. If the causal mutations are very rare than these approaches will not succeed (cf. common variant common disease model, [12]). While the generality of this very simple pattern of haplotype diversity is not yet known, the results of the recent studies are highly encouraging. So what does this mean for association studies in neurological pharmacology, in the near term?

Imagine that we are interested in identifying the mutations influencing how an individual responds to the anti-convulsant lamotrigine. Obvious candidate genes that might influence response to lamotrigine include the relevant DMEs CYP3A4, CYP2C8, CYP1AC, and UGT1A4. Also, as the presumed major mode of action of lamotrigine is interference with the sodium channels, other candidates include the eight genes that encode the sodium channel and that are expressed in the brain. Although these genes in total represent over a megabase of genomic sequence, the apparent simplicity of haplotype diversity in human populations means that large numbers of cases and controls could be screened with the resources available in typical project grants, once the haplotype structure of the genes has been determined. The haplotype structure for the genes could probably also be determined in the context of a project or program grant, and once done could be used for the analysis of any clinical material. The approach that appears most efficient to us, therefore, is to determine haplotype structure first in random samples from the populations of interest, and then to use this to identify appropriate SNPs that are sufficient to represent the desired amount of haplotype diversity. Then this subset of SNPs would be typed in clinical cases to look for association between particular haplotypes and drug response. An advantage of this approach is that genes such as those encoding the sodium channel and DMEs will be relevant not only to the condition under study, such as variable response to anti-convulsants, but to a broad range of conditions. The question of exactly how much of the total haplotype diversity present in a population would need to be represented by the SNPs selected will depend in large part on the genetic control of variable drug response, and in particular the frequency distribution of causal mutations. As these are largely unknown, the program will clearly need to be re-calibrated regularly as we learn more about these things. Given the relative ease of carrying out such studies now, however, we expect to see a great number them for lamotrigine and other drugs in the coming months.

16.3 Conclusions

We have outlined a framework for genetic association studies in neurological disease that can be systematically applied with current technologies to relatively large sets of genes that are candidates for influencing drug response. Looking slightly farther ahead, the NIH and the Wellcome Trust and other bodies are currently considering plans to define the haplotype structure of the entire genome in multiple populations. This is likely to be achieved within the coming years, and it will obviate a focus on candidate genes, allowing the entire genome to be screened for mutations influencing drug response. In other words, the genetic and population-genetic side of genetic association studies for variable drug response are rapidly coming of age. This means that there is an increasingly urgent need to establish appropriate structures for the identification of large numbers of patients with different response profiles to a broad range of drugs, and for the enrollment of fully consented volunteers into genetic studies.

It will only be by fully combining these clinical and genetic aspects that we will be able to begin delivering the real aims of the human genome project: more effective medicines and treatments for the most common diseases.

16.4 References

1 Phillips K, Veenstra D, Oren E, Lee J, Sedee W. Potential role of pharmacogenomics in reducing adverse drug reactions. JAMA **2001**; 286:2270–2273.
2 Perrucca E, Richens A. Antiepileptic drugs. In: General Principles in Biotransformation (Levy RH, Mattson RH, Eds.). New York: Raven Press; **1995**, 31–50.
3 Catalano M. Psychiatric genetics: the challenges of psychopharmacogenetics. Am J Hum Gent **1990**; 65:606–610.
4 Ratcliffe CF, Qu Y, McCormick KA, Tibbs VC, Dixon JE, Scheuer T, Catterall WA. A sodium channel signaling complex: modulation by associated receptor tyrosine phosphatase beta. Nature Neurosci **2000**; 3:437–444.
5 Exner DV, Dries DL, Domanski MJ, Cohn JN. Lesser response to angiotension-converting enzyme inhibitor therapy in black as compared with white patients with left ventricular dysfunction. N Engl J Med **2001**; 344:1351–1357.
6 Anonymous. American College of Cardiology, Advocacy Weekly. March 19, **2001**.
7 Schwartz RS. Racial profiling in medical research. N Engl J Med **2001**; 344:1392–1393.
8 Cavalli-Sforza LL, Menozzi P, Piazza A. The History and Geography of Human Genes. Princeton: Princeton University Press; **1994**.
9 Wilson JF, Weale ME, Smith AC, Gratrix F, Fletcher B, Thomas MG, Bradman N, Goldstein DB. Population genetic structure of variable drug response. Nature Genet **2001**; 29:265–269.
10 Pritchard JK, Stephens M, Donnelly P. Inference of population structure using multilocus genotype data. Genetics **2000**; 155:945–959.
11 Weiss KM, Terwilliger JD. How many diseases does it take to map a gene with SNPs? Nature Genet **2000**; 26:151–157.
12 Chakravarti A. Population genetics – making sense out of sequence. Nature Genet **1999**; 21:56–60.
13 Wright, AF, Carothers AD, Pirastu M. Population choice in mapping genes for complex diseases. Nature Genet **1999**; 23:397–404.
14 Wilson J, Goldstein DB. Consistent long-range linkage disequilibrium in a Bantu-Semitic hybrid population. Amer J Hum Genet **2000**; 67:926–935.
15 Daly MJ, Rioux JD, Schaffner SF, Hudson TJ, Lander ES. High-resolution haplotype structure in the human genome. Nature Genet **2001**; 29:229–232.
16 Jeffreys AJ, Kauppi L, Neumann R. Intensely punctuate meiotic recombination in the class II region of the major histocompatibility complex. Nature Genet **2001**; 29:217–222.
17 Goldstein, DB. Islands of linkage disequilibrium. Nature Genet **2001**; 29:109–111.

17
Pharmacogenomics of Neurodegenerative Diseases: Examples and Perspectives

Philippe Hantraye, Emmanuel Brouillet and Christian Néri

Abstract

The pathogenesis of human neurodegenerative diseases remains poorly understood. The application of pharmacogenomics to neurodegenerative diseases, therefore, requires a better understanding of the etiopathogenic mechanisms underlying these illnesses. This is expected to foster the identification of novel genetic regulators of disease onset and progression, which may also function as modifiers of drug response. Because the characterization and quantification of human neurodegeneration is difficult, the overall perspective is that the integration of clinical, genetic, and biological information from human patients and model systems should greatly enhance pharmacogenomics of neurodegenerative disorders. Here, we will illustrate these notions by first outlining the cell death mechanisms in human neurodegenerative diseases. We will comment more specifically on Huntington's disease, a model disease for inherited disorders that are primarily associated with expanded polyglutamines in the disease proteins and that may also involve additional and secondary genetic modifiers. We will also comment on Parkinson's disease as a model disease for complex neurodegenerative disorders displaying sporadic and inherited forms. We will review the cellular mechanisms and genetic variations that influence the pathogenesis of each of these two disorders, with the hope of identifying new targets for future therapeutic and pharmacogenomic approaches to these diseases.

17.1
Mechanisms of Neuronal Death in Neurodegenerative Disorders: General Concepts

The mechanisms underlying the pathogenesis of neurodegenerative disorders may be separated in two phases:

1) an "initiation" phase during which a particular event or combination of several events provoke the disruption of cellular homeostasis through a specific molecular mechanism, and
2) an "execution" phase in which specific cell death pathways, such as excitotoxicity/necrosis, apoptosis, and/or autophagy, finally lead to cell demise.

The machinery involved in what could be considered the initiation point is probably quite specific for each neurodegenerative disorder. On the contrary, it is largely believed that neurodegenerative diseases have in common several molecular components of the machinery involved in the later phase of cell death execution.

17.1.1
Apoptosis

Apoptosis is a form of physiological cell death that plays a major role during development and organ maturation. Apoptosis is characterized by structural abnormalities including reduction of the cytoplasm, irreversible degradation of the nucleus and of its genetic material, and the relative preservation of other organelles, notably the mitochondria. It is now firmly established that apoptosis is an energy-dependent cell death process, regulated by an intracellular proteolytic cascade, primarily mediated by members of the caspase family of cystein proteases, which may cleave other caspases as well as various key intracellular target proteins, finally leading to cell destruction. Three prototypical signaling pathways for the induction of apoptosis have been described. One pathway involves death ligands (such as Fas) binding to death receptors of the TNF (tumor necrosis factor) receptor family, which, in turn, through the recruitment of adaptator proteins, lead to the ligation of pro-caspase 8 and its cleavage into three fragments, two of which forming the active caspase 8 (the form with proteolytic activity) [1]. Active pro-caspase 8 can in turn cleave the zymogen form of downstream caspases (named effector caspases) such as caspase 3. Finally, activated effector caspases can degrade a variety of intracellular proteins. Caspase 8 can also cleave proteins other than effector caspases such as Bid, which once cleaved can directly disturb the mitochondrial membrane, further amplifying the apoptotic process. A second important apoptotic pathway is controlled by the mitochondrion itself [2], and involves the apoptosis protease activating factor-1 (Apaf-1), cytochrome c, and the mitochondrially-localized caspase 9. Apoptotic stimuli can either alter the permeability of the outer membrane of the mitochondria, allowing the release of small proteins, or induce a severe loss of mitochondrial membrane potential. Loss of membrane potential (permeability transition) involves alteration of a macromolecular complex, the permeability transition pore. During the early phases of apoptosis, the permeability transition pore will dissipate the proton gradient. Among the most important proteins released from the mitochondria in the early phases of apoptosis, cytochrome c and the apoptosis inducing factor (AIF) play a major role. On one hand, the release of AIF leads to nuclear DNA degradation (diffuse cleavage leading to large breakdown products) without primary involvement of caspases [3]. On the other hand, once released, cytochrome c interacts with Apaf-1, together with ATP, then binds to pro-caspase 9, forming the so-called “apoptosome” complex. In the apoptosome, caspase 9 is first activated (its proteolytic activity is increased by a factor 1,000 as compared to the zymogene form) to activate in turn caspase 3 and other downstream caspases (6 and 7) [4]. Many proteins are substrates of effector caspases, among which various proteins known to be involved in neurodegenerative diseases such as actin, tubulin, presenilin, amyloid precursor protein, huntingtin (htt), poly-ADPribosyl transferase, fodrin, lamin, and the inhibitor of caspase activated DNAse (ICAD). The ICAD protein is an important target of caspase 3 [5]. In normal cells, ICAD sequesters CAD (caspase activated DNAse) within the cytoplasm. Caspase 3 activation during apoptosis leads to cleavage of ICAD, allowing CAD to enter the nucleus, leading to the inter-nucleosomal fragmentation of the DNA molecule, a hallmark of "*bona fide*" apoptosis. More recently, an endoplasmic reticulum apoptotic pathway mediated by caspase 12 has also been described that may contribute to β-amyloid neurotoxicity [6].

The apoptotic cascade is tightly regulated by proteins of the Bcl-2 family [7]. The integrity of the mitochondria can thus be protected by anti-apoptotic members of the Bcl-2 protein family (such as Bcl-2 and Bcl-X_L). In contrast, pro-apoptotic members of the Bcl-2 protein family such as Bax, Bak and Bad may activate apoptosis through pore formation or antagonistic effects towards Bcl-2 and Bcl-X_L at the level of the mitochondria. Inhibitors of apoptosis proteins (IAPs) can also inhibit the rate of apoptosis [8]. Although the biochemical mechanism that underlies the suppression of apoptosis by IAP family proteins remains largely controversial, human IAPs (XIAP, cIAP-1 and cIAP-2) have been reported to interfere directly with cytochrome c-mediated activation of caspases and/or activation of caspase 8. More recently, SMAC/DIABLO, a protein localized in the intermembrane space of the mitochondria, has been shown to trigger apoptosis after release into the cytoplasm by antagonizing the "tonic" antiapoptotic effects of IAP [9, 10]. In neurons, ionic deregulation can also facilitate apoptosis. For example, in certain situations where energy production is partially preserved, Ca^{2+} overload leads to apoptosis [11]. In neuronal death triggered by staurosporine or following trophic factor withdrawal, K^+ currents actively participate in neurodegeneration [12]. In addition to these levels of regulation, transcriptional regulation is also likely to intervene, although the stochiometry of these regulatory phenomena remains to be outlined precisely. Under certain circumstances, it has been shown that increased expression of "pro-apoptotic" proteins such as Bax can be an early triggering event of apoptosis. In this context, immediate early gene (IEG) regulation through cell signaling pathways such as ras/MAP kinase pathways can modulate cell survival through transcriptional activity. Other mechanisms of apoptosis regulation are present at the level of protein phosphorylation. For example, under normal circumstances, the pro-apoptotic protein Bad is sequestrated outside the mitochondrial membrane as a result of its phosphorylation by the kinase Akt and cAMP-dependent protein kinase PKA [13, 14]. Under pathological circumstances, Akt is inactivated, and Bad is no longer phosphorylated and, as a consequence, relocalizes in the outer mitochondrial membrane, initiating apoptosis.

In summary, it appears that the apoptosis pathway can be regulated at various levels (the aforementionned description is far from being exhaustive), while many proteins and cell systems involved in this regulation remain to be discovered. Whereas the cascade of intracellular events implicated in experimentally-induced apoptosis has become more and more elucidated, it is worth noting that only few direct evidences arguing for the presence of an actual apoptosis in the brain of patients with neurodegenerative diseases have been published to date [15]. Therefore, the majority of experimental evidence in favor of a possible involvement of apoptosis in neurodegenerative diseases comes from observations obtained in cell culture systems or transgenic animals in which blockade of caspases can have, under certain circumstances, apparent beneficial effects [16].

17.1.2 Excitotoxicity: Direct and Indirect Activation of Glutamate Receptors

Structurally, excitotoxicity has been generally described as a necrotic process involving initial swelling of the cell and of the endoplasmic reticulum, clumping of chromatin, followed by swelling of mitochondria and vacuolization and eventually disruption of plasma membrane and leakage of the intracellular contents. Excitotoxicity is thought to be the main mechanism of neuronal death in various acute pathological conditions such as hypoxia–ischemia and head trauma [17, 18]. In these conditions, massive increases in extracellular glutamate concentrations are known to produce deleterious effects on neuronal cells through the interaction of the excitatory neurotransmitter with selective membrane receptors. Two main types of glutamate receptors have been described, the metabotropic receptors (receptors coupled to G proteins) and the ionotropic receptors (receptors coupled to cation channels). The ionotropic glutamate receptors can be pharmacologically divided into three major types named after their selective agonists (rigid chemical analogs of glutamate): N-methyl-D-aspartate (NMDA), α-amino-3-hydroxy-5-methyl-4-isoxazolepropionate (AMPA) and kainate receptors. Various experimental observations indicate that overactivation of any of these receptors by high concentrations of glutamate can initiate a series of events called the "excitotoxic cascade", whose main features can be summarized as follows.

- First, an initial swelling of the cell occurs which greatly depends on the extracellular concentrations of Na^+ and Cl^-.
- The second phase of the process involves a massive increase in intracellular calcium concentrations. Interestingly, the calcium entry through NMDA receptors is particularly toxic, with one immediate pathogenic target of calcium entry being the mitochondrion [19].
- The last phase of the excitotoxic cascade involves the activation of various biochemical pathways, among which phospholipases, proteases (in particular calpain), kinases and calmodulin-regulated enzymes such as nitric oxide synthase (NOS) play a prominent role.

Most likely, increased production of free radicals may also contribute to a cell's demise.

The source of free radicals is multiplied under these circumstances, arachidonic acid metabolism, activation of xanthine oxidase, perturbation of electron flow within the respiratory chain, and NOS activation. Structurally, excitotoxicity is generally described as a necrotic process involving initial swelling of the cell and of the endoplasmic reticulum, clumping of chromatin, followed by swelling of the mitochondria, vacuolization and eventually disruption of plasma membrane with leakage of intracellular contents.

A number of *in vitro* studies have shown that impairment in energy metabolism can result in a secondary excitotoxic insult without necessarily being accompanied by changes in extracellular glutamate concentrations [20] (for review, see [21]). Thus, energy compromise could indirectly activate the excitotoxic cascade *in vitro*. Several animal studies examining the mechanism of toxicity of a variety of mitochondrial inhibitors hampering oxidative metabolism have demonstrated that this phenomenon can also occur *in vivo* [22]. One hypothesis explaining this phenomenon is that the partial membrane depolarization resulting from energy depletion can release the voltage-dependent magnesium block of the NMDA receptor, leading to increased probability of calcium channel opening in the presence of physiological extracellular glutamate concentrations.

Involvement of direct or indirect excitotoxicity in neurodegenerative disorders such as Huntington's disease, Parkinson's disease, and Alzheimer's disease has long been postulated on the basis of several lines of circumstantial evidence [23]. In particular, it has been demonstrated that experimental overactivation of NMDA receptors using glutamate agonists or mitochondrial toxins in laboratory animals strikingly reproduces neurochemical, histological and behavioral abnormalities highly reminiscent of Huntington's disease (HD) (for review, see [21]). The hypothesis that a mitochondrial defect may have a causal role in HD is also supported by robust studies demonstrating mitochondrial abnormalities in tissue samples from HD patients [24–27]. Interestingly, the preferential susceptibility of the striatum to mitochondrial toxins acting at complex II in both rodents and primates supports the hypothesis that a mitochondrial function specific to the striatum may underlie the preferential vulnerability of this anatomical region in HD [28–30].

Similarly, abnormalities in the mitochondrial machinery and resulting oxidative stress may also intervene in Parkinson's disease (PD) [31, 32]. The decreased activity of mitochondrial complex I in PD patients [33], and the preferential toxicity of the complex I inhibitor rotenone [34] and MPP+ (the active metabolite of MPTP) [32] towards substantia nigra compacta in animal models of the disease, support this view. The involvement of a phenomenon highly reminiscent of "indirect excitotoxicity" in MPP+ induced neurotoxicity has been suggested almost 10 years ago [35] and has been confirmed recently [32], suggesting that an excitotoxic cascade could also play a role in PD's neurodegenerative process.

Even in Alzheimer's disease (AD), the possible involvement of a weak excitotoxic process cannot be ruled out. Indeed, mitochondrial abnormalities (such as cytochrome oxidase alterations) [36, 37] and increased levels of markers of oxidative stress [38] have been reported in AD. This has been the rationale for testing the NMDA antagonist memantine in Alzheimer's dementia [39].

17.1.3
Autophagy

Autophagy is a process which involves the bulk degradation of cytoplasmic components by the lysosomal/vacuolar system. This mechanism of cell degeneration appears to be highly conserved from yeast through mammalian cells [40]. When autophagy is induced under nutrient starvation conditions in yeast, an autophagosome is formed in the cytosol. The outer membrane of the autophagosome fuses with the vacuole, releasing the inner membrane structure, an autophagic body, into the vacuole. The autophagic body is subsequently degraded by vacuolar hydrolases. An increasing number of yeast and mammalian homolog proteins playing a role in the regulation of this process has been identified. One critical initial cellular alteration known to initiate autophagy in mammalian cells is the phosphorylation of the ribosomal protein S6. The S6 phosphorylation can, in turn, trigger the inactivation of the upstream kinase mTor. The subsequent orchestrated autophagic cascade relayed by complex sequences of protein–protein interactions and protein dephosphorylations (e.g., Apg13) will finally control the formation of autophagosomes and regulate their sizes. As a result, autophagy can be seen as a pathological mechanism whereby cells digest themselves from within. Little is known about autophagy in neurons but some studies suggest that this cell death process could be involved in pathological conditions such as PD [41], bovine spongiform encephalopathy [42, 43] and possibly in the etiology of AD [44] and HD [45].

17.1.4
Pharmacogenomics of Cell Death Pathways: Mechanisms from Cell to Brain

As mentioned above, apoptosis, necrosis and/or autophagy can theoretically be involved in the cell death process of many neurodegenerative disorders through alteration of cellular functions such as transcription, intracellular routing, protein folding and degradation, intracellular organelle trafficking, energy metabolism, free radicals scavenging, ion homeostasis and regulation of enzymatic activities (e.g., by phosphorylation). In addition to these "molecular" components taking place at the cellular level, neurodegeneration may also involve more complex functional alterations, at the level of neuronal circuitry. For example, several models of striatal degeneration reminiscent of HD demonstrate that cerebral cortex (i.e., corticostriatal glutamatergic afferents to the striatum) participates in the excitotoxic death of striatal neurons. A similar involvement of nigrostriatal dopaminergic neurons of the substantia nigra compacta has been described in the excitotoxic-like striatal lesion induced by the mitochondrial toxin 3NP [46]. Paradoxically, the presence of striatal cells is also necessary for the long-term survival of cerebral cortex. Experimental lesions of the striatum can produce delayed cell death within selected areas of the cerebral cortex known to project axons to the striatum. The mechanisms of orthograde and trans-synaptic degeneration are not fully elucidated but likely involve the machinery related to synaptic transmission, synaptic plasticity, and trophic factors. Such mechanisms of neuronal death specifically associated with a progressive loss of neuronal connections are likely to play a role in the interindividual variability of disease phenotypes, adding a level of complexity when designing new potential therapeutics for these neurodegenerative disorders.

Our knowledge of the cell death machinery possibly implicated in the pathogenesis of neurodegenerative diseases has significantly improved in the past two decades and consequently, the number of new potential therapeutic targets implicated in the cell death cascade has increased proportionally. In the context of pharmacogenomics, it is highly likely that many of these molecular components or, at a larger scale, cellular compartments and structural or functional neuronal networks involved in neurodegeneration would show a certain degree of polymorphism. These variations are likely to play a role in the inter-individual differences in susceptibility to various toxic insults as well as in the efficacy of new therapeutic strategies. In this context, pharmacogenomics has to identify the presence of polymorphisms in the various components of these cell death cascades (Fig. 17.1) in order to be effective.

17.2 Neurodegenerative Diseases

Pharmacogenomics is a general concept that may strongly apply to several neurodegenerative diseases, from the most common forms of illnesses such as AD or PD, to the less represented disorders such as amylotrophic lateral sclerosis, HD, and Creutzfeld–Jakob's disease. The number of individuals affected by common neurodegenerative diseases such as AD is expected to increase significantly. In addition, if a drug would be available to slow down disease progression, individuals affected with neurodegenerative diseases that occur relatively late in life are expected to need a chronic treatment. Therefore, drug response is likely to be a significant aspect of neurodegenerative disease therapy [47–49]. We elected to comment on a relatively rare disease, namely HD, since progresses recently made on understanding its pathogenesis qualify this illness as a "model disease". In addition, we selected PD as an example of a common neurodegenerative disorder involving genetic and non-genetic factors.

17.2.1 Huntington's Disease

Expanded polyglutamine repeats have been proposed to cause neuronal degeneration in Huntington's disease (HD) and in other disorders such as spinocerebellar ataxia I, dendatorubral pallidoluysian atrophy (DRPLA), and spinal and bulbar muscular atrophy [50]. HD is a dominant neurodegenerative disorder characterized clinically by motor abnormalities, cognitive impairment, and psychiatric disturbances [51], and caused by polyglutamine (polyQ) expansion tract in huntingtin (htt), a primarily cytoplasmic and ubiquitously expressed protein of unknown

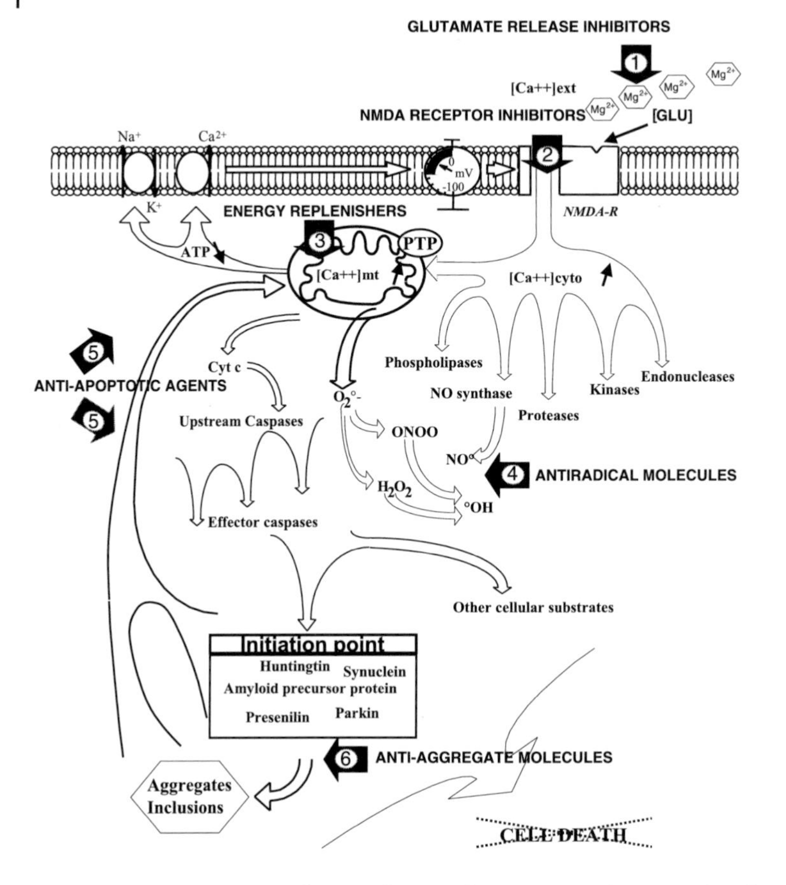

Fig. 17.1 Pharmacogenomics of neuronal death mechanisms. Important pathways of neuronal death are schematically represented with potential therapeutic targets. In many neurodegenerative disorders, mutations in specific sets of proteins (huntingtin, parkin, amyloid precursor protein) trigger initial perturbation of cell homeostasis ("Initiation point"). Abnormal processing, trafficking and degradation of these proteins or some of their partner proteins generate the formation of abnormal aggregates or inclusions which may also participate in initiating cell demise. One interesting working hypothesis, is that initiation points lead to dysregulation of mitochondrial function (e.g., through transcription effects possibly) which would in turn generate a number of self-amplifying vicious cycles within the cell and, finally, activation of the "execution" phase. The aim of this comprehensive picture (however, far from being exhaustive) is to show examples of molecular components involved in three prototypical pathways of neuronal death, downstream of initiation points, including (from left to right): apoptosis, oxidative stress, and excitotoxicity. The "scenario" of events leading to cell death is schematically shown by empty arrows. The three pathways represented are not clearly separated in purpose, since cross talks can occur between these cascades of deleterious events. For instance, overactivation of glutamate receptors can trigger the excitotoxic cascade where Ca^{++} overload activates a number of enzymes which produce cell demise. One target of excitotoxicity is the mitochondrion, where perturbation may in certain circumstances trigger the activation of the cystein protease caspases, the main effectors of the apoptotic cascade. Mitochondria can be seen as a major source of free radical species in cell death. It is noticeable that, once the ex-

function [52, 53]. The protein htt contains a proline-rich region adjacent to the polyQ tract and is thought to interact with a large number of other proteins such as SH3 domain and WW domain proteins [54, 55], as well as components of the neuronal cytoskeleton [56, 57], components of protein complexes involved in intracellular transport [58], and transcription factors (see below). As shown by gene inactivation studies in mice, htt is required for neurogenesis during development [59–61] and for neuronal function and survival in the adult brain [62]. The polyQ size in HD patients inversely correlates with the age-of-onset and severity of symptoms [51], resulting in selective neuronal loss within the basal ganglia, notably within the striatum and various cortical areas [63]. Neuronal intranuclear inclusions, immunopositive for the mutated disease protein, have become the neuropathological hallmark of polyglutamine neurodegenerative diseases. However, the actual cytotoxicity of intranuclear inclusions remains highly controversial. As illustrated by studies of HD patients, nuclear inclusions might not be essential to the occurrence of cell death [64, 65]. In transgenic mice models, the situation is still unclear. In transgenic mice that express htt exon 1 product [66, 67], the appearance of ubiquitinated neuronal intranuclear inclusions (NIIs) containing truncated polyQ-expanded htt before the onset of neurological symptoms has suggested that NIIs may be toxic to neurons [68]. However, a cellular model for HD has suggested that the translocation of soluble polyQ-expanded htt cleavage products in the nucleus might be required in order to produce neuronal death [69]. Studies on transgenic mice that express full-length htt [70, 71] have also suggested that polyQ aggregates may not be essential to the initiation of neuronal death. Neuropil aggregates may be cytotoxic as they are detected in HD patients and mice models [64, 72], and not in normal subjects. It has been hypothesized that a primary mechanism leading to cell death in HD patients might be the alteration of htt conformation when the protein is mutated [73]. The importance of a misfolding of polyQ-expanded proteins has been illustrated by the ability of mo-

---

Fig. 17.1 (continued) ecution phase is triggered, it may amplify the starting points. Large black numbered arrows show identified therapeutic targets, which may be of interest for neurodegenerative diseases. Some of these targets are the rationale for experimental studies in animal models or clinical trials in patients (from 1 to 4). Other targets (5 and 6) remain purely experimental. The various strategies can be summarized as follows: **1**, chronic and partial blockade of glutamate release by using compounds such as riluzole and lamotrigine; **2**, direct blockade of NMDA receptors with antagonists such as remacemide or memantine; **3**, improvement of energy metabolism by using substrate supplementation with agents such as creatine; **4**, decrease of free radical by using scavenging compounds or inhibitors of enzymes (e.g., nitric oxide syntase) involved in radical species production; **5**, anti-apoptotic molecules blocking upstream or downstream apoptotic events; **6**, anti-aggregates strategies. It is highly probable that polymorphisms may exist in the molecular components involved in cell death pathways. This may influence the vulnerability of individuals to neurodegenerative diseases. Similarly, the efficacy of agents acting at the various therapeutic targets depicted above may show variability on a patient-to-patient basis. In this context, pharmacogenomics may help to optimize future treatments for neurodegenerative disorders.

lecular chaperones in reversing polyQ-induced neuronal toxicity in transgenic *Drosophila* models [74] as well as ameliorating muscular toxicity in a transgenic *C. elegans* model [75]. Besides aggregate formation, the altered conformation of soluble polyglutamine-expanded disease proteins may also result in the abnormal interaction of htt with proteins essential for neuronal survival. These essential proteins may in turn show an abnormal cellular distribution, further contributing to cellular toxicity. One example of a putative direct toxicity of soluble mutated htt is provided by abnormal interactions of htt with WW domain proteins [76]. Another interesting example is provided by the decreased ability of mutated htt to bind the postsynaptic protein PSD-95 and inhibit glutamate-mediated excitotoxicity [77]. Recent data suggest that expanded polyQ may also lead to aberrant transcriptional regulation [78] through unappropriate interaction with, and subsequent abnormal expression (cellular depletion, accumulation, mislocalization) of cellular transcription factors such as the TATA-binding protein (TBP) [79], the co-repressor N-Cor [80], mSin3a, the cAMP-responsive element-binding protein (CREB)-binding protein (CBP), p53 [81, 82], TAFII-130 [83], and the co-activator CA150 [84]. In some instances, these transcription factors contain a normal polyglutamine stretch (TBP, CBP) that may interact with expanded polyQ, resulting in the redistribution of these transcription factors to nuclear and cytoplasmic aggregates. Studies conducted on CBP [85] have provided detailed insight into the molecular basis for the cellular depletion of transcription factors containing a normal polyglutamine stretch, and shown the relevance of this mechanism to several polyglutamine expansion disorders. Expanded polyQ present in htt and the DRPLA (atrophin-1) protein can interact with the short polyglutamine tract of CBP and lead to the redistribution of this co-activator away from its normal location in the nucleus into aggregates, resulting in abnormal transcription. The redistribution of CBP is dependent on the presence of CBP's polyglutamine tract since a mutated form of CBP without polyQ was not sequestered from the nucleus.

Data reported in cases of htt and atrophin-1 highlight a unifying mechanism of indirect toxicity for nuclear inclusions in polyglutamine expansion diseases through effects on the localization of transcription factors that contain a short polyglutamine repeat. However, it remains to be determined whether this represents an early or late pathogenic event in different polyglutamine diseases. This observation raises another question: how can an indirect toxicity of nuclear inclusions be reconciled with an apparent lack of a strong correlation between nuclear inclusions and the occurrence of cell death? The formation of nuclear inclusions is likely to be a progressive phenomenon modulated by several parameters such as the length of the polyglutamine tract in the mutated protein, and the length of the disease protein as it is processed by the neuronal cell [86]. One possibility is that multiple (transcription factor) proteins with a different susceptibility to the sequestering effect of expanded polyQ are likely to be affected during this process, some of them not or partially trapped into inclusions. The model of indirect loss-of-function of transcription factors points to interactions between expanded polyQ and proteins with short polyQ as targets for potential therapeutics. Validation studies in animal models of polyglutamine toxicity will tell if the overexpression of CBP, or the manipulation of other transcriptional modulators with an abnormal expression pattern in polyglutamine expansion disorders such as N-Cor [80] or CA150 [84] might constitute a viable therapeutic approach. It will also be interesting to learn about the phenotypes observed in adult mice with loss-of-function of these proteins in the nervous system. Gene expression monitoring using microarrays for animal models of polyglutamine toxicity may also reveal up- and down-regulated genes as potentially interesting targets. The consequence of polyQ expansion on transcription in HD appears to be 2-fold: relocalization of transcription factors to unappropriate cellular compartments [80, 84, 85], and loss-of-transcription for genes essential to striatal and cortical neurons as normally mediated by wild-type htt [87]. An additional implication of transcription factor abnormalities in polyglutamine expansion disorders might be the genetics and pharmacogenomics of these disorders. Aggregate formation is likely to be a progressive process tightly regulated [88] and reversible [89]. Sequence polymorphims of proteins that regulate or are affected by aggregation might modify the cytotoxicity of expanded polyQ and translate into a genetic effect [84]. Thus, whereas 70% of the variability in HD onset age can be attributed to the size of the CAG repeat in the htt gene [90], various studies have also shown that rare polymorphisms can significantly affect HD disease phenotype from patient to patient. Two independent studies have pointed to the association of a rare, untranslated repeat allele of the kainate receptor gene GluR6 with younger onset age of HD [90, 91]. A third study has shown that a rare repeat allele corresponding to a shortly extended $(\text{Gln-Ala})_{38}$ region of the transcriptional coactivator CA150 – carrying three additional amino acids that may be Gln or Ala – may account for a small proportion of the variability in HD onset age [84]. The small magnitude of the CA150 effect may be in part attributable to the low frequency (4.4%) of genotypes with an extended allele, and may correspond to a significant variation in CA150 properties. In human HD brain tissues, CA150 appears to accumulate not only within, but also around nuclear inclusions. In this case, very short expansions of polyalanines are suspected to be more toxic to cells than polyQ expansions [92], and one to three additional alanines in the Gln-Ala repeat of CA150 may significantly modify the biochemical properties of this protein, leading to a more severe relocalization of CA150 in the presence of mutated htt. The influence of the extended CA150 allele on HD onset age would, therefore, be consistent with a more severe accumulation of CA150 as mediated by stronger binding to mutated htt.

It is likely that genetic modifiers of HD will result in multiple small effects such as the ones of GluR6 [90, 91]. In addition, small genetic effects may actually correspond to significant biological effects such as CA150 accumulation in HD brain tissues [84]. It may be of interest to test whether the genes that encode proteins with an abnormal localization following their interaction with mutated htt may constitute a useful source of genetic markers (SNP or microsatellite markers) for understanding the pharmacogenomics of HD in relation to anti-aggregate therapeutic strategies.

17.2.2 Parkinson's Disease

PD is a common neurodegenerative disorder that affects 1% of the population above age 65 and is clinically characterized by tremor, rigidity, and bradykinesia. The cause of sporadic PD is unknown but is likely to involve several pathogenic components including genetic susceptibility [93, 94] as well as environmental factors [95, 96]. Neuropathologically, PD is associated with a selective loss of neurons of the substantia nigra that connect mesencephalic dopaminergic nuclei to the caudate–putamen complex, which in turn results in a striatal loss of dopamine, affecting dopaminergic neurons that connect to the striatum and resulting in a loss of dopamine, dopamine metabolites, tyrosine hydroxylase (the biosynthetic enzyme of dopamine), and dopamine transporters located on striatal dopaminergic afferents [97]. PD is also characterized by the presence of cytoplasmic, round, and filamentous inclusions – the Lewy bodies – that are positive for ubiquitin [98]. Lewy neurites are also observed, and nuclear inclusions are rare.

The discovery of mutations that underlie autosomic dominant or recessive forms of PD has shed light on the pathogenesis of this complex disorder. Rare missense mutations in the *α*-synuclein gene (A53T, A30P) have been associated with an autosomic dominant form of PD [99, 100]. *α*-Synuclein is a protein of unknown function that appears to be limited to vertebrates, and is found in nerve terminals at the level of synaptic vesicles [101]. While *α*-synuclein –/– mice are viable and fertile with no Parkinsonian phenotype, they show functional deficit in the nigrostriatal dopamine system [102], suggesting that soluble wild-type *α*-synuclein is a presynaptic protein directly involved in the regulation of dopaminergic neurotransmission. Mutant *α*-synuclein is not able to bind vesicles [103], and some neurons of the substantia nigra of PD brain tissues are immunopositive for the microtubule-associated protein-2, a cytoskeleton protein primarily localized in neuronal dendrites [104], suggesting that impaired neuronal transport may contribute to the progression of neuronal loss in the brains of PD patients. The *α*-synuclein protein contains N-terminal KTKEGV consensus repeats, which may mediate binding to lipid membranes [101] and bind to and inhibit phospholipase D2 that localizes to the plasma membrane [105]. These studies suggest a function of *α*-synuclein in the partitioning of membranes between the cell surface and intracellular stores. The *α*-synuclein protein also binds to and inhibits protein kinase C [106], and the microtubule-associated proteins Tau [107], suggesting additional toxic effects of *α*-synuclein when this protein is overexpressed or mutated. Besides a partial loss-of-function of soluble *α*-synuclein when this protein carries missense mutations, autosomal dominant PD may also involve a strong gain of toxic properties resulting from the abnormal accumulation of mutated *α*-synuclein. The A53T and A30P mutations oligomerize faster than wild-type *α*-synuclein [108], and Lewy bodies appear to contain several proteins that normally bind to *α*-synuclein. For example, *α*-synuclein binds to synphylin-1 [109], a protein that co-localizes with *α*-synuclein in Lewy bodies [110], and to the microtubule-associated protein 1B, also a component of Lewy bodies [111]. Interestingly, Lewy bodies are immunopositive for α-synuclein in both sporadic and autosomal dominant PD. Studying the toxicity resulting from α-synuclein accumulation may provide insight into the mechanisms of PD in general. In autosomal dominant PD, α-synuclein aggregation appears to result directly from mutations in this protein. In sporadic PD, a consistent feature delineated by dopaminergic toxicity, induced by acute administration of 1-methyl-4-phenyl-1,2,3,4-tetrahydropyridine (MPTP) in non-human primates and by analysis of *post mortem* PD brain tissues, is a decrement of complex I and subsequent increased oxidative stress as mediated by the inducible NOS [31], which may in turn promote α-synuclein aggregation. The molecular and genetic dissection of mechanisms underlying α-synuclein accumulation and toxicity will likely benefit from the development of transgenic animal models. Overexpression of wild-type α-synuclein in mice affects dopaminergic nerve terminals [112]. The pattern of α-synuclein accumulation in these mice is atypical of PD since both nuclear and cytoplasmic inclusions are observed, and since no loss of dopaminergic neurons is detected. However, this model may be representative of an early step of α-synuclein toxicity inducing neuronal dysfunction. Interestingly, a *Drosophila* model has suggested that both overexpression of wild-type and mutated α-synuclein can cause adult-onset loss of dopaminergic neurons, with formation of filamentous intraneuronal inclusions containing α-synuclein, associated with locomotor dysfunction [113]. This model is highly amenable to the genetic dissection of intracellular pathways mediating α-synuclein toxicity. It will be interesting to know if overexpression of α-synuclein can produce highly-penetrant neuronal phenotypes in transgenic *C. elegans* since this model organism is suitable in rapid genetic and pharmacological screens.

In addition to mutations in α-synuclein, homozygous deletions and point mutations in the parkin gene are associated with early-onset autosomal recessive parkinsonism (before the age of 40), slow progression of the disease, and severe levodopa-induced dyskinesia in families of Japanese, European, and Middle East origin [114, 115]. The mechanisms that underlie this autosomal recessive form of PD remain to be understood. However, parkin has several features that make this protein relevant to Parkinsonism in general. Parkin contains two RING-finger motifs and one IBR domain and functions as a E3 ubiquitin–protein ligase [116–118]. Consistently, a significant number of point mutations in parkin are localized at the level of RING-finger motifs [115]. Interestingly, parkin and α-synuclein were recently found to interact functionally. Normal parkin is part of a protein complex that includes UbcH7 as an associated E2 ubiquitin conjugating enzyme, and a 22 kDa glycosylated form of α-synuclein (alphaSp22) as its substrate. In contrast to normal parkin, mutant parkin fails to bind alphaSp22, and alphaSp22 accumulates in a non-ubiquitinated form in parkin-deficient PD brains [119]. From this observation, a critical link between two PD-linked gene products as well as a unified mechanism for the accumulation of α-synuclein in conventional PD can be delineated.

The following also points to a dysregulation of the ubiquitin–proteasome pathway as an important aspect of PD pathogenesis: rare mutations in ubiquitin carboxy-terminal hydrogenase L1 (UCH-L1) associated with autosomal dominant PD in a German family [120], a UCH-L1 polymorphism associated with the development of sporadic PD [121], and the impairment of proteasome activity by expression of mutant α-synuclein in neuronal cells [122]. In familial forms of PD, the dysregulation of the ubiquitin–proteasome pathway may constitute a critical mechanism that occurs as a direct consequence of harmful mutations in α-synuclein, parkin, and maybe other biologically-related genes that remain to be identified on chromosomes 2p13 and 4p [123–125]. In sporadic PD, abnormal degradation of ubiquitin-tagged proteins may be only one of the major cellular mechanisms that contribute to neuronal degeneration. In both cases, the effector mechanisms downstream of the dysregulation of the ubiquitin–proteasome system are not clear. What are the biological links between Lewy bodies (or their upstream toxic intermediates) and cell death in PD? Which of the cellular mechanism(s) that may be associated with neurodegeneration in PD (oxidative stress, excitotoxicity, altered dopamine metabolism, deficient detoxification of metabolites) is the most critical to the onset and progression of PD? Investigating the role of transcription factors and their transcriptional targets in dopaminergic neurons may provide important clues in order to better answer these questions, to identify potentially interesting therapeutic points of intervention, and to select for physiological or genetic markers of therapeutic response. For example, the stress-inducible transcription factor NF-kappaB is activated in neurons in response to excitotoxic, metabolic, and oxidative stress, has been associated with neurodegeneration in HD [126] and dopamine toxicity in PD models [127, 128]. Whether NF-kappaB activation is part of the neurodegenerative process or the hallmark of neuroprotective mechanisms remains unclear. Another example comes from several *in vitro* studies suggesting that appropriate combinations of certain growth and transcription factors are able to promote the induction of a dopaminergic phenotype in neural stem cells. The Nurr1 nuclear orphan receptor together with the homeodomain transcription factor ptx-3 [129] are required for tyrosine hydroxylase expression [130] and for terminal differentiation of dopaminergic neurons [131–133]. The overexpression of Nurr1 together with growth factor stimulation in stem cells result in a phenotype indistinguishable from that of endogenous dopaminergic neurons [134]. Neurotrophic factors in combination with survival-promoting factors such as interleukin-1beta or glial cell line-derived neurotrophic factor, significantly enhance Nurr1 and tyrosine hydroxylase mRNA levels in pluripotent mouse embryonic stem cells [135]. The engrailed genes also appear to be involved in controlling the survival of midbrain dopaminergic neurons [136]. Altogether, these studies indicate several directions that could be explored to further understand PD pathogenesis. One path may be to characterize the regulation of transcription of PD-associated genes since their expression levels may modulate the time course of the disease (α-synuclein, parkin, others). It may be interesting to further explore the transcriptional events essential to dopaminergic function and to compare the localization of transcription factors in normal dopamine neurons with potential abnormal relocalization in dopaminergic neurons affected by the PD-associated cellular processes.

As illustrated above, there appears to be multiple perspectives on the pharmacogenomics of PD since this complex disease seems to involve several pathogenic components, and several potentially interesting genetic susceptibilities. However, as for complex neurological diseases other than PD, strong conclusions have often been impaired by the lack for replication, or the small size of the cohorts tested. One example is the polymorphism in the promoter region of the α-synuclein gene (NACP-Rep1) that revealed significant differences in the allelic distributions between PD patients and the control group [137]. In this study, a combined α-synuclein/apolipoprotein E-ε4 genotype increased the relative risk of developing PD around 12-fold. However, this observation was not confirmed in another study involving a large sample of PD cases [138]. To approach the problem of variability and validity of association studies in PD, a meta-analysis of individual gene polymorphism reported in the literature has been performed and, from 84 studies on 14 selected genes, polymorphisms in N-acetyltransferase 2 (NAT2), monoamine oxidase B (MAOB), glutathione transferase GSTT1, and a mitochondrial gene (tRNAGlu) has been identified as potential factors influencing PD development [139]. Undoubtedly, additional studies are needed to confirm the pathophysiological significance of these polymorphisms in NAT2, MAOB, and GSTT1 and their relevance to PD pathogenesis [140].

17.3 Perspectives on the Pharmacogenomics of Neurodegenerative Processes

Understanding the early biological steps of neurodegenerative disorders (the "upstream initiation phases" as defined in Sect. 17.1) may help define the molecular basis of several poorly understood aspects of these illnesses. There remains several major and unanswered questions on the pathogenesis of neurodegenerative disorders. What is the molecular basis for selective neuronal cell loss in these disorders, and what is the pathological impact, if any, of neuronal cell dysfunction (sick neurons) occurring before actual cell death? Are there critical molecular denominators (changes in protein processing, protein trafficking, and transcription regulation) common to several diseases? Are altered neuron-to-neuron signals (e.g., cortico-striatal projections in HD) critical to disease development? What is the order of appearance of intracellular changes between two different forms of the same disease or between two different diseases? Answering these questions may result in the identification of genetic variations and new therapeutics more effective in targeting a part of the disease mechanism or a particular neurodegenerative disorder.

Genes that appear to significantly influence the risk of developing a disease are usually considered good potential markers of response to future drugs that may block their activities, or act on the pathways to which they belong [49]. This is expected to be fully exact only if, within a group of affected individuals showing homogeneous clinical and neuropathological phenotypes, the biological impact and genetic penetrance of a given polymorphism are of equal strengths. In other words, when there is a strong difference between the genetic and biological influence of gene sequence polymorphisms (weak genetic penetrance and strong biological effect or strong genetic penetrance and weak biological effect) the prediction on potentially interesting pharmacogenomic and therapeutic targets will likely be under- or over-estimated. For example, weak genetic penetrances (because of a rare allele, or sample heterogeneity) may correspond to a relatively strong biological consequence as illustrated by studies on the protein α-synuclein in PD. Therefore, predicting the pharmacogenomics of neurodegenerative diseases might best rely on the biology rather than on the current knowledge of genetic associations.

The properties of a drug (mode of action, metabolizers) is also an important parameter, and it seems difficult to speculate on the polymorphisms that might control drug response "in theory", without referring to a given compound and its specific cellular targets. In the present case, PD is an interesting example because it differs from other neurodegenerative diseases as it has a specific treatment, namely levodopa. Levodopa, which is the most effective drug available for the symptomatic treatment of PD, is metabolized by both decarboxylase and catechol-O-methyl transferase enzymes, and is associated with adverse effects such as dyskynesias [141]. The relative contribution of dopamine D(1) and D(2) receptor function to the pathophysiology of levodopa-induced dyskinesias has not been distinguished [142]. However, genetic variation in the dopamine D(2)receptor (DRD2) gene may influence the risk of developing dyskinesias in levodopa-treated PD patients, suggesting that the DRD2 gene is a susceptibility locus for PD [143]. These studies illustrate again that understanding the mode of action of a drug is one of the strongest indicators used to understand the pharmacogenomics of a disease.

17.4 Conclusions

The ability to cure neurodegenerative diseases and to use "brain pharmacogenomics" will both require knowledge of the cellular mechanisms that cause human neuronal dysfunction and cell loss, which relies on integrating information such as processing of disease-associated proteins by the neuronal cell, pharmacological and genetic data from model systems, brain imaging data, definition of clinical and neuropathological endophenotypes, and postgenomic data. Over the past 20 years, data has been accumulating on cell death pathways in neurodegenerative diseases, providing a stronger scientific rationale for the patient-specific selection of medications. Phenomenological denominators (oxidative stress, excitotoxicity, energy deprivation) common to several neurodegenerative diseases remain putative. In contrast, common molecular denominators are evaluated such as changes in protein trafficking and cleavage, saturation or decrement in efficiency of the ubiquitin–proteasome system, and, more recently, abnormal regulation of transcription. The order of appearance and time course of critical cellular abnormalities are two important parameters of neurodegenerative disease processes, which may greatly differ when comparing inherited and sporadic forms of the same disease, or two different diseases. Studies of the changes in neuron-to-neuron and glial cell-to-neuron signaling may be essential. Pharmacogenomics of neurodegenerative diseases also relies on technologies such as using microarrays for gene expression profiling, searching for polymorphic markers in human candidate genes as they are identified, screening for therapeutics as based on current knowledge of neurodegeneration and brain development, and using bioinformatics to build and mine "*in silico* surrogates" of disease-associated pathways. Research on neuronal cell biology, studies of disease-associated protein toxicity in neuronal cells, and preclinical neuropharmacology therefore appear as three separate but complementary entities that, when integrated, will allow scientists to better test for genetic associations with disease occurrence and progression and to develop therapeutics on a patient-to-patient basis.

17.5 References

1 Ashkenazi A, Dixit VM. Science **1998**; 281:1305–1308.
2 Kroemer G, Reed JC. Nature Med **2000**; 6:513–519.
3 Susin SA, Zamzami N, Castedo M, Hirsch T, Marchetti P, Macho A, Daugas E, Geuskens M, Kroemer G. J Exp Med **1996**; 184:1331–1341.
4 Li P, Nijhawan D, Budihardjo I, Srinivasula SM, Ahmad M, Alnemri ES, Wang X. Cell **1997**; 91:479–489.
5 Enari M, Sakahira H, Yokoyama H, Okawa K, Iwamatsu A, Nagata S. Nature **1998**; 391:43–50.
6 Nakagawa T, Zhu H, Morishima N, Li E, Xu J, Yankner BA, Yuan J. Nature **2000**; 403:98–103.
7 Adams JM, Cory S. Science **1998**; 281:1322–1326.
8 Deveraux QL, Reed JC. Genes Dev **1999**; 13:239–252.
9 Du C, Fang M, Li Y, Li L, Wang X. Cell **2000**; 102:33–42.
10 Verhagen AM, Ekert PG, Pakusch M, Silke J, Connolly LM, Reid GE, Moritz RL, Simpson RJ, Vaux DL. Cell **2000**; 102:43–53.
11 Nicotera P, Orrenius S. Cell Calcium **1998**; 23:173–180.
12 Yu SP, Yeh CH, Sensi SL, Gwag BJ, Canzoniero LM, Farhangrazi ZS, Ying HS, Tian M, Dugan LL, Choi DW. Science **1997**; 278:114–117.
13 Datta SR, Dudek H, Tao X, Masters S, Fu H, Gotoh Y, Greenberg ME. Cell **1997**; 91:231–241.
14 Harada H, Becknell B, Wilm M, Mann M, Huang LJ, Taylor SS, Scott JD, Korsmeyer SJ. Mol Cell **1999**; 3:413–422.
15 Hartmann A, Hunot S, Michel PP, Muriel MP, Vyas S, Faucheux BA, Mouatt-Prigent A, Turmel H, Srinivasan A, Ruberg M, Evan GI, Agid Y, Hirsch EC. Proc Natl Acad Sci USA **2000**; 97:2875–2880.
16 Ona VO, Li M, Vonsattel JP, Andrews LJ, Khan SQ, Chung WM, Frey AS, Menon AS, Li X. J, Stieg PE, Yuan J, Penney JB, Young AB, Cha JH, Friedlander RM. Nature **1999**; 399:263–267.
17 Choi DW. J Neurosci **1990**; 10:2493–2501.
18 Lipton SA, Rosenberg PA. N Engl J Med **1994**, 330, 613–622.
19 Stout JG, Zhou Q, Wiedmer T, Sims PJ. Biochemistry **1998**; 37:14860–14866.
20 Novelli A, Henneberry RC. Eur J Pharmacol **1985**; 118:189–190.
21 Brouillet E, Conde F, Beal MF, Hantraye P. Prog Neurobiol **1999**; 59:427–468.
22 Beal MF. Ann Neurol **1995**; 38:357–366.
23 Beal MF. Trends Neurosci **2000**; 23:298–304.
24 Gu M, Gash MT, Mann VM, Javoy-Agid F, Cooper JM, Schapira AH. Ann Neurol **1996**; 39:385–389.
25 Browne SE, Bowling AC, MacGarvey U, Baik MJ, Berger SC, Muqit MM, Bird ED, Beal MF. Ann Neurol **1997**; 41:646–653.

26 Sawa A, Wiegand GW, Cooper J, Margolis RL, Sharp AH, Lawler JF Jr., Greenamyre JT, Snyder SH, Ross CA. Nature Med **1999**; 5:1194–1198.

27 Tabrizi SJ, Cleeter MW, Xuereb J, Taanman JW, Cooper JM, Schapira AH. Ann Neurol **1999**; 45:25–32.

28 Beal MF, Brouillet E, Jenkins BG, Ferrante RJ, Kowall NW, Miller JM, Storey E, Srivastava R, Rosen BR, Hyman BT. J Neurosci **1993**; 13:4181–4192.

29 Brouillet E, Jenkins BG, Hyman BT, Ferrante RJ, Kowall NW, Srivastava R, Roy DS, Rosen BR, Beal MF. J Neurochem **1993**; 60:356–359.

30 Brouillet E, Hantraye P, Ferrante RJ, Dolan R, Leroy-Willig A, Kowall NW, Beal MF. Proc Natl Acad Sci USA **1995**; 92:7105–7109.

31 Zhang Y, Dawson VL, Dawson TM. Neurobiol Dis **2000**; 7:240–250.

32 Beal MF. Nature Rev Neurosci **2001**; 2:325–334.

33 Schapira AH. Mov Disord **1994**; 9:125–138.

34 Betarbet R, Sherer TB, MacKenzie G, Garcia-Osuna M, Panov AV, Greenamyre JT. Nature Neurosci **2000**; 3:1301–1306.

35 Turski L, Stephens DN. Synapse **1992**; 10:120–125.

36 Mutisya EM, Bowling AC, Beal MF. J Neurochem **1994**; 63:2179–2184.

37 Hirai K, Aliev G, Nunomura A, Fujioka H, Russell RL, Atwood CS, Johnson AB, Kress Y, Vinters HV, Tabaton M, Shimohama S, Cash AD, Siedlak SL, Harris PL, Jones PK, Petersen RB, Perry G, Smith MA. J Neurosci **2001**; 21:3017–3023.

38 Mecocci P, MacGarvey U, Beal MF. Ann Neurol **1994**; 36:747–751.

39 Jain KK. Expert Opin Investig Drugs **2000**; 9:1397–1406.

40 Klionsky DJ, Emr SD. Science **2000**; 290:1717–1721.

41 Anglade P, Vyas S, Javoy–Agid F, Herrero MT, Michel PP, Marquez J, Mouatt-Prigent A, Ruberg M, Hirsch EC, Agid Y. Histol Histopathol **1997**; 12:25–31.

42 Jeffrey M, Scott JR, Williams A, Fraser H. Acta Neuropathol (Berl) **1992**; 84:559–569.

43 Boellaard JW, Kao M, Schlote W, Diringer H. Acta Neuropathol (Berl) **1991**; 82:225–228.

44 Cataldo AM, Hamilton DJ, Barnett JL, Paskevich PA, Nixon RA. J Neurosci **1996**; 16:186–199.

45 Kegel KB, Kim M, Sapp E, McIntyre C, Castano JG, Aronin N, DiFiglia M. J Neurosci **2000**; 20:7268–7278.

46 Reynolds DS, Carter RJ, Morton AJ. J Neurosci **1998**; 18:10116–10127.

47 Price DL, Sisodia SS, Borchelt DR. Science **1998**; 282:1079–1083.

48 Goedert M. Nature Rev Neurosci **2001**; 2:492–501.

49 Maimone D, Dominici R, Grimaldi LM. Eur J Pharmacol **2001**; 413:11–29.

50 Zoghbi HY, Orr HT. Annu Rev Neurosci **2000**; 23:217–247.

51 Harper PS. Hum Genet **1992**; 89:365–376.

52 THsDCR Group Cell **1993**; 72:971–983.

53 DiFiglia M, Sapp E, Chase K, Schwarz C, Meloni A, Young C, Martin E, Vonsattel JP, Carraway R, Reeves SA et al. Neuron **1995**; 14:1075–1081.

54 Faber PW, Barnes GT, Srinidhi J, Chen J, Gusella JF, MacDonald ME. Hum Mol Genet **1998**; 7:1463–1474.

55 Sittler A, Walter S, Wedemeyer N, Hasenbank R, Scherzinger E, Eickhoff H, Bates GP, Lehrach H, Wanker EE. Mol Cell **1998**, 2:427–436.

56 Kalchman MA, Koide HB, McCutcheon K, Graham RK, Nichol K, Nishiyama K, Kazemi-Esfarjani P, Lynn FC, Wellington C, Metzler M, Goldberg YP, Kanazawa I, Gietz RD, Hayden MR. Nature Genet **1997**; 16:44–53.

57 Wanker EE, Rovira C, Scherzinger E, Hasenbank R, Walter S, Tait D, Colicelli J, Lehrach H. Hum Mol Genet **1997**, 6:487–495.

58 Li SH, Gutekunst CA, Hersch SM, Li XJ. J Neurosci **1998**; 18:1261–1269.

59 Duyao MP, Auerbach AB, Ryan A, Persichetti F, Barnes GT, McNeil SM, Ge P, Vonsattel JP, Gusella JF, Joyner AL et al. Science **1995**; 269:407–410.

60 Zeitlin S, Liu JP, Chapman DL, Papaioannou VE, Efstratiadis A. Nature Genet **1995**; 11:155–163.

61 White JK, Auerbach W, Duyao MP, Vonsattel JP, Gusella JF, Joyner AL, MacDonald ME. Nature Genet **1997**; 17:404–410.

62 Dragatsis I, Levine MS, Zeitlin S. Nature Genet **2000**; 26:300–306.

63 Vonsattel JP, Myers RH, Stevens TJ, Ferrante RJ, Bird ED, Richardson EP Jr. J Neuropathol Exp Neurol **1985**; 44:559–577.

64 Gutekunst CA, Li SH, Yi H, Mulroy JS, Kuemmerle S, Jones R, Rye D, Ferrante RJ, Hersch SM, Li XJ. J Neurosci **1999**; 19:2522–2534.

65 Kuemmerle S, Gutekunst CA, Klein AM, Li XJ, Li SH, Beal MF, Hersch SM, Ferrante RJ. Ann Neurol **1999**; 46:842–849.

66 Mangiarini L, Sathasivam K, Seller M, Cozens B, Harper A, Hetherington C, Lawton M, Trottier Y, Lehrach H, Davies SW, Bates GP. Cell **1996**; 87:493–506.

67 Turmaine M, Raza A, Mahal A, Mangiarini L, Bates GP, Davies SW. Proc Natl Acad Sci USA **2000**; 27:27.

68 Davies SW, Turmaine M, Cozens BA, DiFiglia M, Sharp AH, Ross CA, Scherzinger E, Wanker EE, Mangiarini L, Bates GP. Cell **1997**; 90:537–548.

69 Saudou F, Finkbeiner S, Devys D, Greenberg ME. Cell **1998**; 95:55–66.

70 Reddy PH, Charles V, Williams M, Miller G, Whetsell WO Jr., Tagle DA. Philos Trans R Soc Lond B Biol Sci **1999**; 354:1035–1045.

71 Hodgson JG, Agopyan N, Gutekunst CA, Leavitt BR, LePiane F, Singaraja R, Smith DJ, Bissada N, McCutcheon K, Nasir J, Jamot L, Li XJ, Stevens ME, Rosemond E, Roder JC, Phillips AG, Rubin EM, Hersch SM, Hayden MR. Neuron **1999**; 23:181–192.

72 Li H, Li SH, Cheng AL, Mangiarini L, Bates GP, Li XJ. Hum Mol Genet **1999**; 8:1227–1236.

73 Sisodia SS. Cell **1998**; 95:1–4.

74 Warrick JM, Chan HY, Gray-Board GL, Chai Y, Paulson HL, Bonini NM. Nature Genet **1999**; 23:425–428.

75 Satyal SH, Schmidt E, Kitagawa K, Sondheimer N, Lindquist S, Kramer JM, Morimoto RI. Proc Natl Acad Sci USA **2000**; 97:5750–5755.

76 Passani LA, Bedford MT, Faber PW, McGinnis KM, Sharp AH, Gusella JF, Vonsattel JP, MacDonald ME. Hum Mol Genet **2000**; 9:2175–2182.

77 Sun Y, Savanenin A, Reddy PH, Liu YF. J Biol Chem **2001**; 276:24713–24718.

78 Lin X, Antalffy B, Kang D, Orr HT, Zoghbi HY. Nat Neurosci **2000**; 3:157–163.

79 Huang CC, Faber PW, Persichetti F, Mittal V, Vonsattel JP, MacDonald ME, Gusella JF. Somat Cell Mol Genet **1998**; 24:217–233.

80 Boutell JM, Thomas P, Neal JW, Weston VJ, Duce J, Harper PS, Jones AL. Hum Mol Genet **1999**; 8:1647–1655.

81 McCampbell A, Taylor JP, Taye AA, Robitschek J, Li M, Walcott J, Merry D, Chai Y, Paulson H, Sobue K, Fischbeck H. Hum Mol Genet **2000**; 9:2197–2202.

82 Steffan JS, Kazantsev A, Spasic-Boskovic O, Greenwald M, Zhu YZ, Gohler H, Wanker EE, Bates GP, Housman DE, Thompson LM. Proc Natl Acad Sci USA **2000**; 97:6763–6768.

83 Shimohata T, Nakajima T, Yamada M, Uchida C, Onodera O, Naruse S, Kimura T, Koide R, Nozaki K, Sano Y, Ishiguro H, Sakoe K, Ooshima T, Sato A, Ikeuchi T, Oyake M, Sato T, Aoyagi Y, Hozumi I, Nagatsu T, Takiyama Y, Nishizawa M, Goto J, Kanazawa I, Davidson I, Tanese N, Takahashi H, Tsuji S. Nature Genet **2000**; 26:29–36.

84 Holbert S, Denghien I, Kiechle T, Rosenblatt A, Wellington C, Hayden MR, Margolis RL, Ross CA, Dausset J, Ferrante RJ, Neri C. Proc Natl Acad Sci USA **2001**; 98:1811–1816.

85 Nucifora FC Jr., Sasaki M, Peters MF, Huang H, Cooper JK, Yamada M, Takahashi H, Tsuji S, Troncoso J, Dawson VL, Dawson TM, Ross CA. Science **2001**; 291:2423–2428.

86 Kazantsev A, Preisinger E, Dranovsky A, Goldgaber D, Housman D. Proc Natl Acad Sci USA **1999**; 96:11404–11409.

87 Zuccato C, Ciammola A, Rigamonti D, Leavitt BR, Goffredo D, Conti L, MacDonald ME, Friedlander RM, Silani V, Hayden MR, Timmusk T, Sipione S, Cattaneo E. Science **2001**; 14:14.

88 Diamond MI, Robinson MR, Yamamoto KR. Proc Natl Acad Sci USA **2000**; 97:657–661.

89 Yamamoto A, Lucas JJ, Hen R. Cell **2000**; 101:57–66.

90 MacDonald ME, Vonsattel JP, Shrinidhi J, Couropmitree NN, Cupples LA, Bird ED, Gusella JF, Myers RH. Neurology **1999**; 53:1330–1332.

91 Rubinsztein DC, Leggo J, Chiano M, Dodge A, Norbury G, Rosser E, Craufurd D. Proc Natl Acad Sci USA **1997**; 94:3872–3876.

92 Brais B, Bouchard JP, Xie YG, Rochefort DL, Chretien N, Tome FM, Lafreniere RG, Rommens JM, Uyama E, Nohira O, Blumen S, Korczyn AD, Heutink P, Mathieu J, Duranceau A, Codere F, Fardeau M, Rouleau GA, Korcyn AD. Nature Genet **1998**; 18:164–167.

93 Payami H, Larsen K, Bernard S, Nutt J. Ann Neurol **1994**; 36:659–661.

94 Piccini P, Burn DJ, Ceravolo R, Maraganore D, Brooks DJ. Ann Neurol **1999**; 45:577–582.

95 Masalha R, Herishanu Y, Alfahel-Kakunda A, Silverman WF. Brain Res **1997**; 774:260–264.

96 Thiruchelvam M, Richfield EK, Baggs RB, Tank AW, Cory-Slechta DA. J Neurosci **2000**; 20:9207–9214.

97 Dunnett SB, Bjorklund A. Nature **1999**; 399:A32–39.

98 Forno LS. J Neuropathol Exp Neurol **1996**; 55:259–272.

99 Polymeropoulos MH, Lavedan C, Leroy E, Ide SE, Dehejia A, Dutra A, Pike B, Root H, Rubenstein J, Boyer R, Stenroos ES, Chandrasekharappa S, Athanassiadou A, Papapetropoulos T, Johnson WG, Lazzarini AM, Duvoisin RC, Di Iorio G, Golbe LI, Nussbaum RL. Science **1997**; 276:2045–2047.

100 Munoz E, Oliva R, Obach V, Marti MJ, Pastor P, Ballesta F, Tolosa E. Neurosci Lett **1997**; 235:57–60.

101 Clayton DF, George JM. J Neurosci Res **1999**; 58:120–129.

102 Abeliovich A, Schmitz Y, Farinas I, Choi-Lundberg D, Ho WH, Castillo PE, Shinsky N, Verdugo JM, Armanini M, Ryan A, Hynes M, Phillips H, Sulzer D, Rosenthal A. Neuron **2000**; 25:239–252.

103 Jensen PH, Nielsen MS, Jakes R, Dotti CG, Goedert M. J Biol Chem **1998**; 273:26292–26294.

104 D'Andrea MR, Ilyin S, Plata-Salaman CR. Neurosci Lett **2001**; 306:137–140.

105 Jenco JM, Rawlingson A, Daniels B, Morris AJ. Biochemistry **1998**; 37:4901–4909.

106 Ostrerova N, Petrucelli L, Farrer M, Mehta N, Choi P, Hardy J, Wolozin B. J Neurosci **1999**; 19:5782–5791.

107 Jensen PH, Hager H, Nielsen MS, Hojrup P, Gliemann J, Jakes R. J Biol Chem **1999**; 274:25481–25489.

108 Conway KA, Lee SJ, Rochet JC, Ding TT, Williamson RE, Lansbury PT Jr. Proc Natl Acad Sci USA **2000**; 97:571–576.

109 Engelender S, Kaminsky Z, Guo X, Sharp AH, Amaravi RK, Kleiderlein JJ, Margolis RL, Troncoso JC, Lanahan AA, Worley PF, Dawson VL, Dawson TM, Ross CA. Nature Genet **1999**; 22:110–114.

110 Wakabayashi K, Engelender S, Yoshimoto M, Tsuji S, Ross CA, Takahashi H. Ann Neurol **2000**; 47:521–523.

111 Jensen PH, Islam K, Kenney J, Nielsen MS, Power J, Gai WP. J Biol Chem **2000**; 275:21500–21507.

112 Masliah E, Rockenstein E, Veinbergs I, Mallory M, Hashimoto M, Takeda A, Sagara Y, Sisk A, Mucke L. Science **2000**; 287:1265–1269.

113 Feany MB, Bender WW. Nature **2000**; 404:394–398.

114 Kitada T, Asakawa S, Hattori N, Matsumine H, Yamamura Y, Minoshima S, Yokochi M, Mizuno Y, Shimizu N. Nature **1998**; 392:605–608.

115 Abbas N, Lucking CB, Ricard S, Durr A, Bonifati V, De Michele G, Bouley S, Vaughan JR, Gasser T, Marconi R, Broussolle E, Brefel-Courbon C, Harhangi BS, Oostra BA, Fabrizio E, Bohme GA, Pradier L, Wood NW, Filla A, Meco G, Denefle P, Agid Y, Brice A. Hum Mol Genet **1999**; 8:567–574.

116 Imai Y, Soda M, Takahashi R. J Biol Chem **2000**; 275:35661–35664.

117 Shimura H, Hattori N, Kubo S, Mizuno Y, Asakawa S, Minoshima S, Shimizu N, Iwai K, Chiba T, Tanaka K, Suzuki T. Nature Genet **2000**; 25:302–305.

118 Zhang Y, Gao J, Chung KK, Huang H, Dawson VL, Dawson TM. Proc Natl Acad Sci USA **2000**; 97:13354–13359.

119 Shimura H, Schlossmacher MG, Hattori N, Frosch MP, Trockenbacher A, Schneider R, Mizuno Y, Kosik KS, Selkoe DJ. Science **2001**; 28:28.

120 Lincoln S, Vaughan J, Wood N, Baker M, Adamson J, Gwinn-Hardy K, Lynch T, Hardy J, Farrer M. Neuroreport **1999**; 10:427–429.

121 Wintermeyer P, Kruger R, Kuhn W, Muller T, Woitalla D, Berg D, Becker G, Leroy E, Polymeropoulos M, Berger K, Przuntek H, Schols L, Epplen JT, Riess O. Neuroreport **2000**; 11:2079–2082.

122 Tanaka Y, Engelender S, Igarashi S, Rao RK, Wanner T, Tanzi RE, Sawa A, Dawson TM, Ross CA. Hum Mol Genet **2001**; 10:919–926.

123 Farrer M, Gwinn-Hardy K, Muenter M, DeVrieze FW, Crook R, Perez-Tur J, Lincoln S, Maraganore D, Adler C, Newman S, MacElwee K, McCarthy P, Miller C, Waters C, Hardy J. Hum Mol Genet **1999**; 8:81–85.

124 Gasser T. J Neural Transm (Suppl) **2000**:31–40.

125 Gwinn-Hardy K, Mehta ND, Farrer M, Maraganore D, Muenter M, Yen SH, Hardy J, Dickson DW. Acta Neuropathol (Berl) **2000**; 99:663–672.

126 Yu Z, Zhou D, Cheng G, Mattson MP. J Mol Neurosci **2000**; 15:31–44.

127 Grilli M, Memo M. Biochem Pharmacol **1999**; 57:1–7.

128 Weingarten P, Bermak J, Zhou QY. J Neurochem **2001**; 76:1794–1804.

129 Cazorla P, Smidt MP, O'Malley KL, Burbach JP. J Neurochem **2000**; 74:1829–1837.

130 Sakurada K, Ohshima-Sakurada M, Palmer TD, Gage FH. Development **1999**; 126:4017–4026.

131 Le W, Conneely OM, Zou L, He Y, Saucedo-Cardenas O, Jankovic J, Mosier DR, Appel SH. Exp Neurol **1999**; 159:451–458.

132 Wallen A, Zetterstrom RH, Solomin L, Arvidsson M, Olson L, Perlmann T. Exp Cell Res **1999**; 253:737–746.

133 Witta J, Baffi JS, Palkovits M, Mezey E, Castillo SO, Nikodem VM. Brain Res Mol Brain Res **2000**; 84:67–78.

134 Wagner J, Akerud P, Castro DS, Holm PC, Canals JM, Snyder EY, Perlmann T, Arenas E. Nature Biotechnol **1999**; 17:653–659.

135 Rolletschek A, Chang H, Guan K, Czyz J, Meyer M, Wobus AM. Mech Dev **2001**; 105:93–104.

136 Simon HH, Saueressig H, Wurst W, Goulding MD, O'Leary DD. J Neurosci **2001**; 21:3126–3134.

137 Kruger R, Vieira-Saecker AM, Kuhn W, Berg D, Muller T, Kuhnl N, Fuchs GA, Storch A, Hungs M, Woitalla D, Przuntek H, Epplen JT, Schols L, Riess O. Ann Neurol **1999**; 45:611–617.

138 Khan N, Graham E, Dixon P, Morris C, Mander A, Clayton D, Vaughan J, Quinn N, Lees A, Daniel S, Wood N, de Silva R. Ann Neurol **2001**; 49:665–668.

139 Tan EK, Khajavi M, Thornby JI, Nagamitsu S, Jankovic J, Ashizawa T. Neurology **2000**; 55:533–538.

140 Lee M, Hyun D, Halliwell B, Jenner P. J Neurochem **2001**; 76:998–1009.

141 Koller WC. Neurology **2000**; 55:S2–7; discussion S8–12.

142 Rascol O, Nutt JG, Blin O, Goetz, CG, Trugman JM, Soubrouillard C, Carter JH, Currie LJ, Fabre N, Thalamas C, Giardina WW, Wright S. Arch Neurol **2001**; 58:249–254.

143 Oliveri RL, Annesi G, Zappia M, Civitelli D, De Marco EV, Pasqua AA, Annesi F, Spadafora P, Gambardella A, Nicoletti G, Branca D, Caracciolo M, Aguglia U, Quattrone A. Mov Disord **2000**; 15:127–131.

18
Psychiatric Pharmacogenetics: Prediction of Treatment Outcomes in Schizophrenia

Mario Masellis, Vincenzo S. Basile, Alex Gubanov and James L. Kennedy

Abstract

In the era of evidence-based and molecular medicine, genetics is playing an increasingly important role in the management of psychiatric disease. Psychiatric pharmacogenetics combines the fields of pharmacology and genetics in order to predict inter-individual outcomes (i.e., therapeutic response and adverse effect profile) to a psychotropic drug. In the last decade, there has been an explosion in research examining variable responsiveness and adverse effects to antipsychotic medications used to treat schizophrenia. Research has focused on pharmacogenetic studies of clozapine response, typical antipsychotic-induced tardive dyskinesia, and, more recently, antipsychotic-induced weight gain. This chapter provides a review of studies conducted to date in these areas, and also provides some discussion as to future directions.

18.1 Introduction

The aim of both pharmacogenetic and pharmacogenomic studies is to determine the impact of genetic variation or polymorphisms on the inter-individual differences in drug outcomes, with the ultimate goal of predicting the patients' response to medication and/or propensity to develop side effects [1]. Although there is a rich history of pharmacogenetic research, the area of psychiatric pharmacogenetics is a relatively young field, with most research conducted in the last decade. Psychiatric pharmacogenetics seeks to merge the fields of genetics and pharmacology to predict the clinical effects of the prescribed psychiatric medication. It is hoped that the final outcome of the research will lead to a stronger scientific basis for selecting the optimal drug therapy and dosages for each individual patient based on their own specific genetic, environmental, clinical and demographic characteristics. Treatment can then be provided based on these client-centered characteristics to maximize efficacy and minimize the risk of adverse events – getting the right medicine to the right patient.

A meta-analysis of drug safety and pharmacoepidemiology studies from 1966 to 1996 found that the incidence of serious and fatal adverse drug reactions in the

U.S. was 6.7 and 0.32%, respectively, ranking between the fourth and sixth leading cause of death, ahead of pneumonia and diabetes [2]. These adverse drug reactions occurred during treatment with standard doses of drugs, and did not include reports due to intentional or accidental overdose, errors in drug administration or noncompliance [2]. Antidepressant- and antipsychotic-associated falls in the elderly are of great concern because they may result in fractured hips, pneumonia and subsequent death. Although the morbidity and mortality from illnesses such as hypertension, diabetes and many infectious diseases have been substantially reduced through the availability of therapeutic agents, optimal personalized pharmacological therapy for major illnesses remains elusive.

It has been established through clinical observations, population, biochemical and molecular research, that there is significant heterogeneity in the efficacy and toxicity of most therapeutic agents within and among different populations. Unfortunately, prospective identification of those patients who are most likely to benefit from a specific therapy is not routinely possible for many diseases and medications. This is particularly important in the current health care environment, where cost containment and evidence-based initiatives are having a significant influence on patient care [3]. Potential causes for variability in drug effects include the nature and severity of the disease being treated, the individual's age and race, organ function, drug interactions, and concomitant illnesses. Although these factors are often important, inherited differences in the metabolism and disposition of drugs, as well as genetic polymorphisms in the targets of drug therapy can have an even greater influence on the efficacy and toxicity of medications.

18.2 Schizophrenia and its Pharmacotherapy: An Example of Major Mental Illness

Schizophrenia is a chronic, complex psychiatric disorder affecting approximately 1% of the population worldwide. The chronic nature of the illness, in addition to the early age of onset, results in direct and indirect health care expenditures in the U.S., which amount to approximately $ 30 to $ 64 billion dollars per year [4]. It is perhaps the most devastating of psychiatric disorders, with approximately 10% of patients committing suicide. The dopamine hypothesis of schizophrenia postulates that overactivity at dopaminergic synapses in the central nervous system (CNS), particularly the mesolimbic system, causes the psychotic symptoms (hallucinations and delusions) of schizophrenia. Roth and Meltzer [5] have provided a review of the literature and have concluded a role for serotonin as well in the pathophysiology and treatment of schizophrenia. The basic premise of their work stems from the known interaction between the serotonergic and dopaminergic systems.

With the introduction of chlorpromazine in 1952, there was a small revolution in psychiatry; patients suffering from psychosis were able to be de-institutionalized. Chlorpromazine and other "typical" antipsychotics (e.g., haloperidol) demonstrate high *in vitro* binding affinities for the dopamine D2 receptor (D2). Specifically, their binding potential for D2 correlates well with their clinical potencies [6]. The ability of these drugs to antagonize D2 receptors in the mesolimbic system is thought to be central to their antipsychotic properties. The re-introduction of clozapine, the prototype of "atypical" antipsychotics, in the late 1980s led to further advances in the pharmacological management of schizophrenia. Clozapine has a better tolerability profile than typical antipsychotics, particularly with respect to the extrapyramidal symptoms, a heterogeneous group of movement abnormalities including pseudoparkinsonism (tremor, rigidity, bradykinesia/akinesia, and postural instability) and tardive dyskinesia (TD). The more diverse binding profile of clozapine across several CNS receptors (e.g., serotonergic, dopaminergic, histaminergic, adrenergic and cholinergic) is thought to be responsible for these therapeutic advantages. Since then, there has been a rapid development of novel "atypical" antipsychotics that have been pharmacologically modeled, to a certain extent, after clozapine. Although both classes of antipsychotics, typical and atypical, offer some degree of efficacy, it is clear that they do not accommodate all symptoms of the disease.

The pharmacologic treatment of schizophrenia continues to present a therapeutic challenge for clinicians. A "trial and error" approach to prescribing antipsychotics is often adapted resulting in changes to the type of prescribed antipsychotic or titrating the dose to maximize efficacy and minimize side effects. In spite of the wide array of medicines available, 10–20% of the patients initially do not respond to treatment with typical antipsychotic therapy [7]. An additional 20–30% who do respond, eventually relapse on their maintenance programs and some develop serious adverse reactions, which cause them to discontinue the medication [7]. It is evident through clinical observations that there is a considerable variability of patient responses to the same recommended dose of a particular antipsychotic, with some responding adequately to treatment, others showing little response to treatment, and others developing toxic adverse reactions. It is likely that this variability is determined by a combination of genetic and environmental factors. This chapter presents a model for pharmacogenetic studies in psychiatry based on its application to schizophrenia. Both antipsychotic response and side effects have been investigated from a pharmacogenetic perspective. In terms of genetic studies of antipsychotic efficacy, we will review and discuss the paradigm of clozapine response. TD and weight gain will be reviewed as examples of pharmacogenetic approaches to antipsychotic side effects.

18.3 Pharmacodynamics of Clozapine Response

Clozapine is the prototype of atypical antipsychotic drugs, and it has been used effectively to treat patients with schizophrenia who are unresponsive or intolerant to typical antipsychotics [7]. Clozapine is characterized as "atypical" by its preferential binding to serotonin (5-HT2) and dopamine D4 receptors (D4) relative to dopamine D2 receptors [8]. A recent body of work also suggests that "atypicality" may be defined by the rate at which clozapine dissociates from D2 receptors. Specifically, clozapine binds loosely to and dissociates 100 times more rapidly from D2 receptors in the presence of endogenous dopamine as compared to typical antipsychotics [9]. Clozapine is effective in up to 60% [10] of treatment-refractory patients with schizophrenia with 6 months or longer of treatment. It improves negative symptoms and some cognitive functions and it has a reduced rate of extrapyramidal side effects (EPS), especially TD [11]. Nevertheless, some patients are refractory to clozapine treatment, and up to 0.38% develop potentially fatal agranulocytosis with regular blood monitoring [12].

Psychiatric pharmacogenetic studies to date have focused on using polymorphism to predict response to antipsychotics. The extent of response to clozapine across individuals clearly varies significantly and, as a result, several pharmacogenetic studies have focused on determining factors that contribute to the variance in clozapine response. Pharmacogenetic studies of clozapine response have, for the most part, focused on the pharmacodynamic paradigm assessing candidate genes that encode receptors from two major neurotransmitter systems: serotonin and dopamine. Among the four main groups worldwide, studies have focused on the impact of genetic polymorphism in serotonin (5-HT) receptors such as 5-HT2A, 5-HT2C, 5-HT6, and 5-HT7 [13–20], as well as dopamine (D) receptors including D3 and D4 [21–25] as they relate to the therapeutic efficacy of clozapine. In summary, although there are conflicting results, there are only two studies with sufficient statistical power to demonstrate a role for the structural His452Tyr 5-HT2A polymorphism in predicting clozapine response [14, 20]. For a comprehensive review, please refer to Masellis et al. [26]. Although in its early stages, the study of pharmacogenetic variability in antipsychotic response may lead to significant improvements in the clinical management of complex psychiatric disorders, such as schizophrenia.

18.4
Pharmacokinetics of Clozapine Response

Variable absorption and excretion rates and, in particular, variable amounts and activities of the liver enzymes may account for the large inter-individual variation in plasma levels of clozapine [27]. Clozapine is largely metabolized in the liver by the cytochrome P450 1A2 (CYP1A2) enzyme. This enzyme's activity can be induced by cigarette smoke and exposure to other polycyclic aromatic hydrocarbons. Conversely, caffeine intake and the antidepressant fluvoxamine may lead to a decrease in clozapine metabolism *in vivo* [28, 29]. A recent case report of patients who were non-responsive to clozapine had very high levels of CYP1A2, and the addition of fluvoxamine to the treatment regime resulted in increased plasma concentrations of clozapine [10].

The role of CYP1A2 enzyme polymorphisms may be an important factor affecting inter-individual response to clozapine [30]. Recently, a (C→A) polymorphism in the first intron of the CYP1A2 gene was found to be associated with variation in CYP1A2 inducibility in healthy volunteer smokers. Sachse and coworkers [31] have shown that the (C/C) genotype confers low inducibility for CYP1A2 in smokers. A (G→A) polymorphism in the 5'-flanking region of the CYP1A2 gene at position –2964 was associated with a significant decrease of CYP1A2 activity in Japanese smokers [32].

Pharmacogenetic studies of response to clozapine from a pharmacokinetic perspective have been limited to date. Arranz et al. [33] investigated the role of genetic variation in the cytochrome P450 2D6 (CYP2D6) enzyme in the trait of clozapine response and found no significant association. However, the role of CYP2D6 in the metabolism of clozapine is not likely to contribute to a substantial proportion of the variance [34]. Given the recent discovery of these two functional CYP1A2 variants [31, 32], our group has evaluated the role of CYP1A2 polymorphism in inter-individual variation in response to clozapine and found that CYP1A2 variation was not directly associated with clozapine response [35].

18.5 Tardive Dyskinesia

More recently, pharmacogenetic studies in psychiatry have begun to examine the genetics of predisposition to antipsychotic side effects such as EPS (akathisia, pseudoparkinsonism and TD), which are often seen with typical antipsychotic treatment. Antipsychotic-induced TD continues to be a serious and common problem in the psychopharmacology of schizophrenia, affecting 20–30% of patients suffering from psychotic disorders. TD is a motor disorder characterized by involuntary movements of the orofacial musculature and may involve the trunk and extremities as well. It is frequently chronic and has no definitive treatment. It occurs in predisposed individuals during or following cessation of prolonged antipsychotic therapy. While the mechanism of TD remains unknown, it has been postulated that an overactivity of dopaminergic neurotransmission, thought to be secondary to chronic blockade of D2-like receptors in the basal ganglia, may play a crucial role in the manifestation of TD [36]. Although the reasons for variable susceptibility to the development of TD are unknown, early animal [37–39] and human studies [40, 41] indicate a familial/genetic predisposition to this syndrome. This has stimulated a search for pharmacogenetic predictors of TD.

Following our initial report of association between genetic variation at the dopamine D3 receptor gene and TD [42], research has focused on a role for this receptor in predicting TD [43–47]. A recent collaborative effort by several groups (nine centers) in a combined analysis confirms association between the Ser9Gly dopamine D3 receptor gene polymorphism and TD [47]. Another interesting pharmacodynamic candidate is the 5-HT2A receptor gene, which was found to be associated with TD in two independent studies [48, 49], although we did not replicate this finding in a relatively large prospective North American sample [50]. We have also reported an association between a pharmacokinetic candidate gene, CYP1A2, and typical antipsychotic-induced TD [51]. However, a recent study in a German sample did not replicate this finding [52].

18.6 Weight Gain

Although atypical antipsychotics have a lower incidence of motor adverse effects, the use of these drugs is hindered by weight gain. Of the atypical drugs, clozapine appears to have the greatest potential to induce weight gain [52a]. Reviewing the literature, Leadbetter et al. [53] found that 13–85% of patients treated with clozapine had an associated increase in weight. Umbricht et al. [54] found that the cumulative incidence of all patients reaching 20% or more overweight, representing a significant long-term health risk, was greater than 50%. This adverse effect can undermine compliance inclining patients to relapse and may also lead to significant psychological distress. Considerable weight gain is associated with health risks such as type II diabetes mellitus, hypertension, cardiovascular disease, respiratory dysfunction and some types of cancer, which are linked to significant morbidity and mortality in our society [55]. There appears to be considerable variability among individuals with respect to the ability of an antipsychotic to induce weight gain. Hence, the side effect of weight gain occurs in only a proportion of treated patients that are predisposed. It is likely that this variability in patient propensity to gain weight is determined, in part, by genetic factors. The genetic factors may include pharmacokinetic and pharmacodynamic elements. Genetic predisposition to clozapine's ability to induce weight gain has been suggested [11, 54] and ample evidence exists demonstrating that body weight and feeding behavior are influenced by genetic factors [56, 57].

Atypical antipsychotic-induced weight gain is likely to be due to a combination of disturbances and alterations in satiety control mechanisms, energy expenditure, metabolism and lipogenesis. Collectively, data from several research paradigms converge and suggest that atypical antipsychotic-induced weight gain and obesity result from multiple neurotransmitter/receptor interactions with resultant changes in appetite and feeding behavior. Patients treated with clozapine generally complain that they have an inability to control their appetite even after eating a full meal. Satiety signals arise in a variety of areas including the olfactory and gustatory tracts, the esophagus, stomach, liver, intestines and are processed in the hypothalamus. These signals serve as CNS feedback mechanisms in the regulation and maintenance of an individual's homeostatic body weight. It is possible that antipsychotics may disturb satiety processing in the hypothalamus by binding to receptors involved in weight and satiety regulation. As such, genetic differences in these receptors that possess affinity for clozapine are prime candidates for investigation of genetic determinants of clozapine-induced weight gain.

More recently, we have extended our investigations of the pharmacogenetics of adverse reactions to the study of clozapine-induced weight gain. Common factors among the antipsychotics that induce weight gain are that they all exhibit 5-HT2C receptor antagonism, 5-HT1A receptor agonism, as well as histamine H1 receptor (H1) antagonism [58]. Numerous other candidate genes may play a role and for a comprehensive review, see Basile et al. [59]. To our knowledge, we are the first to have conducted association analyses investigating the role of genetic factors in clozapine-induced weight gain [59]. We tested this hypothesis for 10 genetic polymorphisms across 9 candidate genes including the serotonin 2C, 2A and 1A receptor genes (HTR2C/2A/1A); the histamine H1 and H2 receptor genes (H1R/H2R); the cytochrome P450 1A2 gene (CYP1A2); the β_3 and α_{1A} adrenergic receptor genes (ADRB3/ADRA1A) and tumor necrosis factor α (TNFα). Trends were observed for ADRB3, ADRA1A, TNFα and HTR2C, however, replication in larger, independent samples is required [59].

18.7 Conclusions and Future Directions

The main findings thus far in pharmacogenetics of antipsychotic drug effects have been several replications of an association between tardive dyskinesia and the Ser9Gly polymorphism in the dopamine D3 receptor, as well as evidence that polymorphism within the 5-HT2A receptor confers risk to non-response to clozapine. Given these interesting leads in the relatively young field of psychiatric pharmacogenetics, the future looks bright for a steady stream of findings in the genetic prediction of psychotropic drug response and adverse effects. Recent advances in gene chip array technology have demonstrated promise in terms of identifying novel candidate gene targets, which are differentially expressed in the presence of a particular drug. While microarray-based studies of gene expression are becoming more successful, the utility of this technology in mass scale genotyping has seen several technical difficulties and their role in psychiatric pharmacogenetics is still only speculative at present. With respect to pharmacogenetic phenotype, productive areas of investigation will be the dissection of the behavioral phenotype into particular symptoms, and the incorporation of objective biological correlates, such as EEG and positron emission tomography (PET) imaging, as endophenotypes.

Acknowledgements

We are grateful for the technical assistance from Mary Smirniw and Stephanie Care. Some of the work reviewed in this article was supported by grants to JLK from the Canadian Institutes of Health Research, the Ontario Mental Health Association, and the National Alliance for Research in Schizophrenia and Depression.

18.8 References

1 Regalado A. Inventing the pharmacogenomics business. Am J Health Syst Pharm **1999**; 56(1):40–50.

2 Lazarou J, Pomeranz BH, Corey PN. Incidence of adverse drug reactions in hospitalized patients: a meta-analysis of prospective studies. JAMA **1998**; 279(15):1200–1205.

3 McLeod HL, Evans WE. Pharmacogenomics: unlocking the human genome for better drug therapy. Annu Rev Pharmacol Toxicol **2001**; 41:101–121.

4 Williams R, Dickson RA. Economics of schizophrenia. Can J Psychiatry **1995**; 40:60–67.

5 Roth BL, Meltzer HY. The role of serotonin in schizophrenia. Raven Press, New York; **1995**.

6 Seeman P, Lee T, Chau-Wong M, Wong K. Antipsychotic drug doses and neuroleptic/dopamine receptors. Nature **1976**; 261:717–719.

7 Kane J, Honigfeld G, Singer J, Meltzer H. Clozapine for the treatment-resistant schizophrenic. A double-blind comparison with chlorpromazine. Arch Gen Psychiatry **1988**; 45(9):789–796.

8 Stockmeier CA, DiCarlo JJ, Zhang Y, Thompson P, Meltzer HY. Characterization of typical and atypical antipsychotic drugs based on *in vivo* occupancy of serotonin2 and dopamine2 receptors. J Pharmacol Exp Ther **1993**; 266(3):1374–1384.

9 Kapur S, Seeman P. Does fast dissociation from the dopamine d(2) receptor explain the action of atypical antipsychotics? A new hypothesis. Am J Pyschiatry **2001**; 158(3):360–369.

10 Bender S, Eap CB. Very high cytochrome P4501A2 activity and nonresponse to clozapine. Arch Gen Psychiatry **1998**; 55(11):1048–1050.

11 Meltzer HY. Role of serotonin in the action of atypical antipsychotic drugs. Clin Neurosci **1995**; 3(2):64–75.

12 Lieberman JA. Maximizing clozapine therapy: managing side effects. J Clin Psychiatry **1998**; 59(Suppl 3):38–43.

13 Masellis M, Paterson AD, Badri F, Lieberman JA, Meltzer HY, Cavazzoni P et al. Genetic variation of 5-HT2A receptor and response to clozapine [letter; comment]. Lancet **1995**; 346(8982):1108.

14 Masellis M, Basile V, Meltzer HY, Lieberman JA, Sevy S, Macciardi FM et al. Serotonin subtype 2 receptor genes and clinical response to clozapine in schizophrenia patients. Neuropsychopharmacology **1998**; 19(2):123–132.

15 Masellis M, Basile VS, Meltzer HY, Lieberman JA, Sevy S, Goldman DA et al. Lack of association between the T to C 267 serotonin 5-HT6 receptor gene (HTR6) polymorphism and prediction of response to clozapine in schizophrenia. Schizophr Res **2001**; 47(1):49–58.

16 Malhotra AK, Goldman D, Ozaki N, Breier A, Buchanan R, Pickar D. Lack of associaton between polymorphisms in the 5-HT2A receptor gene and the antipsychotic response to clozapine. Am J Psychiatry **1996a**; 153(8):1092–1094.

17 Malhotra AK, Goldman D, Ozaki N, Rooney W, Clifton A, Buchanan RW et al. Clozapine response and the 5HT(2C)Cys(23)Ser polymorphism. Neuroreport **1996b**; 7(13):2100–2102.

18 Nothen MM, Rietschel M, Erdmann J, Oberlander H, Moller HJ, Naber D, et al. Genetic variation of the 5-HT2A receptor and response to clozapine. Lancet **1995**; 346:908–909.

19 Arranz M, Collier D, Sodhi M, Ball D, Roberts G, Price J et al. Association between clozapine response and allelic variation in 5-HT2A receptor gene. Lancet **1995b**; 346:281–282.

20 Arranz MJ, Munro J, Sham P, Kirov G, Murray RM, Collier DA et al. Meta-analysis of studies on genetic variation in 5-HT2A receptors and clozapine response. Schizophr Res **1998**; 32(2):93–99.

21 Malhotra AK, Goldman D, Buchanan RW, Rooney W, Clifton A, Kosmidis MH et al. The dopamine D3 receptor (DRD3) Ser9Gly polymorphism and schizophrenia: a haplotype relative risk study and association with clozapine response. Mol Psychiatry **1998**; 3(1):72–75.

22 Ozdemir V, Masellis M, Basile VS, Kalow W, Meltzer HY, Lieberman JA et al. Variability in response to clozapine: Potential role of cytochrome P450 1A2 and the dopamine D4 receptor gene. CNS Spectrums **1999b**; 4(6):30–56.

23 Shaikh S, Collier D, Kerwin RW, Pilowsky LS, Gill M, Xu WM et al. Dopamine D4 receptor subtypes and response to clozapine [letter]. Lancet **1993**; 341(8837):116.

24 Shaikh S, Collier DA, Sham P, Pilowsky L, Sharma T, Lin LK et al. Analysis of clozapine response and polymorphisms of the dopamine D4 receptor gene (DRD4) in schizophrenic patients. American Journal of Medical Genetics (Neuropsychiatric Genetics) **1995**; 60:541–545.

25 Shaikh S, Collier DA, Sham PC, Ball D, Aitchison K, Vallada H et al. Allelic association between a Ser-9-Gly poly-

morphism in the dopamine D3 receptor gene and schizophrenia. Hum Genet **1996**; 97(6):714–719.

26 Masellis M, Basile VS, Ozdemir V, Meltzer HY, Macciardi FM, Kennedy JL. Pharmacogenetics of antipsychotic treatment: lessons learned from clozapine. Biol Psychiatry **2000**; 47(3):252–266.

27 Byerly M, DeVane L. Pharmacokinetics of Clozapine and Risperidone: A Review of Recent Literature. J Clin Psychopharmacol **1996**; 16(2):177–187.

28 Bertilsson L, Carrillo JA, Dahl ML, Llerena A, Alm C, Bondesson U et al. Clozapine disposition covaries with CYP1A2 activity determined by a caffeine test. Br J Clin Pharmacol **1994**; 38(5):471–473.

29 Jerling M, Lindstrom L, Bondesson U, Bertilsson L. Fluvoxamine inhibition and carbamazepine induction of the metabolism of clozapine: evidence from a therapeutic drug monitoring service. Ther Drug Monit **1994**; 16(4):368–374.

30 Ozdemir V, Posner P, Collins EJ, Walker SE, Roy R, Walkes W et al. CYP1A2 activity predicts clozapine steady state concentration in schizophrenic patients [abstract]. Clin Pharmacol Therapeutics **1999a**; 65:175.

31 Sachse C, Brockmoller J, Bauer S, Roots I. Functional significance of a C→A polymorphism in intron 1 of the cytochrome P450 CYP1A2 gene tested with caffeine. Br J Clin Pharmacol **1999**;47(4):445–449.

32 Nakajima M, Yokoi T, Mizutani M, Kinoshita M, Funayama M, Kamataki T. Genetic polymorphism in the 5'-flanking region of human CYP1A2 gene: effect on the CYP1A2 inducibility in humans. J Biochem (Tokyo) **1999**;125(4):803–808.

33 Arranz MJ, Dawson E, Shaikh S, Sham P, Sharma T, Aitchison K et al. Cytochrome P4502D6 genotype does not determine response to clozapine. Br J Clin Pharmacol **1995a**; 39:417–420.

34 Fang J, Coutts RT, McKenna KF, Baker GB. Elucidation of individual cytochrome P450 enzymes involved in the metabolism of clozapine. Naunyn Schmiedebergs Arch Pharmacol **1998**; 358(5):592–599.

35 Masellis M, Basile VS, Macciardi FM, Meltzer HY, Lieberman JA, Nothen MM et al. Genetic prediction of antipsychotic response following switch from typical antipsychotics to clozapine. In: XXIst Collegium Internationale Neuro-Psychopharmacologicum (CINP) Congress 1998; Glasgow, Scotland; **1998**.

36 Tarsy D, Baldessarini RJ. The pathophysiologic basis of tardive dyskinesia. Biol Psychiatry **1977**; 12(3):431–450.

37 Rosengarten H, Schweitzer JW, Friedhoff AJ. A mechanism underlying neuroleptic induced oral dyskinesias in rats. Pol J Pharmacol **1993**; 45(4):391–398.

38 Hashimoto T, Ross DE, Gao XM, Medoff DR, Tamminga CA. Mixture in the distribution of haloperidol-induced oral dyskinesias in the rat supports an animal model of tardive dyskinesia. Psychopharmacology (Berlin) **1998**; 137(2):107–112.

39 Casey DE. Dopamine D1 (SCH 23390) and D2 (haloperidol) antagonists in drug-naive monkeys. Psychopharmacology **1992**; 107(1):18–22.

40 Muller DJ, Ahle G, Alfter D, Krauss H, Knapp M, Marwinski K et al. Familial occurrence of tardive dyskinesia. In: 6th World Congress on Psychiatric Genetics, 1998; Bonn, Germany; **1998**.

41 O' Callaghan E, Larkin C, Kinsella A, Waddington JL. Obstetric complications, the putative familial-sporadic distinction, and tardive dyskinesia in schizophrenia. Br J Psychiatry **1990**;157:578–584.

42 Badri F, Masellis M, Petronis A, Macciardi FM, Van Tol HHM, Cola P et al. Dopamine and serotonin system genes may predict clinical response to clozapine. In: 46th Annual Meeting of the American Society of Human Genetics; 1996; San Francisco: Am J Hum Genet; **1996**. p. A247.

43 Basile VS, Masellis M, Badri F, Paterson AD, Meltzer HY, Lieberman JA et al. Association of the MscI polymorphism of the dopamine D3 receptor gene with tardive dyskinesia in schizophrenia. Neuropsychopharmacology **1999**; 21(1):17–27.

44 Steen VM, Lovlie R, MacEwan T, McCreadie RG. Dopamine D3-receptor gene variant and susceptibility to tardive dyskinesia in schizophrenic patients. Mol Psychiatry **1997**; 2(2):139–145.

45 Segman R, Neeman T, Heresco-Levy U, Finkel B, Karagichev L, Schlafman M et al. Genotypic association between the dopamine D3 receptor and tardive dyskinesia in chronic schizophrenia. Mol Psychiatry **1999**; 4(3):247–253.

46 Rietschel M, Krauss H, Muller DJ, Schulze TG, Knapp M, Marwinski K et al. Dopamine D3 receptor variant and tardive dyskinesia. Eur Arch Psychiatry Clin Neurosci **2000**; 250(1):31–35.

47 Lerer B, Segman RH, Fangerau H, Daly AK, Basile VS, Aschauer HN et al. Pharmacogenetics of tardive dyskinesia: Combined analysis of 780 patients supports association with dopamine D3 receptor gene Ser9Gly polymorphism. Neuropsychopharmacology (submitted).

48 Segman RH, Heresco-Levy U, Finkel B, Goltser T, Shalem R, Schlafman M et al. Association between the serotonin 2A receptor gene and tardive dyskinesia in chronic schizophrenia. Mol Psychiatry **2001**;6(2):225-229.

49 Tan EC, Chong SA, Mahendran R, Dong F, Tan CH. Susceptibility to neuroleptic-induced tardive dyskinesia and the T102C polymorphism in the serotonin type 2A receptor. Biol Psychiatry **2001**; 50(2):144–147.

50 Basile VS, Ozdemir V, Masellis M, Meltzer HY, Lieberman JA, Potkin SG et al. Lack of association between serotonin-2A receptor gene (HTR2A) polymorphisms and tardive dyskinesia in schizophrenia. Mol Psychiatry **2001**; 6(2):230–234.

51 Basile VS, Ozdemir V, Masellis M, Walker ML, Meltzer HY, Lieberman JA et al. A functional polymorphism of the cytochrome P450 1A2 (CYP1A2) gene: association with tardive dyskinesia in schizophrenia. Mol Psychiatry **2000**;5(4):410-417.

52 Schulze TG, Schumacher J, Muller DJ, Krauss H, Alfter D, Maroldt A, et al. Lack of association between a functional polymorphism of the cytochrome P450 1A2 (CYP1A2) gene and tardive dyskinesia in schizophrenia. Am J Med Genet **2001**;105(6):498-501.

52a Allison DB, Mentore JL, Heo M, Chandler LP, Cappelleri JC, Infante MC, Weiden PC. Antipsychotic-induced weight gain: a comprehensive research synthesis. Am J Psychiatry **1999**; 156(11): 1686–1696.

53 Leadbetter R, Shutty M, Pavalonis D, Vieweg V, Higgins P, Downs M. Clozapine-induced weight gain: prevalence and clinical relevance. Am J Psychiatry **1992**; 149(1):68–72.

54 Umbricht DS, Pollack S, Kane JM. Clozapine and weight gain. J Clin Psychiatry **1994**; 55(Suppl B):157–160.

55 Henderson DC, Cagliero E, Gray C, Nasrallah RA, Hayden DL, Schoenfeld DA et al. Clozapine, diabetes mellitus, weight gain, and lipid abnormalities: A five-year naturalistic study. Am J Psychiatry **2000**; 157(6):975–981.

56 Wade J, Milner J, Krondl M. Evidence for a physiological regulation of food selection and nutrient intake in twins. Am J Clin Nutr **1981**; 34(2):143–147.

57 Comuzzie AG, Allison DB. The search for human obesity genes. Science **1998**; 280(5368):1374–1377.

58 Baptista T. Body weight gain induced by antipsychotic drugs: mechanisms and management. Acta Psychiatr Scand **1999**; 100(1):3–16

59 Basile VS, Masellis M, McIntyre RS, Meltzer HY, Lieberman JA, Kennedy JL. Genetic dissection of atypical antipsychotic-induced weight gain: Novel preliminary data on the pharmacogenetic puzzle. J Clin Psychiatry **2001**; 62(Suppl 23):45–66.

19
Pharmacogenomics of Major Depression and Antidepressant Treatment

Ma-Li Wong, Israel Alvarado and Julio Licinio

Abstract

Hippocrates described depression 2400 years ago and throughout history depression has been a substantial public health problem worldwide. Major depression is a central nervous system disorder with psychological and physical manifestations that cause severe impairment. This disorder is the major cause of suicide, which is in turn the eighth cause of death in the United States. Additionally, depression is an independent risk factor for cardiovascular illness, the largest cause of death in developed countries. Even though considerable progress has been achieved in neuroscience, the biological substrates underlying depression and the response to antidepressant treatment are still unknown. All existing treatments are symptomatic, not curative. Moreover, we do not yet know why some individuals respond to specific treatments and others do not. The potential benefits of pharmacogenomics in depression are the discovery of innovative, effective, and individualized treatments. The process of identifying the genomic substrates underlying treatment may lead to the definition of new nosologic entities within the spectrum of what we currently define as major depression. Current research has been focused on a pharmacodynamic targets, such as the serotonin transporter gene, and on pharmacokinetic targets, namely the cytochrome P450 superfamily. This chapter reviews emerging work on the use of genomic tools to identify predictors of antidepressant treatment and novel therapeutic targets.

19.1
Introduction

Pharmacogenomics is a new area of medical science that is based on large information databases that are emerging from the sequencing of the human genome. High-throughput technologies have been developed to make it possible for researchers to harness the overwhelming amount of genomic data towards the goal of improving therapeutics.

In this chapter we examine the role of genomics in the treatment of depression. We discuss how the abundance of data that has emerged from the human genome project can be applied to enhance understanding of the fundamental biology of major depression, which is one of the oldest medical mysteries yet to be solved. Melancholia (a subtype of depression) was described over two thousand years ago by Hippocrates (460–337 BC); currently, depression is conceptualized as a common and highly complex disorder that is the second major cause of disability in the United States, and costs the American economy in excess of 50 billion dollars annually [1–4].

19.2 Clinical Aspects

The American Psychiatric Association diagnostic criteria are listed in Table 19.1. Table 19.2 describes the clinical features and Table 19.3 lists medical illnesses and drugs that can cause depressive symptoms.

There are no biological criteria to assist in the diagnosis, which is made solely based on clinical presentation. Various clusters of clinical symptoms are classified into distinct depression subtypes, such as melancholic, atypical, postpartum, psychotic, catatonic, and seasonal. Such descriptive classification is highly complex and of questionable value. It is hoped that progress in genomics and pharmacogenomics will lead to the elucidation of the underlying biology resulting in a more rational classification system that is based on distinct mechanisms, leading to the development of specific treatment strategies.

In the past several decades there has been increased incidence of depression, which motivated Gerald Klerman to describe this era as the "age of melancholia" [5]. The lifetime prevalence of depression in the U.S. is higher in women (21.3%) than in men (12.7%). Although the rates of major depression vary across the world, data from fifty countries support the notion that this disease is the fourth leading cause of disability worldwide (second in developed countries) [3]. Longitudinal studies verify that the typical course of the disease is recurrent, with periods of recovery and periods of depression symptoms; however, approximately 17% of patients have a chronic unremitting disease [6]. Depression is the major cause of suicidal behavior and the rate of suicidal attempts has been estimated to be around 56% in depressed patients [7].

Psychosocial factors contributory to depression have been well documented and accepted. For example, stress may precede the onset of a depressive episode. The psychological effects of loss are also relevant to depression. The work of Lerer and colleagues suggests that the loss of a parent prior to age 9 (early parental loss) is highly associated with episodes of depression in adulthood [8]. In that study, divorce impacted more significantly than death, and loss of a mother more than father. Even though psychosocial factors can contribute to depression and psychosocial treatments can be effective [9, 10], clinical experience supports a biological basis for treatment response. For instance, in clinical trials patients who are treated exclusively with medication report full remission following several weeks of treatment. Results such as these indicate the presence of a biological substrate

Tab. 19.1 DSM-IV criteria for major depression

A diagnosis of major depression can be made when items A–F are fulfilled:

A. At least one of the following three abnormal moods which significantly interfered with the person's life:
   1. Abnormal depressed mood most of the day, nearly every day, for at least 2 weeks.
   2. Abnormal loss of all interest and pleasure most of the day, nearly every day, for at least 2 weeks.
   3. If 18 or younger, abnormal irritable mood most of the day, nearly every day, for at least 2 weeks.

B. At least five of the following symptoms have been present during the same 2-week depressed period:
   1. Abnormal depressed mood (or irritable mood if a child or adolescent) [as defined in criterion A].
   2. Abnormal loss of all interest and pleasure [as defined in criterion A2].
   3. Appetite or weight disturbance, either:
      – abnormal weight loss (when not dieting) or decrease in appetite.
      – abnormal weight gain or increase in appetite.
   4. Sleep disturbance, either abnormal insomnia or abnormal hypersomnia.
   5. Activity disturbance, either abnormal agitation or abnormal slowing (observable by others).
   6. Abnormal fatigue or loss of energy.
   7. Abnormal self-reproach or inappropriate guilt.
   8. Abnormal poor concentration or indecisiveness.
   9. Abnormal morbid thoughts of death (not just fear of dying) or suicide.

C. The symptoms are not due to a mood-incongruent psychosis.

D. There has never been a manic episode, a mixed episode, or a hypomanic episode.

E. The symptoms are not due to physical illness, alcohol, medication, or street drugs.

F. The symptoms are not due to normal bereavement.

By definition, major depressive disorder cannot be due to:

- Physical illness, alcohol, medication, or street drug use
- Normal bereavement
- Bipolar Disorder
- Mood-incongruent psychosis (e.g., schizoaffective disorder, schizophrenia, schizophreniform disorder, delusional disorder, or psychotic disorder not otherwise specified).

From [38] with permission

in depression – and support the concept that pharmacological intervention at the biological level can produce symptom remission. However, the ultimate biological therapeutic targets of antidepressants treatment are still unknown.

Tab. 19.2 Clinical features of major depression

Major depressive disorder causes the following *mood* symptoms:

- **Abnormal depressed mood:**
  Sadness is usually a normal reaction to loss. However, in major depressive disorder, sadness is abnormal because it:
  - persists continuously for at least 2 weeks
  - causes marked functional impairment
  - causes disabling physical symptoms (e.g., disturbances in sleep, appetite, weight, energy, and psychomotor activity)
  - causes disabling psychological symptoms (e.g., apathy, morbid preoccupation with worthlessness, suicidal ideation, or psychotic symptoms).

  The sadness in this disorder is often described as a depressed, hopeless, discouraged, "down in the dumps," "blah," or empty. This sadness may be denied at first. Many complain of bodily aches and pains, rather than admitting to their true feelings of sadness.
- **Abnormal loss of interest and pleasure mood:**
  - The loss of interest and pleasure in this disorder is a reduced capacity to experience pleasure which in its most extreme form is called anhedonia.
  - The resulting lack of motivation can be quite crippling.
- **Abnormal irritable mood:**
  - This disorder may present primarily with irritable, rather than depressed or apathetic mood. This is not officially recognized yet for adults, but it is recognized for children and adolescents.
  - Unfortunately, irritable depressed individuals often alienate their loved ones with their cranky mood and constant criticisms.

Major depressive disorder causes the following *physical* symptoms:

- **Abnormal appetite:** Most depressed patients experience loss of appetite and weight loss. The opposite, excessive eating and weight gain, occurs in a minority of depressed patients. Changes in weight can be significant.
- **Abnormal sleep:** Most depressed patients experience difficulty falling asleep, frequent awakenings during the night or very early morning awakening. The opposite, excessive sleeping, occurs in a minority of depressed patients.
- **Fatigue or loss of energy:** Profound fatigue and lack of energy usually is very prominent and disabling.
- **Agitation or slowing:** Psychomotor retardation (an actual physical slowing of speech, movement and thinking) or psychomotor agitation (observable pacing and physical restlessness) often are present in severe major depressive disorder.

Major depressive disorder causes the following *cognitive* symptoms:

- **Abnormal self-reproach or inappropriate guilt:**
  - This disorder usually causes a marked lowering of self-esteem and self-confidence with increased thoughts of pessimism, hopelessness, and helplessness. In the extreme, the person may feel excessively and unreasonably guilty.
  - The "negative thinking" caused by depression can become extremely dangerous as it can eventually lead to extremely self-defeating or suicidal behavior.

Tab. 19.2 (continued)

- **Abnormal poor concentration or indecisiveness:**
  - Poor concentration is often an early symptom of this disorder. The depressed person quickly becomes mentally fatigued when asked to read, study, or solve complicated problems.
  - Marked forgetfulness often accompanies this disorder. As it worsens, this memory loss can be easily mistaken for early senility (dementia).
- **Abnormal morbid thoughts of death (not just fear of dying) or suicide:**
  - The symptom most highly correlated with suicidal behavior in depression is hopelessness.

Reproduced from [38, 39] with permission

Tab. 19.3 Differential diagnosis of depression

Medical illnesses – *organic mood syndromes caused by*
- Acquired immune deficiency syndrome (AIDS)
- Adrenal (Cushing's or Addison's diseases)
- Cancer (especially pancreatic and other GI)
- Cardiopulmonary disease
- Dementias (including Alzheimer's disease)
- Epilepsy
- Fahr's syndrome
- Huntington's disease
- Hydrocephalus
- Hyperaldosteronism
- Infections (including HIV and neurosyphilis)
- Migraines
- Mononucleosis
- Multiple sclerosis
- Narcolepsy
- Neoplasms
- Parathyroid disorders (hyper- and hypo-)
- Parkinson's disease
- Pneumonia (viral and bacterial)
- Porphyria
- Postpartum
- Premenstrual syndrome
- Progressive supranuclear palsy
- Rheumatoid arthritis
- Sjogren's arteritis
- Sleep apnea
- Stroke
- Systemic lupus erythematosus
- Temporal arteritis
- Trauma
- Thyroid disorders (hypothyroid and "apathetic" hyperthyroidism)
- Tuberculosis
- Uremia (and other renal diseases)
- Vitamin deficiencies (B12, C, folate, niacin, thiamine)
- Wilson's disease

Tab. 19.3 (continued)

Drugs:
- Acetazolamine
- Alphamethyldopa
- Amantadine
- Amphetamines
- Ampicillin
- Azathioprine (AZT)
- 6-Azauridine
- Baclofen
- Beta blockers
- Bethanidine
- Bleomycin
- Bromocriptine
- C-Asparaginase
- Carbamazepine
- Choline
- Cimetidine
- Clonidine
- Cycloserine
- Cocaine
- Corticosteroids (including ACTH)
- Cyproheptadine
- Danazol
- Digitalis
- Diphenoxylate
- Disulfiram
- Ethionamide
- Fenfluramine
- Griseofulvin
- Guanethidine
- Hydralazine
- Ibuprofen
- Indomethacin
- Lidocaine
- Levodopa
- Methoserpidine
- Methysergide
- Metronidazole
- Nalidixic acid
- Neuroleptics (butyrophenones, phenothiazines, oxyindoles)
- Nitrofurantoin
- Opiates
- Oral contraceptives
- Phenacetin
- Phenytoin
- Prazosin
- Prednisone
- Procainamide
- Procyclidine

Tab. 19.3 (continued)

- Quanabenzacetate
- Rescinnamine
- Reserpine
- Sedatives/hypnotics (barbiturates, benzodiazepines, chloral hydrate)
- Streptomycin
- Sulfamethoxazole
- Sulfonamides
- Tetrabenazine
- Tetracycline
- Triamcinolone
- Trimethoprim
- Veratrum
- Vincristine

Reproduced from [39] with permission

19.3 Pharmacology of Depression

For over 50 years investigators have tried to use antidepressant treatment response as a probe into the biology of depression. After having been discovered serendipitously, antidepressants were shown to have central monoaminergic systems as their initial and acute targets. Drugs that initially modulate one or more of the monoamines (serotonin, norepinephrine, and dopamine) may be completely effective in the symptomatic treatment of depression [11, 12]. The elucidation of the neurobiology of norepinephrine was one of the great advances of 20th century neuroscience and it was recognized in 1970 by the Nobel Prize awarded to Ulf von Euler and Julius Axelrod. Von Euler discovered that norepinephrine serves as a neurotransmitter at the nerve terminals of the sympathetic nervous system; Axelrod discovered the mechanisms which regulate the formation of this important transmitter in the nerve cells and the mechanisms which are involved in the inactivation of norepinephrine [13]. The role of norepinephrine in depression is confirmed by the findings that dysregulation of central norepinephrine systems is a hallmark of depression [14] and that selective inhibition of norepinephrine reuptake represents an effective treatment strategy [15].

Even though it has been well recognized that monoamines have a role in antidepressant treatment response, the amine hypothesis is insufficient to fully explain the biology and therapeutics of depression. An important clue to the existence of additional substrates is the clinical observation that the effect of antidepressants drugs on one or more of the monoaminergic systems occurs rapidly, generally within hours; however, the clinical response to antidepressants occurs after chronic treatment lasting several weeks. This indicates that some targets which are still unknown and common to various drug categories are elicited after chronic treatment to cause the therapeutic effect. Intriguingly, decades of psychopharmacological research have not yet revealed the identity of those targets.

Patients respond variably to the more than 20 FDA-approved antidepressants: Only 60–70% of patients show significant response to any specific antidepressant, and there is no predictor of response to those drugs. Thus, the development of novel therapies should be geared to solve two important issues in treatment: treatment resistance or refractoriness to current antidepressants, and medication compliance.

Approximately 30–40% of patients will not respond to a given antidepressant and 60–75% may fail to achieve complete remission [16]. Consequently, in its least restricted definition, treatment resistance could be detected in the majority of depressed patients under treatment. Moreover, prior treatment failure negatively influences the response to subsequent antidepressant treatment, decreasing the odds of treatment response by a factor of 15–20% for each failed treatment [17]. The delayed onset of symptom relief (which takes three to eight weeks to occur) and the presence of adverse drug reactions contribute significantly to low treatment compliance.

19.4 Treatment Targets

A major challenge to the development of new drugs is the discovery of new therapeutic targets. For example, the phenomenal success of fluoxetine (Prozac®) has been due to the fact that it was the first selective serotonin re-uptake inhibitor approved for world market release, combined with its improved adverse drug reaction profile. However, no new classes of antidepressants have emerged in recent years.

Promising new targets for the treatment of depression include neuropeptides, which are molecules thought to play a significant role in the biology of depression. For this reason, pharmaceutical companies and academic researchers have promoted the development of drugs that modify neuropeptidergic function. Drugs that target the receptor of neuropeptides, such as corticotropin-releasing hormone (CRH) and substance P, have been reported to be effective in the treatment of depression in initial trials; however, they have yet to make their way to market [18, 19]. Other targets for the treatment of depression are emerging, but clinical research trials have yet to conclusively to substantiate their validity.

19.5 Pharmacogenomics of Depression: The Potential for Drug Discovery

Genomics gives us the tools to search for new therapeutic targets by providing new avenues of research. The application of advanced molecular biology techniques in conjunction with a complete understanding of genomic sequences will

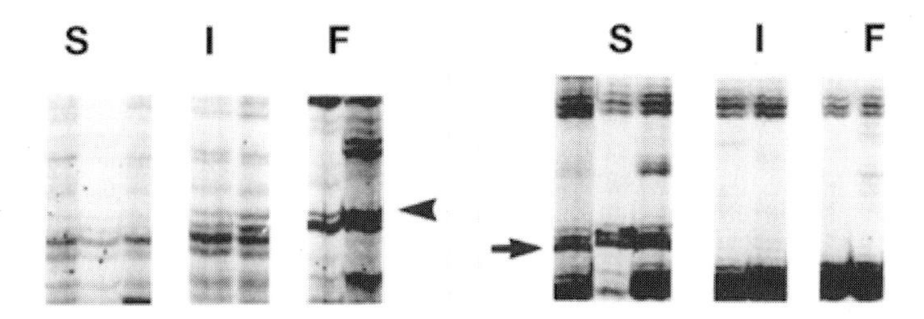

Fig. 19.1 Differential displays comparing RNAs from saline (S)-, imipramine (I)- or fluoxetine (F)-treated rats. Total RNA was extracted from hypothalami of animals treated with the different drugs for two months. Autoradiograms of amplified -[35S]-dATP-labeled PCR (polymerase chain reaction) products after electrophoresis in 6% polyacrylamide gels are shown for two different primer combinations that identified one upregulated (arrowhead) and one downregulated (arrow) fragment in the groups treated with antidepressants (from [4] with permission).

facilitate the search for novel target sites for antidepressants. As an example, our laboratory has used methods including differential mRNA display and DNA microarrays (DNA chips) to isolate previously unidentified transcripts expressed in the central nervous system (CNS) after chronic depression treatment using both fluoxetine and imipramine (see Figure 19.1). Such transcripts may be involved in pathways modulated by antidepressants. At the moment, we are confirming these results through various independent techniques, in addition to characterizing those genes. Following the full characterization of these genes, they would serve as likely new targets for antidepressant drug development.

19.6 Pharmacogenomics of Depression: Treatment Tailored to the Individual

Those who practice medicine are fully aware that patients respond in rather diverse ways to identical treatments. Generally, the best predictor of drug response is the patient's personal or familial treatment response history. This suggests an inherited, hence genetic, basis for treatment response. However, that type of information is seldom available. The treating clinician often observes that following the same dose of the same drug certain patients will show a full remission of symptoms while others report only severe side effects or no response at all. Many times physicians feel the need to balance a patient's health with potentially adverse drug interaction when prescribing various antidepressants; however, this may fail to predict drug response.

Research underway by our group and others on the clinical pharmacogenomics of depression is focused on identifying genetic markers as positive and negative predictors of treatment response. It is necessary that rigorous clinical studies be performed in order to examine closely the relation between genotype and the phenotype of drug response. In well-conducted clinical trials, positive clinical responses and adverse events can be related to specific genetic polymorphisms. Such research is designed to identify markers significantly linked with treatment responses. This line of investigation offers enormous promise for the individualization of drug therapy. Obviously, it would be to the benefit to all if we knew the likelihood of favorable or adverse responses *before* administering a drug. In spite of the potential for enormous return, a myriad of problems emerge in the conduction of such work. Clinical factors that can be problematic include the confounding variable presented by the placebo response, as well as the issues of sample size, genetic background of the patients, ethnic stratification, use of continual versus categorical outcome measures, drug choice and treatment strategy, treatment regime compliance and the impact of the environment on treatment outcome [4].

The dilemmas raised by the genetics of the work are also considerable and are discussed elsewhere [4]. A major difficulty is the choice of the appropriate genetic polymorphisms to associate to drug response. A statistical conundrum is created by the need to assess multiple variables that are partially related to one another (without *a priori* knowledge of the exact nature of such interactions) and that contribute to small effects in clinical trials that are highly costly and often cannot be as large as desired.

It must be kept in mind that pharmacogenetic findings are of a correlational nature. Positive results should lead to further investigation into the fundamental biological mechanisms by which specific gene sequences may impact on drug response.

19.7 Serotonin Transporter Gene

The serotonin transporter gene is an obvious candidate for clinical pharmacogenomic studies in depression. Following release in the synaptic cleft, the serotonin transporter (5-HTT) brings serotonin back into the pre-synaptic neuron, and thus reduces the amount of serotonin that is available in the synapses. Blocking transporters with selective serotonin re-uptake inhibitors (SSRIs) causes elevated synaptic concentrations of serotonin [20]. The serotonin transporter displays a polymorphism in its regulatory region, the presence or absence of a 44 base-pair insertion. The short variant of the polymorphism reduces the transcriptional efficiency of the 5-HTT gene promoter, resulting in decreased 5-HTT expression and 5-HT uptake in lymphoblasts [21]. The group at the San Raffaele Hospital in Milano has shown that patients with the “long” form of the 5-HTT regulatory region show a higher response to fluvoxamine and paroxetine [22, 23]. Furthermore, a not uncommon problem with the treatment of bipolar (manic-depressive) disorder with antidepressants is the precipitation of a manic phase; that risk is higher if the patient has the “short” form of the 5-HTT regulatory region [24]. Additionally, the group from Milano also showed that antidepressant response to a non-pharmacologic intervention, namely sleep deprivation, was more likely to occur in patients with the "long" genotype [25]. The "short" and "long" forms of the 5-HTT regulatory region can also influence the occurrence of extrapyramidal side effects and akathisia that can be induced by SSRIs. These side effects are known to affect compliance [26]. Kim et al. in Korea also studied the effects of the "long" and "short" forms of the 5-HTT regulatory region and examined the association of treatment response to the presence of an insert in the second intron of that gene [27].

One should note that although the Milano group reported that the gene long (l/l or l/s) form was linked with better treatment response to an SSRI, the Korean team found quite the opposite – that the "short" (s/s) form is associated with improved treatment response. These data indicate that factors apart from a specific genotype may actually play critical roles in measuring the effect of gene-drug interactions. Differences between these two studies, including as culture, diet, medical care, psychosocial support and genetic background are so numerous as to make it nearly impossible to attribute a specific cause for such disparities in the data. These discrepancies remind us of the critical importance for genomic and pharmacogenomic studies to be carried out identically in a minimum of two dissimilar populations, to document replicability.

Genetic variations in single targets of antidepressant action (the serotonin transporter) are associated with treatment response; however, they do not completely estimate the response to a particular therapy. It is quite probable that no single genetic marker will fully define the genetic makeup of antidepressant treatment. In order for that to be realized, a number of markers in multiple genes that contribute to treatment responses will have to be identified. Translational research efforts intended to isolate genomic targets of antidepressant action may be of help: Polymorphisms in antidepressant-regulated genes will be obvious candidates for future clinical pharmacogenomic studies [28].

19.8 Additional Targets

The targets for clinical pharmacogenomic associations are not necessarily the same as those for genetics. A gene may have a role in the response to antidepressant treatment and not be necessarily involved in the causation of the disorder. Biological systems that are thought to be involved in the response to antidepressant treatment include the following:

- Monoamines
  - Serotonin
  - Norepinephrine
  - Dopamine
- Neuropeptides and steroids
  - CRH and HPA axis
  - Substance P
  - TRH
  - Growth hormone

  - Sex hormones
  - Arginine vasopressin
  - NPY and PYY
- Opioids
- Amino acids
  - Glutamate
  - GABA
- Cytochrome P450 genes

The genes encoding synthesizing enzymes, transporters, receptors, and signal transduction pathways for the above cited systems are logical candidates for association studies. The challenge to the field is a rational selection of candidates that does not exclude interesting targets but that is not so overinclusive that it will dilute the statistical power of a study. The genes encoding metabolizing pathways are of great relevance as they determine drug availability and can, therefore, impact on treatment outcome.

19.9 Cytochrome P450 System – Drug-metabolizing Enzymes

Research has been conducted on the relation between the cytochrome P450 (CYP450) superfamily (see [29] for a review) and antidepressant response because most antidepressants are metabolized by this system. The CYP450 is a group of related enzymes located in the endoplasmic reticulum. These enzymes are expressed primarily in the liver but are also present in the intestine and brain, and use oxygen to transform endogenous (e.g., steroids) or exogenous (e.g., drugs) substances into more polar products that can be released in the urine. Electrons are supplied by NADPH-cytochrome P450 reductase, a flavoprotein that transfers electrons from NADPH (the reduced form of nicotinamide-adenine dinucleotide phosphate) to cytochrome P450. The CYP superfamily is divided into 14 families and 17 subfamilies of enzymes defined on the basis of similarities in their amino acid sequences. The enzymes transforming drugs in humans belong to the CYP families 1–4. Antidepressants are widely metabolized by these enzymes. Consequently genetic variations that affect enzyme activity will impact on the metabolism of antidepressant drugs and will affect clinical responses to treatment. Among the CYP450 superfamily, CYP2D6 has an important role in the metabolism of various antidepressants, as well as other commonly used drugs (see Table 19.4).

The activity of CYP2D6 is bimodal, some people (6% of Caucasians) have no copy of the gene while others have gene duplication. The frequency of gene duplication varies according to ethnicity – e.g., one-third of Ethiopians possess such duplication. Overall, the CYP2D6 cluster has 48 mutations and 50 alleles [30–34]. A dramatic case report illustrates the clinical relevance of this gene cluster. A 9-year-old diagnosed with attention-deficit hyperactivity disorder, obsessive-compulsive

Tab. 19.4 Antidepressants metabolized by CYP450 enzymes

Drug	*CYP1A2*	*CYP2D6*	*CYP2C19*	*CYP3A4*
Amitriptyline	✓	✓	✓	✓
Nortriptyline		✓		
Imipramine	✓	✓	✓	✓
Desipramine		✓		
Citalopram			✓	✓
Fluoxetine		✓		
Fluvoxamine	✓	✓		
Paroxetine		✓		
Sertraline				✓
Venlafaxine		✓		✓

Reproduced from [28, 32] with permission

disorder and Tourette's disorder was treated with a combination of methylphenidate, clonidine, and fluoxetine. After treatment was initiated he had generalized seizures that evolved to status epilepticus followed by cardiac arrest and death. The medical examiner's report indicated death caused by fluoxetine toxicity. At autopsy, blood, brain, and other tissue concentrations of fluoxetine and norfluoxetine were several-fold higher than expected based on literature reports for overdose situations. This led authorities to charge the parents with murder and prompted juvenile authorities to remove the remaining two children in the household, pending the outcome of the homicide investigation. Subsequent testing of autopsy tissue revealed the presence of a gene defect at the cytochrome P450 CYP2D locus, which is known to result in poor metabolism of fluoxetine [35]. Criminal charges to the parents were subsequently dismissed. The fact that the population frequency of such clinically relevant mutant alleles and duplicated genes is dependent on ethnicity raises critical ethical considerations.

19.10 Ethical Considerations

There are multiple ethical considerations in pharmacogenomics. A detailed review of this topic by Robertson discusses issues of confidentiality (which is always a problem in any type of genetics test), and labeling patients as "non-responders" [36]. Such a label could affect the patient's perception of self, future medical care and ability to obtain insurance or employment. An important possible complication of drugs that are tested and approved for people with specific genetic markers, is the issue of how to treat those who do not have the markers that are associated with favorable outcome. For instance, if a new antidepressant is approved for patients with specific genetic polymorphisms, it may prove difficult to use that drug in individuals lacking specifically that genotype. A chronically depressed patient who is resistant to existing therapies might not possess those particular genetic markers. In addition, his or her physician may still think that a new drug is best for that patient, despite the risk of adverse reactions or low efficacy. Will "off-label" drug administration such as this be covered by the patient insurance? Furthermore, what would the physician's liability costs be if patients experience severe adverse reactions to such non-recommended drug therapy, even if the patient gives informed consent.

Because allele frequencies can vary across ethnic groups, it is sometimes necessary to conduct studies in ethnically identified groups. The inclusion of a person in a research protocol based not only on diagnosis, but also on ethnicity raises a multitude of complex questions that require much thought and consideration. This is well discussed by Weijer and Emanuel [37], who raise important questions: How do community protections relate to individual informed consent? Is it more appropriate to conceive of a community as a vulnerable group protected by current regulations? Might a community use added protections for research to legitimize the oppression of groups within the community? Who counts as the community leader? What if the community wants to suppress adverse or undesirable research findings? Those authors recommend precision in distinguishing different types of communities in research, their characteristics, and protections appropriate for each, and discuss the distinction between community consent and community consultation.

For all groups to benefit from progress in pharmacogenomics it is crucial to include members of various ethnically identified communities in such studies. However, as we do that, additional ethical issues emerge. Future research will determine whether the concepts of race or ethnicity are relevant to pharmacogenomics. Even if they are not, it is necessary to conduct studies to achieve that conclusion. Those studies are themselves fraught with ethical issues. Careful consideration of the interplay of genomics, psychiatry and ethics should guide a conscientious effort to advance the science of pharmacogenomics in a manner that maximizes the translation of scientific advances into better health care for all, without furthering cultural stigmatization, ethnic stereotyping and racism.

19.11
Conclusions

Pharmacogenomics is a new area of investigation that integrates genomics and therapeutics. It has much to offer to the treatment of depression. While the promise of individualized therapeutics is considerable, the obstacles cannot be overlooked. Those include clinical, technical, and ethical issues that are only now being fully addressed. Current work is aimed at identifying the genomic targets of antidepressant action and genomic markers of clinical antidepressant treatment response. Such work may give us new insight into the biology of major depression and may facilitate biologically based classification systems. The identification of genomic markers for depression or for antidepressant treatment response will be a major accomplishment in the field of psychiatry.

Acknowledgements

Ma-Li Wong is supported by NIH grants MH/NS62777, GM61394, AT00151, and HG/CA02500 and by an award from NARSAD. Israel Alvarado is supported by NIH grants GM 61394, and HG/CA02500. Julio Licinio is supported by NIH grants GM61394, DK58851, HL04526, HG/CA02500, RR16996 and by an award from the Dana Foundation. Parts of this text were previously published, with modification, in Refs. [4, 28, 40].

19.12 References

1 Weissman MM, Bland RC, Canino GJ, Faravelli C, Greenwald S, Hwu HG et al. Cross-national epidemiology of major depression and bipolar disorder. JAMA **1996**; 276:293–299.

2 Greenberg PE, Stiglin LE, Finkelstein SN, Berndt ER. The economic burden of depression in 1990. J Clin Psychiatry **1993**; 54:405–418.

3 Murray CJL, Lopez AD. The Global Burden of Disease: A Comprehensive Assessment of Mortality and Disability from Diseases, Injuries, and Risk Factors in 1990 and Projected to 2020. Cambridge, MA: Harvard University Press; **1996**.

4 Wong M-L, Licinio J. Research and treatment approaches to depression. Nature Rev Neurosci **2001**; 2:343–351.

5 Klerman GL. Affective Disorders. In: The Harvard Guide to Modern Psychiatry (Nicholi J, Ed.). Cambridge, MA: Belknap/Harvard University Press; **1978**, 253–281.

6 Zis AP, Grof P, Goodwin FK. The Natural Course of Affective Disorders: Implications for Lithium Prophylaxis. Amsterdam: Excerpta Medica **1979**, 391–398.

7 Goodwin FK, Jamison KR. Bipolar Disorder. New York: Oxford University Press; **1990**.

8 Agid O, Shapira B, Zislin J, Ritsner M, Hanin B, Murad H et al. Environment and vulnerability to major psychiatric illness: a case control study of early parental loss in major depression, bipolar disorder and schizophrenia. Mol Psychiatry **1999**; 4:163–172.

9 Weissman MM, Markowitz J, Klerman GL. Comprehensive Guide to Interpersonal Psychotherapy. New York: Basic Books; **2000**.

10 Beck JS. Cognitive Therapy: Basics and Beyond. New York: Guilford Press; **1995**.

11 Charney DS. Monoamine dysfunction and the pathophysiology and treatment of depression. J Clin Psychiatry **1998**; 59(Suppl 14):11–14.

12 Lecrubier Y, Boyer P, Turjanski S, Rein W. Amisulpride versus imipramine and placebo in dysthymia and major depression. Amisulpride Study Group. J Affective Disord **1997**; 43(2):95–103.

13 http://www.nobel.se/medicine/laureates/1970/press.html.

14 Wong ML, Kling MA, Munson PJ, Listwak S, Licinio J, Prolo P et al. Pronounced and sustained central hypernoradrenergic function in major depression with melancholic features: relation to hypercortisolism and corticotropin-releasing hormone. Proc Natl Acad Sci USA **2000**; 97:325–330.

15 Versiani M, Mehilane L, Gaszner P, Arnaud-Castiglioni R. Reboxetine, a unique selective NRI, prevents relapse and recurrence in long-term treatment of major depressive disorder. J Clin Psychiatry **1999**; 60:400–406.

16 Amsterdam JD, Hornig-Rohan M. Treatment algorithms in treatment-resis-

tant depression. Psychiatr Clin North Am **1996**; 19:371–386.

17 Amsterdam JD, Maislin G, Potter L. Fluoxetine efficacy in treatment resistant depression. Prog Neuropsychopharmacol Biol Psychiatry **1994**; 18:243–261.

18 Kramer MS, Cutler N, Feighner J, Shrivastava R, Carman J, Sramek JJ et al. Distinct mechanism for antidepressant activity by blockade of central substance P receptors. Science **1998**; 281:1640–1645.

19 Zobel AW, Nickel T, Kunzel HE, Ackl N, Sonntag A, Ising M et al. Effects of the high-affinity corticotropin-releasing hormone receptor 1 antagonist R121919 in major depression: the first 20 patients treated. J Psychiatr Res **2000**; 34:171–181.

20 Iversen L. Neurotransmitter transporters: fruitful targets for CNS drug discovery. Mol Psychiatry **2000**; 5:357–362.

21 Lesch KP, Bengel D, Heils A, Sabol SZ, Greenberg BD, Petri S et al. Association of anxiety-related traits with a polymorphism in the serotonin transporter gene regulatory region. Science **1996**; 274:1527–1531.

22 Smeraldi E, Zanardi R, Benedetti F, Di Bella D, Perez J, Catalano M. Polymorphism within the promoter of the serotonin transporter gene and antidepressant efficacy of fluvoxamine. Mol Psychiatry **1998**; 3:508–511.

23 Zanardi R, Benedetti F, Di Bella D, Catalano M, Smeraldi E. Efficacy of paroxetine in depression is influenced by a functional polymorphism within the promoter of the serotonin transporter gene. J Clin Psychopharmacol **2000**; 20:105–107.

24 Mundo E, Walker M, Cate T, Macciardi F, Kennedy JL. The role of serotonin transporter protein gene in antidepressant-induced mania in bipolar disorder: preliminary findings. Arch Gen Psychiatry **2001**; 58:539–544.

25 Benedetti F, Serretti A, Colombo C, Campori E, Barbini B, di Bella D et al. Influence of a functional polymorphism within the promoter of the serotonin transporter gene on the effects of total sleep deprivation in bipolar depression. Am J Psychiatry **1999**; 156:1450–1452.

26 Lane RM. SSRI-induced extrapyramidal side-effects and akathisia: implications for treatment. J Psychopharmacol **1998**; 12:192–214.

27 Kim DK, Lim SW, Lee S, Sohn SE, Kim S, Hahn CG et al. Serotonin transporter gene polymorphism and antidepressant response. Neuroreport **2000**; 11:215–219.

28 Licinio J, Wong M-L. The pharmacogenomics of depression. Pharmacogenomics J **2001**; 1:175–177.

29 Raucy JL, Allen SW. Recent advances in P450 research. Pharmacogenomics J **2001**; 1:178–186.

30 Bertilsson L, Aberg-Wistedt A, Gustafsson LL, Nordin C. Extremely rapid hydroxylation of debrisoquine: a case report with implication for treatment with nortriptyline and other tricyclic antidepressants. Ther Drug Monit **1985**; 7:478–480.

31 Johansson I, Oscarson M, Yue QY, Bertilsson L, Sjoqvist F, Ingelman-Sundberg M. Genetic analysis of the Chinese cytochrome P4502D locus: characterization of variant CYP2D6 genes present in subjects with diminished capacity for debrisoquine hydroxylation. Mol Pharmacol **1994**; 46:452–459.

32 Bertilsson L, Dahl ML, Tybring G. Pharmacogenetics of antidepressants: clinical aspects. Acta Psychiatr Scand Suppl **1997**; 391:14–21.

33 Bertilsson L, Lou YQ, Du YL, Liu Y, Kuang TY, Liao XM et al. Pronounced differences between native Chinese and Swedish populations in the polymorphic hydroxylations of debrisoquine and S-mephenytoin. Clin Pharmacol Ther **1992**; 51:388–397.

34 Aklillu E, Persson I, Bertilsson L, Johansson I, Rodrigues F, Ingelman-Sundberg M. Frequent distribution of ultrarapid metabolizers of debrisoquine in an Ethiopian population carrying duplicated and multiduplicated functional CYP2D6 alleles. J Pharmacol Exp Ther **1996**; 278:441–446.

35 Sallee FR, DeVane CL, Ferrell RE. Fluoxetine-related death in a child with cytochrome P-450 2D6 genetic deficiency. J Child Adolesc Psychopharmacol **2000**; 10:27–34.

36 Robertson JA. Consent and privacy in pharmacogenetic testing. Nature Genet **2001**; 28:207–209.
37 Weijer C, Emanuel EJ. Ethics. Protecting communities in biomedical research. Science **2000**; 289:1142–1144.
38 American Psychiatric Association. Diagnostic and Statistical Manual of Mental Disorders, 4th Edn. Washington, DC: American Psychiatric Association; **1994**.
39 http://www.psychologynet.org.
40 Licinio J, Hannestad J, Wong M-L. Pharmacogenomics of antidepressants: Drug discovery, treatment, and ethical considerations. TEN: Trends in Evidence-Based Neuropsychiatry 2002 (in press).

20
Pharmacogenomics of Bipolar Disorder

Husseini K. Manji, Jing Du and Guang Chen

Abstract

Bipolar disorder (BD) is a common, severe, chronic and often life-threatening illness. Despite well-established genetic diatheses and extensive research, the biochemical abnormalities underlying the predisposition to and the pathophysiology of this disorder remain to be clearly established. In this chapter, we discuss the exciting recent progress being made in elucidating the role of therapeutically relevant mood stabilizer-regulated genes. In particular, a concerted series of mRNA reverse transcriptase-polymerase chain reaction (RT-PCR) studies has identified novel, completely unexpected targets. These targets include major cytoprotective protein B-cell lymphoma protein-2 (bcl-2), a human mRNA-binding (and -stabilizing) protein, A+U-binding protein (AUH), and a Rho kinase. This adds to the growing body of data suggesting that mood stabilizers may bring about some of their long-term benefits by enhancing neuroplasticity and cellular resilience. These results are quite noteworthy in the light of recently conducted morphometric brain imaging and postmortem studies demonstrating that bipolar disorder is associated with the atrophy and/or loss of neurons and glia. The development of novel treatments that more directly target the molecules involved in critical central nervous system (CNS) cells, selectively tailor treatments to individual patients, and thereby modulate the long-term course and trajectory of these devastating illnesses.

20.1
Introduction

Bipolar disorder (BD), also known as manic-depressive illness, is a common, severe, chronic, and often life-threatening illness that represents one of the leading causes of disability worldwide [1, 2] (Table 20.1). Despite the devastating impact that this illness has on millions of lives, there is still a dearth of knowledge concerning its etiology and/or pathophysiology. Increasingly, it is being appreciated that a true understanding of the pathophysiology of an illness as complex as BD must involve clearly addressing its neurobiology at different physiological levels, i.e., molecular, cellular, systems, and behavioral [3]. Undoubtedly, abnormalities in

Tab. 20.1 Bipolar disorder: scope the problem

- Lifetime prevalence ~1.2%; sex ~equal
- One of the most heritable of all psychiatric disorders; likely multiple susceptibility genes of small effect
- Recurrence almost invariable; ~1 episode/year on average
- For many patients, ↓ interepisode recovery and ↓ long-term outcome
- Global burden of disease study – BD ist one of the leading causes of disability worldwide
- Markedly elevated mortality (if untreated, mortality is higher than most types of heart disease, and many types of cancer)
- Suicide risk is higher than any other psychiatric disorder
- Increased substance abuse and cardiovascular disorders
- $ 45 Billion annual economic cost in the USA

gene expression underlie the neurobiology of the disorder at the molecular level. This will become evident as we identify the various susceptibility (and potentially protective) genes for BD in the coming years. However, following identification lies the more difficult work of examining the impact that the faulty expression of these gene products has on integrated cell function. As we discuss in greater detail below, it is at these levels that critical genes are discernable as novel targets for the actions of mood stabilizers, and may thus provide important new clues about the pathophysiology and optimal treatment of BD. The pathophysiology of this illness must account not only for the profound changes in mood, but also for the constellation of neurovegetative and psychomotor features which likely derive from dysfunction in interconnected limbic, striatal and fronto-cortical neuronal circuits. Several research laboratories have recently been focusing extensively on delineating the molecular and cellular mechanisms of mood-stabilizing agents in both preclinical and clinical studies. Such an experimental strategy may prove to be most promising, and has the potential to lead to the development of truly novel treatments. Furthermore, it may provide data derived from the "overall physiological response of the system" and address the critical dynamic interaction with pharmacological agents that effectively modify the clinical expression of the pathophysiology of BD [3]. Although a number of acute, *in vitro* effects of mood stabilizers have previously been identified, the clinical effects in the treatment of BD are only seen after chronic administration, thereby precluding any simple mechanistic interpretations based on its acute biochemical effects. Patterns of effects requiring such prolonged administration of the drug suggest that the therapeutic effects involve the strategic regulation of gene expression in critical neuronal circuits [4–7]. In this context, it is worth noting the substantial progress recently made in identification of genes responsive to trans-synaptic stimulation, as well as in determining the processes that convert often/occasionally ephemeral second messenger-mediated events into long-term cellular phenotypic alterations. These developments have been particularly important with respect to our attempts to understand the mechanisms by which short-lived events (e.g., stressors) can have profound, long-term (perhaps lifelong) behavioral consequences [8]. More importantly for the present discussion, these findings help to unravel the processes by which seemingly "simple" molecules including monovalent cations (e.g., lithium) and fatty acids (e.g., valproic acid) may produce a long-term stabilization of mood in individuals vulnerable to BD.

However, several factors impede our attempts to fully understand the molecular and cellular mechanisms of action of mood stabilizers. For instance, a suitable experimental model of BD is currently not available. Thus, many studies are of necessity conducted on "normal" rodents. This is done with the view that any targets identified may have functions conserved by evolution, lending therapeutic relevance to the human treatments. In this context, the animal models of drug dependence have been very instrumental in accelerating the pace of research on their molecular mechanisms [4]. Another inherent problem in the identification of therapeutically relevant target genes for the actions of mood stabilizers is the relative paucity of easily detectable phenotypic changes induced by these agents [9]. This makes the task of ascribing functional significance to the multiple treatment-induced changes at the genomic level quite daunting. Moreover, the genetic basis of mood as a quantitative trait is still tentative [10]; therefore, we cannot focus on a group of already known genes. Finally, as alluded to already, there is a real lack of knowledge concerning the underlying etiology and pathophysiology of what is likely a group of complex, heterogeneous disorders that show overlap of symptom clusters, and are subsumed under the rubric of "manic-depressive illness" or "bipolar disorder" (Table 20.2).

Nevertheless, despite these significant obstacles, there is currently considerable excitement about the progress that *is* being made, using two fundamental strategies to identify changes in gene expression that may have therapeutic relevance in the long-term treatment of mood disorders:

(1) Firstly, investigators have been focusing on the *known primary biochemical targets* for the actions of mood stabilizers (e.g., inositol monophosphatases), and subsequently identifying alterations in downstream signaling cascades, transcription factors, and ultimately the expression of genes known to be regulated by these primary biochemical targets.
(2) Second, several technological advances are allowing more "black box" screening approaches to be increasingly utilized; these approaches attempt to focus

Tab. 20.2 Pathophysiology of bipolar disorder: constraints for experimental design

- Complex diseases – etiologically heterogeneous and polygenic
- Disease evolves over time – neurobiology not static
- Neurobiology of recurrence distinct from that of specific symptom clusters?
- Dynamic interaction between primary disease neurobiology and adaptations – measuring overall "system response"
- Circadian factors – single time-point studies adequate?
- Lack of suitable animal models
- Characterization of "mood" as a quantitative trait (QTL analysis) not yet accomplished

directly on changes in gene expression produced by the administration of mood stabilizers in therapeutically meaningful paradigms, without necessarily focusing upon the "initiating biochemical events" (i.e., the medication's primary biochemical target). Using screening methods like subtractive hybridization, microarrays, and mRNA differential display, this strategy usually attempts to simultaneously identify treatment-induced changes in multiple genes often numbering in the thousands without any *a priori* focus on specific "candidate genes". However, as we will discuss in greater detail, these methodologies can, if necessary, be biased towards the detection of certain classes of candidate genes.

Both of these strategies require an initial reductive step, which attempts to isolate the specific genes and proteins that are the targets of mood-stabilizing agents. Also included, ideally, is a subsequent integrative step that attempts to establish the relationship between the molecular/cellular changes and certain facets of the therapeutic response [10]. In this chapter, we describe recent research endeavors utilizing both of the strategies outlined above, which have led to the identification of novel, hitherto completely unexpected targets for the long-term actions of mood stabilizers [9]. The identification of these targets may not only lead to the development of improved therapeutics that can be selectively tailored for individual patients, but may also facilitate our understanding about the pathophysiology of a very complex neuropsychiatric disorder.

20.2
Effects of Mood Stabilizers on Immediate Early Genes

Several independent laboratories have now demonstrated that both lithium and valproate (VPA) exert complex, isozyme-specific effects on the PKC (protein kinase C) signaling cascade (reviewed in [3, 5, 11–13]). Not surprisingly, considerable research has recently attempted to identify changes in the activity of transcription factors known to be regulated (at least in part) by the PKC signaling pathway – in particular the activator protein 1 (AP-1) family of transcription factors. In the CNS, the genes that are regulated by AP-1 include those for various neuropeptides, neurotrophins, receptors, transcription factors, enzymes involved in neurotransmitter synthesis, and proteins that bind to cytoskeletal elements [14].

Recent studies have demonstrated that lithium (and to a lesser extent VPA) produces, at therapeutically relevant concentrations, complex alterations in basal and/or stimulated DNA-binding of 12-*o*-tetradecanoyl-phorbol 13-acetate (TPA) response element (TRE) to AP-1 transcription factors. These alterations are produced not only in human SH-SY5Y cells *in vitro*, but also in rodent brain following chronic, *in vivo* administration [5, 7, 15–21]. Corresponding to an increase in basal AP-1 DNA-binding activity, lithium and VPA have been shown to increase the expression of a luciferase reporter gene driven by an SV40 promoter that contains TREs in a time- and concentration-dependent fashion. Mutations in the TRE sites of the reporter gene promoter markedly attenuate lithium's effects [16, 17, 21]. In order to ascribe potential therapeutic relevance to the changes in AP-1-regulated gene expression, it is necessary to demonstrate that they occur in the CNS *in vivo*. It is well established that the expression of tyrosine hydroxylase (TH) is mediated largely by the AP-1 family of transcription factors [22]. The effects of acute and chronic lithium on the levels of TH have, therefore, been investigated in brain areas that have been implicated in the pathophysiology of mood disorders [23]. It has been found that chronic lithium significantly increases the levels of TH in all three areas examined: frontal cortex, hippocampus, and striatum [24]. Recent research has also revealed important roles for the different nitric oxide synthases (NOS) in mediating various aspects of CNS function [25–28]; the expression of endothelial NOS (eNOS) is known to be regulated by AP-1 sites. Thus, the effects of chronic lithium and VPA on eNOS levels in the rat frontal cortex have been investigated; chronic lithium or VPA has been found to produce a 2- to 3-fold increase in the levels of eNOS [29]. Importantly, independent laboratories have also recently demonstrated lithium-induced increases in the levels of proteins whose genes are known to be regulated by AP-1 sites [30, 31]. These results clearly show that, in addition to increasing basal AP-1 DNA-binding activity and the expression of the luciferase reporter gene *in vitro*, chronic lithium increases the levels of several endogenous proteins whose genes are known to be regulated by AP-1 sites, in rat brain *ex vivo* [21, 24]. Together, these data suggest that lithium and VPA, via their effects on the AP-1 family of transcription factors, may bring about strategic changes in gene expression in critical neuronal circuits, effects that may ultimately underlie their efficacy in the treatment of bipolar disorder.

However, while many specific genes which are the targets of long-term lithium and/or VPA action have indeed been identified, it has been estimated that $\sim$10 000–15 000 genes may be expressed in a given cell at any time. Clearly, additional, novel methodologies are required to study the complex pattern of gene expression changes induced by chronic drug treatment [4–6, 32–34]. In recent years, new methodologies have evolved to identify the *differential expression* of multiple genes (e.g., in pathological vs. normal tissue, or in control vs. treated tissue); we now turn to a discussion of recent neuropharmacological studies using such methodologies.

20.3 The Use of a Concerted RT-PCR mRNA Differential Display Screening Strategy to Identify Genes whose Expression is Regulated by Mood-Stabilizing Agents

As alluded to already, a major problem inherent in neuropharmacologic research is the lack of *phenotypic changes* that are clearly associated with treatment response; this difficulty is particularly evident for mood-stabilizing agents [9]. In the absence of suitable animal models, studies have attempted to overcome this experimental hurdle by employing paradigms that involve the identification of com-

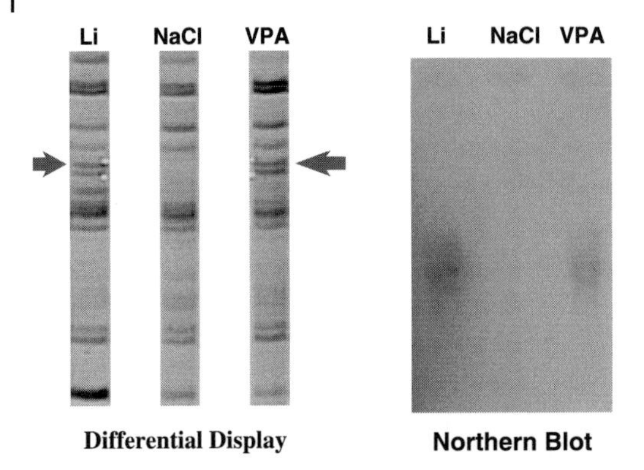

Fig. 20.1 mRNA RT-PCR differential display. The figure on the left provides a representative example of the mRNA RT-PCR DD strategy utilized to identify genes whose expression is regulated in concert by both chronic lithium and VPA. The arrows indicate two bands whose levels are markedly increased by both lithium and VPA compared to saline-treated animals. The figure on the right depicts a representative Northern blot verifying that the differential display results do not represent false positives. It should be noted that, for many rare genes, RNase protection assays were used for the verification stage.

mon long-term molecular targets of structurally dissimilar mood-stabilizing agents when these agents are administered chronically *in vivo*. Thus, in order to identify changes in gene expression that are likely to be associated with components of the therapeutic efficacy of mood stabilizers, reverse transcriptase-polymerase chain reaction differential display (RT-PCR DD) studies have investigated concurrently the effects of lithium and VPA in the CNS following chronic treatment of rodents *in vivo* [35]. These are two highly dissimilar agents structurally; although they likely do not exert their therapeutic effects by *precisely* the same mechanisms, identifying the genes that are regulated in concert by these two agents when administered in a therapeutically relevant paradigm may provide important indication of the molecular mechanisms underlying mood stabilization.

Inbred male Wistar Kyoto rats were used for the mRNA RT-PCR DD studies in order to reduce potential false positives due to individual differences. From the initial ~300 reactions, it was found that 12 bands showed markedly greater levels in the frontal cortical samples obtained from the rats treated chronically with *both* lithium and VPA, whereas 7 bands showed markedly lower levels in both treatment groups (Figure 20.1). A BLAST (Basic Local Alignment Search Tool) search revealed that 11 of these cDNAs have considerable sequence homology to known sequences in the GenBank database. Studies are currently underway to elucidate the identity and potential functional significance of the transcripts regulated in concert by lithium and VPA. Here, we discuss two unexpected target genes for the action of chronic lithium and VPA.

20.4
Identification of the Major Cytoprotective Protein bcl-2 as a Target for Mood Stabilizers

Clone 12, which is 355 bp long and contains a poly-A tail, shows very high sequence homology to a mouse and human transcription factor: polyomavirus enhancer-binding protein 2 (PEBP2) β subunit [also known as core-binding factor (CBF) β and acute myelogenous leukemia 1 (AML1) β] (E-values are e-106 and 3e-58, respectively). BESTFIT comparison indicates that clone 12 has 92% identical sequences to the mouse 3′-end of PEBP2β mRNA [36], and 85% identical sequences to human CBFβ mRNA [37]. The homology between the mouse and human sequences in the same region is 83%. In the mouse, there are at least three forms of PEBP2β mRNAs arising from alternative splicing [36]; all three have the same 3′-end. To verify the existence of three subtypes of PEBP2β in the rat, rat PEBP2β cDNAs were cloned from pooled rat brain cDNAs using the PCR method with primers corresponding to the putative ends of the splicing region. These rat cDNA fragments (Genbank accession numbers for βI, βII, and βIII AF087437) are identical to the mouse sequence and βII differs from the human CBFβ sequence by 16 bases [37]. To verify the DD results, RNase protection assays were conducted with a ^{32}P-RNA probe generated using this clone as the template. Consistent with the DD results, chronic treatment with both lithium and VPA increased the levels of PEBP2β mRNA in the frontal cortex (FCx). In the absence of available antibodies to PEBP2β, we next sought to determine if the treatments induced *functional changes* in PEBP2 transcription factor activity. Although PEBP2β does not directly bind to DNA, the binding of PEBP2β to PEBP2a (the DNA-binding subunit of the PEBP2 transcription factor) forms a high-affinity DNA-binding complex, which results in a dramatic increase in transcription [38, 39]. Chronic treatment of rats with either lithium or VPA significantly increased the DNA-binding activity of PEBP$a\beta$ in frontal cortex [40]. Chronic treatment of rats with either d-amphetamine sulfate (5 mg/kg/day) or chlordiazepoxide (5 mg/kg/day) did not produce any detectable changes in the DNA-binding activity of PEBP2$a\beta$ in frontal cortex [40]. The putative targets of the PEBP2 transcription factor, which may be of therapeutic relevance in the treatment of BD were next investigated. As with any transcription factor, PEBP2 undoubtedly regulates the expression of a number of known and unknown genes. Most of the genes whose expression is currently known to be regulated by PEBP2 are in the hematopoietic and immune system. However, the promoter of the human B-cell lymphoma protein-2 (bcl-2) gene has a PEBP2 binding site, and this site has been clearly demonstrated to increase the expression of a reporter gene driven by the bcl-2 promoter. In view of the growing body of data suggesting that mood disorders are associated with cell atrophy and/or loss in the frontal cortex (discussed in [6, 23, 41]), as well as the robust neurotrophic and neuroprotective effects of bcl-2 [42–44], the effects of lithium and VPA's on bcl-2 were next investigated. It was found that chronic treatment of rats with lithium or VPA resulted in a *doubling* of bcl-2 levels in FCx. Furthermore, immunohistochemical studies showed that chronic treatment of rats with lithium or VPA resulted in a marked increase in the number of bcl-2 immunoreactive cells in layers II and III of FCx (Figure 20.2). Interestingly, the im-

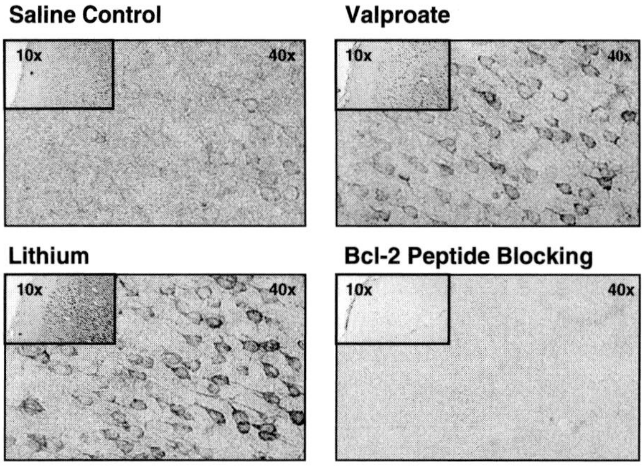

Saline Control 10x 40x

Valproate 10x 40x

Lithium 10x 40x

Bcl-2 Peptide Blocking 10x 40x

Fig. 20.2 Chronic lithium and valproate robustly increase bcl-2 immunoreactive neurons in the frontal cortex. Male Wistar Kyoto rats were treated with either Li_2CO_3, valproate or saline by twice daily i.p. injections for four weeks. Rats brains were cut at 30 μm; serial sections were cut coronally through the anterior portion of the brain, mounted on gelatin-coated glass slides and were stained with thionin. The sections of the second and third sets were incubated free-floating for 3 d at 4 °C in 0.01 M PBS containing a polyclonal antibody against bcl-2 (N-19, Santa Cruz Biotechnology, Santa Cruz, CA 1 : 3000), 1% normal goat serum and 0.3% Triton X-100 (Sigma, St. Louis, MO). Subsequently, the immunoreaction product was visualized according to the avidin-biotin complex method. The figure shows immunohistochemical labeling of bcl-2 in layers II and III of frontal cortex in saline-, lithium- or valproate-treated rats. Blocking peptide shows the specificity of the antibody. Photographs were obtained with 40× magnification. Modified and reproduced, with permission, from [40] (see color plates, p. XXXVI).

portance of neurons in layers II–IV of the FCx in mood disorders has recently been emphasized, since primate studies have indicated that these are important sites for connections with other cortical regions, and major targets for subcortical input [41]. Chronic lithium was subsequently also shown to markedly increase the number of bcl-2 immunoreactive cells in the dentate gyrus and striatum [32], and to robustly increase bcl-2 levels in C57BL/6 mice [45], in human neuroblastoma SH-SY5Y cells [33] and in rat cerebellar granule cells [46] *in vitro*. Consistent with the robust upregulation of bcl-2, lithium has now been demonstrated to exert neurotrophic/neuroprotective effects in a variety of preclinical paradigms both *in vitro* and *in vivo* (Table 20.3).

In view of bcl-2's major neuroprotective and neurotrophic roles, a study was undertaken to determine if lithium, administered at therapeutically relevant concentrations, affects neurogenesis in the adult rodent brain. To investigate the effects of chronic lithium on neurogenesis, mice were treated with "therapeutic" lithium (plasma levels 0.97 ± 0.20 mM), for ~4 weeks. After treatment with lithium for 14 days, the mice were administered single doses of BrdU (bromodeoxyuridine, a thymidine analog which is incorporated into the DNA of dividing cells) for 12 consecutive days. Lithium treatment continued throughout the duration of the

Tab. 20.3 Neurotrophic and neuroprotective effects of lithium

• **Protects cultured cells of rodent and human neuronal origin *in vitro* from** – Glutamate, NMDA (N-methyl-D-aspartate) – Calcium – MPP$^+$ (1-methyl-4-phenylpyridinium) – β-Amyloid – Aging – Growth factor deprivation	• **Protects rodent brain *in vivo* from** – Cholinergic lesions – Radiation injury – MCA (middle cerebral artery) occlusion – Quinolinic acid • Enhances hippocampal neurogenesis in adult mice • **Human effects** – No subgenual prefrontal cortical changes – Increases NAA levels – Increases gray matter volumes

BrdU administration. Following BrdU immunohistochemistry [45], 3-D cell counting was performed using a computer-assisted image analysis system, and it was found that chronic lithium administration *does*, indeed, result in an increase in the number of BrdU-positive cells in the dentate gyrus [45] (Figure 20.3). Moreover, approximately 2/3 of the BrdU-positive cells also double-stained with the neuronal marker NeuN (neuronal nuclear antigen), confirming their neuronal identity (Figure 20.3). Double staining of BrdU and bcl-2 was also observed, and studies using bcl-2 transgenic animals are currently underway to delineate the role of bcl-2 overexpression in the enhanced hippocampal neurogenesis observed.

20.5 Human Clinical Research Studies Arising Directly from the Rodent mRNA RT-PCR Studies

While the body of preclinical data demonstrating neurotrophic and neuroprotective effects of lithium is striking (discussed in [5, 6, 32, 46–48]), considerable caution must clearly be exercised in extrapolating to the clinical situation with humans. A longitudinal clinical study was, therefore, recently undertaken to determine if lithium also exerts neurotrophic/neuroprotective effects in the *human brain in vivo*. Proton magnetic resonance spectroscopy (MRS) was utilized to quantitate N-acetyl-aspartate (NAA) levels longitudinally. NAA is one of the many neurochemical compounds which can be quantitatively assessed via MRS. NAA is a putative neuronal marker, localized to mature neurons and not believed to be present in appreciable levels in mature glial cells, cerebral spinal fluid (CSF), or blood [49]. A number of studies have now shown that initial abnormally low brain NAA measures may increase and even normalize with remission of CNS symptoms in disorders such as demyelinating disease, amyotrophic lateral sclerosis, mitochondrial encephalopathies, and HIV dementia [49]. NAA is synthesized within mitochondria, and inhibitors of the mitochondrial respiratory chain decrease NAA

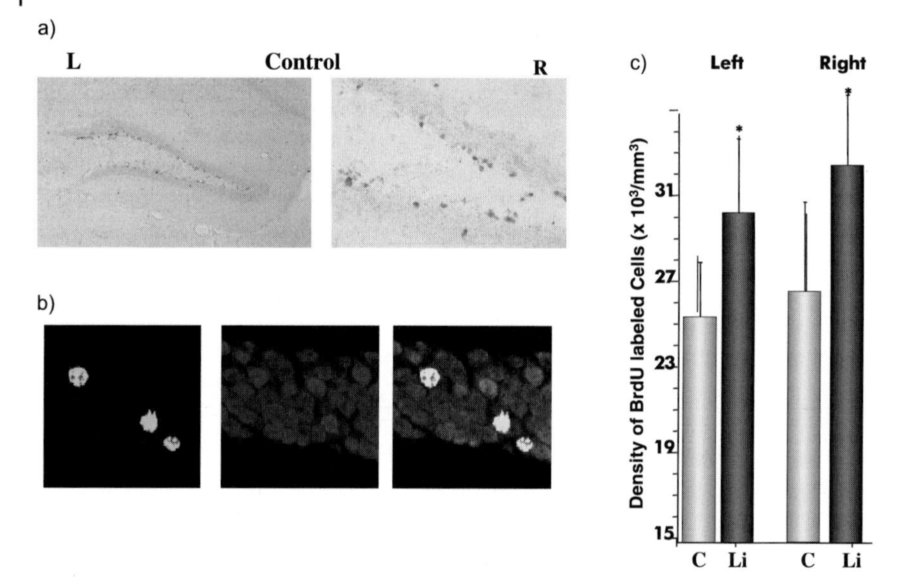

Fig. 20.3 Chronic lithium increases hippocampal neurogenesis. C57BL/6 mice were treated with lithium for 14 days, and then received once daily BrdU injections for 12 consecutive days while lithium treatment continued. 24 hours after the last injection, the brains were processed for BrdU immunohistochemistry. Cell counts were performed in the hippocampal dentate gyrus at three levels along the dorsoventral axis in all the animals. BrdU-positive cells were counted using unbiased stereological methods. Chronic lithium produced a significant 25% increase in BrdU immunolabeling in both right and left dentate gyrus (* $p<0.05$). (**a**) BrdU immunohistochemistry; (**b**) quantitation of BrdU-positive cells; (**c**) double labeling with BrdU and NeuN (neuron-specific nuclear protein, a neuronal marker). Modified and reproduced with permission from [45] (see color plates, p. XXXVII).

concentrations, effects that correlate with reductions in ATP and oxygen consumption [50]. Thus, NAA is now generally regarded as a measure of *neuronal viability and function*, rather than strictly a marker for neuronal loss, *per se* (for an excellent overview of NAA see [51]). After extensive validation of this method for longitudinal *in vivo* measurement, regional NAA concentrations were measured in BD patients at baseline and again after four weeks of lithium at therapeutic doses. Chronic Li administration was found to significantly increase NAA concentration, with all the brain regions investigated, demonstrating an increase in NAA over the course of the study. When regional brain NAA increases were examined together with the regional voxel image segmentation data, a striking ~0.97 correlation between lithium-induced NAA increases and regional voxel gray matter content was observed [52]. Examining data on a regional basis again, together with voxel gray matter content segmentation results in BD patients, revealed an NAA increase per voxel gray matter content two-fold higher in both the frontal and temporal lobe regions compared to the parietal and occipital regions. These results suggest that chronic lithium may not only exert robust neuroprotective

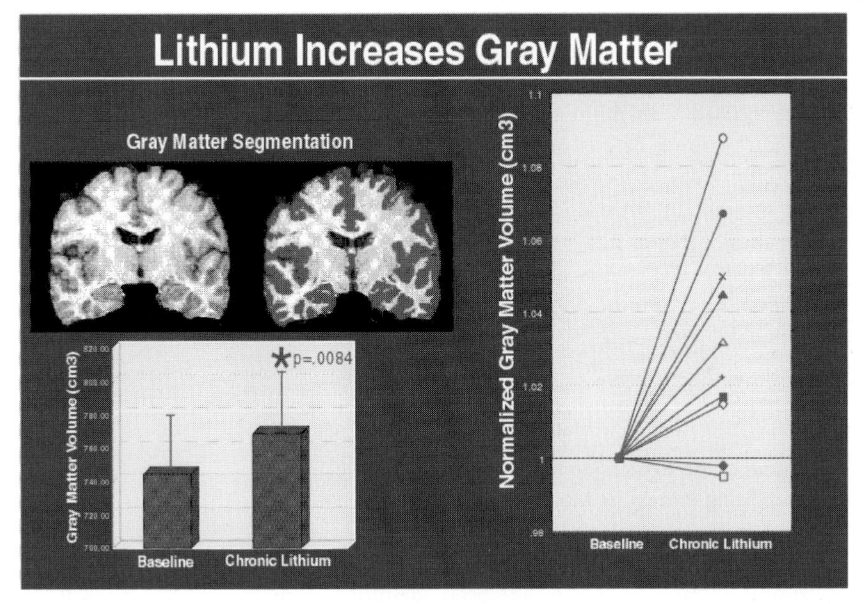

Fig. 20.4 Brain gray matter volume is increased following four weeks of lithium administration at therapeutic levels in BPD patients. Inset illustrates a slice of the three-dimensional volumetric MRI data which was segmented by tissue type using quantitative methodology to determine tissue volumes at each scan time point. Brain tissue volumes using high-resolution three-dimensional MRI (124 images, 1.5 mm thick Coronal T1 weighted SPGR images) and validated quantitative brain tissue segmentation methodology to identify and quantify the various components by volume, including total brain white and gray matter content. Measurements were made at baseline (medication free, after a minimum 14 day washout) and then repeated after four weeks of lithium at therapeutic doses. Chronic lithium significantly increases *total gray matter content* in the human brain of patients with BPD. No significant changes were observed in brain white matter volume, or in quantitative measures of regional cerebral water. Modified and reproduced with permission, from [53] (see color plates, p. XXXVIII).

effects (as has been demonstrated in a variety of preclinical paradigms), but also exerts *neurotrophic effects* in humans.

In a follow-up study to the NAA findings, it was hypothesized that, in addition to increasing functional neurochemical markers of neuronal viability, lithium-induced increases in bcl-2 would also lead to neuropil increases, and thus to increased brain gray matter volume in BD patients. In this clinical research investigation, brain tissue volumes were examined using high-resolution three-dimensional magnetic resonance imaging (MRI) and validated quantitative brain tissue segmentation methodology to identify and quantify the various components by volume, including total brain white and gray matter content. Measurements were made at baseline and then repeated after four weeks of lithium at therapeutic doses. This study revealed an extraordinary finding that chronic lithium significantly increases *total gray matter content* in the brains of patients with BD [53] (Figure 20.4). No significant changes were observed in brain white matter volume, or in quantitative measures of regional cerebral water content, thereby providing strong evidence that the observed increases in gray matter content are likely due to neurotrophic effects as opposed to any possible cell swelling and/or osmotic effects associated with lithium treatment. A finer grained sub-regional analysis of this brain imaging data is ongoing. Since it is believed that the majority of neuron-specific NAA is localized to the neurites rather than the cell body, the observed increase in NAA is likely due to expansion of neuropil content. Taken together, these exciting new results – following up on the identification of bcl-2 in rodent RT-PCR DD studies and other preclinical studies (Table 20.3) – support the contention that lithium does indeed exert neurotrophic effects in the human brain *in vivo*.

20.6 Regulation of the Expression of an mRNA-Binding and Stabilizing Protein by Mood Stabilizers

In addition to bcl-2, another hitherto completely unexpected target for the actions of chronic lithium and VPA has been identified from the mRNA RT-PCR DD study described above. Another clone, also derived from a transcript whose levels were increased by both lithium and VPA, shows very strong homology to a human mRNA-binding protein, the AUH protein ([54, 55]; Genbank accession number X79888). BESTFIT analysis revealed 83.2% sequence homology between this rodent clone and the human AUH protein [54–56].

These findings are of considerable interest since, as we discuss in detail below, the AUH protein is a protein known to bind to AU-rich motifs in the 3′-untranslated region (3′-UTR) of various transcripts, thereby stabilizing them [54–56]. Thus, via their effects on the expression of the AUH protein, lithium and VPA could potentially *regulate the temporal and spatial patterns of the expression of multiple genes in the CNS*, effects which could potentially underlie their therapeutic effects in an illness as complex as BD.

20.6.1 Regulation of mRNA Stability by RNA-Binding Proteins

Increasing recent evidence suggests that far from being "passive way-stations" of encoded information and simple intermediates in the pathway from gene to protein, mRNA molecules may exhibit markedly distinct properties based on structural features embedded in discrete regions of the molecule (see [57, 58]). The regulation of mRNA stability has now emerged as a critical control step in determining cellular mRNA levels, with individual mRNAs displaying a wide range of stability that has been linked to discrete sequence elements and specific RNA-protein interactions (for excellent recent reviews see [58, 59]). Evidence is accumulating that strongly implicates the 3′-UTR of mRNA in the regulation of transcript stability, and thus steady-state mRNA levels. A common *cis* element found in the 3′-UTR of rapidly decaying mRNA is an AU-rich element (ARE), containing various numbers of AUUUA pentamers, sometimes associated with a general AU richness with a surplus of uridylic residues [57–59]. In hybrid constructs, AREs are able to confer rapid degradability to otherwise stable reporter transcripts.

Most relevant for the present discussion, it is noteworthy that various hormones and signaling pathways, including glucocorticoids and MAP (mitogen-activated protein) kinases, are now known to control the levels of certain mRNAs in part by regulating transcript stability [57–63]. Furthermore, the expression of a number of proteins that are important for CNS function is known to be markedly regulated by alterations in transcript stability. These proteins include the glucose transporter 1 (GLUT1), nerve growth factor (NGF), granulocyte-macrophage colony-stimulating factor (GM-CSF), tumor necrosis factor (TNF), interferons (INF), interleukins (IL1, IL3, IL6), tyrosine hydroxylase, growth cone associated protein-43 (GAP-43), the period (per) protein (a circadian rhythm regulator), c-fos, c-myc, and even bcl-2 [64].

Several classes of RNA-binding proteins have been implicated in regulating mRNA stability and turnover [58–60, 62, 63]. More recently, a subset of smaller (ranging from M_r 30000 to 40000) mRNA-binding proteins have been identified that display recognition of AU-rich domains in the 3′-UTR [58–60, 62, 63]. A human A+U-binding protein (AUH) has recently been cloned, and a growing body of data suggests that the regulation of mRNA stability by AUH plays an important role in a variety of physiological and pathophysiological processes. This has generated considerable excitement about the possibility that certain disorders of neuronal plasticity may also arise from pathological processes mediating the proper expression and targeting of genes by 3′-UTR-mediated processes.

The unexpected observation that chronic lithium and VPA increase the expression of a human mRNA-binding (and -stabilizing) protein, raises the possibility that these mood stabilizers may regulate the expression of other transcripts whose levels are critically regulated by AU-rich motifs in their 3′-UTRs (Figure 20.5). To identify these specific putative transcripts, we have undertaken a second round of mRNA RT-PCR DD studies, this time utilizing primers specifically designed to pick up transcripts containing AU-rich regions in the 3′-UTR (Figure 20.6).

mRNA RT-PCR DD, TA Cloning and Sequence Analysis were conducted as described in Section 20.3. The second round of DD studies is still ongoing; however, it has already led to the identification of a novel target. BESTFIT analysis revealed 92.6% identity between clone C2-1 (1 to 167) and the end of *Rattus norvegicus* Rho-associated kinase-alpha (ROK-alpha) mRNA, complete coding sequence (cds) (U38481; 4300 to the end). Furthermore, there was 99.8% identity between clone C2-1 (1–428) and a rat EST (expressed sequence tag) 5′-end sequence (AW918832; whole sequence), and 100% identity between clone C2-1 (344 to poly-A end) and a rat EST sequence (BE104475; whole sequence). ROK-alpha represents a Rho-associated, coiled-coil-containing kinase, and is known as ROCK-II [64a, 64b]. ROCK-II is a serine/threonine kinase known to play a major role as an effector for the Rho family of small G-proteins [65–67]. Rho family members have been demonstrated to regulate dendritic remodeling, neurite initiation, elongation,

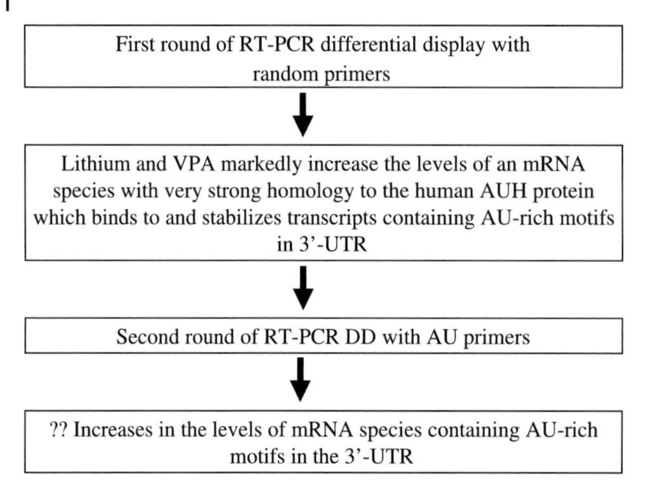

Fig. 20.5 Schematic representation of the second round of differential display studies undertaken to identify transcripts containing AREs in the 3′-UTR that are regulated by chronic lithium and VPA.

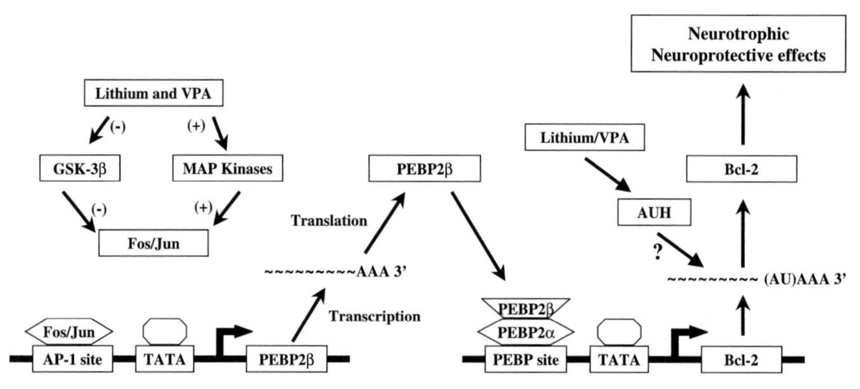

Fig. 20.6 Regulation of bcl-2 levels by chronic lithium and VPA: the role of MAP kinases and GSK-3. This figure depicts a mechanism by which lithium and VPA may regulate bcl-2 levels. Both lithium and VPA are known to regulate MAP kinases and GSK-3β, both of which should result in increases in AP-1 DNA-binding activity. PEBP2β is known to be regulated by AP-1 sites, and, thus, lithium and VPA produce an increase in PEBP2β transcript and protein levels, and ultimately in the function of PEBP2β. PEBP2β serves as a transcription factor in concert with PEBP2α and regulates the expression of many genes, including bcl-2. Bcl-2 is also known to contain AU-rich elements in its 3′-UTR, raising the possibility that lithium- and VPA-induced increases in AUH expression may also regulate bcl-2 mRNA stability and thus bcl-2 levels. It should be noted that this figure simply depicts two potential mechanisms by which lithium and VPA regulate bcl-2 levels. Additional mechanisms, most notably the regulation of bcl-2 by CREB (cAMP Response Element-binding protein), may also play a role.

and regeneration, axonal guidance, axonal outgrowth, cytoskeletal remodeling in neuronal growth cones in response to extracellular cues, and to play important roles in the maintenance of dendritic spines and branches in hippocampal pyramidal neurons [65–71]. While it is clear that several distinct effectors are involved in mediating these diverse functions of activated Rho [72–74], a growing body of evidence suggests that the Rho kinases (ROCK-I and ROCK-II) play critical roles in many of these functions.

In view of the important role of ROCK-II in mediating various neuroplastic events in the CNS, we undertook a study to determine if chronic lithium or VPA regulates the protein levels of ROCK-II. These studies showed that 3–4 weeks of lithium or VPA administration brings about an ~25–45% increase in the levels of particulate ROCK-II in FCx, with the lithium-induced increases reaching statistical significance.

20.7 The Pharmacogenomics of Bipolar Disorder: A Synthesis

We clearly still have much to learn about the mechanisms of action of mood-stabilizing agents, but the rate of progress in recent years has been exciting indeed. The behavioral and physiological manifestations of the recurrent mood disorders are complex and are likely mediated by a network of interconnected neurotransmitter pathways; thus, regulation of signal transduction and gene expression within critical regions of the brain represents an attractive target for psychopharmacological interventions. As discussed in the introduction, there are a number of impediments in our attempts to fully understand the molecular and cellular mechanisms of action of mood stabilizers. Foremost among these are the lack of suitable animal models, the relative paucity of easily detectable phenotypic changes induced by mood stabilizers, and the dearth of knowledge concerning the precise neuronal circuits and pathways underlying the etiology pathophysiology of a complex and heterogeneous group of disorders. Nevertheless, it is our strong conviction that it is at the cellular and molecular level that some of the most exciting advances in our understanding of the long-term therapeutic action of lithium and other mood stabilizers will take place in the coming years. Current studies of the long-term treatment-induced changes in signaling pathways and gene expression regulation are most promising avenues for investigation. The rapid technological advances in both biochemistry and molecular biology have greatly enhanced our ability to understand the complexities of the regulation of neuronal function; these advances hold much promise for the development of novel improved therapeutics for mood disorders, as well as for our understanding of the pathophysiology of these life-threatening illnesses. Several laboratories are starting to utilize the power of genomics and bioinformatics strategies to identify novel targets for the actions of mood stabilizers. Thus, other mRNA RT-PCR DD studies have shown that lithium increases 2′,3′-cyclic nucleotide 3′-phosphodiesterase mRNA levels in C6 glioma cells [34] and regulates the expression of a novel gene which shows significant homology to the yeast nitrogen permease regulator 2 (NPR2) [75], as well as aldolase A expression in the rodent brain [76]. Fewer studies have examined VPA's effects, but in addition to the studies described in this chapter, a recent study demonstrated that VPA increases the expression of the molecular chaperone GRP78 (glucose regulated protein 78) [77].

Interestingly, several of the novel targets recently identified, most notably bcl-2, ROCK-II, and GRP78 are known to play critical roles in neuroplasticity and cellular resilience. As discussed already, bcl-2 is widely regarded as a major cytoprotective and neurotrophic protein, whereas the endoplasmic reticulum (ER) chaperone protein GRP78 suppresses elevations of intracellular Ca^{2+} following exposure of neurons to glutamate, effects which appear to occur via suppression of Ca^{2+} from ryanodine-sensitive stores [78]. Although not as extensively studied as lithium, a growing body of data does indeed suggest that VPA also exerts neurotrophic and neuroprotective effects [79, 80]. These findings are noteworthy since a growing body of data suggests that while the severe mood disorders are *clearly* not classical neurodegenerative diseases, they are, in fact, associated with impairments of neuroplasticity and cellular resilience [6, 81, 82]. It is presently unclear to what extent the cell death and atrophy that occurs in mood disorders arises due to the magnitude and duration of the biochemical perturbations (e.g., glucocorticoid elevations), an enhanced vulnerability to the deleterious effects of these perturbations (due to genetic factors and/or early life events), or a combination thereof. While some data suggests that hippocampal atrophy in major depression is related to illness duration, it is presently not clear if the volumetric and cellular changes that have been observed in other brain areas (most notably frontal cortex) are related to affective episodes *per se*. This raises the intriguing possibility that the cell death and atrophy that occurs in BD may arise more from an endogenous impairment of cellular resiliency, whereas that observed in major depression may be more a manifestation of the neurotoxic sequelae of repeated affective episodes *per se*. It is, thus, noteworthy that a variety of strategies to enhance neurotrophic factor signaling are currently under investigation, efforts that are underway in large part due to recent pharmacogenomic studies.

20.8 Conclusions

In summary, the studies outlined in this chapter have demonstrated the utility of identifying both gene cluster categories as well as individual genes, which may represent therapeutically relevant targets for the actions of mood-stabilizing agents. Interestingly, several of the novel targets recently identified, most notably bcl-2, ROCK-II, and GRP78, are known to play critical roles in neuroplasticity and cellular resilience. These observations have generated considerable excitement among the clinical neuroscience community, and are reshaping views about the neurobiological underpinnings of bipolar disorder. Thus, it is now clear that regionally selective impairments of structural plasticity and cellular resiliency, which have been postulated to contribute to the development of classical neurodegenerative disorders, may also exist in bipolar disorder. The intracellular signaling cascades which are involved in regulating neuroplastic events and cell survival also affect the signal generated by multiple neurotransmitter and neuropeptide systems; alterations in these signaling pathways may, therefore, account for the findings of dysfunction in diverse neurochemical and neurophysiological systems in bipolar disorder. In sum, an increasing number of strategies have been utilized to identify genes whose expression in critical neuronal circuits may underlie the therapeutic actions of mood stabilizers; these developments hold much promise for the development of novel individualized therapeutics for the long-term treatment of severe mood disorders, and for improving the lives of millions.

Acknowledgements
The authors wish to acknowledge the support of the NIMH, NARSAD, and the Theodore and Vada Stanley Foundation. Outstanding editorial assistance was provided by Ms. Sarah Tsou and Ms. Kerri R. Gibala.

20.9 References

1 Goodwin FK. Manic-Depressive Illness. NewYork. Oxford University Press, New York; **1990**.
2 Murray CJ. Global mortality, disability, and the contribution of risk factors: Global Burden of Disease Study. Lancet **1997**; 349:1436–1442.
3 Manji HK. Signaling: Cellular insights into the pathophysiology of bipolar disorder. Biol Psych **2000**; 48:518–530.
4 Hyman SE, Initiation and adaptation: A paradigm for understanding psychotropic drug action. Am J Psych **1996**; 153:151–162.
5 Jope RS. Anti-bipolar therapy: Mechanism of action of lithium. Mol Psych **1999**; 4:117–128.
6 Manji HK, Rajkowska G, Chen G. Neuroplasticity and cellular resilience in mood disorders. Millennium article. Mol Psych **2000**; 5:578–593.
7 Wang JF, Li PP et al. Signal transduction abnormalities in bipolar disorder in: Bipolar Disorder: Biological Models and their Clinical Application (Joffe RT and Young LT, Eds), Marcel Dekker, New York; **1997**, pp 41–79.
8 Kandel ER. A new intellectual framework for psychiatry. Am J Psychiatry **1998**; 155:457–469.
9 Ikonomov O, Manji HK. Molecular Mechanisms Underlying Mood-Stabilization in Manic-Depressive Illness: The Phenotype Challenge. Am J Psychiatry **1999**; 156:1506–1514.
10 Flint J. Do animal models have a place in the genetic analysis of quantitative human behavioral traits? J Mol Med **1996**; 74:515–521.
11 Hahn CG. Abnormalities in protein kinase C signaling and the pathophysiology of bipolar disorder. Bipolar Disorders **1999**; 1:81–86.
12 Jope RS. Lithium and brain signal transduction systems. Biochem Pharm **1994**; 77:429–441.
13 Manji HK. Protein kinase C signaling in the brain: Molecular transduction of mood stabilization in the treatment of bipolar disorder. Biol Psychiatry **1999**; 46:1328–1351.
14 Hughes P. Induction of immediate-early genes and the control of neurotransmitter-regulated gene expression within the

nervous system. Pharmacol Rev **1995**; 47:133–178.

15 Asghari V, Reiach JS, Young LT. Differential effects of mood stabilizers on Fos/Jun proteins and AP-1 DNA binding activity in human neuroblastoma SH-SY-5Y cells. Mol Brain Res **1998**; 58:95–102.

16 Chen G, H.K., Bebchuk JM et al. Regulation of signal transduction pathways and gene expression by mood stabilizers and antidepressants. Psychosomat Med **1999**; 61:599–617.

17 Chen G, Y.P., Jiang Y et al. Valproate robustly enhances AP-1 mediated gene expression. Mol Brain Res **1999**; 64:52–58.

18 Ozaki N. Lithium increases transcription factor binding to AP-1 and cyclic AMP-responsive element in cultured neurons and rat brain. J Neurochem **1997**; 69:2336–2344.

19 Unlap MT. Lithium attenuates nerve growth factor-induced activation of AP-1 DNA binding activity in PC12 cells. Neuropsychopharmacology **1997**; 17:12–17.

20 Williams MB. Circadian variation in rat brain AP-1 DNA binding activity after cholinergic stimulation: modulation by lithium. Psychopharmacology (Berlin), **1995**; 122:363–368.

21 Yuan PX, Manji HK. Lithium Stimulates Gene Expression Through the AP-1 Transcription Factor Pathway. Mol Brain Res **1998**; 58:225–230.

22 Kumer SC. Intricate regulation of tyrosine hydroxylase activity and gene expression. J Neurochem **1996**; 67:443–461.

23 Drevets WC, Krishnan KR. Neuroimaging studies of mood disorders in: Neurobiology of Mental Illness (Charney DS, Nestler EJ, Bunney BS, Eds) Oxford University Press, New York; **1999**, pp 394–418

24 Chen G, Y.P., Jiang Y et al. Lithium increases tyrosine hydroxylase levels both *in vivo* and *in vitro*. J Neurochem **1998**; 70:1768–1771.

25 Haley JE. Gases as neurotransmitters. Essays Biochem **1998**; 33:79–91.

26 Haul S, Schrader J, Haas HL, Luhmann HJ. Impairment of neocortical long-term potentiation in mice deficient of endothelial nitric oxide synthase. J Neurophysiol **1999**; 81:494–497.

27 Lipton SA. Neuronal protection and destruction by NO. Cell Death Differ **1999**; 6:943–951.

28 Murphy S. Production of nitric oxide by glial cells: regulation and potential roles in the CNS. Glia **2000**; 29:1–13.

29 Chen G, Manji HK. Lithium regulates PKC-mediated intracellular cross-talk and gene expression in the CNS *in vivo*. Bipolar Disorders **2000**; 2:217–236.

30 Feinstein DL. Potentiation of astroglial nitric oxide synthase type-2 expression by lithium chloride. J Neurochem **1998**; 71:883–886.

31 Zigova T, Tedesco EM, Borlongan CV, Saporta S, Snable GL, Sanberg PR. Lithium chloride induces the expression of tyrosine hydroxylase in hNT neurons. Exp Neurol **1999**; 157:251–258.

32 Manji HK, Chen G. Lithium at 50: Have the neuroprotective effects of this unique cation been overlooked? Biol Psych **1999**; 46:929–940.

33 Manji HK, Chen G. Lithium upregulates the cytoprotective protein bcl-2 *in vitro* and in the CNS *in vivo*: a role for neurotrophic and neuroprotective effects in manic-depressive illness. J Clin Psychiatry **2000**; 61:82–96.

34 Wang JF. Differential display PCR reveals increased expression of 2′,3′-cyclic nucleotide 3′-phosphodiesterase by lithium. FEBS Lett **1996**; 386:225–229.

35 Chen G, Jiang Y et al. The mood stabilising agent valproate inhibits the activity of glycogen synthase kinase 3. J Neurochem **1999**; 72:1327–1330.

36 Ogawa E, Maruyama M et al. Molecular cloning and characterization of PEBP2 beta, the heterodimeric partner of a novel *Drosophila* runt-related DNA binding protein PEBP2 alpha. Virology **1998**; 194:314–331.

37 Liu P, Hajra A et al. Fusion between transcription factor CBF beta/PEBP2 beta and a myosin heavy chain in acute myeloid leukemia. Science **1993**; **261**:1041–1044.

38 Kagoshima H, Ito Y, Shigesada KJ. Functional dissection of the alpha and beta subunits of transcription factor PEBP2 and the redox susceptibility of its

DNA binding activity. J Biol Chem **1996**; 271:33074–33082.

39 Wang S, Crute BE, Melnikova IN, Keller SR, Speck NA. Cloning and characterization of subunits of the T-cell receptor and murine leukemia virus enhancer core-binding factor. Mol Cell Biol **1993**; 13:3324–3339.

40 Chen G, Jiang L et al. The mood stabilizing agents lithium and valproate robustly increase the expression of the neuroprotective protein bcl-2 in the CNS. J Neurochem **1999**; 72:879–882.

41 Rajkowska G. Postmortem studies in mood disorders indicate altered numbers of neurons and glial cells. Biol Psychiatry **2000**.

42 Adams JM. The Bcl-2 protein family: arbiters of cell survival. Science **1998**; 281:1322–1326.

43 Bruckheimer EM, Sarkiss M, Herrmann J, McDonnell TJ. The Bcl-2 gene family and apoptosis. Adv Biochem Eng Biotechnol **1998**; 62:75–105.

44 Merry DE. Bcl-2 gene family in the nervous system. Ann Rev Neurosci **1997**; 20:245–267.

45 Chen G, Du F, Seraji-Bozorgzad N, Manji HK. Enhancement of Hippocampal Neurogenesis by Lithium. J Neurochem **2000**; 75:1729–1734.

46 Chen RW, Chuang DM. Long term lithium treatment suppresses p53 and Bax expression but increases bcl-2 expression. J Biol Chem **1999**; 274:6039–6042.

47 Nonaka S, Chuang DM. Chronic lithium treatment robustly protects neurons in the central nervous system against excitotoxicity by inhibiting N-methyl-D-aspartate receptor-mediated calcium influx. Proc Natl Acad Sci USA **1998**; 95:2642–2647.

48 Post RM, Clark M, Chuang DM, Hough C, Li H. Lithium Carbamazepine, and Valproate in Affective Illness: Biological and Neurobiological Mechanisms in: Bipolar Medications: Mechanisms of Action. Manji HK, Bowden CL, Belmaker RH Eds. Washington DC, Am Psychiatric Press **2000**; 219–248.

49 Tsai KC, C.V., Kohn DT, Neve RL, Perrone-Bizzozero NI. Post-transcriptional regulation of the GAP-43 gene by specific sequences in the 3′ untranslated region of the mRNA. J Neurosci **1997**; 17:1950–1958.

50 Bates TE, S.M., Keelan J, Davey GP, Munro PM, Clark JB. Inhibition of N-acetylaspartate production: implications for 1H MRS studies *in vivo*. Neuroreport **1996**; 7:1397–1400.

51 Tsai G. N-acetylaspartate in Neuropsychiatric Disorders. Prog Neurobiol **1995**; 46:531–540.

52 Moore GJ, Hasanat K, Chen G, Seraji-Bozorgzad N, Wilds IB, Faulk MW, Koch S, Jolkovsky L, Manji HK. Lithium Increases N-Acetyl-Aspartate in the Human Brain: *In Vivo* Evidence in Support of bcl-2's Neurotrophic Effects. Biol Psychiatry **2000**; 48:1–8.

53 Moore GJ, Wilds IB, Chen G, Manji HK. Lithium-induced increase in Human Brain Gray Matter. Lancet **2000**; 356:1241–1242.

54 Nakagawa J, Nakagawa J, Moroni C. A 20-amino-acid autonomous RNA-binding domain contained in an enoyl-CoA hydratase. Eur J Biochem **1997**; 244:890–899.

55 Nakagawa J, Meyer-Monard S, Hofsteenge J, Jeno P, Moroni C. AUH, a gene encoding an AU-specific RNA binding protein with intrinsic enoyl-CoA hydratase activity. Proc Natl Acad Sci USA **1995**; 92:2051–2055.

56 Brennan LE, Egger D, Bienz K, MoroniC. Characterisation and mitochondrial localisation of AUH, an AU-specific RNA-binding enoyl-CoA hydratase. Gene **1999**; 228:85–91.

57 Conne B, Vassalli JD. The 3′ untranslated region of messenger RNA: A molecular 'hotspot' for pathology? Nature Med; 6:637–641.

58 Staton JM, Leedman PJ. Hormonal regulation of mRNA stability and RNA-protein interactions in the pituitary. J Mol Endocrinol **2000**; 25:17–34.

59 Wilson KF. Signal transduction and post-transcriptional gene expression. Biol Chem **2000**; 381:357–365.

60 Chen CY, Wu Z, Karin M. Stabilization of interleukin-2 mRNA by the c-Jun NH2-terminal kinase pathway. Science **1998**; 280:1945–1949.

61 Chen CY, Schotland P, Sehgal A. Alterations of per RNA in noncoding regions affect periodicity of circadian behavioral rhythms. J Biol Rhythms **1998**; 13:364–379.
62 Chen CY. AU-rich elements: characterization and importance in mRNA degradation. Trends Biochem Sci **1995**; 20:465–470.
63 Derrigo M, Savettieri G, Di Liegro I. RNA-protein interactions in the control of stability and localization of messenger RNA (review). Int J Mol Med **2000**; 5:111–123.
64 Chen G, Manji HK. Mood stabilizers regulate cytoprotective and mRNA binding proteins in the brain: Long term effects on cell survival and transcript stability. Int J Neuropsychopharmacol **2001**; 4:47–64.
64a Leung T, Chen XQ, Manser E, Lim L. The p160 RhoA-binding kinase ROK alpha is a member of a kinase family and is involved in the reorganization of the cytoskeleton. Mol Cell Biol **1996**; 16(10): 5313–5327.
64b Nagakawa O, Fujisawa K, Ishizaki T, Saito Y, Nakao K, Narumiya S. ROCK-I and ROCK-II, two isoforms of Rho-asociated coiled-coil forming protein serine/threonine kinase in mice. FEBS Lett **1996**; 392(2):189–193.
65 Bito H, Ishihara H, Shibasaki Y, Ohashi K, Mizuno K, Maekawa M, Ishizaki T, Narumiya S. A critical role for a Rho-associated kinase, p160ROCK, in determining axon outgrowth in mammalian CNS neurons. Neuron **2000**; 26:431–441.
66 Katoh H, Ichikawa A, Negishi M. p160 RhoA-binding kinase ROKalpha induces neurite retraction. J Biol Chem **1998**; 273:2489–2492.
67 Takahashi N, Saya H, Kaibuchi K. Localization of the gene coding for ROCK II/Rho kinase on human chromosome 2p24. Genomics **1999**; 55:235–237.
68 Gallo G. Axon guidance: GTPases help axons reach their targets. Curr Biol **1998**; 8:R80–82.
69 Luo L, Jan YN. Small GTPases in axon outgrowth. Perspect Dev Neurobiol **1996**, 4:199–204.
70 Luo L, Jan YN. Rho family GTP-binding proteins in growth cone signalling. Curr Opin Neurobiol **1997**; 7:81–86.
71 Nakayama AY, Luo L. Small GTPases Rac and Rhoin the maintenance of dendritic spines and branches in hippocampal pyramidal neuronsi. J Neurosci **2000**; 20:5329–5338.
72 Bishop AL. Rho GTPases and their effector proteins. Biochem J **2000**; 348:241–255.
73 Evers EE, Malliri A, Price LS, ten Klooster JP, van der Kammen RA, Collard JG. Rho family proteins in cell adhesion and cell migration. Eur J Cancer **2000**; 36:1269–1274.
74 Kaibuchi K, Amano M. Regulation of the cytoskeleton and cell adhesion by the Rho family GTPases in mammalian cells. Annu Rev Biochem **1999**; 68:459–486.
75 Wang JF, Young LT. Identification of a novel lithium regulated gene in rat brain. Mol Brain Res **1999**; 70:66–73.
76 Hua LV, G.M., Warsh JJ, Li PP. Lithium regulation of aldolase A expression in the rat frontal cortex: identification by differential display. Biol Psychiatry **2000**; 48:58–64.
77 Wang JF, Young LT. Differential display PCR reveals novel targets for the mood-stabilizing drug valproate including the molecular chaperone GRP78. Mol Pharmacol **1999**; 55:521–527.
78 Mattson MP, Chan SL, Leissring MA, Shepel PN, Geiger JD. Calcium signaling in the ER: Its role in neuronal plasticity and neurodegenerative disorders. Trends Neurosci **2000**; 23:222–229.
79 Bruno V, Scapagnini U, Nicoletti F, Canonico P. Antidegenerative effects of Mg(2+)valproate in cultured cerebellar neurons. Funct Neurol **1995**; 10:121–130.
80 Mora A, Fuentes JM, Soler G, Centeno F. Different mechanisms of protection against apoptosis by valproate and Li+. Eur J Biochem **1999**; 266:886–891.
81 Duman RS, Nestler EJ. A molecular and cellular theory of depression. Arch Gen Psychiatry **1997**; 54:597–606.
82 Manji HK, Charney DS. The cellular neurobiology of depression. Nature Med **2001**; 7:541–547.

21
Pharmacogenomics of Alcoholism

Thomas D. Hurley, Howard J. Edenberg and Ting-Kai Li

Abstract

Alcoholism is a complex behavior that is affected by both environmental and genetic factors. Predisposition to alcohol dependence is influenced by personality characteristics that include sensation seeking, behavioral disinhibition and poor decision-making. The progression to dependence is influenced by additional factors; including sensitivity to the positive and negative effects of alcohol and the capacity for tolerance to alcohol's aversive effects. To date, only the genes encoding the alcohol and aldehyde dehydrogenases have been firmly linked to vulnerability to alcoholism. Certain of these alcohol and aldehyde dehydrogenase genes also affect risk for complications associated with alcohol abuse, including alcoholic liver disease, digestive tract cancer, heart disease and fetal alcohol syndrome. A large genomic survey of over 11,000 people is underway as part of the Collaborative Study on the Genetics of Alcoholism (COGA). These studies have identified five regions of the genome that affect risk for alcoholism on chromosomes 1, 2, 3, 4 and 7. Analysis of the COGA population using more specific and homogeneous endophenotypes has identified additional genomic regions associated with these traits. This review discusses the current state of genetic research in alcoholism and summarizes recent advances in our understanding of this complex disorder.

21.1
Introduction

The term "alcoholism" as a disease entity was coined by the Swedish physician, Magnus Huss, in the mid-19th century to describe the harmful physical and mental effects of chronic excessive alcohol consumption. This strictly medical model held sway for almost a century before it became apparent that a variety of psychosocial factors also influence the onset and course of the disorder. Indeed, drinking behavior and the problems attributable to excessive drinking, including alcoholism, vary widely within and across different cultures and population groups, and even within the same person across the life span. In the last 30 years, basic and clinical research have shown that drinking behavior is a complex trait influenced by both biological and environmental factors.

Twin, family and adoption studies have provided compelling evidence that there is genetic as well as environmental risk for developing alcoholism. Genetic and environmental factors contribute approximately equal proportions to risk in both men and women [1].

The study of the natural history of alcoholism has shown that certain personality/temperament characteristics predispose to alcohol abuse and dependence. They include sensation seeking, behavioral disinhibition, and poor executive functioning in cognition and decision-making. There is substantial genetic influence on these traits; however, these traits are not specific to alcohol abuse and alcoholism. They affect deviant behavior in general and appear to be particularly important in the early initiation of drinking. Thereafter, alcohol-specific risk factors become salient in the progression to habitual drinking and alcoholic drinking. These include: sensitivity to the positive (hedonic) and the negative (aversive) effects of alcohol, the capacity to develop tolerance to the aversive effects of alcohol, and susceptibility to alcohol dependence. Studies in human twin samples and in animal models have shown that there is strong genetic influence and wide variation across individuals on all of these alcohol-specific traits, thereby qualifying alcoholism as a pharmacogenetic disorder [2].

This chapter reports the advances in our understanding of the pharmacogenetics of alcoholism and responses to alcohol and the search for genes that confer susceptibility to this common complex disease.

21.2 Definitions and Diagnostic Criteria

Today, alcoholism is recognized as a disease characterized by impaired regulation of alcohol consumption that, over time, leads to: (1) impaired control over drinking; (2) tolerance; (3) psychological dependence (craving); and (4) physical dependence (withdrawal signs upon cessation). The terms alcohol dependence and alcohol addiction are used synonymously with alcoholism. The neuroadaptive changes to chronic, excessive alcohol exposure, viz, tolerance and dependence, are thought to contribute to impaired control over drinking and the relapsing nature of the disorder. These cardinal features of the so-called “alcohol dependence syndrome” (3) form the basis of the current DSM (4) and ICD (5) criteria for diagnosis of alcoholism. The diagnosis of alcohol dependence requires meeting three of the seven DSM IV criteria and three of the six ICD10 criteria. These include one or more items on: tolerance, withdrawal, impaired control, more time spent in drink-related activities, drinking despite problems, and compulsion to drink. Alcohol drinking leading to recurring social problems, but not meeting three or more of the alcoholism criteria is termed alcohol abuse.

Based upon DSM diagnostic criteria, the lifetime prevalence of alcoholism in the adult American population is estimated to be 12%, and that of alcohol abuse,

5%. The one-year (current) prevalence for dependence is 5% and for abuse, 3%. Although there are biomarkers of excessive alcohol consumption (e.g., γ-glutamyl transferase, carbohydrate-deficient transferrin), quantity and frequency of drinking are not used as part of the diagnostic criteria. Because vulnerability to alcohol abuse and alcohol dependence varies greatly among individuals, it is difficult to assess the risk for abuse and dependence in relation to how much a person drinks. There are no other surrogate markers for alcoholism. The best predictor for alcoholism remains family history.

21.3 Alcohol Pharmacokinetics and Metabolism

The effects of ethanol on bodily functions, e.g., those of the brain, heart, and liver, are dependent upon the systemic concentrations of ethanol over time. Therefore, the pharmacokinetics of ethanol play a pivotal role in the pharmacodynamic actions of ethanol and of its metabolic product acetaldehyde [6].

After oral ingestion, ethanol pharmacokinetics must take into account: (1) Absorption from the gastrointestinal tract. Since ethanol is absorbed most efficiently from the small intestines, the rate of gastric emptying is an important factor that governs the rate of rise of blood alcohol concentration (BAC), i.e., the slope of the ascending limb of the BAC–time curve, and the extent of first pass metabolism of ethanol by the liver and stomach. (2) Distribution of ethanol in the body. Ethanol distributes equally in total body water, which is related to the lean body mass of the person, and (3) the elimination of ethanol from the body, which occurs primarily by metabolism in the liver, first to acetaldehyde and then to acetate [7].

All three processes are influenced by both genetic and environmental factors. For example, gastric emptying and ethanol absorption vary with the concentration of ethanol in the beverage, the rate of ingestion of ethanol, and the presence of food in the stomach or its concomitant ingestion. The peak BAC and the time to reach peak BAC are influenced by genetic factors. Total body water is related to height and body weight, both of which are influenced by genetic as well as environmental factors. Alcohol elimination rate varies as much as 4-fold from person to person. Studies in monozygotic and dizygotic twins have shown that the heritability of alcohol elimination rate (i.e., the genetic component of variance) is about 40–50% [7]. Ethanol metabolic rate is influenced by the genetic variations in the principal alcohol metabolizing enzymes, cytosolic alcohol dehydrogenase (ADH) and mitochondrial aldehyde dehydrogenase (ALDH2), discussed below.

Both ADH and ALDH use NAD^+ as cofactor in the oxidation of ethanol to acetaldehyde. The rate of alcohol metabolism is determined not only by the amount of ADH and ALDH2 enzyme in tissue and by their functional characteristics, but also by the concentrations of the cofactors NAD^+ and NADH and of ethanol and acetaldehyde in the cellular compartments (i.e., cytosol and mitochondria). Environmental influences on elimination rate can occur through changes in the redox ratio of NAD^+/NADH and through changes in hepatic blood flow. The equilibrium of the ADH reaction is poised toward the formation of ethanol and NAD^+, whereas that of the ALDH2 reaction is very strongly directed toward the oxidation of acetaldehyde to acetate. Accordingly, acetaldehyde exerts strong product inhibition on the ADH reaction and the elimination of acetaldehyde is the most critical factor for ethanol to be oxidized rapidly. The total activity of ALDH2 (amount of enzyme and its functional properties) becomes a key determinant of the rate of ethanol metabolism [7]. In agreement with this scheme, the usual concentrations of ethanol and acetate in the circulation during ethanol oxidation are millimolar, whereas that of acetaldehyde is less than 10 micromolar.

21.4 Pharmacodynamic Effects of Ethanol on the Brain

The acute central nervous system (CNS) and behavioral effects of ethanol are biphasic. At low ethanol concentrations and upon the rising limb of the alcohol–time curve after oral ingestion, the effect is stimulatory. At moderate to high concentrations and upon the descending limb of the alcohol–time curve, the effect is inhibitory. A number of neurotransmitter and neuro-modulatory systems have been shown to be directly or indirectly influenced by ethanol: γ-aminobutyric acid (GABA), glutamate, serotonin (5HT), dopamine, (DA), and neuropeptides, including endogenous opioids, corticotrophin releasing hormone (CRF) and neuropeptide Y (NPY). Ethanol has been shown to bind directly to GABAergic and glutamatergic receptors. With chronic ethanol exposure, behavioral- and neuro-adaptions to the sedative-hypnotic/inhibitory effects of ethanol occur, as evidenced by behavioral tolerance and a shift of the ethanol dose–response curve to higher ethanol concentrations. Upon cessation of chronic exposure, there is a rebound excitation, or a withdrawal reaction, indicative of physical dependence.

Studies in human twins have shown that there is a genetic component of variance for drinking behavior as well as for the acute CNS response to the sedative/hypnotic effects of ethanol as measured by the electroencephalogram. Acute response is defined here as the reaction to alcohol within a single session of exposure to ethanol-experienced subjects and encompasses the development of acute or within session tolerance [8]. Studies in rodent animal models of alcoholism have demonstrated genetic influence on not only drinking behavior, but also on susceptibility to physical dependence, and acute tolerance development in ethanol naïve animals (see below). Recent studies further show that selectively bred alcohol-preferring animals are more likely to develop loss of control drinking and more severe loss of control drinking after bouts of repeated voluntary exposure and forced abstinence. Therefore, there appears also to be a genetic component of variance to loss of control drinking [9, 10].

21.5
Structural and Kinetic Features of Alcohol and Aldehyde Dehydrogenase

The majority of ingested alcohol is metabolized in the liver by the action of two enzymatic systems: alcohol dehydrogenase (ADH) and aldehyde dehydrogenase (ALDH). In the liver, there are a number of ADH and ALDH isozymes encoded by separate genes. The ADH isozymes responsible for the metabolism of the majority of beverage ethanol in the liver are the *ADH1A, ADH1B, ADH1C* and *ADH4* gene products (Table 21.1). There is additional complexity within the ADH isozyme system: functional polymorphisms at both the *ADH1B* and *ADH1C* loci give rise to the *ADH1B*1, ADH1B*2* and *ADH1B*3* variants and the *ADH1C*1* and *ADH1C*2* variants (Table 21.2) [7]. Only two aldehyde dehydrogenase isozymes, the products of the *ALDH1A1* (formerly *ALDH1)* and *ALDH2* genes, are thought to contribute to the metabolism of acetaldehyde [7]. The well known functional polymorphism of the *ALDH2* gene gives rise to two common genetic variants in the human population, the *ALDH2*1* and the *ALDH2*2* alleles [7]. This section will focus on the structural and kinetic properties of these isozymes and their polymorphic variants.

21.5.1
Alcohol Dehydrogenase

All forms of human ADH are dimeric zinc-metalloenzymes comprised of subunits with molecular masses of approximately 40,000, and are located in the cytoplasm. The different ADH isozymes are grouped into five classes based on their

Tab. 21.1 Kinetic properties of the human ADH isozymes

New gene name [1)]	*Old gene name*	*Subunit*	*Class*	*K_m [ethanol] [mM]*	*Turnover rate [min^{-1}]*	*% liver contribution at 22 mM ethanol* [2)]
ADH1A	*ADH1*	α	I	4.0	30	8.1
*ADH1B*1*	*ADH2*	β_1	I	0.05	4	21.8
*ADH1C*1*	*ADH3*	γ_1	I	1.0	90	41.5
ADH4	*ADH4*	π	II	30	20	28.6
ADH5	*ADH5*	χ	III	>1000	100	<1
ADH6	*ADH6*	ADH6	V	?	?	<1
ADH7	*ADH7*	σ	IV	30	1800	<1

1) Official gene nomenclature, HUGO Gene Nomenclature Committee, 25 April 2001 (*http://www.gene.ucl.ac.uk/nomenclature/ADH.shtml*)

2) Estimated based on published protein expression (19) and converted to total number of subunit active sites per g of liver tissue. The Michaelis–Menton parameters for ethanol oxidation at pH 7.5 were utilized to calculate individual isozyme activities as a total activity per g of tissue and then normalized as a percent of total activity per μg of tissue following summation of all isozyme activities

Tab. 21.2 Characteristics of the human class I isozymes and their polymorphic variants

Gene	*Subunit*	*Amino acid substitution in polymorphic variants*	*K_m [Ethanol] [mM]*	*Turnover rate [min^{-1}]*	*Three-dimensional structure*
ADH1A	α	None	4.0	20	[12]
*ADH1B*1*	β_1	Arg47, Arg369	0.05	4.0	[12]
*ADH1B*2*	β_2	His47, Arg369	0.9	350	[89]
*ADH1B*3*	β_3	Arg47, Cys369	40	300	[90]
*ADH1C*1*	γ_1	Arg271, Ile349	1.0	90	ND [1)]
*ADH1C*2*	γ_2	Gln271, Val349	0.6	40	[12]

The kinetic constants are based on published values at pH 7.5 [7, 21]

1) The three-dimensional structure of this isozyme has not yet been determined

amino acid sequence identity and enzymatic properties (Table 21.1). The human Class I isozymes share >93% sequence identity and can form both homodimers and heterodimers, while the Class II, Class III and Class IV isozymes share 60–70% sequence identity with the Class I isozymes or each other and have only been observed to form homodimers. Due to their tissue distribution and kinetic properties, only the Class I and Class II isozymes contribute significantly to ethanol metabolism in the liver. One can estimate the relative contributions of the human Class I and II isozymes toward liver ethanol metabolism based on isozyme content [11] and kinetic properties (Table 21.1). The Class I isozymes account for approximately 70% of the total ethanol oxidizing activity at 22 mM ethanol (100 mg dL^{-1}). Surprisingly, due to high expression levels in the human liver, the Class II isozyme could account for fully 29% of the ethanol oxidation activity, even though the isozyme is only operating at about 50% of its maximal rate at this concentration. It is significant that over 50% of the total ethanol oxidation activity is catalyzed by those isozymes that are polymorphic, the products of *ADH1B* and *ADH1C* genes, since changes in the kinetic properties of these isozymes will have the largest impact.

The polymorphic variants of the *ADH1B* and *ADH1C* genes have been studied in great detail [7]. Their kinetic properties are known and the three-dimensional structures for four of the five isozymes have been determined by X-ray crystallography (Table 21.2). While the overall structures of the Class I isozymes are essentially identical, specific localized differences in the structures account for the large variations in their kinetic properties [12]. One feature that emerges from the relationships between the Class I polymorphic variants is that mutations within the coenzyme-binding site of the enzyme structure are a prevalent theme (Figure 21.1). This leads to relatively large changes in the kinetic constants primarily because the release of NADH is the slowest (rate-limiting) step in the enzymatic reaction [13]. In all cases, those polymorphic variants that possess faster turnover rates do so because weaker binding of NADH leads to faster release.

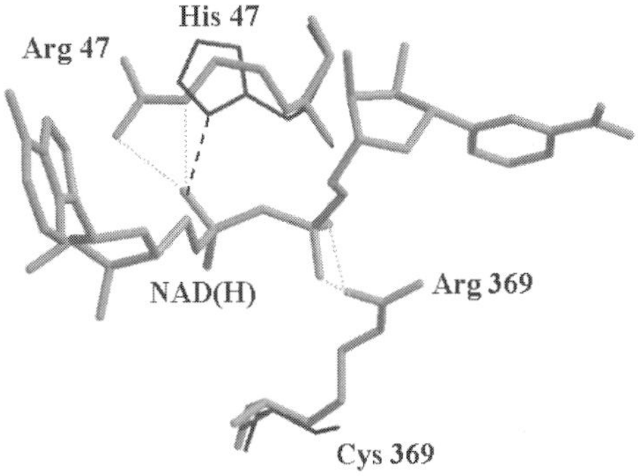

Fig. 21.1 The interactions between the bound coenzyme molecule and the amino acids at positions 47 and 369 in the β_1, β_2, and β_3 polymorphic variants as observed in their respective structures determined by X-ray crystallography. The dashed lines indicate possible hydrogen-bonds between the amino acids and the phosphate oxygens of the bound coenzyme molecule, NAD(H). Arg47 is substituted by a His residue in the β_2 isozyme and Arg369 is substituted by a Cys residue in the β_3 isozyme. In each case, the substitution results in a net loss of hydrogen-bonding interactions and weaker affinity for the coenzyme.

Although the majority of ingested ethanol is metabolized by the liver, a small but significant fraction of the total metabolism takes place during the so-called first pass metabolism prior to ethanol's entry into the systemic circulation. The location of this metabolism is still somewhat controversial, but it can include the initial pass through the liver on the way into the systemic circulation, as well as the epithelial tissues lining the stomach that contain relatively high levels of the Class IV alcohol dehydrogenase encoded by *ADH7*. This enzyme possesses a high catalytic turnover rate, but a relatively high K_m for ethanol. Therefore, although it is not saturated with ethanol as a substrate at typical blood alcohol concentrations, it is likely to be saturated at the much higher concentrations present in stomach during ingestion [14]. In addition, the Class IV isozyme is particularly efficient at oxidizing retinol to retinal as part of the pathway to generate the potent developmental hormone, retinoic acid [14]. *ADH7* is expressed in the appropriate tissues and at the appropriate times during embryonic development to participate in developmental progression and has led to a hypothesis that competition between the substrates ethanol and retinol for this particular ADH isozyme may ultimately lead to the developmental changes associated with fetal alcohol syndrome [15].

21.5.2 Aldehyde Dehydrogenase

Both the *ALDH1A1* and *ALDH2* gene products are tetrameric enzymes with subunit molecular masses of approximately 55,000. The product of the *ALDH1A1* gene is found in the cytosol; the product of the *ALDH2* gene is imported into the mitochondria [7]. The three-dimensional structures for both the ALDH1A1 and ALDH2 enzymes have been determined by X-ray crystallography and, as their 70% sequence identity would suggest, their structures are very similar [16, 17]. The major differences between the isozymes are localized to the substrate-binding site and correlate well with their known preferences for different substrates [17]. These two isozymes possess similar, but not identical, kinetic properties. Most studies agree that the bulk of the acetaldehyde is metabolized in the mitochondria by ALDH2, based on the extremely low K_m for acetaldehyde (2 micromolar) exhibited by ALDH2 and the low circulating levels of acetaldehyde, which are generally 5-fold lower than the K_m for acetaldehyde exhibited by ALDH1A1 (30 micromolar). Thus, in order for the ALDH1A1 enzyme to function significantly in acetaldehyde oxidation, the concentration of acetaldehyde must rise to a level approximately 5 to 10-fold higher than that normally maintained by the cell. This is precisely what happens when the activity of ALDH2 is reduced due to the presence of the *ALDH2*2* allele. The kinetic properties of ALDH2*2 are known; its affinity for NAD^+ and its turnover rate are reduced by approximately 200-fold and 10-fold, respectively [18]. ALDH2*2 is essentially inactive *in vivo* because the intracellular concentrations of NAD^+ are approximately 15-fold lower than its K_m value, thereby reducing its already 10-fold lower turnover rate (defined at full substrate saturation) by an additional 30-fold.

The inactive enzyme results from the substitution of a lysine for the normal glutamate at position 487 of the 500 amino acid chain [19]. Studies on the level of enzyme activity indicated that the inactive (*ALDH2*2*) allele was dominant, with both homozygous and heterozygous individuals having undetectable enzyme activity in liver extracts [20]. It was further shown that heterozygous individuals expressed all five combinations of heterotetramers possible based on random association between two different types of subunits (4:0, 3:1, 2:2, 1:3, 0:4) [18]. It is perhaps more precise to define the *ALDH2*2* allele as semi-dominant, because more recent and precisely defined phenotypic measurements show significant differences between hetero- and homozygotes in facial flushing and risk for alcoholism.

Based on the chemical characteristics of the two amino acids, it was understood that this is a non-conservative change and was likely to lead to substantial changes in enzymatic function. However, it was not clear until the structure of the enzyme was determined how the substitution might actually lead to the observed changes in the enzyme's behavior and to the dominance of the inactive allele in the tetramer. The substituted lysine at position 487 (Lys487) occurs at the interface between two subunits that make up the tetramer of ALDH2 (Figure 21.2) [16]. The normal acidic glutamate (Glu487) forms two important ionic interactions with surrounding basic amino acids that help to stabilize the subunit interface. This interface is located in very close proximity to the active site of each subunit (Figure 21.2). The substitution of Lys487 at this critical position places three basic residues (the two original and the additional Lys487, all positively charged) in very close proximity. These like charges repel each other and will lead to a disruption of this crucial interface and consequently the active site of the enzyme. The dominance of the inactive allele in a tetramer is also due to its location at the subunit interface and the unique nature of this interface. The mutation actually occurs at the interface between dimers in the tetramer (the tetramer can be thought of as a dimer of dimers), but the opposite side of this interface region also forms the interface between dimers in the tetra-

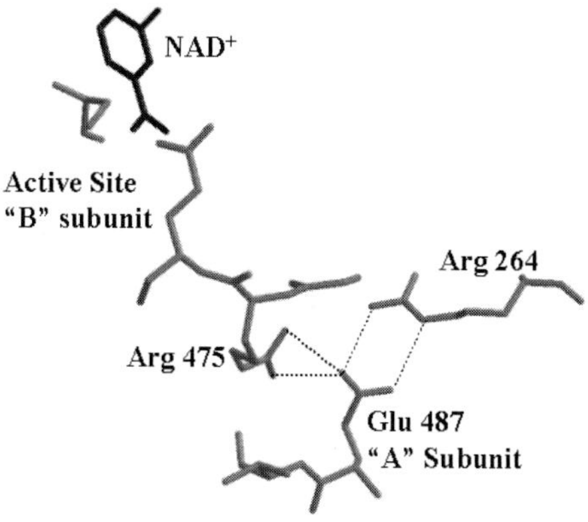

Fig. 21.2 The interactions between the glutamate residue at position 487 in human ALDH2 with the surrounding protein structure as determined by X-ray crystallography. Glu487, a negatively charged amino acid, interacts closely with two positively charged amino acids, Arg264 and Arg475, as indicated by the dashed lines. The interaction with Arg475 is across the interface between subunits in the ALDH2 tetramer. The interaction takes place very close to the active site in the second subunit. Thus, structural changes induced by the substitution of the positively charged lysine at position 487 could disrupt the active site structure and thereby inactivate the enzyme.

mer. Thus, the structural perturbation of one subunit can propagate changes to all subunits in the tetramer [18].

21.6 Genomic Structure and Regulation of the *ADH* and *ALDH* Genes

21.6.1 *ADH* Genes

All seven *ADH* genes are clustered together in a head-to-tail array on chromosome 4, in the order *ADH7, ADH1C, ADH1B, ADH1A, ADH6, ADH4, ADH5.* The individual genes range between 14 kb and 23 kb. The spacing between them ranges from 15 kb between the class I genes to about 60 kb flanking the class I genes. The entire set of seven genes spans 365 kb. Each of the individual genes has 8 introns, and these are located in the same positions relative to the coding sequence [21]. It had been thought that *ADH6* was an exception with only 7 introns [22], but recently it was shown that alternatively spliced forms exist, and a "full-length" transcript containing the missing last exon is the predominant form [23]. The ADH genes each have a different pattern of expression, although nearly all are expressed in the liver. The ADHs together comprise approximately 3% of the total soluble liver protein. The expression patterns of ADH and the *cis*-acting promoter elements that regulate them have been reviewed recently in greater detail [24].

All three class I ADHs are expressed in adult liver. There is a temporal pattern of expression during development, with *ADH1A* expressed first, *ADH1B* by midgestation, and *ADH1C* beginning some months after birth [25, 26]. Class I ADHs are also highly expressed in adrenal glands, and at lower levels in kidney (predominantly *ADH1B* in adults), lung (*ADH1B* at all stages of development), skin, and other tissues [24]. *ADH1B* has also been found in blood vessels. *ADH1C* is expressed in kidney, where it predominates in the fetus, and stomach mucosa. Interindividual differences in the relative levels of expression of *ADH1C* in kidney and stomach were reported early, but have not been further investigated. *ADH1A* is the predominant form in fetal liver and in hepatomas [27].

The three class I ADH promoters are very similar. Prominent among the *cis*-acting elements that contribute to promoter function are the TATA box, a pair of C/EBP sites (that can also be bound by DBP) flanking the TATA box, an E-box sequence (CACGTG) just upstream at which USF can bind, and a G3T sequence (that binds Sp1) one helical turn further upstream from the E-box [28, 29]. Further upstream are CTF/NF-1 and HNF-1 sites, and some elements that are specific to only some of these genes [24]. Differences among the class I genes in these and other sites affect the tissue distribution and amount of expression. Sequence differences among individuals could well affect the level and site(s) of expression, and thereby the effects of alcohol.

ADH4, the class II enzyme, is expressed primarily in the liver and at lower levels in the lower gastrointestinal tract and spleen [24]. The promoter functions in both hepatoma cells and fibroblasts, but several of the *cis*-acting elements have different effects in the different cell types [29]. There are several important C/EBP sites and an AP1 site in the proximal promoter. The C/EBP sites appear to play the predominant role in allowing high-level expression in the liver [29]. Recently, an *ADH4* polymorphism was shown to have a major effect on gene expression. Promoters with an A at bp –75 are twice as active as those with a C [30]. This would be expected to affect alcohol metabolism, particularly at intoxicating levels of alcohol.

Curiously, the ancestral class III ADH isozymes are least like the others in expression pattern. Class III ADHs, including the human *ADH5*, are ubiquitously expressed and have CG-rich promoters characteristic of many "housekeeping" genes. Despite this apparent simplicity, the *ADH5* promoter has many tissue-specific *cis*-acting elements [31]. Another anomaly is that although the class III genes are by far the most conserved in amino acid sequence, the promoters of the human and mouse class III genes share nearly no identity, other than both being very GC rich [32].

The transcription factor Sp1 plays the predominant role in the expression of *ADH5* [31, 33]. There is a pair of Sp1 sites flanking the transcription start site, and a minimal promoter containing these sites is a strong promoter in several different cell types [31]. This region of the promoter is essentially inactive in *Drosophila* cells that lack Sp1, and are strongly activated by coexpression of Sp1 [33].

Upstream AUGs are uncommon in mammalian genes, but present in *ADH5*. Two upstream AUGs encode overlapping peptides of 10 and 20 amino acids, with a common stop codon located just upstream of the AUG that encodes the χ-polypeptide. Altering the length of the upstream ORF affects the level of gene expression [31]. Recent experiments demonstrate that mutating one or both of these upstream AUGs increased the level of gene expression about 2- to 5-fold in different cell lines, as measured either by transient transfection or *in vitro* transcription [34].

ADH7 is the only member of this gene family that is not expressed in liver. It is the major ethanol-active form present in the stomach. It is also found at high levels in the upper gastrointestinal tract, including esophagus, gingiva, mouth, and tongue, and in the cornea and epithelial tissues [24].

The *ADH7* promoter has been examined in several cell lines [35]. The first 232 bp function in all cells tested, and no cell-specific elements were found out to 800 bp. A proximal AP1 site plays a crucial role in gene expression. A nearby site that can be bound by C/EBP also plays a strong role, although paradoxically the coexpression of C/EBP inhibits transcription [35]. A 9 kb fragment of the orthologous mouse gene (now called *Adh7*) directed tissue-specific expression in transgenic mice [15]. There have been reports of interindividual differences in expression in stomach, with about 70% of Chinese and Japanese not showing expression in stomach.

Recently, seven polymorphisms have been identified in *ADH7*, one of which is in the promoter. These polymorphisms were grouped into five alleles in a Swedish population, and one of the alleles was associated with Parkinson's disease [36]. The potential association of polymorphisms with alcoholism or its sequelae would be important.

ADH6 was discovered by nucleotide cross-hybridization, and has not yet been demonstrated as a functional protein. The mRNA is found in liver (both adult and fetal) and stomach. *ADH6* has a fully functional promoter, active in both hepatoma cells and fibroblasts [37]. There are several positive *cis*-acting elements in the proximal promoter, several of which are bound by C/EBP. There is a compound cell-specific regulatory element about 2 kb upstream, that is a positive element in the hepatoma cells and a negative element in fibroblasts [23].

21.6.2
ALDH Genes

The two *ALDH* genes most important in ethanol metabolism and alcoholism are *ALDH1A1* (cytosolic ALDH, previously known as *ALDH1*) and *ALDH2* (mitochondrial, low-K_m ALDH). These two genes are very closely related, with approximately 70% identity at the coding level. Evolutionary analysis of the ALDH gene family shows that these are more closely related to each other than to any other class of ALDH (*www.uchsc.edu/sp/sp/alcdbase/aldhcov.html*). They are located on different chromosomes. *ALDH1A1* is located on chromosome 9q21, and is a member of a family of related genes. It is encoded in 13 exons spanning 53 kb [38]. *ALDH2* is located on chromosome 12q24.2. It is encoded in 13 exons spanning 44 kb; the largest intron is 15 kb [39]. The overall structure and position of many of the introns of *ALDH2* is the same as in *ALDH1A1;* 9 of the 12 introns are located in homologous positions in the coding region [38]. The first exon differs most between the two genes; it encodes the mitochondrial import sequence and first 21 amino acids of the mature ALDH2 polypeptide [39].

Both *ALDH1A1* (UniGene HS.195432) and *ALDH2* (UniGene HS.76392) are very widely expressed. The highest level of ALDH1A1 mRNA is in liver, kidney, muscle and pancreas [40]. *ALDH2* (UniGene HS.195432) is ubiquitously expressed, and a large number of ESTs have been sequenced from both adult and embryonic tissues. It is expressed at highest levels in liver, kidney, muscle and heart [40].

The *ALDH1A1* promoter was active in Hep3B hepatoma cells but not in K562 erythroleukemia cells or in LTK$^-$ fibroblasts [41]. A minimal promoter containing a CCAAT box has been defined, and shown to bind NF-Y. IL-1 and TNFα both increase *ALDH1A1* mRNA levels in human bone marrow cells, but not in several leukemic cells [42]. This is interesting, because ALDH1A1 can be protective against toxicity of 4-hydroperoxycyclophosphamide, a chemotherapeutic agent [42, 43, 44]. ALDH1A1 also plays a role in retinoic acid synthesis; its expression is reduced in the presence of retinoic acid [45].

Two important *cis*-acting elements in the *ALDH2* promoter have been studied. A site located from 79 to 116 bp upstream of the ATG initiating translation is bound by nuclear factor(s) present in all cells tested; the CCAAT box in this region is important for transcriptional activity, and appears to be bound primarily by the transcription factor NF-Y/CP1 [46]. There is a site, approximately 300 bp upstream of the ATG, at which HNF-4 and retinoid X receptors can bind, as can the apolipoproteins regulatory protein (ARP-1) [47, 48]. Transcription from this promoter can be activated by HNF-4 and RXRs [47, 48].

A promoter polymorphism at bp –361 [49, 50] affects promoter function. The G-allele is more active in H4IIE-C3 cells than the A-allele [49]. In the Japanese population, this polymorphism is in linkage disequilibrium with the structural polymorphism; the inactive variant *ALDH2*2* allele is more frequently associated with the G promoter allele [50]. Unlike the structural polymorphism, this promoter polymorphism is found in a wide variety of populations, including North-American Caucasians and African Americans [49,50]. This makes it of possible pharmacogenetic importance in affecting the risk for alcoholism or alcohol effects.

21.7
Genome Screens for Alcohol Related Phenotypes in Humans

Alcoholism, defined as alcohol dependence, is an important, common disease. It is a member of the large group of complex genetic diseases without simple inheritance patterns of risk, including diabetes, hypertension, many psychiatric disorders and cancers. The lack of a clear pattern of inheritance is presumed to result from the contribution of multiple genes, no one of which is essential, and from environmental factors. It is important to remember that while the genetic contribution to risk is important, neither any single gene nor the entire genotype of an individual is deterministic. Individuals can dramatically alter their risk by choices of environment and lifestyle.

The Collaborative Study on the Genetics of Alcoholism (COGA) is a large, multi-center, family-based study of the general U.S. population Approximately 11,000 people have been interviewed and extensively characterized, and genomic surveys have been carried out on two separate subsamples (initial and replication). The initial data provided evidence for four regions that contain genes affecting the risk for alcoholism. Regions on chromosomes 1, 2, and 7 were detected in analyses of allele sharing among sibling pairs affected by alcoholism; a region on chromosome 4 was detected in unaffected sibling pairs, and might be related to lowered risk. The region on chromosome 4 was near the *ADH* gene cluster, for which there is evidence of a protective effect (above). The replication study supported the loci on chromosomes 1 and 7. The locus on chromosome 2 was more significant in the initial data set when new markers were added, but there was little evidence in the replication data set and, therefore, reduced evidence in the combined analysis. A new locus on chromosome 3 was found in the replication data set [51].

Another large study focused upon a Native American population in which alcoholism is extremely prevalent. Three regions provided evidence for genes that affect the risk for alcoholism: a site on chromosome 11 near the dopamine D4 receptor and tyrosine hydroxylase genes, and two sites on chromosome 4, one near the GABA-receptor gene cluster and one near the *ADH* gene cluster [52]. It is interesting that both this and the COGA study have evidence for genes near the GABA and *ADH* clusters on chromosome 4.

The phenotype of alcoholism is complex; diagnosis can result from many different combinations of symptom groups. For that reason, it is useful to seek genes that affect endophenotypes, traits related to alcoholism that might be more homogeneous, closer to biological processes, and affected by fewer genes. COGA measured many endophenotypes, including the event related brain potential (ERP). Analyses of ERP (the visual P300) showed evidence for an effect of regions on chromosomes 2, 6, 5 and 13 [53]. COGA analyzed a phenotype based on latent-class analysis, which provided a grouping that resembles a severity scale. There was evidence for a locus on chromosome 16p that affected the more severe groups. Another phenotype that provided interesting results was Maxdrinks, the largest number of drinks consumed within 24 h at any time in one's life. Although this seems unlikely to have a large genetic component, it is, in fact, relatively heritable. A single region gave strong evidence that it contained a gene affecting this phenotype in both initial and replication samples. This region was on chromosome 4, near the *ADH* cluster [54]; this is consistent with the data on a protective gene in that region.

Comorbid conditions can also provide a way to either narrow or extend the phenotype. An analysis of the COGA data showed a very strong signal for the broadened phenotype of alcoholism or major depression; this was located in the same region of chromosome 1 in which the alcoholism phenotype gave a signal [55]. The data for the combined phenotype was much stronger than that for the alcoholism-only phenotype.

21.8
Genome Screens for QTLs Affecting Alcohol Related Phenotypes in Animal Models

Two standard methods in genetic animal model research have been employed in the search for genes contributing to alcoholism: the study of inbred strains and selective breeding. Inbred strains of animals are genetically identical at all loci to all other members of that strain; the remaining variation of a quantitative trait within that strain is non-genetic or environmental in origin. If the mean variation among different inbred strains exceeds that within the strains, the difference demonstrates the presence of significant genetic control of the trait. Recombinant inbred strains can be developed from intercrossing the inbred strains that differ quantitatively across the strains in the phenotype under study. If genomic screens are performed on these animals and their parental strains, Quantitative Trait Loci (QTLs) for the measured traits can be identified on different chromosomes. Selective breeding systematically mates animals with high and low measures of a quantitative trait to increase the frequencies of alleles that affect the responses in the desired directions. The alcoholism-related traits in laboratory mice and rats that have been studied in this manner are: high and low acute tolerance to ethanol; high and low sensitivity to ethanol withdrawal; and high and low alcohol preference or voluntary alcohol drinking. Selectively bred lines can also be inbred and the inbred strains crossed to generate F2 offspring that have different quantitative measures of the selection phenotype. Genome screens in these animals and their parental strains can similarly yield QTLs for the selected phenotype [56].

21.9
Genome Screens for Alcohol-Related Phenotypes in Rodents

A large amount of research has been performed in the last decade to discern QTLs associated with ethanol preference, acute sensitivity to ethanol, and withdrawal seizure severity in mouse inbred strains and recombinant inbred strains. A lesser amount of work has been done in mice selectively bred for differences in ethanol withdrawal seizure severity, acute ethanol tolerance and in rats selectively bred for differences in free-choice alcohol consumption. All studies have shown that each of these intermediate or endophenotypes of alcoholism is influenced by multiple genes. For example, the studies on withdrawal seizure severity in mice have consistently shown QTLs on chromosomes 1, 4, and 11 that together account for about 70% of the genetic variance [57]. Potential candidate genes include: those for the GABA receptor subunits, glutamic acid decarboxylase and Na^+, K^+-ATPase [58]. The studies of acute tolerance development in mice have identified as many as four significant QTLs that account for over 50% of the genetic variance. Potential candidate genes include the high-affinity neurotensin receptor gene, the acetylcholine receptor subunit genes and the GABA-A receptor subunit genes [59]. For acute locomotor response to ethanol in mice, a major QTL was identified on chromosome 2, implicating the catalase gene underlying this acute stimulatory effect of ethanol, presumably through the action of acetaldehyde [60].

Ethanol preference or voluntary consumption has also been examined in inbred and recombinant inbred mouse strains. A large number of QTLs have been suggested, but confirmation across 4 or more studies has been seen on chromosomes 2, 3, 4, and 9. Interesting candidate genes in the QTL regions include those for the voltage-sensitive sodium channel proteins, alcohol dehydrogenase, the 5HT1A serotonin receptor, the dopamine D2 receptor, and the 5HT1B serotonin receptor [61].

Recent studies in genetically selected rat lines bred for high and low preference and voluntary ethanol consumption have yielded additional candidate genes. A genome screen comparing alcohol-preferring and non-preferring rats revealed a major effect size QTL accounting for about 35% of the genetic variance on chromosome 4. The gene for neuropeptide Y, a neuropeptide known to have anxiolytic and orexigenic properties, is located near the peak [62]. Other QTLs include the DRD2 receptor and the 5HT1b receptor genes [63].

In the last five years, there have been a number of studies using transgenic and knock-out technology in mice to test hypotheses that 5HT, DA, GABA, endogenous opioids and other neurotransmitter/modulators underlie the actions of ethanol. These candidate gene approaches buttress the findings from the QTL studies described above and generally support an inverse relationship between voluntary alcohol consumption and sensitivity to the sedative/hypnotic effects of ethanol, i.e., the more the animals are impaired by ethanol, the less are they able to drink.

21.10
Clinical Correlations of the Pharmacogenomics of ADH and ALDH2, Alcohol Drinking Behavior, Alcoholism and the Systemic Effects of Alcohol

Unlike that for all other drugs of abuse and addiction, the pharmacologically effective concentrations of ethanol are in the millimolar range. The central nervous system response to ethanol is biphasic, with behavioral and motor stimulation occurring at the lower blood alcohol concentrations, e.g., 10–50 mg % (2.2–13 mM), impaired cognitive and cortical executive functioning occurring at moderate blood alcohol concentrations, and depression of behavioral and motor activity occurring at higher concentrations. Legal intoxication is 80–100 mg % (17 mM to 22 mM) and death due to respiratory depression is seen at 500–600 mg %, in non-habituated or non-tolerant individuals.

In subjects who have normal ALDH2 enzyme activity, i.e., those with *ALDH2*1/*1* genotype, the systemic concentrations of acetaldehyde are usually less than 5 micromolar (μM). When the ALDH2 activity is decreased, as in subjects who have the heterozygous *ALDH2*1/*2* and the homozygous *ALDH*2/*2* genotypes, or in patients taking medications that inhibit ALDH2 activity, e.g., disulfiram (Antabuse®), acetaldehyde is increased in the circulation and organ systems to levels as high as 60–100 μM, even with low doses of ethanol, 0.2 g kg^{-1}

[64]. Elevated acetaldehyde levels are associated with a decrease in ethanol elimination rate, consistent with the product inhibition of ADH activity.

Acetaldehyde is a very reactive substance that is capable of forming stable adducts through reaction with amino groups of small organic molecules (e.g., biogenic amines and amino acids) and of proteins and DNA. At the cellular level, it reacts with chromaffin cells to release catecholamines and with mast cells to release kinins and histamine [65]. These in turn produce a physiological reaction called the alcohol flush reaction, characterized by facial flushing, vasodilatation, increased heart rate, nausea and vomiting. When the reaction is severe, there may be hypotension, bronchial asthma, syncope and even death. These reactions are generally aversive even when mild and tend to deter the rapid ingestion of alcoholic beverages and the ingestion of large amounts over time. The alcohol flush reaction is seen in Asians who have the *ALDH2*2/*2* and *ALDH2*2/*1* genotypes [2].

Functional polymorphisms of the *ADH1B* and the *ADH1C* genes do not appear to influence systemic blood acetaldehyde levels and ethanol elimination rates as clearly as does the *ALDH2* polymorphism. A higher elimination rate has been reported in African Americans who carry the *ADH1B*3* allele as compared with those who do not [66]. Although it has been reported that Chinese and Japanese have higher alcohol elimination rates than do Caucasians, a study in Japanese subjects did not find a difference in ethanol elimination rate nor in blood acetaldehyde levels among subjects with the *ADH1B*1/*1*, *ADH1B*1/*2*, and *ADH1B*2/*2* genotypes [67]. However, the presence of the *ADH1B*2* allele has been reported to exacerbate the alcohol flush reaction in Japanese subjects heterozygous for the *ALDH2*2* allele. Studies in Caucasian Jewish populations have shown that there is lower frequency and quantity of drinking in subjects harboring the *ADH1B*2* allele [68], but these effects are also influenced strongly by environmental factors, peer relationships and developmental stages of life. The mechanism underlying the *ADH1B*2* effect on drinking behavior is, therefore, not yet established. A possibility is elevation of acetaldehyde levels from increased first pass metabolism in the stomach and liver leading to the release of kinins, histamine and other reactive compounds from the gastrointestinal tract and liver.

21.10.1 Alcoholism

To date, *ADH2* and *ALDH2* are the only genes that have been firmly established to influence vulnerability to alcohol dependence or alcoholism [2]. There is consensus in the literature that the allele frequencies of *ADH1B*2, ADH1C*1* and *ALDH2*2* are significantly decreased in Japanese, Chinese, and Korean subjects diagnosed with alcohol dependence by DSM III R criteria, as compared with the general population controls. These association studies are consistent with the hypothesis that higher rates of production of acetaldehyde and/or reduced metabolism of acetaldehyde play protective roles against developing alcoholism. The protective effect of *ALDH2*2* is greater than that of *ADH1B*2*; the association of *ADH1C*1* is accountable entirely by linkage disequilibrium with *ADH1B*2* [69]. Association between reduced risk for alcoholism and the *ADH1B*2* variant allele has been found also in other ethnic groups that do not carry the *ALDH2*2* allele, including Caucasians [70], Mongolians in China [71], and the Atayal natives of Taiwan [72]. These observations are consistent with the finding that *ADH1B* polymorphism affects vulnerability to alcoholism independently of *ALDH2*.

The deficiency of ALDH2 enzyme activity, as seen in the Asian populations, follows the pattern of a classical "inborn error of metabolism", with semi-dominant inheritance and high phenotypic penetrance. The heterozygous *ALDH2*1/*2* is very protective, and the homozygous *ALDH2*2/*2* genotype provides almost complete protection against alcoholism. While 5–10% of East Asians are of this genotype, only one alcoholic person homozygous for *ALDH2*2* has been reported in the extant literature [73]. This person also had psychiatric comorbidities of bipolar illness and anxiety disorder. Approximately 30–40% of East Asian populations are heterozygous for *ALDH2*2* (i.e., *ALDH2*1/*2* genotype), and they are about 5-times less likely to becoming alcoholic. However, with changing cultural norms and drinking practices across time, the percentage of alcoholics with the *ALDH2*1/*2* genotype has been increasing among the Japanese and Chinese populations [73, 74].

21.10.2 Alcoholic Liver Disease

The development of alcoholic liver disease (ALD) is clearly related to the amount and duration of alcohol intake. It is equally clear that there are host susceptibility factors that are genetic in origin, since not everyone exposed to equivalent amounts of alcohol develops ALD. Because acetaldehyde forms adducts with proteins that are immunogenic and may contribute to liver injury, the relationships of *ADH1B* and *ALDH2* polymorphisms to ALD have been explored. There appears to be a positive association of the *ADH1B*2* allele as well as the *ALDH2*2* allele with ALD susceptibility, but the relationships are weak [75, 76], indicating the coexistence of other genetic factors. Some studies have seen an association of the *ADH1C* and the *CYP2E1* polymorphisms with ALD [77], whereas others have not.

21.10.3 Cancer

High levels of chronic alcohol consumption are associated with an increased risk for upper aerodigestive tract (oral cavity, pharynx, larynx, esophagus), stomach and colorectal cancers. There is convincing evidence that acetaldehyde has direct mutagenic and carcinogenic effects. Acetaldehyde causes point mutations in DNA and induces sister chromatid exchanges and chromosomal aberrations [78]. The formation of stable DNA adducts represents one mechanism whereby acetaldehyde can contribute to replication errors and impair DNA repair [79]. In support of acetaldehyde having a major role in the genesis of these cancers, it has been shown that salivary acetaldehyde levels are increased in heavy drinkers and smokers. The increased acetaldehyde levels derive not only from the bacterial flora in the oral cavity and the digestive tract of smokers and drinkers, but also from a high rate of ethanol oxidation to acetaldehyde or a low rate of acetaldehyde metabolism by the parotid salivary gland [80].

Many excellent epidemiological studies have shown that the risk of ethanol-associated digestive tract cancers is markedly increased in Asians who carry the *ALDH2*2* allele: increased risks (odds ratios) are 3- to 10-fold for stomach, colon and lung and 11- to 15-fold for oropharyngeal and esophageal cancer [81]. The risk for esophageal cancer occurring concomitantly with oropharyngeal and/or stomach cancer is increased more than 50-fold in this population.

The relationship of *ADH* polymorphisms to oropharyngeal cancer is less clear. Whereas some studies have found an association of *ADH1B*1* and *ADH1C*2* with increased risk for oropharyngeal cancer, others have not [82].

21.10.4
Heart Disease

Epidemiological studies have consistently demonstrated that moderate consumption of alcohol is associated with a reduced risk of myocardial infarction; however, the mechanism underlying this association is unclear. A recent study found that moderate drinkers who are homozygous for the ADH1C*2 had higher HDL (high-density lipoprotein) levels and a substantially decreased risk of myocardial infarction [83]. Whether the protective effect is mediated through the action of the ADH1 gene or another gene in linkage disequilibrium with this gene in unclear.

Chronic heavy alcohol consumption produces alcoholic heart muscle disease (AHMD). Apart from the history of alcoholism, the pathological and clinical course and pathological features are consistent with other dilated cardiomyopathies. The mechanisms by which chronic ethanol exposure produces myocardial damage are unknown, but acetaldehyde has been implicated by studies in animal models. A recent study used a transgenic mouse model of acetaldehyde overproduction (overexpression of ADH) to test the acetaldehyde hypotheses [84]. Overexpression of ADH by 40-fold did not produce deleterious effects to the heart in the absence of ethanol. In the presence of alcohol, acetaldehyde content was 4-fold higher in transgenic hearts than in control hearts. Chronically, alcohol exposed transgenic hearts developed lesions similar to those seen in AHMD more rapidly and to a greater extent than control hearts. The study would suggest that *ADH1B* and *ALDH2* polymorphism in humans might influence the incidence of AHMD.

21.10.5
Fetal Alcohol Syndrome

Fetal alcohol syndrome (FAS) is a pattern of birth defects caused by maternal consumption of ethanol during pregnancy. It is recognized by growth deficiency, a characteristic set of craniofacial features and neurodevelopmental abnormalities leading to cognitive and behavioral deficits [85]. FAS is considered to be the most common non-hereditary cause of mental retardation.

While alcohol is clearly the environmental teratogen, it is unclear whether ethanol itself or acetaldehyde is the principal agent that triggers the developmental abnormalities in brain during gestation. A number of mechanisms have been identified, e.g., apoptosis, damage from free-radical formation, interference with growth factor functions, among others. Retinoic acid, derived from vitamin A (retinol) is essential for controlling the normal patterns of development of tissues and organs. Retinol and ethanol are competing substrates for oxidation by ADH to retinal and acetaldehyde [86]. Accordingly, the relationship of *ADH1B* polymorphism to FAS risk has been examined. The frequencies of the *ADH1B*2* and *ADH1B*3* alleles are significantly lower in the mothers with FAS offspring than in case control subjects [87, 88]. The isozymes encoded by these alleles exhibit higher ADH activity than do the isozyme encoded by *ADH1B*1*, suggesting that a more rapid metabolism of ethanol tends to lessen the teratogenic effect of ethanol consumption. On the other hand, individuals possessing *ADH1B*2* and *ADH1B*3* alleles may also drink less. Thus the pathway of reduced risk associated with the *ADH1B* polymorphism is still unclear.

21.11
Conclusion

Alcoholism and alcohol abuse are complex traits with both environmental and genetic influences. Much more work is necessary before we can understand the interplay and relative strength of these different factors. At present, only the functional polymorphisms associated with the alcohol and aldehyde dehydrogenase genes are firmly linked to risk for alcoholism and complications associated with ethanol's abuse. The *ALDH2*2* allele is associated with a strong protective effect toward alcoholism, but has been associated with increased risk for alcoholic liver disease and gastrointestinal cancers. The functional polymorphisms in the ADH locus are associated with changes in risk for other complications related to alcohol consumption. The *ADH1C*2* locus would appear to be linked to decreased risk for myocardial infarction and both the *ADH1B*2* and *ADH1B*3* loci would seem to be protective toward fetal alcohol syndrome. In the search for additional genes that affect risk for alcoholism, the ongoing COGA study has identified several broad genomic regions that contain genes that affect risk for alcoholism on chromosomes 1, 2, 3, 4 and 7. Consistent with its location near the ADH locus, the region on chromosome 4 would appear to be protective and is found with higher frequency in unaffected individuals. The regions on chromosomes 1 and 7 were found to be significant in both the initial and replicate study. In parallel with the above work, the COGA data was also subjected to analysis using so-called endophenotypes, such as event related brain potential and a latent-class analysis that resembles a severity scale. These analyses identified additional regions of interest. Importantly, work is now in progress to narrow down the regions of interest, before we can search for candidate genes in these regions.

Acknowledgements
The authors acknowledge the help of Dieter Sasse and Brian Gibbons for providing the information for the ADH activity measurements and calculation of contribution to total metabolism. The authors also wish to acknowledge grant support from the following grants: AA10399, AA11982, AA06460, AA02342, AA10722 and AA07611.

21.12
References

1 Heath AC, Bucholz KK, Madden PA et al. Genetic and environmental contributions to alcohol dependence risk in a national twin sample: Consistency of findings in women and men. Psychol Med **1997**; 27:1381–1396.
2 Li T-K.Pharmacogenetics of responses to alcohol and genes that influence alcohol drinking. J Stud Alcohol **2000**; 61:5–12.
3 Edwards G, Gross MM. Alcohol dependence: Provisional description of a clinical syndrome. Br Med J **1976**; 1:1058–1061.
4 American Psychiatric Association. Diagnostic and Statistical Manual of Mental Disorders, Fourth Edition. Washington, DC: The Association, **1994**.
5 World Health Organization. The ICD-10 Classification of Mental and Behavioral Disorders: Clinical Descriptions and Diagnostic Guidelines. Geneva: the Organization, **1992**.
6 Eckardt MJ, File SE, Gessa GL et al. Effects of moderate alcohol consumption on the central nervous system. Alcohol Clin Exp Res **1998**; 22:998–1040.
7 Bosron WF, Ehrig T, Li T-K. Genetic factors in alcohol metabolism and alcoholism. Semin Liver Dis **1993**; 13:126–135.
8 O'Connor S, Sorbel J, Morzorati S, Li T-K, Christian JC. A twin study of genetic influences on the acute adaptation of the EEG to alcohol. Alcohol Clin Exp Res **1999**; 23:494–501.
9 McBride WJ, Li T-K. Animal models of alcoholism: Neurobiology of high alcohol-drinking behavior. Crit Rev Neurobiol **1998**; 12:339–369.
10 Rodd-Hendricks ZA, McKinzie DL, Shaikh SR et al. Alcohol deprivation effect is prolonged in the preferring (P) rat after repeated deprivations. Alcohol Clin Exp Res **2000**; 24:8–16.
11 Maly IP, Toranelli M, Sasse D. Distribution of alcohol dehydrogenase isoenzymes in the human liver acinus. Histochem Cell Biol **1999**; 111:391–397.
12 Niederhut MS, Gibbons BJ, Perez-Miller S, Hurley TD. Three-dimensional structures of the three human class I alcohol dehydrogenases. Protein Sci **2001**; 10:697–706.

13 Stone CL, Bosron WF, Dunn MF. Amino acid substitutions at position 47 of human beta 1 beta 1 and beta 2 beta 2 alcohol dehydrogenases affect hydride transfer and coenzyme dissociation rate constants. J Biol Chem **1993**; 268:892–899

14 Kedishvili NY, Bosron WF, Stone CL, Hurley TD, Peggs CF, Thomasson HR, Popov KM, Carr LG, Edenberg HJ, Li T-K. Cloning and expression of a human stomach alcohol dehydrogenase: Comparison of structure and catalytic properties with the liver isoenzymes. J Biol Chem **1995**; 280:3625–3630.

15 Haselbeck RJ, Duester G. ADH4-lacZ transgenic mouse reveals alcohol dehydrogenase localization in embryonic midbrain/hindbrain, otic vesicles, and mesencephalic, trigeminal, facial, and olfactory neural crest. Alcohol Clin Exp Res **1998**; 22:1607–1613.

16 Steinmetz CG, Xie P, Weiner H, Hurley TD. Structure of mitochondrial aldehyde dehydrogenase: The genetic component of alcohol aversion. Structure **1997**; 5:701–711.

17 Moore SA, Baker HM, Blythe TJ, Kitson KE, Kitson TM, Baker EN. Sheep liver cytosolic aldehyde dehydrogenase: the structure reveals the basis for the retinal specificity of class 1 aldehyde dehydrogenases. Structure **1998**; 6:1541–1551.

18 Zhou J, Weiner H. Basis for half-of-the-sites reactivity and the dominance of the K487 oriental subunit over the E487 subunit in heterotetrameric human liver mitochondrial aldehyde dehydrogenase. Biochemistry **2000**; 39:12019–12024

19 Yoshida A, Huang I-Y, Ikawa M. Molecular abnormality in an inactive aldehyde dehydrogenase variant commonly found in Orientals. Proc Natl Acad Sci USA **1984**; 81:258–261.

20 Crabb DW, Edenberg HJ, Bosron WF, Li T-K. Genotypes for aldehyde dehydrogenase deficiency and alcohol sensitivity. The inactive ALDH2*2 allele is dominant. J Clin Invest **1989**; 83:314–316.

21 Edenberg HJ, Bosron WF. Alcohol dehydrogenases in: Biotransformation Vol. 3, Comprehensive Toxicology. Pergamon Press, New York. **1997**; pp 119–131.

22 Yasunami M, Chen CS, Yoshida A. A human alcohol dehydrogenase gene (ADH6) encoding an additional class of isozyme. Proc Natl Acad Sci USA **1991**; 88:7610–7614.

23 Stromberg P, Hoog JO. Human class V alcohol dehydrogenase (ADH5): a complex transcription unit generates C-terminal multiplicity. Biochem Biophys Res Commun **2000**; 278:544–549.

24 Edenberg HJ. Regulation of the mammalian alcohol dehydrogenase genes. Prog. Nucleic Acid Res Mol Biol **2000**; 64:295–341.

25 Smith M, Hopkinson DA, Harris H. Developmental changes and polymorphism in human alcohol dehydrogenase. Ann Hum Genet (London) **1971**; 34:251–271.

26 Smith M, Hopkinson DA, Harris H. Alcohol dehydrogenase isozymes in adult human stomach and liver: evidence for activity of the ADH3 locus. Ann Hum Genet (London) **1972**; 35:243–253.

27 Smith M. Genetics of human alcohol and aldehyde dehydrogenases. Adv Human Genet **1986**; 15:249–290.

28 Brown CJ, Zhang L, Edenberg HJ. Gene expression in a young multigene family: tissue-specific differences in the expression of the human alcohol dehydrogenase genes *ADH1*, *ADH2* and *ADH3*. DNA Cell Biol **1996**; 15:187–196.

29 Li M, Edenberg HJ. Function of *cis*-acting elements in human alcohol dehydrogenase 4 (*ADH4*) promoter, and role of C/EBP proteins in gene expression. DNA Cell Biol **1998**; 17:387–397.

30 Edenberg HJ, Jerome RE, Li M. Polymorphism of the human alcohol dehydrogenase 4 (*ADH4*) promoter affects gene expression. Pharmacogenetics **1999**; 9:25–30.

31 Hur M-W, Edenberg HJ. Cell-specific function of cis-acting elements in the regulation of human alcohol dehydrogenase 5 gene expression and effect of the 5′-nontranslated region. J Biol Chem **1995**; 270:9002–9009.

32 Edenberg HJ, Ho W-H, Hur M-W. Promoters of the mammalian class III alcohol dehydrogenase genes. Adv Exp Med Biol **1995**; 372:295–300.

33 Kwon H-S, Kim M-S, Edenberg HJ, Hur M-W. Sp3 and Sp4 can repress transcription by competing with Sp1 for core *cis*-elements on the human *ADH5/FDH* minimal promoter. J Biol Chem **1999**; 274:20–28

34 Kwon H-S, Lee D-K, Edenberg HJ, Lee J-J, Ahn Y-H, Hur M-W. Post-transcriptional regulation of human ADH5/FDH and Myf6 gene expression by upstream AUG codons. Arch Biochem Biophys **2001**; 386:163–171.

35 Kotagiri S, Edenberg HJ. Regulation of human alcohol dehydrogenase gene *ADH7*: importance of an AP-1 site. DNA Cell Biol **1998**; 17:583–590.

36 Buervenich S, Sydow O, Carmine A, Zhang Z, Anvret M, Olson L. Alcohol dehydrogenase alleles in Parkinson's disease. Mov Disord **2000**; 15:813–818.

37 Zhi X, Chan EM, Edenberg HJ. Tissue-specific regulatory elements in the human alcohol dehydrogenase 6 (ADH6) gene. DNA Cell Biol **2000**; 19:487–497.

38 Hsu LC, Chang WC, Yoshida A. Genomic structure of the human cytosolic aldehyde dehydrogenase gene. Genomics **1989**; 5:857–865.

39 Hsu LC, Bendel RE, Yoshida A. Genomic structure of the human mitochondrial aldehyde dehydrogenase gene. Genomics **1988**; 2:57–65.

40 Stewart M J, Malek K, Crabb DW. Distribution of messenger RNAs for aldehyde dehydrogenase 1, aldehyde dehydrogenase 2, and aldehyde dehydrogenase 5 in human tissues. J Invest Med **1996**; 44:42–46.

41 Yanagawa Y, Chen JC, Hsu LC, Yoshida A. The transcriptional regulation of human aldehyde dehydrogenase I gene. The structural and functional analysis of the promoter. J Biol Chem **1995**; 270:17521–17527.

42 Moreb JS, Turner C, Sreerama L, Zucali JR, Sladek NE, Schweder M. Interleukin-1 and tumor necrosis factor alpha induce class 1 aldehyde dehydrogenase mRNA and protein in bone marrow cells. Leuk Lymphoma **1995**; 20:77–84.

43 Moreb JS, Schseder M, Gray B, Zucali J, Zori R. *In vitro* selection for K562 cells with higher retrovirally mediated copy number of aldehyde dehydrogenase class-1 and higher resistance to 4-hydroperoxycyclophosphamide. Hum Gene Ther **1998**; 20:611–619.

44 Tsukamoto N, Chen J, Yoshida A. Enhanced expressions of glucose-6-phosphate dehydrogenase and cytosolic aldehyde dehydrogenase and elevation of reduced glutathione level in cyclophosphamide-resistant human leukemia cells. Blood Cells Mol Dis **1998**; 24:231–238.

45 Elizondo G, Corchero J, Sterneck E, Gonzalez FJ. Feedback inhibition of the retinaldehyde dehydrogenase gene ALDH1 by retinoic acid through retinoic acid receptor alpha and CCAAT/enhancer-binding protein beta. J Biol Chem **2000**; 275:39747–39753.

46 Stewart MJ, Dipple KM, Stewart TR, Crabb DW. The role of nuclear factor NF-Y/CP1 in the transcriptional regulation of the human aldehyde dehydrogenase 2-encoding gene. Gene **1996**; 173:155–161.

47 Stewart MJ, Dipple KM, Estonius M, Nakshtri H, Everett LM, Crabb DW. Binding and activation of the human aldehyde dehydrogenase-2 promoter by hepatocyte nuclear factor 4. Biochim Biophys Acta **1998**; 1399:181–186.

48 Pinaire J, Hasanadka R, Fang M, Chou WY, Stewart MJ, Kruijer W, Crabb DW. The retinoid X receptor response element in the human aldehyde dehydrogenase 2 promoter is antagonized by the chicken ovalbumin upstream promoter family of orphan receptors. Arch Biochem Biophys **2000**; 280:192–200.

49 Chou W-Y, Stewart MJ, Carr LG, Zheng D, Stewart TR, Williams A, Pinaire J, Crabb DW. An A/G polymorphism in the promoter of mitochondrial aldehyde dehydrogenase (ALDH2): effects of the sequence variant on transcription factor binding and promoter strength. Alcohol Clin Exp Res **1999**; 26:963–968.

50 Harada S, Okubo T, Nakamura T, Fujii C, Nomura F, Higuchi S, Tsutsumi M. A novel polymorphism (–357G/A) of the ALDH2 gene: linkage disequilibrium and

an association with alcoholism. Alcohol Clin Exp Res **1999**; 23:958–962.

51 Foroud T, Edenberg HJ, Goate A, Rice J, Flury L, Koller DL, Bierut LJ, Conneally PM, Nurnberger JI Jr., Bucholz KK, Li T-K, Hesselbrock V, Crowe R, Schuckit M, Porjesz B, Begleiter H, Reich T. Alcoholism susceptibility loci: confirmation studies in a replicate sample and further mapping. Alcohol Clin Exp Res **2000**; 24:933–945.

52 Long JC, Knowler WC, Hanson RL, Robin RW, Urbanek M, Moore E, Bennett PH, Goldman D. Evidence for genetic linkage to alcohol dependence on chromosomes 4 and 11from and autosome-wide scan in an American Indian population. Am J Med Genet (Neuropsych Genet) **1998**; 81:216–221.

53 Begleiter H, Porjesz B, Reich T, Edenberg HJ, Goate A, Blangero J, Almasy L, Foroud T, VanEerdewegh P, Polich J, Rohrbaugh J, Kuperman S, Bauer LO, O'Connor SJ, Chorlian DB, Li T-K, Conneally PM, Hesselbrock V, Rice JP, Schuckit MA, Cloninger R, Nurnberger JJr., Crowe R, Bloom, FE. Quantitative trait loci analysis of human event-related brain potentials: P3 voltage. Electroenceph Clin Neurophysiol **1998**; 1–7.

54 Saccone NL, Kwon JM, Corbett J, Goate A, Rochberg N, Edenberg HJ, Foroud T, Li T-K, Begleiter H, Reich T, Rice JP. A genome screen of maximum number of drinks as an alcoholism phenotype. Am J Med Genet (Neuropsych Genet) **2000**; 96:632–637.

55 Nurnberger JI Jr., Foroud T, Flury L, Su J, Meyer ET, Hu K, Crowe R, Edenberg H, Goate A, Bierut L, Reich T, Schuckit M, Reich W. Evidence for a locus on chromosome 1 that influences vulnerability to alcoholism and affective disorder. Am J Psychiatry **2001**; 158:718–724.

56 Crabbe JC, Li T-K. Genetic strategies in preclinical substance abuse research in: Psychopharmacology. The Fourth Generation of Progress. (Bloom FE, Kupfer DJ, Eds). New York Raven Press, **1995**; pp 799–813.

57 Buck KJ, Metten P, Belknap JK, Crabbe JC. Quantitative trait loci involved in genetic predisposition to acute alcohol withdrawal in mice. J Neuroscience **1997**; 17:3946–3955.

58 Hood HM, Buck KJ. Allelic variation in the GABA-A receptor $\gamma 2$ subunit is associated with genetic susceptibility to ethanol-induced motor incoordination and hypothermia, conditioned taste aversion, and withdrawal in BxD/Ty recombinant inbred mice. Alcohol Clin Exp Res **2000**; 24:1327–1334.

59 Markel PD, Bennett B, Beeson M, Gordon L, Johnson TE. Confirmation of quantitative trait loci for ethanol sensitivity in long-sleep and short sleep mice. Genome Res **1997**; 7:92–99.

60 Xu Y, Demarest K, Hitzemann R, Sikela J. Gene coding variant in Cas1 between the C57BC/6J and DBA/2J inbred mouse strains: Linkage to a QTL for ethanol-induced locomotor activation. Alcohol Clin Exp Res **2001**, in press.

61 Foroud T, Li T-K. Genetics of alcoholism. A review of recent studies in human and animal models. Am J Addict **1999**; 8:261–278.

62 Carr LG, Foroud T, Bice P et al. Mapping of a quantitative locus for alcohol consumption in selectively bred rat lines. Alcohol Clin Exp Res **1998**; 22:884–887.

63 Bice P, Foroud T, Bo R et al. Genomic screen for QTLs underlying alcohol consumption in the P and NP rat lines. Mamm Genome **1998**; 9:949–955.

64 Peng GS, Wang MF, Chen CY et al. Involvement of acetaldehyde for full protection against alcoholism by homozygosity of the variant allele of mitochondrial aldehyde dehydrogenase gene in Asians. Pharmacogenetics **1999**; 9:463–476.

65 Koivisto T, Kaihovaara P, Salaspuro M. Acetaldehyde induces histamine release from purified rat peritoneal mast cells. Life Sci **1999**; 64:183–190.

66 Thomasson HR, Beard JD, Li T-K. ADH2 gene polymorphisms are determinants of alcohol pharmocokinetics. Alcohol Clin Exp Res **1995**; 19:1494–1499.

67 Mizoi Y, Yamamoto K, Ueno Y, Fukunaga T, Harada S. Involvement of genetic polymorphism of alcohol and aldehyde dehydrogenases in individual variation of

alcohol metabolism. Alcohol Alcohol **1994**; 29:707–710.

68 Neumark YD, Friedlander Y, Thomasson HR, Li T-K. Association of the ADH2*2 allele with reduced alcohol consumption in Jewish men in Israel: A pilot study. J Stud Alcohol **1998**; 22:202–210.

69 Osier M, Pakstis AJ, Kidd JR, Lee J-F, Yin S-J, Ko H-C, Edenberg HJ, Lu R-B, Kidd KK. Linkage disequilibrium at the *ADH2* and *ADH3* loci and risk of alcoholism. Am J Hum Genet **1999**; 64:1147–1157.

70 Borras E, Coutelle C, Rosell A et al. Genetic polymorphism of alcohol dehydrogenase in Europeans: The ADH2*2 allele decreases the risk of alcoholism and is associated with ADH3*1. Hepatology **2000**; 31:984–989.

71 Shen Y-C, Fan J-H, Edenberg HJ et al. Polymorphism of ADH and ALDH genes among four ethnic groups in China and effects upon the risk for alcoholism. Alcohol Clin Exp Res **1997**; 21:1272–1277.

72 Thomasson HR, Crabb DW, Edenberg HJ et al. Low frequency of the ADH2*2 allele among Atayal natives of Taiwan with alcohol use disorders. Alcohol Clin Exp Res **1994**; 18:640–643.

73 Chen Y-C, Lu R-B, Peng G-S et al. Alcohol metabolism and cardiovascular response in an alcoholic patient homozygous for the ALDH2*2 variant gene allele. Alcohol Clin Exp Res **1999**; 23:1853–60.

74 Higuchi S, Matsushita S, Imazeki H et al. Aldehyde dehydrogenase genotypes in Japanese alcoholics. Lancet **1994**; 343:741–742.

75 Yamauchi M, Maezawa Y, Mizuhara Y et al. Polymorphisms in alcohol metabolizing enzyme genes and alcoholic cirrhosis in Japanese patients. Hepatology **1995**; 22:1136–1142.

76 Whitfield JB. Meta-analysis of the effects of alcohol dependence and alcoholic liver disease. Alcohol Alcohol **1997**; 32:613–619.

77 Monzoni A, Masutti F, Saccoccio G et al. Genetic determinants of ethanol-induced liver damage. Mol Med **2001**; 7:255–262.

78 Dellarco VL. A mutagenicity assessment of acetaldehyde. Mutat Res **1988**; 195:1–20.

79 Fang JL, Vaca CE. Detection of DNA adducts of acetaldehye in peripheral white blood cells of alcohol abusers. Carcinogenesis **1997**; 18:627–632.

80 Vakevainen S, Tillonen J, Agarwal DP, Srivastava N, Salaspuro M. High salivary acetaldehyde after a moderate dose of alcohol in ALDH2-deficient subjects: Strong evidence for the local carcinogenic action of acetaldehyde. Alcohol Clin Exp Res **2000**; 24:873–877.

81 Yokoyama A, Muramatsu T, Ohmori T et al. Alcohol-related cancers and aldehyde dehydrogenase-2 in Japanese alcoholics. Carcinogenesis **1998**; 8:1383–87.

82 Olshan AF, Weissler MC, Watson MA, Bell DA. Risk of head and neck cancer and the alcohol dehydrogenase 3 genotype. Carcinogenesis **2001**; 22:57–61.

83 Hines LM, Stampfer MJ, Ma J et al. Genetic variation in alcohol dehydrogenase and the beneficial effect of moderate alcohol consumption on myocardial infarction. N Engl J Med **2001**; 344:549–555.

84 Liang Q, Carlson EC, Borgerding AJ, Epstein PN. A transgenic model of acetaldehyde overproduction accelerates alcohol cardiomyopathy. J Pharmacol Exp Ther **1999**; 291:766–772.

85 Stratton K, Howe C, Battaglia F. Fetal Alcohol Syndrome: Diagnosis, Epidemiology, Prevention and Treatment. National Academy Press, Washington, DC **1996**.

86 Deltour L, Foglio MH, Deuster G. Metabolic deficiencies in alcohol dehydrogenase Adh1, Adh3, and Adh4 null mutant mice: Overlapping roles of Adh1 and Adh4 in ethanol clearance and metabolism of retinol to retinoic acid. J Biol Chem **1999**; 274:16796–16801.

87 McCarver DG, Thomasson HR, Martier SS, Sokol RJ, Li T-K. Alcohol dehydrogenase 2*3 allele protects against alcohol-related birth defects among African Americans. J Pharmacol Exp Ther **1997**; 283:1095–1101.

88 Ericksson CJP, Fukunaga T, Sarkda T et al. Functional relevance of human ADH polymorphism. Alcohol Clin Exp Res **2001**; 25:157S–163S.

89 Hurley TD, Bosron WF, Stone CL, Amzel LM. Three-dimensional structures of three human alcohol dehydrogenase variants: Correlations with their functional differences. J Mol Biol **1994**; 239:415–429.

90 Davis GJ, Bosron WF, Stone CL, Owusu-Dekyi K, Hurley TD. X-ray structure of human $\beta_3\beta_3$ alcohol dehydrogenase. The contributions of ionic interactions to coenzyme binding. J Biol Chem **1996**; 271:17057–17061.

22
Pharmacogenomics of Tobacco Addiction

Elaine Johnstone, Marcus Munafò, Matt Neville, Siân Griffiths,
Mike Murphy and Robert Walton

Abstract

The precise mechanism of nicotine addiction and the influence of genetics on smoking behavior are beginning to be elucidated. A major challenge for the new science of pharmacogenomics is to utilize recent discoveries in genetics to improve existing smoking cessation therapies. Traditional candidate gene studies show that genetic polymorphisms affecting nicotine metabolism and dopaminergic transmission increase the susceptibility to tobacco dependence. Chromosomal regions in which other relevant genes may be located have recently been identified by genome scans. Combining these techniques may open the door to large-scale rationally designed studies to identify new genes that are important in the development of nicotine dependence. Animal models have also been used to unravel genetic influences on the behavioral effects of nicotine and highlight in particular the important role that the acetylcholine receptor plays in nicotine action. This chapter reviews these various approaches to elucidating the genetic basis of nicotine dependence. We explore the potential benefits of classifying smokers according to the molecular etiology of their habit in order to plan individually targeted cessation strategies.

22.1
The Neurophysiological Basis for Nicotine Addiction

The pleasure derived from using tobacco is linked to the stimulation of dopamine-dependent neurotransmitter pathways in the brain, particularly in the mesolimbic system. The precise nature of this link remains controversial, but many of the neurophysiological processes underlying nicotine addiction are common to other addictive drugs with diverse pharmacological actions such as opiates, cannabis, alcohol and cocaine.

Drugs stimulate receptors on the cell bodies of dopaminergic neurons causing dopamine release and stimulating postsynaptic dopamine receptors in the nucleus accumbens, supposedly resulting in the perception of pleasure [1]. Other hypotheses suggest that these mesolimbic dopaminergic pathways are necessary for the associative learning necessary to link perception of pleasure (or the subjective experience of "wanting", discussed below) with particular external stimuli [2, 3]. The distinction between drug "liking" (i.e., the pleasurable effects derived from its use) and drug "wanting" (i.e., the cravings experienced in addiction) is conceptually important, and Robinson and Berridge [3, 4] suggest that it is the latter subcomponent of reward, rather than hedonic pleasure, which results from critical neuroadaptations (see below) in dopaminergic systems. These theories may in fact be complementary, but hypotheses implicating dopamine alone cannot give the whole picture, when heavily abused drugs such as benzodiazepines, which have rewarding properties, cause no dopaminergic activation. Other neurotransmitters implicated in the development of addiction include serotonin, GABA, glutamate [5] and noradrenaline [6].

A fundamental and interesting observation is that many people experience the pleasurable effects of addictive drugs but only a few persistently abuse them. This supports the distinction between drug liking and drug craving, and suggests that the positive reinforcement aspects of drug use are not sufficient for addiction to develop. The molecular mechanisms underlying the enhanced drug craving seen in addicts in withdrawal may be key to explaining why only some people become dependent [3] and to informing strategies to combat addiction (Figure 22.1). It is also worth noting that the negative reinforcement aspects of addiction (i.e., the suggestion that drug use persists to counter the effects of drug withdrawal) do not offer a sufficient explanation either, as some drugs which result in tolerance and withdrawal do not result in dependence, such as tricyclic antidepressants.

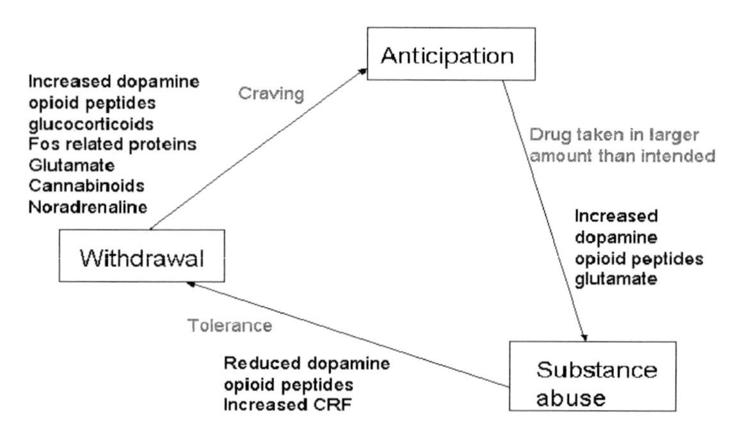

Fig. 22.1 Mechanisms underlying substance abuse. Addictive behavior is seen as a cyclical process where anticipation of the pleasure resulting from use of the drug leads to failure of the psychological mechanisms that usually control consumption. Excess quantities of drug are taken and an unpleasant withdrawal state arises when drug levels fall. This leads to excessive desire to seek out new supplies of the drug and a cycle of excessive use is established. In time tolerance occurs to the pleasurable effects of intoxication. However, in withdrawal addicts may be sensitized to pleasurable effects of the drug leading to a "positive feedback" loop.

A range of cellular changes occurs following repeated drug administration and these processes may help to give rise to the withdrawal state [7]. New proteins that are synthesized include transcription factors in the Fos family, which increase responsiveness to the beneficial effects of drugs [8]. These cellular changes coincide with the withdrawal state characterized by depression, irritability and anxiety.

In addition to changes in cellular biochemistry, neuroadaptations at the synaptic level may also play an important part in establishing the cycle of addiction [1]. Long-term drug use results in impairment of dopaminergic function that may be related to dopamine receptor downregulation. Adaptations to the glutaminergic system also seem to be important in the development of the negative affective state, and noradrenaline in the ventral forebrain plays a key role in the changes associated with drug withdrawal [6]. Linking the unpleasant effects of drug withdrawal with environmental stimuli may occur in the basolateral amygdala, which has been shown in animal experiments to be responsible for conditioned responses to stimuli linked with acute withdrawal [9]. These neuroadaptive changes may be responsible for associative learning processes (i.e., the classical or Pavlovian conditioned responses) that occur in addicts. Such learning processes could play a major part in the development of cravings (i.e., excessive "wanting"), for the addictive substance, which results in the drug user seeking out new drug supplies and either persisting in drug use (i.e., dependence) or beginning again the cycle of addiction (i.e., relapse).

22.2 Genetic Variation and Smoking Predisposition

Two complementary approaches have been used to investigate the relationship between genetic variation and tobacco use. The first type of investigation we shall describe here are "candidate gene" or "association" studies, where a gene is identified for investigation from a knowledge of the underlying biological mechanisms in the disease process. Allelic variants of the gene are defined and the allele frequencies are compared in two or more groups with different smoking phenotypes. These groups might be smokers and ex-smokers in a case-control design or heavy smokers and light smokers in a study to investigate effects of genotype on level of consumption.

The second type of investigation is the "linkage study" or "genome scan" in which there is no predetermined biological hypothesis for the genetic basis of the phenotype. This type of study measures the degree to which the trait is linked to markers which are placed randomly across the whole genome. These studies are discussed further in Section 22.3.

A summary of positive association studies and linkage studies in tobacco addiction is given in Figure 22.2. Many studies are small and the significance of the results is therefore uncertain. The strongest evidence linking genetic polymorphisms to nicotine addiction comes from the investigation of candidate genes in the dopaminergic neurotransmitter system, particularly studies of alleles affecting the dopamine D2 receptor and the dopamine transporter. However, a variety of poly-

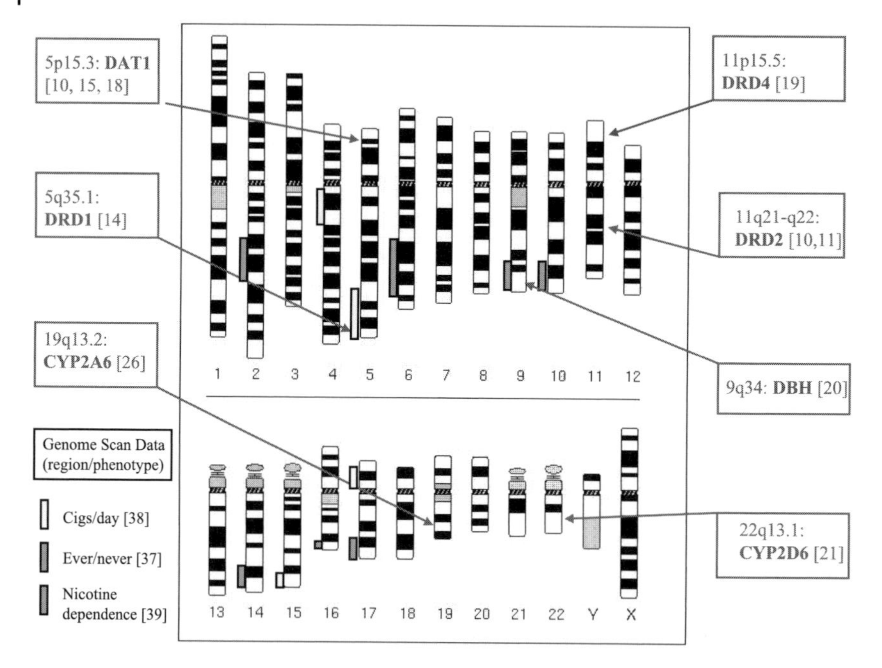

Fig. 22.2 Existing studies of links between polymorphisms in genes and their relationship to smoking behavior. *DAT1*, dopamine transporter; *DRD1*, dopamine D1 receptor; *DBH*, dopamine β-hydroxylase; *DRD2*, dopamine D2 receptor; *DRD4*, dopamine D4 receptor; *CYP2A6*, cytochrome P450 2A6; *CYP2D6*, cytochrome P450 2D6 (see color plates, p. XXXIX).

morphisms affecting proteins involved in synthesis, re-uptake and breakdown of other neurotransmitters have also been studied. Links with smoking have also been found with polymorphisms that reduce the activity of cytochrome P450 enzymes involved in nicotine metabolism and neurotransmitter synthesis.

22.2.1
Alleles Affecting Central Dopaminergic Function

In a large study, Caporaso found that a polymorphism in the 3′-untranslated region of the *DRD2* gene was about twice as common in smokers compared to non-smokers [10]. Originally defined as a restriction fragment length polymorphism (Taq 1A RFLP), the polymorphism results from a C to T change at position 32806 in *DRD2*. These findings, linking the polymorphism to smoking, confirmed earlier work [11] and recent studies suggest the same link although the sizes of the effects are smaller [11–15].

The exact mechanism by which the allele exerts its effects on predisposition to tobacco addiction is not known. People with one or more of the variant alleles are believed to have reduced numbers of dopamine receptors in the corpus striatum

[16]. If these changes are also present in central dopaminergic reward pathways, it may be that the allele is linked to impaired perception of reward. It has been suggested that an inherited dopamine deficit could be overcome by nicotine, which stimulates dopamine release thereby restoring dopamine function to normal levels [17]. In this way the polymorphism could confer susceptibility to tobacco use.

Several studies suggest a similar link between the 9 repeat allele of the dopamine transporter VNTR (variable number of tandem repeats) and smoking behavior [10, 15, 18]. In this case the variant allele, which is related to low scores for novelty seeking and extraversion in personality questionnaires, seems to protect people from persistent smoking. The mechanism of action on a molecular level for this polymorphism has not yet been determined but it is thought to enhance dopaminergic transmission and, therefore, reduce the need to use nicotine to augment dopaminergic function.

A polymorphism in the dopamine D4 receptor, which results in reduced cAMP formation when the receptor is stimulated, may also be linked with smoking. The strongest evidence comes from a small study in only one ethnic group and needs to be confirmed in larger studies [19]. Other studies on enzymes important in dopamine metabolism such as monoamine oxidase give further weight to the argument that dopaminergic pathways are important in tobacco dependence [20].

22.2.2 The Relationship Between Genetically Determined Variants of Cytochrome P450 Enzymes and Tobacco Dependence

Cytochrome P450 enzymes are responsible for the breakdown and inactivation of nicotine. Since genetically determined variations in these enzymes resulting in three distinct phenotypes – extensive, fast and slow metabolizers – are well-described, it seems reasonable to postulate that these variations would lead to differences in susceptibility to developing tobacco addiction. Slow metabolizers might be less likely to take up smoking because they experience more adverse effects such as nausea from nicotine and they may also be less likely to persist because they do not experience the rapid fall in nicotine levels, seen in extensive metabolizers, that stimulates the desire for another cigarette. However, these enzymes are also expressed in the brain and it is possible that any effects on tobacco dependence are more complicated since they may also be involved in central metabolism of monoamine neurotransmitters. The cytochromes P450 are also important in metabolizing many of the drugs used in therapy for tobacco dependence (see Section 22.5.3).

22.2.2.1 Effects of Genetically Determined Variation in Nicotine Metabolism on the Development of Nicotine Dependence

Several studies have looked at cytochrome P450 enzymes in relation to smoking behavior. Originally *CYP2D6* was thought to be responsible for hydroxylation of nicotine to cotinine and an early study looking at *CYP2D6* phenotype showed a striking association with smoking status [21], but this has not been confirmed in studies using molecular typing [22, 23]. It may be that current genetic typing methods for *CYP2D6* do not predict phenotype accurately enough to detect the association. However, current opinion suggests that *CYP2D6* does not play a major part in nicotine metabolism in most people and that a related enzyme *CYP2A6* is largely responsible [24]. In fact, both *CYP2A6* and *CYP2D6* catalyze the formation of cotinine from nicotine by N-oxidation but the *Km* for *CYP2D6* is an order of magnitude greater than for *CYP2A6* and hence the reaction proceeds more slowly [25]. This finding led to the search for an association between *CYP2A6* polymorphisms and smoking with conflicting results [26, 27].

There are considerable structural similarities between members of the *CYP2* subfamily. Figure 22.3 shows an alignment of the amino acid sequences of human *CYP2D6* and *CYP2A6* generated using multiple sequence alignment software (Clustal X). There are areas of substantial homology around the cysteine

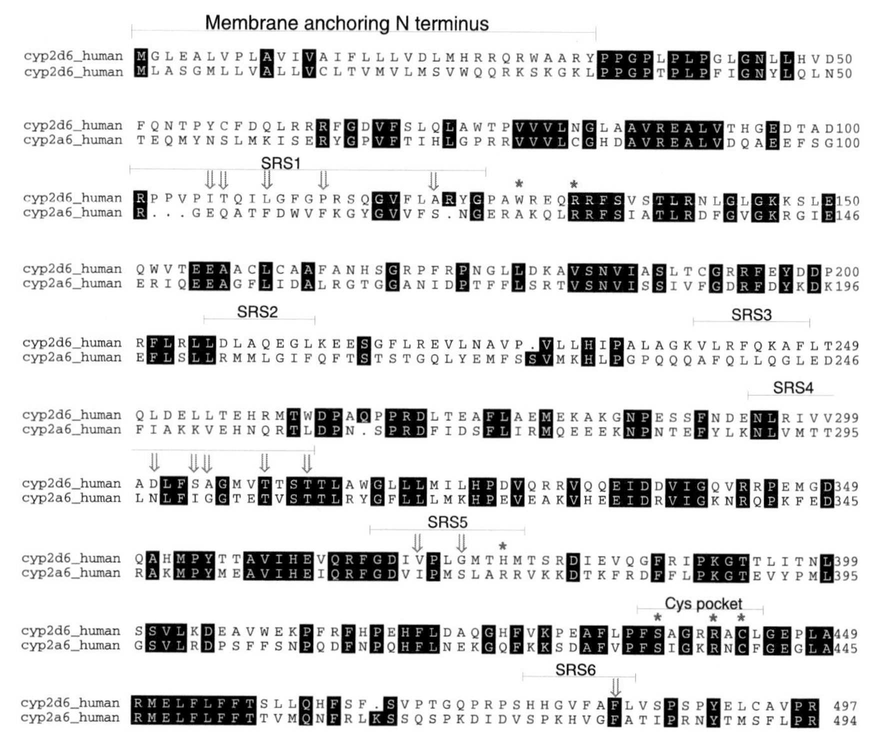

Fig. 22.3 Cytochrome P450 monooxygenases and nicotine metabolism. An alignment of the amino acid sequences of the enzymes 2A6 and 2D6. Occurrences of the same amino acid residue at the same position are shown in black. Putative substrate recognition sites (SRS1–SRS6) are shown by horizontal lines. Vertical arrows indicate amino acid residues predicted from modeling studies on cytochromes to bind to the enzyme substrate. The cysteine pocket contains key residues that bind to the heme cofactor, which is essential for enzyme activity. Inactivating amino acid changes for both enzymes are shown in red (see color plates, p. XL).

pocket, and the forth substrate recognition site (SRS4). Each enzyme has a 33 amino acid N-terminal sequence which attaches it to the membrane. The active site responsible for substrate oxidation is close to the membrane surface thereby allowing access to lipid soluble compounds [28]. A wide variety of compounds are oxidized in this way with substrate specificity being given by amino acid residues in six substrate recognition sites (SRS). These sites are highly polymorphic and genetic mutations that change amino acids occur more frequently in the substrate recognition sites than would be expected by chance [29].

Differences in the amino acid sequences at the substrate recognition sites responsible for binding nicotine may be responsible for functional differences between the two enzymes. Because of the close similarities between the enzymes we suggest that the minority of Caucasians who have inactive alleles of *CYP2A6* will metabolize nicotine to cotinine using *CYP2D6*. If this is the case the composite genotype for both enzymes will better reflect nicotine metabolic phenotype than the genotype for either enzyme alone. People who are slow metabolizers with *CYP2A6* and who also have inactive *CYP2D6* alleles are likely to oxidize nicotine extremely slowly.

Smoking researchers may have to wait for more robust genetic assays before the issue is finally settled [30], but pharmacological inhibitors of *CYP2A6* reduce smoking [31] and it seems likely that an inherited defect in nicotine metabolism would have the same effect.

22.2.2.2 Metabolism of Other Pharmacologically Active Compounds by Cytochrome P450 Enzymes

Differences in specificity between the two cytochromes are more marked for compounds other than nicotine with *CYP2D6* deactivating many drug molecules and *CYP2A6* relatively few. *CYP2D6* is particularly important in metabolizing drugs that affect the central nervous system including tricyclic antidepressants such as nortriptyline, selective serotonin re-uptake inhibitors (fluoxetine, fluvoxamine, paroxetine) and the monoamine oxidase inhibitor tranylcypromine [32].

CYP2D6 also converts tyramine, a ubiquitous dietary amine, to dopamine [33]. Although the major biosynthetic pathway for dopamine is from tyrosine using tyrosine hydroxylase, this alternative pathway may still be important and is likely to be affected by genetically determined variations in *CYP2D6* activity. Assuming a "dopamine deficiency" hypothesis of addiction, poor metabolizers with inactivating mutations in *CYP2D6* would have less available dopamine and, therefore, be more likely to smoke to return their dopamine levels to normal [17].

Perhaps because of these effects on metabolism of brain amines *CYP2D6* status has been linked to personality traits [34]. These personality traits are themselves associated with tobacco dependence and hence provide another possible mechanism linking allelic variation at the CYP2D6 locus to smoking behavior.

22.2.3
Future Targets for Candidate Gene Studies

Since nicotine has wide ranging effects on the central nervous system it seems likely that pharmacogenomic effects on the development of nicotine dependence will span several neurotransmitter systems. One study found an association between a polymorphism in dopamine β-hydroxylase and level of tobacco consumption [20]. This enzyme is important in noradrenaline synthesis and it is tempting to speculate that genetically regulated variations in activity might influence susceptibility to nicotine withdrawal symptoms mediated by noradrenergic pathways, but more information is required on the molecular effects of the polymorphism.

Given that underlying mechanisms of addiction to different drugs are likely to be similar it is surprising that more studies have not looked for links between smoking and neurotransmitters known to be involved in addiction to other substances. Possible candidates might be opioid peptides and GABA since evidence derived from animal studies suggests that these neurotransmitters are implicated in alcohol and opiate dependence and preliminary evidence suggests that they are also involved in nicotine addiction [35].

Perhaps the largest gap in the evidence surrounds the primary site of action of nicotine in the brain – the acetylcholine receptor itself. Studies on links with tobacco dependence in humans will await closer definition of the complex interaction between nicotine and its receptor and the identification of the receptor subtypes that are important in addiction pathways (see Section 22.4).

22.3
Genome Scans to Investigate Tobacco Dependence

Typically genome-wide studies look within families for co-transmission of a disease trait with one of a set of polymorphic markers. These markers are simple sequence variants which are evenly spaced throughout the genome and in the past have taken the form of variable numbers of CA repeats (microsatellites), although more recently interest has focused on single nucleotide polymorphisms (SNPs) as biallelic markers. Linkage studies can support the candidacy of known genes and also suggest chromosomal regions where new genes that contribute to phenotype may be located. Complex traits such as smoking may require a combined linkage and association strategy to unravel genetic influences.

Addiction is extremely unlikely to derive from a single major gene; rather we expect a large number of susceptibility genes to be involved, none fully penetrant. Nicotine dependence is a multi-factorial process involving initiation and maintenance of the smoking habit along with cessation and relapse, and each aspect is likely to come under separate genetic control. Until recently candidate gene studies have been carried out on a gene-by-gene basis. Genome scans may allow a more extensive and unbiased search and suggest new genes that may be involved in tobacco dependence.

Scans identify areas of the genome which are usually too large for positional cloning and may contain several genes. In order to narrow the candidate region, linkage disequilibrium (LD) mapping can be carried out. This relies on the non-random allelic association of closely neighboring markers of a gene variant. Through linkage disequilibrium association studies it is possible to identify a region that may harbor a susceptibility gene, without a prior biological hypothesis. LD mapping has been done in a limited fashion in families [36] and also has the potential to be carried out in large cohorts of non-related individuals.

22.3.1 Genomic Areas Linked with Susceptibility to Nicotine Dependence

Three genome-wide screens have been reported, two of which use the COGA families (Collaborative Study on the Genetics of Alcoholism). Bergen et al. [37] looked at two smoking related traits in sib-pairs, a dichotomous "ever never smoker" trait and also a quantitative measure of the numbers of cigarettes smoked (pack-year history). Some evidence for linkage to areas of chromosomes 6, 9 and 14 was reported. Stronger evidence (a lod score of 3.2) for linkage between smoking behavior and a genetic location on 5q (marker D5S1354) was also found in these COGA families [38]. This region contains the dopamine D1 receptor gene (DRD1), a biologically plausible candidate gene, variations in which have already been implicated in smoking behavior [14]. Weaker evidence also implicated chromosomes 4, 15 and 17 in this study.

Genetic linkage to nicotine dependence was examined in families from New Zealand, and regions on chromosomes 2, 4, 10, 16, 17 and 18 yielded small but positive lod scores [39]. These findings were not replicated in a study on families from the United States. All these studies were relatively small and it is probable they did not provide sufficient power to detect genes of small effect or those which influence risk in only a proportion of the families.

22.3.2 Biallelic or Multiallelic Markers for Linkage Studies?

Microsatellite markers have been extensively used in family studies. These are regions of variation, mostly $(CA)_n$ repeats, which are distributed throughout the genome. There are potentially as many as 10^5 loci, each of which has many alleles and these markers are, therefore, highly informative. Microsatellites can be typed by automated multiplex PCR.

Current interest is, however, focusing on panels of SNPs to provide similar information about genome-wide variation. SNPs are biallelic and, therefore, less informative so more markers are needed, but this potential disadvantage is offset by the improvements in high-throughput technology which means SNP typing can be carried out on a large scale.

22.3.3
Genome Scans in the Future

The number of SNPs needed for a genome scan is currently the subject of some debate and depends in part on the degree to which nearby markers in the genome are inherited together (linkage disequilibrium). The extent of linkage disequilibrium (LD) depends on population admixture along with genetic drift, mutation and natural selection. LD generally declines as the distance between markers increases and the extent of linkage disequilibrium varies with position in the genome. The theoretical average extent of LD has been estimated to be between three and 100 kb [40, 41], this is useful for estimating the number of markers required in a SNP map. To cover the 3 million base pairs of the genome, a minimum of 30000–500000 evenly spaced markers are needed to have a marker within the average range of LD. The density of current maps ranges from 60000–300000 [42] which is sufficient for regions of extensive LD, but may miss susceptibility loci in areas of small LD.

The strength of LD around the locus linked to phenotype will also determine the magnitude of the association. In regions of high LD the marker will be in more complete linkage with the susceptibility SNP and the relative risk will be higher. Differences in frequency between the marker SNP and SNP of interest further increase the need for a greater number of markers [43]. This fact along with the large sample sizes, which will be needed to detect small effect sizes from the many genes responsible for tobacco addiction, emphasizes the need for inexpensive high-throughput genotyping technology.

Nicotine addiction is very complex and pharmacological evidence alongside structure and function analysis of candidate genes may clarify the mechanisms involved. The challenge will be to translate known genetic variation into therapeutic strategies to reduce the prevalence of smoking in the population.

22.4
Evidence for the Genomic Basis of Nicotine Addiction from Animal Models

Likelihood of developing dependence to nicotine will involve specific functional changes in the brain. Examining the detailed genetic basis for these functional changes is difficult in humans, so animal models are needed. Three approaches have been taken to examining genetic influences of the effects of nicotine in rodents namely inbred lines, selectively bred lines and knockout mice.

Nicotine has a wide range of effects on behavior in humans and individual response to nicotine may predict predisposition to addiction. Some individuals may be genetically more likely to be hypersensitive to nicotine and, therefore, find it aversive; others may be more positively reinforced by nicotine and seek to repeat the stimulus. Genetically modified animals are increasingly important tools for elucidating the molecular mechanisms involved in addiction.

Inbred Lines

In order to obtain homozygosity of virtually all genetic loci, inbred strains are created by mating close relations over several generations (> 20). The genetic influence of the inter-individual variation of a variety of related phenotypes has been clearly demonstrated. These include sensitivity to nicotine, first demonstrated by Collins and coworkers to be related to genotype [44], also extent of self-administration [45], and ability to develop tolerance to nicotine [46] and other abused substances [47, 48]. Future work on mapping quantitative trait loci of inbred rodent strains by differential expression studies using high-density DNA microarrays will enable a high-throughput approach to the genetic influences on nicotine dependence.

Selectively Bred Lines

Mating animals that display a required trait and selecting offspring that also display the trait reveals genetic differences in behavioral responses such as cognition. Rats have been selectively bred for high or low emotionality on the basis of defecation rates and these two strains have been found to differ in their sensitivity to the stimulant/depressant effects of nicotine [49].

Knockout Mice

Recent advances in transgenic technology enable inactivating mutations in specific genes to be expressed in mice. This will greatly aid the complete elucidation of the molecular and cellular basis of nicotine action. There are many lines of evidence to suggest the reinforcing properties of nicotine occur principally through the mesolimbic dopaminergic system [50]. Nicotine acts by binding to nicotinic acetylcholine receptors (nAChR) in the nucleus accumbens and ventral tegmental area, regions of the brain known to be implicated in reward. This stimulates an increase in synaptic dopamine levels which is thought to mediate reward. Ten subunits of neuronal nicotinic receptors have been identified so far and these combine to form a variety of receptor subtypes that are sensitive to nicotine and some of which are expressed in dopaminergic neurons [51]. Null mice can be used to unravel the contribution of the different subunits. Knockout mice that lacked the β-2 subunit of the nAChR lacked the high-affinity binding of nicotine in the brain [52] and showed attenuated self-administration of nicotine [53].

Other knockout models that could be used to validate candidate genes include mice that lack monoamine oxidase A (MAO-A), which have demonstrated altered behavior and alcohol tolerance [54]. Transgenic mice in which the dopamine transporter gene has been deleted show striking hyperactivity via enhanced persistence of dopamine which is not altered by cocaine or amphetamine administration [55]. Knockouts of the serotonin 1B receptor are also available and are best used as models of vulnerability to drug abuse [56].

22.5
Applying Pharmacogenomics to Therapy for Nicotine Addiction

Given that there is strong evidence for a genetic component to tobacco addiction the identification of genes that may be responsible for this link leads to exciting new opportunities to help people to withdraw from nicotine and to prevent relapse.

22.5.1
Mechanism of Action of Existing Treatments and New Directions for Drug Therapy

The most effective therapies that we have available to aid smoking cessation are antidepressant drugs and nicotine replacement therapy [57]. Antidepressants such as nortriptyline and buproprion probably work by inhibiting the action of transporter proteins that remove monoamine neurotransmitters from the synaptic cleft. In this way they potentiate the action of the transmitter. The actions of the drugs are not completely specific to particular transporters but nortriptyline is thought to act preferentially on noradrenaline and buproprion on the dopamine transporter. The mechanism of action of nicotine replacement therapy is not clear but the steady delivery available from the nicotine patch may alleviate withdrawal symptoms while faster acting preparations such as sprays control craving.

Dissection of the molecular mechanisms underlying tobacco addiction should lead to new and better treatments to achieve nicotine withdrawal. It seems clear that the dopamine D2 receptor is involved in nicotine dependence and drugs that block this receptor, such as tiapride, could be useful in the treatment of tobacco dependence. Tiapride has been shown to be successful in alcohol withdrawal [58] but would represent a new avenue for tobacco addiction therapy.

Similarly opioid peptides are important in nicotine addiction and may have a role in causing nicotine withdrawal symptoms in some smokers [35]. Opioid antagonists such as naltrexone are licensed treatments for dependence syndromes arising from other addictive drugs and could also be of use in some smokers to aid nicotine withdrawal [59] although there is no definitive evidence overall that they are beneficial [60].

While we have treatments in current clinical use to help smokers to withdraw from nicotine there are none that are routinely used to maintain abstinence. Perhaps the most interesting potential therapeutic developments come from a consideration of molecular mechanisms involved in smoking relapse where glutaminergic systems play an important part. Drugs such as acamprosate that block NMDA (N-methyl-D-aspartate) receptors and have been shown to be effective in preventing relapse in alcoholism [61] may, therefore, also be effective in treating tobacco addiction. It also seems reasonable to suppose that drugs modifying dopaminergic activity associated with the perception of reward such as buproprion and tiapride would also be useful in the prevention of smoking relapse.

22.5.2 Classifying Smokers According to the Molecular Basis for their Habit

Only a small proportion of people respond to the best treatments that are currently available to aid long-term smoking cessation – perhaps about 20% [57]. It may be that genetics has a part to play in determining which patient responds to a particular treatment. For example, we know that polymorphisms in the dopamine transporter predispose to persistent smoking. Might these same polymorphisms also predict therapeutic response to buproprion which acts, in part, by binding to this same protein? Similarly it seems reasonable to suppose that smokers with the *DRD2* C32806T polymorphism will respond well to drugs that augment dopaminergic function and those with inactive alleles of *CYP2A6* would be unlikely to respond to nicotine replacement therapy. Identifying these people in advance with a simple genetic test would be a major step forward in the treatment of tobacco addiction but more basic research is needed before this hypothesis becomes clinical reality.

22.5.3 Accurate Determination of Dosage for Therapeutic Interventions

Perhaps nearer to clinical application is the use of DNA analysis to predict the most appropriate dose of drug. While *CYP2D6* is important in metabolizing drugs such as nortriptyline, determining the phenotype for *CYP2D6* is relatively time-consuming and unlikely to be feasible in clinical practice. Assays for *CYP2D6* genotype predict drug metabolizing phenotype relatively accurately [62]. Since side effects often limit the usefulness of antidepressant drugs such as nortriptyline, speedy and accurate determination of phenotype could be helpful in dose adjustment and perhaps in helping a person to avoid a particular treatment when the chance of adverse effects is high.

Accurate prediction of the rate of nicotine metabolism based on *CYP2A6/D6* genotype could make adjusting the dose of nicotine replacement to avoid side effects and ensure effectiveness much simpler. It may be that information technology could help in making decisions about the choice [63] and dose [64] of drug based on genetic data.

22.5.4 "Minimum SNP Set" for Tobacco Dependence and Need for High-Throughput Genotyping

It seems likely that a small set of single nucleotide polymorphisms (SNPs) could be used in the future to classify smokers more accurately according to the molecular basis for their addiction. Although existing experimental data are still limited, it is easy to see how smokers might fall into different categories. For example, some with polymorphisms affecting dopaminergic function, perhaps *DRD2* and *DAT1*, might be particularly sensitive to the rewarding aspects of consuming tobacco. Others may become addicted because they need to avoid withdrawal symptoms that are made more severe by genetic variation in central pathways affecting neurotransmission with excitatory amino acids or noradrenaline. Perhaps a third group who are more likely to develop long-term adaptations in cellular biochemistry will experience the enduring craving for nicotine often described by heavy smokers which lasts for many years.

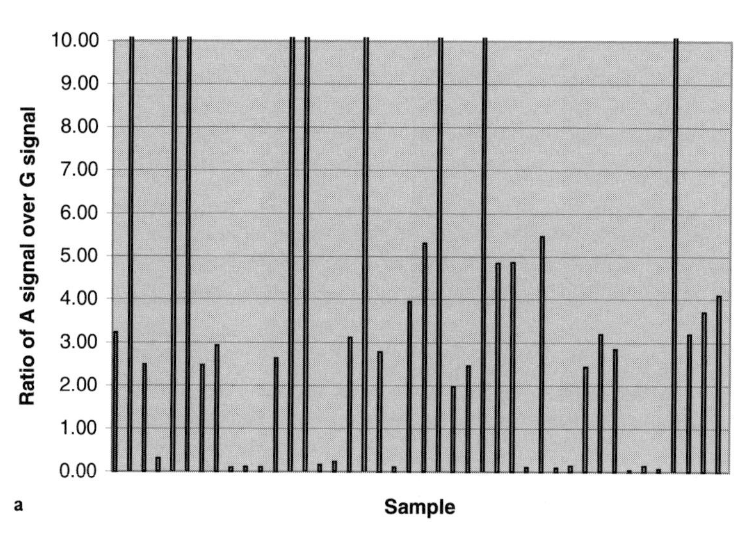

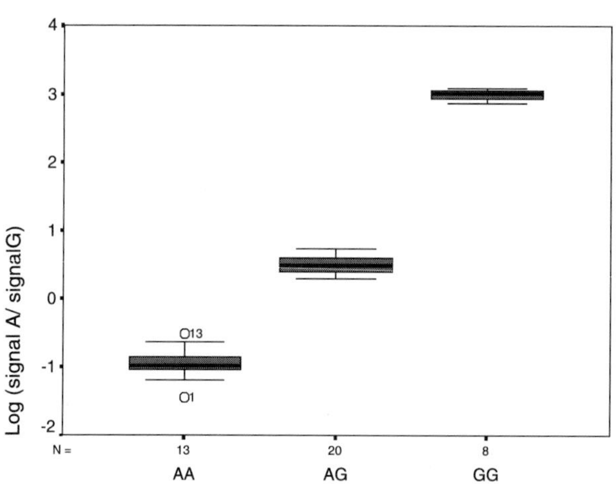

Fig. 22.4 (**a**) Invader genotyping for the DBH 1368 polymorphism. The figure shows results of typing 42 people. The bars represent the fluorescent signal from the reaction detecting the G-allele divided by the signal from the A-allele. Upper extreme of the linear scale has been omitted. (**b**) Discrimination between genotypes with Invader. In the box plot the black bars represent medians, whiskers interquartile range and circles outliers.

Smokers falling into these different groups are likely to respond to different therapeutic approaches both behavioral and pharmacological. The set of polymorphisms used to classify smokers might also predict response to therapy with a particular drug. Another overlapping set, perhaps encompassing common genetic variants in *CYP2A6*, *CYP2D6* and other cytochromes, could suggest the most appropriate dose for drugs such as antidepressants or nicotine replacement therapy. Computers can be an important aid to interpreting this complex genetic data in a clinical setting [65].

One certain need for the future will be for accurate, high-throughput genotyping systems. The process of developing a genotyping assay using the polymerase chain reaction with sequence specific primers [66] is relatively simple for single nucleotide polymorphisms. Another option is to use the Invader assay [67] in a fluorescence detection format as shown in Figure 22.4.

22.6 Conclusion

A deeper understanding of the molecular mechanisms underlying tobacco addiction will lead to the identification of different types of smokers. Classifying smokers according to the underlying biological processes involved in their addiction will lead to new treatments for tobacco dependence. Patient-specific therapy with both choice of treatment and dose of drug informed by DNA analysis seems likely to be more effective than conventional therapy with fewer unwanted effects.

Acknowledgements
Some material in this chapter was previously published in Trends in Molecular Genetics.

22.7 References

1 Koob GF, Le Moal M. Drug abuse: hedonic homeostatic dysregulation. Science **1997**; 278(5335):52–58.
2 Spanagel R, Weiss F. The dopamine hypothesis of reward: past and current status. Trends Neurosci **1999**; 22(11):521–527.
3 Robinson TE, Berridge KC. The neural basis of drug craving: an incentive-sensitization theory of addiction. Brain Res Brain Res Rev **1993**; 18(3):247–291.
4 Robinson TE, Berridge KC. The psychology and neurobiology of addiction: an incentive-sensitization view. Addiction **2000**; 95(Suppl 2):91–117.
5 Wickelgreen, I. Teaching the brain to take drugs. Science **1998**; 280(26 June): 2045–2047.
6 Delfs JM et al. Noradrenaline in the ventral forebrain is critical for opiate withdrawal-induced aversion. Nature **2000**; 403(6768):430–434.

7 Nestler EJ, Aghajanian GK. Molecular and cellular basis of addiction. Science **1997**; 278(5335):58–63.

8 Kelz MB et al. Expression of the transcription factor deltaFosB in the brain controls sensitivity to cocaine. Nature **1999**; 401(6750):272–276.

9 Schulteis G et al. Conditioning and opiate withdrawal. Nature **2000**; 405(6790):1013–1014.

10 Caporaso N et al. The genetics of smoking: the Dopamine receptor (DRD2) and transporter polymorphisms in a smoking cessation study. Proc Am Assoc Cancer Res **1997**; 38(March 1997):168.

11 Noble EP et al. D2 dopamine receptor gene and cigarette smoking: a reward gene? Med Hypotheses **1994**; 42(4):257–260.

12 Bierut LJ et al. Family-based study of the association of the dopamine D2 receptor gene (DRD2) with habitual smoking. Am J Med Genet **2000**; 90(4):299–302.

13 Comings DE et al. Exon and intron variants in the human tryptophan 2,3-dioxygenase gene: potential association with Tourette syndrome, substance abuse and other disorders. Pharmacogenetics **1996**; 6(4):307–318.

14 Comings DE et al. Studies of the potential role of the dopamine D1 receptor gene in addictive behaviors. Mol Psychiatry **1997**; 2(1):44–56.

15 Lerman C et al. Evidence suggesting the role of specific genetic factors in cigarette smoking. Health Psychol **1999**; 18(1):14–20.

16 Thompson J et al. D2 dopamine receptor gene (DRD2) Taq1 A polymorphism: reduced dopamine D2 receptor binding in the human striatum associated with the A1 allele. Pharmacogenetics **1997**; 7(6):479–484.

17 Blum K et al. The D2 dopamine receptor gene as a determinant of reward deficiency syndrome. J R Soc Med **1996**; 89(7):396–400.

18 Sabol SZ et al. A genetic association for cigarette smoking behavior. Health Psychol **1999**; 18(1):7–13.

19 Shields PG et al. Dopamine D4 receptors and the risk of cigarette smoking in African-Americans and Caucasians. Cancer Epidemiol Biomarkers Prev **1998**; 7(6):453–458.

20 McKinney E et al. Association between polymorphisms in dopamine metabolic enzymes and tobacco consumption in smokers. Pharmacogenetics **2000**; 10:1–9.

21 Turgeon J et al. Debrisoquine metabolic ratio (DMR) distribution differs among smokers and non-smokers. Am Soc Clin Pharmacol Ther **1995**; PI61:150.

22 Saarikoski ST et al. CYP2D6 ultrarapid metabolizer genotype as a potential modifier of smoking behavior. Pharmacogenetics **2000**; 10(1):5–10.

23 Cholerton S et al. CYP2D6 genotypes in cigarette smokers and non-tobacco users. Pharmacogenetics **1996**; 6(3):261–263.

24 Nakajima M et al. Role of human cytochrome P4502A6 in C-oxidation of nicotine. Drug Metab Dispos **1996**; 24(11):1212–1217.

25 Yamazaki H et al. Roles of CYP2A6 and CYP2B6 in nicotine C-oxidation by human liver microsomes. Arch Toxicol **1999**; 73(2):65–70.

26 Pianezza ML, Sellers EM, Tyndale RF. Nicotine metabolism defect reduces smoking. Nature **1998**; 393:750.

27 London SJ et al. Genetic variation of CYP2A6, smoking, and risk of cancer. Lancet **1999**; 353(9156):898–899.

28 Williams PA et al. Mammalian microsomal cytochrome P450 monooxygenase: structural adaptations for membrane binding and functional diversity. Mol Cell **2000**; 5(1):121–131.

29 Gotoh O. Substrate recognition sites in cytochrome P450family 2 (CYP2) proteins inferred from comparative analyses of amino acid and coding nucleotide sequences. J Biol Chem **1992**; 267(1):83–90.

30 Oscarson M et al. Identification and characterisation of novel polymorphisms in the CYP2A locus: implications for nicotine metabolism. FEBS Lett **1999**; 460(2):321–327.

31 Sellers EM, Kaplan HL, Tyndale RF. Inhibition of cytochrome P450 2A6 increases nicotine's oral bioavailability and

decreases smoking. Clin Pharmacol Ther **2000**; 68(1):35–43.

32 Daly AK. Molecular basis of polymorphic drug metabolism. J Mol Med **1995**; 73(11):539–553.

33 Hiroi T, Imaoka S, Funae Y. Dopamine formation from tyramine by CYP2D6. Biochem Biophys Res Commun **1998**; 249(3):838–843.

34 Llerena A et al. Relationship between personality and debrisoquine hydroxylation capacity. Suggestion of an endogenous neuroactive substrate or product of the cytochrome P4502D6. Acta Psych Scand **1993**; 87(1):23–28.

35 Corrigall WA et al. Response of nicotine self-administration in the rat to manipulations of mu-opioid and gamma-aminobutyric acid receptors in the ventral tegmental area. Psychopharmacology **2000**; 149(2):107–114.

36 Houwen RH et al. Genome screening by searching for shared segments: mapping a gene for benign recurrent intrahepatic cholestasis. Nature Genet **1994**; 8(4):380–386.

37 Bergen AW et al. A genome-wide search for loci contributing to smoking and alcoholism. Genet Epidemiol **1999**; 17(Suppl 1):55–60.

38 Duggirala R, Almasy L, J. Blangero J. Smoking behavior is under the influence of a major quantitative trait locus on human chromosome 5q. Genet Epidemiol **1999**; 17(Suppl 1):139–144.

39 Straub RE et al. Susceptibility genes for nicotine dependence: a genome scan and followup in an independent sample suggest that regions on chromosomes 2, 4, 10, 16, 17 and 18 merit further study. Mol Psychiatry **1999**; 4(2):129–144.

40 Jorde LB et al. Linkage disequilibrium predicts physical distance in the adenomatous polyposis coli region. Am J Hum Genet **1994**; 54(5):884–898.

41 Lai E et al. A 4-Mb high-density single nucleotide polymorphism-based map around human APOE. Genomics **1998**; 54(1):31–38.

42 Marshall E. Drug firms to create public database of genetic mutations. Science **1999**; 284(5413):406–407.

43 McCarthy JJ, Hilfiker R. The use of single-nucleotide polymorphism maps in pharmacogenomics. Nature Biotechnol **2000**; 18(5):505–508.

44 Hatchell PC, Collins AC. The influence of genotype and sex on behavioural sensitivity to nicotine in mice. Psychopharmacology **1980**; 71(1):45–49.

45 Robinson SF, Marks MJ, Collins AC. Inbred mouse strains vary in oral self-selection of nicotine. Psychopharmacology **1996**; 124(4):332–339.

46 Marks MJ et al. Genotype influences the development of tolerance to nicotine in the mouse. J Pharmacol Exp Ther **1991**; 259(1):392–402.

47 Deitrich RA, Bludeau P, Erwin VG. Phenotypic and genotypic relationships between ethanol tolerance and sensitivity in mice selectively bred for initial sensitivity to ethanol (SS and LS) or development of acute tolerance (HAFT and LAFT). Alcohol Clin Exp Res **2000**; 24(5):595–604.

48 Tolliver BK et al. Genetic analysis of sensitization and tolerance to cocaine. J Pharmacol Exp Ther **1994**; 270(3):1230–1238.

49 Fleming JC, Broadhurst PL. The effects of nicotine on two-way avoidance conditioning in bi-directionally selected strains of rats. Psychopharmacologia **1975**; 42(2):147–152.

50 Pontieri FE et al. Effects of nicotine on the nucleus accumbens and similarity to those of addictive drugs. Nature **1996**; 382(6588):255–257.

51 Le Novere N, Zoli M, Changeux JP. Neuronal nicotinic receptor alpha 6 subunit mRNA is selectively concentrated in catecholaminergic nuclei of the rat brain. Eur J Neurosci **1996**; 8(11):2428–2439.

52 Picciotto MR et al. Abnormal avoidance learning in mice lacking functional high-affinity nicotine receptor in the brain. Nature **1995**; 374(6517):65–67.

53 Picciotto MR et al. Acetylcholine receptors containing the beta2 subunit are involved in the reinforcing properties of nicotine. Nature **1998**; 391(6663):173–177.

54 Popova NK et al. Altered behavior and alcohol tolerance in transgenic mice lacking MAO A: a comparison with effects of

MAO A inhibitor clorgyline. Pharmacol Biochem Behav **2000**; 67(4):719–727.

55 Giros B et al. Hyperlocomotion and indifference to cocaine and amphetamine in mice lacking the dopamine transporter. Nature **1996**; 379(6566):606–612.

56 Scearce-Levie K et al. 5-HT receptor knockout mice: pharmacological tools or models of psychiatric disorders. Ann NY Acad Sci **1999**; 868:701–715.

57 Lancaster T et al. Effectiveness of interventions to help people stop smoking: findings from the cochrane library. Br Med J **2000**; 321(7257):355–358.

58 Shaw GK et al. Tiapride in the prevention of relapse in recently detoxified alcoholics. Br J Psychiatry **1994**; 165(4):515–523.

59 King AC, Meyer PJ. Naltrexone alteration of acute smoking response in nicotine-dependent subjects. Pharmacol Biochem Behav **2000**; 66(3):563–572.

60 Wong GY et al. A randomized trial of naltrexone for smoking cessation. Addiction **1999**; 94(8):1227–1237.

61 Tempesta E et al. Acamprosate and relapse prevention in the treatment of alcohol dependence: a placebo-controlled study. Alcohol **2000**; 35(2):202–209.

62 Sachse C et al. Correctness of prediction of the CYP2D6 phenotype confirmed by genotyping 47 intermediate and poor metabolizers of debrisoquine. Pharmacogenetics **1998**; 8(2):181–185.

63 Walton RT et al. Evaluation of computer support for prescribing (CAPSULE) using simulated cases. Br Med J **1997**; 315(7111):791–795.

64 Walton R et al. Computer support for determining drug dose: systematic review and meta-analysis. Br Med J **1999**; 318(7189):984–490.

65 Emery J. Computer support for genetic advice in primary care. Br J Gen Pract **1999**; 49(444):572–575.

66 Bunce M et al. Phototyping: comprehensive DNA typing for HLA-A, B, C, DRB1, DRB3, DRB4, DRB5 & DQB1 by PCR with 144 primer mixes utilizing sequence-specific primers (PCR-SSP). Tissue Antigens **1995**; 46(5):355–367.

67 Hessner MJ, Budish MA, Friedman KD. Genotyping of factor V G1691A (Leiden) without the use of PCR by invasive cleavage of oligonucleotide probes [In Process Citation]. Clin Chem **2000**; 46(8 Pt 1):1051–1056.

23
Pharmacogenomics of Opioid Systems

Terry Reisine

Abstract

Opiates are one of the major neurotransmitter systems in the body involved in the control of pain sensation. The cloning of the opioid receptors has provided the tools to identify the molecular basis of the analgesic action of endogenous opioid peptides and opiate drugs. The cloning will also provide the means to discover a new family of therapeutics which are powerful analgesics with few if any of the limiting side-effects of drugs used today.

23.1
Introduction

Opium and its derivatives have been employed for centuries for the treatment of pain. Morphine was first synthesized in 1805 and has proven to be one of the most effective analgesic agents available [1]. Morphine and its analogs are particularly useful because they diminish pain sensation while maintaining consciousness. However, opiates induce severe side-effects including respiratory depression, nausea, bradycardia and constipation and long-term use of opiates can cause addiction [2].

Opiates induce these diverse pharmacological actions by stimulating membrane-bound receptors. Opiate receptors were first characterized in pharmacological assays. While the synthetic alkaloid derivatives of morphine and the endogenous peptides all interact with opiate receptors, pharmacological studies showed heterogeneity in the responses to these opiates which suggested that subclasses of opiate receptors were expressed in the body [3–7]. Morphine and its derivatives are more potent than the enkephalins in binding to guinea pig ileum receptors whereas the enkephalins are more potent than morphine in binding to receptors in the mouse vas deferens. The morphine-sensitive receptors were named μ [7] and the enkephalin preferring receptors have been referred to as δ receptors [4]. Extensive pharmacological analysis as well as cloning studies in the 1990s demonstrated that morphine and enkephalin preferring receptors were in fact distinct molecular entities [8].

Additional pharmacological studies identified a third opiate receptor with high affinity for ketocyclazocine and its derivatives which has been referred to as the κ receptor [3]. Dynorphin A selectively binds to this receptor. Stimulation of κ receptors induces pharmacological actions distinct from those associated with μ or δ receptor activation. κ agonists cause diuresis in contrast to the constipation induced by μ agonists. They also induce dysphoria in humans in contrast to the euphoria of μ agonists [2]. Cloning studies have established that the κ receptor is distinct from δ and μ receptors and like the other opiate receptors is encoded by a unique gene [8, 9].

The three major classes of opiate receptors have been further discriminated into subtypes based on pharmacological analysis of the characteristics of each receptor. The most clearly established subtypes are for the δ receptor. All δ receptors respond to enkephalins and beta-endorphin equally well and show no affinity differences. Synthetic peptide agonists at the δ receptor discriminate subtypes, although the affinity differences are minimal. For example, the specific δ agonist DPDPE has been proposed to preferentially bind to δ_1 receptors whereas deltorphin II has been reported to be δ_2 receptor selective [10]. However, these peptides have only 5- to 10-fold differences in affinity for the subtypes. δ receptor subtypes can be distinguished by the antagonists BNTX and NTB [11, 12]. These compounds show over 50-fold selectivity for δ receptor subtypes with BNTX being δ_1 selective and NTB being δ_2 selective. Pharmacological studies in rodents have suggested that the spinal analgesic effect of δ agonists is primarily via the δ_2 receptor [13]. In contrast, supraspinal analgesia induced by intracerebrally administered δ agonists is believed to be mediated by both δ_1 and δ_2 receptors.

Subtypes of μ and κ receptors have also been proposed, based on pharmacological evidence [2, 14, 15]. The antagonist naloxonazine has been suggested to be a selective ligand for μ_1 receptors. Similarly, spiradoline and U69, 593 have been reported to be κ_1 receptor-selective agonists while nor-BNI is a κ_1 selective antagonist. Few drugs, however, have been identified as selective for the other μ or κ receptor subtypes, which has made it difficult to access their distinct functional properties.

23.2 Cloning of the Opiate Receptors

The identification of the genetic basis of opiate receptor subclasses was determined when the opiate receptor cDNAs were cloned. Cloning of an opiate receptor was first accomplished by two independent groups [16, 17] who used expression cloning procedures to identify δ receptor cDNA from a mouse-derived NG-108 neuroblastoma cell line. The cloning procedures involved identifying populations of cells transfected with cDNA generated from the mRNA of the NG-108 cell library for clones expressing the receptor using receptor binding assays. Through an iterative process of screening for expression and isolation of putative receptor clones, the δ receptor cDNA was finally isolated and sequenced.

Cloning of the rat μ opioid receptor was somewhat easier than the cloning of the δ receptor since it was based on the knowledge of what the δ receptor sequence was and on the assumption that δ and μ receptors had high amino acid sequence similarity. Chen et al. [18] used probes directed against conserved regions of the δ receptor to screen a rat brain cDNA library. cDNAs identified by this approach were cloned into expression vectors, transfected into COS cells and μ receptor expression detected with radioactive μ receptor selective ligands.

The κ receptor was cloned by a completely different approach. Yasuda et al. [9] employed probes against conserved regions of somatostatin receptors to screen a mouse brain cDNA library. Mouse κ and δ receptor cDNAs were isolated using this procedure that established that opiate and somatostatin receptors have high amino acid sequence similarity, consistent with their ability to bind some common ligands, such as Sandostatin.

Comparison of the predicted amino acid sequences of the three opiate receptors revealed that they had a high degree of overall amino acid sequence similarity [2, 8, 19–21] (approximately 60% overall). Highest amino acid sequence similarities were found in the transmembrane segments and the intracellular loops of the opiate receptors. The amino and carboxy termini and the extracellular loops have the greatest divergence in amino acid sequence between the opiate receptors. These regions of amino acid sequence divergence may contain functional domains of the receptors responsible for their distinct properties. In fact, mutagenesis studies have revealed that these regions contain ligand binding domains of the receptors that confer onto the receptors the ability to interact with selective drugs.

The opiate receptor genes have been identified and are located on different chromosomes. In the mouse, the δ receptor gene is located on chromosome 4 (locus 4D) [22], the κ receptor gene is located on chromosome 1 while the μ receptor gene is localized to chromosome 10 [23]. In humans, the δ receptor gene is on chromosome 1 [24], the κ receptor gene is on the proximal arm of chromosome 8 [25] and the μ receptor gene is on the distal arm of chromosome 6 [26]. There is no evidence of multiple genes for each opiate receptor.

The genes have multiple exons which encode distinct fragments of the receptor proteins [27]. This exon/intron structure suggests that subtyping of receptors could be created by splicing of different exons especially since the potential splice junctions correspond to critical sites within the receptors involved in differential ligand binding. Zimprich et al. [28] and Bare et al. [29] have found splice variants of the μ receptor. The splice variants differ in amino acid sequence in the C-terminal region of the receptors. The splicing is created by a unique exon that is inserted between the previously described exons 3 and 4 of the μ receptor gene [30]. Expression of the two splice variants in CHO cells show that they have similar pharmacological specificities and are capable of coupling to adenylyl cyclase and phospholipase C [31]. However, they differ in their ability to be desensitized by agonists suggesting an important role of the C-terminus of the μ receptor for desensitization.

Gaveriaux-Ruff et al. [32] have shown evidence that δ and κ receptors undergo alternative splicing. Their evidence suggests that splicing could generate truncated forms of the receptors. This could also be functionally important since the C-terminus of the δ and κ receptors has been proposed to be important in receptor internalization and, as with the μ receptor, splice variants may have different sensitivities to agonists induced regulation [8, 19].

Lu et al. [33] have also found that the κ receptor has multiple promoters and have suggested that this could result in differential regulation of the expression of the κ receptor.

The cloning of the opiate receptors provided tools to test whether the cloned and native receptors are similar proteins. Homologous knock out of the μ opiate receptor gene in mice resulted in animals that did not express μ receptor although they expressed δ and κ receptors at normal levels [34, 35]. These findings suggest that only one gene encodes all of the μ receptors in the body. The knock out animals were insensitive to the analgesic effects of morphine and the animals did not become dependent on morphine demonstrating that both the therapeutic and side effects of morphine are specifically mediated by the μ receptor.

23.3 Distribution of the Opiate Receptors

The cloning of the opiate receptors has provided tools to investigate the distribution and expression of the receptor genes and the receptor protein. Northern analysis using RNA probes identified a single κ receptor mRNA in mouse brain of 5.2 kb [9]. Multiple, large δ receptor transcripts were identified in mouse of 11 and 8.5 kb and transcripts of 11 and 4.5 kb were detected in rat brain [17]. Large μ receptor transcripts of 16 and 10.5 kb were detected in rat brain and multiple transcripts of 13.5, 11, 4.3 and 2.8 kb were detected in human brain [36, 37]. Alternative splicing of primary transcripts could account for the multiple species detected in the RNA blotting.

23.3.1 μ Receptor

Distribution of μ receptor mRNA in rat brain has been investigated by RNA blotting and *in situ* hybridization [37, 38]. μ receptor mRNA was found to be expressed in neurons in the ascending and descending pain pathways. Highest levels were detected in the thalamus. mRNA was expressed in the spinal trigeminal nucleus, raphe nucleus and periaqueductal gray. Functional studies have shown that administration of morphine or other opiates to these regions blocks pain perception and that μ receptors are highly expressed in these areas.

The areas expressing μ receptor mRNA corresponded to those found to express μ receptor binding sites and μ receptor immunoreactivity [37–41]. In most brain and spinal cord regions, μ receptor immunoreactivity was detected in cell bodies and dendrites of neurons [39–41]. Immunoreactivity was also detected in superficial layers of the dorsal horn, which contain primary afferent sensory input. μ receptor immunoreactivity was detected in dorsal root ganglia and dorsal rhizotomy caused a reduction of μ receptor immunoreactivity in the dorsal horn suggesting that some terminal afferent inputs to the spinal cord have presynaptic μ receptor, as previously suggested from lesion and receptor binding studies [39].

A major side-effect of morphine is respiratory depression. Opiates are believed to cause this effect via actions in brainstem nuclei. μ receptor immunoreactivity and mRNA were detected in neurons of the nucleus of the solitary tract, nucleus ambiguous, and parabrachial nucleus. mRNA was detected in the bed nucleus of the stria terminalis which projects to the nucleus of the solitary tract. μ receptor immunoreactivity is found in the nucleus of the solitary tract and dorsal rhizotomy reduced receptor immunoreactivity in the nucleus suggesting a presynaptic localization of the receptor.

Central loci believed to be involved in morphine dependence are the ventral tegmental area and nucleus accumbens. μ receptor mRNA and immunoreactivity were also detected in these regions as well as the hippocampus and amygdala, other regions that may be involved in the euphoric effects of μ agonists.

The locus coeruleus is a noradrenergic nucleus well characterized for its expression of μ receptor binding sites. Electrophysiological studies have clearly shown a role of μ agonists in reducing firing activity of locus coeruleus neurons [42] possibly to reduce arousal since this nucleus contains most of the noradrenergic neurons in the brain and norepinephrine is the critical transmitter involved in arousal. The locus coeruleus has also been proposed as a site for morphine to induce physical dependence [43]. Both μ receptor immunoreactivity and mRNA was detected in the locus coeruleus, confirming the expression of μ receptors in this area.

A number of studies have suggested that enkephalins may have a role as endogenous ligands at the μ receptor [44] and enkephalins have high affinity in binding to the cloned μ receptor [36, 45]. In general, the expression of enkephalins paralleled that of μ receptors. This was particularly striking in the globus pallidus, which receives a large enkephalinergic input and has a high expression of μ receptor mRNA [37]. Since dynorphin A does not bind to μ receptors [45] and there is very little β-endorphin in extrahypothalamic structures [46], enkephalins may be endogenous transmitters at μ receptors. However, recent studies have suggested that another class of endogenous opiates, the endomorphans may be endogenous transmitters at the μ receptor [47].

23.3.2
δ Receptors

δ receptors are expressed at a much lower level in the central nervous system than μ receptors [48, 49]. Highest levels are found in the striatum, nucleus accumbens and cerebral cortex. Relatively few δ receptors are expressed in the brainstem, which may explain the general lack of autonomic side effects by δ agonists.

Immunohistochemical studies have shown that δ receptors are closely associated with descending serotoninergic neurons [49]. δ receptor-positive fibers were also found apposed to tyrosine hydroxylase-positive neurons in the locus coeruleus, suggesting that δ receptors may be presynaptic to neurons impinging on amine containing neurons.

Ultrastructural studies showed δ receptor immunoreactivity was presynaptically localized to sensory inputs to the spinal cord [48]. Dorsal rhizotomy caused a dramatic decrease in δ receptor immunoreactivity in spinal cord and δ receptors were also expressed in the peripheral ganglia which send inputs to the spinal cord. These findings support the notion that δ receptors are presynaptic to sensory inputs and are involved in the presynaptic inhibition of the release of transmitters involved in mediating nociceptive transmission.

23.3.3 κ Receptors

κ receptor mRNA was expressed at high levels in the hypothalamus [50, 51] – a region in which κ agonists are known to regulate hormone secretion. κ agonists induce diuresis by inhibiting vasopressin release. The presence of κ receptor mRNA in the supraoptic nucleus is consistent with an action of κ agonists on vasopressin neurons.

Consistent with receptor binding studies, κ receptor mRNA is expressed in the claustrum and interpeduncular nucleus as well as the ventral tegmental area. High levels of κ receptor mRNA levels are expressed in the substantia nigra pars compacta, suggestive that κ receptors may be expressed in dopaminergic neurons with a presynaptic location.

κ receptors are expressed in hippocampus and are known to mediate dynorphin A's inhibition of glutamate release. Electron microscopy showed a presynaptic location of κ receptor immunoreactivity, particularly in the CA3 region and dentate gyrus [52]. These findings are consistent with electrophysiological studies which have shown that κ receptors couple with an N-type Ca^{++} channel in glutaminergic neurons to mediate dynorphin's inhibition of calcium influx to block glutamate release and long-term potentiation [53].

κ receptor transcripts have also been detected in the immune system [54]. κ receptor transcripts were found in human lymphocytes and monocytes. Neither δ nor μ receptor transcripts were detected in these tissues, as assessed by reverse transcriptase-polymerase chain reaction (RT-PCR). These findings suggest a potential important and selective function of κ ligands in regulating immune system function.

23.4 Pharmacological Properties of the Cloned Opiate Receptors

The pharmacological properties of the cloned opiate receptors are similar to the characteristics of the endogenously expressed receptors [9, 36, 45]. The binding of opiates to the cloned receptors is stereoselective and the antagonist naloxone interacts with all three of the cloned receptors, although naloxone had much lower affinity for δ receptors than μ or κ, consistent with the lower potency of naloxone to block mouse vas deferens δ receptors than guinea pig ileum μ or κ receptors. The endogenous ligands for the μ and δ receptors, the enkephalins and endorphins, potently bind to the cloned δ and μ receptors but do not bind to the cloned κ receptor. In contrast, dynorphin A potently binds to the cloned κ receptor but has much less affinity for the other cloned opiate receptors.

Specific agonists at the κ receptor such as spiradoline and U50,488 bind to the cloned κ receptor with high affinity but do not interact with either the cloned μ or δ receptors. Similarly, the antagonist nor-BNI selectively binds to the cloned κ receptor. A comparison of the rank order of affinities of a large number of opiates at binding to the cloned and endogenously expressed μ and κ receptors revealed a very high correlation [45], suggesting that the ligand binding characteristics of the cloned and native receptors were similar.

In contrast, there was very little correlation of the binding affinities of opiates to the cloned δ receptor and native δ receptors in rat brain [45]. While the δ selective agonists DPDPE and deltorphin II bound selectively to the cloned δ receptor, deltorphin II had over 10-fold higher affinity for the cloned receptor than DPDPE. Similarly, the antagonist NTB had over 50-fold higher affinity for the cloned receptor than BNTX. These pharmacological characteristics suggested that the cloned receptor was similar to the δ_2 receptor. In fact, opiate receptors in NG-108 cells, from which the δ receptors were cloned, have the pharmacological characteristics of a δ_2 receptor subtype. The rat brain, from which most receptor binding data on native δ receptors has been obtained, consists of both δ_1 and δ_2 subtypes. Binding affinities from that tissue reflect a mixture of affinities at the two subtypes which may explain the different rank order of potencies of drug binding affinities to the cloned δ receptor compared to the brain receptors.

The hypothesis that the cloned δ receptor corresponded to one subtype was further supported by *in vivo* studies using antisense derived from the δ receptor cDNA to knock down δ receptor mRNA in rodents [13]. Intrathecal administration of δ receptor antisense blocked the analgesic effects of all δ agonists. This finding is consistent with pharmacological studies suggesting that only the δ_2 receptor is expressed in spinal cord. Antisense administration intracerebrally only blocked analgesic effects of δ_2 agonists. δ_1 agonists such as DPDPE were unaffected. Efforts to clone the δ_1 receptor subtype have so far been unsuccessful and, therefore, the molecular identity of this other δ receptor is not known. However, if the δ receptor gene undergoes splicing, sub-typing of the receptor could be generated by different splice variants, as has been identified for the μ receptor [29–31].

23.5 Functional Properties of the Cloned Opiate Receptors

In addition to their well-known action to inhibit adenylyl cyclase activity, opiates inhibit Ca^{++} conductance in neurons by modulating Ca^{++} channel activity [42]. Inhibition of Ca^{++} influx is a major mechanism by which opiates inhibit neurotransmitter release. In particular, κ opiate receptors in the hippocampus couple to an N-type Ca^{++} channel to reduce Ca^{++} conductance and mediate dynorphin A's inhibition of glutamate release to block long-term potentiation [53]. Like the endogenously expressed receptor, the cloned κ receptor expressed in PC12 cells also coupled to an N-type Ca^{++} channel and the κ selective agonist U50, 488 was found to inhibit Ca^{++} conductance in these cells [55]. Furthermore, the cloned κ and μ receptors expressed in oocytes can couple to Ca^{++} channels [56].

The cloned opiate receptors, like the endogenously expressed receptors, can couple to phospholipase C to increase Ca^{++} mobilization in transfected cells [31, 57, 58]. Thus, stimulation of opiate receptors can lead to inhibition of Ca^{++} influx via voltage-sensitive channels as well as increase Ca^{++} levels in cells released from intracellular stores. This dual role of opiates may explain the opposing excitatory and inhibitory effects they may have on some cells in the nervous system [59].

In general, opiates depress neuronal activity by stimulating K^+ conductance to reduce firing activity [42]. μ and δ receptors in brain neurons [42] and κ receptors in substania gelantinosa neurons [60] have been shown to potentate K^+ currents via an inwardly rectifying K^+ channel. The cloned μ receptor co-expressed in oocytes along with the recently cloned inward rectifier K^+ (GIRK1) mediates morphine stimulation of K^+ conductance [61–64]. Similarly, the cloned κ receptor expressed in AtT-20 cells coupled with an endogenously expressed inwardly rectifying K^+ channel indicating that the cloned receptors are capable of coupling to the same cellular effector systems as the endogenously expressed receptors [65].

23.6
G Protein Coupling to the Opiate Receptors

The cloned opiate receptors expressed in homogeneous cell lines associate with guanine nucleotide binding (G) proteins which couple the receptors to cellular effector systems. G proteins consist of hetero-trimeric complexes of α, β and γ subunits [66]. Multiple subtypes of each subunit have been identified and cloned. In particular, three subtypes of G_i have been cloned that are approximately 90% identical in amino acid sequence. Furthermore, two splice forms of G_o have been identified.

Both the recombinant and native opiate receptors have the capability of coupling to more than one cellular effector system at a time [67–73]. In PC12 cells expressing the cloned κ receptor, opiates inhibit cAMP accumulation and Ca^{++} conductance [55]. The cloned κ and μ receptors expressed in AtT-20 cells are capable of coupling to adenylyl cyclase and K^+ channels [65]. Tsu et al. [57] reported that δ receptors expressed in HEK 293 cells can mediate agonist stimulation of phosphoinositol turnover and Ca^{++} mobilization and Bot et al. [74] reported that in these same cells δ receptor can mediate agonist inhibition of cAMP accumulation. Similarly, δ receptors endogenously expressed in NG-108 cells mediate agonist modulation of cAMP accumulation, ionic conductance and phosphoinositol turnover.

The multiplicity of cellular responses induced by opiates involve G proteins coupling the opiate receptors to distinct cellular effector systems. The independent regulation of distinct effector systems may be mediated by different G proteins. G_o has been shown to couple δ receptors to Ca^{++} channels [75, 76]. This was first reported in reconstitution experiments in which neurons expressing the δ receptors were treated with pertussis toxin to inactivate G_i/G_o [75]. Pertussis toxin uncoupled δ receptors from Ca^{++} channels in these cells. Purified G_o was perfused into the pertussis toxin pretreated cells and found to reconstitute the coupling of the δ receptor to the Ca^{++} channel. Similarly, Taussig et al. [76] transfected NG-108 cells with either pertussis toxin insensitive G_i or G_o. The transfected cells were treated with pertussis toxin to inactive endogenous G proteins. Only the cells transfected with pertussis toxin insensitive G_o responded to δ opiate modulation of Ca^{++} conductance. These findings not only demonstrated that G_o coupled the δ receptor to the Ca^{++} channel but also showed that G_i did not couple the receptor to this channel, indicating specificity in the δ receptor/Ca^{++} channel coupling.

Law and Reisine [73] reported that the cloned δ receptor physically associated with G_o. They solubilized the δ receptor with a mild detergent which allowed solubilized δ receptors to remain associated with G proteins. They then showed that antisera directed against G_o co-immunoprecipitated δ receptor/G protein complexes.

G_i has been proposed to couple δ receptors to adenylyl cyclase. McKenzie and Milligan [77] reported that G_{i2} primarily couples δ receptors in NG-108 cells to adenylyl cyclase. They showed that antisera directed against G_{i2}-blocked δ receptor agonist stimulated GTPase activity and inhibition of adenylyl cyclase activity. Behavioral studies have also revealed an important role of G_{i2} in δ receptor signaling, since knock down of G_{i2} in brain with G_{i2} antisense administered to mice blocked the ability of δ agonists to induce analgesia [68].

However, G_{i2} is unlikely to be the only G protein that couples δ receptors to adenylyl cyclase since the cloned δ receptor expressed in HEK 293 cells effectively couples to adenylyl cyclase via a pertussis toxin sensitive G protein [74], yet these cells lack immunologically detectable G_{i2} and G_o [78] suggesting that either G_{i1} or G_{i3} in these cells couples the δ receptor to adenylyl cyclase. Prather et al. [69] and Law and Reisine [73] have reported that δ receptors can associate with G_{i1} and G_{i3}. Furthermore, Sanchez-Blazquez et al. [68] have reported that antisense against G_{i3} mRNA administered to the nervous system blocked δ agonist-induced analgesia.

Tsu et al. [57] have reported that G_z can reconstitute δ receptor coupling to adenylyl cyclase in HEK 293 cells after pertussis toxin treatment and studies by Law and Reisine [73] have shown that G_z can physically associate with the cloned δ receptor. Tsu et al. [57] have also shown that G_z can couple the cloned δ receptor to phospholipase C, providing a potential dual role of this pertussis toxin-insensitive G protein in δ receptor signaling.

In addition to G protein α subunits, β subunits of G proteins also associate with δ receptors and may be involved in δ receptor signaling. Antisera directed against the β_1 and β_2 subunits co-immunoprecipitated the δ receptor/G protein complex suggesting that these two β subunits physically associate with the cloned receptor [73]. β/γ complexes have been reported to directly interact with K^+ channels and type II adenylyl cyclase, two cellular effectors regulated by δ agonists [79].

The multiplicity of G proteins coupled to opiate receptors may explain how different opiates can bind to the same receptor yet induce different cellular responses. For example, morphine binds to the cloned rat μ receptor expressed in HEK 293, CHO and COS-7 cells and inhibits cAMP accumulation [80–82]. Morphine can be continuously applied to the cells for up to 16 h, and the potency and magnitude of morphine inhibition of adenylyl cyclase does not diminish [80, 81]. In contrast, the opiate sufentanil can bind to the same cloned μ receptor in HEK 293 cells to inhibit cAMP accumulation. However, sufentanil's actions rapidly desensitize [83]. Since both compounds bind to the same receptor, and the μ receptor is the only receptor these drugs can interact with in these cells, the ability of these two full agonists to differentially regulate the μ receptor must be due to their abilities to affect separate adaptive processes in these cells.

Recent antisense studies have shown that knock down of G_{i2} in mice-blocked morphine-induced analgesia, suggesting that morphine binds to μ receptors to activate G_{i2} to modulate neuronal circuits involved in analgesia [68, 84]. Sufentanil-induced analgesia was not diminished by G_{i2} knock down suggesting that a different G protein mediated its behavioral effects [84].

Mutagenesis studies have shown that morphine and sufentanil bind differently to the μ receptor [83, 85]. Mutation of an aspartic acid at residue 114 of the μ receptor to an asparagine resulted in a mutant that did not bind morphine and morphine was ineffective in inhibiting adenylyl cyclase via that receptor. In contrast, sufentanil bound to the mutant and wild-type receptors equally well and it effectively inhibited cAMP accumulation via the mutant receptor. These findings demonstrate that morphine and sufentanil have different requirements for binding to the μ receptor. By binding differentially, these two agonists may induce the μ receptor to interact with different G proteins to induce distinct cellular effects.

23.7
Regulation of Opiate Receptors

Acute stimulation of opiate receptors can lead to rapid changes in biological responses from changes in ionic conductance to pain relief. Chronic use of μ agonists can induce two serious side effects, tolerance and dependence. Tolerance involves the diminished ability of a given dose of opiate to induce its biological response, following repeated administration of the opiate. Dependence results in a physical requirement of the organism for the opiate to maintain normal physiology. Tolerance develops to μ, κ and δ agonists. However, dependence is primarily associated with drugs that selectively bind to the μ receptor.

The molecular basis of tolerance is believed to involve the gradual desensitization of opiate receptors [86, 87]. The desensitization can be manifest as a loss of receptor from cells normally responsive to opiate or an uncoupling of the receptor from cellular effector systems critical for the pharmacological actions of opiates. The cloned opiate receptors have been particularly useful in gaining insights into the molecular events involved in opiate-induced tolerance.

23.7.1 μ Receptor

A critical cellular response to opiates is the potentiation of K^+ currents [42]. Stimulation of μ receptors in neurons causes an increase in K^+ conductance and a reduction in cell firing. Prolonged administration of μ agonists diminishes the ability of the opiates to increase K^+ conductance to inhibit neuronal firing and pain transmission is no longer attenuated.

The molecular basis of μ receptor desensitization has been investigated by co-expressing the μ receptor with the cloned inward rectifying K^+ channel GIRK1 in oocytes [61–64]. Stimulation of the cloned μ receptor increases K^+ conductance through GIRK1. Repeated application of morphine or other μ agonists causes a gradual uncoupling of the receptor from the channel such that morphine is no longer able to stimulate K^+ conductance.

Parallel studies by Tallent et al. [65] have employed AtT-20 cells transfected with the μ receptor which couples to an endogenously expressed inwardly rectifying K^+ channel. Prolonged application of DAMGO to these cells also desensitizes the ability of opiates to potentate the K^+ current. The desensitization of the μ receptor in AtT-20 cells did not involve changes in the ability of the K^+ channel to be activated since GTP analogs perfused into the opiate-treated cells increased K^+ currents to a similar extent as in drug naive cells.

Chen et al. [62] have proposed that protein kinase C is involved in the μ receptor desensitization since activators of this enzyme potentate μ receptor uncoupling from the K^+ channel. The μ receptor has consensus protein kinase C phosphorylation sites in its intracellular domains [8]. Arden et al. [88] have reported that morphine can induce the phosphorylation of the cloned μ receptor and Zhang et al. [64] have reported that activators of protein kinase C can induce the phosphorylation of the μ receptor.

The desensitization of the μ receptor was heterologous. In oocytes cotransfected with μ and serotonin receptors, chronic morphine treatment abolished morphine and serotonin potentiation of the K^+ current [63]. Similarly, in AtT-20 cells transfected with the cloned μ receptor, chronic DAMGO treatment abolished the ability of opiates and somatostatin, acting via endogenous somatostatin receptors in these cells, to stimulate K^+ conductance [65].

The ability of morphine to desensitize other neurotransmitter receptors coupled to K^+ channels may cause long-term consequences in the activity of neurons. The uncoupling of K^+ channel from non-opioid receptors that normally tonically inhibit cell firing could result in an increase in the basal firing of the cells. Changes in the set point of neuronal firing could influence gene expression in the cells and alter the molecular properties of the neurons.

While chronic morphine treatment uncouples the μ receptor from K^+ channels, it did not affect the coupling of μ receptors to adenylyl cyclase. Pretreatment of the cloned μ receptor expressed in HEK 293, AtT-20, CHO and COS cells with morphine or DAMGO for up to 16 h did not alter the subsequent ability of μ agonists to inhibit cAMP accumulation [25, 65, 80–82]. These findings suggest that morphine treatment induces a selective desensitization of the coupling of the μ receptor to K^+ channels.

The maintenance of μ receptor/adenylyl cyclase coupling indicates that chronic morphine treatment does not downregulate the μ receptor to the extent of abolishing function. This has been confirmed by immunohistochemical studies which have shown that morphine treatment does not internalize or downregulate the cloned μ receptor [80, 89].

If μ receptors couple to K^+ channels and adenylyl cyclase via different G proteins, it is possible that chronic morphine treatment uncouples the receptor from those G proteins linked to the K^+ channel and not those coupling μ receptors to adenylyl cyclase. Such a hypothesis would require that G proteins couple to different intracellular domains of the μ receptor so that interaction of G proteins with some domains could be blocked by post-translational events, such as phosphorylation, whereas binding of G proteins to other μ receptor domains would not be affected.

While chronic morphine or DAMGO treatments do not alter the ability of μ agonists to inhibit adenylyl cyclase, chronic treatment does induce other adaptive responses in adenylyl cyclase. Chronic morphine or DAMGO treatments of HEK 293, CHO and COS-7 cells transfected with the μ receptor caused a 2- to 3-fold increase in the ability of forskolin to stimulate cAMP formation [80–84, 90]. Avidor-Reiss et al. [90] have reported that opiate treatment of cells expressing the cloned μ receptor induces a selective increase in type I, V, VI and VII cyclase. The heterogeneous distribution of the adenylyl cyclase subtypes in different tissues and cell types provides a basis for chronic morphine treatment to induce long-term changes in some cells but not others. Avidor-Reiss et al. [81, 82, 90] have proposed that the superactivation of adenylyl cyclase may be an underlying basis of dependence as first suggested by Sharma et al. [91] in the 1970s.

Recent studies by Nestler and associates have suggested that long-term increases in adenylyl cyclase activity in the locus coeruleus may also be an adaptive response to morphine that causes dependence [43, 92, 93]. Chronic treatment of rats with morphine increases levels of adenylyl cyclase and protein kinase A in the locus coerleus. This has been suggested to increase the firing of locus coeruleus neurons in dependent animals [94, 95]. Administration of inhibitors of protein kinase A into the locus coeruleus of morphine-dependent animals precipitated withdrawal which further suggests that the heighten cAMP pathway is critical for maintaining dependence [43].

Not all opiates induce an increase in forskolin stimulation of adenylyl cyclase. Buprenorphine, which is used to treat morphine dependence, does not cause a compensatory rise in cyclase activity [80]. Co-treatment of morphine with buprenorphine prevented morphine from increasing cyclase activity suggesting that buprenorphine's therapeutic efficacy may be related to its ability to block adaptive responses induced by morphine.

Buprenorphine does not cause dependence in humans [96]. Unlike morphine, buprenorphine desensitizes the μ receptor coupling to adenylyl cyclase [80]. The desensitization occurs in the absence of any receptor internalization or downregulation [80]. The desensitization of the μ receptor may be the underlying basis for why buprenorphine does not cause a heightened adenylyl cyclase activity in μ receptor-responsive cells. Buprenorphine's unique cellular regulation of the μ receptor may explain its ability to be a non-addictive analgesic as well as its usefulness in treating opiate dependence.

Both morphine and buprenorphine are agonists at μ receptors. However, buprenorphine is not μ selective and is a partial agonist, showing less efficacy than morphine [80]. This property may be due to the distinct manner in which buprenorphine binds to the μ receptor. Mutagenesis studies have shown that morphine and buprenorphine have different determinants for binding to the cloned μ receptor [85]. Buprenorphine appears to have a mixture of binding properties. It associates with the μ receptor in a manner somewhat similar to antagonists, such as diprenorphine but in contrast to antagonists is able to activate the receptor. These unique binding properties of buprenorphine may allow it to activate different signaling pathways than morphine so as to not to induce the same long-term adaptive responses as morphine and other addictive opiates.

Buprenorphine is a weak analgesic [91], which precludes its ability to replace morphine in the treatment of chronic pain. However, development of compounds that interact with μ receptors in a similar manner as buprenorphine but that are more effective agonists and analgesics could lead to the development of drugs that can be used for the treatment of chronic pain but which have little or no abuse potential.

23.7.2
δ Receptors

δ selective agonists also undergo tolerance development [23]. Chronic exposure of δ receptors to peptide agonists has been reported to uncouple the receptor from adenylyl cyclase [74, 97, 98]. δ receptor desensitization involves the phosphorylation of the receptor by β-adrenergic receptor kinase (BARK) an enzyme that catalyzes the phosphorylation and homologous desensitization of a number of G protein-linked receptors [97].

Prolonged δ agonist treatment causes homologous desensitization in contrast to the heterologous nature of μ receptor desensitization. Furthermore, δ agonists do not cause adaptive increases in adenylyl cyclase activity [74]. The lack of these compensatory responses may be one reason that δ agonists do not cause addiction [99].

Since δ agonists have few of the long-term side effects of μ agonists, δ agonists that could overcome tolerance development may be useful drugs in the treatment of chronic pain. In fact, the non-peptide δ selective agonist SIOM [100] did not desensitize the cloned δ receptor expressed in HEK 293 cells [74] and there is no evidence that this compound can induce tolerance *in vivo* in animals. SIOM or its analogs may be future non-addicting, long acting analgesics.

23.7.3
κ Receptors

Like δ selective agonists, κ agonists have few of the side effects of morphine and recent studies have suggested that κ selective agonists may be effective analgesics [2]. However, a limitation to the clinical use of κ agonists is tolerance development.

κ receptors coupled to K^+ channels rapidly desensitize [65, 101–103]. In contrast to μ receptors, κ receptor desensitization is homologous [65]. κ receptor desensitization also involves an uncoupling of the receptor from adenylyl cyclase [104] as well as Ca^{++} channels [56] and phospholipase C [105]. Like the δ receptor [97], κ receptor desensitization involves G protein receptor kinases [103]. These enzymes are predicted to catalyze the phosphorylation of cytosolic sites within the receptor to cause G protein uncoupling. Since the intracellular loops of the δ and κ receptors are almost identical in amino acid sequence [8], it is possible that common domains in the two receptors may be involved in their regulation. Recent studies have suggested that phosphorylation of residues in the C-terminus of the κ receptor is involved in the desensitization because truncation of the C-terminus attenuated κ receptor desensitization [103].

While the κ selective agonists U50,488 and dynorphin A desensitize the receptor, non-selective opiates such as etorphine did not desensitize the human κ receptor [104]. This divergence in the ability of selective versus non-selective agonists to desensitize the κ receptor may be due to their abilities to bind differently to the receptor to activate distinct cellular adaptive response pathways. Mutagenesis studies have suggested that selective and non-selective agonists bind to different domains of the κ receptor [106, 107]. Development of novel κ selective agonists that bind to the receptor in a similar manner as etorphine may result in the generation of longer acting opiates which would be especially useful in treating chronic pain.

23.8
Structure-Function Analysis of the Cloned Opiate Receptors

Changing the amino acid sequence of the cloned receptors by mutating nucleotides within the receptor cDNAs has proven to be an effective mechanism by which to identify structural features of the receptors responsible for their unique functional properties. In particular, site-directed mutagenesis has been employed to determine the ligand binding domains of each opiate receptor.

23.8.1
Point Mutations of the Opiate Receptors

Site-directed mutagenesis has involved either the deletion of portions of the receptor, the changing of a single amino acid in a receptor or the large-scale transposition of a fragment of one receptor for the corresponding fragment of another receptor to generate chimeric receptor mutants. The first mutagenesis done on an opiate receptor was the simple substitution of an aspartate residue (Asp) in transmembrane (TM2) of the δ receptor for an asparagine (Asn) residue [108]. Aspartic acid residues are negatively charged, while Asn residues are not charged. A vast number of G protein-linked receptors have a conserved Asp residue in their second TM spanning region [109]. Mutations of those conserved Asp have been found to diminish Na^+ regulation of agonist binding [108–111]. Na^+ was first shown to affect G protein linked receptors by Pert et al. [112] who reported the ability of the ion to reduce agonist binding to brain opiate receptors. Na^+ is believed to influence agonist binding by dissociating G proteins from receptors so as to convert the receptor into a low-affinity state for agonists. The positive charge of the Na^+ is believed to interact with the negatively charged Asp in TM2 via electrostatic interactions to induce conformational changes in the receptor. Removal of the negative charge in TM2 of the α_2-adrenergic receptors [110], somatostatin receptors [111] and the δ opiate receptor [108] by site-directed mutagenesis abolishes sodium regulation of agonist binding demonstrating that this conserved Asp is the selective site of Na^+ regulation of receptors.

Mutation of Asp^{95} to Asn in the δ receptor greatly reduced the affinity of the receptor for selective agonists such as DPDPE, DSLET and SIOM [108, 113]. In contrast, the binding of antagonists and the non-selective agonists such as bremazocine and buprenorphine was not altered by the mutation. This finding indicates that Asp^{95} of the δ receptor has a critical role in the binding of selective agonists.

The conserved Asp in TM2 of the μ receptor also has a critical role in the binding of agonists [85, 114, 115]. Mutation of Asp^{114} in the μ receptor to non-charged amino acids greatly diminished the binding affinities of DAMGO and non-peptide μ agonists such as morphine. In contrast, antagonist binding was not reduced by the mutation. Interestingly, the binding of partial agonists, such as buprenorphine, bremazocine and nalorphine to the mutant receptor was not significantly different than their binding to the wild-type receptor, suggesting that these compounds have similar determinants for binding as antagonists [85].

Buprenorphine is a partial agonist at the opiate receptors [96]. The partial agonist characteristic of the opiate may be due to its ability to interact with opiate receptors with a mixture of agonist and antagonist characteristics. The agonist and antagonist binding properties of buprenorphine can be separated out through mutagenesis of the δ opiate receptor. Mutation of Asp^{95} of the δ receptor to an asparagine resulted in a receptor that bound buprenorphine as well as the wild-type receptor, but buprenorphine was unable to inhibit cAMP accumulation [113]. In fact, buprenorphine antagonized DSLET inhibition of cAMP accumulation in cells expressing the Asn^{95} mutant. Thus, the mutation in the δ receptor abolished the agonism of buprenorphine and left buprenorphine as a pure antagonist at the mutant receptor.

Buprenorphine has a similar structure as the opiate antagonist diprenorphine [96]. Both buprenorphine and diprenorphine are N-cyclopropylmethylnordihydro-orvinol derivatives of thebaine-methyl vinyl ketone adducts. They differ in structure at carbon atom 19 with diprenorphine having a methyl group and buprenorphine has a tertiary butyl substitution. The larger butyl group of buprenorphine may provide the agonistic properties of this opiate. The remaining structure may be critical for buprenorphine to bind to the opiate receptors as an antagonist. Understanding the structural requirements for this non-addicting analgesic to bind to μ receptors should allow for the rational development of buprenorphine analogs with improved analgesic properties.

A second conserved Asp residue found in most G protein-linked receptors is present in TM3. Mutation of this charged residue to neutral amino acids in the α_2- and β-adrenergic receptors was found to greatly reduce agonist binding affinities [116, 117]. From these studies it was proposed that the Asp serves as a counter-ion to basic regions of ligands to facilitate binding through electrostatic interactions.

Consistent with this hypothesis all opiate agonists have a positively charged region that has been found to be necessary for binding to opiate receptors and mutation of Asp^{128} or Asp^{147} of the δ and μ receptors, respectively, to neutral amino acids greatly reduced the affinity of the receptors for agonists [74, 114, 118]. In contrast, the binding of antagonists to the mutant and wild-type receptors was similar demonstrating that the Asp in TM3 is not necessary for antagonist binding.

These studies indicate that the conserved Asp in TM3 as well as the conserved Asp in TM2 of the opiate receptors contributes to ligand binding. They may play a role in reducing dissociation rates of agonist interaction with the receptor. This may explain why some agonists were able to activate the mutant Asn^{128} δ receptor and Asn^{114} mutant μ receptor to inhibit cAMP accumulation while having very low binding affinity. For example, methadone has an affinity of greater than 1 μM at the Asn^{114} mutant μ receptor yet effectively inhibits adenylyl cyclase activity via this receptor [85]. Similarly, DSLET has over a 500-fold lower affinity for interacting with the Asn^{128} mutant δ receptor yet inhibits adenylyl cyclase to the same magnitude as it does via the wild-type receptor. Efficacy of agonists at the receptor may be primarily dependent on association rates of ligand interaction with the receptor whereas binding affinity detected in a receptor binding assay is primarily dependent on dissociation rates of the ligand.

The lack of effect of the mutations on antagonist binding suggests that these compounds bind in a fundamentally different manner to the opiate receptors than do full agonists. Testing of chimeric opiate receptors have further established that agonists and antagonists bind to different domains of the opiate receptors.

23.9 Chimeric Opiate Receptors

23.9.1 κ Receptor

Chimeric opiate receptors have been generated to identify ligand binding domains of the opiate receptors. This mutagenesis is more complex than point mutations because it requires deleting corresponding regions of different receptors and then sewing the fragments back on to the host receptor. Kong et al. [119] were among the first investigators to use chimeric receptors to identify regions of the κ receptor involved in naloxone binding. The N-terminus and TM1 of the mouse κ and δ receptors were exchanged between the two receptors and the chimeric receptors were tested for their ligand binding properties. Naloxone binds with much higher affinity to the κ receptor than to the δ receptor. Pharmacological analysis of the chimeric receptors showed that the N-terminus and TM1 of the κ receptor was essential for the high-affinity binding of naloxone to the κ receptor.

The N-terminus is not the only region of the κ receptor involved in antagonist binding. Hjorth et al. [120] reported that Glu^{297} at the juncture of TM6 and the third extracellular loop of the κ receptor is a recognition site for the κ selective antagonist nor-BNI. These authors have suggested that nor-BNI expresses lower affinity at the μ and δ receptors because these other opiate receptors have a lysine (μ receptor) or tryptophan (δ receptor) at residues corresponding to Glu^{297} of the κ receptor which serve to repel nor-BNI from interacting with the binding pockets. They have suggested that the negative charge of the Glu^{297} may serve as a counter-ion to attract nor-BNI to the κ receptor whereas the positive charge of Lys^{303} of the μ receptor may repel nor-BNI.

Chimeric receptors were also used to identify agonist binding domains of the κ receptor. The extracellular loops of the opiate receptor have largely divergent amino acid sequences and are potential sites for peptide agonist binding. Transposition of the second extracellular loop of the κ receptor onto either the δ or μ receptors conferred onto those receptors high affinity for the κ selective agonist dynorphin A [106, 107]. The second extracellular loop of the κ receptor has a much greater number of negatively charged amino acids than the corresponding regions of either the μ or δ receptors. Dynorphin A has within its N-terminal region leucine enkephalin which binds to both μ and δ receptors. However, the C-terminal region of dynorphin A is rich in positively charged amino acids that might be attracted to the anionic regions of the second extracellular loop of the κ receptor. Smaller fragments of dynorphin A with less charged residues in the C-terminus have less specificity for the κ receptor and tend to bind with higher affinity to μ receptors. The electrostatic interactions between the C-terminal fragment of dynorphin A and the second extracellular loop of the κ receptor may form the basis of the selective binding of dynorphin A to this opiate receptor type.

The finding that extracellular loop 2 of the κ receptor is a ligand binding domain is not unique among G protein-linked receptors. Recent mutagenesis studies on the somatostatin receptor subtype SST1 have shown that extracellular loop 2 is the binding domain for selective peptides [121]. In fact, extracellular loops have been identified as binding domains for a number of peptide ligands.

However, previous mutagenesis studies on the adrenergic receptors have suggested that small molecule binding domains usually reside in the hydrophobic pockets formed by the transmembrane spanning regions [116, 117]. In fact, the small synthetic agonists at the κ receptor, such as U50,488, have been proposed not to bind to the second extracellular loop of the κ receptor but instead may associate further into the receptor binding pocket [106]. This finding suggests that synthetic κ agonists have binding domains distinct from those that bind the endogenous peptides. The much larger size of the peptides and their more hydrophilic nature make the extracellular loops more likely targets for their binding.

Identifying the second extracellular loop as the selective binding domain of dynorphin A may have important implications in the design and development of new κ agonists. In general, κ agonists have several therapeutically beneficial properties. They do not cause dependence nor respiratory depression and they do not cause constipation, like most clinically used opiates. In fact, they are potent diuretics, and as such could be useful in treating congestive heart failure. Most importantly, they are effective analgesics. However, the U50,488 analog spiradoline was found to cause dysphoria and psychosis in humans which stopped its further development as clinical drugs.

In human and animal studies, dynorphin A is not known to induce the same side effects as spiradoline. This may be due to it binding differently to the κ receptor to induce distinct biological responses. Since dynorphin A can induce analgesia, its lack of side effects makes it a potentially desirable pharmaceutical agent. However, peptides in general do not make good drugs because they are easily degraded in the bloodstream and do not cross diffusion barriers such as the gut wall or blood-brain barrier. Identification of the dynorphin A binding domain of the κ receptor should allow for the development of a new generation of non-peptide dynorphin A-like drugs that could have the analgesic properties of the parent compounds but lack the serious side effects of previously developed κ selective drugs. In fact, recent modeling studies by Paterlini et al. [122] describe a structural model by which dynorphin A binds to the second extracellular loop of the κ receptor. Such models provide the basis for designing non-peptides that would simulate the same conformation of dynorphin A in binding to the κ receptor.

23.9.2 δ Receptor

Chimeric receptors have also been useful in identifying ligand binding domains in the δ receptor [123, 124]. Studies using chimeras in which fragments of δ and μ receptors have been exchanged suggested that δ selective agonists bind to a region encompassed by TM5–TM7. Varga et al. [125] have further specified that the third extracellular loop contains the binding domain of δ selective agonists. Valiquette et al. [126] have further specified that three amino acids in this extracellular loop, Trp^{284}, Val^{296} and Val^{297}, are necessary for selective agonist binding.

Mutagenesis studies have also suggested that hydrophobic amino acids in transmembrane spanning regions of the δ receptor are essential for peptide agonist binding but that the binding of the δ selective non-peptide agonist BW373U86 was not dependent on these residues [127]. This finding suggests that subtle differences may exist in how peptide and non-peptide selective agonists bind to the δ receptor. Such subtle differences could cause different cellular responses and may explain why peptides such as DPDPE and DSLET rapidly desensitize the δ receptor while the non-peptide selective agonist SIOM does not [74].

23.9.3
μ Receptor

Initial mutagenesis studies on the μ receptor showed that neither the N-terminal 64 amino acids nor the C-terminal 33 amino acids were essential for the receptor to bind ligands [114]. Based on studies on a series of chimeric receptors generated by exchanging regions of the μ receptor with either the κ or δ receptor, peptides such as DAMGO were found to bind to regions encompassing the first and third extracellular loops while morphine binding was primarily localized to the third extracellular loop and the surrounding transmembrane spanning regions [123, 128–130]. Neither of the extracellular loops was essential for antagonist binding indicating that agonists and antagonists have different binding domains in this receptor.

Mutagenesis studies have established that the C-terminal region of the μ and δ receptors is not essential for the receptors to couple to adenylyl cyclase [131, 132]. The remaining intracellular domains of the opiate receptors have almost identical amino acid sequences. As a consequence, it is likely that the intracellular loops are the main regions of the opiate receptor involved in G protein coupling and effector system regulation.

23.10
The C-Terminus of the Opiate Receptors

Major differences in the amino acid sequences of the intracellular domains of the opiate receptors reside in the C-terminal tail [8]. While this domain is not essential for coupling to G proteins it may be important for the desensitization and internalization of the opiate receptors.

The carboxy terminus of the μ receptor was essential for agonist-induced desensitization [83, 132] since truncation of the receptor prevented desensitization. Like those findings with the κ receptor, the enzyme G protein receptor kinase (GRK) appears to be involved in the desensitization process, since blockade of GRK prevented the desensitization process. Wang [132] has proposed that GRK catalyzes the phosphorylation of a series of serine/threonine residues in the C-terminus of the μ receptor to desensitize the receptor.

Cvejic et al. [133] and Trapaidze et al. [134] have reported that the C-terminus of the δ receptor is essential for the internalization and downregulation of the receptors. Truncation of the δ receptor attenuates receptor internalization. Residue Thr^{353} seems to be selectively involved in the internalization of the receptor, since mutation of this amino acid blocks the internalization process. Recent studies by Chu et al. [135] not only confirm the role of the C-terminus in internalizing the δ receptor but have shown that clathrin-coated pits are involved in the internalization since a K44I mutant of dynamin I blocks the rapid internalization of the δ receptor. These studies have identified a structural basis for the differential regulation of the three opiate receptors.

The *in vitro* studies have shown an important role of G protein receptor kinases in opiate receptor desensitization. These kinases have been proposed to catalyze the phosphorylation of the receptors and this post-translational event is believed to attract a family of protein referred to as β-arrestins to the receptor to uncouple the receptors from G proteins and effector systems. Studies by Bohn et al. [136] have shown that the β-arrestins are critical for desensitization and tolerance development to morphine since in β-arrestin knock out mice, morphine-induced analgesia was potentiated and greatly prolonged compared to wide-type mice. In follow-up studies, Bohn et al. [137] showed that in β-arrestin knock out mice, morphine's analgesic effects did not desensitize. However, morphine-induced dependence was not affected. Similarly, prolonged morphine treatment increased basal adenylyl cyclase activity, just like in *in vitro* studies. These studies indicate a dissociation between the development of tolerance and dependence and suggest that they arise via distinct molecular mechanisms. Tolerance occurs with all the opiate receptors and involves an uncoupling of the receptor from G proteins and effector systems via the actions of protein kinases and β-arrestin. Dependence primarily involves the μ receptor, does not involve β-arrestin and is associated with an upregulation of adenylyl cyclase activity. These provide insight into the possibility to develop drugs that could block the dependence induced by opiates without affecting the analgesia-producing effects of these drugs.

23.11 Future Directions

The cloning of the opiate receptors has provided valuable tools to identify the structural basis by which drugs interact with and regulate these receptors. Information gained from such structural analysis should provide the basis for the development of a new class of opiates that may produce many of the desired pharmacological effects of opiates, such as analgesia, but with few of the side effects. Such rational design of drugs may lead to the development of long acting opiates which do not cause dependence.

In fact, several pharmaceutical companies, such as Adolor, are developing peripheralized μ and κ analgesics which have been found to relieve pain but because they do not cross the blood-brain barrier do not cause the side effects of clinically used opiates such as respiratory depression, nausea, affective behaviors and dependence. Furthermore, centrally acting κ analgesics which mimic the actions of dynorphin A might be developed to treat severe forms of chronic pain, such as neuropathetic pain, but have none of the side effects of μ agonists nor cause the dysphoria associated with U-50,488 and other previously developed κ agonists. Such drugs could be particularly beneficial to the millions of chronic pain patients who have few options presently to manage their debilitating disorders.

23.12 References

1 SERTURNER FWA. Darstellung der reinen Mohnasäure (Opiumsäure); nebst einer chemischen Untersuchung des Opiums, mit vorzüglicher Hinsicht auf einen darin neu entdeckten Stoff. J Pharm Ärzte Apoth Chem **1805**; 14:47–93.

2 REISINE T, PASTERNAK G. Opioid analgesics and antagonists. In: The Pharmacological Basis of Therapeutics, 9th edn (HARDMAN JG, LIMBIRD LE, MOLINOFF, PB et al., eds). New York: McGraw-Hill, **1996**; 521–555.

3 MARTIN WR, EADES CG, THOMPSON JA et al. The effects of morphine and nalorphine-like drugs in the non-dependent and cyclazocine-dependent chronic spinal dog. J Pharmacol Exp Ther **1976**; 197:517–532.

4 LORD J, WATERFIELD A, HUGHES J et al. Endogenous opioid peptides: multiple agonists and receptors. Nature **1977**; 267:495–498.

5 PORTOGHESE PS. A new concept on the mode of interaction of narcotic analgesics with receptors. J Med Chem **1965**; 8:609–616.

6 SIMON EJ, GIOANNINI TL. Opioid receptor multiplicity: isolation, purification and chemical characterization of binding sites. In: Opioids I. Handbook of Experimental Pharmacology. Vol. 104 (HERZ A, ed). Springer, Heidelberg, **1993**; 3–26.

7 MARTIN WR. Opioid antagonists. Pharmacol Rev **1967**; 19:463–521.

8 REISINE T, BELL GI. Molecular biology of opioid receptors. Trends Neurosci **1993**; 16:506–510.

9 YASUDA K, RAYNOR K, KONG H, BREDER C, TAKEDA J, REISINE T, BELL GI. Cloning and functional comparison of kappa and delta opioid receptors from mouse brain. Proc Natl Acad Sci USA **1993**; 90:6736–6740.

10 VANDERAH T, TAKEMORI A, SULTANA M, PORTOGHESE P, MOSEBERG H, HRUBY V, HAASETH R, MATSUNAGA T PORRECA F. Interaction of [D-Pen2, D-Pen5]enkephalin and [D-Ala2, Glu4]deltorphin with delta opioid receptor subtypes *in vivo*. Eur J Pharmacol **1994**; 252:133–137.

11 PORTOGHESE P, SULTANA M, NAGASE H et al. A highly selective delta-1 opioid receptor antagonist: 7-benzylidenenaltrexone. Eur J Pharmacol **1992**; 218:195–196.

12 SOFUOGLU M, PORTOGHESE P, TAKEMORI A. Differential antagonism of delta opioid agonists by naltrindole and its benzofuran analog (NTB) in mice: evidence for delta opioid receptor subtypes. J Pharmacol Exp Ther **1991**; 257:676–680.

13 STANDIFER K, CHIEN C, WAHLSTEDT C, BROWN G, PASTERNAK G. Selective loss of delta opioid analgesia and binding by antisense oligodeoxynucleotide to a delta opioid receptor. Neuron **1994**; 12:805–810.

14 PASTERNAK G, WOOD P. Multiple κ opiate receptors. Life Sci **1986**; 38:1889–1898.

15 CLARK JA, LUI L, PRICE B et al. Kappa opiate receptor multiplicity: evidence for two U50,488-sensitive kappa1 subtypes and a novel kappa3 subtype. J Pharmacol Exp Ther **1989**; 251:461–468.

16 KIEFFER BL, BEFORT K, GAVRIAUX-RUFF C et al. The δ-opioid receptor: isolation of a cDNA by expression cloning and phar-

macological characterization. Proc Natl Acad Sci USA **1992**; 89:12048–12052

17 Evans CJ, Keith DE Jr, Morrison H et al. Cloning of a delta opioid receptor by functional expression. Science **1992**; 258:1952–1955.

18 Chen Y, Mestek A, Liu J, Hurley J, Yu L. Molecular cloning and functional expression of a mu opioid receptor from rat brain. Mol Pharmacol **1993**; 44:8–12.

19 Blake AD, Bot G, Reisine T. Structure-function analysis of the cloned opiate receptors: peptide and small molecule interactions. Chem Biol **1996**; 3:967–972.

20 Kieffer BL. Recent advances in molecular recognition and signal transduction of active peptides: receptors for opioid peptides. Cell Mol Neurobiol **1995**; 15:615–635.

21 Satoh M, Minami M. Molecular pharmacology of the opioid receptors. Pharmacol Ther **1995**; 68:343–365.

22 Kaufman D, Xia J, Keith D, Newam D, Evans C, Lusis A. Localization of the delta opioid receptor to mouse chromosome 4 by linkage analysis. Genomics **1994**; 19:405–406.

23 Giros B, Pohl M, Rochelle J, Seldin M. Chromosomal localization of opioid peptide and receptor genes in the mouse. Life Sci **1995**; 56:PL369–PL375.

24 Befort K, Mattei M, Roeckei N, Kieffer B. Chromosomal localization of the delta opioid receptor gene to human 1p 34.3–p36.1 and mouse 4D bands by *in situ* hybridization. Genomics **1994**; 20: 143–145.

25 Yasuda K, Espinosa R, Takeda J, Le Beau M, Bell GI. Localization of the kappa opioid receptor gene to human chromosome band 8q11.2. Genomics **1994**; 19:596–597.

26 Wang J, Johnson P, Persico Am Hawkins A, Griffin C, Uhl G. Human mu opiate receptor cDNA and genomic clones, pharmacological characterization and chromosomal assignment. FEBS Lett **1994**; 338:217–222.

27 Min B, Augustin L, Felsheim R, Fuchs J, Loh HH. Genomic structure and analysis of promoter sequence of a mouse mu opioid receptor gene. Proc Natl Acad Sci USA **1994**; 91:9081–9085.

28 Zimprich A, Simon T, Hollt V. Cloning and expression of an isoform of the rat mu opioid receptor (rMOR1B) which differs in agonist induced desensitization from rMOR. FEBS Lett **1994**; 359:142–146

29 Bare LA, Mansson E, Yang D. Expression of two variants of the human κ opioid receptor mRNA in SK-N-SH cells and human brain. FEBS Lett **1994**; 354:213–216.

30 Mayer P, Schulzeck S, Kraus J, Zimprich A, Hollt V. Promoter region and alternative spliced exons of the rat mu opioid receptor gene. J Neurochem **1996**; 66:2272–2278.

31 Zimprich A, Simon T, Hollt V. Transfected rat mu opioid receptors (rMOR1 and rMOR1B) stimulate phospholipase C and Ca^{++} mobilization. Neuroreport **1995**; 7:54–56.

32 Gaveriaux-Ruff C, Peluso J, Befort K, Simonin F, Zilliox C, Kieffer B. Detection of opioid receptor mRNA by RT-PCR reveals alternative splicing for the delta- and kappa-opioid receptors. Mol Brain Res **1997**; 48:298–304.

33 Lu S, Loh HH, Wei L. Studies of dual promoters of mouse kappa-opioid receptor gene. Mol Pharmacol **1997**; 52:415–420

34 Matthes HWD, Maldonado R, Simonin et al. Loss of morphine-induced analgesia, reward effect and withdrawal symptoms in mice lacking the μ-opioid-receptor gene. Nature **1996**; 383:819–823.

35 Sora I, Takahashi N, Funada M et al. Opiate receptor knockout mice define μ receptor roles in endogenous nociceptive responses and morphine-induced analgesia. Proc Natl Acad Sci USA **1997**; 94:1544–1549.

36 Raynor K, Kong H, Mestek A et al. Characterization of the cloned human μ receptor. J Pharmacol Exp Ther **1995**; 272:423–428.

37 Delfs J, Kong H, Mestek A, Chen Y, Yu L, Reisine T, Chesselet MF. Expression of mu opioid receptor mRNA in rat brain: An *in situ* hybridization study at the single cell level. J Comp Neurol **1994**; 345:46–68.

38 Mansour A, Fox C, Burke S, Meng F, Thompson R, Akil H, Watson S. Mu, delta, and kappa opioid receptor mRNA expression in the rat CNS: an *in situ* hybridization study. J Comp Neurol **1994**; 350:412–438.
39 Arvidsson U, Riedl M, Chakrabarti S, Lee J, Nakano A, Dado R, Loh HH, Law P, Wessendorf M, Elde R. Distribution and targetting of a mu opioid receptor in brain and spinal cord. J Neurosci **1995**; 15:3328–3341.
40 Ding Y, Kaneko T, Nomura S, Mixuno N. Immunohistochemical localization of mu opioid receptors in the central nervous system of the rat. J Comp Neurol **1996**; 367:375–402.
41 Moriwaki A, Wang J, Svingos A, van Bockstaele E, Cheng P, Pickel V, Uhl GR. Mu opiate receptor immunoreactivity in rat central nervous system. Neurochem Res **1996**; 21:1315–1331.
42 North RA. Opioid receptor types and membrane ion channels. Trends Neurosci **1986**; 9:114–117.
43 Punch L, Self D, Nestler E, Taylor J. Opposite modulation of opiate withdrawal behaviors on microinfusion of a protein kinase A inhibitor versus activator into the locus coeruleus or periaqueductal gray. J Neurosci **1997**; 17:8520–8527.
44 Mansour A, Hoversten MT, Taylor LP et al. The cloned μ, δ and κ receptors and their endogenous ligands: Evidence for two opioid peptide recognition cores. Brain Res **1995**; 700:89–98.
45 Raynor K, Kong H, Chen Y, Yasuda K, Yu L, Bell GI, Reisine T. Pharmacological characterization of cloned kappa, delta and mu opioid receptors. Mol Pharmacol **1994**; 45:330–334.
46 Young E, Bronstein D, Akil H. Proopiomelanocortin biosynthesis, processing and secretion: functional implications. In: Opioids I. Handbook of Experimental Pharmacology, Vol. 104 (Herz A, ed). Springer, Heidelberg, **1993**; 393–721.
47 Zadina JE, Hackler L, Ge L-J et al. A potent and selective endogenous agonist for the μ-opiate receptor. Nature **1997**; 386:499–502.
48 Dado R, Law P, Loh HH, Elde R. Immunofluorescent identification of a delta-opioid receptor on primary afferent nerve terminals. Neuroreport **1993**; 5:341–344.
49 Arvidsson U, Dado R, Riedl M. Lee J, Law P, Loh HH, Elde R, Wessendorf W. Delta-opioid receptor immunoreactivity: distribution in brainstem and spinal cord, and relationship to biogenic amines and enkephalin. J Neurosci **1995**; 15:1215–1235.
50 DePaoli A, Hurley K, Yasuda K, Reisine T, Bell GI. Distribution of kappa opioid receptor mRNA in adult mouse brain: An *in situ* hybridization histochemistry study. Mol Cell Neurobiol **1994**; 5:327–335.
51 Mansour A, Fox C, Meng F, Akil H, Watson S. Kappa 1 receptor mRNA in the rat CNS: comparison of kappa receptor binding and prodynorphin mRNA. Mol Cell Neurosci **1994**; 5:124–144.
52 Drake C, Patterson T, Simmons M, Chavkin C, Milner T. Kappa opioid receptor-like immunoreactivity in guinea pig brain: ultrastructural localization in presynaptic terminals in hippocampal formation. J Comp Neurol **1996**; 370:377–395.
53 Weisskopf M, Zalutsky R, Nicoll RA. The opioid peptide dynorphin mediates heterosynaptic depression of hippocampal mossy fibre synapses and modulates long-term potentiation. Nature **1993**; 365:188–190.
54 Gaveriaux C, Peluso J, Simonin F, Laforet J, Kieffer B. Identification of kappa- and delta-opioid receptor transcripts in immune cells. FEBS Lett **1995**; 369:272–276.
55 Tallent M, Dichter M, Bell GI, Reisine T. The cloned kappa opioid receptor couples to an N-type Ca^{++} current in undifferentiated PC 12 cells. Neuroscience **1994**; 63:1033–1040.
56 Kaneko S, Yada N, Fukuda K, Kikuwaka M, Akaike A, Satoh M. Inhibition of Ca^{++} channel current by mu- and kappa-opioid receptors coexpressed in *Xenopus* oocytes: desensitization dependence on Ca^{++} channel alpha 1 subunits. Br J Pharmacol **1997**; 121:806–812.
57 Tsu R, Chan J, Wong Y. Regulation of multiple effectors by the cloned delta opioid receptor: stimulaton of phospholi-

pase C and type II adenylyl cyclase. J Neurochem **1995**; 64: 2700–2707.

58 Smart D, Lambert DG. The stimulatory effects of opioids and their possible role in the development of tolerance. Trends Pharmacol Sci **1996**; 17:264–269.

59 Dhawan BN, Cesselin F, Raghubir R et al. International Union of Pharmacology XII. Classification of Opioid Receptors. Pharmacol Rev **1997**; 48:567–592.

60 Grudt T, Williams JT. Kappa opioid receptors also increase potassium conductance. Proc Natl Acad Sci USA **1993**; 90:11429–11432.

61 Mestek A, Hurley JH, Bye LS et al. The human κ opioid receptor: modulation of functional desensitization by calcium/calmodulin-dependent protein kinase and protein kinase C. J Neurosci **1995**; 15:2396–2406.

62 Chen Y, Yu L. Differential regulation by cAMP-dependent protein kinase and protein kinase C of the κ opioid receptor coupling to a G protein-activated K^+ channel. J Biol Chem **1994**; 269:7839–7842.

63 Kovoor A, Henry DJ, Chavkin C. Agonist-induced desensitization of the mu opioid receptor-coupled potassium channel (GIRK1). J Biol Chem **1995**; 270:589–595.

64 Zhang L, Yunkai Y, Mackin S et al. Differential μ opiate receptor phosphorylation and desensitization induced by agonists and phorbol esters. J Biol Chem **1996**; 271:11449–11454.

65 Tallent M, Dichter M. Reisine T. Coupling of the cloned kappa and mu opioid receptors to cellular effector systems is differentially regulated. Neuroscience **1998**; 85:873–885.

66 Simon M, Strathmann M, Gautam N. Diversity of G proteins in signal transduction. Science **1991**; 252:802–808.

67 Prather PL, McGinn TM, Claude PA et al. Properties of a kappa-opioid receptor expressed in CHO cells: interaction with multiple G-proteins is not specific for any individual Ga subunit and is similar to that of other opioid receptors. Mol Brain Res **1995**; 29:336–346.

68 Sanchez-Blazques P, Garcia-Espana A, Garzon J. *In vivo* injection of antisense oligonucleotides to G alpha subunits and supraspinal analgesia evoked by mu and delta opioid agonists. J Pharmacol Exp Ther **1995**; 275:1590–1596.

69 Prather PL, Loh HH, Law PW. Interaction of δ-opioid receptors with multiple G proteins: A non-relationship between agonist potency to inhibit adenylyl cyclase and to activate G proteins. Mol Pharmacol **1994**; 45:997–1003.

70 Gintzler AR, Xu H. Different G proteins mediate the opioid inhibition or enhancement of evoked methionine] enkephalin release. Proc Natl Acad Sci USA **1991**; 88:4741–4745.

71 Laugwitz K-L, Offermanns S, Spicher K et al. mu and delta opioid receptors differentially couple to G protein subtypes in membranes of human neuroblastoma SH-SY5Y cells. Neuron **1993**; 10:233–242.

72 Offermanns S, Schultz G, Rosenthal W. Evidence for opioid receptor-mediated activation of the G-proteins, Go and Gi2, in membranes of neuroblastoma x glioma (NG108-15) hybrid cells. J Biol Chem **1991**; 266:3365–3368.

73 Law SF, Reisine T. Changes in the association of G protein subunits with the cloned mouse delta opioid receptor upon agonist stimulation. J Pharmacol Exp Ther **1997**; 281:1476–1486

74 Bot G, Blake A, Li X et al. Regulation of the mouse δ-opioid receptor expressed in HEK 293 cells by opiates. Mol Pharmacol **1997**; 52:272–281.

75 Hescheler J, Rosenthal W, Trautwein W, Schultz G. The GTP-binding protein, Go, regulates neuronal Ca^{++} channels. Nature **1987**; 325:445–447.

76 Taussig R, Sanchez S, Rifo M, Gilman AG, Delardetti F. Inhibition of the omega conotoxin-sensitive calcium current by distinct G proteins. Neuron **1992**; 8:799–809.

77 McKenzie FR, Milligan G. Delta-opioid-receptor-mediated inhibition of adenylyl cyclase is transduced specifically by the guanine-nucleotide-binding protein G_{i2}. Biochem J **1990**; 267:391–398

78 Law SF, Yasuda K, Bell GI, Reisine T. Gia3 and Goa selectively associate with the cloned somatostatin receptor subtype

SSTR2. J Biol Chem **1993**; 268:10721–10727.

79 Lustig KD, Conklin BR, Herzmark P et al. Type II adenylyl cyclase integrates coincident signals from GS, Gi and Gq. J Biol Chem **1993**; 268:13900–13905.

80 Blake A, Bot G, Li X, Freeman J, Reisine T. Differential opioid agonist regulation of the mouse mu opioid receptor. J Biol Chem **1997**; 272:782–790.

81 Avidor-Reiss T, Bayewitch M, Levy R, Matus-Leibovitch N, Nevo I, Vogel Z. Adenylylcyclase supersensitization in mu opioid receptor transfected chinese hamster ovary cells following chronic opioid treatment. J Biol Chem **1995**; 270:29732–29738.

82 Avidor-Reiss T, Nevo I, Levy R, Pfeuffer T, Vogel Z. Chronic opioid treatment induces adenylyl cyclase V superactivation: Involvement of G beta/gamma. J Biol Chem **1996**; 271:21309–21315.

83 Bot G, Blake A, Li X, Reisine T. Fentanyl and its analogs desensitize the cloned mouse μ-opioid receptor. J Pharmacol Exp Ther **1998**; 285:1207–1218.

84 Raffa RB, Martinez RP, Connelly CD. G-protein antisense oligodeoxyribonucleotides and mu-opioid supraspinal antinociception. Eur J Pharmacol **1994**; 258:R5–R7.

85 Bot G, Blake A, Li S, Reisine T. Mutagenesis of a single amino acid in the rat mu opioid receptor discriminates the binding of full agonists from partial agonists and antagonists. J Neurochem **1998**; 70:358–365.

86 Loh H, Smith AP. Molecular characterization of opioid receptors. Ann Rev Pharmacol Toxicol **1990**; 30:123–147.

87 Childers SR. Opioid receptor-coupled second messenger systems. Life Sci **1991**; 48:1991.

88 Arden JR, Segredo V, Wang Z et al. Phosphorylation and agonist-specific intracellular trafficking of an epitope-tagged κ-opioid receptor expressed in HEK 293 cells. J Neurochem **1995**; 65:1636–1645.

89 Keith DE, Murray SR, Zaki PA et al. Morphine activates opioid receptors without causing their rapid internalization. J Biol Chem **1996**; 271:19021–19024.

90 Avidor-Reiss T, Nevo I, Saja D, Bayewitch M, Vogel Z. Opiate-induced adenylyl cyclase superactivation is isozyme-specific. J Biol Chem **1997**; 272:5040–5047.

91 Sharma S, Klee W, Nirenberg M. Dual regulation of adenylate cyclase accounts for narcotic tolerance and dependence. Proc Natl Acad Sci USA **1975**; 72:3092–3096.

92 Nestler E. Under seige: the brain and opiates. Neuron **1996**; 16:897–900.

93 Nestler E, Tallman J. Chronic morphine treatment increases cAMP dependent protein kinase activity in the rat locus coeruleus. Mol Pharmacol **1988**; 33:127–132.

94 Alreja M, Aghajanian G. Pacemaker activity of locus coeruleus neurons: whole-cell recordings in brain slices show dependence on cAMP and protein kinase A. Brain Res **1991**; 556:339–343.

95 Shiekhattar R, Aston-Jones G. Modulation of opiate responses in brain noradrenergic neurons by cAMP cascade: changes with chronic morphine. Neuroscience **1993**; 57:879–885.

96 Cowan A, Lewis J. Buprenorphine: Combating Drug Abuse with a Unique Opioid. Wiley-Liss, New York, **1995**.

97 Pei G, Kieffer BL, Lefkowitz RJ et al. Agonist-dependent phosphorylation of the mouse κ-opioid receptor: involvement of G protein-coupled receptor kinases but not protein kinase C. Mol Pharmacol **1995**; 48:173–177.

98 Malatynska E, Wang Y, Knapp RJ et al. Human delta opioid receptor: Functional studies on stably transfected chinese hamster ovary cells after acute and chronic treatment with the selective non-peptidic agonist SNC-80. J Pharmacol Exp Ther **1996**; 278:1083–1089.

99 Rapka R, Porreca F. Development of delta opioid peptides as non-addicting analgesics. Pharmaceut Res **1991**; 8:1–8.

100 Portoghese P, Moe S, Takemori A. A selective delta1 opioid receptor agonist derived from oxymorphone: evidence for separate recognition sites for delta1 opioid receptor agonists and antagonists. J Med Chem **1993**; 36:2572–2574.

101 Jin W, Terman G, Chavkin C. Kappa opioid receptor tolerance in the guinea pig hippocampus. J Pharmacol Exp Ther **1997**; 281:123–128.

102 Ma GH, Miller R, Kuznestov A et al. Kappa-opioid receptor activates an inwardly rectifying K^+ channel by a G protein-linked mechanism: coexpression in *Xenopus* oocytes. Mol Pharmacol **1995**; 47:1035–1040.

103 Appleyard S, Celver J, Pineda V, Kovoor A, Wayman G, Chavkin C. Agonist-dependent desensitization of the kappa opioid receptor by G protein receptor kinase and beta-arrestin. J Biol Chem **1999**; 274:233802–233807.

104 Blake A, Bot G, Li X et al. Differential agonist regulation of the human κ opioid receptor. J Neurochem **1997**; 68:1846–1850.

105 Kaneko S, Nakamura S, Adachi K, Akaike A, Satoh M. Mobilization of intracellular Ca^{++} and stimulation of cAMP production by kappa opioid receptors in *Xenopus* oocytes. Mol Brain Res **1994**; 27:258–264.

106 Xue J-C, Chen C, Zhu J et al. Differential binding domains of peptide and nonpeptide ligands in the cloned rat κ-opioid receptor. J Biol Chem **1994**; 269:30195–30199.

107 Wang J, Johnson P, Wu J, Wang W, Uhl G. Human kappa opiate receptor second extracellular loop elevates dynorphin's affinity for human mu/kappa chimeras. J Biol Chem **1994**; 269:25966–25969.

108 Kong H, Raynor K, Yasuda K et al. A single residue, aspartate 95, in the δ opioid receptor specifies high affinity agonist binding. J Biol Chem **1993**; 268:23055–23058.

109 Dohlman H, Caron M, Stader C, Amlaiky N, Lefkowitz R. A family of receptors coupled to guanine nucleotide regulatory proteins. Biochemistry **1988**; 27:1813–1817.

110 Horstman D, Brandon S, Wilson A, Guyer C, Cragoe E, Limbird L. An aspartate among G protein receptors confers allosteric regulation of alpha2-adrenergic receptorys by sodium. J Biol Chem **1990**; 265:21590–21595.

111 Kong H, Raynor K, Yasuda K, Bell GI, Reisine T. Mutation of an aspartate at residue 89 in somatostatin receptor subtype 2 prevents Na^+ regulation of agonist binding but does not alter receptor/G protein association. Mol Pharmacol **1993**; 44:380–384.

112 Pert C, Snyder SH. Opiate receptor binding of agonists and antagonists affected differentially by sodium. Mol Pharmacol **1974**; 10:868–879.

113 Bot, G, Blake, A, Li, S, Reisine, T. Mutagenesis of the mouse delta opioid receptor converts – buprenorphine from a partial agonist to an antagonist. J Pharmacol Exp Ther **1998**; 284:283–290.

114 Surratt CK, Johnson PS, Moriwaki A et al. μ opiate receptor. Charged transmembrane domain amino acids are critical for agonist recognition and intrinsic activity. J Biol Chem **1994**; 269:20548–20553.

115 Blake AD, Bot G, Reisine T. Molecular pharmacology of the opioid receptors. In: Molecular Neurobiology of Pain. Progress in Pain Research and Management, Vol. 9 (Borsook D, ed), International Association for the Study of Pain Press, USA, **1997**; 259–273.

116 Strader C, Sigal I, Register R, Candelore M, Rands E, Dixon R. Identification of residues required for ligand binding to the beta-adrenergic receptor. Proc Natl Acad Sci USA **1987**; 84:4384–4388.

117 Savarese TM, Fraser CM. *In vitro* mutagenesis and the search for structure-function relationships among G protein-coupled receptors. Biochem J **1992**; 283:1–19.

118 Befort K, Tabbara L, Bausch S et al. The conserved aspartate residue in the third putative transmembrane domain of the δ-opioid receptor is not the anionic counterion for cationic opiate binding but is a constituent of the receptor binding site. Mol Pharmacol **1996**; 49:216–223.

119 Kong H, Raynor K, Yano H et al. Agonists and antagonists bind to different domains of the cloned κ opioid receptor. Proc Natl Acad Sci USA **1994**; 91:8042–8046.

120 Hjorth S, Thirstrup K, Gandy D et al. Analysis of selective binding epitopes for

the κ-opioid receptor antagonist nor-binaltorphimine. Mol Pharmacol **1995**; 47:1089–1094.

121 Liapakis G, Fitzpatrick D, Hoeger C, Rivier J, Vandlen R, Reisine T. Identification of ligand binding determinants in the somatostatin receptor subtypes 1 and 2. J Biol Chem **1996**; 271:20331–20339.

122 Paterlini G, Portoghese P, Ferguson D. Molecular simulation of dynorphin A-(1–10) binding to extracellular loop 2 of the kappa opioid receptor. A model for receptor activation. J Med Chem **1997**; 40:3254–3262.

123 Wang WW, Shahrestanifar M, Jin J et al. Studies on μ and δ opioid receptor selectivity utilizing chimeric and site-mutagenized receptors Proc Natl Acad Sci USA **1995**; 92:12436–12440.

124 Meng F, Ueda Y, Hoversten MT et al. Mapping the receptor domains critical for the binding selectivity of δ-opioid receptor ligands. Eur J Pharmacol **1996**; 311:285–292.

125 Varga EV, Li X, Stropova D et al. The third extracellular loop of the human δ-opioid receptor determines the selectivity of the δ-opioid agonists. Mol Pharmacol **1996**; 50:1619–1624.

126 Valiquette M, Vu HK, Yue SY et al. Involvement of Trp-284, Val-296 and Val-297 of the human δ-opioid receptor in binding of δ-selective ligands. J Biol Chem **1996**; 271:18789–18796.

127 Befort K, Tabbara L, Kling D et al. Role of aromatic transmembrane residues of the δ-opioid receptor in ligand recognition. J Biol Chem **1996**; 271:10161–10168.

128 Onogi T, Minami M, Katao Y et al. DAMGO, a μ-opioid receptor selective agonist, distinguishes between the μ- and δ-opioid receptors around their first extracellular loops. FEBS Lett **1995**; 357:93–97.

129 Fukuda K, Kato S, Mori K. Location of regions of the opioid receptor involved in selective ligand binding. J Biol Chem **1995**; 270:6702–6709.

130 Xue J-C, Chen C, Zhu J et al. The third extracellular loop of the μ opioid receptor is important for agonist selectivity. J Biol Chem **1995**; 270:12977–12979.

131 Zhu X, Wang C, Cheng Z et al. The carboxyl terminus of mouse δ-opioid receptor is not required for agonist-dependent activation. Biochem Biophys Res Commun **1997**; 232, 513–516.

132 Wang HL. A cluster of Ser/Thr residues at the C-terminus of mu-opioid receptor is required for G protein-coupled receptor kinase 2-mediated desensitization. Neuropharmacology **2000**; 353–363.

133 Cvejic S, Trapaidze N, Cyr C et al. Thr353, located within the COOH-terminal tail of the δ-opioid receptor, is involved in receptor down-regulation. J Biol Chem **1996**; 271:4073–4076.

134 Trapaidze N, Keith DE, Cvejic S et al. Sequestration of the δ opioid receptor. J Biol Chem **1996**; 271:29279–29285.

135 Chu P, Murray S, Lissin D, von Zastrow M. Delta and kappa opioid receptors are differentially regulated by dynamin-dependent endocytosis when activated by the same alkaloid agonist. J Biol Chem **1997**; 272:27124–27130.

136 Bohn L, Lefkowitz R et al. Enhanced morphine analgesia in mice lacking beta-arrestin 2. Science **1999**; 286:2495–2498.

137 Bohn L, Gainetdinov R et al. Mu-opioid receptor desensitization by beta-arestin-2 determine morphine tolerance but not dependence. Nature **2000**; 408:720–723.

24
Ethnicity and Pharmacogenomics

Howard L. McLeod and Margaret M. Ameyaw

Abstract

There is great heterogeneity in the way humans respond to medications, often requiring empirical strategies to define the appropriate drug therapy for each patient. Genetic polymorphisms in drug metabolizing enzymes, transporters, receptors, and other drug targets have been linked to inter-individual differences in the efficacy and toxicity of many medications. These DNA variations provide putative markers for predicting which patients will experience extreme toxicity or treatment failure. Both quantitative (allele frequency) and qualitative (specific allele) differences in polymorphic genes have been observed between different population groups. For example, the frequency of mutations in thiopurine methyltransferase is lower in Chinese than in Caucasian populations. In addition, the predominate mutation responsible for deficient enzyme activity differs between the two populations (*TPMT*3C* vs *TPMT*3A*). Understanding the influence of ethnicity on pharmacogenomics will lead to comprehensive strategies for using the genome to optimize therapy for patients throughout the world.

24.1
Genetic Variation in Drug Metabolism and Disposition

Inter-individual genetic variability is a potential problem in clinical practice because it may result in idiosyncratic drug reactions or lack of predictable response to normal doses of drugs. Causes for such variability include the type of disease being treated and its severity, other concomitant illnesses, drug interactions, age of the patient, nutritional status, as well as the health status of the kidney and liver. In addition to the above variables, genetic polymorphism in drug metabolizing enzymes, transporters, receptors, and other drug targets is increasingly being recognized as sources of treatment failure and drug toxicity [1].

Clinical observations of such inherited differences in drug effects were first documented in the 1950s, as shown by the relationship between prolonged muscle relaxation after suxamethonium and an inherited deficiency of plasma choline esterase [2]. Some psychiatric patients were found to be unusually susceptible to suxamethonium and had low esterase activity without any obvious cause. Further studies of family members of these patients and normal volunteers revealed that there were two alleles for pseudocholine esterase giving rise to two homozygous extremes, with either very high or very low enzyme levels, and heterozygous intermediates [2].

Studies after the World War II showed that hemolysis, which affected black soldiers in the U.S. Army who were given the antimalarial therapy primaquine, was due to a genetic deficiency of erythrocyte glucose 6-phosphate dehydrogenase (G6PD) activity [3]. Later, the high frequency of G6PD deficiency in black populations was shown to be associated with the ability to survive falciparum malaria. The deficiency conveyed a biological advantage in malaria-infested countries and, therefore, a higher frequency of G6PD-deficient individuals was present in populations originating from such countries [3].

A study of 484 tuberculosis patients on isoniazid showed that the development of peripheral neuropathy in a subgroup of patients was due to inherited differences in the acetylation of this medication [4]. Individuals could be divided into rapid or slow acetylators of isoniazid. Family studies showed that rapid or slow acetylation status was inherited, with rapid acetylation being dominant and slow acetylation recessive. Polyneuritis was found to occur in 4 out of 5 slow acetylators, while only 2 out of 10 rapid acetylators developed polyneuritis. The rapid acetylators also tolerated longer courses of the drug [4].

24.2 Ethnic Variation in Drug Disposition

The rate and manner in which an individual metabolizes drugs is partly determined by the inheritance of alleles encoding therapeutically important genes. All pharmacogenetic polymorphisms studied to date differ in frequency among ethnic and racial groups. Clinical drug trials have historically been conducted in Caucasians, and, based on the data obtained in these clinical trials, dose recommendations have been made. However, as different ethnic groups have begun to be incorporated into clinical research studies, it has become clear that ethnic groups may differ in their response to drugs. Inter-ethnic differences in response to medications have been observed since the 1920s. In 1921, Paskind investigated the effect of atropine sulfate on 20 Caucasians and 20 African American men in Cook County Hospital, Chicago, USA [5]. Initial slowing of the heart rate, reaching a maximum in 10–15 min, was observed frequently in Caucasians but not in African American subjects. Chen and Poth (1929) measured the change in the transverse diameter of the pupil after the instillation of various mydriatics [6]. The increase in the diameter of the pupil was greatest in Caucasians, intermediate in Chinese and least in African Americans.

Thus it was clear that drug metabolism can differ between ethnic groups, and data generated in one population cannot be directly extrapolated to another population. When such differences exist, one ethnic group may be at an increased risk of therapeutic failure or toxicity because of differences in drug metabolism.

24.3 Ethnic Variation in Polymorphic Genes

There are both quantitative (allele frequency) and qualitative (specific allele) differences in polymorphic genes among different population groups. For example, separation of individuals into either rapid or slow acetylators of isoniazid constitutes one of the first discovered genetic polymorphisms of drug and carcinogen metabolism [7]. The proportion of rapid and slow acetylators varies remarkably in different ethnic or geographic populations. For example, 5% of Canadian Eskimos are slow acetylators, whereas, this phenotype rises to over 80% among Egyptians and 90% among Moroccans. Most populations in Europe and North America are 40–70% slow acetylators, whereas in Asian populations only 10–20% are slow acetylators [8]. The molecular basis for these differences has been elucidated and there are at least 20 alleles of the *NAT2* gene, which correlate with variable acetylation phenotype [9].

One consequence of uneven world distribution of different alleles is that clinical response to different drugs will vary widely between populations. If a gene that influences response to a particular compound has been identified, then one needs to assess the frequency of variant alleles in different populations. If such studies are not pursued, genotyping for a specific polymorphism may not adequately identify populations likely to respond to a drug or those at risk for excessive toxicity. Evaluating ethnic variation in drug response is important for drug manufacturers as a means to better characterize drug behavior in individuals who obviously differ from each other in terms of clinical phenotype.

24.3.1 Reasons for Ethnic Variation in Allele Frequencies

The variation in allele frequencies seen between different ethnic groups may reflect differences in the inheritance of original balanced polymorphisms throughout evolution. For example, individuals heterozygous for the sickle cell mutant allele resist malaria better than wild-type homozygotes. The Duffy null allele which confers resistance to *Plasmodium vivax* malaria is found in 100% of persons from Papua New Guinea and 0% of other populations living in non-endemic regions [10].

Other factors providing selective pressure for a particular allele may not be so readily ascertained. Drug metabolizing enzymes (DMEs) and their receptors are thought to have first evolved for critical life functions [11]. For example, mutations and deletions in the microsomal fatty aldehyde dehydrogenase (*FALDH*) gene have been shown to be the cause of Sjögren-Larsson syndrome, characterized by mental retardation, spasticity and ichthyosis. This implies that absence of FALDH activity during an important phase of differentiation leads to this neurocutaneous disorder [12]. Mutations in the *CYP1B1* gene are responsible for primary congenital glaucoma, implying that failure of CYP1B1 to metabolize some endogenous substrate in the anterior chamber of the eye leads to abnormal differentiation [13].

The function of DMEs is also thought to include the detoxification of dietary products and the evolution of plant metabolites, including drugs [11]. The selective forces responsible for the maintenance of different alleles in different populations may include the fact that one allele may enable improved rates of implantation, improved prenatal growth and development, improved postnatal health in response to dietary or environmental selective pressures or improved resistance to bacteria, viruses or parasites [11, 14]. Allele frequencies may also reflect ethnic dietary differences that have evolved over thousands of years [15].

24.3.2
Tracing the Molecular History of Genetic Polymorphisms

Knowing the frequencies of various polymorphic gene alleles in diverse populations is useful for understanding the history of genetic polymorphisms. Human genetics and archaeological research have provided the out-of-Africa model of human evolution [16]. The out-of-Africa model proposes that *Homo sapiens* originated in Southern and Eastern Africa 100,000–200,000 years ago and that all contemporary human populations are descended from this single African population.

The hypothesis on the origin of modern humans assumes that anatomically modern humans advanced geographically from Africa to West Asia then toward East Asia, Europe, America and Australia. The three continents entered last, the Americas and Australia, were occupied by expanding populations originally located in Northeast Asia and Southeast Asia, respectively [17]. The Negroid and the Caucasoid and Mongoloid groups are said to have diverged about 110,000 years ago, whereas Caucasoid and Mongoloid diverged about 40,000 years ago [17].

By tracing the molecular history and evolution of these polymorphisms, most analyses of allele frequencies in different populations reveal major ethnic differences. For example, CYP2D6*4 and *5 alleles occur in all populations studied to date and thus are very ancient mutations [18]. In contrast, CYP2D6*3, *6, *7 and *8 have not been detected in African populations studied so far, but occur in Caucasians and so may have emerged after divergence of Caucasians and Mongoloids from the Negroid race [18, 19]. On the other hand, CYP2D6*17 has been found only in Africans, and so must have also occurred in Africans after the divergence of Caucasians and Mongoloids from the Negroid race [18–20].

24.4
Variability Pharmacogenetic Polymorphisms Within a Population

Great heterogeneity in the frequency of genetic polymorphisms has been observed within a geographic population, such as in Africa. Genetic polymorphisms of drug metabolizing enzymes have been poorly characterized in ethnic African populations compared to Western countries. Furthermore, within Africa, genetic and cultural diversity is substantial and may thus affect the distribution of alleles in different African populations. For example, the prevalence of poor metabolizers of CYP2D6 varies considerably among different African populations ranging from 2–19% for debrisoquine and 2–4% for sparteine [20–25]. The frequency of CYP2C19 poor metabolizers in populations of African descent has also been reported to range from 1–35.4% [26]. These data suggest a pronounced heterogeneity among different African populations possibly as a result of different diseases, dietary preferences and cultural practices throughout the course of microevolution.

24.4.1 Interpretation of African American Data with Respect to Africans Living in Africa

Very few studies have focused on ethnic African groups. Therefore, very little is known about the molecular basis of ethnic differences in disease incidence and drug response in native Africans. Unfortunately, previous studies in African American subjects cannot be generalized to include native African populations, due to the great heterogeneity of racial ancestry of African American subjects, as well as other factors.

African slaves were brought to the United States beginning in 1619 and the movement of Africans lasted for over 250 years [27]. More than 98% came from an extensive area of West Africa and West-central Africa from both coastal and inland areas, with the contribution from East Africa being negligible [27]. At some early point in American slavery, intermarriages between the Africans and Caucasians began to occur. It is estimated that there is 10–50% Caucasian admixture in African Americans [17]. By the 1920 census in the United States, an African American was defined as anyone with even one black ancestor (the "one-drop rule"). This rule is the commonly accepted definition of an African American in the United States [28]. Many African Americans, therefore, have both African and Caucasian genes and will not show the true genetic picture of Africans living on the African continent.

24.5 Relevance of a Pharmacogenomic Approach to Therapeutics in Different Ethnic Groups

Individualized tailoring of treatments is becoming increasingly important in medicine. Predicting the response of a patient to a drug before its administration is both beneficial for the patient and economically sound. Pharmacogenetic studies focus on how known genetic variations affect people's responses to a particular drug treatment for a disease. With such knowledge, researchers can determine the appropriate dose of a specific drug to obtain the maximum effect while avoiding serious toxicity. For physicians, patients susceptible to adverse responses could be rapidly identified, and they can avoid expensive patient monitoring and treatment for drug toxicity. For patients, they could be assured that they receive more effective and safer treatments.

To date several genetic polymorphisms of therapeutic relevance have been identified and characterized, and studies are now underway testing their effects on drug response.

24.6
Ethnic Variation of Thiopurine Methyltransferase Alleles

Thiopurine methyltransferase (TPMT) catalyzes the S-methylation of thiopurine drugs, such as 6-mercaptopurine (6-MP), 6-thioguanine and azathioprine, to inactive metabolites [29–32]. Thiopurines form part of the routine treatment for patients with acute lymphoblastic leukemia, rheumatoid arthritis, and autoimmune diseases such as SLE and Crohn's disease, and are used as an immunosuppressant following organ transplantation.

TPMT enzyme activity and immunoreactive protein levels in human tissues are controlled by a common genetic polymorphism [32, 33]. Variation in TPMT activity determines thiopurine toxicity and therapeutic efficacy of thiopurine drugs. Approximately 1 in 300 white subjects have low activity, 6–11% have intermediate activity, and 89–94% have high activity [29, 32, 34]. Patients with low or undetectable levels of TPMT activity develop severe myelosuppression when treated with "standard" doses of thiopurines, while patients with very high TPMT are more likely to have a reduced clinical response to these agents [35–40]. 6-MP and the other thiopurines, azathioprine and thioguanine, are all inactive prodrugs, requiring metabolism to thioguanine nucleotides (TGNs) in order to exert cytotoxicity [29]. The principal mechanism by which these drugs exert cytotoxicity is thought to be the result of the incorporation of TGNs into DNA and RNA.

Interethnic variability in RBC TPMT activity has been reported in several populations. RBC TPMT was 29% higher in Saami subjects in Northern Norway compared to white subjects from the same geographic region [41]. African American subjects have 17–33% lower RBC TPMT activity than American white subjects [34, 42]. The TPMT activity in African and white Americans was substantially lower than that reported in 119 Chinese subjects [43].

TPMT is encoded by a 28 kb gene consisting of 10 exons and 9 introns and has been localized to chromosome 6p22.3 [29, 44]. Molecular pharmacogenetic studies have revealed a series of single nucleotide polymorphisms within the cDNA open reading frame (ORF) or at certain splice junctions of the *TPMT* gene [44–47]. These have been associated with significantly decreased levels of TPMT activity. To date, eight polymorphic *TPMT* alleles have been identified, including three alleles (*TPMT*2, TPMT*3A, and TPMT*3C*), accounting for approximately 80–95% of low or intermediate TPMT activity in Caucasians (Figure 24.1) [33, 47, 50].

The association between low TPMT activity and excessive hematological toxicity has been recognized [31, 35, 37]. Molecular analysis of the TPMT genotype is able to identify patients at risk for acute toxicity from thiopurines. A recent study involving 180 children identified that the TPMT genotype plays an important role in a patient's tolerance to 6-MP therapy [51]. Two of the patients, who were TPMT-de-

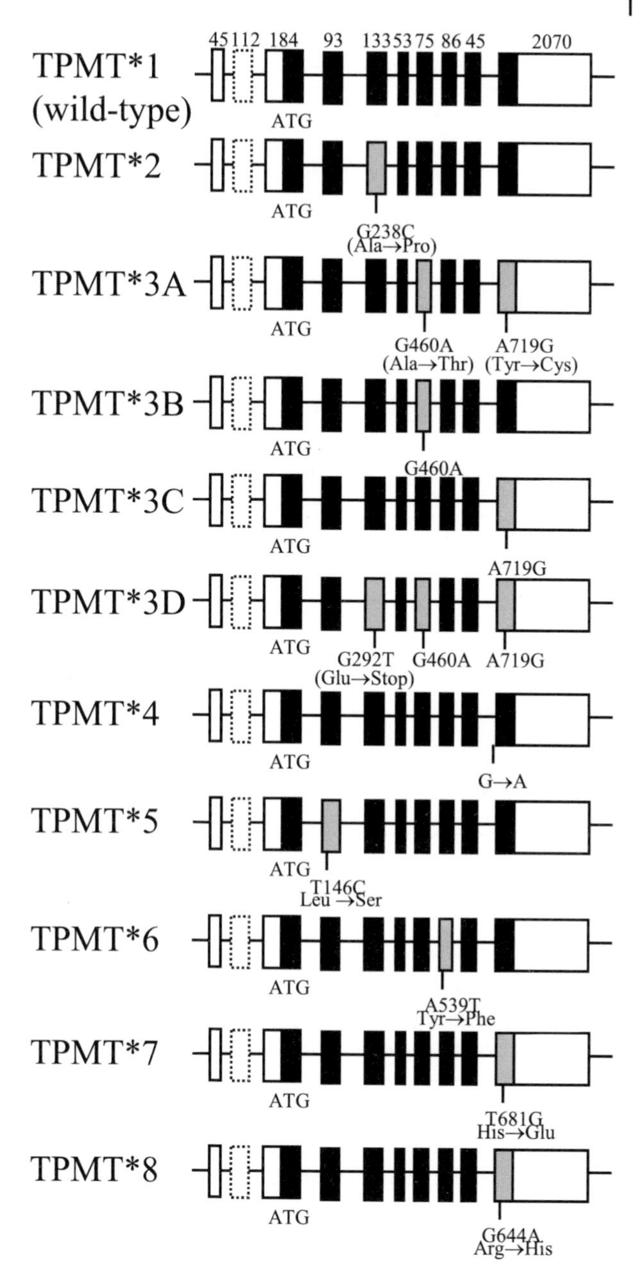

Fig. 24.1 Variant alleles at the human TPMT locus. Grey boxes are exons containing mutations. White boxes are untranslated regions and black boxes represent exons in the ORF. Dashed box represents exon 2, which was detected in one of 16 human liver cDNAs (adapted from [30]).

Tab. 24.1 *TPMT* genotype and allele frequencies (%) in different ethnic populations. The relevant reference is shown in paretheses

Population (Reference)	*n*	*Allele frequency*			*Genotype frequency*		
		*TPMT *2*	*TPMT *3A*	*TPMT *3C*	*wt/wt*	*wt/mut*	*mut/mut*
Ghanaians ([53] and unpublished data)	975	0	0	6.3	87.6	12.2	0.2
Kenyans [54]	101	0	0	5.4	89.1	10.9	0
Sudanese (unpublished)	52	0	0	2.9	94.2	5.8	0
African American [55]	NA	0.4	0.8	2.4	90.7	9.2	0.2
Caucasian, USA [50]	NA	0.2	3.2	0.2	92.5	7.4	0.14
Caucasian, UK [53]	199	0.5	4.5	0.3	89.9	9.6	0.5
Caucasian French [49]	191	0.5	5.7	0.8	85.9	13.6	0.5
Southwest Asians [52]	99	0	1	0	98	2	0
Chinese [52]	192	0	0	2.3	95.3	4.7	0
Filipino (unpublished)	74	0	0	1.4	97.3	2.7	0
Japanese (30)	553	0	0	1.5	97.3	2.4	0.4

n=number of subjects; genotype frequencies include mutant frequencies for *TPMT*2*, *TPMT*3A*, *TPMT*3C*; NA=not applicable.

ficient, tolerated a full dose of 6-MP for only 7% of the planned therapy. Heterozygous and homozygous wild-type patients tolerated full doses for 65 and 84% of the 2.5 years of treatment, respectively. The percentage of time in which 6-MP dosage had to be decreased to prevent toxicity was 2, 16 and 76% in wild-type, heterozygous and homozygous mutant individuals, respectively [51].

Based on the population genotype–phenotype studies performed to date, assays for the molecular diagnosis of TPMT deficiency have focussed on alleles *TPMT*2*, *TPMT*3A* and *TPMT*3C*, as these represent 80-95% of all mutant alleles of this gene in Caucasians [46, 50]. However, the frequency and pattern of mutant alleles of this gene is different among various ethnic populations. For example, Southwest Asians (Indian, Pakistani) have a lower frequency of mutant *TPMT* alleles and all mutant alleles identified to date are *TPMT*3A* (Table 24.1) [52]. This is in contrast to Kenyans and Ghanaians where the frequency of mutant alleles is similar to Caucasians, and all mutant alleles are *TPMT*3C* (Table 24.1) [53, 54]. Among African Americans, *TPMT*3C* is the most prevalent allele, but *TPMT*2*

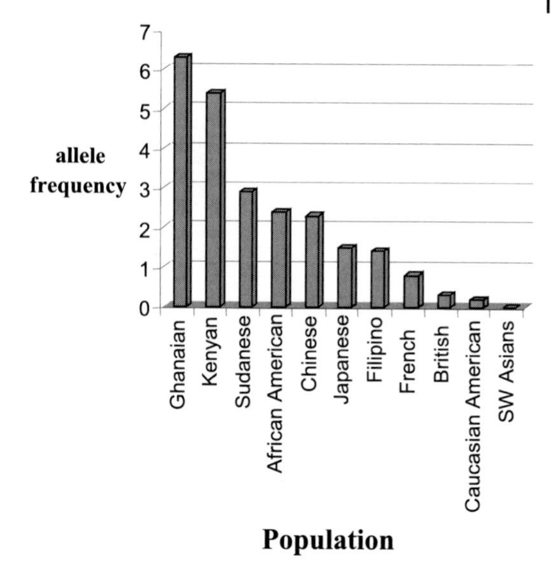

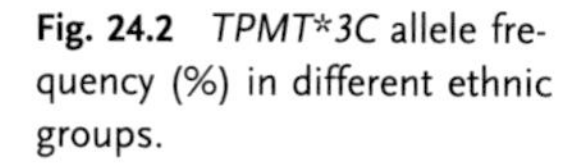
Fig. 24.2 *TPMT*3C* allele frequency (%) in different ethnic groups.

and *TPMT*3A* are also found [55]. In contrast, in Caucasians, *TPMT*3A* is the prevalent allele, but *TPMT*2* and *TPMT*3C* are also found in this population [52].

*TPMT*3C* accounted for 100% of the mutant alleles observed in the Ghanaian subjects. This is similarly found in 101 Kenyans and 192 Chinese subjects, as well as in Sudanese and Filipino subjects (Table 24.1; Figure 24.2) [52, 54]. This contrasts with the Caucasian (British, American, French) subjects, where 5.7, 5.5 and 11.4% of variant alleles were *TPMT*3C*, respectively (Table 24.1). *TPMT*3A* was not detected in the African or Asian populations, but accounted for 84.9, 81.4 and 88.9% of variant alleles in British, American, and French Caucasians, respectively. Therefore, mutations at nucleotide 719 (TPMT*3C) is common in all populations studied to date, but occurs most often in the presence of a simultaneous mutation at nucleotide 460 (TPMT*3A) in Caucasian subjects (Table 24.1).

Yates et al. suggested that *TPMT*3C* may be more prevalent in black subjects than white subjects, as four out of nine African Americans with a heterozygous phenotype had the *TPMT*3C* allele [50]. In that study, the *TPMT*3C* allele was associated with a loss of RBC TPMT activity, and was subsequently shown to be associated with the loss of immunodetectable TPMT protein in the RBCs of humans inheriting this allele [50]. More extensive analysis in African American subjects with an intermediate or low RBC TPMT activity phenotype revealed that *TPMT*3C* represented 66.7% of variant alleles, with the remaining alleles being *TPMT*3A*, *TPMT*2*, and *TPMT*8* [55].

The presence of *TPMT*3A* alleles in the African American population is consistent with the genetic mixing that has been identified through historical and molecular analysis [12, 27]. These data, compared with that of different ethnic populations, reveal that the pattern of variant *TPMT* alleles differs significantly between ethnic groups [53]. There is a decrease in *TPMT*3C* allele frequency from Africans to Asians to Caucasians ($p<0.05$ between African/Asian populations and Caucasian populations; Figure 24.2). The difference in allele frequency between the African populations and Asian populations, excluding the Southwest Asians, was not significant ($p>0.05$). The similarity of Southwest Asian *TPMT* allele frequencies to that of Caucasians is not surprising as it has been postulated that the origin of East Asians is different from the rest of Asia. South and West Asians are thought to originate from Northeast Africa where populations are more Caucasoid than Negroid [17].

The *TPMT*2* allele (G238C) accounted for 9.4, 7.1 and 5.5% of the mutant alleles in the British, French and American populations, respectively. However, this allele was not found in any of the Ghanaian subjects and was not detected in the other African or Asian subjects. These findings are consistent with those of a recent study, which found that the *TPMT*2* allele was present in British Caucasians and not in Chinese or Southwest Asian subjects [52]. In addition, the *TPMT*2* allele was not detected in Kenyan subjects [54]. This suggests that *TPMT*2* is either very rare in non-Caucasian populations or specific to Caucasians, or both. The recently identified *TPMT* mutant alleles *TPMT*4–*8* appear to be relatively rare in Caucasian subjects. Their contribution to variant alleles in other ethnic groups has yet to be defined.

Phylogenetic analyses estimate the divergence of Africans and non-Africans to be at least 100,000 years ago [17]. From gene evolution studies it is considered that the most common allele in all populations is usually the ancestral allele. Mutation and recombination then give rise to the other genotypes. Genotype analysis in this study suggests that the A719G mutation may be the ancestral *TPMT* mutant allele, as it was present in both Caucasian and African subjects and has been described in Southwest Asian and Chinese populations [52]. This further indicates that the G460A allele was acquired later and added to form *TPMT*3A*. Since the *TPMT*2* allele appears to be confined to Caucasians, it may be a more recent allele of this polymorphic enzyme.

TPMT genotyping prior to thiopurine drug administration may soon become a routine molecular diagnostic test in many centers. Because the pattern of variant *TPMT* alleles differs significantly between ethnic groups, it is important to fully characterize which alleles are more prevalent in particular populations. The more prevalent alleles should be analyzed to identify the majority of patients at risk and will depend on the ethnic origin of the patient about to receive thiopurine drugs. Analysis of all thiopurine drug recipients to identify the population at risk for TPMT-mediated toxicity, would not only be beneficial to these patients but will also be cost-effective. Moreover, patients with wild-type alleles associated with high TPMT activity, may not respond to this drug therapy. Such patients will be identified, and higher doses of thiopurines may then be administered.

24.6.1
Influence of Ethnicity on Drug Transport Pharmacogenetics

The human multidrug-resistance (*MDR1; ABCB1*) gene encodes an integral membrane protein, P-glycoprotein (PGP), a member of the ATP-binding cassette family of membrane transporters [56, 57]. PGP was originally identified by its ability to confer multidrug resistance to tumor cells against a variety of structurally unrelated anticancer agents. PGP limits the bioavailabilty of several commonly prescribed drugs such as cyclosporine A, paclitaxel, colchicine, doxorubicin, vinblastine, ivermectin, digoxin, and HIV-1 protease inhibitors.

The *MDR1* gene is located on the long arm of chromosome 7 and consists of a core promoter region and 28 exons [56, 57]. PGP protein level is highly variable between subjects [59]. However, the molecular basis for inter-patient variation in PGP is not clear. Recently, fifteen different single nucleotide polymorphisms (SNPs) were detected in the *MDR1* gene. One of these SNPs, resulting in a C to T transition in exon 26 (C3435T), showed a correlation with PGP protein levels and function. The homozygous T allele was associated with more than 2-fold lower duodenal PGP protein levels compared with C3435 homozygotes [59]. Further analysis has confirmed the influence of this SNP on *in vivo* PGP function [60].

The frequency of the MDRI C3435T SNP in various world populations has recently been described [61]. The allele frequency of the mutant T allele ranged between 0.16–0.27 in the four African populations, compared with 0.41–0.66 in the Caucasian and Asian groups (Table 24.2, Figure 24.3). The genotype frequency of the homozygous TT genotype ranged between 0 and 6% in the four African groups and 20–47% in the Caucasian and Asian subjects (Table 24.2) [61]. The distribution of C and T genotype and allele frequencies were significantly different between the African/African American populations and the Caucasian/Asian populations (Table 24.2). The Southwest Asian subjects were significantly different from all the other populations except the Portuguese [61]. Within the African group, the Ghanaian and Kenyan subjects had an identical allele frequency. There was a significant difference between the Sudanese and the Ghanaian/Kenyan subjects ($p=0.009$). The Portuguese were also significantly different from the Filipino population ($p=0.02$), but similar to the other Caucasian and Asian subjects [61].

The mutant T allele, which results in decreased PGP levels, is relatively rare in populations with African ancestry, but exists at higher frequencies in Caucasian, Chinese, Filipino, Portuguese and Saudi populations. The TT genotype was not detected in the 206 Ghanaians studied, but accounted for 1, 4, and 6% of individuals in the African American, Kenyan and Sudanese populations, respectively [61]. From the previous functional studies this data implies that populations of African ancestry will have higher PGP protein levels and drug efflux.

Differences in environmental and dietary factors may contribute to the differences in allele frequencies in the African/African American, Caucasian and Asian populations. PGP exists in several normal tissues where it probably has the physiological role of excreting xenobiotics and protecting important tissues from such compounds when they are present in the blood [62, 63]. Allelic differences of sev-

Tab. 24.2 Inter-ethnic differences in the genotype and allele frequencies (%) of the *MDR1* exon 26 C3435T polymorphism. The relevant reference is shown in parentheses

Population	*n*	*Allele frequency*		*Genotype frequency*		
		C (95%CI)	*T (95%CI)*	*wt/wt*	*wt/mut*	*mut/mut*
Ghanaian [61]	206	83 (78–88)	17 (12–22)	67	34	0
Kenyan [61]	80	83 (75–91)	17 (9–25)	70	26	4
African American [61]	88	84 (76–92)	16 (8–24)	68	31	1
Sudanese [61]	51	73 (61–85)	27 (15–39)	52	43	6
Caucasian, U.K [61]	190	48 (41–55)	52 (45–59)	24	48	28
Caucasian, German [59]	188	52 (45–59)	48 (41–55)	28	48	24
Portuguese [61]	100	43 (33–53)	57 (47–67)	22	42	36
Southwest Asians [61]	89	34 (24–44)	66 (56–76)	15	38	47
Chinese [61]	132	53 (44–62)	47 (38–56)	32	42	26
Filipino [61]	60	59 (47–71)	41 (29–53)	38	42	20
Saudi [61]	96	55 (45–65)	45 (36–54)	37	38	26

C=wild type allele, T=mutant allele, *n*=number of subjects, 95% CI=95% confidence interval.

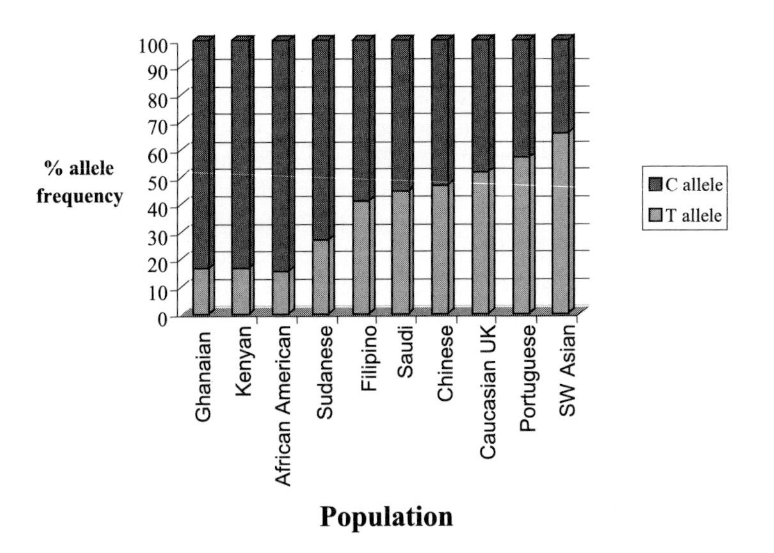

Fig. 24.3 Allele frequency of MDR1 C3435T polymorphism in different ethnic populations.

eral xenobiotic/drug-metabolizing enzymes are thought to sometimes reflect differences in diet and environmental pollutants among ethnic groups that have evolved over thousands of years [11]. The T allele was also relatively low in frequency in the African Americans, which favors dietary influence over environmental effects. Cultural preferences for a diet consisting principally of tropical plants and foodstuffs may have lead to the maintenance of the C allele in persons of African descent.

Furthermore, it has been observed that placental PGP is of great importance in limiting the fetal penetration of various potentially harmful or therapeutic compounds [64]. High PGP levels may have a role in the protection of fetuses from harmful tropical plant metabolites (from which many drugs are derived). An originally balanced polymorphism may have been preferentially selected by improved prenatal growth and development as well as improved postnatal health. This selective pressure may maintain the C allele in populations of African descent.

PGP limits the bioavailability of many commonly prescribed medications including digoxin, ivermectin, several anticancer agents, antipsychotics, antidepressants as well as HIV-1 protease inhibitors. With a population of about 600 million (approximately 10% of the world total), sub-Saharan Africa accounts for over two-thirds of the world's HIV-infected persons and 80% of the world's HIV-infected women and children [65]. HIV infection is already the leading cause of adult death in many cities on the continent and has also increased child mortality in many countries. HIV-1 protease inhibitors are largely inaccessible in most of sub-Saharan Africa, but this may soon change. Most of the approved HIV-1 protease inhibitors are PGP substrates. Bioavailability of these drugs may be limited in African patients as a result of high PGP levels, making the MDR1 genotype an important public health issue for health providers in Africa. Prospective studies are now required to determine the utility of the *MDR1* C3435T genotype for optimizing therapy for HIV, cancer, and other common diseases.

Several new drugs in the developing stages are being designed to reverse or prevent the multidrug resistance mechanism caused by the expression of the *MDR1* gene. Such drugs may be important in populations of African descent in order to improve the bioavailability of drugs that are PGP substrates. Information on the allele distribution of this functional *MDR1* SNP will, therefore, be a valuable tool for drug manufacturers to optimize the efficacy of commonly prescribed drugs.

The high frequency of the C allele in the African group may also contribute to the high incidence of drug resistance and the prevalence of more aggressive tumors, in diseases such as breast cancer, in individuals of African origin [66].

24.7
Ethnic Variation in a Target for Drug Therapy

Single-base substitutions, which affect the amino acid sequence of proteins and lead to altered protein function, are the most frequent type of polymorphisms associated with many disease phenotypes as well as with variation in drug response

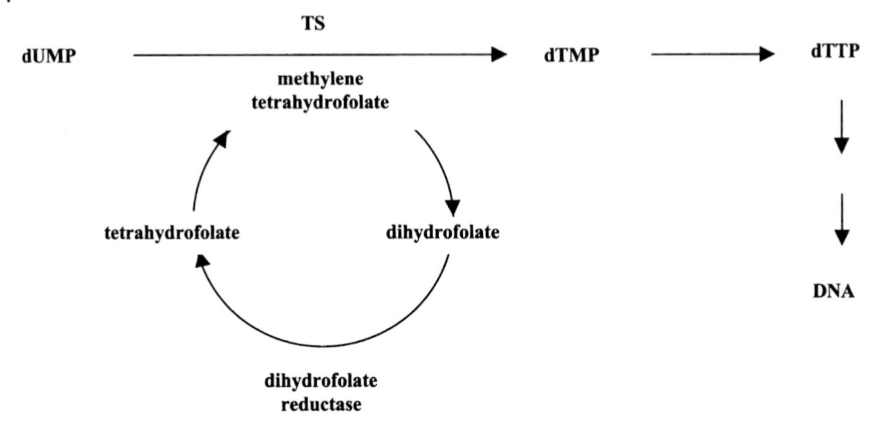

Fig. 24.4 Thymidylate synthase (TS) biochemical pathway. dUMP-deoxyuridine monophosphate, dTMP-deoxythymine monophosphate, dTTP-deoxythymine triphosphate.

[67]. Ethnic variation in allele frequencies can lead to important differences in disease susceptibility, outcome and drug metabolism [68, 69]. In addition to single nucleotide polymorphisms, variable number of tandem repeat (VNTR) regions have been shown to have functional significance.

In 1995, Horie et al. described a polymorphic tandem repeat found in the 5′-untranslated region of the thymidylate synthase gene [70]. Thymidylate synthase (TS; TYMS) catalyzes the intracellular transfer of a methyl group to deoxyuridine-5-monophosphate (dUMP) to form deoxythymidine-5-monophosphate (dTMP), which is anabolized in cells to the triphosphate (dTTP). This pathway is the only *de novo* source of thymidine, an essential precursor for DNA synthesis and repair. The methyl donor for this reaction is the folate cofactor 5,10-methylenetetrahydrofolate (CH2-THF) (Figure 24.4).

The 16 kb human TS gene has been localized to chromosome 18p11.32 and consists of 7 exons and 6 introns [71]. Thymidylate synthase has been of considerable interest as a target for cancer chemotherapeutic agents such as 5-fluorouracil and Raltitrexed [72, 73]. Fluoropyrimidine resistance in several tumors, including colorectal cancer, has been shown to be mediated through increased mRNA and TS protein levels [74]. High levels of TS expression have been correlated with poor prognosis in breast cancer, gastric cancer and colorectal cancer [75–78]. This may be due to increased tumor cell proliferation as a result of increased TS levels [79]. The human TS promoter has recently been characterized, revealing several important mechanisms for gene regulation.

In vitro studies have shown that increasing the number of repeats leads to stepwise increases of TS gene expression with the presence of a triple repeat resulting in a 2.6-fold greater TS expression than a double repeat [70, 80]. *In vivo* studies in human gastrointestinal tumors have shown a significant increase in TS protein levels and functional activity in patients with *TSER*3* compared to individuals with *TSER*2* [81]. As TS tumor levels are important for resistance and survival prediction, this may have important implications for TS-based chemotherapy.

Tab 24.3 TSER allele frequencies (%) for the different ethnic groups. The relevant reference is shown in parentheses

Ethnic group	*n*	*TSER*2 (95%CI)*	*TSER*3 (95%CI)*	*TSER*4 (95%CI)*	*TSER*9 (95%CI)*
Ghanaian [83]	496	41 (33–49)	55 (47–63)	3 (0.3–6)	1 (0–3)
Kenyan [83]	196	44 (34–54)	49 (39–59)	7 (2–12)	ND
Sudanese (unpublished)	104	53 (45–70)	47 (30–56)	ND	ND
African American [83]	184	46 (36–56)	52 (42–62)	2 (0–5)	ND
Caucasian American [83]	208	46 (36–56)	54 (44–64)	ND	ND
Caucasian, UK [82]	194	45 (35–55)	54 (44–64)	1 (0–3)	ND
Southwest Asian [82]	190	38 (28–48)	62 (52–72)	ND	ND
Filipino (unpublished)	148	14 (6–22)	86 (78–94)	ND	ND
Chinese [82]	192	18 (10–26)	82 (74–90)	ND	ND
Japanese [70]	42	19 (7–30)	81 (69–93)	ND	ND

(95%CI) = 95% confidence intervals, *n* = number of alleles, ND = not detected.

An important factor in the resistance to chemotherapy drugs such as 5-fluorouracil and Raltitrexed is an increase in TS expression [72, 37]. The TS enhancer region polymorphism may be one mechanism responsible for increasing TS gene expression.

The allele frequency of *TSER*2* ranged between 0.38–0.53 in Caucasian and African populations (Table 24.3) [82, 83]. However, Chinese and Filipino populations had a lower frequency of the *TSER*2* allele (0.18 and 0.14, respectively) (Table 24.3). The frequency for the *TSER*3* allele ranged between 0.47–0.62 in all the populations except Chinese and Filipino, where it was 0.82 and 0.86, respectively (Table 24.3). *TSER*4* allele was found predominantly in the Kenyan, Ghanaian, and African-American subjects at low frequencies [7, 3, and 2%, respectively) and in one British Caucasian subject (Table 24.3) [82, 83]. *TSER*9* allele was found only in the Ghanaian population studied and accounted for 1% of all the alleles (Table 24.3). Both *TSER*4* and *TSER*9* were always heterozygous with either *TSER*2* or *TSER*3* (Table 24.3) [82, 83].

The allele frequencies for *TSER*2* and *TSER*3* in the Chinese, Japanese and Filipino groups were significantly different from all the other populations in the study ($p<0.001$] (Figure 24.5) [82, 83]. The Sudanese were significantly different from the Southwest Asians ($p=0.033$). No other significant difference in *TSER*2* and *TSER*3* allele frequency between populations were observed. No significant difference in *TSER*2* and *TSER*3* allele frequency was observed between the Caucasian and African populations ($p>0.05$ in all cases) [83].

There was no significant difference in the *TSER*2* and *TSER*3* allele frequencies between the African, Caucasian and Southwest Asian subjects [82, 83]. There was, however, a significant difference between the Asian populations (Chinese, Filipino and Japanese) and all other populations for both *TSER*2* and *TSER*3* ($p<0.001$] (Table 24.4) [82]. This is in contrast to previous population studies where a shift in the predominant allele is usually observed during migration from

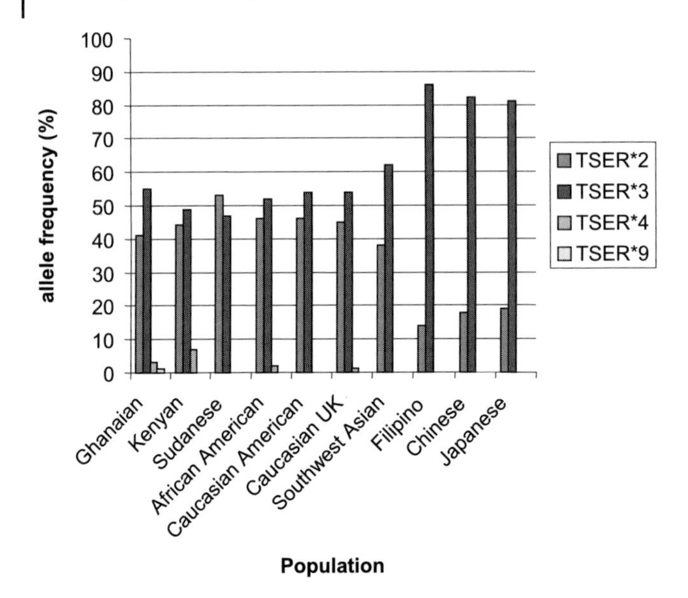

Fig. 24.5 *TSER* allele frequencies in different ethnic groups.

Tab. 24.4 Overview of SNP frequencies in three ethnic groups

Gene	***Mutant allele frequency (%)***			***P value***		
	African (Ghanaian)	***Asian (Chinese)***	***Caucasian (Scottish)***	p_1	p_2	p_3
B2AR	10	10	47	1.00	<0.001	<0.001
*TPMT*3C*	6	2	0.3	0.149	<0.05	0.155
MDR1	17	47	52	<0.001	<0.001	0.479
CYP3A4	69	0	5	<0.001	<0.001	<0.05
CCND1	20	63	42	<0.001	0.001	0.003
HER-2	0	11	20	0.001	<0.001	0.079
COMT	26	18	54	0.172	<0.001	<0.001

(p_1=African vs. Asian; p_2=African vs. Caucasian; p_3=Asian vs. Caucasian).

Africa to Asia and Europe. For example, African populations were significantly different from Caucasians and Asians for the *COMT, MDR1*, and *CYP3A4*, polymorphisms [61, 84–88]. Most models of population genetics define clear differences between African and Caucasian subjects [17]. In contrast, there was no significant difference in TSER allele frequency between the Africans and Caucasians.

These results would imply external influences, e.g., dietary intake of thymidine, are stabilizing the TSER allele frequency among populations where differences would be expected. In the Asian populations, *TSER*3* is about 5- to 6-fold more common than *TSER*2* [82]. Higher levels of TS induced by the presence of

*TSER*3* may have been a selective advantage during migration to Asia through diet or other environmental factors. This advantage may not have been so necessary in other areas of the world, where a roughly equal frequency of *TSER*2* and *TSER*3* are found. This pattern is also found in TPMT, where 9.4% of African-American [55], 10.1% Caucasians and 10.9% of Kenyans [54], and 14.8% of Ghanaian subjects [53], showed variant alleles compared to 4.7% of Chinese subjects [17]. This supports the hypothesis that multiple selective pressures will have distinct influences on different genes within populations (Cavalli-Sforza et al., 1996). The *TSER*4* and *TSER*9* alleles, found at low frequencies mainly in the African populations studied, may have been lost during migration as no selective pressure has been present to maintain them.

The alleles containing 4 and 9 copies of the TSER repeat were primarily confined to African populations. *TSER*4* accounted for 2–7% of *TSER* alleles in all African populations except the Sudanese. However, *TSER*4* was also found in a British Caucasian subject but not among the American-Caucasian population studied [53]. This suggests that *TSER*4* occurs at a low frequency in Caucasian populations. The absence of the *TSER*4* allele in the Sudanese population may be a result of the small sample size or due to the fact that this allele occurs at very low frequencies in this population. The latter possibility would make sense as this population is an admixture of Negroid and Caucasoid characteristics at both the morphological and molecular levels [17].

Although *TSER*9* appears to be unique to the Ghanaian population, the sample size evaluated in this study cannot deny the presence of this allele in other African populations [83]. Alleles corresponding to 5–8 or >9 tandem repeats were not identified in this study; however, considering the low frequencies of *TSER*4* and *TSER*9* in the populations studied, it is possible that a larger population study may identify novel alleles. In addition, the significance of *TSER*4* and *TSER*9* is less clear. An increase in TSER repeats is associated with increased TS expression *in vitro* and TS protein levels *in vivo* [70, 80, 81]. There is no data in the literature that evaluates the role of ethnicity in response to TS inhibitor chemotherapy.

24.7.1 Comparison of Mutant Alleles Across the Three Major Ethnic Groups (African, Asian and Caucasian)

There have been few similar mutant polymorphisms between the three major ethnic groups. The data for the mutant alleles for seven SNPs evaluated in the same set of normal volunteers clearly demonstrates that Africans are different from Caucasians with respect to therapeutically important genetic polymorphisms (Table 24.4; Figure 24.6). The Africans were significantly different from the Caucasian population in all seven genes evaluated (Table 24.4, p_2-values). Significant differences between Africans and Asians occurred in 57% of the genes listed (Table 24.4, p_1-values). The Asians were also significantly distinct from the Caucasians in 57% of the genes studied (Table 24.4, p_3-values), but in different genes

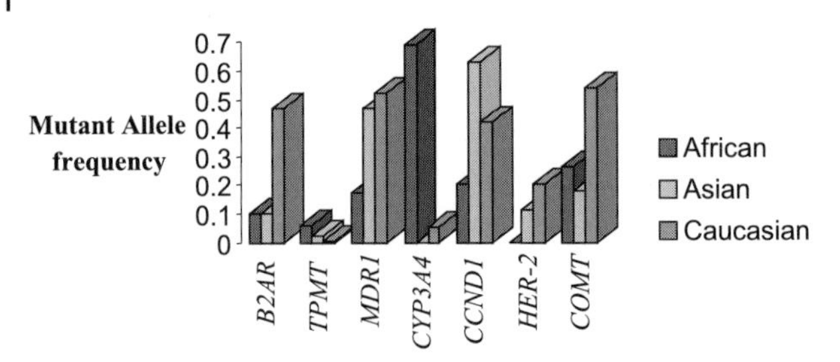

Fig. 24.6 Mutant allele frequencies of SNPs in distinct ethnic populations. B2AR-β2 adrenoreceptor, COMT-catechol-O-methyltransferase. Glu27 and TPMT*3C allele frequencies were used for the *B2AR* and *TPMT* genes, respectively.

than the African/Asian populations. There was a significant difference between the Africans and Asians with respect to the mutant allele frequency for the *MDR1, CYP3A4, CCND1* and *HER-2* genetic polymorphisms (Table 24.4; Figure 24.6). The mutant allele frequencies were significantly different between the Asian and Caucasian populations for the *B2AR, CYP3A4* and *COMT* and *CCND1* genes (Table 24.4; Figure 24.6). The mutant allele frequencies were significantly different for the *CCND1* and *CYP3A4* genes in all three population groups (Table 24.4).

Negroid, Caucasoid and Mongoloid groups are said to have diverged about 100,000 years ago, whereas Caucasoid and Mongoloid diverged about 40,000 years ago [17]. The fact that Africans and Asians are similar for some of the variant allele frequencies, may be explained by a recent hypothesis that there was exchange of genetic information between Asians and Africans before migration to Europe, therefore, maintaining some of the African alleles in Asia [87]. The two African populations (West and East Africans) were almost identical in the allele frequencies for all genes studied. The similarity of the West and East African populations may be explained by their exposure to similar tropical dietary or medicinal plants and flavonoids, which influence the frequency of alleles of genes encoding drug- and xenobiotic-metabolizing enzymes.

24.7.2
CYP3A4 and MDR1 Mutant Genotypes in Ghanaians

There is a striking overlap in PGP and CYP3A4 substrates, including erythromycin, ivermectin, cyclosporine, taxanes, quinidine, steroids and HIV-1 protease inhibitors. PGP has been shown to influence basal expression of CYP3A and also determines the extent of CYP3A metabolism of numerous medications *in vivo* by limiting intracellular substrate availability [88, 89]. The *MDR1* gene encodes PGP protein. *MDR1*-negative mice had higher CYP3A4 activity as shown by their 2-fold higher than average erythromycin breath test $^{14}CO_2$ and increased CYP3A expression [88]. Both genes are located on the short arm of chromosome 7. Individuals with a genetic basis for high PGP levels, i.e., homozygous wild type for the *MDR1* C3435T [61, 86], might therefore be expected to have reduced CYP3A4 ex-

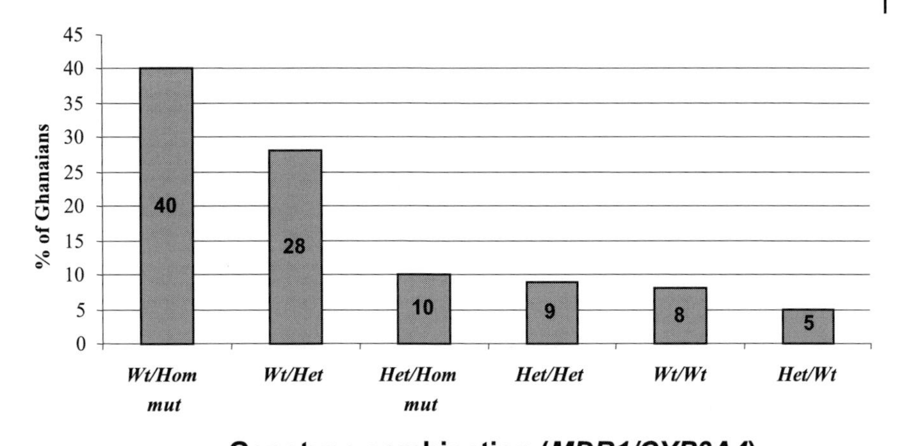

Fig. 24.7 *MDRI/CYP3A4* genotype in 100 Ghanaian subjects. Wt=wild type; Het=heterozygous; Hom mut=homozygous mutant.

pression, thereby influencing metabolism of drugs by CYP3A4. In a series of 100 Ghanaians genotyped for both SNPs, 40% of *CYP3A4* homozygous variant individuals were wild type for *MDR1*, 28% of *CYP3A4* heterozygous individuals were wild type for *MDR1*, 10% homozygous variant *CYP3A4* subjects were heterozygous for *MDR1*, 9% of individuals were heterozygous for both SNPs, 8% were wild type for both SNPs, and only 5% were wild type for *CYP3A4* but heterozygous for *MDR1* (Figure 24.7). There was no homozygous *MDR1* mutant subject in the Ghanaian population.

This data shows that 68% of Ghanaians have wild-type genotypes for *MDR1* and mutant genotypes for *CYP3A4*, and 5% have mutant *MDR1* and wild-type *CYP3A4* genotypes. 73% of Ghanaians may, therefore, have an altered *CYP3A4-MDR1* relationship and be at increased risk of altered response to commonly prescribed medications including erythromycin, steroids, ivermectin, HIV-1 protease inhibitors as well as anticancer agents including taxanes, doxorubicin and cyclosporine A. Considering the fact that HIV in Ghana has a prevalence rate of about 4.6%, this is of significant public health importance. Onchocerciasis is also a major public health problem in Ghana, with ivermectin being widely prescribed for both its treatment and prevention as part of the Onchocerciasis Control Program. Altered response to ivermectin in the Ghanaian population will have a significant public health impact on this developing country. Deficiency of PGP has been associated with hypersensitivity to ivermectin due to drug accumulation in the brain [90]. CYP3A4 is the predominant enzyme responsible for the metabolism of ivermectin [91]. 19% of the Ghanaians studied had mutant alleles for both *MDR1* and *CYP3A4* (Figure 24.7). These individuals may be at increased risk of adverse reactions to ivermectin.

Other factors such as drug interactions that result in inhibition or induction of both CYP3A4 and PGP may influence the bioavailability and metabolism of their drug substrates. Post-transcriptional and translational factors may also influence expression of these genes. Clinical studies are required to assess the impact of these SNPs on the treatment of patients with these drugs.

24.8
Pharmacogenomics as a Public Health Tool

While the promises of pharmacogenomics are enormous, it is likely to have the greatest initial benefit for patients in developed countries, due to expense, availability of technology and the focus of initial research [92]. However, pharmacogenomics should ultimately be useful to all world populations. There is clear evidence for ethnic variation in disease risk, disease incidence, and response to therapy. In addition, many polymorphic drug metabolizing enzymes have qualitative and quantitative differences among racial groups.

One approach to applying pharmacogenomics to public health concerns is through SNP allele frequency analysis in defined populations. For example, TPMT genotype in world populations suggest that TPMT-mediated toxicity from azathioprine or mercaptopurine would be lower in Japanese or Chinese populations than in Caucasians [30]. In contrast, a higher mutant allele frequency was found in the Ghanaian and Kenyan populations [30]. In addition, further analysis of the five major tribes of Ghana found distinct differences in TPMT allele frequency, ranging from 9.9% heterozygotes in the Ewe population to 13.8% in Fanti individuals (Ameyaw and McLeod, submitted). Even greater ethnic differences have been established for other polymorphic drug metabolizing enzymes (e.g., NAT2, CYP2D6, CYP2C19], and this will likely be the case for most pharmacogenomic targets, including drug transporters [69]. This general approach needs to be more extensively evaluated, but does offer the potential for generating information that will have broad application to the development of clinical practice guidelines and national formularies in developing countries.

While using knowledge of ethnic differences may be relevant to much of the world's populations, it is significantly limited in places with extensive genetic mixing. For example, it is well known that the African-American population has a great degree of geographic and social mixing that provide a basis for genetic heterogeneity. This is illustrated in evaluation of TPMT mutations between African-American and West African populations. Although the *TPMT*3C* allele was the most frequently observed variant in both populations, it represented 100% of West African mutant alleles and 52% of African American mutant alleles [53, 55]. The remaining African-American mutant alleles were *TPMT*2* and *TPMT*3A* [55], alleles that are common in Caucasians. Therefore, great care must be made when applying pharmacogenomics to public health issues, and testing at the genetic level in each patient will remain the most definitive approach.

24.9
Non-Scientific Challenges for Pharmacogenomics

There are a number of issues influencing the development of pharmacogenomics, including many that are of a practical or non-scientific nature [92]. An important limitation to the wide application of pharmacogenomics is the availability of gene expression arrays, high-throughput genotyping, and informatics. Currently there is considerable growth in the number of companies offering both genomics analysis on a fee-for-service basis and the equipment for user maintained instruments. As technology and competition bring down the high initial capital costs of array and genotype systems, the potential for general application of these approaches will be further enhanced [92].

A related, and unanswered question, is how much can pharmacogenomics analysis cost to be a viable adjunct to current medical practice [92]? Currently, the technology for gene expression and genotype assessment is only affordable in the research and development setting. Thoughtful economic analysis is needed to justify and direct the further development of pharmacogenomics for rational therapeutics. On the positive side, once a panel of genotypes has been correctly determined for a given individual, they need not be repeated. It is anticipated that a secured, patient-specific database will be established for each person, into which additional results will be deposited as additional genotypes are determined. This potentially web-based compilation of an individual's established genotypes would then be available to authorized health care providers, for the selection of optimal therapy for the treatment or prevention of diseases.

24.10
Conclusions

The ethics of genetic analysis is currently under avid discussion and debate [91]. Previously, a system of trust and internal control was utilized to prevent inappropriate use of genetic information. This approach has been very successful, with breach of trust being a rare event. However, the field of bioethics is now focusing on prevention of potential or theoretical abuses of genetic information against individuals. This has led to questions on what information is needed, who should have access to the data, and how it should be used. Issues such as these are deeply challenging, as the very companies that pay for genetic testing are the same ones that could use the information to restrict future insurance coverage. However, the great potential gains from pharmacogenomics, in terms of both patient well being and cost of health care, heavily outweigh the risks in this field. Putting such powerful information in the hands of knowledgeable health care providers and those involved in the discovery of new approaches to disease treatment or prevention, offers so much promise that society must find a way to ensure that inappropriate exploitation does not preclude the vast public good that will emerge from the burgeoning field of pharmacogenomics.

24.11 References

1 Evans WE, Relling MV. Pharmacogenomics: Translating functional genomics into rational therapeutics. Science **1999**; 286:486–491.

2 Kalow W. Familial incidence of low pseudocholinesterase level. Lancet **1956**; 211:576–577.

3 Carson PE, Flanagan CL, Ickes CE, Alving AS. Enzymatic deficiency in primaquine-sensitive erythrocytes. Science **1956**; 124:484–485.

4 Evans DAP, Manley KA, McKusick VA. Genetic control of isoniazid metabolism in man. Br Med J **1960**; 2:485–491.

5 Paskind HA. Some differences in response to atropine in white and coloured races. J Am Med Assoc **1921**; 76:104–108.

6 Chen KK, Poth EJ. Racial differences as illustrated by the mydriatic action of cocaine, euphthalmine and ephedrine. J Pharmacol Exp Ther **1929**; 36:429–445.

7 Peters JH, Miller KS, Brown P. Studies on the metabolic basis for the genetically determined capacities for isoniazid inactivation in man. J Pharmacol Exp Ther **1965**; 150:298–304.

8 Evans DAP. N-Acetyltransferase. Pharmacol Ther **1989**; 42:157–234.

9 Hein DW, Grant DM, Sim E. Update on consensus arylamine N-acetyltransferase gene nomenclature. Pharmacogenetics **2000**; 10:1–2.

10 Zimmerman PA, Woolley I, Masinde GL, Miller SM, McNamara DT, Hazlett F et al. Emergence of FY*A (null) in a *Plasmodium vivax*-endemic region of Papua New Guinea. Proc Natl Acad Sci USA **1999**; 96:13973–13977.

11 Nebert DW, Ingelman-Sundberg M, Daly AK. Genetic epidemiology of environmental toxicity and cancer susceptibility: human allelic polymorphisms in drug metabolising enzyme genes, their functional importance, and nomenclature issues. Drug Metab Rev **1999**; 31:467–487.

12 De Laurenzi V, Rogers GR, Hamrock DJ, Marekov LN, Steinert PM, Compton JG et al. Sjögren-Larsson syndrome is caused by mutations in the fatty aldehyde dehydrogenase gene. Nature Genet **1996**; 12:52–57.

13 Stoilov I, Akarsu AW, Sarfarazi M. Identification of three different truncating mutations in cytochrome P450 1B1 (CYP1B1] as the principal cause of primary congenital glaucoma (buphthalmos) in families linked to the GLC3A locus on chromosome 2p21. Hum Mol Genet **1997**; 6:641–647.

14 Rotter JI, Diamond JM. What maintains the frequencies of human genetic diseases? Nature **1987**; 329:289–290.

15 Gonzalez FJ, Nebert DW. Evolution of the P450 gene superfamily: animal-plant "warfare", molecular drive, and human genetic differences in drug oxidation. Trends Genet **1990**; 6:182–186.

16 Stringer CB, Andrews P. Genetic and fossil evidence for the origin of modern humans. Science **1988**; 239:1263–1268.

17 Cavalli-Sforza LL, Mennazzi P, Piazza A. The History and Geography of Human Genes. **1996**; Princeton University Press, New Jersey.

18 Griese E-U, Asante-Poku S, Ofori-Adjei D, Mikus G, Eichelbaum M. Analysis of the CYP2D6 gene mutations and their consequences for enzyme function in a West African population. Pharmacogenetics **1999**; 9:715–723.

19 Wennerholm A, Johansson I, Massele AY, Jande M, Alm C, Aden-Abdi Y et al. Decreased capacity for debrisoquine metabolism among black Tanzanians: analyses of the CYP2D6 genotype and phenotype. Pharmacogenetics **1999**; 9:707–714.

20 Masimirembwa C, Persson I, Bertilsson L, Hasler J, Ingelman-Sundberg M. A novel mutant variant of the CYP2D6 gene (CYP2D6*17] common in black African population: association with diminished debrisoquine hydroxylase activity. Br J Clin Pharmacol **1996**; 42:713–719.

21 Masimirembwa C, Hasler J, Bertilsson L, Johansson I, Ekberg O, Ingelman-Sundberg M. Phenotype and genotype

analysis of debrisoquine hydroxylase (CYP2D6] in a black Zimbabwean population. Reduced enzyme activity and evaluation of metabolic correlation of CYP2D6 drugs. Eur J Clin Pharmacol **1996**; 51:117–122.

22 Lennard MS, Iyun AO, Jackson PR, Tucker GT, Woods HF. Evidence for a dissociation in the control of sparteine, debrisoquine and metoprolol metabolism in Nigerians. Pharmacogenetics **1992**; 2:89–92.

23 Sommers de K, Moncrieff J, Avenant J. Metoprolol alpha-hydroxylation polymorphism in the San Bushmen of Southern Africa. Hum Toxicol **1989**; 8:39–43.

24 Sommers de K, Moncrieff J, Avenant J. Non-correlation between debrisoquine and metoprolol polymorphisms in the Vend. Hum Toxicol **1989**; 8:365–368.

25 Simooya OO, Njunju E, Hodjegan AR, Lennard MS, Tucker GT Debrisoquine and metoprolol oxidation in Zambians: a population study. Pharmacogenetics **1993**; 3:205–208.

26 Xie HG, Kim RB, Stein CM, Wilkinson GR, Wood AJJ. Genetic polymorphism of (S)-mephenytoin 4′-hydroxylation in populations of African descent. Br J Clin Pharmacol **1999**; 48:402–408.

27 Reed TE. Caucasian genes in American negroes. Science **1969**; 165:762–768.

28 Freeman HP. The meaning of race in science-considerations for cancer research. Cancer **1998**; 82:219–225.

29 Krynetski EY, Evans WE. Pharmacogenetics as a molecular basis for individualised drug therapy: the thiopurine S-methyltransferase paradigm. Pharmaceut Res **1999**; 16:342–349.

30 McLeod HL, Krynetski EY, Relling MV, Evans WE. Genetic polymorphism of thiopurine methyltransferase and its clinical relevance for childhood acute lymphoblastic leukemia. Leukaemia **2000**; 14:567–572.

31 McLeod HL, Relling MV, Liu Q, Pui CH, Evans WE. Polymorphic thiopurine methyltransferase in erythrocytes is indicative of activity in leukaemic blasts from children with acute lymphoblastic leukaemia. Blood **1995**; 85:1897–1902.

32 Weinshilboum RM, Sladek SL. Mercaptopurine pharmacogenetics: monogenic inheritance of erythrocyte thiopurine methyltransferase activity. Am J Hum Genet **1980**; 32:651–662.

33 Tai HI, Krynetski EY, Schuetz EG, Yashinevski Y, Evans WE. Enhanced proteolysis of thiopurine S-methyltransferase (TPMT) encoded by mutant alleles in human (*TPMT**3A*, TPMT**2*): mechanism for the genetic polymorphism of TPMT activity. Proc Natl Acad Sci USA **1997**; 94:6444–6449.

34 McLeod HL, Lin J-S, Scott EP, Pui C-H, Evans WE. Thiopurine methyltransferase activity in American white subjects and black subjects. Clin Pharmacol Ther **1994**; 55:15–20.

35 McLeod HL, Miller DR, Evans WE. Azathioprine-induced myelosuppression in thiopurine methyltransferase deficient heart transplant recipient. Lancet **1993**; 1341:1151.

36 Evans WE, Horner M, Chu YO, Kalwinsky D, Roberts WM. Altered mercaptopurine metabolism, toxic effects and dosage requirement in a thiopurine methyltranferase-deficient child with acute lymphoblastic leukaemia. J Paediatr **1991**; 119:985–989.

37 Lennard L, Lilleyman JS, Van Loon JA, Weinshilboum RM. Genetic variation in response to 6-mercaptopurine for childhood acute lymphoblastic leukaemia. Lancet **1990**; 336:225–229.

38 Lennard L, Gibson BES, Nicole T, Lilleyman JS. Congenital thiopurine methyltransferase deficiency and 6-mercaptopurine toxicity during treatment for acute lymphoblastic leukaemia. Arch Dis Child **1993**; 69:577–579.

39 Lilleyman JS, Lennard L. Mercaptopurine metabolism and risk of relapse in childhood lymphoblastic leukaemia. Lancet **1994**; 343:1188–1190.

40 Schutz E, Gummert J, Mohr F, Oellerich M. Azathioprine induced myelosuppression in thiopurine methyltransferase deficient heart transplant recipient. Lancet **1993**; 341:436.

41 Klemetsdal B, Tollefsen E, Loennechen T, Johnsen K, Utsi E, Gisholt K, Wist E, Aarbaake J. Interethnic differ-

ence in thiopurine methyltransferase activity. Clin Pharmacol Ther **1992**; 51:24–31.

42 Jones DC, Smart C, Titus A, Blyden G, Dorvril M, Nwadike N. Thiopurine methyltransferase activity in a sample population of black subjects in Florida. Clin Pharmacol Ther **1993**; 53:348–353.

43 Lee EJD, Kalow W. Thiopurine S-methyltransferase activity in a Chinese population. Clin Pharmacol Ther **1993**; 54:28–33.

44 Szulmanski C, Otterness D, Her C, Lee D, Brandiff B, Kelsell D, Thiopurine methyltransferase pharmacogenetics: human gene cloning and characterisation of a common polymorphism. DNA Cell Biol **1996**; 15:17–30.

45 Krynetski EY, Schuetz JD, Galpin AJ, Pui CH, Relling MV, Evans WE. A single point mutation leading to loss of catalytic activity in human thiopurine S-methyltransferase. Proc Natl Acad Sci USA **1995**; 92:949–953.

46 Otterness D, Szulmanski C, Lennard L, Klemestdal B, Aarbaake J, Park-Han H et al. Human thiopurine methyltransferase pharmacogenetics: gene sequence polymorphisms. Clin Pharmacol Ther **1997**; 62:60–73.

47 Otterness DM, Szumlanski CL, Wood TC, Weinshilboum RM. Human thiopurine methyltransferase pharmacogenetics: kindred with a terminal splice junction mutation that results in loss of activity. J Clin Invest **1998**; 101:1036–1044.

48 Tai HL, Krynetski EY, Yates CR, Loennechen T, Fessing MV, Krynetskaia NF et al. Thiopurine S-methyltransferase deficiency: two nucleotide transitions define the most prevalent mutant allele associated with loss of catalytic activity in Caucasians. Am J Hum Genet **1996**; 58:694–702.

49 Spire-Vayron de la Moureyre C, Debuysère H, Sabbagh N, Marez D, Vinner E, Chevalier ED et al. Detection of known and new mutations in the thiopurine S-methyltransferase gene by single-strand conformation polymorphism analysis. Hum Mutat **1998**; 12:177–185.

50 Yates CR, Krynetski EY, Loennechen T, Fessing MY, Tai H-L, Pui CH et al. Molecular diagnosis of thiopurine-S-methyltransferase deficiency: genetic basis for azathioprine and mercaptopurine intolerance. Ann Intern Med **1997**; 126:608–614.

51 Relling MV, Hancock HL, Rivera GK, Sandlund JT, Ribeiro RC, Krymetski EY et al. Intolerance to mercaptopurine therapy related to heterozygosity at the thiopurine methyltransferase gene locus. J Natl Cancer Inst **1999**; 91:2001–2008.

52 Collie-Duguid ESR, Pritchard SC, Powrie RH, Sludden J, Collier DA, Li T, McLeod HL. The frequency and distribution of thiopurine methyltransferase alleles in Caucasian and Asian populations. Pharmacogenetics **1999**; 9:37–42.

53 Ameyaw MM, Collie-Duguid ESR, Powrie RH, Ofori-Adjei D, McLeod HL. Thiopurine methyltransferase alleles in British and Ghanaian populations. Hum Mol Genet **1999**; 8:367–370.

54 McLeod HL, Pritchard SC, Githang'a J, Indalo A, Ameyaw MM, Powrie RH, Booth L, Collie-Duguid ESR. Ethnic differences in thiopurine methyltransferase pharmacogenetics: evidence for allele specificity in Caucasian and Kenyan subjects. Pharmacogenetics **1999**; 9:773–776.

55 Hon YY, Fessing MY, Pui CH, Relling MV, Krynetski EY, Evans WE. Polymorphism of the thiopurine S-methyltransferase (TPMT) gene in African Americans. Hum Mol Genet **1999**; 8:371–376.

56 Borst P, Evers R, Kool M, Wijnholds J. A family of drug transporters: The multidrug resistance-associated proteins. J Natl Cancer Inst **2000**; 92:1295–1302.

57 Ueda K, Cornwell MM, Gottesman MM, Pastan I, Roninson IB, Ling V. The *MDR1* gene, responsible for multidrug resistance, codes for P-glycoprotein. Biochem Biophys Res Commun **1986**; 141:956–962.

58 Chen CJ, Clark D, Ueda K, Pastan I, Gottesman MM, Roninson IB. Genomic organization of the human multidrug resistance (*MDR1*) gene and origin of P-glycoprotein. J Biol Chem **1990**; 265:506–514.

59 Hoffmeyer S, Burk O, vonRichter O, Arnold HP, Brockmoller J, Johne A et al. Functional polymorphisms of the human multidrug-resistance gene: Multiple sequence variations and correlation of one allele with P-glycoprotein expression and activity *in vivo*. Proc Natl Acad Sci USA **2000**; 97:3473–3478.

60 Hitzl M, Drescher S, van der Kuip H, Schaffeler E, Fischer J, Schwab M, Eichelbaum M, Fromm MF. The C3435T mutation in the human MDR1 gene is associated with altered efflux of the P-glycoprotein substrate rhodamine 123 from CD56+ natural killer cells. Pharmacogenetics. **2001**; 11:293–298.

61 Ameyaw MM, Regateiro F, Li T, Liu X, Tariq M, Mobarek A et al. *MDR1* pharmacogenetics: Frequency of the C3435T mutation in exon 26 is significantly influenced by ethnicity. Pharmacogenetics **2001**;11:217–221.

62 Cordon-Cardo C, O'Brien JP, Casals D, Rittman-Grauer L, Biedler JL, Melamed MR et al. Multidrug resistance gene P-glycoprotein is expressed by endothelial cells at blood–brain barrier sites. Proc Natl Acad Sci USA **1989**; 86:695–698.

63 Thiebaut F, Tsumo T, Hamada H, Gottesman MM, Pastan I, Willingham MC. Cellular localization of the multidrug resistance gene product in normal human tissues. Proc Natl Acad Sci USA **1987**; 84:7735–7738.

64 Smit JW, Huisman MT, van Telligen O, Whiltshire HR. Absence or pharmacological blocking of placental P-glycoprotein profoundly increases foetal drug exposure. J Clin Invest **1999**; 104:1441–1447.

65 De Cock KM, Weiss HA. The global epidemiology of HIV/AIDS. Trop Med Int Health **2000**; 5:A3–A9.

66 Elmore JG, Moceri VM, Carter D, Larson EB. Breast carcinoma tumor characteristics in black and white women. Cancer **1998**; 83:2509–2515.

67 Schork NJ, Fallin D, Lanchbury S. Single nucleotide polymorphisms and the future of genetic epidemiology. Clin Genet **2000**; 58:250–264.

68 Rebbeck TR, Jaffe JM, Walker AH, Wein AJ, Malkowicz SB. Modification of clinical presentation of prostate tumours by a novel genetic variant in CYP3A4. J Natl Cancer Inst **1998**; 90:1225–1229.

69 Kalow W, Bertilsson L. Interethnic factors affecting drug response. Adv Drug Res **1994**; 25:1–53.

70 Horie N, Aiba H, Oguro K, Hojo H, Takeishi K. Functional analysis and DNA polymorphism of the tandemly repeated sequences in the 5′-terminal regulatory region of the human gene for thymidylate synthase. Cell Struct Funct **1995**; 20:191–197.

71 Hori T, Takahashi E, Ayusawa D, Takeishi K, Kaneda S, Seno T. Regional assignment of the human thymidylate synthase (TS) gene to chromosome band 18p11.32 by nonisotopic *in situ* hybridization. Hum Genet **1990**; 8:576–580.

72 Rustum YM, Harstrick A, Cao S, Vanhoefer U, Yin MB, Wilke H et al. Thymidylate synthase inhibitors in cancer therapy: direct and indirect inhibitors. J Clin Oncol **1997**; 15:389–400.

73 Leichman CG, Lenz HJ, Leichman L, Danenberg K, Baranda J, Groshen S et al. Quantitation of intratumoral thymidylate synthase expression predicts for disseminated colorectal cancer response and resistance to protracted-infusion fluorouracil and weekly leucovorin. J Clin Oncol **1997**; 15:3223–3229.

74 Berger SH, Jenh CH, Johnson LF, Berger FG. Thymidylate synthase overproduction and gene amplification in fluorodeoxyuridine-resistant human cells. Mol Pharmacol **1985**; 28:461–467.

75 Pestalozzi BC, Peterson RD, Gelber A, Goldhirsch A, Gusterson BA, Trihia H et al. Prognostic importance of thymidylate synthase expression in early breast cancer. J Clin Oncol **1997**; 15:1921–1931.

76 Lenz H-J, Leichman CG, Danenberg KD, Danenberg PV, Grashen S, Cohen H et al. Thymidylate synthase mRNA level in adenocarcinoma of the stomach: a predictor for primary tumour response and overall survival. J Clin Oncol **1995**; 14:176–182.

77 Johnston PG, Fisher ER, Rockette HE, Fisher B, Wolmark N, Drake JC, Chabner BA, Allegra CJ. The role of thymidylate synthase expression in prognosis and outcome of adjuvant chemotherapy in patients with rectal cancer. J Clin Oncol **1994**; 12:2640–2647.

78 Kornmann M, Link KH, Lenz H-J, Pillasch J, Metzger R, Butzer U et al. Thymidylate synthase is a predictor for response and resistance in hepatic artery infusion chemotherapy. Cancer Lett **1997**; 118:29–35.

79 Kaye SB. New antimetabolites in cancer chemotherapy and their clinical impact. Br J Cancer **1998**; 78:1–7.

80 Kawakami K, Salonga D, Omura K, Park JM, Danenberg KD, Watanabe Y et al. Effects of polymorphic tandem repeat sequence on the *in vitro* translation of messenger RNA. Proc Am Assoc Cancer Res **1999**; 40:436–437.

81 Kawakami K, Omura K, Kanehhira E, Morishita M, Watanabe Y. Polymorphic tandem repeats in the thymidylate synthase gene is associated with it's protein expression in human gastrointestinal cancers. Anticancer Res **1999**; 19:3249–3252.

82 Marsh S, Collie-Duguid ESR, Li T, Liu X, McLeod HL. Ethnic variation in the thymidylate synthase enhancer region polymorphism among Caucasian and Asian populations. Genomics **1999**; 58:310–312.

83 Marsh S, Ameyaw M-M, Githang'a J, Indalo A, Ofori-Adjei D, McLeod HL. Novel thymidylate synthase enhancer region alleles in African populations. Human Mut **2000**; 16:528.

84 Ameyaw MM, Syvänen A-C, Ulmanen I, Ofori-Adjei D, McLeod HL. Pharmacogenetics of catechol-O-methyltransferase: Frequency of low activity allele in a Ghanaian population. Hum Mutat **2000**; 16:445–446.

85 McLeod HL, Syvänen A-C, Githang'a J, Indalo A, Ismai D, Dewar K et al. Ethnic differences in catechol O-methyltransferase pharmacogenetics: frequency of the codon 108/158 low activity allele is lower in Kenyan than Caucasian or South-west Asian individuals. Pharmacogenetics **1998**; 8:195–199.

86 Tayeb M, Clark C, Ameyaw MM, Haites NE, Price Evans DA, Tariq M et al. CYP3A4 promoter variant in Saudi, Ghanaian, and Scottish Caucasian populations. Pharmacogenetics **2000**; 10:753–756.

87 Hammer MF, Karafet T, Rasanayagam A, Wood ET, Althiede TK, Jenkins T et al. Out of Africa and back again: Nested cladistic analysis of human Y chromosome variation. Mol Biol Evol **1998**; 15:427–441.

88 Schuetz EG, Umbenhauer DR, Yasuda K, Brimer C, Nguyen L, Relling MV. Altered expression of hepatic cytochromes P450 in mice deficient in one or more mdr1 genes. Mol Pharmacol **2000**; 57:188–197.

89 Lan L-B, Dalton JT, Schuetz EG. Mdr1 limits CYP3A metabolism *in vivo*. Mol Pharmacol **2000**; 58:863–869.

90 Pulliam JD, Seward RL, Henry LT, Steinberg SA. Investigating ivermectin toxicity in collies. Vet Med **1985**; 80:33–40.

91 Zeng Z, Andrew NW, Arison BH, Luffer-Atlas D, Wang RW. Identification of cytochrome P4503A4 as the major enzyme responsible for the metabolism of ivermectin by human liver chromosomes. Xenobiotica **1998**; 28:313–321.

92 McLeod HL, Evans WE. Pharmacogenomics: unlocking the human genome for better drug therapy. Annu Rev Pharmacol Toxicol **2001**; 41:101–121.

25
Pharmacogenomics: Ensuring Equity Regarding Drugs Based on Genetic Difference

Phyllis Griffin Epps and Mark A. Rothstein

Abstract

Certain genetic polymorphisms relevant to drug response occur with varying frequency among different ethnic groups. With its focus on genetic variations among individuals, pharmacogenomics illustrates several challenges regarding the integration of the role of genetic diversity in drug response with ongoing debates over the role of ethnicity in research and medical practice. A gap exists between the decreasing relevance of race as an explanation for physical differences, and the role that racial identity plays in matters pertaining to social equality. The role of diversity in clinical drug trials and the equitable distribution among ethnic groups of drugs produced through pharmacogenomics are only a few examples. From drug manufacturers to health service providers, all bear a responsibility to balance the competing interests, if the knowledge gained through pharmacogenomics is to be put to the greatest possible advantage.

Pharmacogenomics is the name given to the study and development of genomic approaches to drug discovery and response. Similarly, pharmacogenetics is the study of the role of inherited genetic variations in drug response. Pharmacogenomics expands beyond pharmacogenetics to include genomic variations in drug target genes and gene expression differences in health and disease states. The objective of pharmacogenomics is the definition of the pharmacological significance of genetic variations among individuals and the use of this information in drug discovery and development. The application of genomic tools and analysis strategies to therapeutics could ultimately decrease the incidence of adverse drug reactions by tailoring certain drugs to specific genetic forms of a disease. By understanding which genetic factors are particularly relevant to the success or failure of a particular drug therapy, researchers and manufacturers hope to provide better drugs designed to benefit persons of a particular genotype [1]. Increasingly, a diagnostic test may be available to determine the correct pharmaceutical intervention and even the optimum dosage. New technologies such as high-throughput screening and computational biology have facilitated the rapid genotyping of many genomes, which is necessary to achieve the comprehensive representation of diverse groupings of human populations and the different relevant polymorphisms [2].

Advances in pharmacogenetics demonstrate the suggestion that, from the biomedical perspective, genetic research is valuable and important for what it reveals about human variation rather than human sameness [3]. Many of the ethical implications of pharmacogenomics are shared across populations. Pharmacogenomics presents several ethical concerns, however, some of which are related to the fact that certain polymorphisms of pharmacological significance may disproportionately affect certain ethnic groups. Future strategies for disease treatment and drug therapy in clinical medicine will depend not on such imprecise indicators as race or ethnicity [4], but on the individual patient's genotype. The idea, then, is not to eradicate or ignore differences but to redefine or move beyond race to more precise categories of difference with justification for establishing such differences. Newer, more precise categories will not be immune to such social forces as discrimination and stigma. Rather, pharmacogenetics is but one segment of genetic research that promises to change the traditional concept of race as a meaningful biological indicator.

This chapter outlines the implications of pharmacogenomics particular to the concepts of race and ethnicity in medical research. The first section is a brief discussion of pharmacogenomics and population genetics. It includes examples that highlight some inter-ethnic differences in the frequencies of certain phenotypes. The second section explores areas of concern presented by pharmacogenomics to ethnic groups, not by virtue of shared phenotype but by virtue of shared social and political histories. The impact of pharmacogenomics on requirements of diversity in research trials is one example. The ability to fragment the market for pharmaceuticals by genotype and the likelihood that such pharmaceuticals will be expensive could exacerbate current shortages of pharmaceuticals in low-income markets comprised disproportionately of persons of a shared ethnicity.

25.1 Pharmacogenomics and Population Genetics

Current drug therapy is based on the knowledge that individuals vary in how they respond to the same dose of a single drug. Physicians currently employ a trial-and-error stance in monitoring the effectiveness of drug therapy. Pharmacogenomics is the study of how inherited genetic variations affect an individual's ability to respond to a drug and the use of that knowledge in drug discovery and development. The presence of certain patterns in genetic composition can explain why one person may benefit from a drug while another may suffer toxic effects from the same dosage of the same drug. Researchers have identified genetic variations or genetic polymorphisms in alleles of a gene that produce either a higher or a lower expression of proteins, which is associated with drug absorption, or drug-resistant malignancies. Computational biology tools and improvements in molecular biology will benefit the way researchers predict gene function and pharmacology [5].

Genes involved in drug metabolism encode enzymes or receptor proteins that dictate drug response. Variations in gene structure, primarily single nucleotide polymorphisms (SNPs), modify the function of a protein and yield phenotypic differences in response to a particular drug [6]. The presence of a SNP can affect drug response by decreasing catalytic activity of the protein product, increasing activity of the enzyme, or affecting the expression of the gene product. Only a small percentage of the approximately 1.42 million SNPs thus far documented in the literature have an impact on drug response. The challenge to researchers in pharmacogenomics is to identify those SNPs that are closely associated with drug response.

Of the SNPs involved in drug response, a large number vary in the frequency of occurrence in different ethnic groups. For example, cytochrome P4502D6, an enzyme also known as debrisoquine hydroxylase, is central to the metabolism of at least 25 therapeutic drugs. Genetic polymorphism of the CYP2D6 gene may lead to reduced production of an active protein. Persons with this polymorphism are considered to have a poor metabolizer phenotype. Conversely, those with an ultrarapid metabolizer phenotype possess a polymorphism that prompts increased production of active protein. The poor-metabolizer phenotype has a frequency of 5–10% in North American and European white populations, but only 1–2% in African Americans, native Thailanders, Chinese, and native Malay populations [7]. The poor-metabolizer phenotype is hardly present among native Japanese populations. Similarly, the frequency of the ultra-metabolizer phenotype varies from 5-10% among Caucasian Americans to 29% among Ethiopians.

Deficiency of thiopurine S-methyl transferase (TPMT) is another phenotype that exhibits inter-ethnic differences in frequency. TPMT is an enzyme that catalyzes methylation of therapeutic agents used in the treatment of acute lymphoblastic leukemia, rheumatoid arthritis, and autoimmune/inflammatory diseases, as well as in organ transplantation. Patients who have TPMT deficiency experience less efficient methylation and are at greater risk of fatal toxicity when treated with standard doses of thiopurines. TPMT phenotype is defined by erythrocyte 6-mercaptopurine methylation. African American populations exhibit a 20% lower erythrocyte TPMT than Caucasian Americans, and persons of Chinese descent tend to exhibit greater activity than either of these other American subpopulations.

Finally, polymorphisms associated with arylamine N-acetyltransferase (NAT2) may result in slow acetylators. The slow-acetylator phenotype is present in 50–70% of the population in Western countries and is associated with several drug-induced side effects. The frequency of the slow-acetylator phenotype rises to 80% in Egyptian and certain Jewish populations; however, the frequency drops to 10% or 20% among Japanese and Canadian Eskimos.

Inter-ethnic differences in the frequency of phenotypes are relevant because the full development of pharmacogenetics in drug discovery and development will depend upon the application of its principles on a genome-wide basis [8]. The two completed drafts of the human genome represented the genome of a "composite" human [9, 10]. The collective value of these drafts is beyond question, but efforts to integrate inherited genetic difference in drug response require a broader scope of genetic diversity. The validation of candidate pharmacogenetic target genes will rely partly on population genetics [11, 12]. Timely identification of candidates for pharmacogenetic target genes will require the genotyping of clinically relevant populations for the set of relevant alleles and the implementation of genomic technologies – high-throughput screening, for example – to establish linkage between alleles and selected phenotypes. Documentation of the frequency of occurrence of a significant SNP is necessary to this process.

The differences in the frequency of certain pharmacogenetic phenotypes are significant for another reason. The preceding examples demonstrate the existence of pharmacologically meaningful differences among individuals that correlate with traditional ethnic identifiers. The differences are the result of patterns of human migration and reproduction, a full analysis of which lies beyond the scope of this chapter. Nevertheless, inter-ethnic differences in phenotypic frequencies provide a basis for the discussion of questions regarding pharmacogenomics and its social implications.

25.2 Pharmacogenomics and Ethnic Groups

25.2.1 Stigma and the Significance of Difference

One of the challenges presented by inter-ethnic differences in the frequency of certain phenotypes is the use or communication of this knowledge in a way that does not contribute to the tradition of ascribing social value to biological differences described as racial characteristics. Much racism has rested upon the concept of classifying persons by physical traits for the purpose of allocating resources and assigning social privilege. The history of eugenics offers many examples of how the scientific method can be applied to efforts to sort persons by physical traits into categories receiving varying qualities of treatment in society. Recent decades are remarkable for a trend away from according significance to race. Yet the trend toward assigning respect to persons without regard to race is accompanied by demands to respect the differences in social histories that racism has produced. In some instances, a history of racism has shaped the development of the environment in which bloodlines and genetic lineage continued by influencing patterns of behaviors and diet. Contemporary efforts to abandon race altogether are met by accusations of disrespect toward differences shared by persons of a particular ethnicity.

Pursuits in pharmacogenomics illustrate the challenges in the continuing evolution of race as used within medical and other social discourse. To suggest that a drug may be suited for one person but not another is not inherently dangerous. To suggest race as a basis for suitability, rather than diversity within the human genome, is scientifically questionable and socially divisive. The danger lies in the temptation to collapse the two and validate race as a basis for further discrimination in the context of medical care and other contexts of social interaction. Past and present racism is cited as a partial explanation for disparities in health outcomes in persons of different ethnicities [13]. The question is whether and when race can be used as a reliable proxy for genetic characteristics that dictate differences in treatment by physicians and other medical researchers. Pharmacogenomics may well contribute an answer and, in so doing, prompt an analysis of the role of traditional racial classifications in medical research and practice.

In designing and conducting pharmacogenetics studies, researchers must take care in the selection of subject populations. In culling a set of persons which best represents the diversity of an area of the genome, the researcher must consider several questions:

- What do patterns of observed health outcomes or even patterns of human migration suggest for sources of genetic diversity?
- How might geographic origin be a factor in achieving genetic diversity within a sample of subjects?
- What are the ultimate limits of geographic origin as a factor, particularly where mobility is both possible and common?
- If ethnicity is determined to be a factor, how is ethnicity defined and determined? [14]

The language used in research studies suggests that scientists still grapple with labels to describe ethnic groups in a way that does not misinform or cause undue generalization of results. What is a Caucasian? What does it mean to be Asian? Research in pharmacogenomics will drive a re-examination of these labels in search of more meaningful descriptive terms.

25.2.2 Clinical Drug Trials

The recruitment and selection of subjects or participants in clinical drug trials highlights an area of conflict regarding the role of race in medical research. Pharmacogenomics promises to streamline the clinical trial phase of drug development [15]. The ability to predict drug efficacy by genotyping participants during the early stages of clinical trials for a drug would enable researchers to recruit for later trials only those patients who, according to their genotype, are likely to benefit from the drug. Smaller clinical trials will produce a greater quantity and quality of information, but on a smaller segment of the population. A drug could reach the market more quickly, but with less information about the side effects or risk of harm to individuals with genotypes not represented in the clinical trials. A related issue is the concern that the product may be prescribed for "off-label" uses by individuals with genotypes not appropriate for the drug. The effects of pharmacogenomics on the current model of clinical drug trials holds interesting implications for current policies governing the use of human subjects in research. Wherever such policies exist in part to recognize the vulnerabilities of certain ethnic groups and protect against exclusion or exploitation based on ethnicity, science may have to accommodate politics in new ways.

Regulations governing the use of human subjects in federally funded research projects within the United States include a statement encouraging racial diversity in the recruitment of subjects. Generally, researchers are required to include women and members of certain ethnic groups in research projects involving human subjects absent a compelling rationale to the contrary [16, 17]. On one level, the policy of inclusiveness reflects a recognition of the need to abandon the use of homogeneous subject groups where sameness limits the application of the results of the research. On another level, the policy reflects the continuing power of race as an inexact proxy for biological homogeneity. The main value of racial diversity in clinical trials, however, is to ensure the inclusion of individuals from social and ethnic groups that were once systematically excluded based on visible physical traits and politics.

The government policy of inclusiveness in clinical research is not intended to institutionalize a belief that races are necessarily biologically distinct; it demonstrates a respect for the potential value in revealing and accommodating what differences may arise within the context of medical research. Guidelines for the selection of participants in clinical drug trials must allay the suspicions of groups that have been neglected or singled out for injurious treatment in the past. In the United States, many of the infamous instances of such mistreatment have been based on race or perceptions of race. Perhaps the legacy of racism is inextricable from the political environment in which research and scientific inquiry must operate. If the reliance upon racial categories retains some value for public education about legitimate health risks or the successful diagnosis and treatment of disease, the reliance must be shifted to a more accurate and effective vocabulary for communication of complex concepts to avoid the danger of perpetuating harmful and inaccurate stereotypes.

Within the context of pharmacogenomics, clinical trials constructed around a particular polymorphism should not conflict with inclusiveness guidelines as a matter of course. Though the sub-population frequencies of any polymorphism vary across different ethnic groups, the frequency rarely drops to zero percent for any one ethnic group. For a given polymorphism, at least someone will possess the variation in each ethnic group. The end result may be greater difficulty in recruiting sufficient participants who not only share a particular SNP but also represent together the ethnic diversity of the national population in sufficient number. The effect of higher standards for inclusion in clinical drug trials may well add to the cost in time and money needed to steward a drug through the approval process.

25.2.3 Clinical Integration of Pharmacogenomic Medicine

As pharmacogenomic-based drugs increase in prevalence over the next several years, the use of genotyping or genetic testing as a diagnostic tool and the prescription of medications based on genotypic information will become the standard of care. Physicians and pharmacists will be charged with sufficient knowledge of genetics to adequately interpret diagnostic tests and prescribe appropriate drug therapies in proper dosages.

Reports of racial disparities in medical care and health status in the United States are common. The United States government has targeted the elimination of racial disparities in health status as a national priority in health policy [18]. Examples of racial disparities include differences in the aggressiveness of treatment ordered for white patients and black patients with cancer [19] and heart disease [20]. One study documented the paucity of opioid pharmaceuticals in low-income neighborhoods consisting primarily of African Americans and Hispanic Americans as evidence of barriers to adequate pain management among populations [21]. The methodologies in some studies documenting racial disparities have been criticized for their lack of rigor [22]. Flaws in data collection, however, do not entirely explain the conclusions [23]. To the extent that health disparities can be explained by lack of access to care or bias among physicians in pursuing aggressive treatment, pharmacogenomics raises questions regarding the extent to which its benefits will be enjoyed across ethnic groups.

Similarly, where the elimination of racial disparities in health status may be linked with the availability of therapies best suited for persons with a genotypic characteristic that occurs with greater frequency in a particular ethnic group, the question is whether the benefit of designing drugs tailored to common genotypes outweighs the danger of perpetuating partially misleading perceptions of biological heterogeneity between racial classifications and homogeneity within racial classifications.

There is additional evidence that race is a factor in drug development strategies. The development of drugs for one particular racial group over another may become more common as drug companies respond to the existence of relevant polymorphisms that may be more common to what may be more accurately described as an ethnic group. The question is whether a drug proven to serve a need particularly pressing to a certain population should be marketed to that group when the group identifier, race, is politically charged and arguably irrelevant as a predictor of beneficial use. If many in an ethnic group share an increased risk for susceptibility to a disease or resistance to conventional drug therapy for a disease, which is more harmful to society: the pretense that, increased risk notwithstanding, a racial group shares so little in common genetically that the label should be ignored, or the perpetuation of race as an inaccurate but still effective shorthand for communicating information about disease susceptibility or a danger of toxic effects from a drug? Even where the relevant genetic variation is attributable to historical patterns of human migration and is not exclusively or even most accurately defined or predicted by skin color, race is likely to remain as a stand-in, however sloppy, until completely demonstrated to be completely irrelevant as a more accurate nomenclature is developed.

25.2.4 Cost

The differences in the frequencies of certain polymorphisms are relevant because of the potential to fragment markets for pharmaceuticals. Just as pharmacogenomics will enable physicians to tailor drug therapy to suit an individual, so will it enable drug companies to focus efforts on developing drugs to suit the most prevalent genotype. Whether the most prevalent genotype also stands to become the most profitable genotype remains to be seen. The high price of pharmaceuticals is a source of continuing international concern. The substantial resources devoted to the realization of pharmacogenomics through diagnostic tests and relatively specialized drugs are not likely to result in cheaper products. Public financial incentives, like those provided under the Orphan Drug Act, may be necessary to ensure that drug companies invest time and money in the development of pharmacogenomic products for relatively rare genotypes [24]. It makes sense economically for a drug company, before investing millions of dollars to bring to market a drug targeted to a particular genotype, to engage in demographic research to determine whether populations most likely to have such a genotype will be able to purchase the drug.

Alternatively, the current shortage of available therapies in impoverished nations may be exacerbated. Consider the shortage of drug therapies for encephalitis in African countries. By one account, the reluctance to manufacture the drug in sufficient quantities is a function of the poverty in the affected areas, the high cost of drug production, and the absence of a profitable market. Where an impoverished community is identifiable in part by the high occurrence of a polymorphism that is not shared by persons in more profitable markets, drug companies may need incentives to develop effective therapies accessible to impoverished communities.

25.3 The Appropriate Use of Race and Ethnicity

Advancements in pharmacogenomics and population genomics will effect changes to the traditional models of research, drug development, and the practice of medicine. Specifically, the emergence of the gene and the effects of its variations on drug response promise to alter the tradition of race and ethnicity in medical research. Because certain patterns of genetic variations differ in frequency among different ethnic groups, ethnicity and race may retain some value. The challenge is to determine the limits of that value so as not to misinform by perpetuating inaccurate stereotypes.

To avoid the stigma and discrimination that may result from misinformation, one must avoid the casual use of race or ethnicity in research and design. Where ethnicity is relevant to a particular area of research, care must be taken to define and justify the use of race and ethnicity for the purpose of that research. As general public knowledge of the role of the gene in biology and in the practice of medicine increases, the definitions of race and ethnicity are likely to evolve to suit the proper, rather than the superficial, use of such concepts in social discourse.

25.4 Conclusions

Pharmacogenomics presents an opportunity to contribute to discourse regarding race and its proper role, if any, in the research and practice settings. The value of race as a predictor of disease and response to, or compliance with, drug therapy is an area of considerable debate. The varying frequency of certain genetic variations relevant to drug response among ethnic groups offers a perspective that may inform the debate. If nothing else, the uneven distribution of certain polymorphisms should elevate the level of discourse by emphasizing genetic differences over racial differences.

The uneven frequency of relevant polymorphisms does not necessarily suggest that race is a valuable predictor of drug response. Of the polymorphisms relevant to drug response, each one is present in every ethnic group if only rarely. Moreover, social interaction among ethnic groups and the pitfalls of self-identification or "eyeballing" suggest that race is a flawed proxy for genetic make-up.

Nevertheless, the uneven distribution of genetic variations among ethnic groups is further evidence of the diversity within the human genome and such diversity must be considered if knowledge regarding the genome is to be used to full advantage. There may be dissonance between the need to consider genetic diversity and political pressure to acknowledge diversity of ethnic groups as units of society. In the context of drug development and manufacture, for example, care to ensure diversity among human subjects in research may be dictated by politics rather than science. This is appropriate. In some contexts, as in efforts to eliminate health disparities, the end of improving health status is better served by first acknowledging the differences in the treatment and outcome of disease among different ethnic groups. With particular regard to traditions of race, science is never entirely divorced from social values and politics. The challenge is to balance the opportunity to abandon discourse that is not helpful in advancing understanding about difference with the need to acknowledge the social realities that are the legacy of the very same discourse. The issues raised by pharmacogenomics illustrate how the understanding of drug response as partially a function of genes rather than race may not obliterate the need to consider the race or ethnicity of human subjects in research or of the market for the results of efforts to develop certain drugs.

25.5 References

1 Omenn GS. Prospects for Pharmacogenetics and Ecogenetics in the New Millennium. Drug Metabolism & Disposition **2001**; 29:611–614.

2 Shi MM et al. Pharmacogenetic Application in Drug Development and Clinical Trials. Drug Metabolism & Disposition **2001**; 29:591–595.

3 Olson S. Genetic Archaeology of Race. Atl Monthly **2001**; 287:69–80.

4 Lin SS, Kelsey JL. Use of Race and Ethnicity in Epidemiologic Research: Concepts, Methodological Issues, and Suggestions for Research. Epidemiol Rev **2001**; 22:187–202.

5 Hanke J. Genomics and New Technologies as Catalysts for Change in the Drug Discovery Paradigm. Whitehead Policy Symposium **2000**.

6 Linder MW, Valdes R Jr. Fundamentals and Applications of Pharmacogenetics for the Clinical Laboratory. Ann Clin Lab Med Sci **2001**; 29:611–614.

7 Linder MW, Valdes R Jr. Pharmacogenetics in the Practice of Laboratory Medicine. Mol Diagnosis **1999**; 4:365–379.

8 Shi MM et al. Pharmacogenetic Application in Drug Development and Clinical Trials. Drug Metabolism & Disposition **2001**; 29:591–595.

9 International Human Genome Sequencing Consortium. Initial sequencing and analysis of the human genome. Nature **2001**; 409:860–921.

10 Venter JC et al. The Sequence of the Human Genome. Science **2001**; 291:1304–1351.

11 Harrington JJ et al. Creation of Genome-Wide Protein Expression Libraries Using Random Activation of Gene Expression. Nature Biotechnol **2001**; 19:440–445.

12 Lin SS, Kelsey JL. Use of Race and Ethnicity in Epidemiologic Research: Concepts, Methodological Issues, and Suggestions for Research. Epidemiol Rev **2001**; 22:187–202.

13 United States Dept. of Health & Human Serv. The initiative to teliminate racial and ethnic disparities in health (*http://raceandhealth.hhs.gov*).

14 Lin SS, Kelsey JL. Use of Race and Ethnicity in Epidemiologic Research: Concepts, Methodological Issues, and Suggestions for Research. Epidemiol Rev **2001**; 22:187–202.

15 Shi MM et al. Pharmacogenetic Application in Drug Development and Clinical Trials. Drug Metabolism & Disposition **2001**; 29:591–595.

16 National Institutes of Health. Guidelines on the Inclusion of Women and Minorities as Subjects in Clinical Research. 59 Fed. Reg. 14508 (March 28, 1994).

17 Centers for Disease Control. Policy on the Inclusion of Women and Racial and Ethnic Minorities in Externally Awarded Research. 60 Fed. Reg. 47947 (Sep 15, 1995).

18 United States Department of Health & Human Services. The initiative to eliminate racial and ethnic disparities in health (*http://raceandhealth.hhs.gov*).

19 Jun G et al. Ethnic differences in poly CADP-ribose) polymerase pseudogene genotype distribution association with lung cancer risk. Carcinogenesis **1999**; 20:1465–1469.

20 Chen J et al. Racial differences in the use of cardiac catheterization after acute myocardial infarction. N Eng J Med **2001**; 344:1443–1999.

21 JAMA Apr. **2001**

22 Satel S. The indoctrinologists are coming. Atl Monthly **2001**; 287:59–64.

23 Wood AJJ. Racial differences in the response to drugs – pointers to genetic differences. N Eng J Med **2001**; 344:1393–1396.

24 Thamer M et al. A Cross-National Comparison of Orphan Drug Policies: Implications for the U.S. Orphan Drug Act. J Health Politics Policy Law **1998**; 23:265–290.

26
Translation of Vascular Proteomics into Individualized Therapeutics

Renata Pasqualini and Wadih Arap

Abstract

Despite major advances from the Human Genome Project, the molecular diversity of receptors in the human vasculature is still largely unknown. We, among others, have developed a methodology to allow cell-free, *in vitro* and *in vivo* selection of libraries of random peptides to identify ligands that home to specific vascular beds *in vivo*. These strategies revealed a vascular address system that allows targeting of tissue-specific and angiogenesis-related receptors expressed in blood vessels. Targeted delivery of cytotoxics, gene therapy vectors, imaging agents, or proteins to specific receptors has been accomplished in animal models. We are now working towards the definition of a ligand/receptor-based map of human vasculature. High-throughput translation of these technologies into clinical targeting applications may form the basis for the development of a personalized vascular pharmacology.

26.1
Introduction

A key issue in individualizing therapeutics is targeting delivery of drug to specific sites. The diversity of the vascular endothelium may provide an important tool to facilitate site-specific drug delivery. In that regard, the work of our group and others has been aimed at defining tissue-specific and/or angiogenesis-related markers in the vasculature and using them for targeted therapeutics. We define vascular proteomics or "angiomics" as the molecular phenotyping of cells forming blood vessels at the protein-protein interaction level. The translation of the molecular diversity of cell surface receptors expressed in the endothelium may lead to a receptor-based targeting map of the human vasculature [1].

We are developing integrated, combinatorial library-based platform technologies whose goal is to enable the identification, validation, and prioritization of molecular targets in blood vessels. This methodology will allow drug development based on targeting the differential protein expression in the vasculature associated with normal tissues or diseases with an angiogenesis component; these include cancer, arthritis, diabetes, and cardiovascular diseases. Our long-term goal is to create a functional map of molecular targets and biomarkers for imaging and therapeutic applications.

Here we review several approaches used to screen combinatorial libraries of peptides and antibodies to identify ligands that target specific tissues or sites of disease. Validated ligands may be used for targeting diagnostic and/or therapeutic agents. Moreover, the ligands themselves may be used as either drug discovery leads or for therapeutic modulation of their corresponding receptor(s). Finally, another application of the selected targeted ligands is identification of their vascular receptors.

26.2 Cell-Free Screening on Isolated Vascular Ligands or Receptors

Phage display library systems involve the manipulation of the bacteriophage genome in order to express all possible permutations of a short peptide or a large collection of antibodies [2]. Improvements in phage display random peptide libraries screening methodology have led to significant advances in the elucidation of ligand-receptor binding sites and antibody epitope mapping. Ligands that interact with functionally relevant sites within a given receptor can be selected and isolated by “biopanning”, a process in which phages expressing ligands with specific properties are eluted and amplified in a host bacteria. Phage display random peptide libraries were designed to define binding sites of antibodies [2]. Later, many ligands for isolated receptors (including proteases, adhesion molecules, proteoglycans, signaling molecules, among others) were found by this technology. As examples we highlight cell-free panning on isolated receptors: the targeting of gelatinases in the endothelial cell surface.

We have previously described the isolation of specific gelatinase inhibitors from phage display peptide libraries [3]. Such inhibitors belong to a class of cyclic peptides that inhibit activity of matrix metalloproteases (MMP) in a selective manner. Studies aimed at determining the biological properties of such peptides have demonstrated that they suppress cell migration, interfere with MMP function on both tumor cells and endothelial cells *in vitro*, home to tumor vasculature *in vivo*, and prevent the growth and invasion of several tumor types in mice. Peptides displaying the motif CXXHWGFXXC (isolated by phage display on the active form of gelatinases *in vitro*) inhibit only this subclass of metalloproteases, and have minimal or no effect on several other metalloprotease family members. A line of work from our group focuses on the development of HWGF-derived peptidomimetics. Such compounds show promise as potential anticancer leads because they display two levels of specificity. First, they selectively inhibit gelatinases; second, they specifically target tumors because these enzymes are overexpressed in tumor cells and tumor vasculature, and third, they target activated gelatinases that are accessible to circulating ligands administered intravenously, thus representing suitable targets for probes homing to tumor vasculature. Treatment of tumor-bearing mice with a prototype peptide (sequence CTTHWGFTLC) results in delayed tumor growth and, ultimately, in increased survival of tumor-bearing mice, based on the combination of tumor-targeting, antiangiogenic, and anti-invasive properties of this peptide class [3].

26.3 Targeting the Molecular Diversity of Endothelium-Derived Cells

Screening the molecular heterogeneity of receptor expression in endothelial cell surfaces is required for the development of vascular-targeted therapies. First, as opposed to targeting purified proteins as discussed above, membrane-bound receptors are more likely to preserve their functional conformation, which can be lost upon purification and immobilization outside the context of intact cells. Moreover, many cell surface receptors require the cell membrane microenvironment to function so that protein-protein interaction may occur. Finally, combinatorial approaches may allow the selection of cell membrane ligands in a functional assay and without any bias about the cellular surface receptor. Therefore, even as yet unidentified receptors may be targeted.

We have recently developed a novel approach for the screening of cell surface-binding peptides from phage libraries. Biopanning & Rapid Analysis of Selective Interactive Ligands (termed BRASIL) is based on differential centrifugation in which a cell suspension incubated with phage in an aqueous upper phase is centrifuged through a non-miscible organic lower phase [4]. This single-step organic phase separation is faster, more sensitive and more specific than current methods that rely on washing steps or limiting dilution. As a proof-of-concept, we screened human endothelial cells stimulated with vascular endothelial growth factor (VEGF), constructed a peptide-based ligand-receptor map of that VEGFR family, and validated a new targeting ligand [4]. Mapping ligand-receptor interactions by BRASIL may allow an understanding of binding requirements for other endothelial cell surface receptor families and enable isolation of a panel of peptides for endothelium-derived cell targeting applications. The method may also be used in tandem with fine needle aspirates of solid tumors or fluorescence-activated cell sorting of circulating cells obtained directly from patients or clinical samples. We, therefore, expect that BRASIL will prove to be a superior method for probing target cell surfaces with several potential applications.

26.4 *In vivo* Vascular Targeting in Animal Models

We have previously developed an *in vivo* selection method in which peptides that home to specific vascular beds are selected after intravenous administration of a phage display random peptide library [5]. This strategy revealed a vascular address system that allows tissue-specific targeting of normal blood vessels [6–8] and angiogenesis-related targeting of tumor blood vessels [3, 6, 9–12]. While the biological basis for such vascular diversity is still largely unknown, a number of peptides selected by homing to blood vessels in animal models have been used as carriers to guide the delivery of chemotherapeutics [11], recombinant peptides [9], metalloprotease inhibitors [3], cytokines [13], fluorophores [14], and genes [15]. Moreover, vascular receptors corresponding to the selected peptides have been identified in blood vessels of normal organs [16] and in tumor blood vessels [17].

26.5 Vasculature-Targeted Cytotoxic Agents

It is well recognized that tumor growth and metastasis depend on angiogenesis. We have used peptides that home to tumor blood vessels to target the cytotoxic drug doxorubicin [11] and pro-apoptotic peptides [9] to the vasculature of xenografts in mouse models. The endothelial cell surface receptors targeted included vascular integrins [12] and aminopeptidases [17]. Other receptor candidates for vascular targeting include gelatinases [3] and proteoglycans [10]. Other investigators have used the same strategy and vascular ligand-receptor systems to target tumor necrosis factor (TNF) to tumor blood vessels [13]. In both cases, similar results were observed with coupling to homing peptides resulting in targeted compounds that are more effective but less toxic than the parental compound [9, 11, 13]. Taken together, these data suggest that it may be possible to develop therapeutic strategies based on selective expression of vascular receptors [1]. Well-designed clinical trials must determine the final value of this approach in humans.

26.6 *In vivo* Vascular Targeting in Human Subjects

As discussed above, ligands and receptors isolated in animal models have been useful to identify putative human homologs. However, it is unlikely that targeted delivery will always be achieved in humans through such approach. Data from the Mouse and Human Genome Projects indicate that the higher complexity of the human species relative to other mammalian species derives from expression patterns of proteins at different tissue sites, levels, or times rather than from a greater number of genes [18, 19]. Indeed, striking examples of species-specific differences in gene expression within the human vascular network have recently surfaced [20]. Such differences in protein expression patterns and ligand-receptor accessibility caution that vascular proteomics results obtained in animal models must be carefully evaluated before extrapolation to human studies. Therefore, selection of phage display random peptide libraries in humans may reduce costly late-stage clinical trial failures by shifting decisions to earlier stages of the drug development process.

Given this rationale, we have reasoned it would be possible, and have recently reported the direct mapping of the human vasculature by *in vivo* phage display in a patient [20]. This large-scale survey of motifs that localized to different organs showed that the tissue distribution of circulating peptides is non-random. Moreover, a high-throughput analysis of the motifs revealed similarities to ligands for differentially expressed cell surface proteins. Finally, we validated a candidate ligand-receptor pair in the vasculature of the human prostate [20]. This methodology represents a major step towards the ultimate goal of outlining a molecular map of the human vasculature. If successful, a completed receptor-based map will have broad implications for the development of a new vascular-targeted pharmacology.

26.7 Future Directions

A major goal in drug development has long been to develop a technology for targeting therapeutics more effectively to their intended disease site and to improve their therapeutic index by limiting the systemic exposure of other tissues to untoward or toxic effects. The methods described here have two main applications. First, they may identify vascular targeting ligands. Second, they may enable the construction of a molecular map of human vascular receptors.

In theory, targeted delivery of drugs, liposomes, peptide sequences, gene therapy vectors, and biological therapies can be achieved in clinical applications. Ultimately, it may be possible to guide imaging or therapeutic compounds to the target site in real clinical situations. Similarly, ligands that are targeted to a specific vascular bed or specific disease site may themselves have potential as therapeutics. In the future, the determination of molecular profiles of blood vessels in specific conditions may also lead to vascular targets. Early identification of targets, optimized regimens tailored to the molecular profile of individual patients, and identification of new vascular addresses may lead to revisiting or even salvaging of ineffective or toxic drug candidates.

26.8 Acknowledgements and Disclosures

Funded in part by grants from NIH (CA90270 and CA8297601 to R.P., CA90270 and CA9081001 to W.A.) and awards from the Gilson-Longenbaugh Foundation. The V Foundation and CaP CURE (to R.P. and W.A.). The University of Texas M.D. Anderson Cancer Center has licensed the technology to NTTX Biotechnology, Inc., a biotechnology company that will develop the technology and information derived from the studies described here into clinical applications. Drs. Arap and Pasqualini and the University of Texas have equity in NTTX Biotechnology that is subjected to certain restrictions under university policy. The University of Texas manages the terms of these agreements in accordance with its conflict-of-interest policies.

26.9 References

1 Kolonin MG, Pasqualini R, Arap W. Molecular addresses in blood vessels as targets for therapy. Curr Opin Chem Biol **2001**; 5:308–313.

2 Smith GP, Scott JK. Libraries of peptides and proteins displayed on filamentous phage. Methods Enzymol **1993**; 217:228–257.

3 Koivunen E, Arap W, Valtanen H et al. Tumor targeting with a selective gelatinase inhibitor. Nature Biotechnol **1999**; 17:768–774.

4 Giordano RJ, Cardó-Vila M, Lahdenranta J et al. Biopanning and rapid analysis of selective interactive ligands. Nature Med **2001**; 7:1249–1253.

5 Pasqualini R, Arap W, Rajotte D et al. *In vivo* selection of phage display libraries. In: Phage Display: A Laboratory Manual (Barbas CF III, Burton DR, Scott JK, Silverman GJ, Eds.). New York: Cold Spring Harbor Laboratory Press; **2000**, 22.1–22.24.

6 Arap W, Haedicke W, Bernasconi M et al. Targeting the prostate for destruction through a vascular address. Proc Natl Acad Sci USA **2002**; 99:1527–1531.

7 Rajotte D, Arap W, Hagedorn M et al. Molecular heterogeneity of the vascular endothelium revealed by *in vivo* phage display. J Clin Invest **1998**; 102:430–437.

8 Pasqualini R, Ruoslahti E. Organ targeting *in vivo* using phage display peptide libraries. Nature **1996**; 380:364–366.

9 Ellerby HM, Arap W, Ellerby LM et al. Anti-cancer activity of targeted pro-apoptotic peptides. Nature Med **1999**; 5:1032–1038.

10 Burg MA, Pasqualini R, Arap W et al. NG2 proteoglycan-binding peptides target tumor neovasculature. Cancer Res **1999**; 59:2869–2874.

11 Arap W, Pasqualini R, Ruoslahti E. Cancer treatment by targeted drug delivery to tumor vasculature in a mouse model. Science **1998**; 279:377–380.

12 Pasqualini R, Koivunen E, Ruoslahti E. Alpha v integrins as receptors for tumor targeting by circulating ligands. Nature Biotechnol **1997**; 15:542–546.

13 Curnis F, Sacchi A, Borgna L et al. Enhancement of tumor necrosis factor alpha antitumor immunotherapeutic properties by targeted delivery to aminopeptidase N (CD13). Nature Biotechnol **2000**; 18:1185–1190.

14 Hong FD, Clayman GL. Isolation of a peptide for targeted drug delivery into human head and neck solid tumors. Cancer Res **2000**; 60:6551–6556.

15 Trepel M, Grifman M, Weitzman MD et al. Molecular adaptors for vascular-targeted adenoviral gene delivery. Hum Gene Ther **2000**; 11:1971–1981.

16 Rajotte D, Ruoslahti E. Membrane dipeptidase is the receptor for a lung-targeting peptide identified by *in vivo* phage display. J Biol Chem **1999**; 274:11593–11598.

17 Pasqualini R, Arap W, Koivunen E et al. Aminopeptidase N is a receptor for tumor-homing peptides and a target for inhibiting angiogenesis. Cancer Res **2000**; 60:722–727.

18 Lander ES, Linton LM, Birren B et al. Initial sequencing and analysis of the human genome. Nature **2001**; 409:860–921.

19 Venter JC, Adams MD, Myers EW et al. The sequence of the human genome. Science **2001**; 291:1304–1351.

20 Arap W, Kolonin MG, Trepel M et al. Steps toward mapping the human vasculature by phage display. Nature Med **2002**; 8:121–127.

27
Glossary of Key Terms in Molecular Genetics and Pharmacogenomics

Bernhard R. Winkelmann, Michael M. Hoffmann, Markus Nauck
and Winfried März

Abstract

The proliferation in genomic technology and genomic data in recent years has been dramatic. Research in life sciences depends increasingly on genomic approaches. Thus, the use of genomic terms is widespread also in mainstream scientific literature. This glossary of genomic terms should help to clarify the meaning of such terms and is intended to be used as a handy reference.

Note

The collection of terms is partly based on the glossary of molecular genetic terms found in the "Primer on Molecular Genetics" from the June 1992 DOE Human Genome 1991–92 Program Report (see *http://www.ornl.gov/TechResources/Human_Genome/glossary/*). The primer is intended to be an introduction to basic principles of molecular genetics pertaining to the genome project. A visit to this web site is highly recommended.

27.1
Alphabetical Listing of Genomic Terms

If an expression used in the definition of a key term is itself defined as a key term in this glossary such a *term* is highlighted in italics and will be listed at the corresponding position of the alphabetical A–Z listing.

Adenine (A): A nitrogenous purine base, which bonds with *thymine* (*T*) to form the A–T *base pair* in *DNA* and the A-U *base pair* in *RNA*.

Allele: One form of a genetic *locus,* distinguished from other forms (=alleles) by its particular *nucleotide* or its coded *amino acid* sequence. *Single nucleotide polymorphisms* are biallelic, i.e., one *base pair* is exchanged with just one other *base pair.* In such a case, only two different alleles exist in the population for such a *gene locus*. However, several different alleles may exist in the population if the *locus* is defined by more than one *nucleotide* (i.e., microsatellites, *haplotypes*). Thus, alleles are alternative forms of a genetic *locus* whereby each of the two alleles for one *locus* found in a *diploid* individual is inherited separately from the parents.

Amino acid: A group of 20 different small molecules which are linked together to form *proteins* (see also *genetic code*).

Autosome: A *chromosome* not involved in sex determination. The *diploid* human *genome* consists of 46 *chromosomes*, 22 pairs of autosomes, and one pair of sex *chromosomes* (X and Y *chromosomes*).

Base pair (bp): The four *nucleotides* in the *DNA* contain the bases: *adenine* (A), *guanine* (G), *cytosine* (C), and *thymine* (T). Two bases (*adenine* and *thymine* or *guanine* and *cytosine*) are held together by weak bonds to form base pairs. The two strands of human *DNA* are held together in the shape of a double helix by those bonds between *base pairs*. For example, the complementary nucleic acid base sequence to G-T-A-C that forms a double-stranded structure with the matching bases is C-A-T-G.

cDNA: See *complementary DNA*.

Centimorgan (cM): The *recombination* frequency is measured in centimorgans. One cM is equal to a 1% chance that a *marker* at one genetic *locus* will be separated from a *marker* at a second *locus* by *crossing over* (in a single generation). One cM is equivalent, on average, to 1 million *base pairs* in humans.

Chromosomes: Chromosomes are self-replicating genetic structures of cells containing *DNA* that bears in its linear *nucleotide* sequence the set of *genes* (i.e., the *genome*, the *genetic code*). *Prokaryotes* (i.e., bacteria) carry their *genome* in one circular chromosome. Eukaryotic *genomes* consist of a number of chromosomes (e.g., humans, n=46).

cM: See *centimorgan*.

Codon: See *genetic code*.

Complementary (coding) DNA (cDNA): *DNA* that is generated from a *messenger RNA* (*mRNA*) and contains only protein-coding DNA sequences.

Crossing over: The breaking of one maternal and one paternal *chromosome* during *meiosis* with exchange of corresponding sections of *DNA*, and rejoining of the *chromosomes*. The process leads to an exchange of *alleles* between *chromosomes* and is the fundamental principle of human evolution (see also *recombination*).

Cytosine (C): A nitrogenous pyrimidine base, which bonds with *guanine* (G) to form the G-C *base pair*.

Dalton: Atomic mass unit, one Dalton = mass of one hydrogen atom (3.32×10^{-24} g).

Diploid: The full set of paired *chromosomes* (one *chromosome* set from each parent). The diploid human *genome* has 46 *chromosomes* (see also *haploid*).

DNA (deoxyribonucleic acid): A double-stranded molecule held together by weak bonds between *base pairs* of *nucleotides* that encodes genetic information. The base sequence of each single strand can be deduced from that of its partner since *base pairs* form only between the bases *A* and *T* and between *G* and *C*.

DNA sequence: The relative linear order of *base pairs*, whether in a fragment of *DNA*, a *gene*, a *chromosome*, or an entire *genome*.

Domain: A discrete portion of a *protein* with its own function. The combination of domains in a single *protein* determines its overall function.

EST: Expressed sequence tags are *STSs* derived from *cDNAs* (see *sequence tagged site, STS*).

Eukaryote: Cell or organism with membrane-bound, structurally discrete nucleus and other well-developed subcellular compartments. Eukaryotes include all organisms except viruses, bacteria, and blue-green algae (see also *prokaryote*).

Exon: *DNA* sequence portion of a *gene* that codes for the *protein*. Human *genes* consist of several exons that are separated by *introns* (see also *intron*).

Gene: A segment of *DNA* containing all information for the regulated biosynthesis of an *RNA* product, including *promoters, exons, introns,* and other untranslated regions that control expression. Fundamental physical and functional unit of heredity and evolution. A gene is an ordered sequence of *nucleotides* located on a specific chromosomal *locus* that encodes particular products (i.e., *RNA* molecules, *proteins*) (see comment at *amino acids*).

Gene expression: Entire process that translates the information coded in a *gene* into *RNA* and *proteins*. Expressed *genes* are transcribed into *mRNA* and subsequently translated into *protein* or they remain as *RNA* (e.g., *transfer* and *ribosomal RNAs*).

Gene locus: See *locus*.

Gene product: The biochemical material, either *RNA* or *protein*, resulting from *gene expression*.

Genetic code: Sequence of *nucleotides* along the *DNA* and coded in triplets (*codons*) along the *mRNA* that determines the sequence of *amino acids* in *protein* synthesis. The *DNA* sequence of a *gene* can be used to predict the *mRNA* sequence, and subsequently to predict the *amino acid* sequence.

Genome: All genetic material in the *chromosomes* of a particular organism. The size of a genome is generally given as its total number of *base pairs.*

Genomic library: A collection of clones made from a set of randomly generated overlapping *DNA* fragments representing the entire *genome* of an organism.

Genotype: An unphased 5′ to 3′ sequence of *nucleotide* pair(s) found at one or more *polymorphic sites* in a *locus* on a pair of *homologous chromosomes* in an individual.

Guanine (G): A nitrogenous purine base, which bonds with *cytosine* (*C*) to form the G-C *base pair.*

Haploid: A single set of *chromosomes* (in humans the 23 *chromosomes* from either father or mother or the single set of *chromosomes* in their reproductive cells) (see also *diploid*).

Haplotype: A phased 5′ to 3′ sequence of *nucleotides* found at one or more *polymorphic sites* on a single *chromosome* from a single individual. In general genotyping methods determine the presence of individual *single nucleotide polymorphisms (SNPs)* in a *diploid* individual, but cannot distinguish which *chromosome* of a *diploid* pair is associated with each *SNP.* The haplotype of an individual, however, describes specific *alleles* defined by the number of *SNP loci* associated with each *chromosome.* A haplotype can be defined on the level of a *gene,* a region of the *chromosome* or any long *DNA* fragment. In case of n biallelic *SNPs,* the number of theoretically possible haplotypes is 2 to the power of n. However, due to *linkage* in humans, only a few of the theoretically possible haplotypes (in the range of 10 to 50) have empirically been identified in human *genes* [e.g., 18 biallelic *SNPs* could give rise to theoretically 2^{18} (=262,144) haplotypes].

Haplotype pair: The two *haplotypes* found for a *locus* in a single individual. For example, the simplest *haplotype* is determined by two biallelic SNPs (Aa, Bb). Four different *haplotypes* are theoretically possible on each of the two *chromosomes*: AB, Ab, aB, ab (2 to the power of 2). However, 10 different haplotype pairs may be found in *diploid* individuals: (1) AB/AB, (2) AB/Ab, (3) AB/aB, (4) AB/ab, (5) Ab/Ab, (6) Ab/aB, (7) Ab/ab, (8) aB/aB, (9) aB/ab, (10) ab/ab (discounting the origin of a *haplotype,* whether it originates from the father's or mother's *chromosome,* i.e., AB/Ab or Ab/AB count as one pair). Only two of the 10 haplotype pairs will represent *heterozygosity* at both *loci* (AB/ab and Ab/aB).

Heterozygosity: The presence of different *alleles* at one or more *loci* on *homologous chromosomes.*

Homologies: Similarities in *DNA* or *protein* sequences between individuals of the same species or among different species.

Homologous chromosomes: The pair of *chromosomes* with the homologous linear *gene* sequence derived from father and mother.

Hybridization: Process of joining two complementary strands of *DNA* or one each of *DNA* and *RNA* to form a double-stranded molecule.

***In situ* hybridization:** Use of a *DNA* or *RNA probe* to detect the presence of the *complementary DNA* sequence in cloned bacterial or eukaryotic cells.

Intron: *DNA* base sequence between *exons,* the *protein*-coding parts of a *DNA* sequence of a *gene.* Intronic sequences are transcribed into *mRNA* but they are spliced out of the *RNA* molecule before *translation* of *RNA* into *protein* (see also *exon*).

Kilobase (kb): Kilo=thousand; unit of length for *DNA* fragments equal to 1,000 *nucleotides.*

Linkage: The closer two or more *markers* (e.g., *polymorphisms*) on a *chromosome* are together the lower is the probability that they will be separated during *DNA* repair or replication. Therefore, the closer they are linked together, the greater is the probability that they will be inherited together.

Linkage map: A map of the relative positions of genetic *loci* on a *chromosome,* constructed from data how often the *loci* are inherited together. The distance is measured in *centimorgans* (see also *centimorgan*) (*cM*).

Locus: A location on a *chromosome* or *DNA* molecule corresponding to a *gene* or a physical or phenotypic feature.

Marker: Known location on a *chromosome* (e.g., restriction enzyme cutting site, *gene*) whose inheritance can be monitored. Markers are located in or close to coding regions of *DNA* (i.e., *genes*) or in segments of *DNA* with no known coding function but whose pattern of inheritance can be determined (i.e., microsatellites).

Megabase (Mb): Unit of length for *DNA* fragments equal to 1 million *nucleotides* and roughly equal to 1 cM.

Meiosis: Two consecutive cell divisions in the *diploid* progenitors of sex cells that result in four rather than two daughter cells, each with a *haploid* set of *chromosomes.*

Messenger RNA (mRNA): *RNA* that serves as a template for *protein* synthesis.

Mitosis: Nuclear division in cells producing daughter cells that are genetically identical.

Mutation: Permanent change in *DNA* sequence that will be heritable. Common mutations are also called polymorphisms (see *polymorphism*).

Nucleotide: A subunit of *DNA* or *RNA* consisting of a purine (*adenine* and *guanine*) or a pyrimidine base [*thymine* (*DNA* only), *uracil* (*RNA* only) and *cytosine*], a phosphate molecule, and a sugar molecule (deoxyribose in *DNA* and ribose in *RNA*).

Oncogene: A *gene* that is involved in cancer. Most oncogenes are involved, directly or indirectly, in controlling the rate of cell growth.

Polymerase chain reaction (PCR): Process of amplifying a *DNA* base sequence using a heat-stable polymerase and two *primers,* commonly about 20 bases in length, one complementary to the (+)-strand at one end of the sequence to be amplified and the other complementary to the (–)-strand at the other end. After separating (denaturing) the double-stranded *DNA,* the *primers* anneal to the corresponding *DNA* sequence on the isolated *DNA* strands. *DNA* polymerase builds a new double-stranded *DNA* molecule by adding the matching *nucleotides.* Newly synthesized *DNA* strands serve as additional templates for the same *primer* sequences. Therefore, successive cycles of denaturing, *primer* annealing, strand elongation, and dissociation produce a rapid and highly specific amplification of the desired sequence (30 cycles produce 2^{30}=268,435,356 copies of the *DNA* fragment, whose length has been determined by the forward and reverse *primer*).

Polymorphic site: A position within a *locus* at which at least two alternative sequences (*alleles*) are found in a population.

Polymorphic variant: A *gene, mRNA, cDNA,* polypeptide or *peptide* whose *nucleotide* or *amino acid* sequence varies from a reference sequence due to the presence of a *polymorphism* in the *gene.*

Polymorphism: A common (i.e., at least 1% prevalence of the minor *allele* in the population) sequence variation observed in an individual at a *polymorphic site.* Polymorphisms include *nucleotide* substitutions, insertions, deletions and microsatellites. They may be functional or silent, i.e., they do not result in detectable differences in *gene expression* or *protein* function.

Primer: Short synthetic polynucleotide chain (generally about 18–25 bases) to which new deoxyribonucleotides can be added by *DNA* polymerase.

Probe: Single-stranded *DNA* or *RNA* molecules of specific base sequence, labeled either radioactively or immunologically, that are used to detect the complementary base sequence by *hybridization*.

Prokaryote: Cell or organism lacking a membrane-bound, structurally discrete nucleus and other subcellular compartments. Bacteria are prokaryotes (see also *eukaryote*).

Promoter: A site on the *DNA* to which *RNA* polymerase will bind and initiate *transcription*.

Protein: Molecules composed of one or more chains that are built from a set of 20 *amino acids* in humans. The order of *amino acids* is determined by the base sequence of *nucleotides* in the *gene* coding for the protein. Proteins are required for the structure, function, and regulation of the body's cells, tissues, and organs, and each protein has unique functions. Examples are hormones, enzymes, and antibodies. However, several factors may interact during the *transcription* process and result in different *proteins* being generated from the same *genetic code* (i.e., alternative splicing, epigenetic modification, distant control regions). That is to say that in humans, in general, one *gene* (with one single *genetic code*) does code for more than one *protein* (i.e., 3 to 5 *proteins*, many of which have yet to be linked to the corresponding *gene*), and the process can be modified by environmental stimuli.

Recombinant DNA molecules: A combination of *DNA* molecules of different origin that are joined using recombinant *DNA* technologies.

Recombination: The process by which progeny derive a combination of *genes* different from that of either parent. In higher organisms, this is achieved by *crossing over* of *chromosomes*.

Restriction fragment length polymorphism (RFLP): Variation between individuals in *DNA* fragment sizes cut by specific restriction enzymes. RFLPs are usually caused by *mutation* at a cutting site.

Ribonucleic acid (RNA): Molecules including *messenger RNA*, *transfer RNA*, *ribosomal RNA*, or small RNA. RNA serves as a template for *protein* synthesis and other biochemical processes of the cell. The structure of RNA is similar to that of *DNA* except for the base *thymidine* being replaced by *uracil*.

Ribosomal RNA (rRNA): *RNA* found in the ribosomes of cells.

Sequence tagged site (STS): Short (200–500 *base pairs*) *DNA* sequence that has been identified and located as a single occurrence in the human *genome*. Detectable by *polymerase chain reaction*, STSs are useful for localizing and mapping of sequence data reported from different laboratories (see also *expressed sequence tags*).

Single nucleotide polymorphism (SNP): Typically, the specific pair of *nucleotides* observed at a single *polymorphic site*. In rare cases, three or four *nucleotides* may be found.

Southern blotting: Technique used to identify and locate DNA sequences which are complementary to another piece of DNA called probe using electrophorectic gels for separation of DNA and membrane filters with radiolabelled complementary probes.

STS: See *sequence tagged site*.

Thymine (T): A nitrogenous pyrimidine base found in *DNA* but not in *RNA*; it bonds with *adenine* (*A*) to form the A-T *base pair*.

Transcription: The synthesis of an RNA copy from a sequence of DNA (i.e., a gene); the first step in gene expression (see also *translation*).

Transfer RNA (tRNA): *RNA* with a triplet *nucleotide* sequence that is complementary to the triplet *nucleotide* coding sequences of *mRNA*. tRNAs in *protein* synthesis bond with *amino acids* and transfer them to the ribosomes, where *proteins* are assembled according to the *genetic code* carried by *mRNA*.

Translation: Process of synthesizing *proteins* from *amino acids* based on the *genetic code* carried by *mRNA* (see also *transcription*).

Uracil: A nitrogenous base normally found in *RNA* but not in *DNA*; it bonds with adenine to form the A-U base pair.

Yeast artificial chromosome (YAC): Large segments of DNA (up to 1 million bases) from another species spliced into DNA of yeast. The new construct that carries the foreign DNA is called a vector.

Selected genomic resources on the World Wide Web (www)

Many different resources and databases are accessible from these major sites.

Organization	http:// address
Pharm GKB: The Pharmacogenetics Knowledge Base	*http://www.pharmgkb.org*
National Center for Biotechnology Information	*http://www.ncbi.nlm.nih.gov*
Department of Energy (DOE) Biology Information Center	*http://www.ornl.gov/hgmis/*
European Bioinformatics Institute (EBI)	*http://www.ebi.ac.uk*
German Cancer Research Center (DKFZ)	*http://genome.dkfz-heidelberg.de*
Stanford Microarray	*http://www.microarray.org/*
Stanford University Center for Molecular and Genetic Medicine	*http://cmgm.stanford.edu/*
SWISS-PROT protein database	*http://www.expasy.ch/sprot/*
UC at Santa Cruz Genome Browser	*http://genome.ucsc.edu*
Whitehead Institute for Biomedical Research	*http://www.genome.wi.mit.edu*
KEGG Encyclopedia of Genes and Genomes	http://*www.genome.ad.jp/kegg/*
WIT pathway database	*http://wit.mcs.anl.gov/WIT2/*
Human Genome Nomenclature Database	*http://www.gene.ucl.ac.uk/*
Human Genome Project by DOE and NIH	*http://www.science.doe.gov/ober/hug_top.html*
Primer on Molecular Genetics (DOE 1992)	*http://www.ornl.gov/hgmis/publicat/primer/intro.html*
Glossary of molecular genetic terms (from Primer of Molecular Genetics internet site, DOE)	*http://www.ornl.gov/TechResources/Human_Genome/glossary/*

Index

b

d